Developments in Petroleum Science, 17B

enhanced oil recovery, II

processes and operations

DEVELOPMENTS IN PETROLEUM SCIENCE
Advisory Editor: G.V. Chilingarian

Volumes 1, 3, 4, 7 and 13 are out of print.

2. W.H. FERTL
ABNORMAL FORMATION PRESSURES

5. T.F. YEN and G.V. CHILINGARIAN (Editors)
OIL SHALE

6. D.W. PEACEMAN
FUNDAMENTALS OF NUMERICAL RESERVOIR SIMULATION

8. L.P. DAKE
FUNDAMENTALS OF RESERVOIR ENGINEERING

9. K. MAGARA
COMPACTION AND FLUID MIGRATION

10. M.T. SILVIA and E.A. ROBINSON
DECONVOLUTION OF GEOPHYSICAL TIME SERIES IN THE EXPLORATION FOR OIL AND NATURAL GAS

11. G.V. CHILINGARIAN and P. VORABUTR
DRILLING AND DRILLING FLUIDS

12. T.D. VAN GOLF-RACHT
FUNDAMENTALS OF FRACTURED RESERVOIR ENGINEERING

14. G. MOZES (Editor)
PARAFFIN PRODUCTS

15A O. SERRA
FUNDAMENTALS OF WELL-LOG INTERPRETATION
1. THE ACQUISITION OF LOGGING DATA

15B O. SERRA
FUNDAMENTALS OF WELL-LOG INTERPRETATION
2. THE INTERPRETATION OF LOGGING DATA

16. R.E. CHAPMAN
PETROLEUM GEOLOGY

17A E.C. DONALDSON, G.V. CHILINGARIAN and T.F. YEN (Editors)
ENHANCED OIL RECOVERY, I
FUNDAMENTALS AND ANALYSES

18A A.P. SZILAS
PRODUCTION AND TRANSPORT OF OIL AND GAS
A. FLOW MECHANICS AND PRODUCTION
second completely revised edition

18B A.P. SZILAS
PRODUCTION AND TRANSPORT OF OIL AND GAS
B. GATHERING AND TRANSPORTATION
second completely revised edition

19A G.V. CHILINGARIAN, J.O. ROBERTSON Jr. and S. KUMAR
SURFACE OPERATIONS IN PETROLEUM PRODUCTION, I

19B G.V. CHILINGARIAN, J.O. ROBERTSON Jr. and S. KUMAR
SURFACE OPERATIONS IN PETROLEUM PRODUCTION, II

20. A.J. DIKKERS
GEOLOGY IN PETROLEUM PRODUCTION

21. W.F. RAMIREZ
APPLICATION OF OPTIMAL CONTROL THEORY TO ENHANCED OIL RECOVERY

22. E.C. DONALDSON, G.V. CHILINGARIAN and T.F. YEN (Editors)
MICROBIAL ENHANCED OIL RECOVERY

Developments in Petroleum Science, 17B

enhanced oil recovery, II
processes and operations

Edited by

ERLE C. DONALDSON

School of Petroleum and Geological Engineering, University of Oklahoma, Norman, Oklahoma, U.S.A.

GEORGE V. CHILINGARIAN

Department of Petroleum Engineering, University of Southern Calfornia, Los Angeles, California, U.S.A.

and

TEH FU YEN

Department of Civil Engineering, University of Southern California, Los Angeles, California, U.S.A.

ELSEVIER — Amsterdam — Oxford — New York – Tokyo 1989

ELSEVIER SCIENCE PUBLISHERS B.V.
Sara Burgerhartstraat 25
P.O. Box 211, 1000 AE Amsterdam, The Netherlands

Distributors for the United States and Canada:

ELSEVIER SCIENCE PUBLISHING COMPANY INC.
655, Avenue of the Americas
New York, NY 10010, U.S.A.

Library of Congress Cataloging-in-Publication Data
(Revised for volume 2)

Enhanced oil recovery.

(Developments in petroleum science ; 17A-17B)
Includes bibliographies and indexes.
Contents: [1] Fundamentals and analysis -- 2.
Processes and operations.
1. Secondary recovery of oil. 2. Oil field flooding.
I. Donaldson, Erle C. II. Chilingarian, George V.,
1929- . III. Yen, Teh Fu, 1927-
IV. Series.
TN871.E527 1985 622'.3382 83-25288
ISBN 0-444-41625-0 (set)
ISBN 0-444-42206-4 (v. 1)
ISBN 0-444-42933-6 (v. 2)

ISBN 0-444-42933-6 (Vol. 17B)
ISBN 0-444-41625-0 (Series)

Printed in The Netherlands

Dedicated to

Victor E. Grijalda, President of Schlumberger, Inc.,
Howard Dingman, President of Western Atlas International
J. Robert Fluor, Fluor Corp., and
Dan K. Adamson, Executive Director of the Society of Petroleum Engineers

for their outstanding contributions and leadership in the petroleum industry.

PREFACE

Enhanced oil recovery (EOR) is an engineering activity concerned with increasing the recovery of hydrocarbons from various types of petroleum reservoirs. Shallow reservoirs containing very viscous oils require special applications of heat; and this has developed into the special branch of EOR known as thermal recovery. Deeper reservoirs that contain less viscous hydrocarbon reserves, are amenable to enhancement of recovery by various methods that depend on the chemistry of the reservoir fluids and the petrophysical properties of the reservoir rocks. The EOR methods that apply under specific circumstances usually involve flooding the reservoir with surfactants, caustics, micellar–polymer solutions, polymers, microbes and gases. The mechanisms of hydrocarbon displacement by each of these methods are different, resulting in differences of equipment design, personnel training, operating procedures, and overall field development economics.

Enhanced oil recovery, therefore, presents a host of questions, decisions, designs and responsibility to select the best technology for a given situation and to solve the myriad of problems that arise in actual practice. The purpose of this book is to provide the managers and engineers with information to form the basis for their decisions, designs, and operating procedures.

Volume I of this two-volume treatise is devoted to the fundamental physical and chemical principles and analyses of EOR processes, whereas Volume II is designed to present standard processes and operations of EOR. Thus the two volumes complement each other by presenting both the theory and practice of EOR.

This is principally a reference book. It can, however, serve as a text for graduate studies and intensive short courses in EOR. The emphasis of this book has been placed on design and operating strategies for EOR, which was maintained by the fruitful cooperation between the authors and the editors. This book contributes to the expanding theory and practice of EOR. It provides a better understanding of EOR theory and processes, and serves as a guide for selection, design, and operation of enhanced oil recovery technology.

ERLE C. DONALDSON,
GEORGE V. CHILINGARIAN
TEH FU YEN

LIST OF CONTRIBUTORS

R.S. BRYANT	*NIPER, P.O. Box 2128, Bartlesville, OK 74005, U.S.A.*
D.B. BURNETT	*3109 Sunset Oaks, Arlington, TX 76016, U.S.A.*
R. CAENN	*4371 Margarita, Irvine, CA 92714, U.S.A.*
G.V. CHILINGARIAN	*Department of Petroleum Engineering, School of Engineering, University of Southern California, University Park, Los Angeles, CA 90089-1211, U.S.A.*
F. CIVAN	*School of Petroleum and Geological Engineering, University of Oklahoma, Norman, OK 73019, U.S.A.*
A.T. COSCIA	*Chemicals Group, American Cyanamid Company, One Cyanamid Plaza, Wayne, NJ 07470, U.S.A.*
P.A. DICKEY	*8050 N. Yale, Owasso, OK 74055, U.S.A.*
E.C. DONALDSON	*School of Petroleum and Geological Engineering, University of Oklahoma, Norman, OK 73019, U.S.A.*
W.H. FERTL	*Atlas Oilfield Services Group, Dresser Industries, Inc., P.O. Box 1407, Houston, TX 77251, U.S.A.*
K.L. GOYAL	*Reservoir and Production Division, Oil and Natural Gas Commission, Dehradun, India*
S.I. KOHEN	*Chemicals Group, American Cyanamid Company, One Cyanamid Plaza, Wayne, NJ 07470, U.S.A.*
A.R. KUKRETI	*School of Civil Engineering and Environmental Science, University of Oklahoma, Norman, OK 73019, U.S.A.*
S. KUMAR	*Petroleum Engineering Department, University of Southern California, Los Angeles, CA 90089-1211, U.S.A.*
G.L. LANGNES	*Mobil Oil Corporation, P.O. Box 900, Dallas, TX 75221, U.S.A.*
C.S. MATTHEWS	*Shell Oil Company, One Shell Plaza, P.O. Box 2463, Houston, TX 77001, U.S.A.*
A. MEHDIZADEH	*Mechanical Engineering Department, California Polytechnic State University, San Luis Obispo, CA 93407, U.S.A.*
M. PRATS	*2834 Bellefontaine, Houston, TX 77025, U.S.A.*

Y. RAJAPAKSA	*School of Engineering and Environmental Science, University of Oklahoma, Norman, OK 73019, U.S.A.*
J.O. ROBERTSON, Jr.	*Earth Engineering, Inc., 4244 Live Oak St., Cudahy, CA 90201, U.S.A.*
D.O. SHAH	*Department of Chemical Engineering, University of Florida, Gainesville, FL 32611, U.S.A.*
M.K. SHARMA	*Chemical Laboratories, Eastman Chemical Division, P.O. Box 1972, Kingsport, TN 37662, U.S.A.*
J. TORABZADEH	*Petroleum Engineering Department, University of Southern California, University Park, Los Angeles, CA 90089-1211, U.S.A.*
T.F. YEN	*School of Engineering, Department of Civil Engineering, University of Southern California, University Park, Los Angeles, CA 90089-0242, U.S.A.*
W.S. YEN	*Chemicals Group, American Cyanamid Company, One Cynanamid Plaza, Wayne, NJ 07470, U.S.A.*

CONTENTS

Chapter 1

INTRODUCTION

E.C. DONALDSON, G.V. CHILINGARIAN, T.F. YEN and M.K. SHARMA

The overall objective of this two-volume treatise on emerging technology for enhancing the recovery of oil beyond that attainable by secondary recovery, is the development of an in-depth review of the state-of-art of the new technology which has become known as *enhanced oil recovery* (EOR). The first volume was devoted to presentation of the fundamentals and analyses of petrophysics, reservoir characterization and secondary recovery. This second volume follows directly with in-depth development of the processes and operations of EOR. Thus, discussions in Volume II are based on the assumption that a thorough technical evaluation of the oil reservoir has indicated that an EOR process is feasible and has been initiated.

No single EOR process has yet emerged which is applicable to all petroleum reservoirs as a general method for enhancement of oil recovery; however, specific processes that are quite distinct from each other have been developed to address reservoirs with special characteristics. The criteria for selection of a particular EOR process are complex because of the large number of petrophysical, chemical, geologic, environmental, and fluid properties that must be considered for each individual case.

A petroleum reservoir may be defined as a porous and permeable medium containing brine, oil, and gas which move toward production wells under either existing or applied pressure. Interconnecting pores of different sizes and shapes contribute to the formation of the reservoir. The oil and gas reservoirs consist of many different rocks, mainly semiconsolidated sands, sandstones, conglomerates, limestones, and dolomites containing pore spaces for fluids. The porosity of most commercially productive petroleum reservoirs ranges from 10 to 25%.

The permeability of reservoir rocks governs the ability of the fluids to move through the pore spaces of the petroleum reservoir. The standard unit of the permeability is μm^2, or darcy. The darcy is defined as: the rate of flow in $cm^3\ s^{-1}$ of a fluid having 1 centipoise viscosity through a cross-section of 1 cm^2 of the porous medium under a pressure gradient of 1 atm cm^{-1}. The permeability of most of the reservoirs containing oil varies from 0.5 millidarcy (mD) to 3500 mD (Foley, 1976).

Petroleum reservoirs contain water which may be classified as either *connate, formation*, or *interstitial* water. The water present in a reservoir essentially contains dissolved solids, primarily sodium chloride. Oilfield water, therefore, also is called brine. Subsurface reservoirs are completely saturated with fluids (gas, oil and water) in various proportions varying from 0 to 100%. The reservoir water contains anions such as SO_4^{2-}, Cl^-, HCO_3^-, CO_3^{2-}, NO_2^-, S^{2-}, Br^-, and I^-, and cations, such as K^+, Mg^{2+}, and Ca^{2+}. The salt concentration depends on the characteristics of the petroleum reservoir, such as ease of dissolution of mineral salts, chemical reactions, and exchange capacity of cations with clay minerals, etc. (McCain, 1973). Salt concentrations range from 200 to 300,000 ppm.

The crude oil mobility in a petroleum reservoir is influenced by the viscosity of the crude oil. The oil viscosity is temperature sensitive, reducing with an increase of temperature. Any analysis of in-situ fluid flow must, therefore, consider the actual temperature of the petroleum reservoir. This temperature may be estimated from the mean surface temperature of the location and the geothermal gradient, e.g., 1.8°C/100 m. Thus, considering a mean surface temperature of 25°C and a depth of 2000 m, the reservoir temperature will be approximately 61°C, which will produce a significant reduction of oil viscosity from the measured value at ambient temperature.

OIL RECOVERY PROCESSES

Oil recovery processes may be subdivided into three major categories: primary, secondary and enhanced oil recovery. In the primary process, the oil is forced out of the petroleum reservoir by existing natural pressure of the trapped fluids in the reservoir. When the reservoir pressure is reduced to a point where it is no longer effective as a stress causing movement of hydrocarbons to the producing wells, water or gas is injected to augment or increase the existing pressure in the reservoir. Conversion of some production wells to injection wells and subsequent injection of gas or water for pressure maintenance in the reservoir has been designated as secondary oil recovery. The oil recovered by both primary and secondary processes ranges from 20 to 50% depending on oil and reservoir properties. The goal of enhanced oil recovery processes is to recover at least a part of the remaining oil-in-place. There are several enhanced oil recovery processes that are considered to be promising: (1) chemical flooding; (2) thermal oil recovery; and (3) miscible displacement methods, which are briefly described below along with other processes. Various mechanisms are involved in the different oil recovery processes.

Primary oil recovery

The efficiency of oil displacement in primary oil recovery process depends mainly on existing natural pressure in the petroleum reservoir. This pressure originated in the reservoir from various forces: (1) expanding force of natural gas, (2) gravitational force, (3) buoyancy force of encroaching water, and (4) an expulsive force due to the compaction of poorly consolidated reservoir rocks

(Uren, 1939). Among these forces, the expanding force of high-pressure natural gas contributes mainly to oil production. These forces in the reservoir either can act simultaneously or sequentially, depending on the composition and properties of the reservoir.

The gravitational force is more effective in steeply inclined reservoirs, where it facilitates the drainage of oil. This force alone may not be effective in moving large amounts of oil into a production well. Another, more effective, force for displacement of oil is encroachment of water from the side or bottom of a reservoir. In some fields, edge water encroaching from a side appears to be stationary. The ability of the edge water to encroach depends upon the pressure distribution in the reservoir and the permeability. Compaction of the reservoir as fluids are withdrawn also is a mechanism for movement of oil to production wells. Part of the oil will be expelled due to the decrease in reservoir volume.

Secondary oil recovery

The general mechanism of oil recovery is movement of the hydrocarbons to production wells due to a pressure difference between the reservoir and the production wells. Thus the objectives of EOR are to increase the pressure difference between the reservoir and production wells, or to increase the mobility of the oil by reduction of the oil viscosity or decrease of the interfacial tension between the displacing fluids and oil.

When oil production declines because of hydrocarbon production from the formation, the secondary oil recovery process is employed to increase the pressure required to drive the oil to production wells. The mechanism of secondary oil recovery is similar to that of the primary oil recovery except that more than one wellbore is involved, and the pressure of the petroleum reservoir is augmented or maintained artificially to force oil to the production wells. The process includes the application of a vacuum to a well, the injection of gas, air, water, and/or aqueous solutions of caustic and polymer.

The decrease of pressure in the reservoir during primary oil recovery may be restored partially by injecting a gas into the reservoir to achieve a high pressure. Gas injection methods can be subdivided into three categories: (a) pressure restoration, (b) pressure maintenance, and (c) gas-drive, depending upon the way in which the gas is injected into the reservoir. In the *pressure restoration* method, the gas is injected into the productive formation through one well while other wells are closed until the pressure is restored throughout the reservoir. This may take as long as a year or more. When the desired reservoir pressure is reached, gas injection is stopped and all of the wells start producing oil under the influence of the artificially developed pressure.

In the *pressure maintenance* method, gas from producing wells is recompressed and injected into the selected wells before the reservoir pressure is totally exhausted. In this method, some wells are operated as injection wells, whereas others are operated as production wells. In the *gas-drive method*, gas is

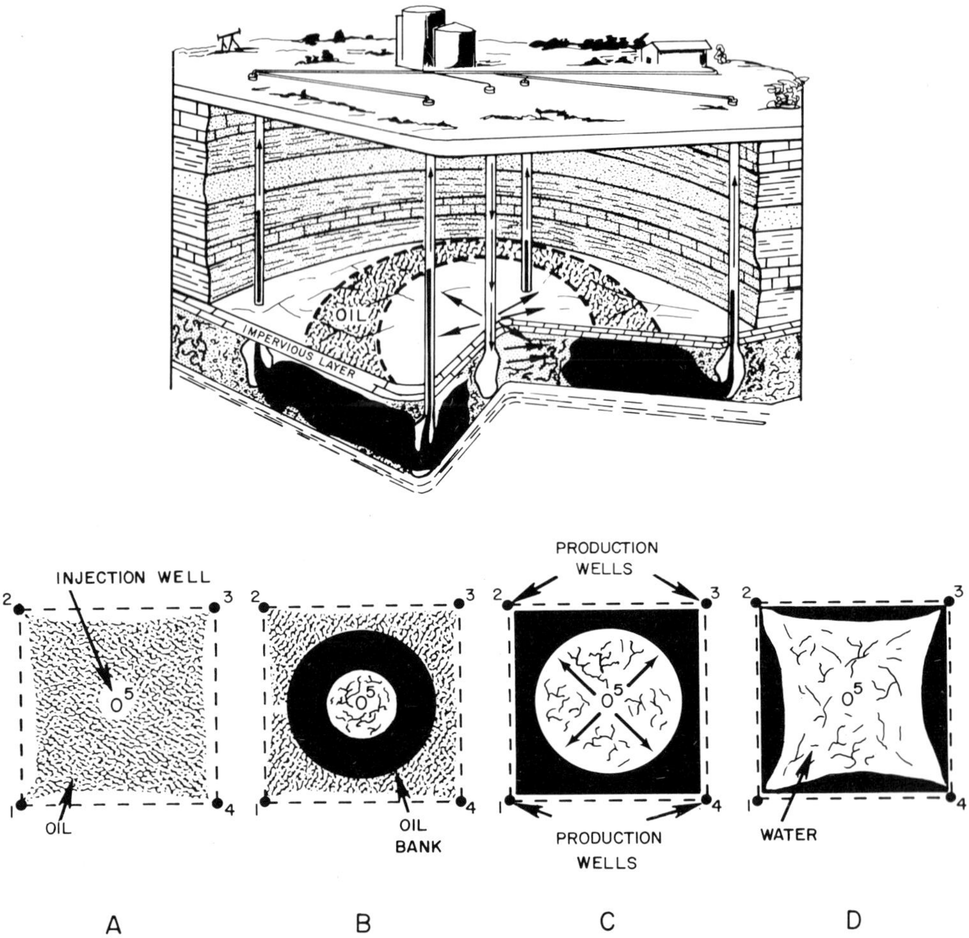

Fig. 1-1. Displacement of oil through reservoir rocks by waterflooding (five-spot pattern).

injected into the reservoir under pressure and a continuous gas flow is maintained from injection wells to the producing wells. The moving gas drives the oil in the form of a film, or gas bubbles ahead of the gas, toward the producing wells.

After primary oil recovery, the pressure of the depleted reservoirs can be restored by *waterflooding*. In the waterflooding process, water is injected into the reservoir through injection wells. The water drives oil through the reservoir rocks toward the producing wells. For waterflooding, the most common pattern of injection and production wells is a *five-spot* configuration, shown in Fig. 1-1, where water is injected in the central well displacing oil to the four surrounding production wells. Fig. 1-1 shows the mechanism of oil displacement through reservoir rocks by waterflooding. To improve the efficiency of the waterflooding process, some chemicals such as caustic or polymers are added to the water.

Some early investigators (e.g., Reisberg and Doscher, 1956) placed primary importance on lowering the interfacial tension by the injection of caustic, which will result in the formation of an oil-in-water emulsion. Thus oil will be produced as a low-viscosity emulsion because an emulsion adopts the viscosity of the external phase.

The efficiency of waterflooding also can be improved by lowering the water–oil mobility ratio, M, which can be defined as $k_w \mu_o / \mu_w k_o$, where k_w = effective water permeability, mD; k_o = effective oil permeability, mD; μ_w = water viscosity, cP; and μ_o = oil viscosity, cP. A mobility ratio greater than unity is unfavorable because water is more mobile than oil. Water would finger through the oil zone and, therefore, reduce the oil recovery efficiency. If the mobility ratio is less than unity, the displacement of oil by water occurs more or less in piston-like displacement. In waterflooding, the mobility ratio can be reduced by decreasing the mobility of water which can be achieved by adding chemicals such as alcohols, sugar, glycerin, polymers, etc., to the injected water (Pye, 1964; Sandiford, 1964; Dauben and Menzie, 1967; Gogarty, 1967; Jennings et al., 1971). Low concentrations (250–1500 ppm) of polymers in water can reduce the mobility ratio dramatically.

Enhanced oil recovery

Enhanced oil recovery processes have as their objective the increase of recovery from reservoirs depleted by secondary recovery with waterflooding or gas injection. The EOR *processes* can be divided into three major categories: (1) chemical, (2) thermal, and (3) miscible. The major enhanced oil displacement processes can be further subdivided into various categories as illustrated in Fig. 1-2. Thermal processes have been used extensively for the displacement of heavy oils, whereas chemical and miscible displacement processes have been employed for the recovery of light oils. Among the various processes for oil recovery, thermal processes have the least uncertainty, and offer a promising approach for about 70% of the world's enhanced oil recovery production. At present, surfactant flooding is the most complex and, therefore, has the highest degree of uncertainty. If the surfactant formulation for oil recovery is properly designed and if the flow of the formulation is properly controlled in the reservoir, it has a high potential for achieving maximum oil recovery.

Chemical flooding processes can be divided into three main categories: (1) surfactant flooding, (2) polymer flooding, and (3) caustic flooding. The mechanism of oil displacement by surfactant and caustic flooding is based on the formation of ultra-low interfacial tension, whereas polymer flooding alone, or injection of surfactant followed by polymer flooding, results in controlling the mobility which in turn enhances oil recovery. Caustic injected into the petroleum reservoir reacts chemically with the fatty acids present in the petroleum derivatives and forms in-situ sodium salts of fatty acids. The formation of these

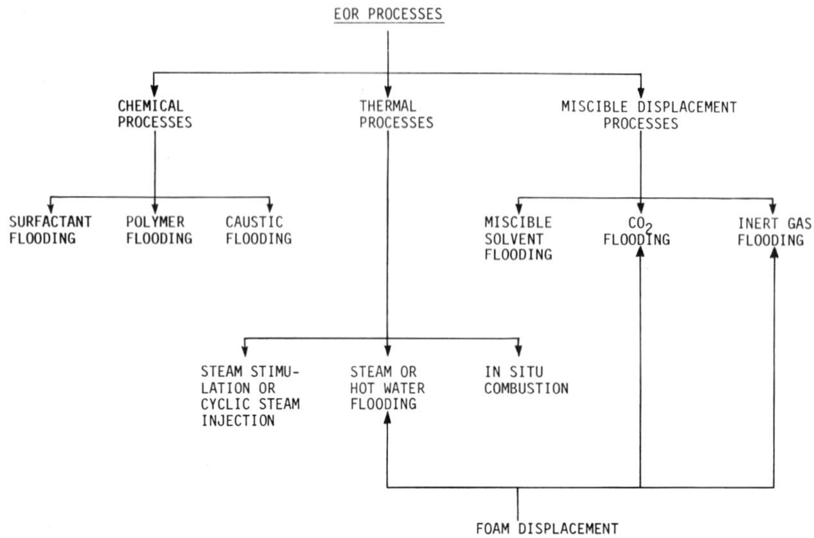

Fig. 1-2. Various EOR processes.

surfactants results in ultra-low interfacial tension. The chemical flooding processes have been schematically illustrated in Fig. 1-3.

When petroleum reservoirs contain a low-gravity (less than 20° API), high-viscosity oil and have a high porosity, secondary recovery methods are not effective for displacement of oil. For such reservoirs, thermal processes have received the most attention. The injection of steam reduces the oil viscosity which causes an increase in the oil mobility. Depending on the way in which the heat is generated in the reservoir, the thermal processes can be divided into three categories: (1) in-situ combustion, (2) steam injection, and (3) wet combustion.

For *in-situ combustion*, the crude oil near the wellbore is ignited using chemicals, downhole electric heaters or downhole gas burners. The mechanism of oil displacement by in-situ combustion was extensively reported by many investigators (e.g., Kuhn and Koch, 1953; Grant and Szasz, 1954; Ramey and Naber,

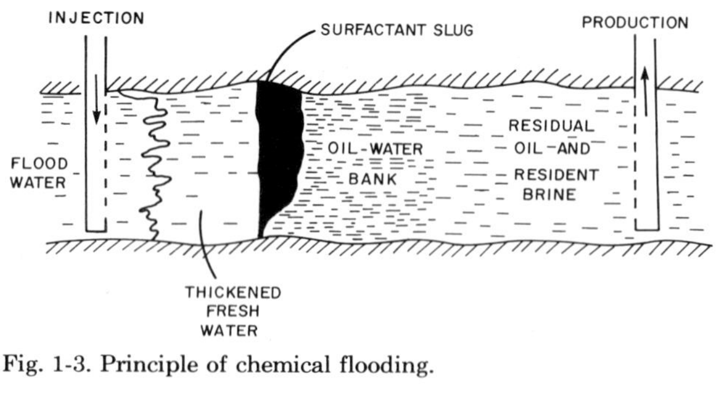

Fig. 1-3. Principle of chemical flooding.

1954; McNeil and Nelson, 1956; Benham and Poettmann, 1958; Martin et al., 1958; McNeil and Moss, 1958; Nelson and McNeil, 1961). After completing ignition in the vicinity of the wellbore, continuous air injection promotes movement in the burning zone toward the producing wells. Propagation of a continuous burning zone results in almost complete removal of all reservoir liquids and leaves behind hot, clean rock, which heats the injected air before it reaches the burning zone.

In the second method, *steam injection*, steam is injected into the reservoir either continuously or in cycles. Continuous steam injection involves both injection and production wells, whereas cyclic injection involves one well only which serves as both injection and production well. Steam floods are easier to control than in-situ combustion. For the same pattern size, the response time is 25–50% lower than the response time for additional production by in-situ combustion.

The third method is called the *wet combustion*. In this in-situ combustion process, a large amount of heat is left behind in the swept formation as waste heat. The heat utilization and efficiency of the process can be improved by water injection. In this process, water is injected with the air. Superheated steam forms in an evaporation front and travels behind the combustion front. The important advantage of this process is that the amount of residual oil left to be burned as fuel by the burning front is considerably decreased (Dietz, 1970; Burger and Sahuquet, 1973), which in turn displaces more oil and less air is required to burn a unit volume of oil in the reservoir.

The *miscible displacement* process involves the injection of a solvent such as alcohol, refined hydrocarbons, condensed hydrocarbon gases, liquefied petroleum gases, or carbon dioxide, which can dissolve in the reservoir oil. The injected solvent reduced the capillary forces that cause oil retention in the pore spaces of the reservoir rocks. In this process, the injected solvent slug is followed by the injection of a liquid or gas to force the solvent–oil mixture out. The miscible displacement process can be subdivided into (1) the miscible slug process; (2) the enriched gas process; (3) the high-pressure, lean gas process; and (4) the mutual solvent and carbon dioxide processes.

The *miscible slug* process involves the injection of a slug of liquid hydrocarbons equal to about one half of the pore volume of the reservoir. The liquid hydrocarbon slug is followed by gas or water injection to drive the slug through the reservoir. In the enriched gas process, the injected slug (10–20% pore volume) of enriched natural gas is followed by lean gas, or lean gas and water. The high-pressure lean gas process consists of injecting a lean gas at a high pressure in order to cause retrograde evaporation of the crude oil and formation of a miscible phase, which consists of C_2–C_6 components, between the reservoir oil and the gas. The main difference between the enriched gas and the high-pressure lean gas processes is that in the enriched gas process the C_2–C_6 components are transferred from the gas to the oil, whereas in the high-pressure lean gas process, the C_2–C_6 components are transferred from the oil to the gas.

The *mutual solvent* process consists of injecting solvents (e.g., alcohols), which are miscible both in the reservoir oil and water. These solvents form a single phase in the petroleum reservoir and improve oil recovery. A very high concentration of these solvents is required to maintain the single phase. In the *carbon dioxide* process, the mechanism of carbon dioxide miscibility in oil is similar to that in the high-pressure lean gas process. Under the right conditions of pressure, temperature and reservoir oil composition, carbon dioxide can create a miscible front, which moves as a single liquid phase and efficiently displaces reservoir oil to the producing wells. The carbon dioxide miscibility can be achieved at pressures as low as 1500 psi at normal reservoir temperatures. The presence of impurities in the carbon dioxide, such as nitrogen and methane, increase the miscibility pressure, whereas impurities such as propane and hydrogen sulfide decrease it.

For example, several companies are successfully using carbon dioxide for enhancement of oil recovery in oilfields in the Permian Basin in western Texas, U.S.A. The supply of carbon dioxide is obtained from the Bravo Dome Formation in northeastern New Mexico (approximately 400 km away), which is estimated to contain 2.8×10^{10} m^3 (12×10^{10} ft^3) of 97% pure carbon dioxide. Subsurface supplies of carbon dioxide also are available in the United States from the Sheep Mountains in southeastern Colorado and the McElmo Dome in southwestern Colorado. The availability of large supplies of carbon dioxide make this process feasible for regional enhanced recovery operations. Operators of the Alaskan north slope oil reserves are considering the application of carbon dioxide processes for enhancement of heavy oil recovery from the West Oak Formation lying at a depth of about 1100 m. The oil reserves here are estimated to be 3 Gm^3 (20×10^9 bbl). Successful application to heavy oil recovery will broaden the utility of carbon dioxide as a stimulant for oil recovery.

The use of micro-organisms and their metabolic products to stimulate oil production is receiving interest worldwide. This technique involves the injection of selected microorganisms into the reservoir and the subsequent stimulation and transportation of their in-situ growth products. Their presence aids in further reduction of residual oil left in the reservoir after secondary recovery. Microbial enhanced oil recovery (MEOR) is unlikely to replace conventional EOR methods because MEOR itself has certain constraints. This unique process seems superior in many respects, however. As the self-duplicating units, namely, the bacteria cells, are injected into the reservoir, their in-situ multiplication magnifies the beneficial effects. Microbial cultures are capable of synthesizing a large variety of biochemical products from crude oil constituents when provided with essential nutrients and proper environmental conditions. In fact, populations of various types of microbes have been found in subsurface petroleum reservoirs down to a depth of about 2000 m (Petzel and Williams, 1986). The prevalent theory is that the microbes were introduced by drilling, well completion, and well workover procedures or by water injection during secondary recovery. The fact that the populations of microbes are present in deep porous formations is positive proof that petroleum reservoirs are not biologically sterile. This offers positive encouragement for fundamental research to discover microbes, procedures and technology that can lead to enhancement of oil recovery. The field trials that have been reported remain to be completely understood. Specific research for field applications will resolve these in the future (e.g., see Donaldson and Clark, 1982; Zajic and Donaldson, 1985; Yen, 1986).

In addition to microbial EOR, which may be considered as an emerging technology, and is currently receiving limited field testing, there is another technique which is in the realm of laboratory experimentation and very limited field testing. This EOR method involves electro-osmotic fluid flow upon application of direct electric current to enhance the injectivity of water and the production of oil. Laboratory experimental results indicate that direct electric current can be used to increase the volumetric flow rate of oil and water during either primary or secondary recovery (Adamson et al., 1963a,b; Anbah et al., 1965a,b; Chilingar et al., 1970). Chilingar et al. (1968) reported a 2- to 32-fold increase in water flow rate through sand cores containing clays on application of direct electric current. In addition, electrochemical changes occur in the porous medium, as evidenced by the higher final permeability upon discontinuation of treatment (hysteresis effect) (Anbah et al., 1965b). Also X-ray analysis of clays after electric treatment shows that their sheet structures are drastically reduced as a result of imposing direct electric current across the cores. Silicates seem to be destroyed upon treatment (Horton et al., 1967). Another concept is to chemically alter the micellar configuration of petroleum in situ based on the principle of membrane-mimetic chemistry; this will lead to the quaternary recovery (all previously discussed EOR can be classified as tertiary recovery) or advanced level of ultimate recovery (Yen, 1986).

Unfortunately, the EOR processes require large initial investments and careful monitoring during application, whereas the rate of incremental oil recovery is low and extended over a long period of time. Thus the EOR projects require an extensive economic evaluation and forecasting because of the low rate of return of investments. Initial reservoir definition with evaluation of oil saturation, stratigraphy and hydrology are necessary to determine the feasibility of any EOR project. Highly trained personnel are, therefore, required for all phases of EOR implementation. Although these constraints to EOR cause hesitancy in the application of any of the processes, two categories of EOR methods are proving to be successful. These are carbon dioxide flooding and thermal recovery methods. The potential exists for significantly increasing oil recovery by other EOR methods; however, continued field experimentation and evaluation is required to remove the economic and technical constraints.

This volume provides a thorough review of EOR processes, included mathematical simulation. It may be used either as a complete text for teaching the EOR processes or as an engineering reference manual for critical evaluation of a specific reservoir.

REFERENCES

Adamson, L.G., Anbah, S.A., Chilingar, G.V. and Beeson, C.M., 1963a. Possible use of electric current for increasing volumetric rate of flow of oil and water during primary or secondary recovery. *Chim. Chron.*, 28(1): 1–4.

Adamson, L.G., Chilingar, G.V. and Beeson, C.M., 1963b. Some data on electrokinetic phenomena and their possible application in petroleum production. *Chim. Chron.*, 28(10): 121–127.

Anbah, S.A., 1963. *Use of direct electrical current for increasing the flow rate of reservoir fluids during petroleum recovery.* Ph.D. Dissertation, University of Southern California, Los Angeles, Calif., 252 pp.

Anbah, S.A., Chilingar, G.V. and Beeson, C.M., 1965a. Application of electrical current for increasing the flow rate of oil and water in porous medium. *J. Can. Pet. Technol.,* 4(2): 1–8.

Anbah, S.A., Chilingar, G.V. and Beeson, C.M., 1965b. Use of direct electrical current for increasing the flow rate of resevoir fluids during petroleum recovery. *J. Can. Pet. Technol.,* 3(1): 8–14.

Benham, A.L. and Poettmann, F.H., 1958. The thermal recovery process — an analysis of laboratory combustion data. *Pet. Trans. AIME.*, 213: 406–408.

Burger, J.G. and Sahuquet, B.C., 1973. Laboratory research on wet combustion. *J. Pet. Technol.*, 25(10): 1137–1146.

Chilingar, G.V., Chang, Kim Sang, Davis, J.E., Farhanghi, H.J., Adamson, L.G. and Sawabini, S., 1968. Possible use of direct electrical current for augmenting reservoir energy during petroleum production. *Compass of Sigma Gamma Epsilon*, 45(4): 272–285.

Chilingar, G.V., El Nasir, Abdulrahman and Stevens, R.G., 1970. Effect of direct electrical current on permeability of sandstone cores. *J. Pet. Technol.*, 22(7): 830–836.

Dauben, D.L. and Menzie, D.E., 1967. Flow of polymer solutions through porous media. *J. Pet. Technol.*, 19(8): 1065–1073.

Dietz, D.N., 1970. Wet underground combustion state of the art. *J. Pet. Technol.*, 22(5): 605–617.

Donaldson, E.C. and Clark, J.B. (Editors), 1982. *Proceedings, 1982 International Conference on Microbial Enhancement of Oil Recovery*. NTIS, Springfield, Va., 219 pp.

Foley, C., 1976. *The Energy Question*. Penguin Books Ltd., Harmondsworth, Middlesex, 210 pp.

Gogarty, W.B., 1967. Mobility control with polymer solutions. *J. Pet. Technol.*, 7(2): 161–173.

Grant, B.F. and Szasz, S.E., 1954. Development of an underground heat wave for oil recovery. *Trans. AIME*, 201: 108–118.

Horton, J.H., Hamid, S., Abi-Chedid, E., and Chilingar, G.V., 1967. Effect of electrochemical treatment on selected physical properties of clayey silt. *Eng. Geol.,* 2(3): 191–196.

Jennings, R.R., Roger, J.H. and West, T.J., 1971. Factors influencing mobility control by polymer solutions. *J. Pet. Technol.*, 23(3): 391–410.

Kuhn, C.S. and Koch, R.L., 1953. In-situ combustion, newest method of increasing oil recovery. *Oil Gas J.*, 52(14): 92–96, 113–114.

Martin, W.L., Alexander, W.L. and Dew, J.N., 1958. Process variables in in-situ combustion. *Trans. AIME*, 213: 28–35.

McCain, W.D., Jr., 1973. *The Properties of Petroleum Fluids*. Petroleum Publishing Company, Tulsa, Okla., 325 pp.

McNeil, J.S. and Moss, J.T., 1958. In-situ combustion. *Oil Gas J.*, 56(37): 235–241.

McNeil, J.S. and Nelson, T.W., 1956. Thermal methods provide three ways to improve oil recovery. *Oil Gas J.*, 57(3): 86–98.

Nelson, T.W. and McNeil, J.S., Jr., 1961. In-situ combustion project. *Oil Gas J.*, 59(23): 59–65.

Petzel, G.A. and Williams, B., 1986. Operations trim basic EOR research. *Oil Gas J.*, 84(6): 41–45.

Pye, D.J., 1964. Improved secondary recovery by control of water mobility. *J. Pet. Technol.*, 16(8): 911–916.

Ramey, H.J., Jr. and Naber, G.W., 1954. A blottertype electrolyte model determination of aereal sweeps in oil recovery by in-situ combustion. *Trans. AIME*, 201: 119–123.

Reisberg, J. and Doscher, T.M., 1956. Interfacial phenomena in crude oil water systems. *Prod. Mon.*, 21(1): 43–50.

Sandiford, B.B., 1964. Laboratory and field studies of waterfloods using polymer solutions to increase oil recovery. *J. Pet. Technol.*, 16(8): 917–922.

Uren, L.C., 1939. *Petroleum Production Engineering, 2*. McGraw Hill, New York, N.Y., 632 pp.

Yen, T.F., 1986. *A State of the Art Review on Microbial Enhanced Oil Recovery*. Report to National Science Foundation, IOR 8405134, 278 pp.

Zajic, J.E. and Donaldson, E.C. (Editors), 1985. *Microbes and Oil Recovery, 1*. Bioresources Publications, El Paso, Texas, 370 pp.

Chapter 2

GEOLOGICAL FACTORS IN ENHANCED OIL RECOVERY

PARKE A. DICKEY

RESERVOIR HETEROGENEITIES (THE MOST FREQUENT CAUSE FOR FAILURE OF ENHANCED RECOVERY PROJECTS)

A very large fraction, probably much higher than half, of the field tests of enhanced recovery projects have been failures because they recovered less oil than had been projected. By far the most common reason for failure was the heterogeneity of the reservoir, which resulted in poor sweep efficiency. (The next most common reason was that there was less remaining oil than had been estimated.) The recovery medium which was injected simply traversed the more permeable layers of reservoir rock bypassing the less permeable layers. Bypassing is also, of course, a serious difficulty with secondary recovery methods, but it is somewhat less fatal. Since water and methane are cheap, it is possible to circulate them to no avail through the permeable beds while waiting for part of the oil to come out of the less permeable layers. All tertiary recovery fluids are expensive. Some of them are viscous in order to control mobility ratio. This can result in no flooding of the less permeable zones.

The reports of the EOR projects funded in part by the DOE often conclude with the statement that the "reservoir heterogeneity was much more dominant than expected". "Geologic and permeability heterogeneities are the most probable causes of the low recovery efficiency realized" (Goodrich and Watson, 1980). "Reservoir characterization is at least as important as process characterization, if not more so, to the overall understanding and interpretation of a field test" (Claridge and Lohse, 1983).

In spite of the fact that reservoir heterogeneity is the most common cause of failure, many lists of "target reservoirs" (Hicks and Foster, 1979; Goodrich and Watson, 1980) and tables of "screening" criteria do not mention it. It is inexcusable to start an EOR project without obtaining as much information as possible on the shape, size, and heterogeneity of the reservoir.

Types of heterogeneity

The most common type of reservoir heterogeneity is simply variations in permeability in the reservoir rock, both lateral and vertical. Besides this, some reservoirs are broken by faults which act as barriers. Often the reservoir consists of a series of channels which pinch out laterally, partially, or completely.

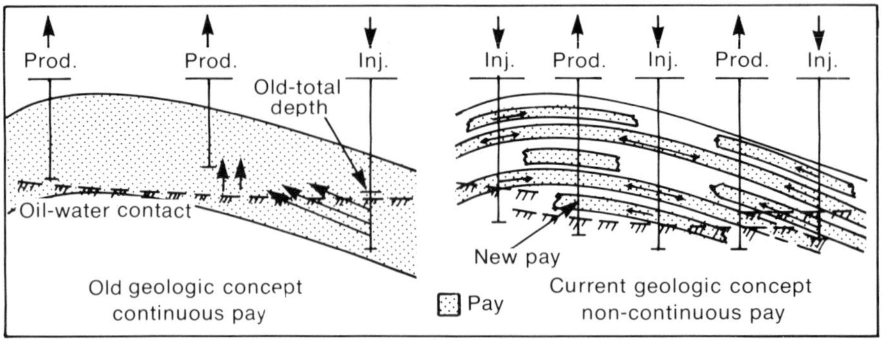

Fig. 2-1. Old and new geologic concepts of pay continuity (After Ghauri et al., 1974, p. 596, fig. 3; courtesy of the Society of Petroleum Engineers of AIME.)

Sometimes there is a system of parallel fractures. These are normally tightly closed, but can easily be opened by the pressure of the injected fluids.

Almost all sandstones and many carbonates are laminated, that is, they consist of horizontal layers of differing permeability. These layers usually are separated by impermeable beds of shale which are often so thin that they are not recorded on wireline logs, and sometimes are not even noticed in cores. They normally divide the reservoir into a series of independent subreservoirs, each with its own porosity and permeability. The presence of the horizontal impermeable barriers is beneficial to secondary and enhanced recovery methods, because it keeps the injected fluids flowing parallel to the stratification. If it were not for them, all the injected fluid would flow across the bedding to the most permeable layer, and the vertical sweep efficiency would be seriously reduced.

The stratification turned out surprisingly effective in the Permian dolomites of West Texas when waterflooding was started (Ghauri et al., 1974; Barrett et al., 1977). It was supposed that the entire thickness was uniform dolomite. When peripheral waterflooding proved ineffective, geological studies showed that the water would not seriously under-run the oil, and pattern floods have been remarkably successful (Fig. 2-1).

The geometry of the pores has an effect on the efficiency of EOR methods. Very often the pores are partly filled with authigenic clays which have a microporosity of their own. This microporosity is filled with immobile water, and is not available for storage of oil. The enormous surface area of these fine flaky crystals adsorbs surfactants.

If the ratio of the diameter of the pore throats to the diameter of the pores is small, the efficiency of the recovery is reduced. The wettability of the pore surface affects recovery efficiency, but this effect is poorly understood because it is almost impossible to determine which portions of the surface are oil-wet and which ones are water-wet in the subsurface.

In carbonates, pore properties sometimes have a major effect. Many carbonate reservoirs, like those in the Permian of West Texas, have quite uniform intergranular porosity, and behave like sandstone reservoirs *. In other cases, carbonates have a dual porosity, with most of the oil in the low-permeability matrix, which is traversed by high-permeability solution channels or fractures. Such rocks have responded poorly to waterflooding and probably will not respond to enhanced recovery. The injected fluids simply go through the permeable channels without penetrating the matrix.

Mathematical attempts to express permeability stratification

When EOR methods began in the late 1940's to be applied to actual reservoirs, the results were disappointing because they were poorer than the recoveries in laboratory experiments, most of which were one-dimensional. The laboratory experimentalists postulated the concept of "conformance", a numerical expression which would quantitatively relate field results to predictions from laboratory tests. This concept illustrated both poor science and poor engineering. In science the mathematics should conform to nature, not the other way around, and any engineering structure should be designed to conform to the local terrain, not to some ideal situation.

Later, other efforts were made to express permeability variations from core analysis in some quantitative way. A good review of these methods is given in Craig (1971). The most commonly used "coefficient of permeability variation" was proposed by Dykstra and Parsons (1950). They defined the permeability variation V as follows:

$$V = \frac{\bar{k} - k_\sigma}{\bar{k}} \tag{2-1}$$

where $\bar{k}$ is the median permeability, and k_σ is the permeability at 84.1% of the cumulative sample. A completely uniform system would have a value of zero.

Permeability variation, like many other geologic properties, usually has a log-normal distribution. The mean of a log-normal distribution is the geometric mean. Thus the average permeability, $\bar{k}$ is:

$$\bar{k} = \sqrt[n]{k_1 \times k_2 \times k_3 \ldots \times k_n} \tag{2-2}$$

None of these mathematical expressions, however, is very useful in predicting the sweep efficiency. The weighted average of the permeability–thickness (kh) product determined from transient well-testing is of questionable value. It is like

* In carbonate reservoir rocks, however, permeability perpendicular to bedding is often about equal to that parallel to the bedding (G.V.C.).

the man who drowned trying to cross a river having a weighted average depth of one meter. In many, perhaps the majority, of reservoirs, most of the primary oil production has come from a small fraction of the total sand thickness as estimated from well logs. Most of the remaining oil is in the tight sand, and is quite unrecoverable. There is, therefore, a tendency to overestimate the effective reservoir volume and thus the volume of residual oil. Possibly the primary and secondary recoveries have been more effective than originally was thought.

The fact is that many oil reservoirs were deposited in the form of either river channels or beaches and are, therefore, limited laterally as well as vertically. In simulating a tertiary recovery project, it will simply be misleading to estimate a coefficient of permeability variation. The actual size and shape of the reservoir and the pattern and manner of completion of injection and producing wells must be designed to sweep the permeable sands.

EXAMPLES OF GEOLOGICAL FACTORS IN ENHANCED RECOVERY PROJECTS

From many descriptions of enhanced recovery, examples have been chosen to show the effects of reservoir heterogeneities. Often these resulted in the failure of the project. Sometimes the project was designed to conform to the geology of the reservoir.

Hydrocarbon solvents

When two fluids are miscible, there is no capillary pressure between them. When a bank of propane and butane (LPG) is passed through a formation containing oil droplets, it will pick up the oil and result in 100% recovery.

The main difficulty with hydrocarbon solvents is their low viscosity and, thus, a poor mobility ratio with respect to the oil and water in the formation. Inasmuch as fingers of solvent penetrate ahead, no proper bank of oil is formed. If gas is used as a fluid to drive the LPG, the mobility ratio is even worse, and the LPG tends to evaporate back into the gas. As long as the LPG is confined in a one-dimensional pipe in the laboratory, it works deceptively well.

The only geological situations where LPG would be successful would be those with steep dip where the LPG slug followed by gas drive would be stabilized by gravity, such as a tall carbonate reef with good vertical permeability, or a steeply dipping stratum on the flank of a salt dome. By injecting gas in a gas cap, a layer of LPG might be forced downward through the reef, dissolving and picking up any oil left behind after gravity drainage, in effect washing the rock. Several attempts have been made to accomplish this in the Devonian reefs of Alberta, Canada.

The Golden Spike pool is one of a line of pinnacle barrier reefs extending from Redwater to Rimbey (Reitzel and Callow, 1977). It has an area of 1385 acres (561 ha) and an average thickness of 480 ft (146 m). It was discovered in 1949.

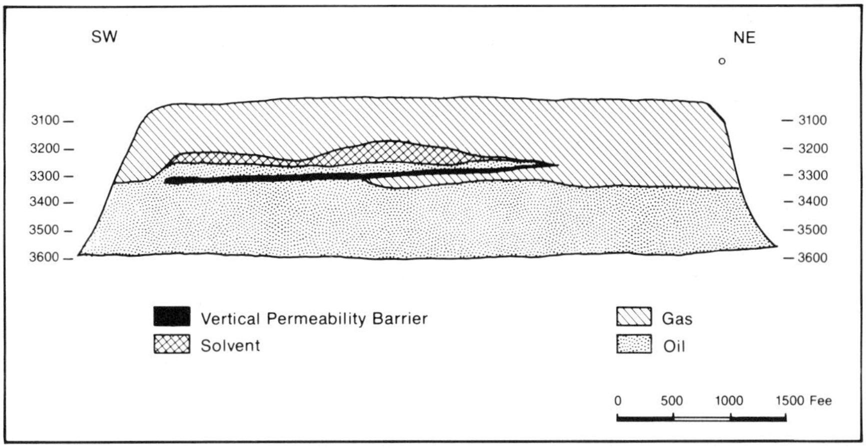

Fig. 2-2. Distribution of gas, solvent, and oil in the Golden Spike reef in 1973. (After Reitzel and Callow, 1977, p. 869, fig. 5; courtesy of the Society of Petroleum Engineers of AIME.)

Pressure maintenance by gas injection began in 1953, which resulted in a secondary gas cap. There was neither any original gas cap nor bottom water. Since the reef was not dolomitized and appeared to be homogeneous, it seemed to be a good candidate for miscible LPG slug injection. It was decided to inject a 7% hydrocarbon pore volume solvent bank between the secondary gas cap and the oil zone. The injection started in 1964.

Additional drilling in 1972, and later, showed that the miscible bank was not performing as expected. In 1973 the oil surface was 130 ft (40 m) deeper on one side of the pool. There was an impermeable zone 325 ft (107 m) above the base of the reservoir. The gas cap had under-run this barrier, bypassing some of the oil and one-half the injected solvent. The remainder of the solvent had dispersed into the gas cap. This situation is shown in Fig. 2-2.

A detailed geological study was then undertaken, which showed that the reef consisted of a variety of lithologic types. There were several other, less extensive, horizontal permeability barriers. Some of these had thin layers of oil on top of them which were flowing off and rejoining the oil column. The 30-million-barrel solvent bank recovered an additional 10 million barrels instead of the 69 million barrels anticipated. The small zone swept by the solvent had recovered more than 90% of the oil in-place. Reitzel and Callow (1977) reported that the final recovery will be 67% of the original oil in-place, although the final results of the project have not yet been reported.

One of the few apparently successful projects using hydrocarbon solvents is being conducted by Texaco (Martin and Young, 1982). The Wizard Lake field is another of the line of tall barrier reefs in Alberta, Canada. The reef was discovered in 1951. The oil column stood 648 ft (197.5 m) above the Cooking Lake aquifer. There was no original gas cap. Production continued from 1951 to

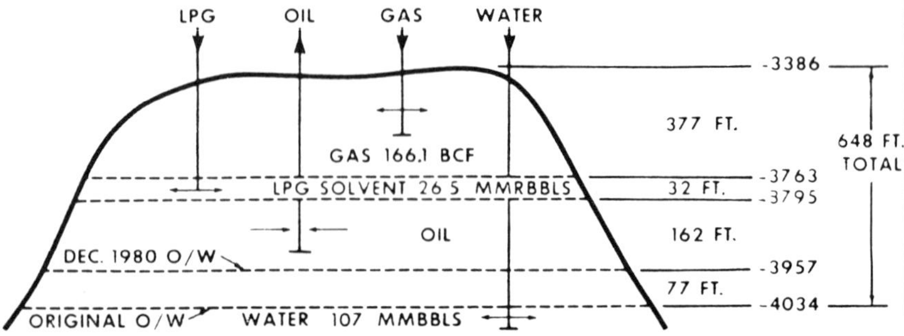

RESERVES STATISTICS	ERCB RECOGNIZED
ORIGINAL OIL-IN-PLACE, MSTB	385,000
EXPECTED PRIMARY, MSTB	254,000
RECOVERY FRACTION	0.66
INC. MISCIBLE GAIN, MSTB	69,000
MISCIBLE RECOVERY FRACTION	0.179
TOTAL RECOVERY, MSTB	323.000
ULTIMATE RECOVERY FRACTION	0.839
PROD. TO 80-12-31, MSTB	234,750
STAGE OF DEPLETION, %	69.7

Fig. 2-3. Wizard Lake reef, showing distribution of oil, gas and solvent in 1980. (After Martin and Young, 1982, p. 550, fig. 8; courtesy of the Society of Petroleum Engineers of AIME.)

1969 by a combination of gas expansion, water drive, and gravity segregation. By this time there was a 78-ft (24 m) secondary gas cap. The pressure had dropped from 2270 psi (15,651 kPa) to 1878 psi (12,948 kPa).

Because of the small areal extent of the pool and its great vertical height it was an ideal candidate for a solvent slug miscible process. There appeared to be no correlatable horizontal layers which might act as permeability barriers. It was expected that primary recovery would be 66% of original oil in-place, whereas with the miscible displacement it would be 84%.

LPG, mostly propane and butane, was injected beginning in 1969. When complete it will constitute 7.4% of the hydrocarbon pore volume. The reservoir pressure will be maintained at 2100 psi (14,479 kPa). The oil–water contact will be raised slightly to −3960 ft (−1207 m) from the original −4034 ft (−1230 m) subsea. The porosity of the dolomitized reef averages 9.5%. Connate water saturation was 7%. Permeability data are uncertain because of the vuggy nature of the reservoir rock, but appear to be high from the drawdown tests. During 1980, 115,534 bbl of solvent, 2,302,771 reservoir barrels of push gas, and 838,169 bbl of water were injected. Oil production was 1,475,528 bbl, whereas gas production was 1,037,861 Mcf. Fig. 2-3 shows the situation in 1980.

The Swan Hills reef of Alberta has a complex pattern of lithologic types. Alternating solvent and water injection parallel to the bedding has been at least partially successful, in spite of severe gas and water breakthrough (Griffin and Cyca, 1980).

Carbon dioxide

The Scurry reef is a mound of bioclastic debris of Pennsylvanian age 23 miles long, 4–8 miles wide, with a maximum thickness of 1535 ft (468 m). The maximum relief above the oil–water contact is 765 ft (233 m); the average relief above the water is 230 ft (70 m) (Goodrich, 1980).

The reservoir rock consists of fragments of calcareous organisms ranging from submicroscopic size to several millimeters in diameter. Its primary porosity was destroyed by recrystallization, and the secondary porosity consists of small interconnecting vugs. The average porosity is 7% and the average permeability is 15 mD. There are two distinct lithologic types: one has an average porosity of more than 4.5%, and the other, which contains some shaly zones, has less than 4.5% porosity. The low-porosity zones constitute vertical permeability barriers. Some can be traced for considerable distances. Fig. 2-4 shows the porosity layering in a particular area of the field.

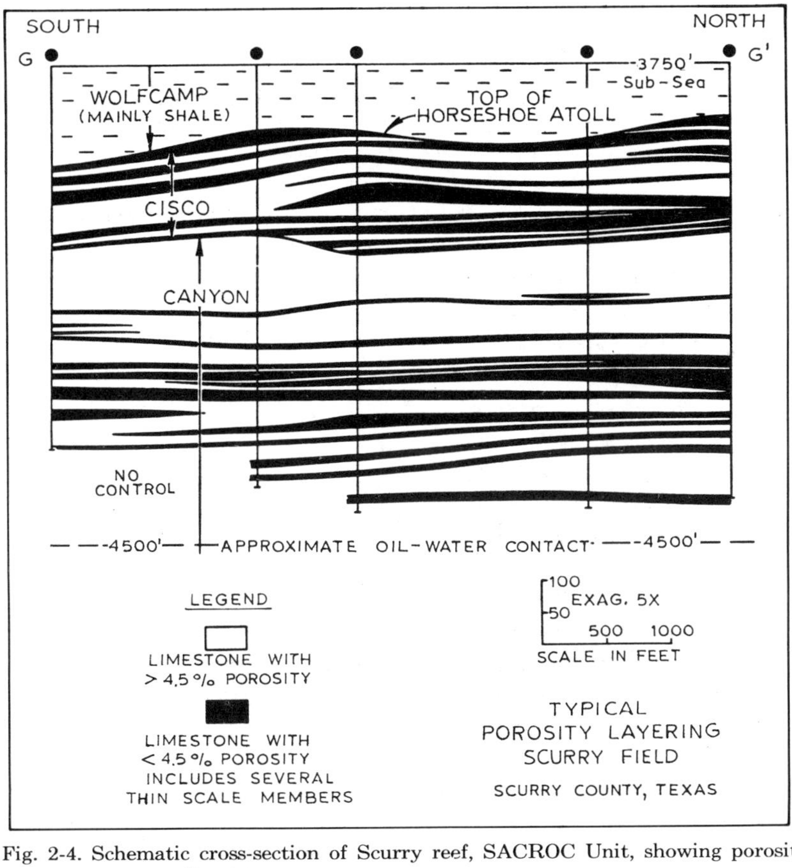

Fig. 2-4. Schematic cross-section of Scurry reef, SACROC Unit, showing porosity layering. (After Vest, 1970, p. 200, fig. 15.)

From the field's discovery in 1948 to 1954, 129 million bbl of oil were produced, but the reservoir pressure dropped from 3137 psi (21,614 kPa) to 1560 psi (10,748 kPa). Water injection was started in 1954. A line of water injection wells was installed right down the center line of the field. By 1971, an additional 410 million bbl of oil had been produced, and much of the reservoir pressure was restored.

In 1971, it was decided to inject alternating slugs of CO_2 and water, mostly into the outside areas that had not been swept by the centerline waterflood. The pattern was an inverted 9-spot, as shown in Fig. 2-5. In Phase 1, there was an

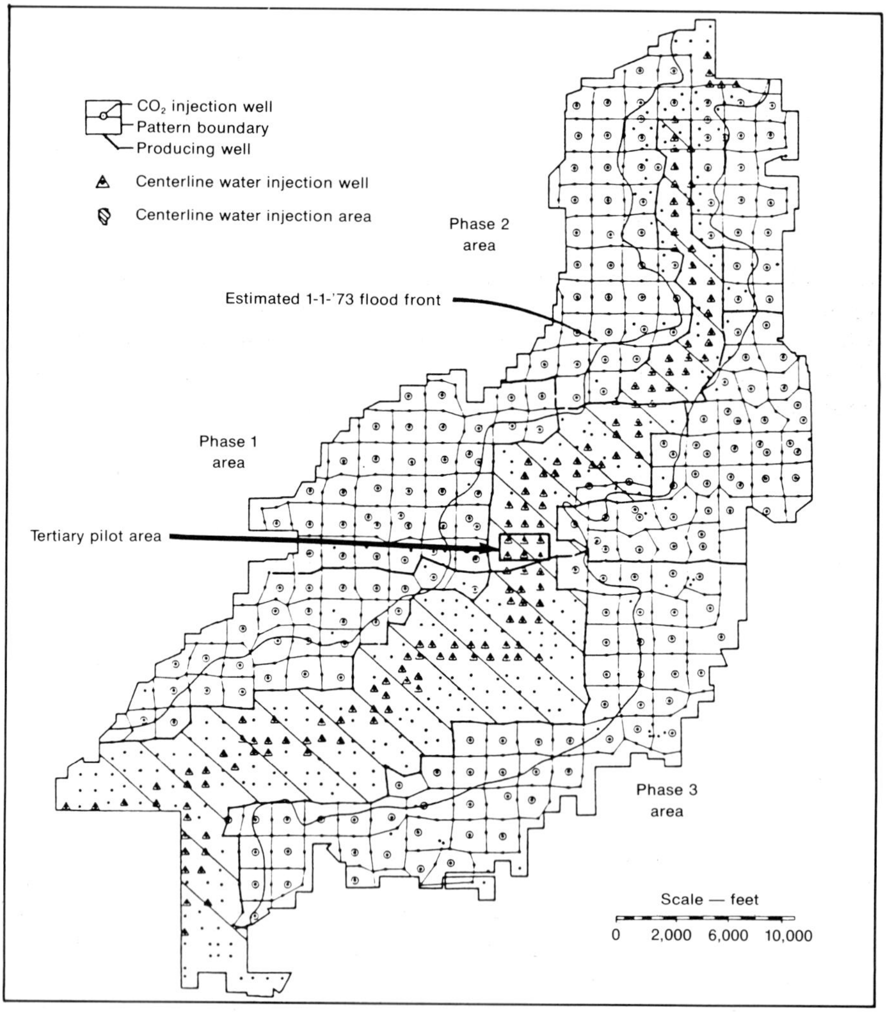

Fig. 2-5. Map of SACROC Unit, showing pattern of water and CO_2 injection wells. (After Goodrich, 1980, p. 71, fig. 7.)

immediate rise in oil production, but also immediate breakthrough of CO_2. To correct the poor injection profiles, about 80 injectors were recompleted with liners and selective injection equipment.

Although oil production increased with the new injection of water and CO_2, it is impossible to tell how much of the increase was due to the CO_2 and how much to the infill drilling and water injection. The disappointing results are ascribed in part to the fact that "reservoir heterogeneity is much more dominant than expected".

Several other CO_2 floods have shown promise, because the CO_2 mobilized residual oil after waterflood; some, however, were failures because of reservoir heterogeneity. In Granny's Creek, West Virginia, the injected CO_2 spread rapidly over some 200 acres (Goodrich, 1980). Geologic and permeability heterogeneities are the most probable causes of the low recovery efficiency which was realized. Calculations show that only 6–10% of the injected CO_2 entered the pattern.

In South Gillock, Texas, on the flank of a salt dome, premature breakthrough of CO_2 was attributed to complicated fault geometry, and the project was terminated (Goodrich, 1980).

In Little Creek, Mississippi, the Tuscaloosa Sand appears to be a point bar accumulation about one mile wide and 40 ft thick (Goodrich and Watson, 1980). Primary production and waterflooding recovered 46.8% of the original oil in-place. After the field was entirely watered out, a single CO_2 injection well (well 110) was installed near the outside of a bend in the channel (Fig. 2-6), so that the sand pinched out on the east and south. Water was injected on the north and west. The CO_2 mobilized and moved the oil to wells 111, 116 and 117, producing the remarkable recovery of 62% of the remaining oil in-place, which was only 21% of pore volume. The cumulative CO_2/oil ratio, however, was 17.1 Mcf/bbl, which was regarded as excessive. Nevertheless, the CO_2 breakthrough did not become unreasonable for a year, so that serious bypassing did not occur. This project may be considered an example where a good geological picture of the reservoir led to an understanding of what was going on. It might be questioned whether or not the injection well should have been located in the thin sand at the edge of the field.

Micellar–polymer

The El Dorado Micellar–Polymer Demonstration Project was conducted by Cities Service Company from June 1974, and was still in operation in September 1983 (Jordan and Tillman, 1982; Claridge and Lohse, 1983; Van Horn, 1983). It was proposed as a side-by-side comparison of two different micellar–polymer processes. The oil recovery was very disappointing: less than half of the original predictions. The failure was attributed primarily to the failure of the preflushes to form the salinity environment required for the micellar systems. The failure occurred in spite of the injection of about 40% of the reservoir volume of preflush

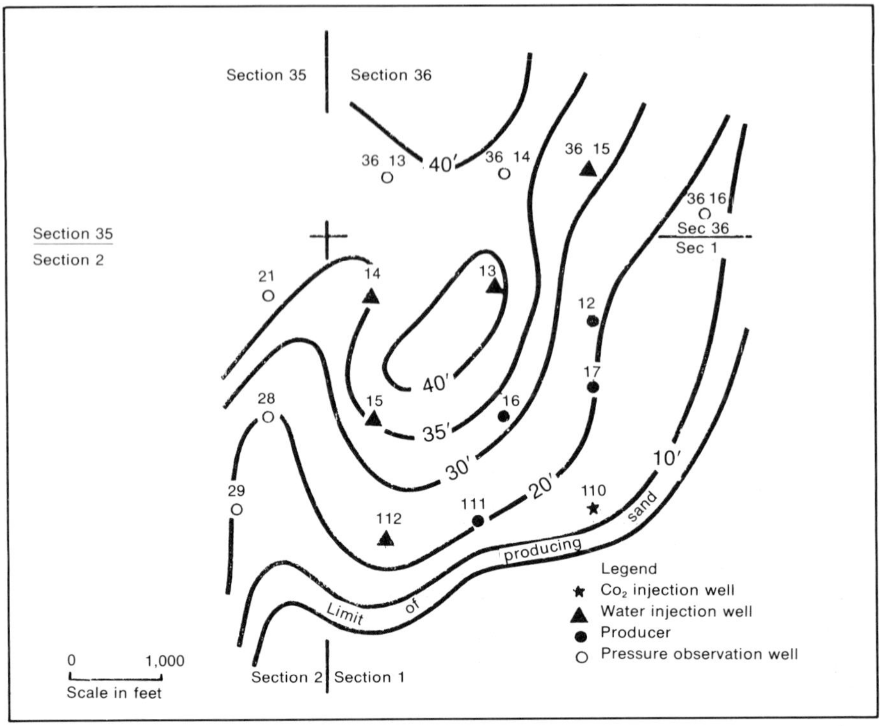

Fig. 2-6. Map of Little Creek CO_2 injection project. The CO_2 injection well (110) is close to the pinchout of the sand, and water injection wells 112, 15, 14, and 13 confine the pattern. (After Goodrich and Watson, 1980, p. 217, fig. 8.)

solution. This failure was ascribed partly to adsorption and reaction of the surfactants on the sand, and partly to reservoir heterogeneity.

An evaluation of the project (Claridge and Lohse, 1983) stated that reservoir heterogeneity gave rise to repeated numerical simulations of doubtful value. It was concluded that "Reservoir characterization is at least as important as process characterization, if not more so, to the overall understanding and interpretation of a field test".

When it was realized that the reservoir was much more heterogeneous than had been supposed, a very detailed geological study was made by Jordan and Tillman (1982). The 650-ft thick Admire sand was deposited in a deltaic system. Its lowest sediments were interdistributary bay muds. Distributary channels prograded into the area, and at their margins spread thin splays of sand. Although porous and permeable they are thin and discontinuous. Eventually distributary channels over-rode the splay deposits. Although these distributary channels contained high-energy sand with good porosity and permeability, they were irregular. There is no distributary channel sandstone in the southeast part of the project. A major channel flowed northward along the western margin of

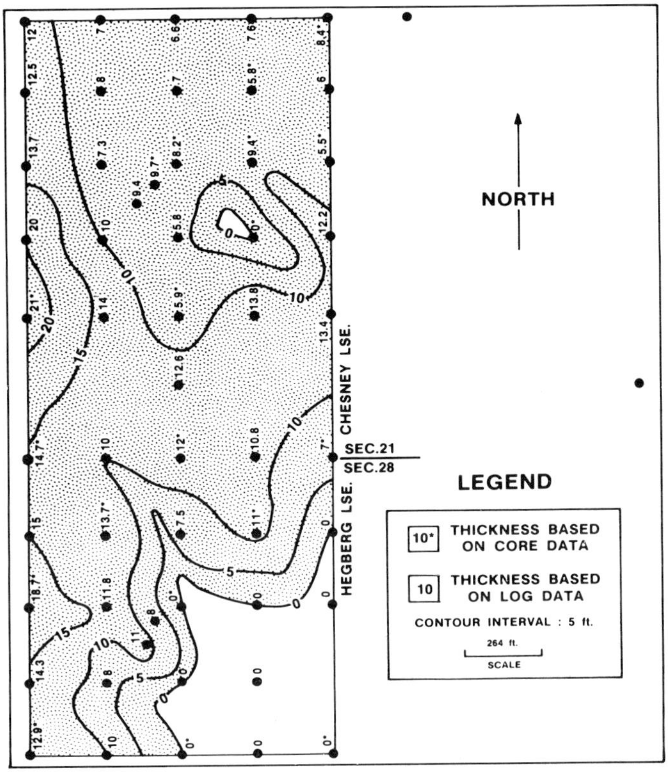

Fig. 2-7. Map showing thickness of channel sand in El Dorado micellar–polymer project. (After Jordan and Tillman, 1982, p. 562, fig. 41.) Scale represents 80 m.

the area and branched to the east near the north edge of the south pattern (Fig. 2-7). This channel contained the bulk of the recoverable oil. Its presence could have been suspected by plotting the cumulative water injected during the secondary waterflood (Fig. 2-8). A fence diagram showing the thickness and extent of the various facies is shown in Fig. 2-9. It is important to note that the channel sand is missing in well 122.

As the currents in the channels decreased later, they were covered with fine-grained silts and shales. Finally, as the water became clear limestone was deposited.

The Bell Creek Field of Montana was recognized as a barrier island shortly after its discovery in 1967 (McGregor and Biggs, 1968). It strikes NE-SW, and is about 15 miles (24 km) long and 1–3 miles (1.6–5 km) wide. It is in the Muddy Sand of Cretaceous age. The sea lay to the northwest, with a lagoon to the southeast. The dip is to the northwest, and there was edge water along the northwest edge of the field.

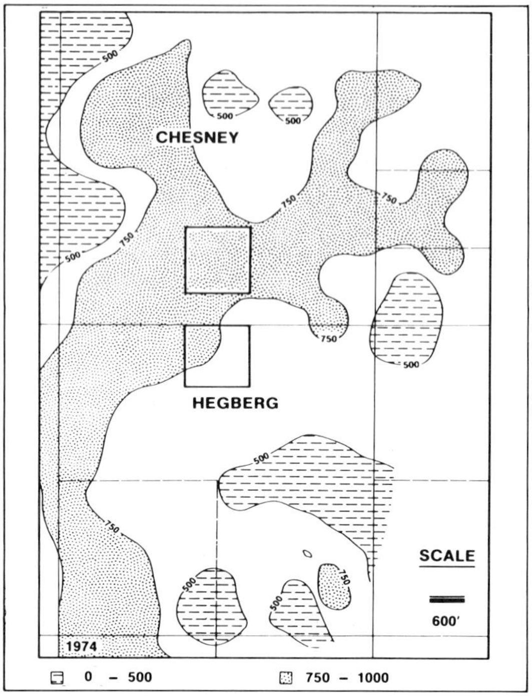

Fig. 2-8. Map showing cumulative water injection in El Dorado micellar–polymer project. The location of the permeable channel is clearly shown. (After Jordan and Tillman, 1982, p. 560, fig. 39.)

A line-drive waterflood was started in August 1970, with water being injected downdip along the northwest edge of the field. In 1977, a micellar–polymer project was developed. It consisted of a 40-acre, 5-spot pilot, with 4 injection wells and an observation well near one of the injection wells. The Muddy Sand in the 40-acre area is 4650 ft (1417 m) deep and has an average thickness of 6.4 ft (1.95 m). Permeability is 1050 mD, porosity is 29.1%, and oil has 32.5° API gravity.

Conventional water injection was started in October 1977, and continued until January 1979 in order to establish the flowlines for a 5-spot instead of a line drive (Goldburg, 1982).

In spite of its apparent uniformity and the success of the line-drive waterflood, the sand body turned out to be very complex. There were several different, highly permeable, narrow channels, striking more or less in the north-south direction (Fig. 2-10). Sometimes they overlapped each other, but the southeast corner of the 5-spot was completely separated from the rest by a permeability barrier (Fig. 2-11). This barrier was first evident from pulse tests and tracer breakthrough times. An Intercomp Chemical Flood Simulation made in 1979 did not match the field performance, apparently because of the complex geology.

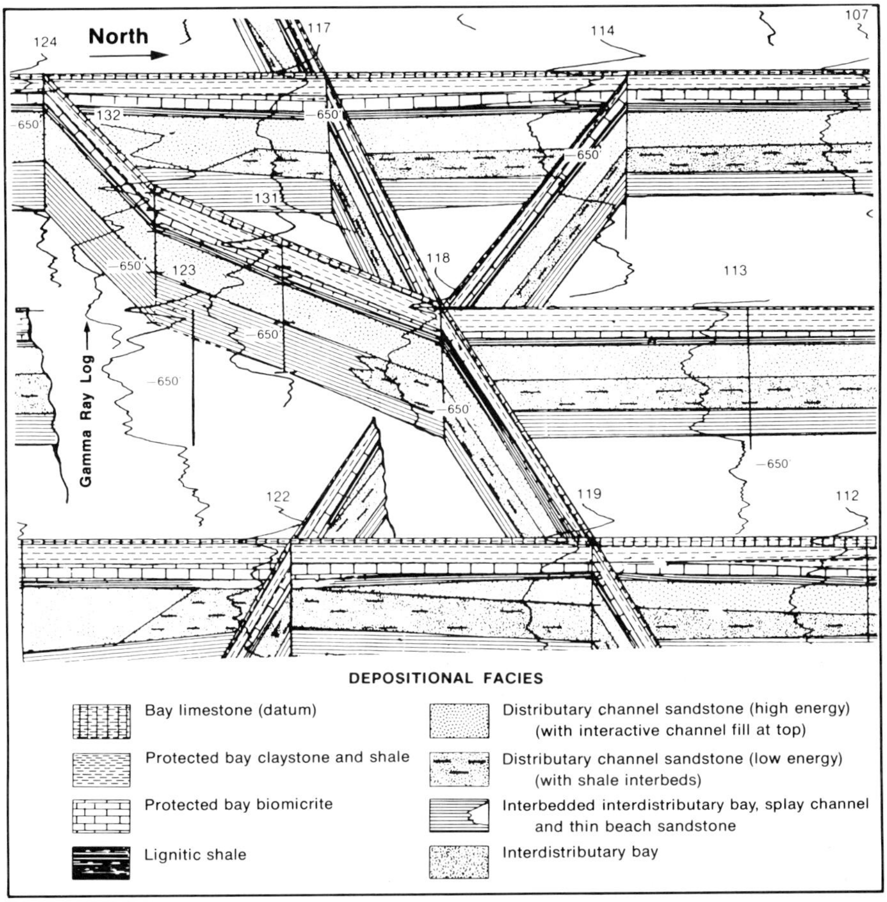

Fig. 2-9. Fence diagram showing different lithologic types making up the Admire Sand in the El Dorado micellar–polymer project. (After Jordan and Tillman, 1982, p. 561, fig. 40.)

Some of the wells showed two or three different increases in the oil cut at different times, suggesting that the fluids came through the different channels at different rates.

Pennzoil's micellar–polymer project No. 7 and No. 8 in the Bradford Field (Pennsylvania) are excellent examples of how the geology of the reservoir was taken into account in the design of an enhanced recovery project (Guckert and Paynter, 1982). The Bradford Field was discovered in the 1870's and produced about 250 million bbl of primary oil. Waterflooding was begun about 1921 and resulted in an additional recovery of 417 million bbl. The field is now mostly watered out.

The sand in the project area is fine-grained, containing abundant rock fragments; it has an average porosity of 15%, permeability is 10 mD, net sand

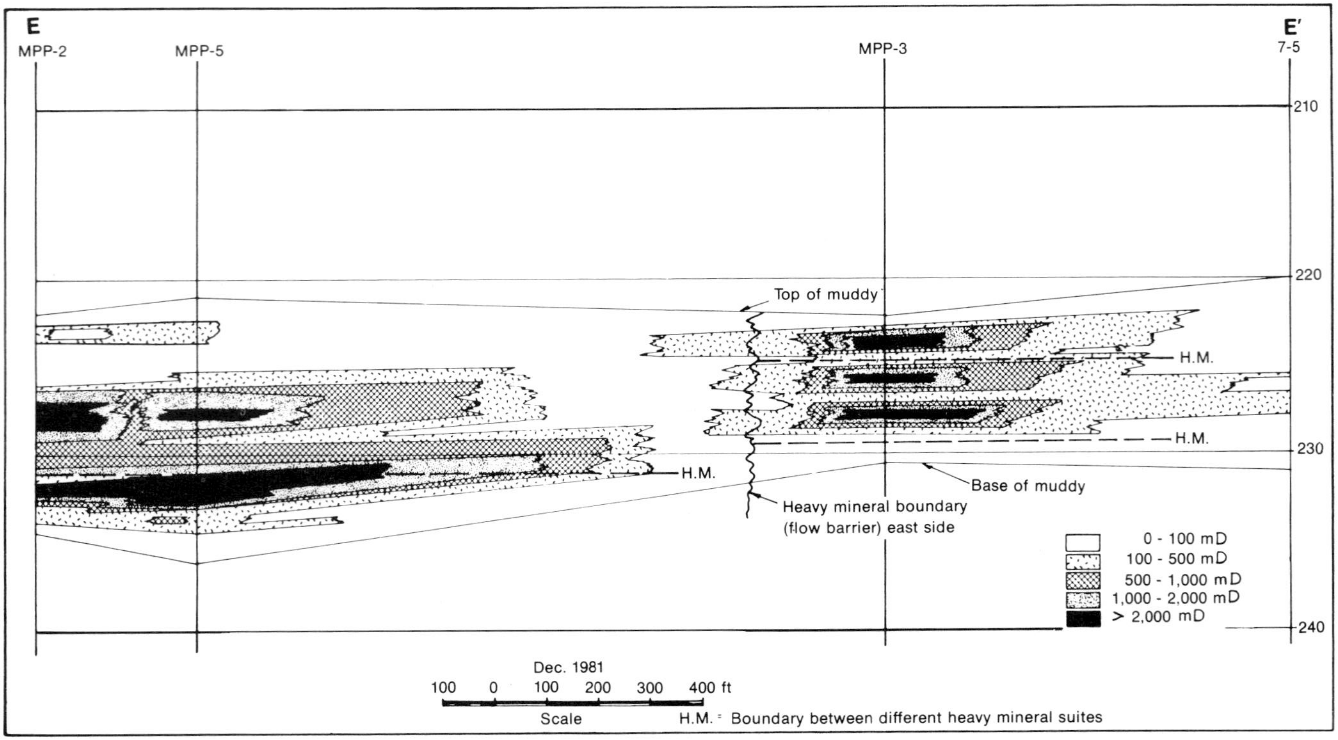

Fig. 2-10. Cross section of Bell Creek Field, Montana, showing layers of permeable sand. (After Goldburg, 1982, p. 62, fig. 5.)

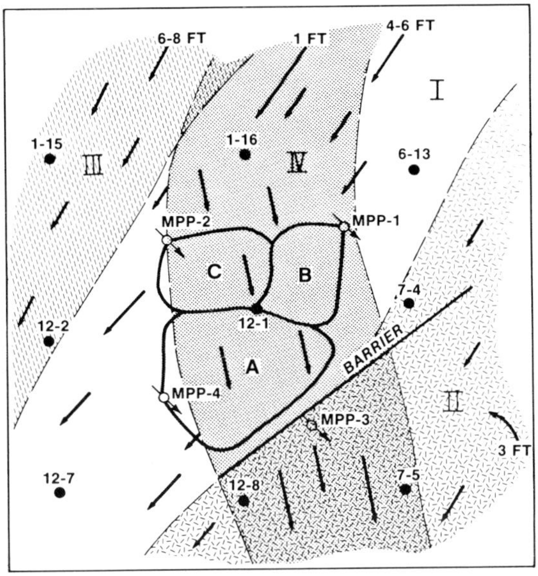

Fig. 2-11. Map of pilot micellar–polymer project in Bell Creek Field, showing channels and barrier in Muddy Sand. (After Goldburg, 1982, p. 63, fig. 6.)

thickness is 30 ft, and residual oil saturation after waterflood is 40%. It is predominantly oil-wet. Apparently it was deposited as a parallel series of shoreline sands striking NE-SW, at the edge of the Catskill delta of Devonian age.

The Bingham leases are in the southeast lobe of the field, comprising 816 acres (330 ha). Primary oil recovery was about 2000 bbl/acre (823 t/ha) and waterflood oil was 6300 bbl/acre (2592 t/ha). The first pilot test was a 3/4-acre, inverted 5-spot pattern with one injection well and 4 producing wells, in a watered-out area. Six percent of the pore volume of micellar solution was injected, beginning in 1968, following by 140% pore volume of polymer solution. The results were encouraging: all the production wells showed increased oil cuts and oil production rates. The project was then extended to a 46.2-acre pilot with 16 injection and 25 production wells in a 3-acre, 5-spot pattern.

The coring and logging program showed that the Bradford sand can be divided into 3 distinct layers (Fig. 2-12). The middle zone ranges from 6 to 25 ft in thickness and has an average porosity of 18% and permeability of 120 mD. The upper and lower zones together have a thickness of 22 to 37 ft with an average porosity of 13% and permeability of 9 mD. It was found that nearly all of the injected fluids entered only the middle zone. Whereas the micellar fluids entered easily, the polymer severely damaged the injectivity of the injection wells, and high pressures were required. However, the results were good. Ultimate oil recovery is estimated to be about 5000 bbl/acre (2000 t/ha) or 340

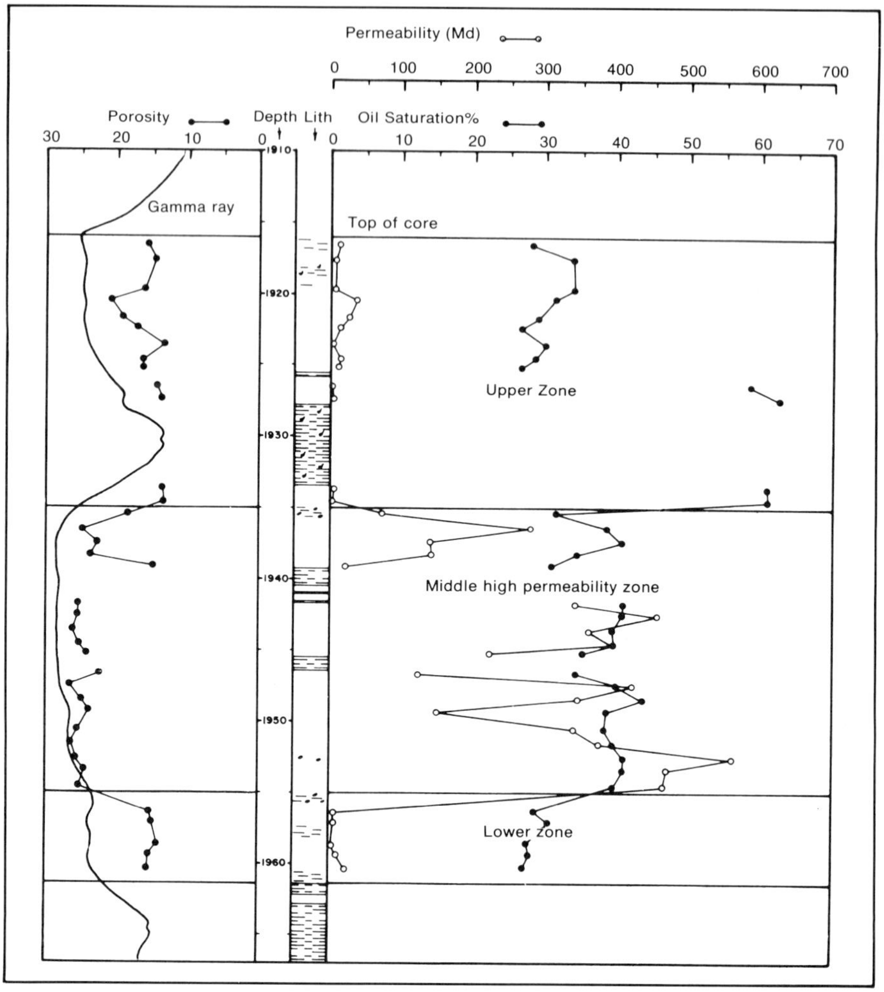

Fig. 2-12. Core analyses of Bradford Third Sand at Bingham micellar–polymer project, showing 3 zones. Only the middle zone is permeable. (After Guckert and Paynter, 1982, p. 567, fig. 5.)

bbl/acre-ft. If the remaining oil was 40% of pore volume, then half of that was recovered.

It was determined that the more permeable middle part of the sand existed in a belt about 2000 ft (600 m) wide, extending NE-SW. The thickness in the center of the channel is 15–20 ft (5–6 m), dropping off to 5 ft or less on the edges. It was decided to extend the project along this belt in the form of a staggered line drive, with a line of injection wells along the axis of the channel and a line of production wells along each edge (Fig. 2-13). The distance between wells in the lines is 140 ft (42 m), and the distance between the lines is 560 ft (170 m).

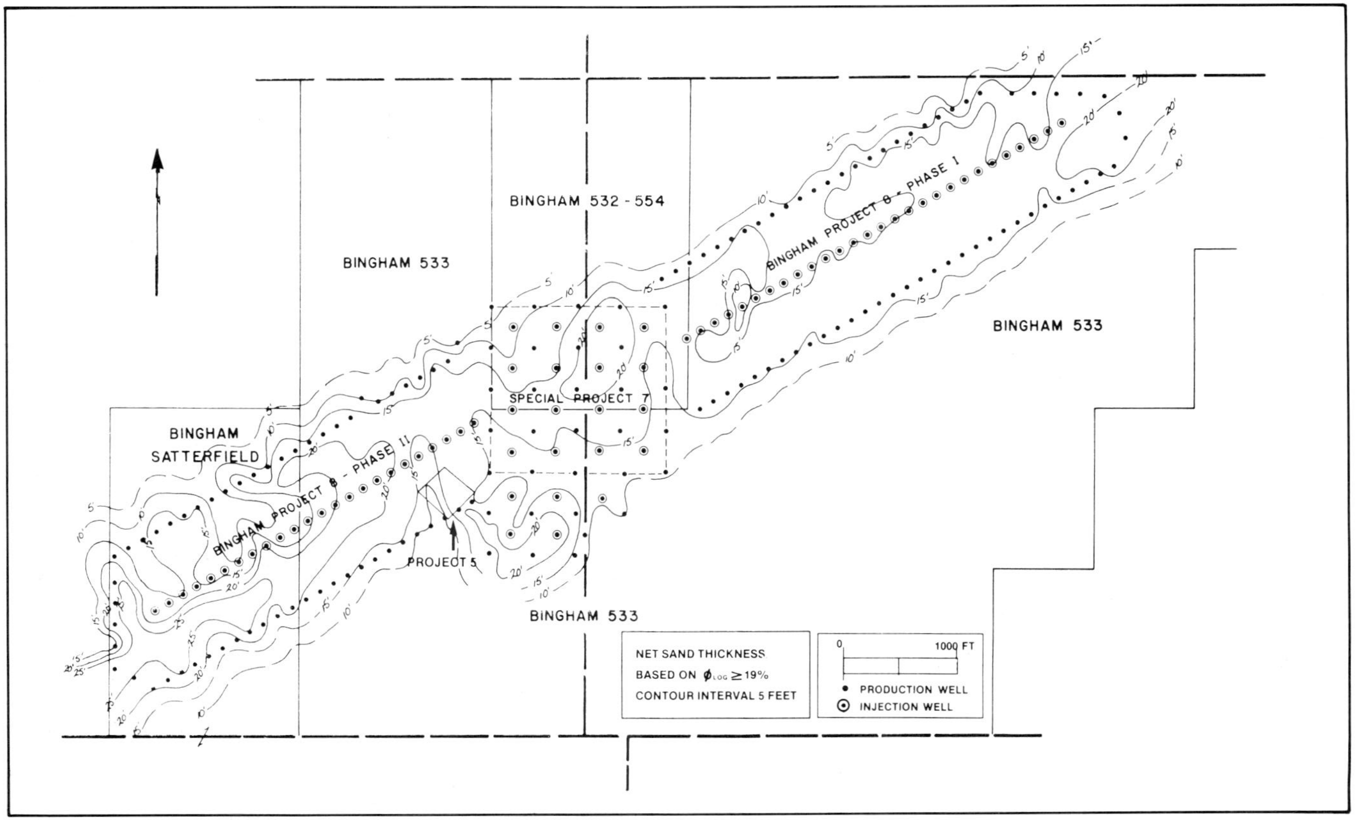

Fig. 2-13. Map showing thickness of permeable zone at Bingham micellar–polymer project and planned pattern. Injection wells will be in the center line of the beach-type sand body. (After Guckert and Paynter, 1982, p. 567, fig. 8.)

Another micellar–polymer project was attempted by Pennzoil on the Lawry lease in the Bradford Field (Ondrusek and Paynter, 1982). The area has been waterflooded, with a recovery of 9000 bbl/acre (3690 t/ha). The net pay sand thickness was 29 ft (8.8 m), with an average porosity of 12.55%. The average permeability was 7.65 mD, with 5 ft (1.5 m) of sand having a permeability of 40 mD and the rest having a permeability of 1 mD or less. The remaining oil saturation after waterflooding was 40%. The more permeable zones could not be correlated from one well to another. Injection rates were less than expected, although the fracture pressure was substantially exceeded in ten of the sixteen wells. It was found that the micellar solution was entering only 5 ft (1.5 m) of the most permeable sand. Because the cumulative recovery was only 5% of the estimated oil in-place, the project was considered a failure. This is a good example of the folly of using average permeability values.

Thermal recovery methods

In thermal methods of recovery, the more permeable bypassing beds are less disastrous. They act as flues, so that the hot fluids can warm the overlying and underlying strata. In the Bodcau in-situ combustion project (Dean, 1982; Trantham, 1982), only 49% of the total pay sand was burned. The lower part of the sand below a limestone layer was burned; however, in the unburned area the saturation of heavy oil was reduced to 32% of pore volume.

The most successful recovery method to date has been the steam injection. In the steam soak (cyclic huff and puff) method, steam is injected into a well for a certain period, after which the well is pumped. In using this method, it is not very important to work out the geology beforehand. The steam simply moves into the more permeable beds and the oil comes back out.

In the continuous steam drive method, on the other hand, formation heterogeneities become very important. In the M-6 project at Tia Juana, Venezuela, the geological data was mapped out before the start of the project (Schenk, 1982a,b) (Fig. 2-14). The Miocene sands of this area contain heavy (11–15° API gravity) oils. Primary production by solution gas and formation compaction drives amounted to 12% of original oil in-place (OOIP). Steam soak production raised this to 23% OOIP. The reservoir (D sands) consists of an intricate stack of medium- to fine-grained channel sands with associated fine-grained natural levee sands. There are many shale beds which are not very continuous laterally. Most of the production during the primary and steam-soak phase is ascribed to formation compaction. The surface over the M-6 area has subsided between 1 and 2 m. If all the compaction took place in the sands, which seems unlikely, the porosity of the sands would have been reduced about 5% of the original value.

The original triangular pattern was on a spacing of 758 ft (230 m) between wells. There are 21 injectors and 130 producers in the project area of 1827 acres (740 ha). The injection wells are all located in areas of thin sand, and the steam is supposed to drive the oil toward the thicker and more permeable zones. After the steam soak and before the steam drive started, gas with tracer was injected

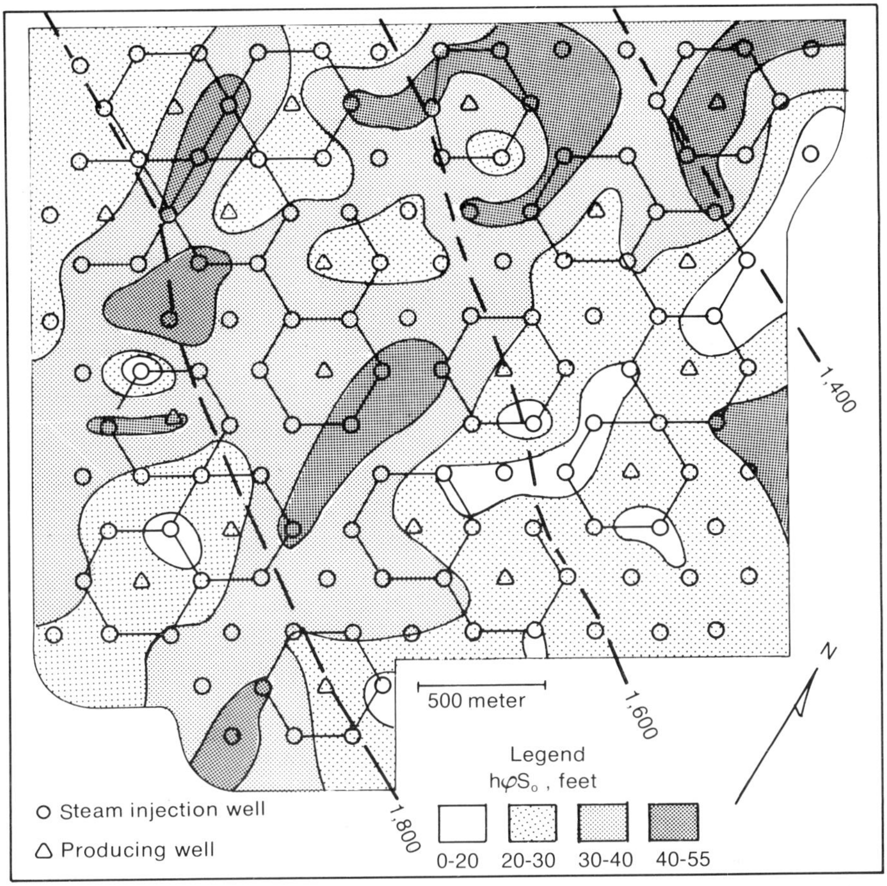

Fig. 2-14. Map of M-6 steam drive project, Tia Juana, Venezuela. (After Schenk, 1982b, p. 504, fig. 4.)

into injection wells. Breakthrough times were short, sometimes only one day at offsetting wells. The channelling was due mainly to the fluvial character of the sand bodies. Of the two pilot hexagons, one had an exaggerated directional flow of steam. After full-scale operation started, it was found that sometimes one or two of the six production wells are likely to react much more strongly than the others. First there is a pronounced increase in oil production, followed by an even larger increase in water production, accompanied by a rapid increase in wellhead temperature. Finally, steam is produced with negligible amounts of oil. The bypassing bed is generally the top 20–30 ft of sand. Efforts to plug it off are difficult, because the producing wells are gravel-packed. As of 1981, 15 producing wells had been recompleted, and the project was performing successfully with about 2 bbl of oil per ton of steam injected.

The Kern River Field of California has been one of the most successful steamfloods. The sand consists of a sequence of braided stream deposits. There are several sets of sands up to 30 ft thick, separated by mudstones. Each of these sets has a well-defined lithologic succession, starting with a conglomerate at the base, then beds of coarse sand and gravel, and finally cross-laminated fine sand (Nicholson, 1980a,b). The individual sand bodies pinch out laterally, sometimes over a distance of as little as 200 ft. The sands are very porous and permeable. The mudstones serve to keep the steam confined to the sandy intervals, which are flooded in succession, starting with the lowest.

There is a tendency for the upper part of the sand to be better flushed than the lower (Bursell and Pittman, 1974) — the residual oil may be reduced to 6 or 7% in the upper part and 20–25% in the lower part. This is probably due to gravity over-ride, but may also be affected by the sand permeability.

A $2\frac{1}{2}$-acre 5-spot pattern has been very successful in spite of the heterogeneity of the reservoir. There are equal numbers of injection and producing wells, which push the steam and oil from every direction toward the producing wells.

According to Restine (1983), the sands above and below the zone which was flooded were "preheated" and tend to flood-out faster, although the ultimate recovery is not improved. The recovery from thin sands may be improved by preheating.

STRUCTURE AND COMPOSITION OF RESERVOIR ROCKS

Important petrophysical properties of petroleum reservoir rock are porosity and permeability. These attributes are commonly conferred on a sediment by the process of washing to remove the fine particles from the coarser ones, leaving pores that are big enough to permit the oil and gas (commonly non-wetting) to enter against the capillary forces which tend to exclude them.

Washing (winnowing) can occur in nature only where there are currents of water or air. Sand in sand dunes is very well sorted, but ancient dune sands seldom contain petroleum. More commonly oil and gas reservoirs were deposited under water. The two principal places where sand can be washed clean are in *river channels* and on *beaches*. The great majority of oilfields consist of sands laid down in one or the other of these environments, which are naturally (a) local and (b) elongated. Most huge oil reservoirs consist of river channel sands lying side by side or stacked on top of each other. Consequently, they are extremely heterogeneous.

The various origins of carbonate reservoir rocks are not as simple as those of sandstones, but most were also deposited in shallow water where waves and tidal currents promoted porosity development. *. In the early 1950's, many geologists

* Diagenetic processes (mainly dolomitization) play a very important role in the development of porosity in carbonate reservoir rocks. Solution porosity is also extremely important. (Editorial comment.)

decided that the key to the understanding of ancient sediments was a study of the modern-day depositional processes. The first comprehensive study was made by F.P. Shepard and his students in the delta of the Mississippi River and the Gulf Coast of Texas (Shepard et al., 1960). Since then many studies of modern depositional environments have been made by both academic institutions and oil companies. As a result, geologists are now able to recognize and characterize the bodies of porous sandstones or carbonates which make good reservoirs. With a good set of cores, or even wireline logs of a single well, one can make guesses as to the form and orientation of the porous rock.

A good summary of this work by Friedman and Sanders (1978) lists almost 8000 references. Another summary edited by Scholle and Spearing (1982) illustrates rock types as seen in cores, in color pictures. A now classical series of papers was published in a special issue of the *Journal of Petroleum Technology*, called "Synergy", in July of 1977. In that issue, R.J. LeBlanc, Sr., one of the pioneers in the study of recent sediments, presented a brief review of the characteristics of sandstone reservoirs which by now should be familiar to all reservoir engineers. In making reservoir simulations it should never be assumed that a reservoir is either homogeneous or isotropic. Channel sands are more permeable in the lower part and tend to be elongated perpendicular to the ancient coast, whereas beach sands are more permeable in the upper part and are parallel to the coast. Many large sand bodies consist of both channel and beach sands.

Sandstone reservoir rocks

Sandstones with porosity and permeability consist of grains of different sizes, most commonly between 0.1 and 0.5 mm in diameter. The pores between the grains often contain some clay-size particles, with diameters less than 0.005 mm. The original sand grains were washed clean, so the clays in the pores have usually been precipitated later by circulating water (authigenic origin). The sand grains in a reservoir are usually quartz, although feldspar and rock fragments are sometimes abundant. The grain composition, however, is not very important. It is the texture, i.e., grain size distribution, that determines the porosity and permeability.

The shape and size of the reservoir are determined by the environment in which the sand was deposited. Fig. 2-15 is a diagram showing the course of an imaginary river from the foot of the mountains to the sea, with the names of the principal depositional environments.

Alluvial fans and braided stream deposits

At the foot of mountains the rushing water during floods carries mud, sand, and boulders, building up a "fan" or conical mound. At the fan's base the water coalesces into a river across the plain. Because it is overloaded with sediments, it continuously changes its course from side to side. There are often multiple

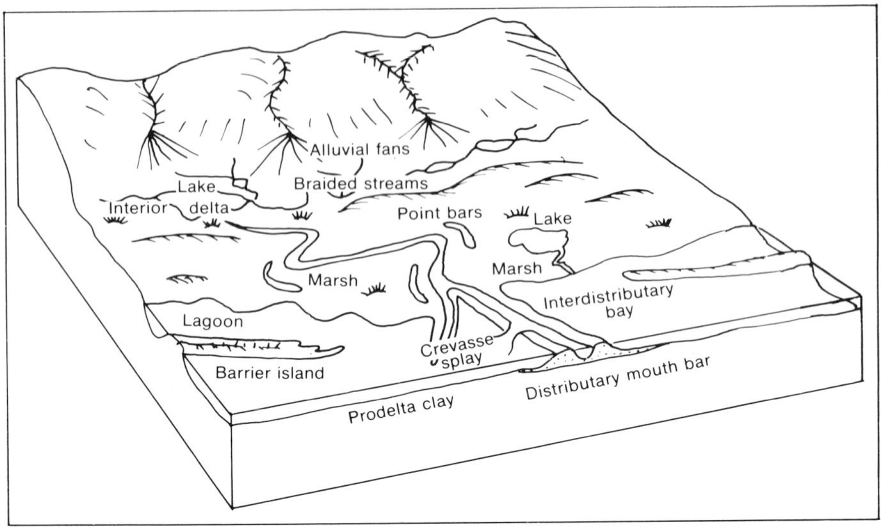

Fig. 2-15. Diagram showing sedimentary environments. (After Dickey, 1981, p. 89, fig. 6-3.)

channels which are continually shifting. Such rivers are common where plains meet mountains. They are especially prominent in the Gangetic Plain at the foot of the Himalayas.

Alluvial fans and braided stream deposits are seldom preserved in the subsurface, because during the processes of erosion and sedimentation mountains are destroyed. The best known alluvial fan oilfield is Quiriquire in eastern Venezuela.

When braided stream deposits are preserved, they make excellent reservoirs; however, they are extremely heterogeneous. As the river shifts from side to side, it deposits coarse sand in one place and clay in another. The best known oilfield in braided stream deposits is Prudhoe Bay in Alaska (Wadman et al., 1979). The 500-ft (150 m) thick sandstones contain four major shale intervals which can be correlated from well to well. There are many other shales which were assumed to extend only a hundred feet or so laterally.

Point bar deposits

When rivers reach the coast, they build deltas. The sediments prograde out to sea and pile on top of each other, so that they are finally buried in the subsurface. Sand bodies in deltas, therefore, are much more likely to be preserved and become petroleum reservoirs. When the stream gradient is low, the rivers wind back and forth in a series of meanders. Upon flooding the flood plain is covered with unsorted sand and clay. After coming back into its bed, the river proceeds to work over the flood plain deposits. It cuts into these deposits on the outside of the bends. Fine clays go into suspension and are swept on downstream, whereas the sand is deposited as a bar on the next inside bend. This sand bar is called a "point bar". Because the meanders are continually moving sideways, the

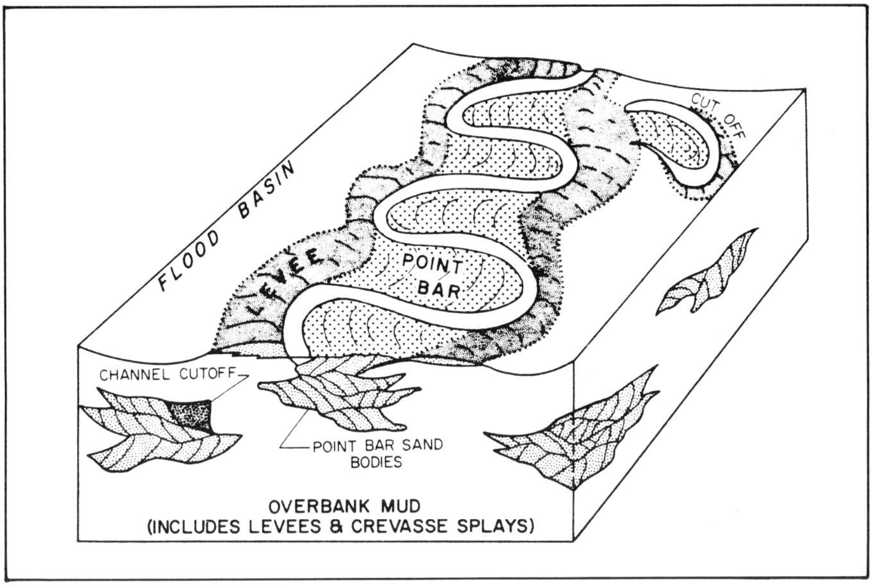

Fig. 2-16. Diagram of the flood plain and meander belt of a river. (After Brown, 1979, p. 56, fig. 17.)

river can clean up a belt of sand along the whole width of the meander belt (Fig. 2-16). Grains at the bottom of the channel on the outside of curves are coarse, whereas on the inside of the curves there is less current and the grains are finer. As the channel migrates sideways, the sand grains deposited in shallow water on the inside of the bend cover the coarse deposits of the outside of the bend (Fig. 2-17). More permeable sand, therefore, is found in the lower part of point bar and other river channel deposits. Fig. 2-18 shows that in a channel sand the permea-

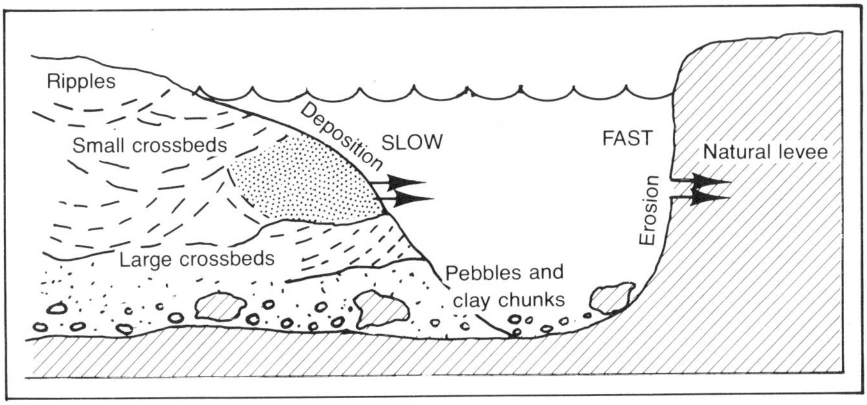

Fig. 2-17. Sediments deposited in river channels. As the river undercuts its bank on the outside of the bend, it moves to the right. The finer sediments deposited in shallow water then cover the coarser sands deposited in deeper water. (After Dickey, 1981, p. 92, fig. 6-6.)

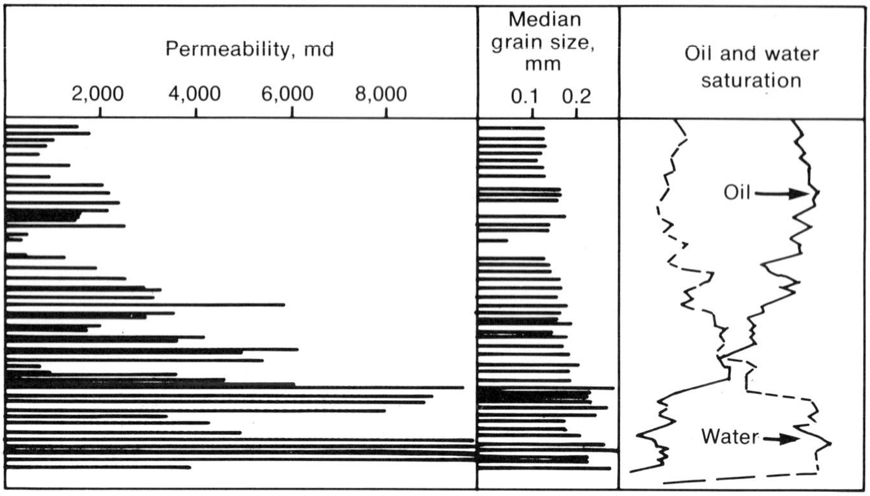

Fig. 2-18. Results of coring a channel sand after waterflooding, Daqing, China. Only the lower, most permeable portion is flushed. (After Jin et al., 1981, p. 26, fig. 15.)

bility is much greater in the lower part. Cores taken in these sands after waterflooding show that all the water had gone through the lower part of the sand, leaving the upper part unswept.

Some oilfields consist of single river channels whereas others are point-bar deposits stacked alongside and on top of each other.

Beach-type deposits

The advance and retreat of waves on a beach (swash) is an excellent way to clean sands. Waves usually hit the beach at an angle, giving rise to a longshore

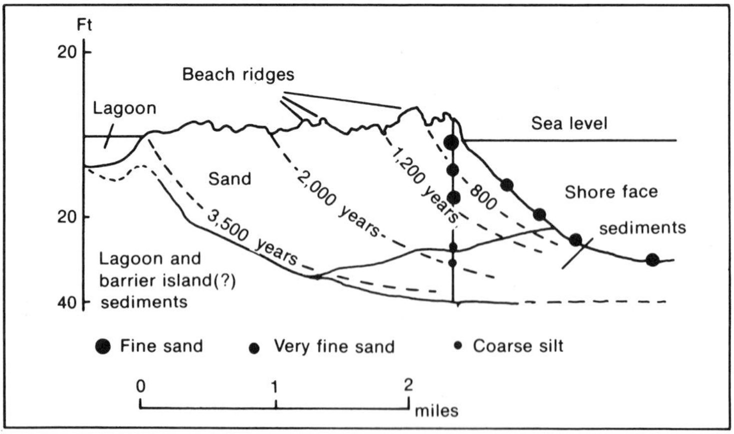

Fig. 2-19. Cross-section of Galveston barrier island. (After LeBlanc, 1977, p. 786, fig. 17; courtesy of the Society of Petroleum Engineers of AIME.)

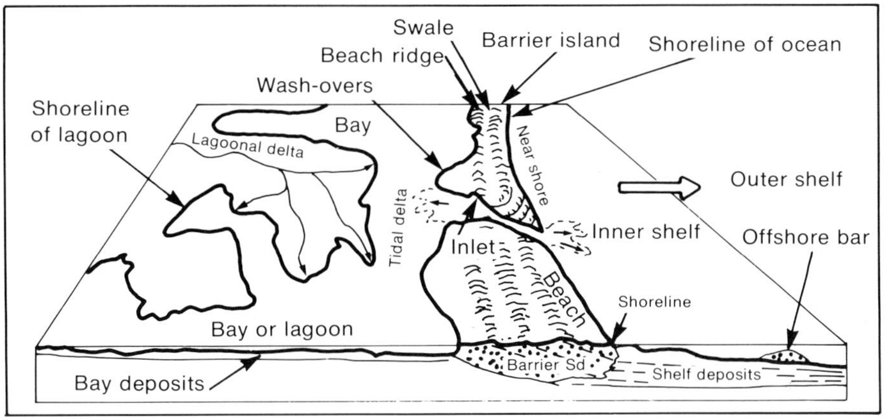

Fig. 2-20. Diagram of barrier island and associated sediments. (After Weimer, 1976, p. 102, fig. 12.3.)

current which extends the beach parallel to the coastline. Beach sands make excellent reservoirs, but they are seldom more than 5 miles (8 km) wide or 30 ft (10 m) thick. The strongest currents affect the sand body above sea level, whereas below sea level the currents are weak. Consequently, beach-type sands are coarser and more permeable in the upper part (Fig. 2-19).

The waves build up barrier islands parallel to the coast, usually separated from the delta proper by a shallow lagoon (Fig. 2-20). The seaward edge of the sand body is straight and abrupt, whereas on the lagoon side the sand gradually loses permeability and becomes interbedded with shale. Many excellent oilfields are present in beach-type deposits. Wind often blows beach sand into dunes, on the lagoon side of the barrier island.

River-mouth bar deposits

When the river enters the sea, the current suddenly stops, so that much of the burden of sediment is piled up right at the mouth, where it forms a shallow bar. Here it is subject to washing by the waves and to the lateral sweeping by longshore currents. A sand body is thus built which is similar to a barrier island beach, but it does not appear above water. As in a barrier island, the grain size and permeability are larger in the upper part where the wave action is more intense.

As the delta builds seaward, the river channel cuts into the river-mouth bar. It is very common to find a hydrocarbon reservoir in a river-mouth bar or barrier island transected at right angles by a channel sand (Fig. 2-21). Such reservoirs are especially common in deltas which advanced rapidly into the sea, like those of the U.S. Gulf Coast and Nigeria.

After a delta has prograded out to sea a few miles, the rivers have a tendency to switch out of their banks many miles (sometimes hundreds of miles) upstream, and start to build another delta alongside the first one. When this happens, the

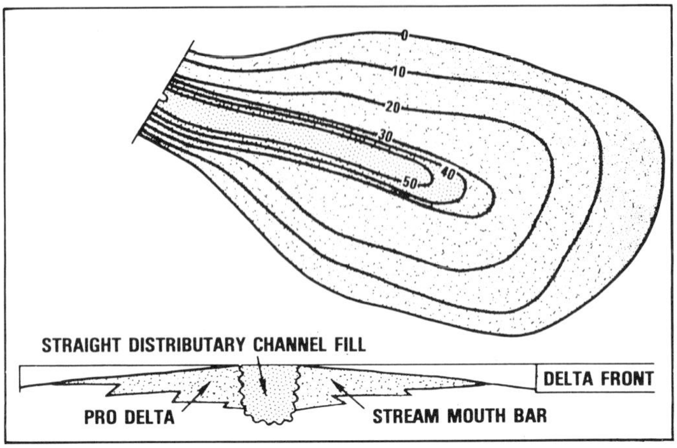

Fig. 2-21. Geometry of a shallow stream-mouth bar, transected at right angles by a channel. (After Swanson, 1979, p. 148, fig. 32.)

abandoned delta subsides below sea level because of the compaction of its unconsolidated sediments. The sea then transgresses over the abandoned delta. Waves at the advancing shore sometimes clean up a small amount of sand which extends over the deltas as a sheet. These sands and those which form in the interdistributary bays (called "crevasse splays") are thin and of low permeability. They do have greater lateral extent than the channel and beach sands, but they are usually poor reservoirs. The subsided delta is finally covered by a marine shale or limestone. At a still later time the river may return and build another delta on top of the first one. Some big oilfields, notably the Bolivar Coastal Fields of Venezuela, have several deltas stacked on top of each other. Usually the new deltas are a little farther out to sea than the old ones.

Turbidite sands

Normally only clays are deposited in the deep water off the front of a delta. In some places, however, large bodies of soft, unsorted clay and sandy material slip down the slope. Sometimes they slip as a coherent mass; occasionally, they disperse and become a body of muddy water, denser than the surrounding water. When this happens, the turbid water rushes down the slope and spreads out at the bottom of the basin. When the motion stops, coarser sand particles settle first, followed by the fine particles and, finally, clay (Fig. 2-22). The resulting sand body is called a turbidite, and can serve as an oil reservoir. It is more permeable on the bottom than on the top. Usually the sands are thin (less than 30 ft or 10 m) and low in permeability. However, there may be dozens of these sands on top of each other, with shale beds in between. Inasmuch as turbidites spread over a large area, they are more isotropic and less directional in their

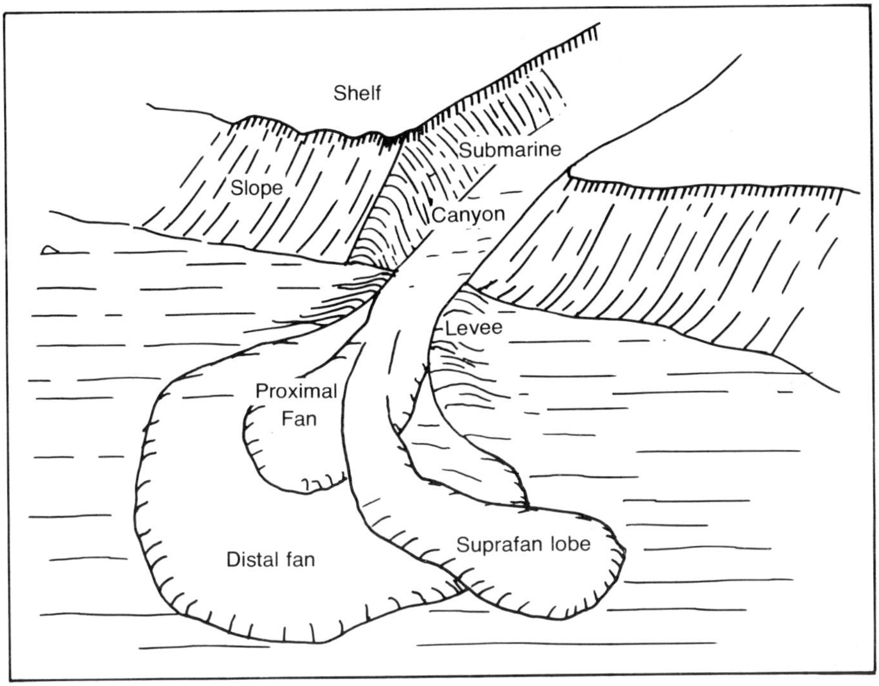

Fig. 2-22. Turbidity current deposits. (After Dickey, 1981.)

reservoir properties. Turbidites are important reservoirs in the valleys of southern California.

Well log expression of depositional environment

Different sand body types can often be recognized by noting their textural features in cores. Much can be also learned from wireline logs. The SP and gamma-ray logs are affected by the clay content of the rock; consequently, the cleaner and coarser sand shows larger deflections. The short-penetration resistivity logs are affected by the invasion of drilling mud filtrate, which is greater in the more permeable sands. In the case of a channel sand where the coarser sand is in the lower part, the electric log will have the shape of a bell. In the case of beach sands and river-mouth bars, where the coarser sand is in the upper part, the electric log will have a funnel shape. Alluvial fans, braided streams, and turbidites show as a thick succession of sands separated by shales. The shales are easily correlated from well to well in turbidites, but are difficult to correlate in alluvial fans and braided streams. Examples of electric logs of different sand bodies are given in Fig. 2-23.

Diagenesis of sandstones

After a sand body is buried it becomes compacted and lithified, resulting in a loss of porosity and permeability. Muds compact and lose porosity by distortion

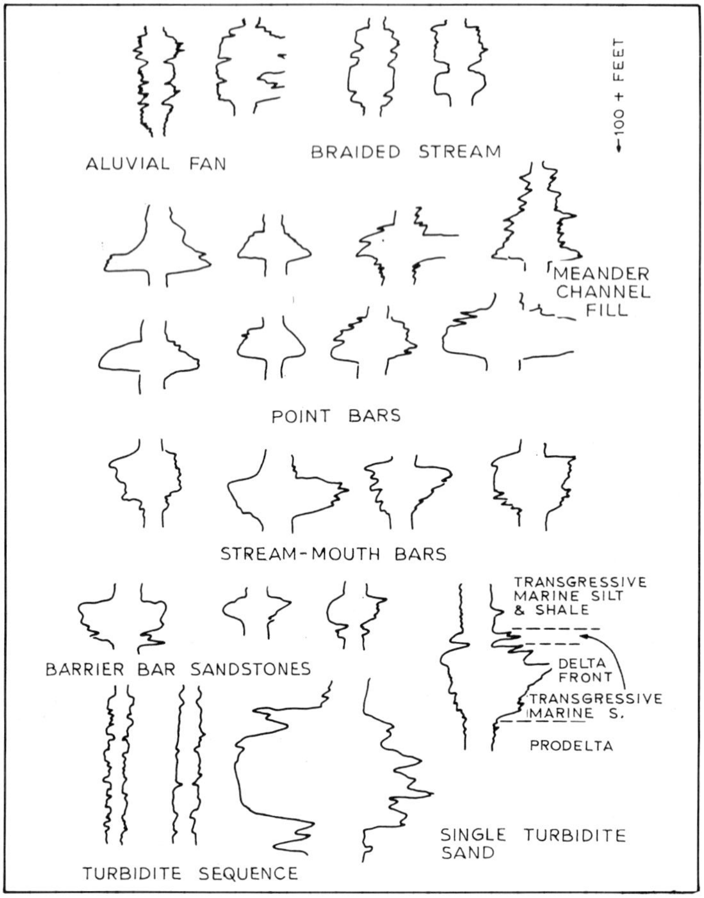

Fig. 2-23. Electric logs of different types of sand bodies. (After Swanson, 1980, pp. 54, 55.)

and flattening of the flaky clay particles. Sandstones also lose porosity due to the precipitation of silica and clay minerals in the pores, brought in by the water which is squeezed out of the adjacent shales.

The silica precipitates on the surface of the quartz grains. At first the grains are barely touching each other at their corners, but the precipitation of silica causes the grains to enlarge and the areas of contact between grains to increase. The additional silica conforms to the optical structure of the original grains, and as the grains grow out into the pores their surfaces become smooth like the surface of a quartz crystal (Fig. 2-24). As the process continues, the pores may become entirely filled with clear quartz, and porosity and permeability may vanish.

The migrating waters sometimes deposit clay minerals in the pores. These minerals are related to the micas, and usually crystallize in flakes (Fig. 2-25). The

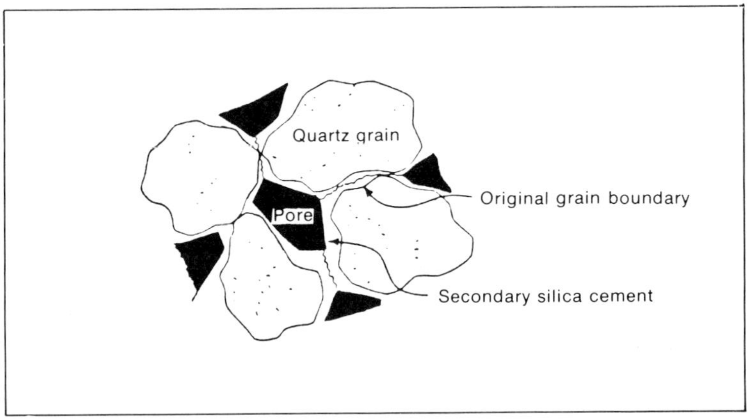

Fig. 2-24. Secondary silica coating quartz grains. (After Dickey, 1981, p. 142, fig. 8-11.)

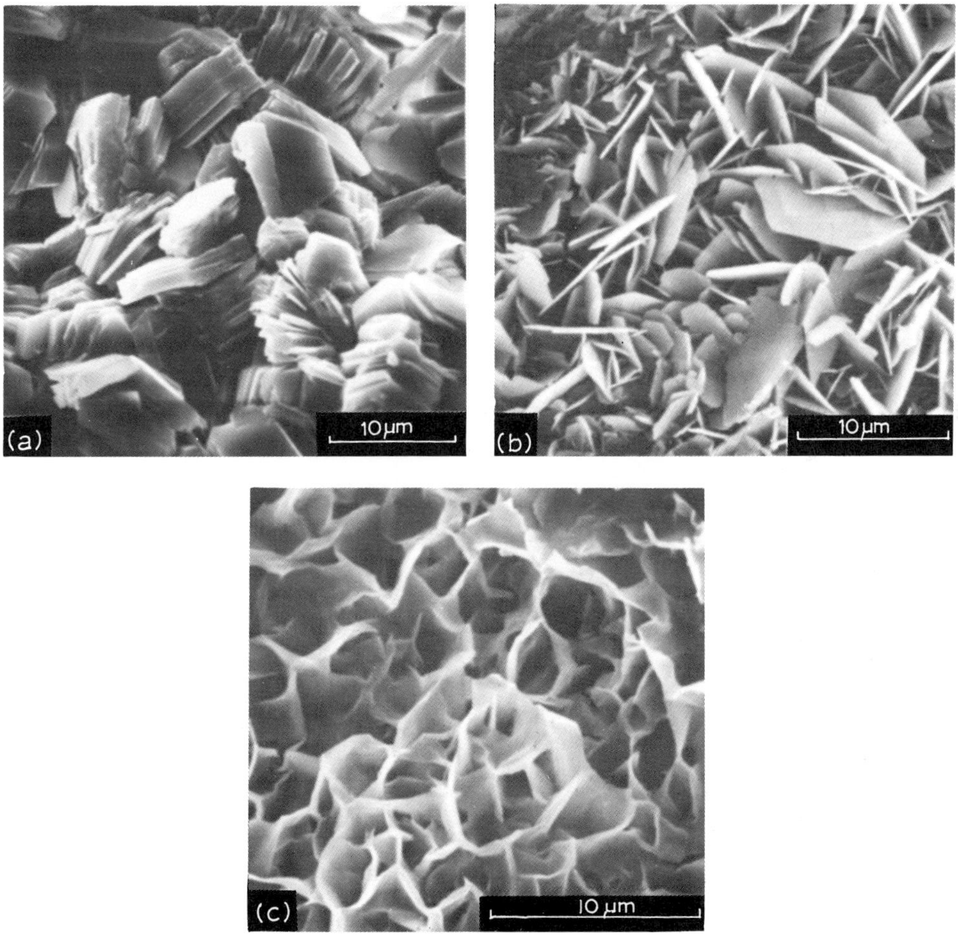

Fig. 2-25. Authigenic clays in the pores of oil reservoir sandstones: (a) kaolinite, (b) chlorite, and (c) smectite. (After Wilson and Pittman, 1977, p. 20, fig. 15.)

flakes partially fill the pores, and they have three very bad effects on the reservoir properties of the sandstones.

(1) Spaces between the clay flakes are so small that they hold the pore water tightly by capillary forces. Oil and gas are unable to enter this portion of the pore space, and fluid flow is blocked. Much of the porosity that is measured in core analysis thus becomes ineffective "microporosity", good neither to store fluids nor to permit them to move.

(2) Some clay flakes are attached only weakly to the sand grains. Any motion of the pore fluids will cause them to break loose and plug the pore throats. This motion may be due to invasion of mud filtrate, primary oil flow, or fluids injected during waterflooding or enhanced recovery. Smectite clays may swell when contacted by injected fluids different from the original pore water.

(3) Chemically attached to the clay surfaces are cations, mostly sodium and calcium. Under the electrical potential of a resistivity survey these ions will move, like the ions in solution. The rock will appear conductive, and the calculated water saturation may be much too high and the oil saturation too low. The calcium cations may damage the surfactants and polymers which are injected for enhanced oil recovery.

During diagenesis there is a tendency for the non-quartz minerals to dissolve. Grains of feldspar and rock fragments and carbonate cement dissolve away (Schmidt and McDonald, 1979). The pores formed by solution are larger than the intergranular pores of the original sandstone (Fig. 2-26). The two pictures of Fig. 2-26 are photomicrographs of resin casts of pores. Resin is injected under pressure until it fills all the pores; then the mineral grains are dissolved away. In Fig. 2-26a, the pores are slightly smaller than the grains, and the pore throats are slightly smaller than the pores. In Fig. 2-26b, certain pores formerly occupied by feldspar (P) are larger than the grains. The pore throats are narrow and sheet-like (lamellar, L).

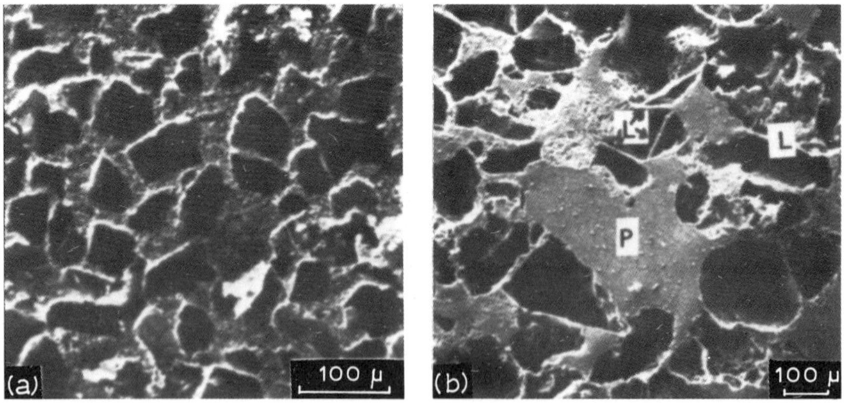

Fig. 2-26. Secondary porosity in sandstone. In (b), large grain of feldspar has been dissolved, leaving an extra large pore (P). The pore throats are lamellar and small. (After Wardlaw and Cassan, 1979, p. 129, fig. 7.)

Wardlaw and Cassan (1979) determined the pore-throat size ratio of a number of samples on which they also had relative permeability data and mercury injection capillary pressure curves. If the oil in a rock is the non-wetting fluid, then the saturation of residual oil (S_{or}) at water breakthrough is a measure of recovery efficiency during waterflooding. Mercury is a non-wetting fluid, so the amount of mercury left in the sample when the injection pressure is removed and the mercury is expelled by capillary forces, correlates with the volume of non-wetting oil held in the rock by capillary forces when water is driven through.

During mercury injection, the pressure is raised incrementally and the amount of mercury which enters the pores is plotted against the pressure. Since mercury is a non-wetting fluid, the surface tension tends to keep it out of the pores. The relation theoretically is:

$$P_c = \frac{2\sigma \cos \theta}{d} \qquad (2\text{-}3)$$

where P_c = capillary pressure, σ = surface tension of mercury, θ = contact angle, and d = width of pore.

As the pressure is increased there is a threshold pressure, that is, an abrupt change in slope (P_t) where the mercury starts to enter the pores (Fig. 2-27). This is a measure of the pore throat width. Then, when the mercury expels itself because of its capillary pressure, the amount left behind compares with the residual oil left behind after a waterflood. The ratio between the maximum amount of mercury injected and the amount left behind may be considered a measure of the recovery efficiency. Wardlaw and Cassan (1979) found that the recovery efficiency was related to the ratio between the size of the pores and the width of the pore throats, measured on resin casts. A small pore to throat-size ratio correlated with both oil and mercury recovery efficiency. A large pore to throat-size ratio causes poor recovery efficiency.

These rules hold when the sand is completely water-wet. Donaldson et al. (1969) showed that the recovery efficiency by waterflooding is greater when the sand is water-wet than if it is oil-wet. If oil in a water-wet sand is expelled by water, all the oil will be recovered by the time the water first breaks through. If the sand is oil-wet, the water breaks through early, and the oil is recovered only after a large amount of water has been injected.

The wettability of a natural core is usually determined by immersing it alternately in oil and water to see which is preferentially imbibed (Amott, 1959). However, no method of determining wettability is really satisfactory. It seems most probable that some of the interior surfaces of a sand are oil-wet and others water-wet. The microporosity in the clays where bound water is tightly held by both capillary and molecular forces is not likely to be oil-wet. Clays, however, form complexes with certain organic compounds. The surfaces of the clean quartz particles might well be oil-wet. Perhaps the electron-scanning microscope could be used to determine which surfaces are oil-wet and which are water-wet.

In spite of laboratory tests showing that water-wet sands have greater recovery efficiency, several natural sands which have responded well to

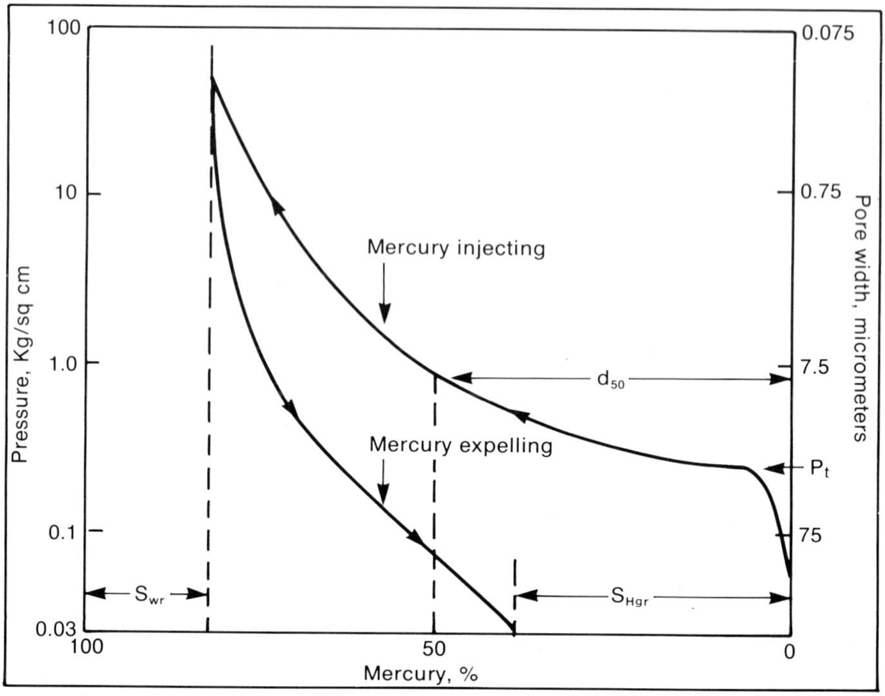

Fig. 2-27. Mercury capillary pressure injection curves. The threshold pressure (P_t) depends on the size of the pore throats, while the mercury retained on imbibition of air (drainage of mercury) (S_{Hgr}) measures the oil held by capillary forces. (After Wardlaw and Cassan, 1979, p. 120, fig. 20.)

waterflooding are apparently oil-wet. These include the Bradford Sand in Pennsylvania and the Bartlesville Sand in many locations in Oklahoma. The only way to explain the extremely low residual oil saturation after water encroachment in the East Texas field is to assume that some of the interior surfaces of the sand are oil-wet (Salathiel, 1973).

Carbonate reservoir rocks

The term "carbonate rock" is generally used to cover both limestones and dolomites, which are closely related. Both are important oil reservoir rocks, depending on their porosity. Like sandstones, the original depositional environment is important in determining the porosity. Unlike sandstones, carbonates are soluble in water, so the porosity in the reservoir is usually somewhat altered from the primary porosity. *

* "Solution porosity" is very significant in many carbonate reservoirs. (Editorial comment.)

Electrical and other wireline logs are not nearly as useful in carbonates as they are in sandstones. They give little information on the nature of the porosity or the depositional environment. Complete cores are indispensable to the prediction of fluid behavior and the shape and size of the reservoir.

General pattern of depositional environments

Practically all carbonates consist basically of ground-up protective and structural parts of calcareous organisms ("sea shells"). Marine organisms which form hard exo-skeletons can grow only where light penetrates in shallow water. Through geologic time the greater bulk of carbonate rocks was formed on the shallow shelves at the margin of seas or oceans (Fig. 2-28).

Fig. 2-29 is a more detailed, but still diagrammatic, sketch showing the usual pattern of depositional environments. Starting at the left, deep basins usually contain only mud, and little carbonate. Sometimes pelagic (open sea), microscopic calcareous organisms flourish on the surface of water. Their tiny shells settle on the bottom, where they form chalk. The shelf slope is usually steep (several degrees), and consists of broken and often finely ground fragments of reef material. The most prolific zone for generation of carbonate is right at the break in slope. Here the currents and breaking waves provide abundant oxygen and nutrients. Certain organisms, nowadays corals and algae, build upstanding structures called reefs. In earlier geologic times other, now extinct, organisms were the principal reef-builders. On the landward side of the reef, there is a broad lagoon where the water is shallow. Because the tides raise and lower the water level in the lagoon twice daily, there are strong currents in channels crossing the reefs and extending for a short distance into the lagoon. These currents are effective in washing particles free of mud. Farther back in the lagoon proper the currents are weak, so that mostly lime mud, that is, shell fragments of only very small size, are deposited.

In arid regions, such as the present-day Persian Gulf, or in West Texas during the Permian, evaporation takes place on the landward side of the lagoon. This zone is now called the "sabkha". As the sea water evaporates, it precipitates first calcium carbonate and then calcium sulfate. It thereby becomes enriched in

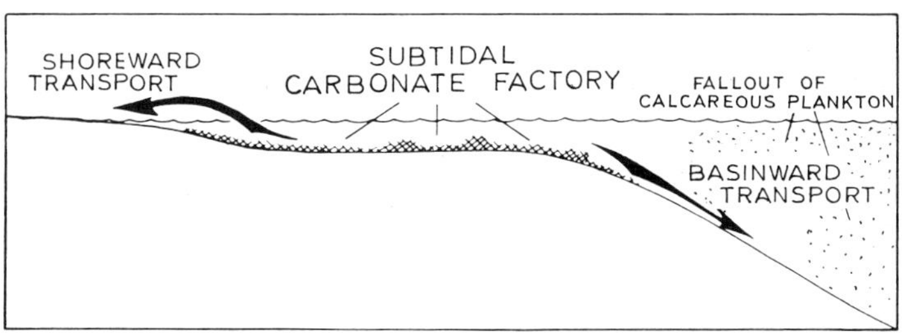

Fig. 2-28. Most carbonate rocks are made by organisms in shallow water on continental shelves, and swept both shoreward and seaward. (After James, 1977, p. 124, fig. 1; courtesy of *Geoscience Canada*.)

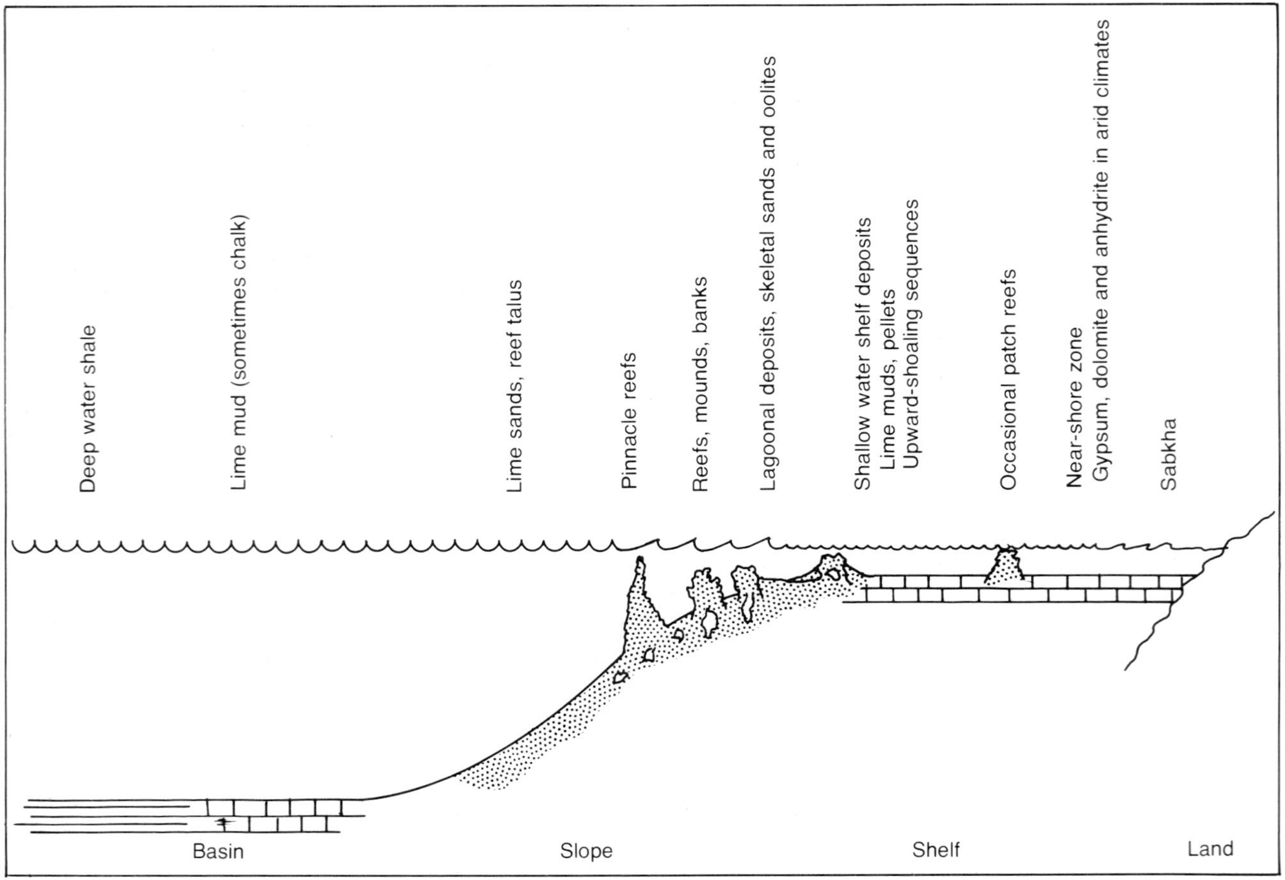

Fig. 2-29. Carbonate depositional environments. (After Dickey, 1981, p. 160, fig. 9-8.)

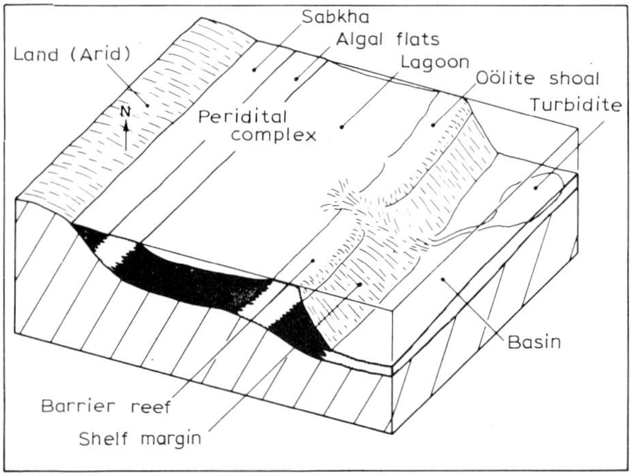

Fig. 2-30. Pattern of carbonate deposition in New Mexico and West Texas during the Permian. (After Friedman and Sanders, 1978, p. 379, figs. 1–4; courtesy of John Wiley and Sons.)

magnesium chloride, which alters the previously deposited lime mud to dolomite. Fig. 2-30 shows the pattern of deposition in the Permian Basin.

Diagenesis of carbonates

Calcium carbonate is slightly soluble in water, and a small change in salinity, acidity, or temperature of the water causes changes in the solubility. Consequently, the carbonate sediments are hardly buried before their texture is altered by migrating waters.

Because carbonates form mostly in shallow water, a slight shallowing of the sea will expose them to the atmosphere. Then rainwater permeates the porous rock, dissolving the carbonate. When water percolates down to where the rock is fully saturated, it redeposits the calcite in the pores, often practically destroying the permeability. The horizontal permeability barrier in the Golden Spike reef was due to a short period of exposure.

Much large-scale solution and deposition of calcite also takes place. The lime mud, when first deposited, is very porous. It does not compact like shale, but its pores are filled by microcrystalline calcite (called "micrite"). This calcite must have come from the solution of a whole layer of lime mud, either above or below. The disappeared bed may be represented by wavy bands of insoluble material, called stylolites. This pressure-solution process is poorly understood.

The reefs (or mounds) built up by organisms, usually at or slightly seaward of the shelf edge, have primary porosity between the frame-building organisms. The spaces, however, fill up with lime mud and secondary calcite. Circular reefs (atolls) have a lagoon in the center where lime mud is deposited. Generally, the best reservoir rock is preserved on the reef flank. Fig. 2-31 shows the depositional

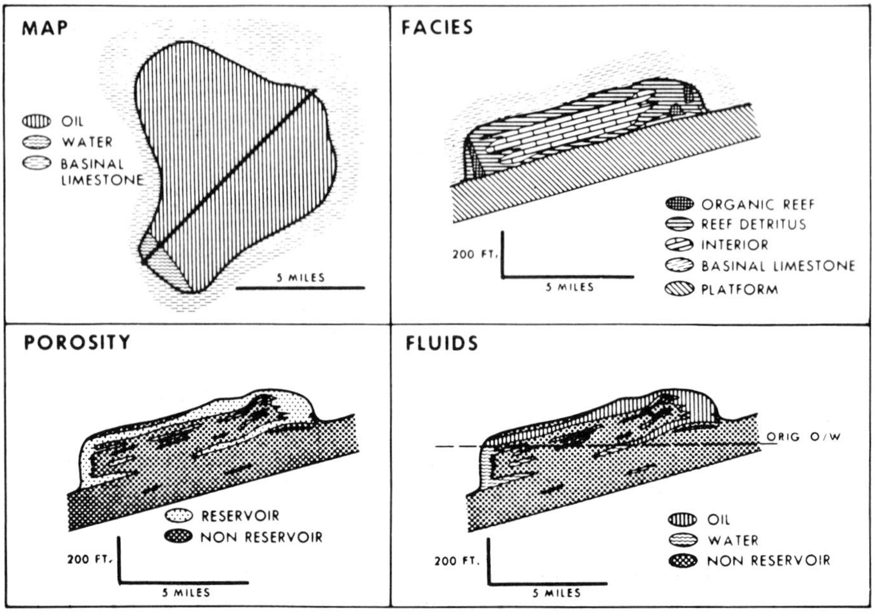

Fig. 2-31. Internal structure of Judy Creek reef, Alberta. (After Jardine et al., 1977, p. 879, fig. 7; courtesy of the Society of Petroleum Engineers of AIME.)

environments, reservoir quality, and fluid content of the Judy Creek reef in Alberta.

In the geological past, more often than today, large areas of the shelf lime muds were converted to dolomite. Beds of pure anhydrite (calcium sulfate) often alternate with the limestone and dolomite, showing that the shelf was, in fact, very shallow and the climate very arid. Lime mud was converted to small dolomite crystals, leaving a rock resembling granulated sugar. The huge reservoirs of West Texas and Saudi Arabia occur in shelf dolomites.

Porosity types in carbonates

The components of a limestone or dolomite consist of (1) micrite, that is, recrystallized lime mud, with particles less than 0.03 mm in diameter; (2) grains, that is, particles usually less than 1.0 mm in diameter; and (3) cement, which is clear calcite deposited by circulating water (Fig. 2-32). Occasionally, there may be large chunks of coral or other frame-building organisms. The different kinds of limestones have differing relative amounts of micrite and grains. There are five principal grain types: (1) fragments of previously cemented limestones; (2) fragments of shells, called skeletal grains: (3) pellets, which are egg-shaped particles believed to be largely the excrement of worms; (4) shapeless lumps of calcite; and (5) oolites, which are spherical grains formed by accretion of calcite on particles which are rolled back and forth by tidal currents (Fig. 2-33).

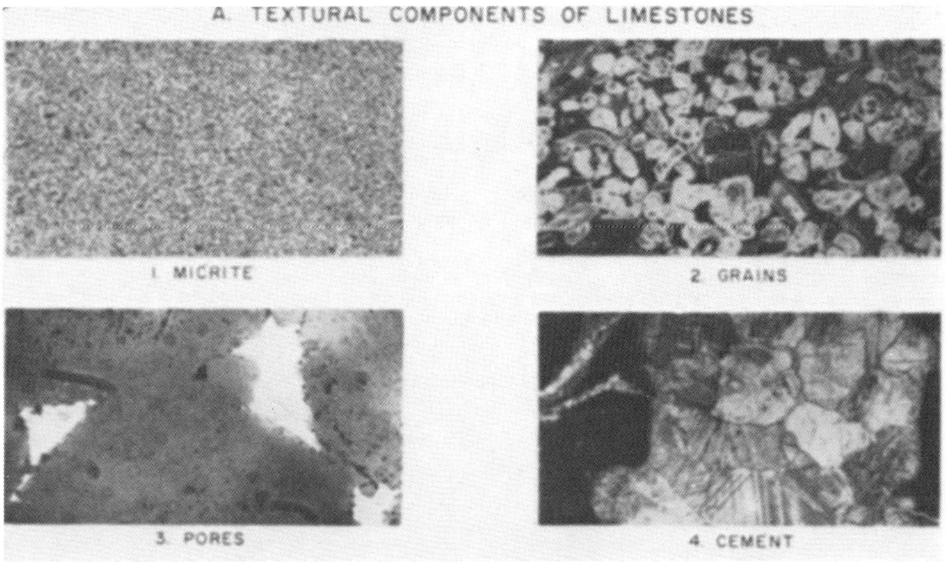

Fig. 2-32. Textural components of limestones. (After Leighton and Pendexter, 1962, p. 35, plate I; courtesy of the American Association of Petroleum Geologists.)

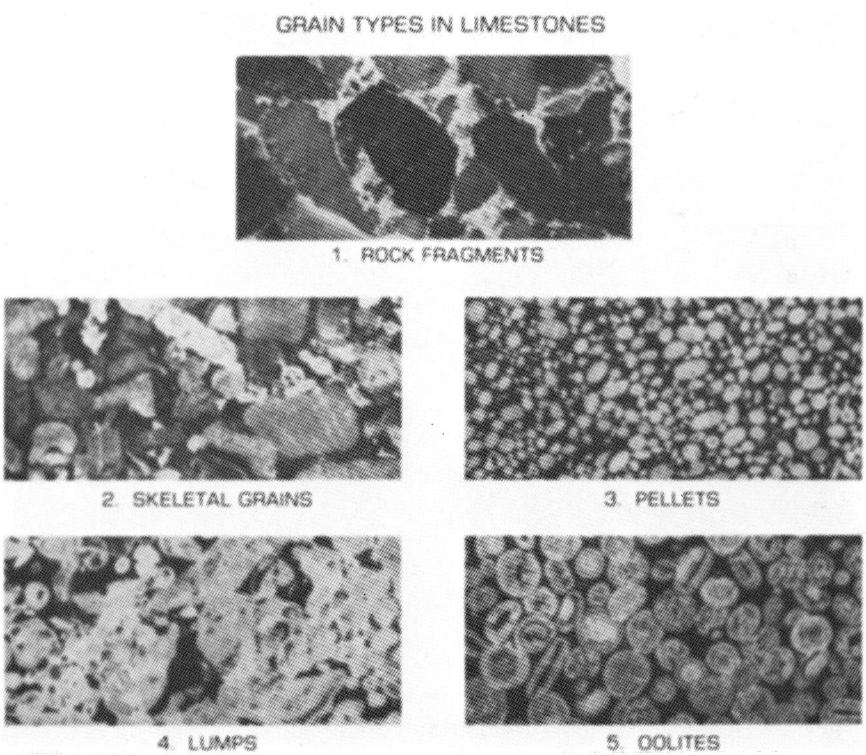

Fig. 2-33. Types of grains in limestones. (After Leighton and Pendexter, 1962, p. 35, plate I; courtesy of the American Association of Petroleum Geologists.)

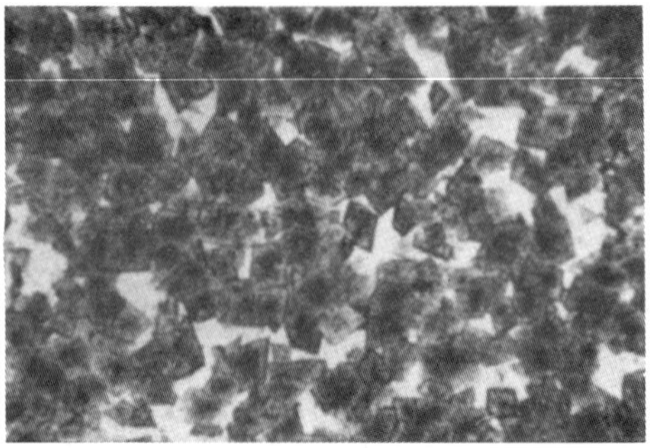

Fig. 2-34. Interparticle porosity in sugary dolomite, Saudi Arabia. (From Powers, 1962, p. 142, pl. V; courtesy of the American Association of Petroleum Geologists.)

Washing by currents removes the lime mud from the grains, and confers the reservoir properties of porosity and permeability. The best limestone reservoirs are those found in localities where there had been currents, which are rather limited zones near the shelf edge. Unfortunately the open pores are frequently found to be cemented by secondary calcite.

Dolomitization causes recrystallization of the lime mud or micrite. Dolomite crystals are small rhombs, and a lime mud with no porosity or permeability may become an excellent reservoir (Fig. 2-34). Continued circulation of dolomitizing water causes the crystals to enlarge, reducing the pore throats and increasing the ratio of pore size to throat size. Eventually the pores are practically closed, and porosity and permeability almost disappear (Fig. 2-35).

Very often the moving water goes through open channels. These pursue irregular paths, diverging and then rejoining. Channels range in size from less than 1.0 mm (pinpoint porosity) to several millimeters (vuggy porosity). Often they are partly filled by later deposits of crystalline calcite. Vugs and channels constitute an important type of carbonate porosity (Fig. 2-36).

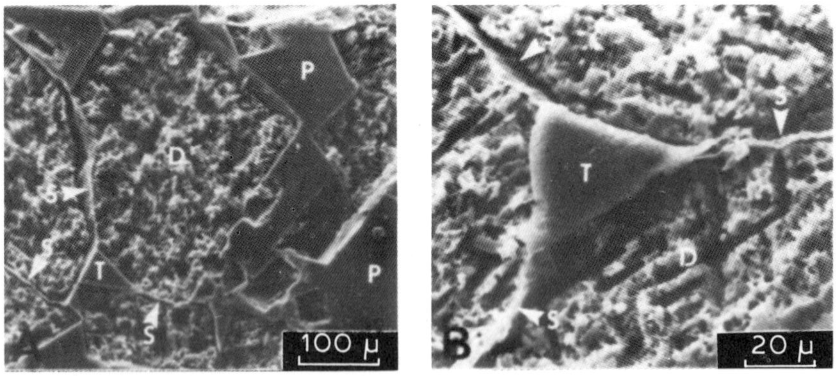

Fig. 2-35. Growth of the dolomite crystals eventually eliminates the porosity. P = polyhedral pores; T = tetrahedral pores. (From Wardlaw, 1976, p. 252, fig. 11; courtesy of the American Association of Petroleum Geologists.)

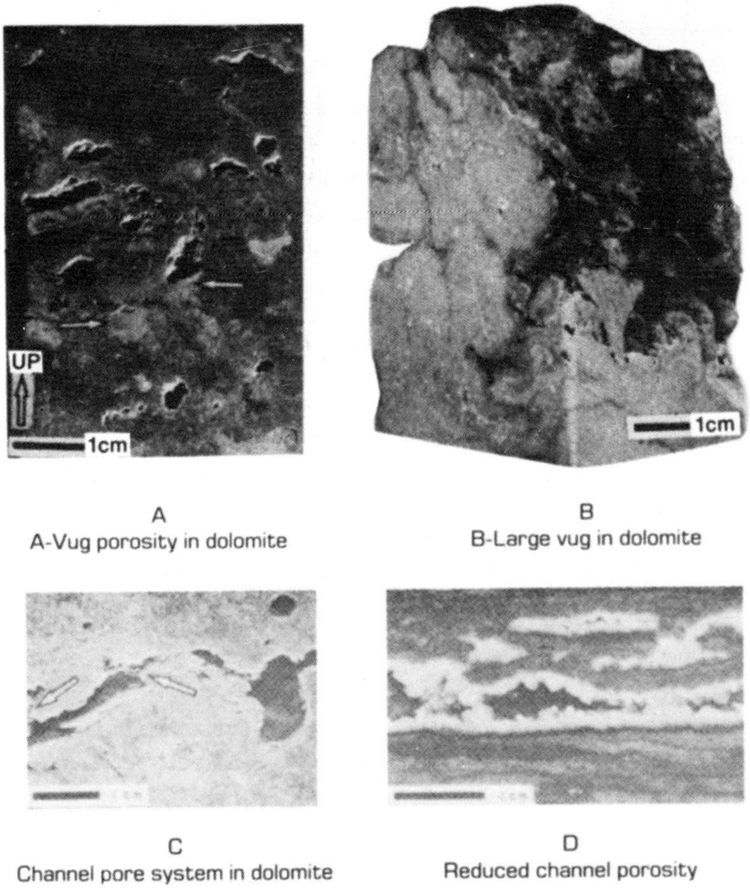

Fig. 2-36. Types of vugs and channels in limestone. (After Choquette and Pray, 1970, p. 236, fig. 10; courtesy of the American Association of Petroleum Geologists.)

When sandstones are bent, the separate layers slide over each other like pages of a bending telephone book. Carbonates, when bent, behave more like marble tombstones, i.e., they fracture. Fig. 2-37 shows the pattern of fractures which developed over the great anticlines of the Iranian oilfields.

Choquette and Pray (1970) classified carbonate porosity into 15 types. For the purposes of enhanced recovery it seems adequate to consider 3 principal types: (1) interparticle; (2) vuggy; and (3) fracture.

Interparticle porosity may be the porosity between the original grains (primary), or between dolomite crystals (secondary). When the original grains are oolites, porosity and permeability are excellent. The Jay Field of Florida and some of the Persian Gulf fields are in oolites. The great fields in the San Andres Limestone of West Texas and Jay Limestone in Florida have responded well to

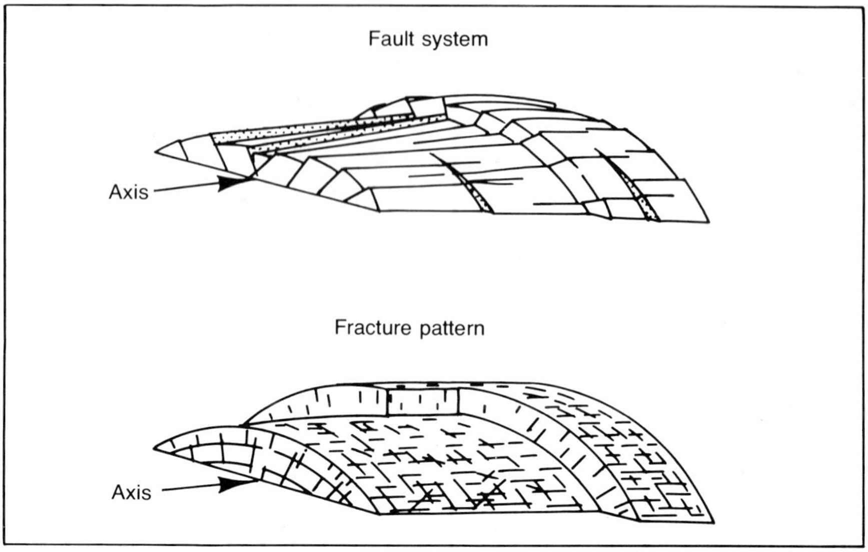

Fig. 2-37. Pattern of fractures in Asmari Limestone reservoirs of Iran. (After Andresen et al., 1963, p. 19, fig. 3.)

waterflooding and are good prospects for enhanced recovery. During waterflooding it became evident that the porous layers were not thick and extensive, as had at first been thought. They do not extend very far laterally. Peripheral water injection was not effective, and infill pattern drilling was necessary. There are also many layers of low permeability which inhibit vertical communication between the porous beds.

Vug and channel porosity can be small (less than 5%), but still the rock can be a good reservoir if the channels are big enough to provide good permeability. Both porosity and permeability are difficult to measure.

Many oilfields with vuggy porosity have not responded well to waterflooding, and may not be candidates for present methods of enhanced recovery. When water is injected, it flows quickly through the more permeable channels. Some oil may be trapped in the smaller channels and cul-de-sacs, but there is no obvious way to displace it. If the reservoir is a tall reef with good vertical permeability, it is possible to inject a layer of miscible fluid (hydrocarbon or CO_2), which will be stable because of its lower density.

The great fields of Iran consist of more or less dolomitized limestone which has been bent sharply over a series of anticlines. An extensive series of fractures developed which have great permeability (Fig. 2-37), so that pressure drops in one well are felt soon in wells more than a kilometer away. The fractures do not extend from one oilfield to another; the fields had different initial pressures. The limestone matrix has substantial porosity — about 15%. Most of the oil is present in the matrix, and the fractures provide such permeable paths that enormous producing rates can be sustained.

In a field with fracture porosity, as the oil is produced the water–oil contact (also present in the fractures) rises. The water thus bypasses the oil still in the matrix. If the pores are water-wet, water will be imbibed into the matrix by capillary pressure, displacing some of the residual oil. It is hard to establish the effectiveness of the imbibition process.

The total pore volume of the fractures is usually less than 1%. If there is no effective porosity in the matrix, oil wells come in with large initial production, which declines in a matter of weeks.

NATURAL FRACTURES

Character of systematic fractures, or "joints"

Nearly all hard rocks, such as consolidated sandstones and carbonates, are broken by sets of regional parallel fractures, usually called joints. They are perpendicular to the bedding. Their surfaces are plane and fairly smooth, although plume-like patterns often appear. There is seldom evidence of sliding motion. Apparently, they are normally tightly closed in the subsurface; if they were open, injected fluids would always pass right through them. The fractures do tend to open on outcrops, where they often contain secondary minerals probably deposited by circulating ground water.

The orientation of fractures is remarkably constant, remaining the same within 5° for hundreds of miles. In western Pennsylvania, it is about 320°; in

Fig. 2-38. Joints as seen in cores from Burbank, Oklahoma. (After Hagen, 1972, p. 15, fig. 4.)

northern Oklahoma it is about 70° (Hagen, 1972). Often there is a conjugate set of joints oriented 90° to the principal set. These are definitely subordinate, usually running only from one principal joint to the next. Joints are especially conspicuous in coal. The principal set is called by the miners "face cleats", and the subordinate set, "butt cleats". The orientation of the joints determines the orientation of the tunnels.

Spacing of the fractures is quite variable, but depends on the thickness of the bed. The thinner the bed, the more closely spaced are the joints; the spacing is of the same order of magnitude as the bed thickness. The joints usually terminate at the top and bottom of the hard bed, not penetrating the soft shale above and below. Sometimes, however, the same joint goes through the shale and cuts the next hard bed above or below.

Cores should be carefully examined for joints. It is sometimes hard to distinguish joints from drilling-induced fractures. Fig. 2-38 shows the joints as seen in cores from the Burbank Field of Oklahoma.

Pressure necessary to open fractures

When fluids are injected into a porous rock that is fractured, the fractures may be opened if a critical pressure is exceeded. The fractures are kept closed by the horizontal stress in the rock. If the vertical stress is σ_v, then the horizontal stress (σ_h) will be:

$$\sigma_h = \sigma_v\left(\frac{\mu}{1-\mu}\right) \tag{2-4}$$

where μ is Poisson's ratio. The vertical stress gradient is about 1 psi/ft (23 kPa/m), but it is reduced by the buoyant effect of the water in the pores, which has a pressure gradient of about 0.5 psi/ft. Poisson's ratios for sedimentary rocks (Daines, 1982) range from 0.2 to 0.4. Thus, the horizontal stress gradient is equal to:

$$\sigma_h = (1.0-0.5)\left(\frac{0.3}{1.0-0.3}\right) = \pm 0.21 \text{ psi/ft.}$$

The pressure in the pore water resists the effect of the pressure in the borehole. If the pore water pressure gradient is 0.5 psi/ft and μ is 0.3, then the pressure gradient necessary to open the fractures (p^*) is:

$$p^* = 0.2 + 0.5 = 0.7 \text{ psi/ft.}$$

If the pore pressure is low because it was drawn down by production, then the pressure gradient necessary to open the joints is much lower. If the pore pressure gradient is 0.2 psi/ft, then the rupture pressure is:

$$p^* = 0.2 + 0.2 = 0.4 \text{ psi/ft.}$$

This is less than the weight of a column of water.

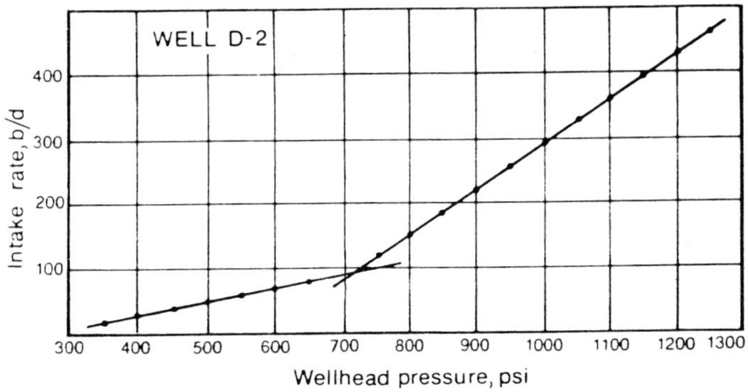

Fig. 2-39. Step-test results. When pressure at wellhead exceeded 700 psi, the fracture opened. (After Dickey and Andresen, 1950, p. 335, fig. 17.)

If the fractures are opened for only a short time they close again. If they are opened for a long time, the passage of the injected fluids erodes the fracture surfaces and they never close.

Many waterflood operations have been irreparably damaged by raising the injection pressure above the rupture pressure. It is customary to determine the rupture pressure by raising the injection pressure step by step, and measuring the intake rate of the well. When one is plotted against the other, a straight line is obtained. When a fracture is opened, the curve breaks to a new straight line (Fig. 2-39). Step tests should be conducted (with caution) before injecting any fluid.

Fractures in the North Burbank Field, Oklahoma

The most detailed study of the effect of fractures on secondary and tertiary oil recovery in the Burbank Field in Oklahoma was made by Trantham et al. (1980). The average reservoir thickness is 42 ft (12.8 m), the average porosity is 16.8%, and average permeability is 50 mD. The petrographic characteristics identify it as fluvial in origin, consisting of a series of river channels about 1000 ft (300 m) wide, superimposed on and cut laterally into each other, providing a wide, apparently massive sand body. There are, however, local areas where the sand is poorly developed, probably between channels. Permeability between beds varies widely.

The field was waterflooded in the 1950's. The water channeled quickly through fractures oriented in an east–west direction from the injection wells, so the pattern was changed to a staggered line drive with line-wells in east–west lines. It is not clear whether the fractures were initially closed or not; they opened soon after water injection started.

In preparing a pilot test for surfactant–polymer injection after the waterflood, pressure fall-off curves showed that natural fractures were formed at an injection rate of 300 to 500 bbl/day (47.7 m^3/day), and they closed when the well was shut in. Estimation of fracture pressures by raising the pressure stepwise in one well showed that the fractures opened at 1465 psia (10,000 kPa) bottomhole pressure. There was considerable variation from well to well, but generally the fractures opened at a pressure gradient of 0.5 psi/ft (11.3 kPa/m).

When serious channeling occurred, the pressure responses were "almost instantaneous", and the tracer response curves were "very steep spikes".

In a final evaluation of the project, Tracy and Dauben (1982) stated that the tracer survey showed a high degree of heterogeneity. As the injectivity declined, the wells were deliberately fractured and then treated with gel to stop the bypassing. This may have caused the polymer to follow different paths from those followed by the micellar solution.

Efforts to correlate the tertiary oil recovery with injection of sulfonate surfactants and polyacrylamides (mobility control solutions) were unsuccessful. This led to the conclusion that "the early tertiary production came from channels or layers that were more permeable and that the sweep of the less permeable layers was limited".

Other fractured sand reservoirs

Fracture pressures were exceeded in the Lawry micellar–polymer project at Bradford, Pennsylvania (Ondrusek and Paynter, 1982), and at a fireflood in West Texas (Anthony et al., 1981). In the Willard CO_2 project in the Wasson Field of West Texas, the fracture pressure was exceeded when the CO_2 injection began, and CO_2 was produced along the axis of fracture orientation. The trouble was quickly identified and pressures were cut back to 1500 psi (10,350 kPa).

In the Street Ranch Field of South Texas, horizontal fractures were formed at the base of the sand which acted as flues to heat the whole sand body (Britton et al., 1982). The sand is 50 ft (15 m) thick and 1500 ft (456 m) deep. The oil is a tar having $-2°$ API gravity. Previous steam injection tests failed to mobilize the oil. In a pilot test, the injection wells were horizontally notched at the base and then fractured with water without proppant. Steam was then injected at higher than fracture pressure (1.08 psi/ft or 24.4 kPa/m). Then the producing wells were perforated, and steam was injected at less than fracture pressure gradient through the sand. It successfully mobilized the oil.

In several pilot steam stimulation projects at Cold Lake, Alberta, it was found that there was a set of fractures oriented NE–SW, which opened at a pressure of about 1300 psi (Fisher, 1982). It was decided to drill the wells in lines oriented at an azimuth of 300°, spaced 200 ft (67 m) between wells and 900 ft (240 m) between lines. Steam is expected to penetrate farther along the channels northeast and southwest, and not so far between the closely spaced wells in the NW–SE lines.

REFERENCES

Amott, E., 1959. Observations relating to the wettability of porous rock. *Trans. AIME*, 216: 156–162.

Andresen, K.H., Baker, R.I. and Raoofi, J., 1963. Development of methods for analysis of Iranian Asmari reservoirs. *6th World Pet. Congr., Part II*, Paper 14: 13–25.

Anthony, M.J., Taylor, T.D. and Gallagher, B.J., 1981. Fireflooding a high gravity crude in a watered-out West Texas Reservoir. In: *Proc., Second Joint SPE-DOE Symp. on Enhanced Oil Recovery, Tulsa, Okla., April 5–8*, SPE 9711.

Barrett, D.D., Harpole, K.J. and Zaaza, M.W., 1977. Reservoir data pays off; West Seminole San Andres Unit, Gaines County, Texas. *Soc. Petr. Eng. Pap.*, 6738.

Britton, M.W., Martin, W.L., Leibrecht, R.J. and Harman, R.L., 1982. The Street Ranch pilot test of fracture-assisted steam flood. In: *Proc., Third Joint SPE-DOE Symp. on Enhanced Oil Recovery, April 4–7, Tulsa, Okla.*, SPE 10707.

Brown, F.L., Jr., 1979. Deltaic sandstone facies of the Midcontinent. In: N.J. Hyne (Editor), *Pennsylvanian Sandstones of the Midcontinent, Tulsa Geol. Soc. Spec. Publ.*, 1: pp. 35–64.

Bursell, C.G. and Pittman, G.M., 1974. Performance of steam displacement Kern River Field. *Soc. Pet. Eng. Pap.*, 5017.

Carroll, H.B. (Chairman), 1982. Impact of geologic parameters on enhanced oil recovery. *Workshop Proc., Camelot Hotel, Tulsa, Okla., October 15–16, 1981.*

Choquette, P.W. and Pray, L.C., 1970. Geologic nomenclature and classification of porosity in sedimentary carbonates. *Bull. Am. Assoc. Pet. Geol.*, 54: 207–250.

Claridge, E.L. and Lohse, A., 1983. *Data Requirements for EOR Surfactant–Polymer Process Simulation and Analysis of El Dorado Pilot Project Simulation, Butler County, Kansas.* DPE/ET/10145–74.

Collins, A.G., 1982. *Enhanced Oil Recovery Injection Waters.* DOE/BETC/R1-82/5.

Craig, F.F., 1971. *The Reservoir Engineering Aspects of Waterflooding. SPE Monograph 3*, Society of Petroleum Engineers, Dallas, Texas, pp. 62–68.

Daines, S.R., 1982. Prediction of fracture pressures for wildcat wells. *J. Pet. Technol.*, 34(4): 863–872.

Dean, D.M., 1982. *Bodcau In Situ Combustion Project, Final Report.* DOE/ET/12057–9.

Dickey, P.A., 1981. *Petroleum Development Geology.* PennWell, Tulsa, Okla., 428 pp.

Dickey, P.A. and Andresen, K.W., 1950. Behavior of water input wells. In: *Secondary Recovery of Oil in the U.S.* API, Dallas, Texas, 2nd ed., pp. 317–341.

Donaldson, E.C., Thomas, R.D. and Lorenz, P.B., 1969. Wettability determination and its effect on recovery efficiency. *Soc. Pet. Eng. J.*, 19(3): 13–20.

Dullien, F.A.L., 1975. Effects of pore structure on capillary and flow phenomena in sandstones. *J. Can. Pet. Technol.*, 14: 48–55.

Dykstra, H. and Parsons, R.L., 1950. The prediction of oil recovery by waterflooding. In: *Secondary Recovery of Oil in the U.S.* API, Dallas, Texas, 2nd ed., pp. 160–174.

Eson, R.L., 1983. *Field Demonstration of the Conventional Steam Drive with Ancillary Materials.* DOE/SF/10762–3.

Fertl, W.H., 1980. Well log analysis concepts in clastic and argillaceous U.S. Gulf Coast sediments. *Trans. Gulf Coast Assoc. Geol. Soc.*, 30: 81–89.

Fisher, W.G., 1982. Enhanced oil recovery. *U.N. Int. Meet. On Oilfield Development Techniques, Daqing, September 7–14.*

Friedman, G.M. and Sanders, J.E., 1978. *Principles of Sedimentology.* John Wiley and Sons, New York, N.Y., 792 pp.

Ghauri, W.K., Osborne, A.F. and Magnuson, W.L., 1974. Changing concepts in carbonate waterflooding, West Texas Denver Unit project: an illustrative example. *J. Pet. Technol.*, 26(6): 595–666.

Gogarty, W.B., 1978. Micellar-polymer flooding — an overview. *J. Pet. Technol.*, 30(8): 1089–1101.

Goldburg, A., 1982. *Bell Creek Field Micellar–Polymer Pilot Demonstration. Final Report.* DOE/SE/01802–61.

Goldburg, A. and Stevens, P., 1981. Micellar–polymer oil recovery demonstration in the Bell Creek Field, Montana. In: *Proc. Int. Energy Workshop Enhanced Oil Recovery, Bartlesville, Okla., April 29, 1980.*

Goodrich, J.H., 1980. *Target Reservoirs for CO_2 Miscible Flooding. Final Report.* DOE/MC/08341-17.

Goodrich, J.H. and Watson, J.A., 1980. CO_2 tertiary recovery field pilot, Little Creek Field, Pike County, Mississippi. In: *Target Reservoirs for CO_2 Miscible Flooding, Final Report, Vol. II.* DOE/MC/08341-17.

Griffin, J.D. and Cyca, L.G., 1980. Performance of Swan Hills miscible flood. In: *Proc., First Joint SPE-DOE Symp. on Enhanced Oil Recovery, April 20–23, Tulsa, Okla.*, pp. 257–266, SPE 8835.

Guckert, L.G. and Paynter, W.T., 1982. Micellar-polymer flooding in the Bradford Field. *Proc., Int. Meet. on Petroleum Engineering, Beijing, China, 19-22 March*, 2: 557–568.

Gulf Universities Research Consortium, 1981. *Survey of Activity in the Area of Instrumentation for Reservoir Description for Enhanced Oil Recovery, Final Report.* DOE/ET/10145-64.

Hagen, K., 1972. *Mapping of Surface Joints in Air Photos Can Help Understand Waterflood Performance Problems at North Burbank Unit.* M.S. Thesis, University of Tulsa, Tulsa, Okla.

Hicks, J.N. and Foster, R.S., 1979. Evaluation of target oil in 50 major reservoirs in the Texas Gulf Coast for enhanced oil recovery. *5th Annu. DOE Symp., Tulsa, Okla., August* 22–24.

James, N.P., 1977. Facies model 8: shallowing upward sequences in carbonates. *Geosci. Can.*, 4: 123–125.

Jardine, D., Andrews, D.P., Wishart, J.W. and Young, J.W., 1977. Distribution and continuity of petroleum reservoir. *J. Pet. Technol.*, 29: 873–885.

Jin Yu-Sun, Wan-li, Yang and Wang, Zhi-wu, 1981. Exploitation of multizones by water flooding in the Daqing oilfield. *Acta Pet. Sinica, Spec. Issue.*

Jordan, D.W. and Tillman, R.W., 1982. *El Dorado Micellar–Polymer Demonstration Project — Reservoir Description.* DOE/ET/13070-79.

Kuzyla, K. and Friedman, G.M., 1981. Geological heterogeneities important to future enhanced recovery in carbonate reservoirs of Upper Ordovician Red River Formation at Cabin Creek Field, Montana. In: *Proc., Second Joint SPE-DOE Symp. on Enhanced Oil Recovery*, Tulsa, Okla., April 5–8, SPE 9802.

LeBlanc, R.J., Sr., 1977. Distribution and continuity of sandstone reservoirs, Part 1 and Part 2. *J. Pet. Technol.*, 29(7): 776–850.

Leighton, M.W. and Pendexter, C., 1962. Carbonate rock types. In: W.E. Ham (Editor), *Classification of Carbonate Rocks. Am. Assoc. Pet. Geol., Mem.*, 1: 33–61.

Martin, W.E. and Young, M.N., 1982. The Wizard Lake D-3 pool miscible flood. *Proc., Int. Meet. on Petroleum Engineering, Beijing, 19–22 March,* 2: 539–555.

McGregor, A.A. and Biggs, C.A., 1968. Bell Creek Field; a rich stratigraphic trap. *Bull. Am. Assoc. Pet. Geol.*, 52:1867–1887.

Morrow, N.R., 1975. Effects of surface roughness on contact angle with special reference to petroleum recovery. *J. Can. Pet. Technol.*, December, pp. 42–53.

Muecke, T.W., 1979. Formation fines and factors controlling their movement in porous rocks. *J. Pet. Technol.*, 32(2): 144–150.

Nicholson, G., 1980a. Geology of the Kern River Field. In: *AAPG Pacific Section, Field Trip Guidebook*, pp. 7–17.

Nicholson, G., 1980b. Kern River Oilfield. *AAPG Pacific Section. Field Trip Guidebook*, 24 pp.

Ondrusek, P.S. and Paynter, W.T., 1982. *Penn Grade Micellar Development Project, Final Report.* DOE/ET/08002-26.

Perry, G.E., 1981. Weeks Island "S" Sand reservoir B gravity stable miscible CO_2 displacement, Iberia Parish, Louisiana. In: *Proc., Int. Energy Agency Workshop on Enhanced Oil Recovery, Tulsa, Okla., September 24.*

Powers, R., 1962. Arabian upper Jurassic carbonate reservoir rocks. In: W.E. Ham (Editor), *Classification of Carbonate Rocks. Am. Assoc. Pet. Geol., Mem.*, 1: 122–192.

Pusch, W. and Dean, M., 1982. *Bodcau In Situ Combustion Project, Bellevue Field, Louisiana.* U.S. Dept. Energy Heavy Oil Contractor's Reports, pp. 59–65.

Reeckmann, A. and Friedman, G.M., 1982. *Exploration for Carbonate Petroleum Reservoirs.* John Wiley and Sons, New York, N.Y., 213 pp.

Reitzel, G.A. and Callow, G.D., 1977. Pool description and performance analysis leads to understanding Golden Spike miscible flood. *J. Pet. Technol.*, 29(7): 867–872.

Restine, J.L., 1983. Effect of preheating on Kern River steam drive. *J. Pet. Technol.*, 35(3): 523–529.

Salathiel, R.A., 1973. Oil recovery by film drainage in mixed wettability rocks. *J. Pet. Technol.*, 25: 1216–1224.

Schenk, L., 1982a. M-6: Venezuela's first step to opening the Orinoco belt. *World Oil*, May, pp. 236–248.

Schenk, L., 1982b. Analysis of early performance of the M-6 steam drive project, Venezuela. In: *Proc., Third Joint SPE-DOE Symp. on Enhanced Oil Recovery, Tulsa, Okla., April 4–7*, p. 497, SPE 10710.

Schmidt, C. and McDonald, D.L., 1979. The role of secondary porosity in sandstone diagenesis. In: P.A. Scholle and Schluger (Editors), *Aspects of Diagenesis. SEPM Spec. Publ.*, 26: 175–207.

Scholle, P.A. and Spearing, D. (Editors), 1982. *Sandstone Depositional Environments*. American Association Petroleum Geologists, Tulsa, Okla., 410 pp.

Shepard, F.P., Phleger, F.B. and van Andel, T.H. (Editors), 1960. *Recent Sediments, Northwest Gulf of Mexico* American Association of Petroleum Geologists, Tulsa, Okla., 394 pp.

Singh, S.P. and Kiel, O.G., 1982. Waterflood design (pattern, rate, and timing). In: *Proc. Int. Meet. on Petroleum Engineering, Beijing, 19–22 March*, 2: 203–219.

Smith, F.W., 1978. Ion-exchange conditioning of sandstones for chemical flooding. *J. Pet. Technol.* 30(6): 959–968.

Somerton, W.H. and Radke, C.J., 1980. Role of clays in enhanced recovery of petroleum. In: *Proc., First Joint SPE-DOE Symp. on Enhanced Oil Recovery, April 20–23, Tulsa, Okla.*, p. 363, SPE 8845.

Swanson, D.C., 1979. Deltaic deposits in the Pennsylvanian Upper Morrow Formation of the Anadarko Basin. In: *Pennsylvanian Sandstones of the Midcontinent*, Tulsa Geological Society, Tulsa, Okla., pp. 115–168.

Swanson, D.C., 1980. *Handbook of Deltaic Facies, Subsurface Clastic Facies Workshop, April 2.* Lafayette Geological Society.

Tracy, G.W. and Dauben, D.L., 1982. *Evaluation of North Burbank Unit Tertiary Recovery Pilot Test.* DOE/BC/10033-2.

Trantham, J.C., 1982. *Evaluation of the Bodcau (Bellevue) In Situ Combustion Project* DOE/BE/10033-4.

Trantham, J.C., Threlkeld, C.B. and Patterson, M.L., Jr., 1980. Reservoir description for a surfactant–polymer pilot in a fractured oil reservoir. *J. Pet. Technol*, 32(9): 1647–1656.

Van Horn, L.E., 1983. *El Dorado Micellar-Polymer Demonstration Project, Eighth Annual Report.* DOE/ET/13070-92.

Vest, E.L., Jr., 1970. Oil fields of the Pennsylvanian-Permian atoll, West Texas. In: *Geology of Giant Petroleum Fields. Am. Assoc. Pet. Geol., Mem.*, 14: 185–203.

Wadman, D.H., Lamprecht, D.E. and Mrosovsky, I., 1979. Geologic/engineering analysis of the Sadlerochit reservoir, Prudhoe Bay field. *J. Pet. Technol.*, 31(7): 933–940.

Wardlaw, N.C., 1976. Pore geometry of carbonate rocks as revealed by pore casts and capillary pressure. *Bull. Am. Assoc. Pet. Geol.*, 60: 245–257.

Wardlaw, N.C., 1980. The effects of pore structure on displacement efficiency in reservoir rocks and in glass micromodels. In: *Proc., First Joint SPE/DOE Symp. on Enhanced Oil Recovery, Tulsa, Okla., April 20–23*, pp. 345–351, SPE 8843.

Wardlaw, N.C. and Cassan, J.P., 1979. Oil recovery efficiency and the rock-pore properties of some sandstone reservoirs. *Bull. Can. Pet. Geol.*, 27(2): 117–138.

Wardlaw, N.C. and Taylor, R.P., 1976. Mercury capillary pressure curves and the interpretation of pore structure and capillary behavior in reservoir rocks. *Bull. Can. Pet. Geol.*, 24(6): 225–262.

Weimer, R.J., 1976. Deltaic and shallow marine sandstones: sedimentation, tectonics and petroleum occurrence. *Am. Assoc. Pet. Geol., Contin. Educ. Course. Not.*, 2.

Wilson, M.D. and Pittman, E.D., 1977. Authigenic clays in sandstones. *J. Sediment. Petrol.*, 47: 3–37.

Chapter 3

DETERMINATION OF RESIDUAL OIL SATURATION BASED ON GEOPHYSICAL WELL LOGGING TECHNIQUES

WALTER H. FERTL and GEORGE V. CHILINGARIAN

INTRODUCTION

The pivotal factor that must be known prior to the initiation of enhanced recovery technique in a reservoir that has reached the economic limit of production is the remaining (or residual) oil saturation. This oil saturation must be known accurately to evaluate the economic feasibility of any enhanced oil recovery technique that may be considered. The remaining oil saturation may be equal to the true residual oil saturation. More frequently, however, the remaining oil saturation is greater than the residual oil saturation of the rock. This is due to the (1) production well damage that restricts oil production near the wellbore, (2) bottom-water coning, (3) extensive fracture porosity, and (4) reservoir heterogeneity. Thus, the term *remaining oil saturation* is used here to define the existing oil saturation of a reservoir at any time after production was initiated, whereas the *residual oil saturation (ROS)* defines the oil saturation of a rock that cannot be reduced further by normal waterflood.

Average primary recovery factors are often low (e.g., solution gas drive: 5–30%; gas cap drive: 20–40%; water drive: 35–75%). Secondary recovery projects may also behave disappointingly due to overall poor sweep and displacement efficiencies, reservoir stratification, directional and/or drastic vertical and lateral permeability variations, etc. For example, in a survey of North American waterflood projects, the principal causes of failure included 45% for poor sweep efficiency, 25% for unusual expenses, 20% for equipment failure, and 10% for other causes (Jackson, 1968).

As a result, major reserves of proven but as yet unrecovered oil, left behind after primary and secondary recovery, exist. Based on conservative estimates, application of tertiary recovery methods (miscible flooding, thermal recovery, chemical flooding, etc.) to already known oil reservoirs could recover from 55 billion additional barrels of oil in the U.S. alone (Geffen, 1973) up to 300 billion barrels (Wesson, 1979). In West Germany, total recoverable oil reserves could be doubled should tertiary recovery projects recover only 10% additional oil from the hydrocarbon resources still left behind in the ground, particularly in heavy oil reservoirs.

Generally speaking, as oil recovery technology advances, more and more of this oil will come within the reach of mankind. Ultimate recovery of this

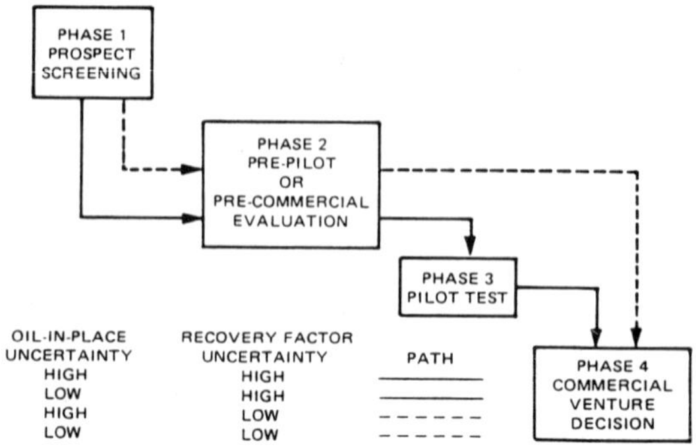

Fig. 3-1. Four phases of enhanced recovery oil development and two decision paths that may be followed, depending on the uncertainty of two factors: oil-in-place and recovery factor. (After Hasiba et al., 1977; courtesy of *World Oil,* Gulf Publishing Company.)

hydrocarbon resource, however, not only depends on available technology but is also largely affected by economic considerations.

As one would expect, geological, petrophysical, reservoir engineering, mechanical, and the economic factors are all interrelated in their superimposed effects on the feasibility, organization, planning, and eventual implementation of any enhanced oil recovery project. On the basis of probabilistic methods, Hasiba et al. (1977) have proposed four sequential steps for corporate planners to reduce the uncertainty and risk involved. These four steps include:

(1) Reservoir prospect screening based on production and injection history, geology, reservoir parameters and fluid properties, etc.

(2) Prepilot and precommercial evaluation based on pressure tests, pulse tests, injectivity tests, and in-fill wells, with emphasis on special coring and logging procedures, tracer studies, reservoir modeling and simulation, etc.

(3) Field pilot test as additional criterion for step 4.

(4) Commercial venture decision based on input from steps 2 and 3 to screen projects by reserve level, project cost, and probability of success.

These concepts (Fig. 3-1) strongly depend upon two key parameters—residual oil-in-place (residual oil saturation and porosity) and recovery factor.

DETERMINATION OF RESIDUAL OIL SATURATION

Reliable evaluation and knowledge of residual oil saturation is a key factor in the selection and proper planning of tertiary recovery methods.

Several techniques are available to determine residual oil saturation. These include (1) reservoir engineering studies, (2) core analysis, (3) single-well tracer tests, (4) pressure transient tests, and (5) geophysical wireline logging techniques.

There are, however, other complicating factors that must be considered, such as the lithology of the reservoir, salinity of the formation waters, composition of the oils, and the oil–water–rock interactions. The large variety of petroleum reservoirs are composed of consolidated sandstones, unconsolidated sands, and carbonates (limestones, dolomites, chalk), which may be clean or may contain varying amounts of silt and shale. The formations also may be uniform (homogeneous) or severely stratified. The intergranular porosity network itself is frequently complicated by fractures and solution vugs that give it a multi-porosity character. Rock–fluid interactions complicate the fluid-flow and electrical conductivity properties of the rock and its fluids by producing variability of preferential wetting. Rocks may exhibit preferential water or oil wettability, or a range of mixed wettability.

The existing oil saturation of petroleum reservoirs may be determined by utilizing:

(1) *Reservoir engineering studies*: material balance techniques based on initial oil-in-place estimates, and production history and modeling (simulation) techniques via analytical models.

(2) *Core analysis data*. The key objective is to bring an actual rock sample from the subject reservoir to the surface without disturbing its fluid content. Hence, ideally, one should maintain the core pressure and minimize loss of the initial core fluid due to flushing effects when core is cut. Resulting effects can be quite significant (Fig. 3-2).

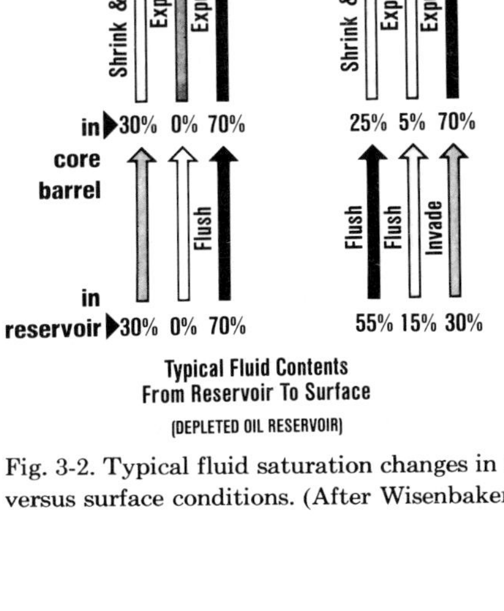

Fig. 3-2. Typical fluid saturation changes in reservoir cores from a depleted oil reservoir under in-situ versus surface conditions. (After Wisenbaker, 1980; courtesy of Core Lab, Dallas, Texas.)

Thus interpretation of core analysis data depends on the type of drilling fluid (high- versus low-fluid loss water-base mud, emulsion mud, or oil-base mud), type of core barrel (conventional, rubber sleeve, pressure core, sponge core, etc.), handling of the cores (wrapping, sealing in wax, freezing, etc.), conventional analyses (distillation, summation of fluids, Dean Stark method) or specialized analyses, such as laboratory waterflood testing and determination of relative permeabilities.

Several mathematical calculations have been developed to correct residual oil saturation estimates from routine core analysis data to in-situ conditions. A method proposed by Kazemi (1977) considers the effects on a core sample taken with water-base mud resulting from reservoir heterogeneity, capillary pressure effects, and the mobility of oil and water. The correction to subsurface conditions is expressed mathematically as follows:

$$ROS = S_{oc} \cdot B_o \cdot E \cdot \frac{M}{1 - V^2} \tag{3-1}$$

where ROS = in-situ residual oil saturation, %; S_{oc} = average residual oil saturation from core analysis, %; B_o = reservoir volume factor for oil; $E = 1.11$, correction factor for core bleeding; M = mobility ratio of the fluids in pore space; and V = coefficient of permeability variation.

(3) *Single-well tracer tests* involve the injection of a bank of primary tracers (e.g., organic ester such as ethyl formate, propyl formate, or ethyl acetate) into a reservoir that is at residual oil saturation (Tomich et al., 1973). This bank is then displaced into the reservoir rock by additional injection of water without the primary tracer. All the injection water, however, is tagged with methanol, which serves as a non-reactive material balance tracer. Following injection, the well is shut in to allow partial hydrolysis of the ester to form an alcohol and equilibrium partitioning of the ester between the oil and water (the alcohol is assumed to be totally insoluble in the oil phase.). Then the well is put on production and all tracer concentrations are measured. Chromatographic separation is used to relate the oil saturation to the difference in arrival time of the primary (ester) and secondary tracers (alcohols) at the wellbore via digital computational modeling techniques. The technique is applicable in open and/or cased wellbores and can analyze the reservoir to large depths (10–20 ft), but has to be corrected for "drift" effects using the methanol tracer.

(4) *Pressure transient (well) testing* includes several methods, such as pressure buildup and draw-down, injectivity, fall-off, and interference tests. The information obtained describes wellbore damage, effective permeability, reservoir drainage volume, etc. Test data can then be analyzed with relative permeability data and core permeability data to estimate oil saturation.

(5) *Geophysical well logging techniques*. Basically, well logging devices measure secondary petrophysical parameters, such as resistivity, nuclear, and acoustic, and radioactive reactions. These can then be related to the primary reservoir properties, such as water saturation (S_w), porosity (ϕ), and formation permeability (k).

Water saturation (S_w) is determined from the basic Archie's (1942) equation or from more advanced shaly sand models (Waxman and Smits, 1968). Residual oil saturation (ROS) values are determined from material balance equations. For example, total reservoir porosity minus water-filled pore space equals the bulk volume of residual oil:

$$\phi - \phi \cdot S_w = \phi \cdot ROS$$

or:

$$1 - S_w = ROS \tag{3-2}$$

Monte Carlo-type simulation techniques can be used to investigate the uncertainty (confidence limits) of log-derived oil saturation values for a given set of optimum reservoir conditions, including consistency in grain-size distribution, cementation, wettability, and absence of unrecognized lithology changes and salinity variations (Walstrom et al., 1967). Similar studies have focused on the accuracy of standard open-hole (Fertl, 1979a, b) and cased-hole logging techniques (Schenewerk et al., 1982).

Archie's equation is one of the very basic petrophysical relationships applied to well log analysis:

$$S_w^n = R_o/R_t = a \cdot \phi^{-m} \cdot R_w/R_t \tag{3-3}$$

where S_w = water saturation, %; n = saturation exponent; R_o = true formation resistivity in water-bearing reservoir ($S_w = 100\%$); R_t = true formation resistivity; a = empirical constant; ϕ = reservoir porosity, %; m = cementation exponent; and R_w = formation water resistivity.
Assuming a hypothetical, shale-free (i.e., clean) reservoir rock for which the parameters ϕ, R_t, and R_w are accurately known (i.e., the uncertainty is zero), with only the saturation exponent (n) and cementation exponent (m) varying, one can calculate the expected uncertainty limits according to the following equation:

$$\Delta S_w \approx \pm \left[\left(\frac{\partial S_w}{\partial m} \cdot \Delta m \right)^2 + \left(\frac{\partial S_w}{\partial n} \cdot \Delta n \right)^2 \right]^{1/2} \tag{3-4}$$

where:

$$\frac{\partial S_w}{\partial m} = -\frac{S_w}{m} \cdot \ln \phi$$

$$\frac{\partial S_w}{\partial n} = -\frac{S_w}{n} \ln S_w$$

$\phi = 20\%$, $m = 2.0 \pm 0.2$, $n = 2.0 \pm 0.2$, and $S_w = 50\%$; then $S_w = 50 \pm 9\%$.

Calculated water saturation values significantly depend upon the value assumed for the saturation exponent n. While many publications tacitly make the

TABLE 3-I

Comparison of data from repeat runs on Frio Sandstone (after Walther, 1968; courtesy of the Society of Petroleum Engineers)

No.	Sample	Runs	1	2	3	4
1	ϕ_{ca}	1	22.0	26.8	27.5	29.3
2		2	22.0	27.2	26.2	28.9
3	ϕ_{ncp}	1	22.0	26.8	26.9	28.5
4		2	21.7	26.5	26.6	28.2
5	F	1	13.8	10.4	10.5	8.6
6		2	14.3	10.7	10.7	8.9
7	m based on ϕ_{ncp}	1	1.74	1.78	1.79	1.72
8		2	1.76	1.82	1.77	1.76
9	S_w by material balance	1	38.1	36.8	43.8	38.4
10	S_w by distillation	1	37.1	36.5	41.1	37.7
11	S_w by material balance	2	38.2	37.6	44.2	36.6
12	S_w by distillation	2	39.4	38.2	44.9	36.7
13	n two-electrode	1	2.08	2.21	1.86	1.81
14		2	2.00	2.11	1.85	1.72
15	n four-electrode	1	2.00	2.08	1.83	–
16		2	1.99	2.06	1.88	1.75
17	n average		2.02	2.11	1.85	1.76

general assumption that n approximates 2.0, the literature furnishes clear evidence that n varies depending upon rock type, wettability, confining pressure, etc. Hence, a laboratory technique was developed by Walther (1968) for measuring m and n under confining pressure and at formation temperature. The technique involved mounting core samples (e.g., Frio, Miocene, and Berea sandstones and Cretaceous limestones) in a Hassler-type sleeve, using an oil–brine flow system, and making two- and four-electrode measurements.

Table 3-I shows data for both m and n from repeat runs on Frio Sandstone samples. According to Walther (1968), "since the accuracy of n factor measurements is probably in the range of $\pm 5\%$, these four samples from the same zone do show significant differences. The major sources of error are probably: (1) saturation measurements and (2) uniform fluid saturation obtained within the core. The m factor measurement generally shows closer agreement between runs. This probably occurs because only one phase is present and the porosity measurements are more accurate than the saturation measurements."

Data by Coates and Dumanoir (1973) listed in Table 3-II comprises experimental values of n for 36 formations of different lithology. The value of n varies from 1.5 to 3.8; only three samples have the "classic" value of $n = 2.0$ and only 12, or 33% of these samples, exhibit an n value within 10% of 2.0.

Log–core study of the low-porosity and rather tight Cotton Valley Sandstone also showed significant variation of n within this formation (Wilson and Hensel,

TABLE 3-II

Values of m and n for various formations (after Coates and Dumanoir, 1973; courtesy of SPWLA)

Formation	Lithology [a]	Average m	Average n
Wilcox, U.S. Gulf Coast	ss	1.9	1.8
Sparta, south Louisiana (Opelousas)	ss	1.9	1.6
Cockfield, south Louisiana	ss	1.8	2.1
Government Wells, south Texas	ss	1.7	1.9
Frio, south Texas	ss	1.8	1.8
Miocene, south Texas	consolidated ss	1.95	2.1
	unconsolidated ss	1.6	2.1
Travis Peak and Cotton Valley	hard ss	1.8	1.7
Rodessa, east Texas	ls	2.0	1.6
Edwards, south Texas	ls	2.0	2.8
Woodbine, east Texas	ss	2.0	2.5
Annona, north Louisiana	chalk	2.0	1.5
Nacatoch, Arkansas	ss	1.9	1.3
Ellenburger, west Texas	ls and dol	2.0	3.8
Ordovician Simpson, west Texas and New Mexico	ss	1.6	1.6
Pennsylvanian, west Texas	ls	1.9	1.8
Permian, west Texas	ss	1.8	1.9
Simpson, Kansas	ss	1.75	1.3
Pennsylvanian, Oklahoma	ss	1.8	1.8
Bartlesville, Kansas	ss	2.0	1.9
Mississippian, Illinois	ls	1.9	2.0
Mississippian, Illinois	ss	1.8	1.9
Pennsylvanian, Illinois	ss	1.8	2.0
Madison, North Dakota	ls	1.9	1.7
Muddy, Nebraska	ss	1.7	2.0
Cretaceous, Saskatchewan, Canada	ss	1.6	1.6
Bradford, Pennsylvania	ss	2.0	1.6
Frio, Chocolate Bayou, Louisiana	ss	1.55–1.94	1.73–2.22
Frio, Agua Dulce, south Texas	ss	1.71	1.66
Frio, Edinburgh, south Texas	ss	1.82	1.47, 1.52
Frio, Hollow Tree, south Texas	ss	1.80, 1.87	1.64, 1.69
Jackson, Cole Sand, south Texas	ss	2.01	1.66
Navarro, Olmos, Delmonte, south Texas	ss	1.89	1.49
Edwards Lime, Darst Creek, Texas	ls	1.94, 2.02	2.04, 2.08
Viola, Bowie field, north Texas	ls	1.77	1.15
Lakota Sand, Crook County, Wyoming	ss	1.52	1.28

[a] ss = sandstone; ls = limestone; dol = dolomite.

1982). It is noteworthy that at 6% porosity, the irreducible water saturation of 26.3% ($n = 1.73$) shows up as 40% + when the "classic" assumption of $n = 2.0$ is applied, which results in a 52% error.

Even in high-porosity reservoirs, the resulting error in water saturation is primarily dependent on the value of n (Dorfman, 1984):

$$\Delta S_w \approx \frac{\partial S_w}{\partial F}\Delta F + \frac{\partial S_w}{\partial R_w}\Delta R_w + \frac{\partial S_w}{\partial R_t}\Delta R_t + \frac{\partial S_w}{\partial n}\Delta n \tag{3-5}$$

$$\frac{\Delta S_w}{S_w} = \left\{\left(\frac{1}{n}\frac{\Delta F}{F}\right)^2 + \left(\frac{1}{n}\frac{\Delta R_w}{R_w}\right)^2 + \left(\frac{1}{n}\frac{\Delta R_t}{R_t}\right)^2 + \left[\frac{1}{n}\ln\left(\frac{FR_w}{R_t}\right)\frac{\Delta n}{n}\right]^2\right\}^{1/2} \tag{3-6}$$

For $\phi = 0.3$, $m = 2.0$, and $a = 0.8$, the formation resistivity factor $F = 8.9$. Also given are $R_w = 0.05$, $R_o = F \cdot R_w = 0.45$, $R_t = 5.0$, and $n = 2.0$. Assuming a standard deviation of 10% in the parameters F, R_w, R_t, and n, then the error analysis for the calculated water saturation ($\Delta S_w/S_w$) yields 14.9%. Even if the parameters, F, R_w, and R_t are known perfectly (i.e., their uncertainty equals zero), there will still be an error of ($\Delta S_w/S_w$) = 12.1%. In order to reduce $\Delta S_w/S_w$, therefore, one must reduce $\Delta n/n$. This statement is a key consideration for log-inject-log techniques using resistivity logs which are discussed in some detail later on.

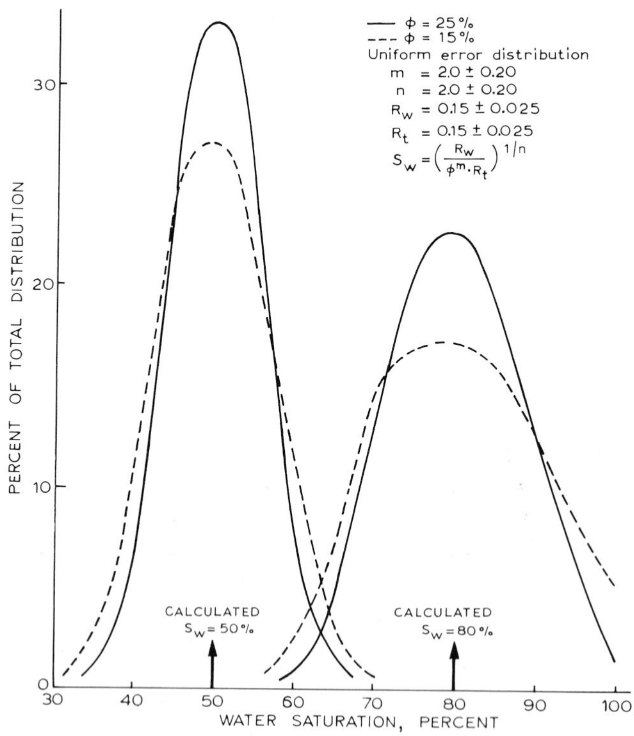

Fig. 3-3. Monte Carlo-type uncertainty distribution for water saturation calculations based on Archie's equation; 90% confidence limits, sample size = 10,000. (After Fertl, 1979b; courtesy of *Oil and Gas Journal.*)

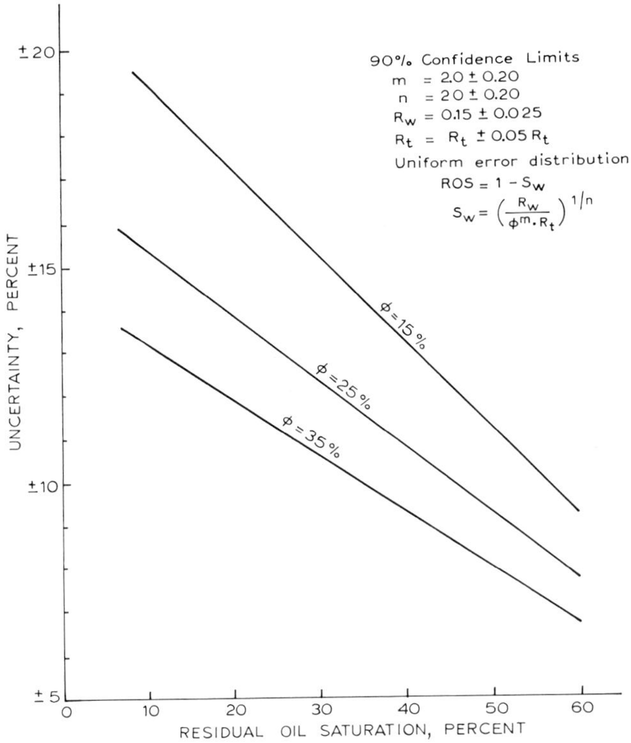

Fig. 3-4. Expected uncertainty (in %) of residual oil saturation calculations as function of reservoir porosity (ϕ) and residual oil saturation (*ROS*); 90% confidence limits, sample size = 10,000. (After Fertl, 1979b; courtesy of *Oil and Gas Journal*.)

In reality, each one of the petrophysical parameters contained in Archie's equation has a specific error distribution (e.g., uniform, triangular, and normal), which in turn significantly affects the accuracy (uncertainty limits) of water saturation values computed by these variables. Monte Carlo-type statistical studies indicate that for a given reservoir porosity (ϕ), the saturation exponent (n) and cementation exponent (m) are responsible for the largest uncertainty in calculated saturation values (S_w, *ROS*), whereas the effect of errors in R_w and R_t is less (Fertl, 1979a). As expected, uncertainty in log-derived saturation values increases with decreasing porosity and decreasing oil saturation.

Results of digital probability distribution studies (Fertl, 1979b) using a sample size of 10,000 for all sets of input variables and postulating uniform error distributions for the latter, are shown in Figs. 3-3 and 3-4 for the 90% confidence limits (i.e., a chance of 9 out of 10). Even under unrealistically optimum reservoir conditions (i.e., $\phi > 30\%$, $ROS = 50\%$), the 90% confidence limits for the error limits listed in Figs. 3-3 and 3-4 exceed the value of $ROS \pm 8\%$.

TABLE 3-III

Comparison of *ROS* estimates from core and log data, El Dorado micellar–polymer project (after Pusch, 1977; courtesy of the Society of Petroleum Engineers)

Well	Core results				Log results		
	permeability (mD)	porosity (%)	oil saturation (%)	thickness (ft)	porosity (%)	oil saturation (%)	thickness (ft)
MP-101	432	26.0	32.7	18	24.3	30.8	21
MP-102	448	25.5	19.0	21	24.5	30.0	24
MP-103	311	25.1	16.4	19	24.7	19.5	20
MP-104	393	26.5	28.3	18	24.3	31.1	19
MP-105	191	21.6	21.2	19	22.3	26.4	20
MP-106	565	27.5	22.2	19	25.3	31.2	18
MP-110	521	27.3	25.0	19	23.8	31.8	19
MP-112	464	27.5	n.m. [a]	19	25.3	27.9	19
MP-114	385	26.2	29.9	17	23.3	42.5	20
MP-116	686	28.3	25.4	19	24.9	31.2	19
MP-120	433	24.5	34.1	18	23.9	34.1	20
MP-122	500	28.0	26.7	15	24.7	30.5	14
MP-126	448	28.3	n.m.	17	24.2	28.2	17
MP-128	403	26.6	33.0	20	25.5	37.2	22
MP-130	379	27.1	26.4	19	24.6	22.6	19
MP-201	405	25.4	27.1	18	23.5	25.9	11
MP-203	331	25.4	28.5	18	24.6	32.7	20
MP-205	294	25.5	26.5	21	23.8	29.9	22
MP-207	301	26.8	14.8 [b]	19	26.0	45.5	20
MP-211	300	25.6	n.m.	18	24.7	27.3	17
MP-213	469	26.6	25.5	11	26.0	28.5	10
MP-215	398	26.5	21.5	21	24.7	30.3	19
MP-221	581	27.2	31.5	17	24.6	34.0	20
MP-223	513	26.0	24.2	18	22.2	28.9	19
MP-225	201	23.4	20.6	21	22.3	25.5	21
Average	414	26.2	26.0		24.3	30.5	

[a] n.m. = not measured

[b] Saturations measured after core was weathered.

Hence, the degree of uncertainty of today's routine logging and interpretive techniques does not offer any promise for their use in determinations of low oil saturations remaining after secondary recovery within acceptable limits for key decisions on whether or not to initiate an enhanced oil recovery project. This is particularly true for reservoir rocks exhibiting medium to low porosity and marginal residual oil saturation ($ROS \approx 20$–30%).

To illustrate his point (with respect to post-secondary recovery oil saturation), the comparison of oil saturation estimates from core and log data is shown in Table 3-III for the El Dorado micellar–polymer flood project, located in Kansas (Pusch, 1977). The well-log-derived oil saturations are consistently higher than those obtained using core analysis.

Log-inject-log techniques

Log-inject-log (LIL) techniques based on resistivity logs, which exhibit various depths of investigation and vertical resolution, provide residual oil saturation estimates within very narrow uncertainty limits. This technique, proposed by Fertl and Reynolds (1973), consists of the following four steps:

(1) Running base resistivity log.

(2) Chemical flooding of an aqueous solution containing a surface-active agent (e.g., non-ionic, cationic, anionic, and ampholytic); sufficient solution is injected into the formation through the borehole to insure that the oil has been displaced from the vicinity of the wellbore to a distance exceeding the radius of investigations of the logging tool.

(3) Resaturation of the chemically "cleaned" reservoir rock with formation water to restore a 100% water-bearing condition ($S_w = 100\%$).

(4) Running a second resistivity log, which provides a measurement designated as R_o because the reservoir rock is now 100% water saturated.

The residual oil saturation (ROS) can readily be determined using the following mathematical relationship:

$$ROS = 1 - (R_o/R_t)^{1/n} \tag{3-7}$$

By employing the log-inject-log method with deep investigating logging devices, one can obtain measurements farther into the formation, thus providing a more representative value of the remaining oil saturations following secondary recovery. Furthermore, this method does not require knowledge of reservoir porosity (ϕ), and the three required parameters (R_o, R_t, n) can all be determined with relatively high accuracy.

Monte Carlo-type statistical studies by the writers have shown that with realistic error limits for the variables R_o, R_t, and n, the residual oil saturation can be determined with an accuracy of ± 3 saturation percent.

Based on the same sequential steps outlined above, other openhole logging responses can be applied in similar fashion. This has also been proposed by Fertl and Reynolds (1973) for acoustic and density measurements.

For example, the difference in the acoustic log responses is due to the amount of residual oil present in the formation:

$$\Delta(\Delta t)_{\log 1,2} = f(\Delta t_{f1})_{ROS, ROS=0} = f(ROS) \tag{3-8}$$

The same concept also holds true for density log measurements such as:

$$\frac{\Delta f_{\log 2,1}}{\phi} = \Delta f_{\text{fluid } 1,2} = f(ROS) \tag{3-9}$$

Inasmuch as the formation contains only residual oil, it is sometimes advantageous to pre-flush the formation with high-salinity brine to enhance the acoustic and/or density response between oil and the brine. When such a pre-flushing step is employed, the water injected subsequently to establish the water-saturated formation is the same as that employed in the pre-flushing step.

Caldwell (1983) suggested quite similar concepts via a log-inject-log operation using the borehole gravity log. From the resulting bulk density, gravity, and porosity indications, water and oil saturation of the formation are determined. Basically, "the formation is firstly flushed with a fluid having a density greater than that of the water present in the formation. A first borehole gravity log is then recorded of the flushed formation. The formation is secondly flushed with a fluid having a density greater than that of the water present in the formation and differing from that of the first fluid. A second borehole gravity log is then recorded of the flushed formation. The porosity of the formation is also measured. From the density, borehole gravity and porosity factors, the water and oil saturation of the formation are determined. However, in the event the pore water density is known or measurable prior to the log-inject-log operation, only a single flushing of the formation will be required with a borehole gravity log being recorded before and after the fluid injection of the single flushing operation. The fluids injected to flush the formation may be brines of higher density than the formation water contents or immiscible hydrocarbons which do not affect the oil-in-place in the formation."

Artificially-induced gamma-ray radiation via radioactive tracers used in log-inject-log techniques should also provide useful residual oil saturation information. Several methods have been proposed which can be applied in both open and cased wellbores. They include the following steps: (1) running a gamma-ray base log; (2) injection of an aqueous solution of a water-soluble radioactive tracer; (3) relogging the interval; (4) injecting a second aqueous solution of a radioactive tracer soluble in both water and oil; and (5) re-logging again.

According to Casad (1978), the data collected from the logging runs determine residual oil saturation as follows:

For the logging run after injection of the first aqueous solution where $R_{o/w} = 0$:

$$\phi_T = \phi_B + \phi_N + \phi_{W1} \tag{3-10}$$

For the logging run after injection of the second aqueous solution, where $R_{o/w} > 1.0$:

$$\phi_T = \phi_B + \phi_{W2} + \phi_o \tag{3-11}$$

where $R_{o/w}$ = equilibrium partition ratio, defined as the concentration of the second compound in the oil divided by the concentration of the second compound in the water at equilibrium; ϕ = gamma-ray flux; ϕ_T = total measured flux; ϕ_B = flux from material in wellbore; ϕ_N = flux from naturally-occurring isotopes; ϕ_{W1} = flux from material in the first aqueous solution; ϕ_{W2} = flux from material in the second aqueous solution; and ϕ_o = flux from material in the oil in the formation.

The ϕ_N is determined by the first logging run prior to the injection of the first aqueous solution and ϕ_B is readily determined by logging a portion of the wellbore above the formation at a time when the wellbore is filled with a material of interest. The ϕ_{W1}, ϕ_{W2}, and ϕ_o are determined from the logging runs after the injection of the first aqueous solution and the second aqueous solution, respectively.

The ϕ_o is proportional to *ROS*, the concentration of tracer material in the oil (C_o); and the formation porosity (*P*), whereas ϕ_W is proportional to the concentration of tracer material in the water. Therefore:

$$\phi_o = K \cdot ROS \cdot P \cdot C_o \text{ and } \phi_W = K \cdot ROS \cdot P \cdot C_W \quad (3\text{-}12)$$

where K is a proportionality constant. Furthermore, ($ROS + S_W$) equals 1.0, and P and K are constants for a given reservoir rock and a given oil–water system. The C_o equals $R_{o/w} \cdot C_W$. Hence, ϕ_o/ϕ_W simplifies, and inasmuch as the sum of the fluid components equals unity (i.e., $ROS + S_W = 1.0$), then:

$$ROS = \left(\frac{\phi_o}{\phi_W R_{o/w}}\right) \cdot \left(1 + \frac{\phi_o}{\phi_W R_{o/w}}\right)^{-1} \quad (3\text{-}13)$$

In another radioactive method proposed by Mitchell et al. (1978), the *ROS* value is determined by using the following steps: (1) injecting radioactive water into the formation; (2) logging the gamma-ray response of the formation; (3) chemically flooding the formation to remove all of the residual oil; (4) injecting radioactive water into the formation; (5) logging the gamma-ray response of the formation; and (6) determining *ROS* from the difference in the two gamma ray log responses. Knowledge of reservoir porosity is not required.

According to Mitchell et al. (1978), their method "is based on the fact that the radiation measured will be directly proportional to the volume of the radioactive aqueous phase present both before and after displacement of residual oil. Therefore, ($GR_2 - GR_1$) is proportional to the volume of residual oil previously present in formation, and:

$$ROS = (GR_2 - GR_1/GR_2) \quad (3\text{-}14)$$

where GR_1 = radiation measurement with residual oil and aqueous phase (1) in formation corrected to the same specific activity as aqueous phase (2), if aqueous phases (1) and (2) have different specific activities; GR_1 is proportional to the volume of water present/unit volume of formation; and GR_2 = radiation measurement with only aqueous phase (2) present in formation; GR_2 is proportional to the pore volume present/unit volume of formation."

These authors suggested a further refinement where conditions seem to warrant it; "a very small amount of radioactive aqueous phase may still exist in essentially film form between the inflatable packer element and the casing of the well. A background measurement, *BG*, of this radioactivity can be made in a borehole containing a radioactive aqueous phase by (1) inflating the packer in an interval where no injection has occurred, and (2) performing logging measurements as before, and (3) correcting the background measurement to the same specific activity as the GR_1 and GR_2 measurements. The expression would then become:

$$ROS = (GR_2 - GR_1)/(GR_2 - BG) \tag{3-15}$$

Thus, we compensate for residual radioactivity in the borehole."

The paramagnetic solution flood technique using nuclear magnetism logging (NML) for determining *ROS* has been investigated by Robinson et al. (1972) and Thomas and Ausburn (1976). The NML is not yet a fully developed standard logging technique, mainly due to the still marginal signal-to-noise ratio.

Basically, in nuclear magnetism logging a magnetic field is being set up within the reservoir rock by means of an energy source and a polarizing coil. The coil is then de-energized, and the voltage induced in the coil by processing hydrogen nuclei is measured. The magnitude of the voltage induced is a measure of the number and geometry of the hydrogen nuclei. The time of realignment of the hydrogen nuclei with the earth's gravitational field is related to the type of fluid containing these nuclei. Thus, the use of nuclear magnetism in well logging is of special interest, because it enables direct measurements on the hydrogen (proton) in the formation fluids and not in the rock matrix.

The NML measurements are possible in open hole only. They exhibit a shallow depth of investigation and require a rather complex stepwise procedure, including:

(1) Preparation of aqueous paramagnetic solution with the desired proton spin-lattice thermal relaxation time.

(2) Displacement of formation water around the borehole by paramagnetic solution.

(3) Multimeasurements of oscillograms for signal-plus-noise and noise alone.

(4) Statistical data analysis to extract hydrocarbon signal to obtain the percent of pore space occupied by oil, such as $ROS = FFI/\phi$, where FFI = free fluid index and ϕ = reservoir porosity.

Table 3-IV summarizes the results reported by Robinson et al. (1972) from a test well located in the Benton Field, Franklin County, Illinois. The probable error results were derived using Monte Carlo techniques. The NML-derived saturation results indicate significantly reduced error limits in residual oil saturation determination ($ROS \pm 3\%$).

Over the last ten years, electromagnetic wave propagation measurements (i.e., dielectric constant well logging) have been developed as a reliable borehole measurement.

Electromagnetic measurements of earth formations basically depend on three main parameters: electrical resistivity, magnetic permeability, and dielectric permittivity.

Today's standard induction-type logging devices use the relatively low frequency of 20 kHz and thus measure the electrical conductivity of formation. Hence, the usefulness of formation conductivity (resistivity) measurements is directly related to the conductivity (resistivity) contrast between water and

TABLE 3-IV

Comparison of probable errors in *ROS* estimates (after Robinson et al., 1972; courtesy of the Society of Petroleum Engineers)

Depth (ft)	Porosity (%)			NML (% bulk volume)		Resistivity [a] (% bulk volume)		
	core analysis	density log	average $\pm 1\sigma$	$S_o\phi \pm 1\sigma$ (FFI)	$S_o \pm 1\sigma$	R_{x_o} (Ω m)	$S_o\phi \pm 1\sigma$	$S_o \pm 1\sigma$
2128	18.0	17.3	17.6 ± 0.5	4.74 ± 0.37	27.0 ± 2.2	17.0 ± 0.5	5.04 ± 0.95	28.5 ± 5.0
2138	18.5	16.9	17.7 ± 0.75	5.33 ± 0.26	30.2 ± 1.9	16.5 ± 0.5	4.95 ± 1.10	27.8 ± 5.5
2164	18.6	16.1	17.3 ± 1.0	4.29 ± 0.27	24.9 ± 2.2	17.0 ± 0.5	4.79 ± 1.25	27.4 ± 6.2
2192	18.2	17.9	18.0 ± 0.5	8.78 ± 0.52 [b]	48.8 ± 3.1	17.2 ± 0.5	5.47 ± 0.94	30.3 ± 4.9

[a] Fixed values and probable errors to one standard deviation are: $R_w = 0.43 \pm 0.02$ Ωm; R_{x_o} as shown, $m = 1.73 \pm 0.070$; $n = 2.00 \pm 0.050$.

[b] Incomplete displacement of paramagnetic complex into formation probably resulted in some measured water signal.

hydrocarbon-bearing formations. The higher the formation water salinity, the better the contrast and differentiation of water and hydrocarbons. In other words, induction logs work poorly to investigate reservoir characteristics in the presence of brackish or very low formation water salinity.

With increasing frequencies of logging devices, however, the real part of the propagation constant is only weakly dependent on conductivity. Several commercial dielectric logging devices are available to the petroleum industry. Instrument frequency cover the megahertz (20, 25, 30, 47, 60, 200 MHz) to gigahertz (1.16 GHz) range. Basically, signal amplitude and phase difference measurements are recorded to determine the apparent dielectric constant in the formation of interest. There are practical trade-offs in selecting a specific frequency logging device. These include:

(1) Attenuation due to conductivity increases with frequency, thus decreasing the depth of investigation in the reservoir.

(2) Salinity and lithology effects decrease with increasing frequency.

(3) Very high frequency logging devices exhibit very shallow depth of investigation and, thus, are significantly affected by borehole rugosity and thick mud cakes.

(4) High frequencies require pad-type logging instrumentation.

These dielectric logging devices can be run in open hole in any type of drilling fluid (including oil-base mud).

Inasmuch as the dielectric constant of water is approximately forty times that of oil, water saturation can be determined even in freshwater environments. Numerous studies and field experiences have shown this to be a practical way of quantitative water saturation evaluation in potential reservoir rocks which contain brackish or fresh formation waters (Friedman and Vogiatzis, 1979; Wharton et al., 1980; AOSG, 1981; Huchital et al., 1981; Xuwen et al., 1982; Janes et al., 1984).

Fertl (1985) proposed a technique based on dielectric logging, which determines water saturation in earth formations independent of lithology. The interval of interest is first logged to obtain a base log. Then the formation is flooded with a sufficient amount of oil-miscible solution containing a surface-active agent to displace substantially all the residual fluids from the reservoir for a radial distance at least equal to the radial zone of investigation. The formation is then flooded with a sufficient amount of water to render the zone of investigation 100% water saturated. The interval is logged a second time to obtain measurements of the water-saturated formation. Combining the information of both logging runs, the water saturation of the reservoir can be determined independent of lithology.

Pulsed neutron logging devices can also be run in open hole, but find their main application in cased wellbores.

RESIDUAL OIL SATURATION DETERMINATION IN CASED WELLBORES

For over fifteen years, pulsed neutron logging devices utilizing several different gating systems (that measure the macroscopic cross-sections (Σ) for thermal neutron capture in a borehole environment) have been highly successful in differentiating high-salinity formation waters from hydrocarbons behind casing on both a qualitative and quantitative basis. Hence, they can be used to monitor time-related variations of water, oil, and gas saturations in reservoirs under primary or enhanced recovery schemes. The resulting information is invaluable input in reservoir engineering and workover projects.

Neutron Lifetime (Serpas et al., 1977; Randall et al., 1981) and Thermal Decay Time (Hall et al., 1980) devices have their optimum range of applications in saline, high-porosity reservoirs. They cannot, however, be used to evaluate and/or monitor hydrocarbon saturations where the macroscopic cross-section of formation water is not significantly different from that of the hydrocarbons, or in cases where salinity is constantly changing and unknown (e.g., in the case of fresh water or steam flood). Logs can be run equally well in fluid- and air-filled boreholes. Response of the Continuous Carbon/Oxygen (C/O) log (Hopkinson et al., 1981; Oliver et al., 1981), which is also a pulsed neutron device, however, is not affected by such salinity constraints. This also holds true for the Gamma-Ray Spectrometer Tool (Westaway et al., 1980).

Log-derived determination of residual oil saturation for enhanced recovery projects requires the best and most accurate *ROS* value obtainable. Special considerations, therefore, have to be given to both the logging operation and interpretive techniques. Inasmuch as the statistical uncertainty limits of the conventional open- *and* cased-hole logging techniques are not acceptable for determining *ROS* values (particularly in reservoirs having medium or low porosity and marginal *ROS* values), a key parameter in the evaluation of enhanced recovery projects, several log-inject-log (LIL) techniques based on

TABLE 3-V

ROS estimates (%) using log-inject-log and core analysis methods (after Murphy et al., 1976; courtesy of the Society of Petroleum Engineers)

Test	Formation and location	Interval analyzed (ft)	Log-inject-log		Pressure core		Native state core flood	
			range	average	range	average	range	average
1	Sims, Oklahoma	120	9–50	33				
2	Muddy "J", Well 1, Nebraska	16	21–42	33			19–24	21
3	Muddy "J," Well 2, Nebraska	13	14–46	31				
4	Grayburg, Texas	115	12–70	34	0–45	32 [a]	12–50	34
5	Morrow, Texas	6		25				
6	San Andres, Field A, Texas	74	20–63	34			18–45	32
7	San Andres, Field B, Texas	72	11–48	36	4–54	31 [a]	15–60	28
8	First Wall Creek, Wyoming	78	13–54	34			20–33	25
9	Second Wall Creek, Wyoming	66	25–53	34			13–36	21
10	Tensleep, Wyoming	29	14–36	25			15–25	20
11	Beaverhill Lake, Well 1, Canada						12–42	33
12	Beaverhill Lake, Well 2, Canada	20	20–41	33				

[a] Core saturation corrected to bottom-hole conditions.

multiple repeat logging runs have been developed to provide more reliable *ROS* values. Numerous articles have been published on this subject, which has been frequently reviewed (Fertl, 1979a, b; Wyman, 1977; Bond, 1978; Wesson, 1979).

Several log-inject-log (LIL) techniques using multiple repeat logging runs allow determination of *ROS* within ± 5 saturation percent. Comparative North American field data for *ROS* determination based on LIL techniques and core analysis data as compiled by Murphy et al. (1976) is given in Table 3-V.

The LIL techniques using pulsed neutron logs have already been used extensively to determine residual oil saturation profiles (distributions) in depleted oil reservoirs. Proper planning of LIL operations, evaluation of reservoir and borehole conditions, and fluid injection procedures are an absolute prerequisite. Important considerations and guidelines are summarized in Table 3-VI (Fertl, 1979a).

Statistical variations are inherent to all pulsed neutron measurements. Averaging of multiple (5–10) repeat logging runs, therefore, are recommended for log quality control (e.g., unreliable logging runs are omitted from the averaging

TABLE 3-VI

Considerations for log-inject-log for residual oil saturation determinations (after Fertl, 1979a; courtesy of *Oil and Gas Journal*)

Reservoir	Logging devices	Well conditions	Injection
1. High porosity, high residual oil saturation and good permeability	1. Properly functioning, calibrated instruments	1. Enough rat-hole so entire zone can be logged	1. Non-uniform injection profiles suggest poor fluid displacement in stratified formations
2. Select uniform reservoir	2. Multiple repeat runs (6–10) at proper logging speed, time constant, etc., to reduce statistics	2. Evaluate a short single zone rather than a too long zone or multiple zones to facilitate control of proper injection procedures	
3. Avoid fractured or fracturing of reservoir which is very detrimental to sweep efficiency	3. Zones investigated by logs must be completely covered by the injection—does not necessarily guarantee complete fluid replacement around the cased wellbore	3. Newly perforated intervals rather than zones with old perforations to avoid formation slumping, sand production, and resulting drastic porosity changes	3. Proper control of injection pressure (versus fracture gradient) and rates
4. Availability of reliable porosity information		4. Avoid tests in old injection wells, since ROS may be drastically reduced due to "stripping effects"	4. Injection fluids prepared under controlled conditions (i.e., batch mixing, calculated and/or measured Σ-values)
5. Gas saturation is zero in subject reservoir		5. Satisfactory well completion and zone isolation	

Note: Do not be concerned about what at the first appears to be conflicting residual oil saturation data obtained from reservoir engineering concepts, single-well tracer tests, core analyses, and log-derived tests. Closely study the valid reasons for apparent discrepancies, which are many. Keep in mind that results may be weighted by permeability, porosity, depth of investigation and vertical resolution of logs, etc. Also note that no single method alone gives totally meaningful results of both the amount and the distribution of residual oil saturation.

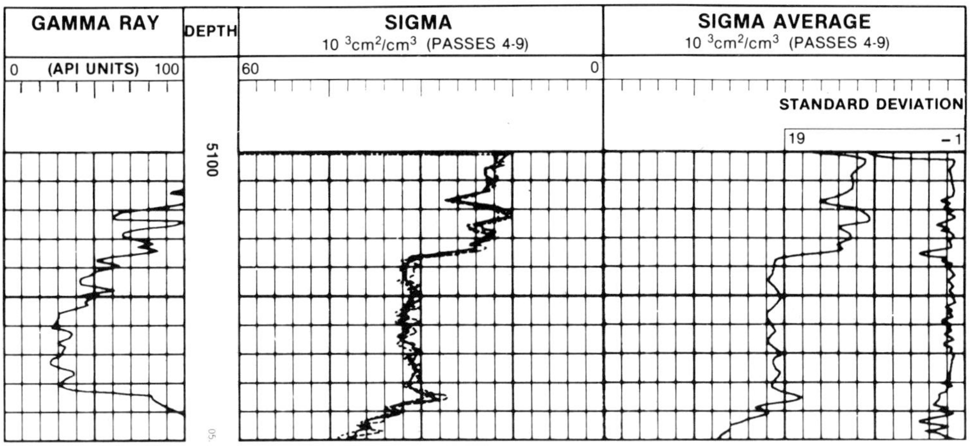

Fig. 3-5. Field example of five pulsed neutron Σ-runs, Σ-average, and standard deviation. (After AOSG, 1982.)

calculations) to provide an improved average log response over zone of interest and its statistically significant standard deviation.

Fig. 3-5 shows five sigma-runs, the average Σ-value, and standard deviation of approximately ±0.7 Σ-unit over the zone of interest (AOSG, 1982). Recent

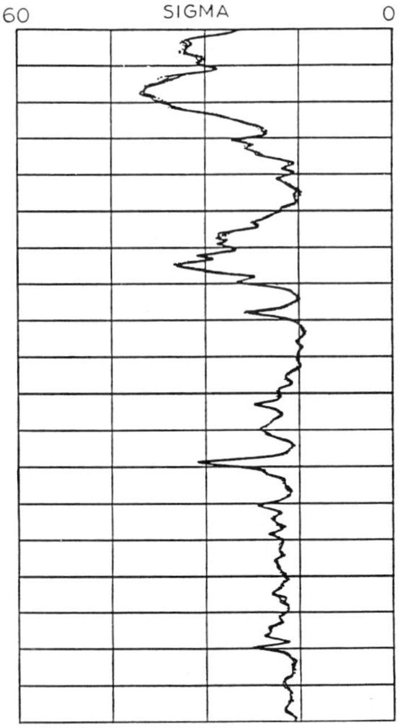

Fig. 3-6. Field example of five pulsed neutron Σ-runs with PDK-100 logging device over 500-ft interval. (After AOSG, 1982.)

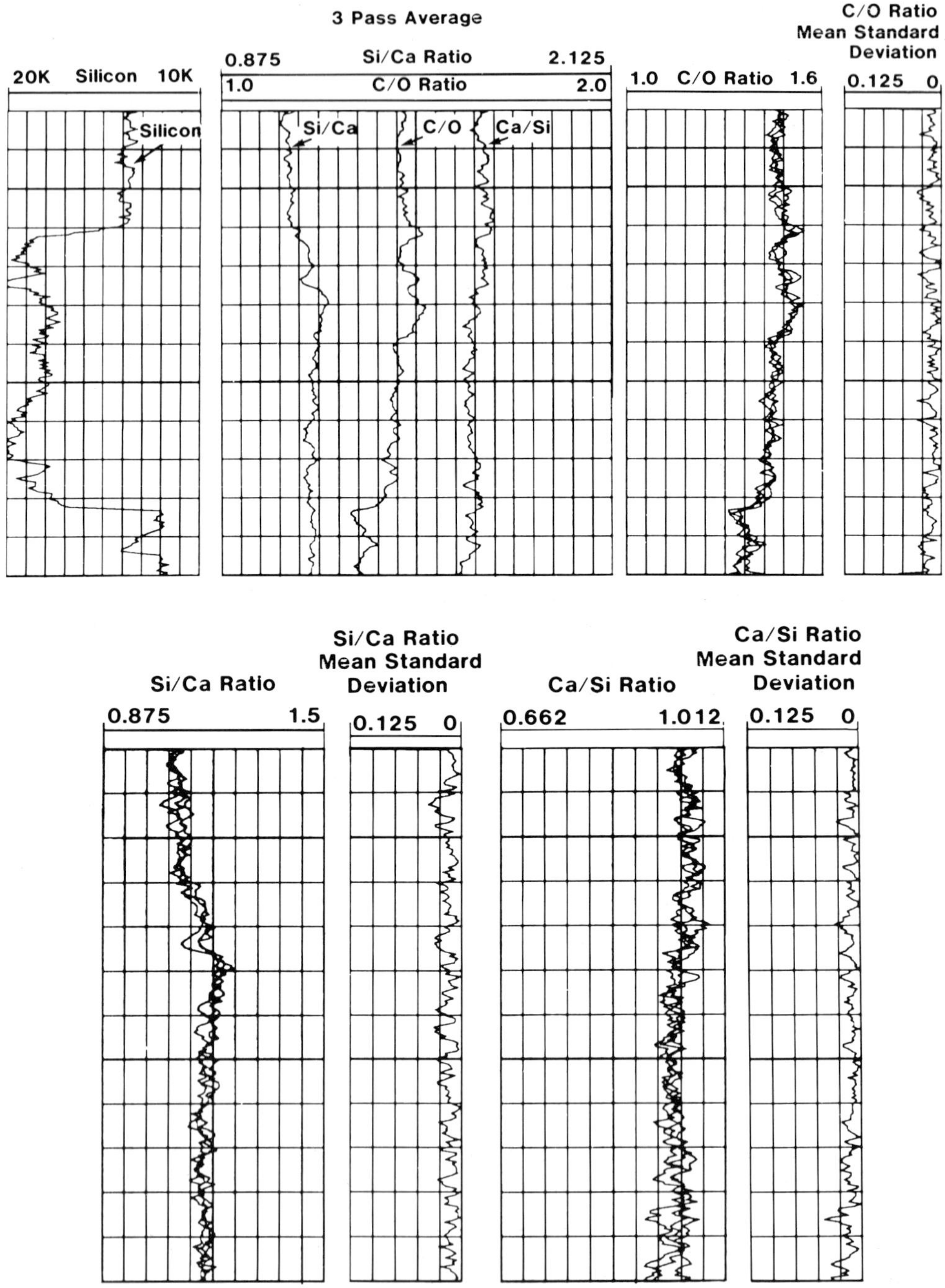

Fig. 3-7. Statistical variations of continuous carbon/oxygen (C/O) logging measurements. Raw data, repeat run overlays, and mean standard deviation of C/O, Si/Ca, and Ca/Si ratios. (After AOSG, 1982.)

instrument improvements in neutron lifetime technology (i.e., the PDK-100 log, now provide accuracies of ± 0.2 Σ-unit on, for example, five repeat runs over a 500-ft interval) (Fig. 3-6). This results in even more reliable *ROS* values.

Another field case presented in Fig. 3-7 shows three Continuous Carbon/Oxygen (C/O) passes over a zone of interest, with the mean standard deviation computed for each of the ratios (Oliver et al., 1981).

Waterflood log-inject-log technique

As a *single-step injection* technique, the waterflood log-inject-log method is applicable only in reservoirs at residual oil saturation conditions. The three operational steps include: (1) base log (Σ_1, Σ_{w1}), (2) injection of brine of pre-selected salinity, (3) repeat log (Σ_2, Σ_{w2}). Then *ROS* is determined (e.g., Youmans et al., 1966; Richardson et al., 1972) as follows:

$$ROS = 1.0 - (\Sigma_2 - \Sigma_1)/[\phi(\Sigma_{w2} - \Sigma_{w1})] \qquad (3\text{-}16)$$

where ϕ = reservoir porosity; Σ_w = capture cross-section of formation water; $\Sigma_{w2} \neq \Sigma_{w1}$ = largest feasible, pre-selected salinity contrast.

This technique is independent of reservoir matrix and the hydrocarbon capture cross-sections. The fluid Σ-values can be calculated provided the chemical compositions are accurately known or can be measured at the wellsite on small fluid samples, e.g., in the Dresser Atlas sigma-fluid cell.

The preferred method of determining the capture cross-sections for fluids is by direct measurement using the sigma-fluid cell, described below. The calculation of Σ for fluids from chemical analysis data is less reliable, because of the disproportionate effect on Σ of some constituents, such as boron. The procedure for this calculation is given below, along with a chart of Σ as a function of concentration for sodium chloride brines.

Sigma-fluid cell. A device capable of measuring the macroscopic cross-section for thermal neutron absorption (Σ) has been developed for use with small fluid samples. This system allows the Σ of water samples to be determined at the wellsite, and can provide a check on salinity variations of fluids used in an LIL program.

The sigma-fluid cell consists of a fluid container, a neutron source, a neutron detector, and associated electronics. It is easily transportable, simple to operate, and provides reliable and accurate measurements of Σ at the wellsite.

Thermal neutron absorption cross-sections are measured using an isotopic (AmBe) neutron source. The fluid sample, collected at the wellsite, is placed between the source and a neutron detector (Fig. 3-8). Neutrons from the AmBe source are emitted at an average energy of approximately 3 MeV, and are rapidly thermalized in the fluid sample. The neutron detector responds primarily to the thermal neutrons which are not absorbed by the moderating material. The number of neutrons absorbed in the fluid is proportional to Σ; therefore, the

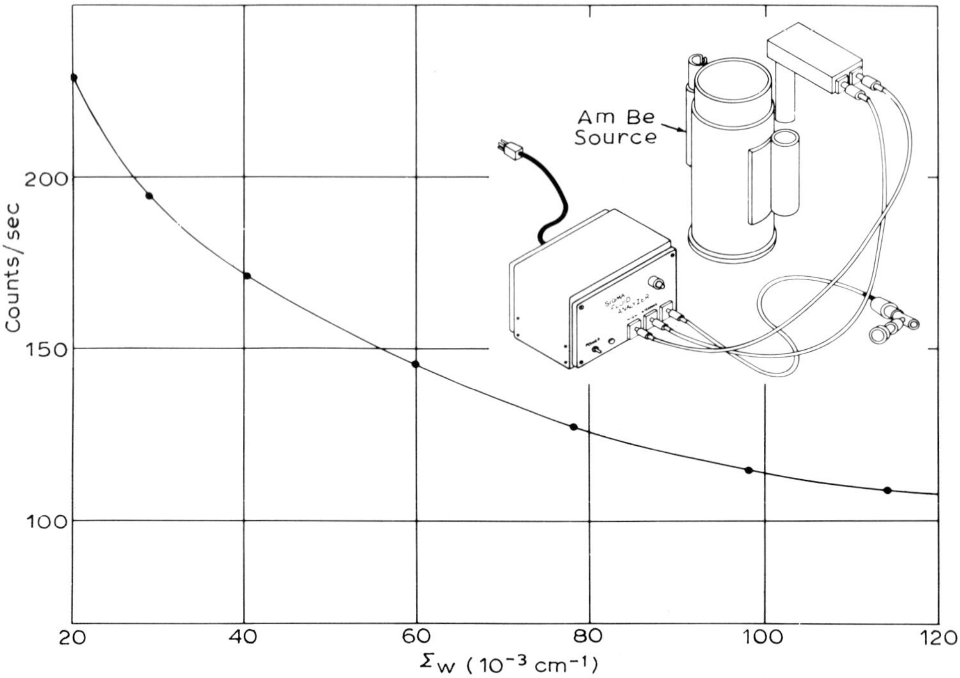

Fig. 3-8. Sigma-fluid cell response versus fluid capture cross-section (Σ_w).

number of thermal neutrons reaching the detector per unit time is proportional to $1/\Sigma$.

Fig. 3-9 depicts the count rate of the detector versus fluid capture cross-section Σ for solutions with varying NaCl concentrations. Measurements of Σ with the device are more accurate than determinations by resistivity or chemical analysis and it can measure contributions from trace elements with high capture cross-sections, such as boron or gadolinium.

Σ-value calculated from fluid composition (chemical analysis). The equation for obtaining Σ using fluid composition data is as follows:

$$\Sigma = \rho AV \left[\frac{W_a(NA_1\sigma_1) + NA_2\sigma_2 + \cdots}{MW_a} + \frac{W_b(NA_1'\sigma_1' + NA_2'\sigma_2' + \cdots)}{MW_b} \right] \tag{3-17}$$

where ρ = density of compound, g/cm^3; AV = Avogadro's number, 6.022×10^{23} molecules/mole; W = weight fraction of each component in compound; MW = molecular weight of compound; NA = number of atoms per molecule; and σ = thermal neutron absorption capture cross-section per atom in barns (10^{-24} cm^2) and the subscripts, a, b, ..., and 1, 2, ..., define atoms in the mixture.

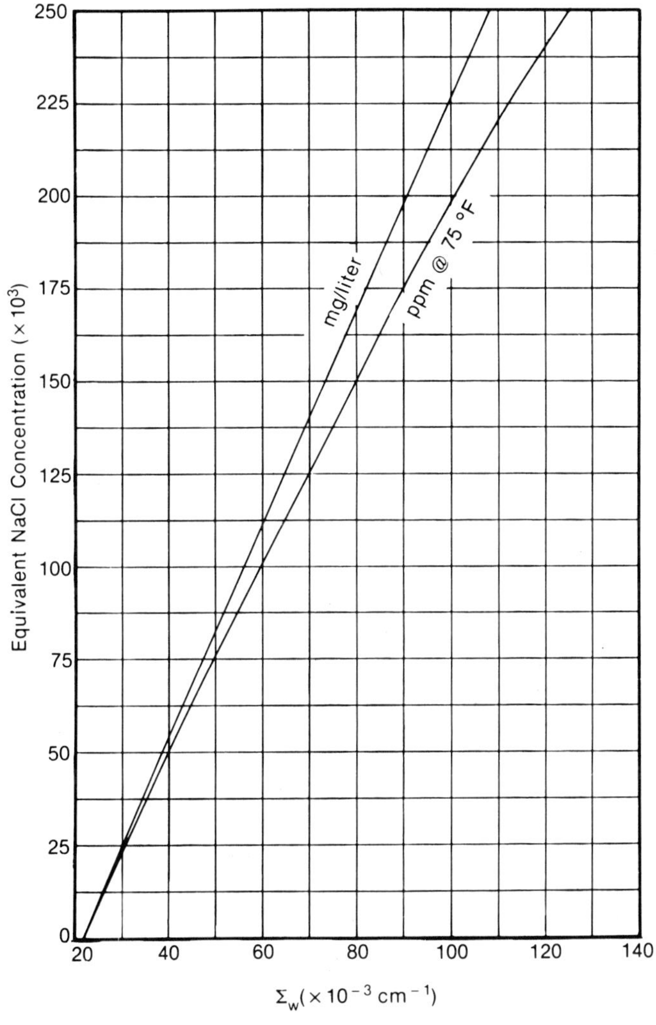

Equation:

Σ_w = [(mg/liter) × 10^{-3} + 64.01]/2.91

For ppm NaCl < 150 k,
$\Sigma_w \cong$ (ppm × $10^{\ 3}$ + 54.31)/2.59

For ppm NaCl ⩾ 150 k,
$\Sigma_w \cong$ (ppm × 10^{-3} + 21.74)/2.17

Fig. 3-9. Neutron capture cross-section (Σ_w) for equivalent sodium chloride solution.

TABLE 3-VII

Cased-hole porosity (ϕ) determinations (after Fertl, 1982; courtesy of the Society of Petroleum Engineers of AIME)

Log type	Instrumentation	Remarks on basic principles, interpretive concepts, constraints
Neutron	Neutron source–capture gamma-ray detection	Capture gamma-ray intensity roughly proportional to thermal neutron density
	Neutron source–thermal neutron detection	Thermal neutrons measured directly
	Neutron source–epithermal neutron detection	Reduction of perturbing effects of thermal neutron absorption characteristics
		ϕ_w = function of rock matrix, porosity, amount and type of fluid in pore space (mud filtrate, oil, gas)
		Environmental effects need to be taken into account when compared to open-hole porosity information
Acoustic (sonic)	Several acoustic transmitter–receiver configurations	*Interval transit time* (Δt) is measured Δt = function of rock matrix, porosity, amount and type of fluid in pore space, compaction, secondary porosity
		Requires good cement bond to both casing and formation
		Detrimental attenuation effects in unconsolidated, gas-bearing clastic formations
Density	Gamma-ray source–scattered gamma-ray detection	*Electron density* (ρ_e, number of electrons per cm^3) is measured, related to true bulk density ($\rho_b = g/cm^3$) $\rho_e = \rho_b \frac{2 \cdot \text{atomic number } Z}{\text{atomic weight } A}$
		ρ_b = function of rock matrix, porosity, amount and type of fluid in pore space (mud filtrate, oil, gas)
		Requires good borehole conditions and reliable cement job
Pulsed neutron	* Neutron lifetime * Thermal decay time	Porosity computed as the ratio (ratio curve from count rates of short-spaced (near) and long-spaced (far) detector
	Several multiple fixed and sliding gate systems	

	High-energy (14 MeV) pulsed neutron source–thermal neutron or gamma-ray detection	Response of ratio curve is similar to porosity index from compensated neutron logs
	* Log-inject-log chemical flood method	Multiple pulsed neutron runs Runs at $S_w = 1 - ROS$ and after chemical flood ($S_w = 1.0$) $\phi = (\Sigma_1 - \Sigma_2)/(\Sigma_{w1} - \Sigma_{w2})$
	Chlorinated oil injection	Runs at $S_w = 1 - ROS$ and after injection of chlorinated oil (*apparent* $S_w = 1.0$)
	* Continuous carbon/oxygen (C/O)	*Silicon count rate* curve is similar in response to compensated neutron log (as a count rate curve it will vary in absolute value for each instrument and source output)
	Inelastic and capture gamma-ray measurements	Ratio of capture gamma ray spectrum to a portion of the inelastic spectrum. This *capture/inelastic* (C/I) ratio is calibrated in porosity units. Independent of instrument and source output variations
	* Gamma-ray spectrometer (GST) tool	Porosity–indicator ratio (PIR): PIR = H/(Si + Ca)
	Inelastic and capture gamma-ray measurement	Pulsed neutron logs exhibit shallow depth of investigation Mud filtrate invasion needs to be dissipated prior to logging in newly cased wells Severe washout effects, poor cement jobs and other environmental effects need to be considered
Transforms	* Mathematical correlations between different log responses to estimate porosity	Resistivity, sigma, natural radioactivity, spontaneous potential ↔ ϕ
	* Estimated maximum porosity (ϕ_{max}) corrected for reservoir shaliness (V_{sh})	$\phi = \phi_{max}(1 - V_{sh})$ Use appropriate shaliness indicators as available (Table 3-III)
	* Depth-related porosity trends with and without shaliness corrections	Requires extensive area experience

Example calculation. Calculate the Σ-value for formation water of 100,000 ppm sodium chloride (NaCl) (i.e., 10% by weight); $\rho = 1.07$ g/cm^3:

$$\Sigma = 1.07 \times 602 \times \left[\frac{0.9(2 \times 0.33 + 1 \times 0.0006)}{18} + \frac{0.1(1 \times 0.44 + 1 \times 33.6)}{58.5} \right] = 58.7$$

The variation of Σ with NaCl concentration is shown in Fig. 3-9.

Injection of chlorinated oil. As an additional injection and logging step is incorporated in the above waterflood LIL technique, the porosity of the reservoir can be determined. Chlorinated oil is injected that has a capture cross-section identical to the one of the formation brine. Hence, to any Σ-measuring device, the reservoir rock appears to be 100% water-bearing (Murphy and Froning, 1974).

Multi-step waterflood injection. Multi-step waterflood injection extends application of the waterflood LIL technique to reservoirs of unknown salinity or still containing significant amounts of movable oil. The initial injection brings the reservoir under residual oil saturation conditions at a controlled defined salinity (Σ_{w2}). The *ROS* is then calculated as follows:

$$ROS = 1.0 - (\Sigma_3 - \Sigma_2)/[\phi(\Sigma_{w3} - \Sigma_{w2})] \tag{3-18}$$

If during fluid injection no complete water displacement is achieved, then the calculated residual oil saturation is too optimistic. Under certain conditions, one can estimate and correct for the effect of incomplete water displacement. The proposed procedure has been developed in a special field test, i.e., LIL in a water sand (Bragg et al., 1978).

Chemical flood log-inject technique. The reservoir does not have to be at residual oil saturation conditions and the rock matrix capture cross-section does not have to be known. The operational steps include: (1) running base log (Σ_1), (2) removing oil (Σ_{HC}) within depth of investigation of logging device (Σ_w = constant), (3) re-saturation with formation brine, and (4) running repeat log (Σ_2). The *ROS* is then calculated as follows:

$$ROS = (\Sigma_2 - \Sigma_1)/[\phi(\Sigma_w - \Sigma_{HC})] \tag{3-19}$$

The Continuous C/O Log (Oliver et al., 1981) can independently evaluate *ROS* at a given porosity:

$$ROS = [(C/O) - (C/O)_w]/[(C/O)_o - (C/O)_w] \tag{3-20}$$

where C/O is the log measurement, whereas $(C/O)_w$ and $(C/O)_o$ represent the water- and oil-saturated reservoir rocks, respectively.

Applied in LIL operations, the C/O log not only determines *ROS*, but also gives a check on fluid injection and possible stripping of hydrocarbons in the vicinity of the wellbore. Equally important, however, is the independence of the C/O ratio measurement from the presence of free gas and unrecognized fluid salinity effects.

Porosity information. The importance of accurate, reliable porosity measurements by well logs cannot be overemphasized. Generally, modern logging suites fill these requirements very adequately.

In many old wells, however, that are now being considered for either recompletion or particularly for enhanced recovery techniques, there is no adequate information for the required evaluation.

Similarly, when highly-deviated wells are drilled, primarily for development purposes, often a complete log suite cannot be obtained in the open hole. In these circumstances, it is often still possible to evaluate adequately the wells in question with a suite of cased-hole porosity devices.

Table 3-VII lists several through-casing porosity determination techniques which are currently available, points out the basic concepts involved, and focuses on possible constraints. Basically, a wide variety of neutron, acoustic, density, and several pulsed neutron logging concepts (including log-inject-log) allow through-casing porosity estimates. In addition, the same also holds true for the judicious selection and application of proper mathematical transforms (Fertl, 1982).

CONCLUSIONS

Residual oil saturation, a key factor in selecting and planning enhanced recovery projects, can be determined on the basis of specialized geophysical well logging techniques.

Uncertainty limits of conventional logging techniques are not too promising. Several log-inject-log techniques, however, applicable in either open- or cased-hole environments and based on multiple repeat logging runs, can determine residual oil saturation within the range of $\pm(\leqslant 5)$ saturation percent.

Proper planning of both logging and injection procedures is a must. Unfortunately, no absolute *ROS* value will ever be available as a yardstick, regardless of the number of residual oil saturation techniques used in a particular reservoir. Monte Carlo-type error analysis techniques, however, can be used to estimate the precision of *ROS* values obtained from a particular method in the oil reservoir under investigation.

REFERENCES

AOSG, 1981. The Dresser Dielectric (47 MHz) Log. *Dresser Atlas Tech. Memo.*, No. 9424, 9 pp.

AOSG, 1982. LIL—Log-Inject-Log measurements of residual oil saturation. *Dresser Atlas Tech. Brochure,* No. PES-0104, 17 pp.

Archie, G.E., 1942. The electrical resistivity log as an aid in determining some reservoir characteristics. *J. Petrol. Technol.,* 5: 165–180.

Bond, D.C., 1978. *Determination of Residual Oil Saturation.* Interstate Oil Compact Commission, 298 pp.

Bragg, J.R., Hoyer, W.A., Lin, C.J., Humphrey, R.A., Marek, J.A. and Kolb, J.E., 1978. A comparison of several techniques for measuring residual oil saturations. *Soc. Pet. Eng. Symp. on Improved Recovery, April 1978, Tulsa, Okla.*, SPE 7074.

Caldwell, R.L., 1983. Method for determining fluid saturation in a subsurface formation. *U.S. Patent,* No. 4,397,181.

Casad, B.M., 1978. Determination of residual oil in a subterranean formation. *U.S. Patent,* No. 4,071,756.

Coates, G.R. and Dumanoir, J.L., 1973. A new approach to log-derived permeability. *Trans. SPWLA,* May.

Dorfman, M.H., 1984. Discussion of reservoir description using well logs. *J. Pet. Technol.,* 36(12): 2195–2196.

Fertl, W.H., 1979a. Find ROS from well logs for enhanced recovery projects, in practical log analysis series, 14. *Oil Gas J.,* 77 (February 7): 120–127.

Fertl, W.H., 1979b. Determination of residual oil saturation from geophysical well logs in tertiary recovery projects. *Energy Sour.,* 4(3): 259–280.

Fertl, W.H., 1982. Well logging and its application in cased holes. *Soc. Pet. Eng. Int. Pet. Exhib. Technol. Symp., Beijing, March 18–26, 1982,* SPE 10034.

Fertl, W.H., 1985. Determination of water saturation in earth formations independent of lithology. *U.S. Patent,* No. 4,494,071.

Fertl, W.H. and Reynolds, E.B., 1973. Determination of residual oil in a formation. *U.S. Patent,* No. 3,878,890.

Friedman, B.R. and Vogiatzis, P.J., 1979. Theory of microwave dielectric constant logging using the electromagnetic wave propagation method. *Geophysics,* 44(5): 969–983.

Geffen, T.M., 1973. Oil production to expect from known technology. *Oil Gas J.,* May, pp. 66–76.

Hall, J.E., Johnstone, C.W., Baldwin, J.L. and Jacobson, L.A., 1980. A new thermal neutron decay logging system—TDT-M. *Soc. Pet. Eng. 55th Annu. Fall Meet., Dallas, Texas, September 21–24, 1980,* SPE 9462.

Hasiba, H.H., Wilson, L.A. and Martinelli, J.W., 1977. How to organize and plan enhanced recovery efforts. *World Oil,* January, pp. 91–95.

Hopkinson, E.C., Fertl, W.H. and Oliver, D.W., 1981. The continuous carbon/oxygen log—basic concepts and recent field experiences. *Soc. Pet. Eng. Reg. Middle East Meet., Bahrain, March 1981,* SPE 9613.

Huchital, G.S., Hutin, R., Thoraval, Y. and Clark, B., 1981. The deep propagation tool (a new electromagnetic logging tool). *Soc. Pet. Eng. 56th Annu. Fall Meet., San Antonio, Texas, October 5–7, 1981,* SPE 10988.

Jackson, R.W., 1968. Why some waterfloods fail. *World Oil,* March, pp. 65–68.

Janes, T.A., Hilliker, D.J. and Carville, C.L., 1984. 200 MHz dielectric logging system. *9th Int. Formation Evaluation Symp., S.A.I.D., Paris, October 24–26, 1984.*

Kazemi, H., 1977. Determination of waterflood residual oil saturation from routine core analysis. *J. Pet. Technol.,* 29(1): 31–32.

Mitchell, F.R., Robinson, J.D. and Vogiatzis, J.P., 1978. Method for determining residual oil saturation of a formation. *U.S. Patent,* No. 4,124,800.

Murphy, R.P. and Froning, H.R., 1974. *U.S. Patent,* No. 3,825,752.

Murphy, R.P., Foster, G.T. and Owens, W.W., 1976. Evaluation of waterflood residual oil saturation using log-inject-log procedures. *Soc. Pet. Eng. Symp. on Improved Recovery, Tulsa, Okla., April 1976,* SPE 5804.

Oliver, D.W., Frost, E. and Fertl, W.H., 1981. Continuous carbon/oxygen (C/O) logging–instrumentation, interpretive concepts and field applications. *Trans. SPWLA,* Paper TT: 28 pp.

Pusch, W.H., 1977. Core analysis study for the El Dorado micellar–polymer project. *J. Pet. Technol.,* 29(8): 1013–1016.

Randall, R.R., Fertl, W.H. and Hopkinson, E.C., 1981. Time-derived sigma for pulsed neutron capture logging. *Soc. Pet. Eng. Reg. Middle East Meet., Bahrain, March 1981,* SPE 9614.

Richardson, J.E., Wyman, R.E., Jordan, J.R. and Mitchell, F.R., 1972. Method for determining residual oil with pulsed neutron capture logs. *Soc. Pet. Eng. Symp. on Improved Recovery, Tulsa, Okla., April 1972,* SPE 3796.

Robinson, J.D., Vajnar, E.A., Loren, J.D. and Hartman, D.E., 1972. Determination of residual oil using the nuclear magnetism log. *Soc. Pet. Eng. Symp. on Improved Recovery, Tulsa, Okla., March 1972,* SPE 3797.

Schenewerk, P.A., Knapp, R.M. and Fertl, W.H., 1982. The accuracy of pulsed neutron capture logs for residual oil saturation determination. *Soc. Pet. Eng. 57th Annu. Fall Meet., New Orleans, La., September 26–27, 1982,* SPE 11148.

Serpas, C.J., Wichmann, P.A., Fertl, W.H., DeVries, M.R. and Randall, R.R., 1977. The dual detector neutron lifetime log—theory and practical applications. *Trans. SPWLA,* Paper CC: 37 pp.

Thomas, E.C. and Ausburn, B.E., 1976. The determination of swept-zone residual oil saturation in a slightly consolidated sandstone reservoir of the Gulf Coast. *Soc. Pet. Eng. Symp. on Improved Recovery, Tulsa, Okla., March 1976,* SPE 5803.

Tomich, J.F., Dalton, R.L., Deans, H.A. and Schallenberger, L.K., 1973. Single well tracer method to measure residual oil saturation. *J. Pet. Technol.,* 25(2): 211–227.

Walstrom, J.E., Mueller, T.D. and McFarlane, R.C., 1967. Evaluating uncertainty in engineering calculations. *J. Pet. Technol.,* 19(12): 1595–1603.

Walther, H.C., 1968. Saturation from logs—laboratory measurements of logging parameters. *J. Pet. Technol.,* 20(3): 251–258.

Waxman, M.H. and Smits, L.J.M., 1968. Electrical conductivities in oil-bearing shaly sands. *Soc. Pet. Eng. J.,* 243(5): 107–122.

Wesson, T.C., 1979. How to determine residual oil saturation. *Pet. Eng.,* 51(12): 25–36.

Westaway, P., Hertzog, R. and Plasek, R.E., 1980. The gamma spectrometer tool inelastic and capture gamma-ray spectroscopy for reservoir analysis. *Soc. Pet. Eng. 55th Annu. Fall Meet., San Antonio, Texas, September 21–24, 1980,* SPE 9461.

Wharton, R.P., Hazen, G.A., Ray, R.N. and Best, D.L., 1980. *Electromagnetic propagation logging: advances in technique and interpretation. Soc. Pet. Eng. 55th Annu. Fall Meet., Dallas, Texas, September 21–24, 1980,* SPE 9267.

Wilson, D.A. and Hensel, W.M., 1982. The Cotton Valley sandstones of east Texas: a log-core study. *Trans. SPWLA,* Paper R: 27 pp.

Wisenbaker, J.D., 1980. *Fundamentals of Core Analysis.* Core Lab, Dallas, Texas, 70 pp.

Wyman, R.E., 1977. How should we measure residual oil saturation? *Bull. Can. Pet. Geol.,* 25(2): 233–270.

Xuwen, G., Yinzhou, Y., Da, L. and Shufang, Z., 1982. A log method for determining oil saturation in a reservoir—phase dielectric log. *Soc. Pet. Eng. Int. Pet. Exhib. Technol. Symp., Beijing, March 18–26, 1982,* SPE 10565.

Youmans, A.H., Hopkinson, E.C. and Wichmann, P.A., 1966. Neutron lifetime logging in theory and practice. *Trans. SPWLA,* Paper Q: 17 pp.

Chapter 4

GAS INJECTION

JALAL TORABZADEH, GERALD L. LANGNES, JOHN O. ROBERTSON, Jr., T.F. YEN, ERLE C. DONALDSON and GEORGE V. CHILINGARIAN

INTRODUCTION

Gas injection is the oldest of the fluid injection processes. The idea of using a gas for the purpose of maintaining reservoir pressure and restoring oil well productivity was suggested as early as 1864, just a few years after the drilling of the Drake well. Since that time, injection of natural gas or other gases has been used successfully in many oil reservoirs throughout the world as a secondary recovery method. The early gas injection projects were designed to increase the immediate productivity and should be classified as pressure maintenance projects. Recent gas injection applications, however, have been intended to increase the ultimate recovery and can be considered as enhanced recovery projects.

As an enhanced recovery technique, the injected fluid must, in addition to increasing the reservoir energy, displace the oil and drive it toward the production wells. The success of an enhanced recovery project depends (1) upon the mechanism by which the injected fluid displaces the oil (displacement efficiency), and (2) on the volume of reservoir which the injected fluid contacts (conformance or sweep efficiency). In this regard, generally, gas because of its adverse viscosity ratio (higher mobility ratio) is inferior to water in recovering oil. Gas may offer economic advantages, however, because of its availability and relative ease of injection. Thus, gas injection is still the preferred secondary recovery method in certain circumstances (Combs and Kuezek, 1971).

Gas injection may be either a miscible or an immiscible displacement process. The character of the oil and gas and the temperature–pressure conditions of the injection determine the type of process. Only the immiscible displacement process is discussed in this chapter. (Miscible displacement is discussed in Chapter 5.) Gas can be injected into a reservoir to maintain the pressure (i.e., dispersed gas injection) or to attempt to bank and sweep oil to the producers (i.e., gas-cap injection). Both systems are considered in this chapter.

Theoretical calculations by Muskat (1946) showed that oil recovery could be greatly increased by pressure maintenance, particularly in the case of oils having a high formation volume factor. In these calculations, however, many physical characteristics of the heterogeneous reservoir were not considered. Elkins and Cooke (1949) noted that the volume of gas injected and the associated change in oil viscosity and formation volume factor appear to be the principal factors in determining the oil recovery. To obtain the maximum benefit from the change in oil properties, the volume of reservoir contacted by the gas must be as high as possible. For massive carbonate reservoirs with good permeability, injection pressure does not appear to have a large effect on final, ultimate, oil recovery (Elkins and Cooks, 1949).

The primary problems with gas injection in carbonate reservoirs are the high mobility of the displacing fluid and the wide variations of permeability. Much greater control over the injection process must be exercised than is necessary with waterflooding. The effect of the reservoir's permeability profile on gas sweep efficiency may be evaluated most satisfactorily by a short-term pilot gas-injection test. The pilot test may also provide necessary data to calculate the required volumes of gas; this, in turn, will aid in the design of compressor equipment and estimating the number of injection wells which will be required. The following four recommendations should be followed in conducting a gas pilot test (Elkins and Cooke, 1949):

(1) The rate of gas injection and its duration should be sufficient to result in significant changes of the produced gas/oil ratio. The daily injection rate should at least equal the volume of fluids produced by the first-row producing wells.

(2) Gauging facilities should be adequate to ensure accurate measurement of the performance of the first-row wells. Continuous monitoring is preferred.

(3) The test period should include a time period prior to and after the injection of gas to establish the trend of the produced gas/oil ratio.

(4) When possible, the pilot test should be conducted early in the life of the field.

Useful design techniques are presented in this chapter followed by summaries of field applications. Analogy must be used with considerable care because of the heterogeneity of carbonate reservoir rocks and the sensitivity of the process.

PREDICTIVE TECHNIQUES

Two major predictive techniques, the Welge (1952) and the Tarner (1944), are reviewed in this chapter. Either technique is applicable to the analysis of both sandstone and carbonate reservoirs.

The Welge method

The Welge technique (1952) is a simplified form of the Buckley-Leverett or frontal advance method. It also can be used in evaluating waterflooding operations. The advantage of Welge's method over the Buckley-Leverett technique is that the curve showing the saturation distribution in the reservoir is not required in the former method. The average gas saturation is used to determine the recovery, whereas the terminal gas saturation, near the outflow face, defines the fractional flow of oil, f_o, and hence the flowing gas/oil ratio at any desired time during production. As a result, a considerable amount of time required for calculation and plotting is saved.

Assumptions

(1) The reservoir is a single homogeneous layer, and the flow is linear.
(2) Gas displaces oil saturated with gas at a constant pressure.
(3) Oil displacement is immiscible.
(4) Flow occurs in one direction only (there is no crossflow).

With these assumptions, the fraction of oil at the outflow face is:

$$f_o = \frac{N_p}{G_p + N_p} \tag{4-1}$$

and:

$$G_i = G_p + N_p \tag{4-2}$$

$$N_p = S_{ga} \tag{4-3}$$

The fraction of oil flowing is thus the slope of the cumulative oil produced versus the cumulative gas injected, expressed in pore volumes:

$$f_o = \frac{dN_p}{dG_i} = \frac{dS_{ga}}{dG_i} \tag{4-4}$$

$$\int dS_g = \int f_o \, dG_i \tag{4-5}$$

$$S_{ga} = S_g + G_i f_o \tag{4-6}$$

At the outflow face:

$$\frac{k_g}{k_o} = \left(\frac{1 - f_o}{f_o}\right)\left(\frac{\mu_g}{\mu_o}\right) \tag{4-7}$$

where G_i = cumulative gas injected (fraction of pore volume); G_p = cumulative gas produced (fraction of pore volume); k_g = gas permeability; k_o = oil permeability; N_p = cumulative oil produced (fraction of pore volume); S_g = gas saturation at the outflow end; S_{ga} = average gas saturation of the reservoir; μ_g = gas viscosity; and μ_o = oil viscosity.

Procedure

(1) Review the relative permeability data obtained from core analysis. If more than one set of data is available, choose a set which appears to be most representative. If stratification exists in the reservoir, a separate calculation for each layer should be considered. (See the discussion of the layered Buckley-Leverett technique, Volume I, p. 279.)

(2) Approximate the irregular shape of the reservoir with a linear model. An idealized length and cross-sectional area of the reservoir to flow must be determined.

TABLE 4-I

Recommended steps in using the Welge (1952) technique of gas-injected flood prediction

(1)	(2)	(3)	(4)	(5)	(6)	(7)	(8)	(9)
S_g	k_g/k_o	k_g	f_o	df/dS [a]	t	S_{ga}	GOR	q_o

[a] $df/dS = df_g/dS_g$.

(3) Obtain the average fluid velocity, v_o, in the hydrocarbon-occupied pore space as follows:

(a) An average cross-sectional area exposed to flow, A_f, is obtained by dividing the original oil-in-place (converted from bbl to scf) by the idealized length, L (ft), of the reservoir.

(b) The v_o is then equal to the average withdrawal rate of oil, q_o, and water, q_w, divided by A_f:

$$v_o = \frac{q_o + q_w}{A_f} \tag{4-8}$$

The units of v_o should be converted from ft/day to cm/s.

(4) The remainder of the steps in the calculations can be presented in tabular form as shown in Table 4-I. The values for the columns in the table can be determined as described here.

(a) Column 1: several selected values of producing gas saturations, S_g, after breakthrough.

(b) Column 2: value of k_g/k_o for the selected S_g as determined from a k_g/k_o versus S_g curve.

(c) Column 3: values of k_g for each value of S_g selected.

(d) Column 4: fractional flow of oil:

$$f_o = \frac{1 + 3.09 \times 10^4 \{ [k k_{ro} (\rho_o - \rho_g) \sin\theta] / \mu_g v_o \}}{1 + (k_{rg}/k_{ro})(\mu_o/\mu_g)} \tag{4-9}$$

where f_o = fraction of oil in flowing stream; k_{rg} = relative permeability to gas with connate water in place; k_{ro} = relative permeability to oil with connate water in place; k = effective permeability to oil with connate water in place, mD; μ_g = gas viscosity, cP; μ_o = oil viscosity, cP; ρ_o = oil density, g/cm³; ρ_g = gas density, g/cm³; θ = angle of structural dip; v_o = average fluid velocity, cm/s. Where dip is negligible, $\sin\theta$ approaches zero and equation (4-9) simplifies to:

$$f_o = \frac{1}{1 + (k_{rg}/k_{ro})(\mu_o/\mu_g)} \tag{4-10}$$

(e) Column 5: the values of df/dS are taken from f_g ($f_g = 1 - f_o$) versus S_g curves similar to the f_w versus S_w curve presented in Vol. I, p. 278, fig. 8-22. The slopes of the curve, df/dS, may be measured directly for the desired saturations.

(f) Column 6: the time, t, may be determined as follows:

$$t = \frac{L}{v_o(\mathrm{d}f/\mathrm{d}S)} \tag{4-11}$$

where L = length of reservoir, cm.

(g) Column 7: average gas saturation, S_{ga}, of the reservoir is:

$$S_{ga} = \left(S_g + \frac{f_o}{(\mathrm{d}f/\mathrm{d}S)} \right) \tag{4-12}$$

(h) Column 8: the producing gas/oil ratio, GOR, for various values of S_g is equal to:

$$GOR = R_s + \left(\frac{B_o}{B_g} \right) \left(\frac{1 - f_o}{f_o} \right) \tag{4-13}$$

where R_s = gas in solution per bbl of oil, scf/STB; $(B_o/B_g) \times [(1 - f_o)/f_o]$ = free gas produced per bbl of oil, scf/STB; B_o = oil formation volume factor, bbl/STB, B_g = gas formation volume factor, bbl/scf.

(i) Column 9: the stock-tank oil producing rate, q_o:

$$q_o = \frac{v_o A_f \phi (1 - S_w) f_o}{5.615 B_o} \tag{4-14}$$

The distance, X, from the injector that the floodfront has traveled can be found at any time, t, by using the following equation:

$$X = v_o t \left(\frac{\mathrm{d}f}{\mathrm{d}S} \right) \tag{4-15}$$

Equation (4-13) is sensitive to the assumed conditions such as gas velocity, permeability, and geometric configuration of the reservoir. It tends to yield GOR values higher than those occurring in the field for the same cumulative recovery. Consequently, if a GOR cutoff is based on this calculation, a conservative reserve estimate is obtained.

This method is sensitive to the throughput velocity, v_o. Inasmuch as the velocity is in the denominator of the gravitational term (column 4, equation (4-9)), rapid displacement reduces the size of the gravitational term and increases the fraction of gas, f_g, flowing through the reservoir. This would indicate a lower displacement efficiency. On the other hand, where the gravitational term is sufficiently large, f_g can approach zero or have negative values; this implies countercurrent flow of gas updip and oil downdip, resulting in gravity drainage which yields the maximum displacement efficiency. If a gas cap overlies the oil zone, drainage can be vertical ($\sin \theta = 1.00$) and the cross-sectional area to flow can be very large. Gravity drainage will yield high recoveries unless the vertical permeability is low.

Frequently a field has been produced for a period of time prior to initiation of gas injection. In this case, the calculation steps in the Welge predictive technique are similar to those previously presented, except that calender time is used. The following procedure, presented in outline form, can be followed:

(1) Determine the residual oil saturation, S_{o1}, at the time of flood:

$$S_{o1} = \left(\frac{B_{o1}}{B_{oi}}\right)\left(S_{oi} - \frac{N_p B_{oi}}{7758 A h \phi (1 - S_w)}\right) \tag{4-16}$$

where B_{o1} = oil formation volume factor at start of injection, bbl/STB; B_{oi} = initial formation volume factor, bbl/STB; N_p = oil produced, STB; Ah = reservoir volume, acre-ft; ϕ = porosity, fraction; S_w = interstitial water saturation, fraction; S_{oi} = initial oil saturation, fraction.

(2) Determine the residual gas saturation, S_{g1}, at start of injection:

$$S_{g1} = (S_{oi} - S_{o1}) \tag{4-17}$$

(3) The time at which this saturation, S_{g1}, is reached, which is computed following the tabulation shown in Table 4-I, is then identified with the calendar date of the start of the injection.

(4) The average gas saturation at later times is calculated as follows:

$$S_{ga} = \left[S_g + \frac{f_o}{\mathrm{d}f/\mathrm{d}S}\left(\frac{B_{oi}}{B_{o1}}\right)\right] \tag{4-18}$$

Tarner's prediction

Tarner (1944) developed a mathematical approach for predicting reservoir performance with and without gas injection. His method employs the simultaneous solution of two gas/oil ratio equations. The first equation is based on material balance concepts, whereas the second equation is based on the relative permeability relationships.

Assumptions

(1) The reservoir can be treated as a single unit.

(2) There are no pressure gradients across the field; the pressure drops uniformly.

(3) All gas must flow through the oil zone in the reservoir; no bypassing occurs.

(4) The size of the oil zone does not change. There is no water influx or gravity drainage.

(5) Representative k_g/k_o versus S_g data are available.

(6) S_w is constant.

(7) Prediction starts when the bubble point is reached. No free gas saturation exists in the reservoir.

Data required

(1) Oil- and gas-in-place initially.

(2) Cumulative oil and gas production until the time when the bubble point pressure is reached, and production to the start of the prediction.

(3) Curves relating the instantaneous produced gas–oil and average reservoir pressure to cumulative production (both curves must extend to the start of the prediction period.)

(4) Curves relating reservoir pressure to (a) formation volume factors (oil, gas, total), (b) solution gas/oil ratio, and (c) viscosity (oil and gas).

(5) Curve relating permeability ratio and total liquid saturation.

Permeability ratio data can be obtained from laboratory work or determined from the reservoir performance data. "Apparent" relationships developed from reservoir performance data, particularly fluctuations caused by changes in the producing rate, enable engineers to predict the performance to be expected with and without gas injection (Elkins, 1950).

Equations

The instantaneous produced gas/oil ratio from the material balance equation is given by the following equation:

$$R_2 = \frac{(B_{t2} - B_{tb}) - \left(\frac{N_{p2}}{N}\right)B_{t2} + \left(\frac{N_{p2}}{N}\right)R_{sb}B_{g2}}{B_{g2}\left(\frac{N_{p2}}{N} - \frac{N_{p1}}{N}\right)} - \frac{\left(\frac{N_{p1}}{N}\right)R_{p1}}{\left(\frac{N_{p2}}{N} - \frac{N_{p1}}{N}\right)} - \frac{R_1}{2} \tag{4-19}$$

where R_2 = instantaneous producing gas/oil ratio at estimated pressure, scf/STB; R_1 = instantaneous producing gas/oil ratio at previous pressure, scf/STB; R_{p1} = cumulative gas/oil ratio at previous pressure, scf/STB; R_{sb} = solution gas/oil ratio at bubble point, scf/STB; B_{tb} = total formation volume factor (two-phase) at bubble point pressure, bbl/STB; B_{t2} = total formation volume factor (two-phase) at estimated pressure, bbl/STB; B_{g2} = gas formation volume factor at estimated pressure, bbl/SCF; N = oil-in-place at bubble point pressure, STB; N_{p2} = cumulative oil produced at estimated pressure, STB; and N_{p1} = cumulative oil produced at previous pressure, STB.

The total liquid saturation at a pressure below the bubble point pressure is given by the following equation:

$$S_{t2} = \left(1 - \frac{N_{p2}}{N}\right)\left(\frac{B_{o2}}{B_{tb}}\right)(1 - S_{wi}) + S_{wi} \tag{4-20}$$

where S_{t2} = total liquid saturation at estimated pressure, fraction; S_{wi} = initial water saturation, fraction; and B_{o2} = oil formation volume factor at estimated pressure, bbl/STB.

The instantaneous produced gas/oil ratio derived from Darcy's equation is given by:

$$R_2 = \left(\frac{k_g}{k_o}\right)_2\left(\frac{\mu_o}{\mu_g}\right)_2\left(\frac{B_o}{B_g}\right)_2 + R_{s2} \tag{4-21}$$

where $(k_g/k_o)_2$ = permeability ratio at estimated pressure; μ_{o2} = oil viscosity at estimated pressure, cP; μ_{g2} = gas viscosity at estimated pressure, cP; B_{g2} = gas formation volume factor at estimated pressure, bbl/SCF; and B_{s2} = solution gas-oil ratio at estimated pressure, scf/STB. The cumulative produced gas is given by the following equation:

$$G_{p2} = \left(\frac{R_1 + R_2}{2}\right)(N_{p2} - N_{p1})(1 - G_i) + G_{p1} \tag{4-22}$$

where G_{p2} = cumulative produced gas at estimated pressure, scf; G_{p1} = cumulative produced gas at previous pressure, scf; and G_i = injected gas (fraction of gas produced which is returned to the reservoir during the incremental period).

The cumulative produced gas/oil ratio, R_{p2} = in scf/STB may be expressed as follows:

$$R_{p2} = \frac{G_{p2}}{N_{p2}} \tag{4-23}$$

Procedure

The two equations for the instantaneous produced gas/oil ratio are solved by use of the following method:

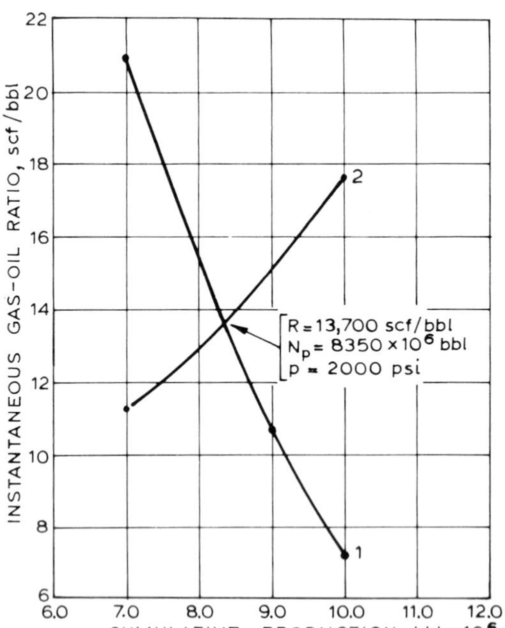

Fig. 4-1. Tarner's prediction plot relating the instantaneous produced gas/oil ratio and the cumulative oil production (curve *1*, performance calculated using the material balance equation; and curve *2*, performance calculated from relative permeability data).

(1) Calculate oil- and gas-in-place at bubble point pressure.

(2) Assume an average reservoir pressure. Select pressure steps on the basis of the change in character of the instantaneous produced gas/oil ratio. (Caution: use smaller pressure increments where the slope of the curve is changing rapidly.)

(3) Make three estimates of cumulative oil production for the assumed pressure (note the trend of the cumulative oil production–pressure plot when selecting values).

(4) Calculate the instantaneous produced gas-oil ratios using equations (4-19) through (4-21) for each of the three estimates of cumulative oil production.

(5) Plot the data as shown in Fig. 4-1. The intersection of the two curves gives the answer.

(6) Calculate the cumulative produced gas using equation (4-22).

(7) Calculate the cumulative produced gas/oil ratio using equation (4-23).

RESERVOIR PERFORMANCE

Gas cycling may be distinguished from gas injection in that the recovery of the crude oil by vaporization is the primary intent of cycling operations. The intent of gas injection is to provide a pressure gradient which moves the oil toward nearby producers.

Gas cycling

Gas cycling has been used successfully in several reservoirs: Pickton in Texas (McGraw and Lohec, 1964), Hermattar-Elkton in Alberta, Canada (Donahoe and Bohannon, 1965), and Opelika in Texas (Clay, 1954). Cook et al. (1967) demonstrated in the laboratory that oil recoveries after cycling operations could range from 15 to 70% of the immobile stock-tank oil depending upon the characteristics of the reservoir fluids. The total (primary plus secondary) recovery from the Pickton reservoir was estimated at near 73% (primary ultimate was estimated at 19.4%) of the oil originally in place (McGraw and Lohec, 1964). The total recovery in the Harmattan-Elkton Field was estimated at 30–40% of the oil-in-place (Donahoe and Bohannon, 1965). The total recovery from the Opelika field was estimated at 27% (Clay, 1954).

Table 4-II and Figs. 4-2 and 4-3 show the theoretical effects of oil gravity, pressure, and temperature and the quantity of gas cycled upon the ultimate volume of oil recovered (Cook et al., 1967). The viscosities for the oil and gas were obtained from data presented by Katz et al. (1959). These data, however, cannot be extrapolated to field performance; they are presented only for the purpose of indicating the impact of various factors.

Jacoby and Berry (1958) showed that the actual vaporization sequence occurring in each reservoir is determined by the production process. They outlined a method for calculating fluid compositions in reservoirs containing low-

TABLE 4-II

Effect of oil gravity, pressure and temperature on vaporization (after Cook et al., 1967; courtesy of the SPE of AIME)

Example	Oil gravity (°API)	Cycling pressure (psia)	Cycling temperature (°F)	Gas/oil viscosity ratio	Vaporization (percent of immobile stock-tank oil [a])	Displacement (percent of original stock-tank oil)
1	45	4100	250	0.338	73.6	76.5
2	35	4100	250	0.113	50.3	62.5
3	22	4100	250	0.020	28.9	54.0
4	35	4100	100	0.047	33.8	60.7
5	35	1100	100	0.007	15.3	48.8
6	45	1100	250	0.050	47.9	65.5

[a] Amount of total gas cycled was 25 Mcf/bbl of immobile stock-tank oil.

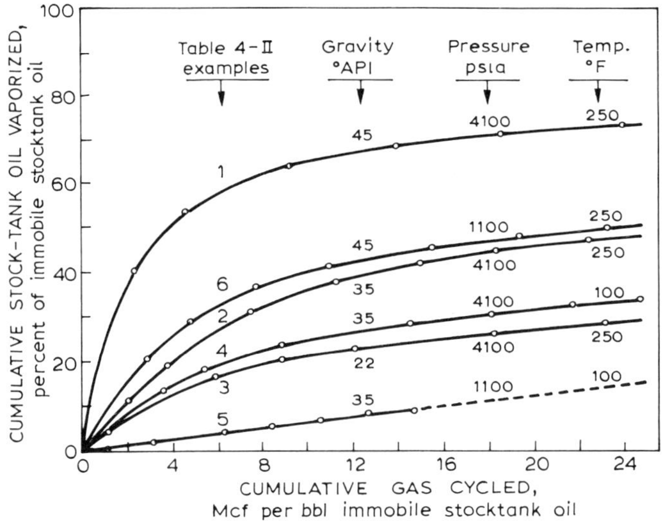

Fig. 4-2. Vaporization as related to gravity of oil, pressure, temperature, and amount of gas cycled. (After Cook et al. 1967; courtesy of the Society of Petroleum Engineers of AIME.)

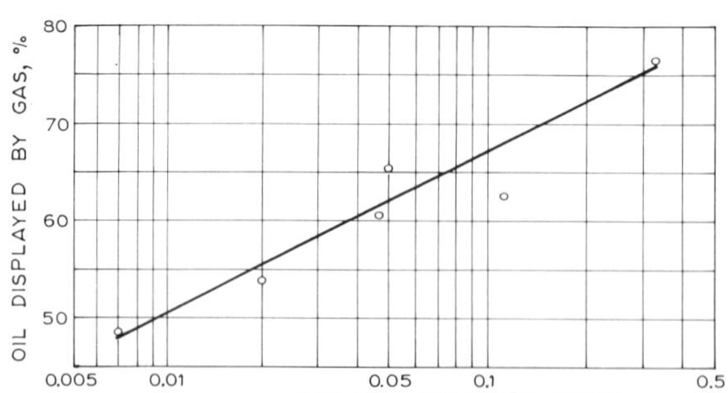

Fig. 4-3. Relationship between gas/oil viscosity ratio and oil displacement. (After Cook et al. 1967; courtesy of the Society of Petroleum Engineers of AIME.)

molecular-weight hydrocarbon liquids. The recovery of stock-tank oil per unit of pressure decline should be predicted from reservoir fluid composition (compositional material balance and oil–gas separator recovery calculations). Jacoby and Berry (1958) noted that for an oil rich in light components, the material balance method indicated a recovery twice that obtained by using conventional calculations.

GAS INJECTION IN CARBONATE RESERVOIRS

Injection of gas in carbonate reservoirs can increase the ultimate recovery of oil. The benefits obtained by the gas injection are dependent upon horizontal and vertical sweep efficiency of the injection gas. The sweep efficiency depends on the type of porosity system present, as discussed below.

Intercrystalline–intergranular porosity systems

According to Longren (1951), insufficient data prohibited a quantitative prediction of the susceptibility to gas injection of the Panhandle dolomite reservoir, Texas, which is classified as an intercrystalline–intergranular porosity system. Secondary recovery predictions for the West Rampa Project in the Panhandle Field, Texas, evolved through pilot testing. Particular attention was paid to the selection and preparation of old producers for gas injection, the determination of maximum injection rates, and the evaluation of injection response (Longren, 1951).

The primary concern in deciding which producers should be used as injectors, was to obtain maximum displacement efficiency. Wells located in the areas having lower permeability were chosen in preference to those located in areas with higher permeability, because reservoir pressure was higher in the former areas and thus helped retard the fingering of gas. Preference was also given to the wells with low gas/oil ratios and those located in areas having lower gas saturations. Producers which penetrated the entire producing zone provided the best vertical coverage. Deepening of producers prior to the initiation of injection helped improve vertical coverage. The injectors located in areas having higher gas saturations required higher quantities of injeced gas. Structural position did not appear to be a major factor affecting the performance of the injectors.

It was found that the maximum benefit from gas injection was derived when the producing wellbores were kept in clean condition and each producer was offset by at least one injection well (Nesluge, 1951). Well cleanout and acidizing programs, performed after gas injection was underway for a year or more, yielded better results than when stimulation work was done in producers prior to gas repressurization.

Fracture–matrix porosity system

Representative data collection is difficult for reservoirs with a well-developed fracture–matrix porosity system consisting of low-permeability matrix and interconnected fractures and solution channels constituting 10% or more of the reservoir's void space (Elkins, 1969). As a result, calculations of recoverable oil by volumetric measurements will seldom yield volumes equal to those determined by material balance calculations. This divergence is greater where the permeability of the reservoir matrix is low.

Factors directly responsible for this difference are (1) significant differences between the pressure in the tight rock matrix and pressure in the interconnected fractures and/or solution channels; (2) difficulties in determining the quantity and distribution of the gas saturation within the rock matrix, fractures, and solution channels (it should be noted that not all of the matrix in a tight reservoir need be saturated with oil) (Elkins, 1969); and (3) the quantity and rate of water influx during repressurization.

A well-developed fracture–matrix porosity system can result in severe channeling of injected gas and/or water influx with little or no benefit to the ultimate recovery of oil. An example of this occurred in the Bois d'Arc Formation pilot in the West Edmond Field, Texas (Elkins, 1969).

To aid in the determination of the reservoir volume swept by gas in the Hunton Lime, West Edmond Field, helium was added for one week to the injected gas as a tracer (1% concentration) (Elkins, 1969). The producing wells were monitored for the daily quantity of produced helium for 5 months. At the end of 5 months, 44% of the injected helium tracer had been produced. The quantity of produced helium was affected by factors such as (1) varying velocities along various flow paths, (2) dispersion of gas along flow paths, and (3) variation in permeability. If the peak helium contents, however, correspond to average travel times of injected gas to producing wells, then the minimum percentage of reservoir volume swept by gas was about 6%. The additional oil recovered by gas injection in the Clearfork Zone dolomite, Fullerton Field, Texas, was around 4% of the total oil-in-place (Nolan and Locker, 1954). For both fields, a well-developed fracture–matrix porosity system reduced the effectiveness of injected gas. It was demonstrated, however, that gas injection into a fractured reservoir could result in recovery of additional oil.

Vugular–solution porosity system

As a gas conservation measure and in order to increase the ultimate oil recovery in a vugular–solution porosity type system, gas was injected into the Westerose D-3 Pool, Alberta, Canada (Hnatiuk and Martinelli, 1967). Based on reservoir pressure behavior, it was determined that all pools on the reef trend are, to a varying degree, in pressure communication with each other. To predict the performance of the Westerose D-3 Pool and other pools on this reef trend, a two-dimensional computer model was developed to simulate the pressure–product performance of the pools. The model parameters were adjusted until there was close agreement between the reservoir model and the 20-year production history.

The future performance was then estimated at various rates of oil production with and without the return of the produced gas to the gas caps. It should be emphasized that the results are only qualitative and not quantitative. The trends determined are likely to be valid, but the quantitative values calculated should not be considered exact. The results of the simulation studies (Hnatiuk and Martinelli, 1967) indicated that the Westerose D-3 Pool had lost, and would continue to lose, considerable oil to the aquifer unless gas injection was stopped. As a result of this investigation, the gas injection project was shut down shortly after the simulation study was completed.

The failure of gas injection in the Westerose D-3 Pool does not mean that it cannot be successful in other reef reservoirs. It does emphasize, however, the need to understand the geology of the reservoir completely when selecting a method of enhanced oil recovery.

INERT GAS INJECTION

In recent years, the ever-rising cost and limited supply of natural gas has prompted operators to search for a substitute. Promising substitutes receiving the most attention are (1) pure nitrogen gas (inert) and (2) inert gas mixture that is predominantly nitrogen. Nitrogen gas injection is rapidly emerging as one of the best enhanced oil and gas recovery methods. Currently, in over thirty fields in the United States nitrogen gas is used to assist the recovery of reservoir fluids. In 1982, over 500 million cubic feet of nitrogen gas were injected daily into some twenty fields (Clancy et al., 1983). Nitrogen is less expensive and more readily available than either natural gas or carbon dioxide.

The injected nitrogen gas can enhance the recovery of oil or gas by one or more of the following mechanisms: (1) pressure maintenance, (2) immiscible displacement, or (3) miscible displacement, depending on the condition of pressure and temperature and the properties of injected gas and reservoir fluids. Pressure maintenance can be quite applicable in a volumetric reservoir to keep pressure above the bubble point or the dew point. Immiscible displacement is effective if mobility ratio is favorable and if structure and density difference between injected and reservoir fluids allow gravity segregation to occur.

Miscible displacement of reservoir fluids by nitrogen (either direct or multiple contact) can occur only in a narrow range of fluid composition, pressure, volume, and temperature. The necessary conditions to achieve miscibility are discussed in the next chapter. Nitrogen miscibility requires high pressure (deep reservoirs) and light oils.

The primary advantage of inert gas is its availability and lower cost. Other advantages include (1) prevention of oil encroachment into the gas cap when gas cap is present, (2) higher recoveries compared to water drive in reservoirs having low permeability, (3) residual inert gas at abandonment rather than saleable natural gas, and (4) reliability of the supply.

CANDIDATES FOR GAS INJECTION

Oil fields that have low oil saturation in either primary or secondary gas cap are prime candidates for improved recovery by gas injection into the gas cap. Further shrinkage of oil is reduced by gas injection. Inasmuch as it maintains a relatively high pressure gradient on the oil phase, relative permeability to oil remains high and oil is produced faster and in greater quantity (Prats and Miller, 1973). In reservoirs having high permeability and high vertical span, gas injection may result in high recovery factors because of gravity segregation.

Richardson and Blackwell (1971) showed that under certain conditions of gas-cap drive, gravity drainage of oil can be highly efficient.

In thick formations or in reservoirs having sizeable formation dip, in the presence of sufficient vertical permeability the injected gas would probably override or move to the higher structural positions. In reservoirs having vertical permeability higher than 200 mD, vertical segregation of gas and oil may occur, allowing the gas cap to displace the oil downwards. If the reservoir lacks either sufficient vertical permeability or relief for gravity segregation to be effective, a frontal drive of the injected gas similar to that used for water injection, can drive the oil towards the production wells.

In reservoirs containing volatile crude oil, vaporization of the lighter components also will result in a higher recovery efficiency (Jacoby and Berry, 1958; Cook et al., 1967). The rich gas generated as a result of contact of the injected dry gas with volatile crude, when produced, contributes to oil production. The oil recovery is substantially greater than that predicted by conventional frontal-drive methods, which do not consider the vaporization equilibrium between the reservoir oil phase and the injected gas (Jacoby and Berry, 1958).

Vaporization of the immobile oil in the flow path can play an important role in higher oil recovery by pressure maintenance and gas cycling. The amount of vaporization is related to volatility of the oil as reflected by the API gravity of the stocktank oil, reservoir pressure, and reservoir temperature during gas cycling. Favorable factors for vaporization are high API gravity, high cycling pressure, and high reservoir temperature (Cook et al., 1967). Additionally, for high recovery by gas injection, the sweep and displacement efficiencies should be high, too. Favorable conditions for high vaporization are also favorable for high sweep and displacement efficiencies.

Barton and Dykes (1958) discussed the successful injection of gas into an oil reservoir having a large amount of dissolved gas and a large shrinkage factor

(Pickton Field). Crestal injection of gas in the Cedar Lake Field to maintain the reservoir pressure above the bubble point was analyzed by Leibrock et al. (1951). The crestal injection effectively delayed adverse shrinkage and relative permeability effects resulting in higher oil recovery. Stewart et al. (1955) discussed the crestal injection of an inert gas in a high-relief sour-crude reservoir leading to enhanced gravity drainage.

REFERENCES

Barton, H.B. and Dykes, F.R., Jr., 1958. Performance of the Pickton Field. *J. Pet. Technol.*, 10(2): 27–31.

Clancy, J.P., Gilchrist, R.E., Cheng, L. and Bywater, D.R., 1983. Analysis of nitrogen injection projects to develop screening guides and offshore design criteria. *Symp. Offshore Europe 83*, SPE 11902.

Clay, T.W., 1954. Pressure maintenance by gas injection in Opelika Field of Henderson County, Texas. In: *Secondary Recovery. Oil Gas J., Repr. Ser.*, 54: 55–58.

Combs, G.D. and Kuezek, R.B., 1971. Gas injection for upstructure drainage. *J. Pet. Technol.*, 23(3): 361–372.

Cook, A.B., Johnson, F.S., Spencer, G.B. and Bayazeed, A.F., 1967. The role of vaporization in high percentage oil recovery by pressure maintenance. *J. Pet. Technol.*, 19(2): 245–250.

Donahoe, C.W. and Bohannon, D.L., 1965. Harmattan Elkton Field—a case for engineered conservation and management. *J. Pet. Technol.*, 17(10): 1171–1178.

Elkins, L.F., 1969. Internal anatomy of a tight, fractured Hunton Lime reservoir revealed by performance—West Edmond Field. *J. Pet. Technol.*, 21(2): 221–232.

Elkins, L.F., 1950. The importance of injected gas as a driving medium in limestone reservoir as indicated by recent gas-injection experiments and reservoir-performance history. In: *Secondary Recovery of Oil in the United States*. API, 2nd ed., pp. 370–382.

Elkins, L.F. and Cooke, J.T., 1949. Pilot gas injection—its conduct and criteria for evaluation. *Trans. AIME*, 148: 180–188.

Hnatiuk, J. and Martinelli, J.W., 1967. The relationship of the Westerose D-3 pool to other pools on the common aquifer. *J. Can. Pet. Technol.*, 6(2): 43–49.

Jacoby, R.H. and Berry, V.J., Jr., 1958. A method for predicting pressure maintenance performance for reservoirs producing volatile crude oil. *Trans. AIME*, 213: 59–64.

Katz, D.L., Cornell, D., Riki, K., Poettmann, F.H., Vary, J.A., Elenbaas, J.R. and Weinany, C.F., 1959. *Handbook of Natural Gas Engineering*. McGraw-Hill, New York, N.Y., pp. 175–179.

Leibrock, R.M., Hiltz, R.G. and Huzarevich, J.E., 1951. Results of gas injection in the Cedar Lake Field. *Trans. AIME*, 192: 357–366.

Longren, H.F., 1951. Increasing oil recovery in the Panhandle. *Proc., 2nd. Conf. of Texas Petroleum Research Committee*, pp. 143–145.

McGraw, J.H. and Lohec, R.E., 1964. The Pickton Field—a review of a successful gas injection project. *J. Pet. Technol.*, 16(4): 399–405.

Muskat, M., 1946. *Flow of Homogeneous Fluids*, J.W. Edwards, Ann Arbor, Mich., 763 pp.

Nesluge, F.J., 1951. Gas injection in dolomite reservoir West Pampa repressuring association project. *Proc., 2nd. Conf. of Texas Petroleum Research Committee*, pp. 119–142.

Nolan, W.E. and Locker, G.R., 1954. How fractured limestone responds to gas injection. In: *Secondary Recovery. Oil Gas J., Repr. Ser.*, 54: 59–63.

Prats, M. and Miller, W.C., 1973. The role of technical publications in the advancement of fluid injection processes for oil recovery. *J. Pet. Technol.*, 25(12): 1361–1370.

Richardson, J.G. and Blackwell, R.J., 1971. Use of simple mathematical models for predicting reservoir behavior. *Trans. AIME*, 255: 1145–1154.

Stewart, F.M., Garthwaite, D.L. and Krebill, F.K., 1955. Pressure maintenance by inert gas injection in the high-relief Elk Basin Field. *Trans. AIME*, 204: 49–57.

Tarner, J., 1944. How different size gas caps and pressure maintenance programs affect the amount of recoverable oil. *Oil Weekly*, pp. 32–34.

Welge, H.J., 1952. Simplified method for computing oil recovery by gas or water drive. *Trans. AIME*, 195: 91–98.

Chapter 5

MISCIBLE FLOODING

AMROLLAH MEHDIZADEH, GERALD L. LANGNES, JOHN O. ROBERTSON, Jr., T.F. YEN, ERLE C. DONALDSON and GEORGE V. CHILINGARIAN

INTRODUCTION

Miscibility exists when two fluids are able to mix in all proportions without any interface forming between them (Clark et al., 1958). Miscibility is controlled by the pressure, temperature, composition of the oil, and composition of the displacing fluid. The pseudo-ternary phase diagram is often used as an aid in understanding the miscibility process for complex hydrocarbon mixtures (Benham et al., 1960). The diagram, however, should not be used for making quantitative predictions. Fig. 5-1 shows a pseudo-ternary composition diagram from the work of Clark et al. (1958). The three-component system shown consists of methane (C_1), the intermediates (C_2 through C_6), and all hydrocarbons heavier than C_6 (C_{7+}). The phase diagram shows whether gas, liquid, or gas–liquid phases exist for various mixtures of components at a given temperature and pressure. Region *A* represents an all-gas phase, whereas region *D* is all liquid (oil). In the critical regions (*B* or *C*) both liquid and gas are present. Region *B* shows the range of compositions for a given temperature and pressure that would be miscible with the mixtures in the dry-gas region. Region *C* contains

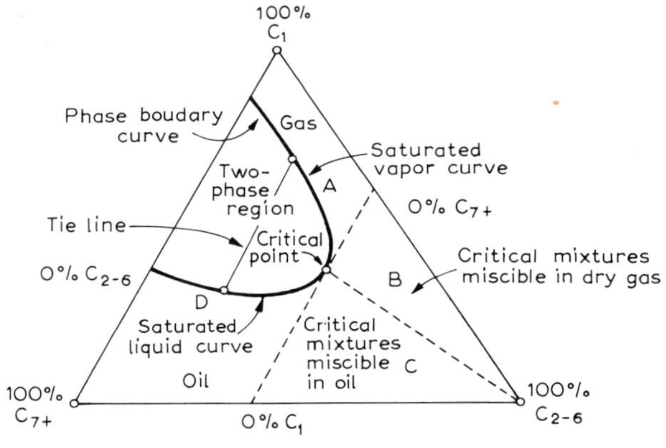

Fig. 5-1. Triangular graph showing physical conditions of hydrocarbon systems at fixed temperature and pressure conditions. (After Clark et al., 1958; courtesy of the Society of Petroleum Engineers of AIME.)

mixtures that are miscible with the mixtures in the liquid (oil) region. The tie lines terminate at points on the saturated vapor curve and the saturated liquid curve. These two points represent a saturated gas and a saturated oil which are in equilibrium (Clark et al., 1958). The sign of the slope for the tie line is determined by the value of the equilibrium ratios (constants) for the intermediates as the system approaches its critical composition (Benham et al., 1960). If the equilibrium ratios total less than one, the slope is negative. If they total more than one, the slope is positive (the tie line in Fig. 5-1 has a positive slope). The limiting tie line passes through the critical point. All fluids having an intermediate composition equal to or greater than that of the critical composition are either immediately miscible or are capable of becoming miscible with the crude in region *D* (Benham et al., 1960).

In a gas–liquid system, a miscible bank is formed by either evaporation or condensation of the intermediate hydrocarbons (C_2 through C_6). If the major transfer of intermediate hydrocarbons occurs by condensation from the gas, the system is known as enriched-gas or condensing-gas drive. If the major transfer of intermediate hydrocarbons is from the reservoir oil, then the system is known as high-pressure or evaporation-gas drive.

SWEEP EFFICIENCY

According to Crosby (1969), many of the unsuccessful miscible projects were improperly engineered. Some of the reasons for failure were improper patterns, insufficient pressure to achieve miscible displacement, and excessive reservoir heterogeneity. When the conditions are suitable for the attainment of miscibility, sweep efficiency determines the success or failure of a project.

The areal coverage or pattern efficiency of a project is controlled by the shape of the reservoir and the location of the injection wells (Clark et al., 1958). The location of the injection wells, injection rates, and timing of fluid production or injection should be evaluated by pilot operations.

In a reservoir, lithology and petrophysical properties usually vary from one area to the other, and the reservoirs frequently tend to be stratified. As variations in the porosity and permeability increase (especially in carbonate reservoirs), the sweep efficiency decreases (see Clark et al., 1958). These variations are less important in waterflooding water-wet reservoirs owing to the benefit received by the water's ability to wet the sides of the pore channels and enter the channels of low permeability. In laboratory tests, Scott and Read (1959) found that miscible displacement sweep efficiency was related directly to pore geometry, i.e., combined effects of tortuosity, pore shape, pore-size distribution, and width of intercommunicating channels.

Steeply dipping reservoirs or reefs with good vertical permeability can be effectively flooded when the impact of gravity is considered and the rates of injection are controlled.

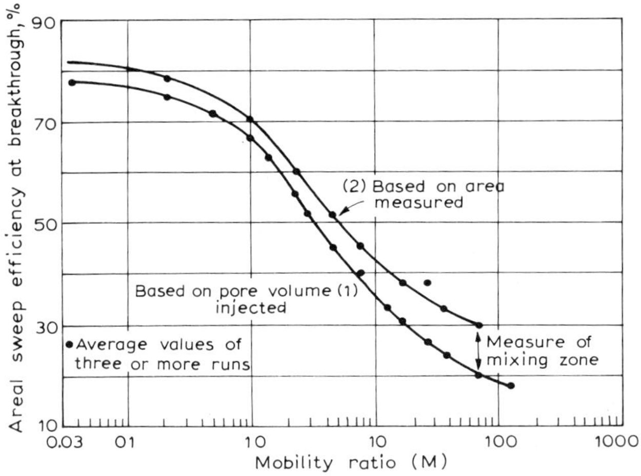

Fig. 5-2. Breakthrough sweep efficiencies for a five-spot well pattern as a function of mobility. The difference between areas contacted and the corresponding volumetric injection is a measure of the size of the mixing zone between the two miscible phases. (After Habermann, 1960, Fig. 1; courtesy of the Society of Petroleum Engineers of AIME.)

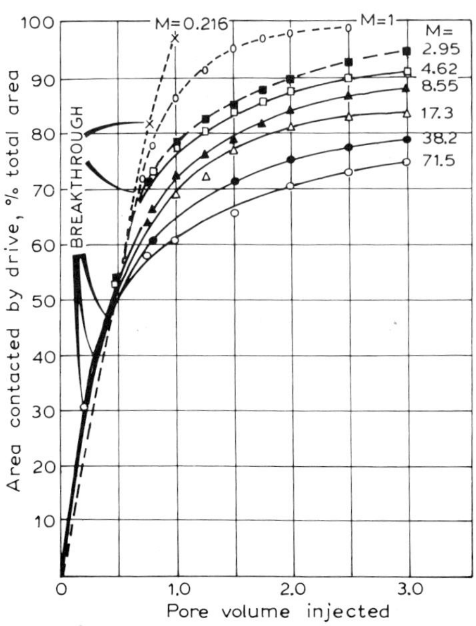

Fig. 5-3. Area contracted by drive after breakthrough, based on a quarter of a five-spot well pattern. (After Habermann, 1960, fig. 3; courtesy of the Society of Petroleum Engineers of AIME.)

In the case of reefs, good vertical drainage can result in oil recovery efficiency as high as 85–95% of the oil-in-place (Editorial, 1965). In massive and fractured reservoirs, the benefits of gravity segregation are generally eliminated by the presence of stratification (Mahaffey et al., 1966).

The impact of mobility ratio, M (= ratio of the mobility of the driving phase to the mobility of the driven phase), on the efficiency of miscible displacement is shown in Figs. 5-2, 5-3, and 5-4 (Habermann, 1960). When a less viscous fluid, such as a liquid petroleum gas (LPG) slug, drives a more viscous reservoir oil, fingers develop which reduce the areal sweep efficiency. For $M \leqslant 1$, the front is essentially radial, and a stable band is formed by the slug between the oil and driving phase until the drawdown of the producer causes a center finger to develop, which moves ahead to the well. For $M > 1$, viscous fingering of the miscible phase occurs sooner with the resultant decrease in oil recovery (Habermann, 1960).

HIGH-PRESSURE GAS INJECTION

In the high-pressure gas injection process, a lean gas (low in the C_2 through C_6 hydrocarbons, i.e., intermediates) is utilized. The injected gas is enriched within the reservoir by a transfer of intermediate hydrocarbons from the oil to the gas. To accomplish this transfer, the oil must contain sufficient quantities of hydrocarbons in the C_2 through C_6 range, and the reservoir pressure should generally be in excess of 2500 psi. If the reservoir oil is low in intermediate hydrocarbons, it will be impossible for an evaporating-gas drive to develop a miscible front (Slobod and Koch, 1953).

The performance of a miscible flood may be approximated with the Buckley-Leverett or Welge predictive techniques when the following assumptions are

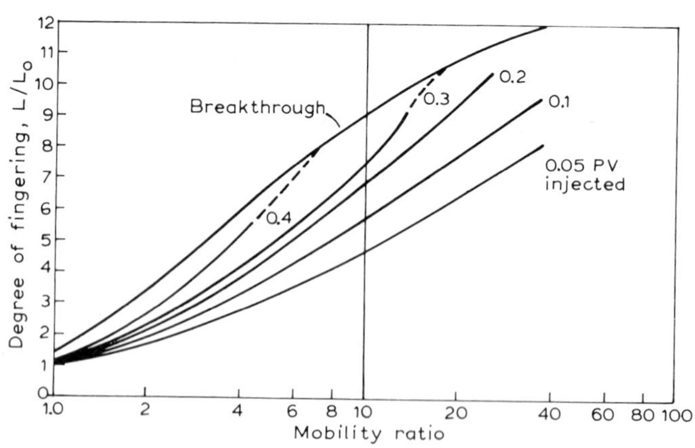

Fig. 5-4. Degree of fingering, based on a quarter of a five-spot well pattern. (After Habermann, 1960, fig. 6; courtesy of the Society of Petroleum Engineers of AIME.)

added to those presented previously in Volume I of this book: (1) Using the viscosity of the reservoir fluid and the viscosity of the displacing fluid, the recovery at breakthrough can be determined using the curves presented for liquid–liquid displacement (Fig. 5-4); (2) the producing gas/oil ratio can be estimated by assuming the fluids are produced in proportion to their saturation and viscosities (Crawford, 1969).

ENRICHED-GAS DRIVE

An enriched-gas-drive system forms a miscible or solvent front by condensation of the intermediates from the injected enriched gas. Under reservoir conditions, the intermediates condense to form the solvent bank. The process is applicable in the general pressure range of 1700–3000 psi.

Benham et al. (1960) prepared a set of curves relating the maximum methane content in the displacing fluid (to support miscibility) to reservoir and injection fluid properties, and to reservoir temperature and pressure (Figs. 5-5 through 5-15). In comparing different reservoir oils at different temperatures and pressures, the C_{2+} content of the reservoir fluid and the C_{5+} content of the displacing fluid were found to be the controlling factors for attainment of miscibility. Figs. 5-5 to 5-15 are not applicable if the displacing gas contains large amounts of

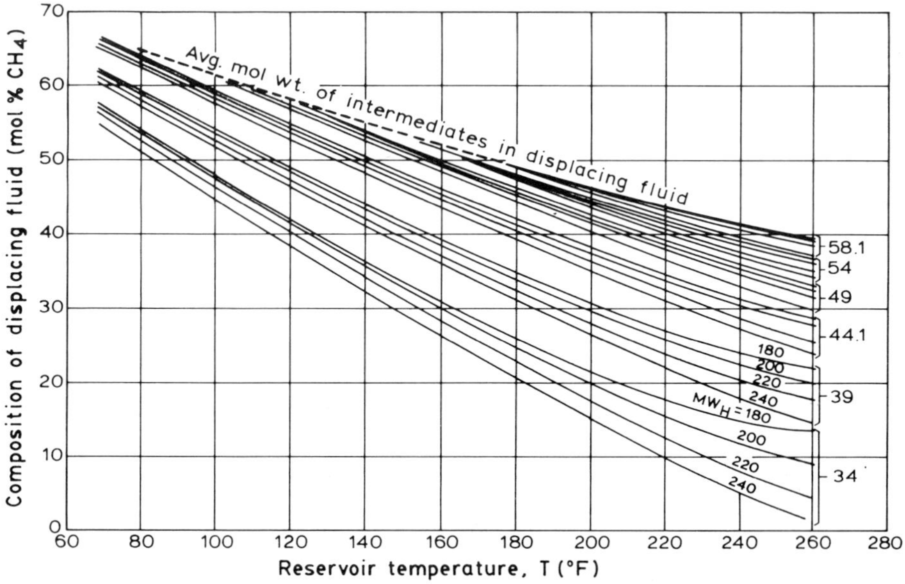

Fig. 5-5. Predicted phase conditions at 1500 psia for a miscible displacement of a reservoir fluid characterized by a C_{5+} molecular weight of 180, 200, 220, and 240. (After Benham et al., 1960, fig. 6; courtesy of the Society of Petroleum Engineers of AIME.)

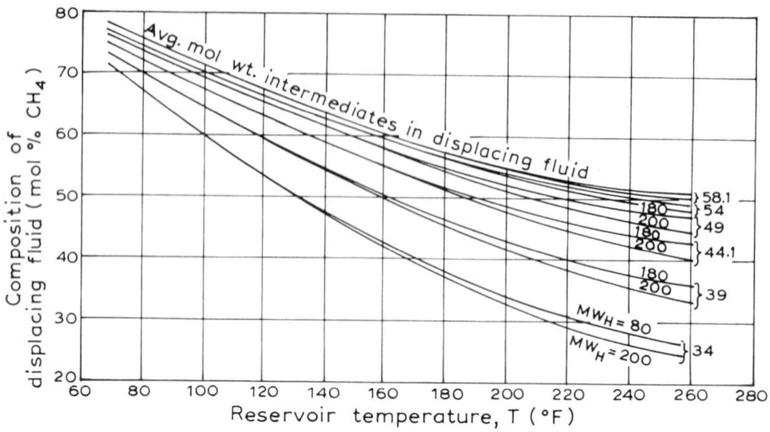

Fig. 5-6. Predicted phase conditions at 2000 psia for a miscible displacement of a reservoir fluid characterized by a C_{5+} molecular weight of 180–200. (After Benham et al., 1960; courtesy of the Society of Petroleum Engineers of AIME.)

non-hydrocarbons. The attainment of miscibility using an enriched-gas-drive process is favored by high pressures, lower temperatures, light reservoir fluids, and presence of light hydrocarbons in displacing fluids (Benham et al., 1960).

The method of predicting maximum methane content to support miscibility for an enriched-gas drive can be summarized as follows (Benham et al., 1960):

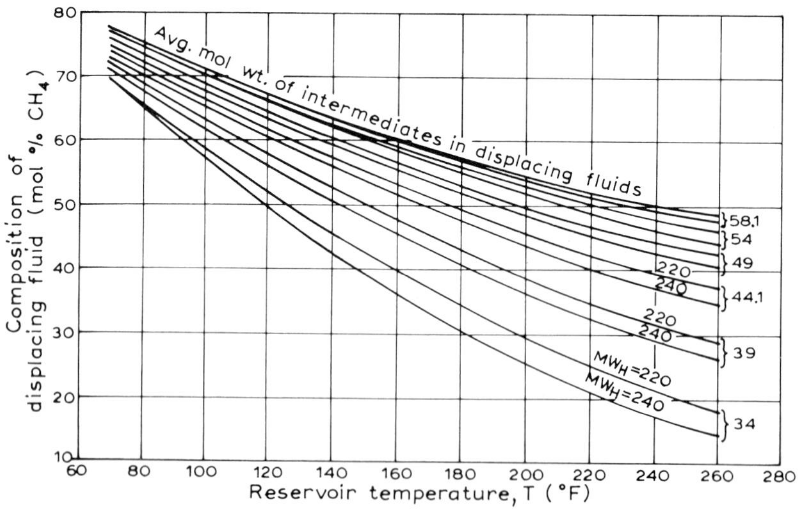

Fig. 5-7. Predicted phase conditions at 2000 psia for a miscible displacement of a reservoir fluid characterized by a C_{5+} molecular weight of 220 and 240. (After Benham et al., 1960, fig. 8; courtesy of the Society of Petroleum Engineers of AIME.)

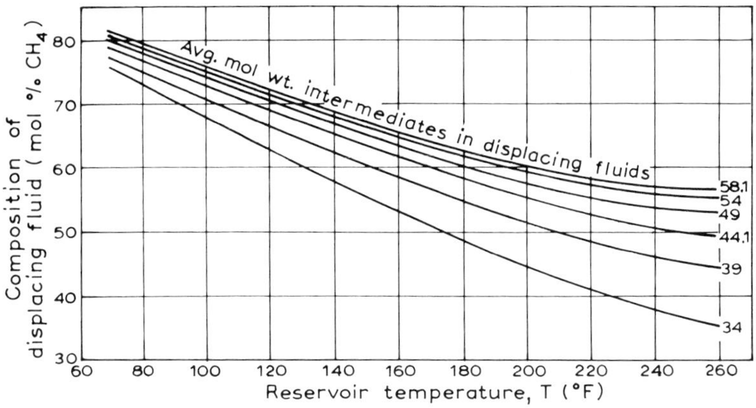

Fig. 5-8. Predicted phase conditions at 2500 psia for a miscible displacement of a reservoir fluid characterized by a C_{5+} molecular weight of 180. (After Benham et al., 1960, fig. 9; courtesy of the Society of Petroleum Engineers of AIME.)

(1) Calculate the molecular weight of the C_{5+} fraction of the reservoir fluid by (a) multiplying the molecular weight of each component, C_5 or heavier, by its mole percent, and (b) summing.

(2) Calculate the molecular weight of the C_{2+} fraction of displacing fluid in the same manner.

(3) Determine the maximum methane content in the displacing fluid from the correlation charts for a particular reservoir pressure and temperature (Figs. 5-5 to 5-15).

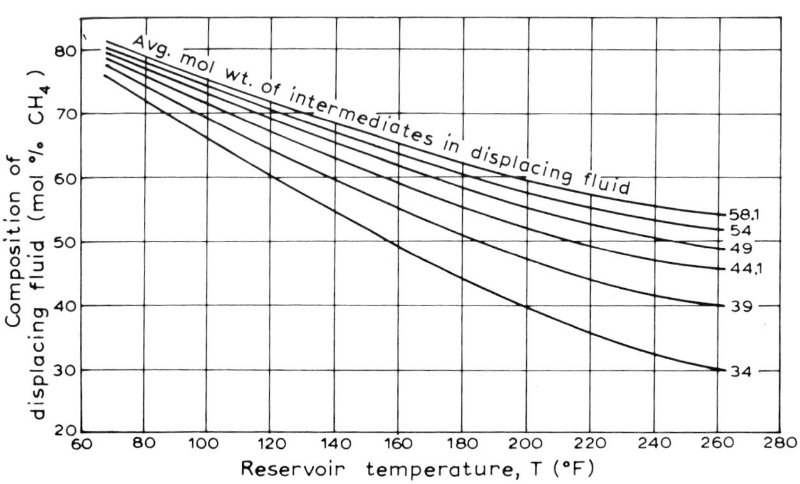

Fig. 5-9. Predicted phase conditions at 2500 psia for a miscible displacement of a reservoir fluid characterized by a C_{5+} molecular weight of 200. (After Benham et al., 1960, fig 10; courtesy of the Society of Petroleum Engineers of AIME.)

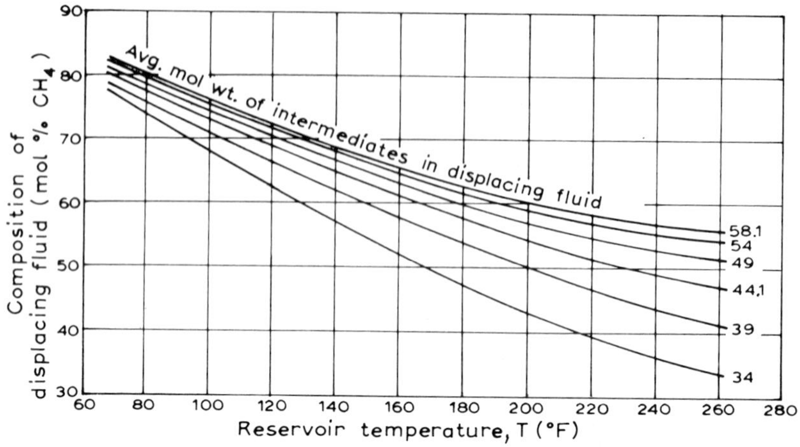

Fig. 5-10. Predicted phase conditions at 2500 psia for a miscible displacement of a reservoir fluid characterized by a C_{5+} molecular weight of 220. (After Benham et al., 1960, fig. 11; courtesy of the Society of Petroleum Engineers of AIME.)

Volumetric prediction methods can be used in predicting oil recovery for enriched-gas-drive floods, if it is assumed that gravity effects may be neglected and that the injected dry gas does not move vertically (Smith, 1966). Methods such as Stiles or Dykstra-Parsons permit calculation of a vertical sweep efficiency, E_v. Areal sweep efficiencies, E_a, may be estimated from Fig. 5-4, for the

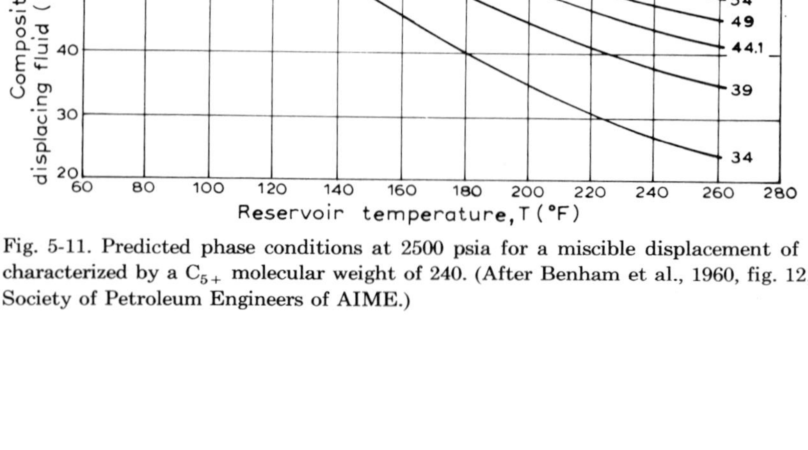

Fig. 5-11. Predicted phase conditions at 2500 psia for a miscible displacement of a reservoir fluid characterized by a C_{5+} molecular weight of 240. (After Benham et al., 1960, fig. 12; courtesy of the Society of Petroleum Engineers of AIME.)

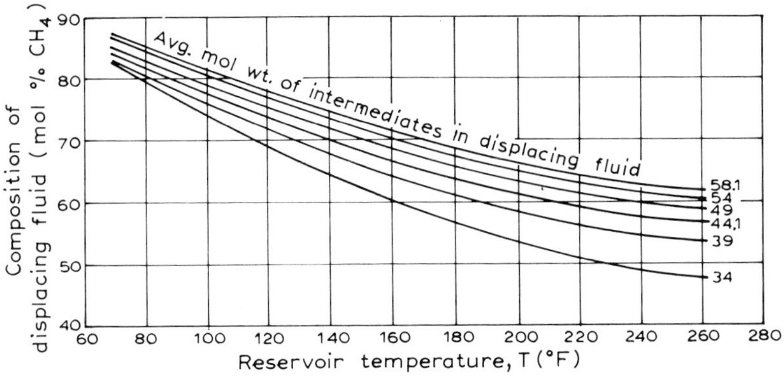

Fig. 5-12. Predicted phase conditions at 3000 psia for a miscible displacement of a reservoir fluid characterized by a C_{5+} molecular weight of 180. (After Benham et al., 1960, fig. 13; courtesy of the Society of Petroleum Engineers of AIME.)

contacted area of the oil-saturated reservoir. The oil recovered, N_p, by miscible drive can be calculated as follows:

$$N_p = 7758Ah\phi(E_v E_a E_d)\left[\frac{(1 - S_{wa} - S_{gi})(B_{or} - B_{oi})}{B_{oi} B_{or}}\right] \tag{5-1}$$

where A = area, acres; B_{oi} = formation volume factor at the beginning of project, bbl/STB; B_{or} = formation volume factor at the end of the project, bbl/STB; h = average reservoir thickness, ft; ϕ = porosity, fraction; S_{wa} = average formation water saturation, fraction; S_{gi} = average gas saturation at the start of the project, fraction; and E_d = displacement efficiency.

Welge et al. (1961) have presented a method of predicting future recovery for a nonmiscible or partially miscible enriched-gas drive. Their method takes into

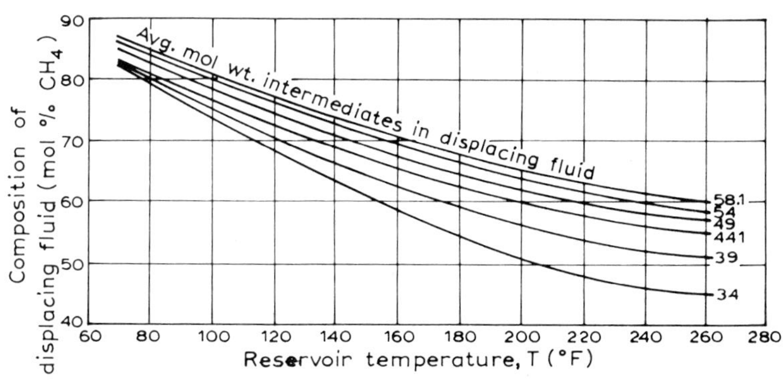

Fig. 5-13. Predicted phase conditions at 3000 psia for a miscible displacement of a reservoir fluid characterized by a C_{5+} molecular weight of 200. (After Benham et al., 1960, fig. 14; courtesy of the Society of Petroleum Engineers of AIME.)

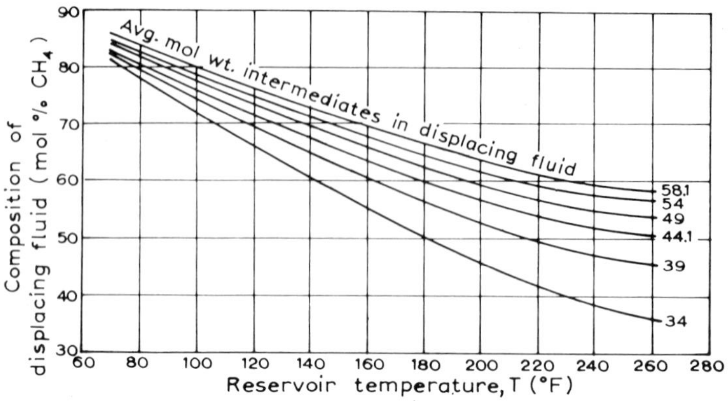

Fig. 5-14. Predicted phase conditions at 3000 psia for a miscible displacement of a reservoir fluid characterized by a C_{5+} molecular weight of 220. (After Benham et al., 1960, fig. 15; courtesy of the Society of Petroleum Engineers of AIME.)

account condensation of some of the intermediate hydrocarbons from the injected gas into the oil, as well as enhanced volatility of the heavier hydrocarbons at elevated temperatures. The recovery of oil is increased by the swelling of oil and the increase of the oil's mobility.

On the basis of laboratory experiments, Arnold et al. (1960) concluded that in reservoirs, which are not subject to gross bypassing of the oil by the injected enriched gas, small banks of enriched gas driven by methane may be used to obtain a recovery comparable to that produced by continuous injection of enriched gas. The minimum bank was estimated at 12% of the hydrocarbon pore volume.

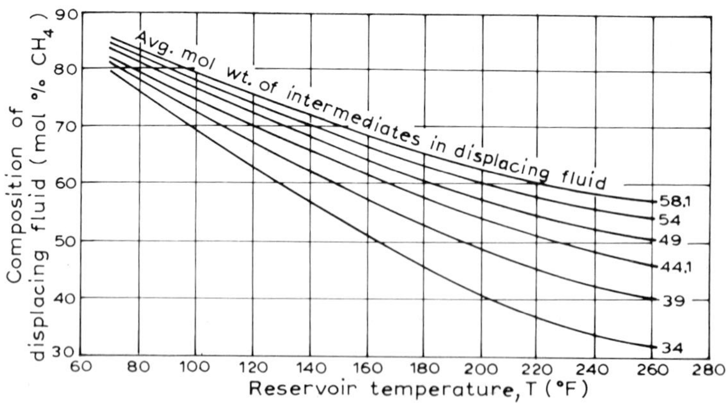

Fig. 5-15. Predicted phase conditions at 3000 psia for a miscible displacement of a reservoir fluid characterized by a C_{5+} molecular weight of 240. (After Benham et al., 1960, fig. 16; courtesy of the Society of Petroleum Engineers of AIME.)

LIQUID PETROLEUM GAS (LPG) SLUG DRIVE

In miscible slug injection, a slug (or band) of LPG or propane is driven by dry gas or water through the reservoir. This slug miscibly displaces the reservoir oil from the swept portions of the reservoir. At pressures above 1100 psi, the LPG is also miscible with the driving gas. The quantity of LPG required to maintain miscible conditions is an important factor in the economics of miscible flooding. In the case of low solvent (LPG) content, miscibility is lost when the bank of LPG deteriorates. At that point, the displacement will become immiscible rather than miscible, and recovery will drop accordingly. On the other hand, an excess of LPG represents an unnecessary capital investment.

Factors such as (1) rock permeability, (2) displacement rate, (3) reservoir oil viscosity, (4) distance traveled, and (5) diffusion rate determine the extent of mixing which occurs at the solvent–crude oil interface and the solvent–driving gas interface (Craig and Owens, 1960). Mixing tends to occur longitudinally in the direction of flow-gas (Craig and Owens, 1960).

The specific quantity of slug required is a subject of controversy. Estimates range from 2–3% of the total pore volume (Koch and Slobod, 1957) to 10–15% of the total pore volume (Habermann, 1960). Field trials indicate that the required slug size for many sandstone reservoirs is 10–15%, whereas carbonates appear to require 4–7% of the total pore volume. Habermann (1960) investigated the effect of slug size (10–25% of the pore volume swept) upon the oil recovery. Using pore volume swept instead of total pore volume requires an estimation of vertical sweep efficiency, E_v, and the areal sweep efficiency, E_a, as previously discussed. It was noted that for reservoirs with an intercrystalline–intergranular porosity system, when the slug size was less than 10% of the pore volume swept, the bank deteriorated prematurely, letting the driving phase penetrate through the bank of solvent. The optimum size for the slug appeared to be about 15% (Habermann, 1960). Habermann's findings were substantiated by the work of Mahaffey et al. (1966).

The sweep efficiency generally improves on using water as the driving fluid. There is, however, a loss of miscibility at the trailing edge of the slug, which leaves expensive solvent behind in the reservoir. To overcome this problem, some operators (Sessions, 1960) inject a slug of propane, followed by a buffer zone of gas, and then water.

PREDICTIVE TECHNIQUES

In addition to the techniques discussed previously, several other more detail predictive techniques such as (1) black-oil simulators (Peaceman and Rachford, 1962; Todd and Longstaff, 1972; Settari et al., 1977), (2) compositional simulators (Roebuck et al., 1969; Nolen, 1973; Kazemi et al., 1978, Nghiem et al., 1981), and (3) scaled physical models (Pozzi and Blackwell, 1963; Mahaffey et al., 1966) have been reported in the literature.

TABLE 5-I

Selected vaporizing-gas drive field projects (after Stalkup, 1984, p. 132, table 7.1; courtesy of the Society of Petroleum Engineers). For references, see Stalkup's classical work

Field	Operator	Year Started	Project Type	Oil Gravity (API)	Viscosity Ratio μ_o/μ_s	Miscibility Pressure (psi)	Depth (ft)	Thickness (ft)	Acreage	Status	Recovery (% OOIP)	References
Block 31 Devonian (TX)	ARCO Oil & Gas Co.	1949 1952	G,N	46	10	3,500	8,400	125 to 150	7,840	NC, succ., prof.	>60*	14,16,23,24,25†
Neale 10,400-ft (LA)	ARCO Oil & Gas Co.	1956	G	47	4	4,000	10,400	10	—	term.	36**	15,24
University Block 9 (TX)	ARCO Oil & Gas Co.	1960	LPG slug by G by G/W	38	29	4,500	5,300	4 to 45, 10	4,000	—	40	26
Raleigh (MS)	Chevron U.S.A. Inc.	1960	G	45.5	3.5	<5,400	12,600	18	812	—	>52	27
Hassi-Massaoud (Algeria)	SONATRACH	1964	G and G/W	44	7	3,700	10,800	300	18,000	succ., prof.	—	28,29,30†
Fairway (TX)	Hunt Oil Co.	1966	G/W	48	5.5	4,800	10,000	55	22,600	HF, succ., prof.	50*	25†,31
Bridger Lake (WY/UT)	Phillips Petroleum Co.	1970	G	40	19	5,300	16,500	20	3,800	NC, succ., prof.	32*	25†,32
East Binger (OK)	Phillips Petroleum Co.	1977	N	38	—	—	10,000	—	13,000	JS, TETT	—	25†,33
Fordoche (LA)	Sun Gas Co.	1971	G+N	45	5	—	13,500	25 to 35	6,100	HF, succ.	50*	25†,34
Jay-Little (FL)	Exxon U.S.A.	1982	N/W (20% HCPV)	51	6	3,600	15,000	350	14,000	—	54* (6.5* incremental)	8
Lake Barre‡ (LA)	Texaco Inc.	1978	N	40-43	—	5,100 (G) 5,800 (N)	17,500	57	1,194	—	53*	9

*Projected.
**Recovery in the east and west sections of the sand where miscible displacement definitely was achieved was estimated by the operator to be approximately 50%.
†Status as given in the *Oil and Gas J.* article identified in Reference column.
‡Full miscibility apparently has not been achieved; reservoir pressure is several hundred pounds per square inch below dynamic miscibility pressure.

Symbol key:
G = natural gas
N = flue gas or nitrogen
G/W = alternate gas-water injection
succ. = successful
prof. = profitable
TETT = to early to tell
NC = nearing completion
term. = terminated
HF = half finished
JS = just started

Stalkup (1984) has summarized some selected field trials in the three following categories: (1) vaporizing-gas drive (high-pressure gas injection); (2) condensing-gas drive (enriched-gas drive); and (3) first-contact miscible drive (liquid petroleum gas slug drive). These are reported in Tables 5-I through 5-IV. Some of these trials have been described in the following section. Stalkup's excellent book belongs on the bookshelf of every petroleum engineer.

Block 31 Field, Texas (intercrystalline–intergranular porosity system)

A high-pressure gas miscible injection project was initiated in the Block 31 Field, Texas, 1949 (Bleakley, 1969). In 1969, it was estimated that 60% of the oil-in-place would be recovered by this project (Herbeck and Blanton, 1961). Several factors contributed to the success of the project: (1) the project was begun early in the life of the reservoir; (2) the formation rock was continuous and homogeneous; and (3) close engineering control over the project ensured miscible displacement and maximum sweep efficiency (Davison, 1967).

In 1966, flue gas was substituted for natural gas after determining its ability to maintain miscibility. The flue gas is cheaper to use than hydrocarbon gas even on considering the amortization of a new flue gas generating plant. On using flue gas, besides economy, more oil can be produced as the reservoir's economic limit will be lower (Davison, 1967). Clark et al. (1964), Davison (1967), and Tittle and From (1967) have provided detailed information on the use of the flue gas in oil field operations. As of 1981, more than 46% of the original oil-in-place has been recovered from the Block 31 Field.

Millican Field, Texas (vugular–solution porosity system)

A miscible pilot flood was started in late 1956 in the Millican Field, Texas (Sturdivant, 1959). The volume of the propane–butane slug was equal to 1.5% of the floodable pore volume. The premature breakthrough of the gas was attributed to the small slug size and an unfavorable permeability distribution (Sturdivent, 1959). An overall recovery of 18% of the original oil-in-place has been observed by the operator.

Golden Spike Field, Alberta, Canada (vugular–solution porosity system)

In the Golden Spike flood, Alberta, Canada, the total volume of the slug was 7% of the total hydrocarbon pore volume (Larson et al., 1967). The data developed by Benham et al. (1960) were used in determining the composition of the solvent bank. The suitability of the estimated composition was then confirmed by laboratory tests. Reservoir simulation was used to determine the optimum method for the placement of the slug. The ultimate recovery was

TABLE 5-II

Condensing-gas drive field projects (after Stalkup, 1984, pp. 120–121, table 6.1; courtesy of the Society of Petroleum Engineers)

Field	Operator	Year Started	Project Type	Oil Gravity (°API)	Oil Viscosity (cp)	Viscosity Ratio μ_o/μ_s	Depth (ft)	Thickness (ft)	Acreage	Slug Size (% HCPV)	Recovery (% OOIP)	Status	References
Haynesville (LA)	Haynesville Operators Comm.	1953	LPG←RG ←G ←W						3,200				11
Bronte (TX)	Exxon U.S.A.	1956	RG						1,300				11
Bronte (TX)	Exxon U.S.A.	1956	RG						460				11
Seeligson Zone 20B-07 (TX)	Exxon U.S.A.	1957	RG←G/W	40		12	6,000		877	52	5 to 10 (incremental) 50 to 54 (ultimate)	NC, disc.	11-21*
Stratton (TX)	Southern Minerals Corp.	1957	RG										11
South Coles Levee (CA)	Ohio	1957	RG										11,22
Elk Basin (WY)	Amoco Production Co.												11
Midland Farms (TX)	Amoco Production Co.	1960	RG←G ←G/W	41	0.3		8,350		600	2		term., disc., prof.	23*-25
Neale Lilliedoll (LA)	ARCO Oil & Gas Co.	1962	tertiary RG←G/W			11	10,100	13.5	640	15		term.	26,†
Golden Spike (Alberta)	Imperial Oil Co.	1964	vertical RG←G	37	0.8		5,672	550	1,305	7	3.1 (incremental)	comp., disc., prof.	27-31*
Pembina Bear Lake Cardium (Alberta)	Cities Service Co.	1968	RG←G ←G/W	37.6	1.5	52	4,857	50	4,160	3			32
Ante Creek (Alberta)	Amoco Production Co.	1968	RG/W	42	0.13	4.7	11,000	29.5	6,000	cont. rich gas		HF, NE	27,31*,33
Intisar D Reef (Libya)	Occidental Petroleum Corp.	1969	vertical RG	40	0.46	20	8,950	888	3,300		70** (ultimate)	HF, succ., prof.	34-37*
Levelland (TX)	Amoco Production Co.	1972	RG←RG/W ←G/W	30	1.9	86	4,900	250	1,190	14	>27** (incremental)	HF, succ., prof.	31*,37,38

South Swan Hills (Alberta)	Amoco Production Co.	1973	RG/W ←G/W ←W	42	0.39	18	8,500	73	830	15	20** (incremental) 65** (ultimate)	HF, prom., prof.	27,31*,39
Central Mallett Slaughter (TX)	Amoco Production Co.	1972		30	2		4,800		12	20 and increasing	6.0 (incremental) and increasing	JS TETT unprof.	31*,41
Camp Sand Haynesville (TX)	Marathon Oil Co.	1966	RG←G←W	46	1.05		8,000	5	~2,200	~10			42
Rainbow area (Alberta)	various		vertical										27,31

*Primary reference reporting status.
**Estimated with simulators.
†Unpublished ARCO Oil & Gas Co. results.

Symbol key:
LPG = liquified petroleum gas
RG = rich gas
G = lean gas
RG←G/W = rich gas displaced by alternate gas-water
NC = nearing completion
disc. = discouraging
term. = terminated
NE = not elevated
TETT = too early to tell
JS = just started
HF = half finished
prof. = profitable

TABLE 5-III

First-contact miscible tertiary recovery tests (after Stalkup, 1984, p. 105, table 5.3; courtesy of the Society of Petroleum Engineers)

Field	Operator	Year Started	Project Type	Oil Gravity (°API)	Oil Viscosity (cp)	Viscosity Ratio μ_o/μ_s	Depth (ft)	Thickness (ft)	Acreage	Slug Size* (% HCPV)	Incremental Recovery* (% OOIP)	Oil/ Gross Slug (STB/RB)	Oil/ Net Slug (STB/RB)	Reference
Burkett Unit (KS)	Phillips Petroleum Co.	1958	LPG←G←W	42	—	25	2,100	30	10	10	7.0	0.67		18,74
Johnson (NE)	Ohio Oil Co.	1958	LPG←G	37	1.1	18	4,600	10	164	5.5	34†	~7		18,75,76
South Ward (TX)	ARCO Oil & Gas Co.	1959	LPG←G←W	35	4	40	2,400	32	10	7.5	11.4	1.5	2.2	18,77
Adena Clar A (CO)	Union Oil Co. of California	1962	LPG←G	44	0.42	6.5	5,500	28	1 injector 2 producers	~17‡		0.36	1.7	78
Adena Hough A (CO)	Union Oil Co. of California	1963	LPG←G/W	44	0.42	6.5	5,500	28	80	~9.0‡		0.47	0.93	78
Hibberd Pool South Cuyama (CA)	ARCO Oil & Gas Co.	1963	LPG←G←W	35	1.7	23	4,300	60	80	7.4	13.5	1.6	>3	79
Phegly Unit (CO)	Mobil Oil Corp.	1964	LPG←G/W	37	3	30	4,900	8	785	3.3	3.5	0.9		80

*Treated area.
**Ultimate estimated from decline curve analysis.
†Nineteen percent OOIP in entire field.
‡Treated area difficult to define because of nature of test.

Symbol key:
LPG←G = LPG-slug driven by gas
LPG←G/W = LPG-slug driven by alternate gas/water injection.

TABLE 5-IV

First-contact miscible projects (after Stalkup, 1984, pp. 103–104, table 5.2; courtesy of the Society of Petroleum Engineers)

Field	Operator	Year Started	Project Type	Oil Gravity (°API)	Oil Viscosity (cp)	Viscosity Ratio μ_o/μ_s	Depth (ft)	Thickness (ft)	Acreage	Slug Size (% HCPV)	Recovery (% OOIP)	Status	Reference
San Pedro Block D (Peru)	Intl. Petroleum Co.	1951	LPG	39	10		2,300	164	20	12	10 incremental	term., succ.	17*
West Panhandle (TX)	Diamond Shamrock Corp.	1952	LPG storage; oil recovery						600			term., good oil response	18*
Seminole City (OK)	Carter Oil Co.	1953	LPG←G	37	3	30	3,500	22		8.5	>7.5 incremental	term., oil response	18*,19
Chicon Lake (TX)	Winthrop Corp.		LPG storage; oil recovery									term.	18*
Eram (OK)	Sunray		tertiary LPG←G										18
Sand Creek (WY)	Mobil Oil Corp.		LPG←G										18
South Leonard (OK)	Amoco Production Co.		LPG←G									term., unsucc.	18*
Spraberry Trend Area (TX)	Phillips Petroleum Co.		LPG									term.	18*
Camp Hill (TX)	Frank Buttram		LPG									term., poor response	18*
Killan (TX)	Petrocel Corp.		LPG storage; oil recovery									term., poor response	18*
West Panhandle (TX)	Diamond Shamrock Corp.	1955	LPG storage; oil recovery										18
Taylor-Ina (TX)	Wherry & Green	1956	LPG←G									term., unsucc., channelling	18*
Millican Reef (TX)	Sun Oil Co.	1956	LPG←G				6,000		300	1.5			18,20–23
North Pembina Cardium Unit (Alberta)	Mobil Oil Corp.	1957	LPG←G	37		14	5,000	16	10	7.1	72 ultimate	term., succ.	24*
Newhall-Potrero (CA)	Sun Oil Co.	1957	LPG←G										18,25,26
Bisti (NM)	Sun Oil Co.	1957	LPG←G		0.9	10	4,900	130	40-pilot 1,600	5.5	55 Pilot ultimate	term., unsucc., prof.	18*,27–37
Short Junction (OK)	Conoco Inc.	1957	LPG←G						640				18,38
Elk City P1 (OK)	Shell Development Co.	1957	LPG←G						pilot				18

(*continued on next page*)

TABLE 5-IV (continued)

Field	Operator	Year Started	Project Type	Oil Gravity (°API)	Oil Viscosity (cp)	Viscosity Ratio μ_o/μ_s	Depth (ft)	Thickness (ft)	Acreage	Slug Size (% HCPV)	Recovery (% OOIP)	Status	Reference
Panhandle (TX)	Cities Service Co.	1957	LPG←G						320				18
Willbarger Co. Regular (TX)	Cities Service Co.	1957	LPG←G						26				18
Three Bar (TX)	Amoco Production Co.	1957	LPG←G						160			term., unsucc.	18*,39
Parks Pennsylvanian (TX)	Mobil Oil Corp.	1957	LPG←G	45	0.38	5	10,400	21	6,400	4	14 incremental	NC, succ., prof.	18,39,40, 41*–48
Tijerina-Canales-Blucher (TX)	Texaco Inc.	1957	LPG←G						pilot				18,49
Meadow Creek Lakota B (WY)	Conoco Inc.	1957	LPG←G ←G/W		0.42	10	7,300	9.5	400	7		term., disc., prof.	18,50,51*
Slaughter Boyd Lease (TX)	ARCO Oil & Gas Co.	1958	LPG←G ←W	30	1.3	13	5,000	40	320	0.6	4 incremental		18,52,53
Rincon (TX)	Conoco Inc.	1958	LPG←G						120				18,54–56
Midway-Sunset 29-D (CA)	Standard Oil Co. of California	1958	LPG←G	31	0.84	13	5,300	20	160	10	51 ultimate		57
Rio Bravo (CA)	Superior Oil Corp.	1958	LPG←G	38	0.28	5	11,200	60	300	12			58
Guijairal Hills (CA)	Union Oil Co. of California	1959	LPG←G										18
Paloma (CA)	Western Oil Corp.	1959	LPG←G										18
Elk City M Zone (OK)	Shell Development Co.	1959	LPG←G						pilot				18
Sunflower (OK)	Tekoil Corp.	1959	LPG←G						40				18,59
East Washunga (OK)	Tekoil Corp.	1959	LPG←G						40				18
La Gloria (TX)	Mobil Oil Corp.	1959	LPG←G						59				18
Panhandle (TX)	Skelly	1959	LPG←G						160				18
Baskington (LA)	Sun Oil Co.	1962	LPG←G		0.4	4	3,800	20	3 wells			term., gravity seg.	60*
Pembina Lobstick Cardium (Alberta)	Amoco Production Co.	1964	LPG←G/W				5,000		2,240			disc.	61,62*
West Ranch Howard Glasscock (TX)	Mobil Oil Corp.	1964	High-relief LPG←G	31	1.4		5,500		480			75% succ., prof.	41*
North Coles Levee (CA)	ARCO Oil & Gas Co.	1967	LPG←G	36	0.41		9,500		624			term., prof.	41*

Pembina Cardium (Alberta)	Texaco Inc.	1972		32	2		5,000		5,540			term., partly succ., unprof.	51,63*
Wasson (TX)	Shell Oil Co.	1959	LPG←G	30	1.2	12	5,000	250	160				64–66
North Pembina Cardium Unit (Alberta)	Mobil Oil Corp.	1959	LPG←G	37		14	5,000	16	13,000	1–1.7		term., disc.	49,67*–69
Wizard Lake (Alberta)	Texaco Inc.	1969	vertical LPG←G	39	0.5		6,200	650	4,800	7.4	84 ultimate projected	HF, succ., prof.	63*,70–71
Rainbow area (Alberta)	various		vertical LPG←G										

*Primary reference reporting status.

Symbol key:
LPG = liquified petroleum gas
G = lean gas
LPG←G = LPG-slug driven by gas
LPG←G/W = LPG-slug displaced by alternate gas/water injection
NC = nearing completion
disc. = discouraging
term. = terminated
HF = half finished
prof. = profitable
succ. = successful.

expected to be near 96% of the stocktank oil-in-place (Sturdivant, 1959). The gas injection project utilizing gravity-controlled rates of injection had been expected to recover 70% of the oil-in-place (Willman, 1966). Upon completion of the project, 67% of the original oil-in-place was recovered.

Slaughter Field, Texas (intercrystalline–intergranular porosity system)

In 1958, a miscible slug project was started in the San Andres Zone of the Slaughter Field, Texas (Sessions, 1960). An irregular-pattern flood was used. The LPG slug was injected first, followed by gas, and, finally, by water. It was estimated that under primary operations, 20% of the oil-in-place would be recovered, whereas the combined slug–gas–water process should result in 62% recovery. The miscible project has recovered 4% more oil (original oil-in-place) as compared with a conventional waterflooding.

Parks Field, Texas (intercrystalline–intergranular porosity system)

In 1957, a miscible slug project was started in the Parks Field, Texas, in the Pennsylvanian Bend reservoir (Marrs, 1961). A slug of propane (4% of the total hydrocarbon pore volume) was injected followed by dry gas. In 1961, Marrs estimated that the 17% recovery by primary means would be increased to 55%. There was also a five-spot pilot waterflood project underway in the field. It has been shown that the incremental recovery by miscible flooding as compared with conventional waterflood recovery has been increased by 14% of the original oil-in-place.

REFERENCES

Agan, J.B. and Fernandes, R.J., 1962. Performance of a miscible slug in a highly-stratified reservoir. *J. Pet. Technol.,* 14(1), 81–86.

Arnold, C.W., Stone, H.L. and Luffel, D.L., 1960. Displacement of oil by rich-gas banks. *Trans. AIME,* 219: 305–312.

Benham, A.L., Dowden, W.E. and Kunzman, W.J., 1960. Miscible fluid displacement--prediction of miscibility. *Trans. AIME,* 219: 229–237.

Bleakley, W.B., 1969. Miscible flood hikes Block 31's oil output. *Oil Gas J.,* October 27, pp. 67–70.

Clark, N.J., Shearin, H.M., Schultz, W.P., Garms, K. and Moore, J.L. 1958. Miscible drive—its theory and application. *J. Pet. Technol.,* 10(6), 11–23.

Clark, N.J., Roberts, T.G. and Lindner, J.D., 1964. Engine exhaust gas boasts heavy oil recovery. *Pet. Eng.,* August, pp. 43–47.

Clay, T.W., 1954. Pressure maintenance by gas injection... In Opelika Field of Henderson County, Texas. In: *Secondary Recovery. Oil Gas J., Repr. Ser.,* 54:. 55–58.

Coats, K.H. 1969. Use and misuse of reservoir simulation models. *J. Pet. Technol.,* 21(11): 1391–1398.

Cook, A.B., Johnson, F.S., Spencer, G.B. and Bayazeed, A.F., 1967. The role of vaporization in high percentage oil recovery by pressure maintenance. *J. Pet. Technol.,* 19(2): 245–250.

Craig, F.F. and Owens, W.W., 1960. Miscible slug flooding—a review. *J. Pet. Technol.,* 12(4): 11–16.

Crawford, P.B., 1969. *Advanced Petroleum Engineering Short Course.* Texas A&M University, College Station, Texas.

Crosby, G.E., 1969. Is miscible flooding dead? *Pet. Eng.,* April, pp. 53–55.

Davison, K., 1967. Inert gas boasts recovery from heavy oil reservoirs. *World Oil,* March, pp. 98–105.

Doepel, G.W. and Sibley, W.P., 1962. Miscible displacement—a multilayer technique for predicting reservoir performance. *J. Pet. Technol.,* 14(1): 73–80.

Donahoe, C.W. and Bohannon, D.L., 1965. Harmattan Elkton Field—A case for engineered conservation and management. *J. Pet. Technol.,* 17(10): 1171–1178.

Editorial, 1965. Miscible flood may capture 95 percent of crude in place. *Oil Gas J.,* July, pp. 164–166.

Editorial, 1968. Rainbow project may hit 96 percent recovery. *Oil Gas J.,* December, pp. 48–49.

Editorial, 1969. Texaco miscible flood program begins for Alberta Pool "soon". *Oil Gas J.,* December, pp. 48–49.

Elkins, L.F., 1950. The importance of injected gas as a driving medium in limestone reservoirs as indicated by recent gas-injection experiments and reservoir-performance history. In: *Secondary Recovery of Oil in the United States.* API, 2nd ed., pp. 370–382.

Elkins, L.F., 1969. Internal anatomy of a tight, fractured Hunton Lime Reservoir revealed by performance—West Edmond Field. *J. Pet. Technol.,* 21(2): 224–232.

Elkins, L.F. and Cooke, J.T., 1949. Pilot gas injection—its conduct and criteria for evaluation. *Trans. AIME,* 148: 180–188.

Enright, R.J., 1961. Are miscible floods worth the cost? *Oil Gas J.,* December, pp. 43–46.

Fitch, R.A. and Griffith, J.D., 1964. Experimental and calculated performance of miscible floods in stratified reservoirs. *J. Pet. Technol.,* 16(11): 1209–1298.

Fitch, R.A. and Holloway, H.D., 1963. Performance of miscible floods with alternate gas-water displacement in the Midland Farms Wolfcamp Field, Andrews County, Texas. *Soc. Pet. Eng. Reg. Meet., New Orleans, La, October 24–25,* SPE 626.

Gardner, A.O., Jr., Peaceman, D.W. and Pozzi, A.L., Jr., 1964. Numerical calculation of multidimensional miscible displacement by the method of characteristics. *Soc. Pet. Eng. J.,* 16(3): 26–36.

Gardner, G.H., Downie, J. and Dendall, H.A., 1962. Gravity segregation of miscible fluids in linear models. *Soc. Pet. Eng. J.,* 14(6): 95–104.

Greenkorn, R.A., Johnson, C.A. and Haring, R.E., 1965. Miscible displacement in a controlled natural system. *J. Pet. Technol.,* 17(11): 1329–1335.

Habermann, B., 1960. The efficiency of miscible displacement as a function of mobility ratio. *Trans. AIME,* 219: 264–272.

Herbeck, E.F. and Blanton, J.R., 1961. Ten years of miscible displacement in Block 31 Field. *J. Pet. Technol.,* 13(6): 543–549.

Hnatiuk, J. and Martinelli, J.W., 1967. The relationship of the Westerose D-3 Pool to other pools on the common aquifer. *J. Can. Pet. Technol.,* April–June, pp. 43–49.

Jacoby, R.H. and Berry, V.J., Jr., 1958. A method for predicting pressure maintenance performance for reservoirs producing volatile crude oil. *Trans. AIME,* 213: 59–64.

Katz, D.L., Cornell, D., Riki, K., Poettmann, F.H., Vary, J. A., Elenbaas, J.R. and Weinany, C.F., 1959. *Handook of Natural Gas Engineering.* McGraw-Hill, New York, N.Y., pp. 175–179.

Kazemi, H.A., Vestal, C.R. and Shank, G.D., 1978. An efficient multicomponent numerical simulator. *Soc. Pet. Eng. J.,* 18(10): 355–368.

Koch, H.A. and Slobod, R.L., 1957. Miscible slug process. *Trans. AIME,* 210: 40–47.

Koval, E.J., 1963. A method for predicting the performance of unstable miscible displacement in heterogeneous media. *Soc. Pet. Eng. J.,* 15(6): 145–154.

Larson, V.C., Peterson, R.B. and Lacey, J.W., 1967. Technology's role in Alberta's Golden Spike miscible project, PDDC-12. In: *Proceedings, Seventh World Petroleum Congress, Vol. 3.* Elsevier, Amsterdam, pp. 533–544.

Longren, H.F., 1951. Increasing oil recovery in the Panhandle. *Proc. 2nd Conf. Texas Pet. Res. Committee,* pp. 143–145.

Mahaffey, J.L., Rutherford, W.M. and Matthews, C.S., 1966. Sweep efficiency by miscible displacement in a five-spot. *Soc. Pet. Eng. J.*, 6(3): 73–80.
Marrs, D.G., 1961. Field results of miscible displacement program using liquid propane driven by gas, Parks Field Unit, Midland County, Texas. *J. Pet. Technol.*, 13(4): 327–332.
McCulloch, R.C., Langton, J.R. and Spivak, A., 1969. Simulation of high relief reservoirs, Rainbow Field, Alberta, Canada. *J. Pet. Technol.*, 21(11): 1339–1408.
McGraw, J.H. and Lohec, R.E., 1961. The Picton Field—a review of a successful gas injection project. *J. Pet. Technol.*, 13(4): 399–405.
Muskat, M., 1946. *Flow of Homogeneous Fluids.* J.W. Edwards, Ann Arbor, Mich.
Nesluge, F.J., 1951. Gas injection in dolomite reservoir, West Pampa Repressuring Association Project. *Proc. 2nd Conf. Texas Pet. Res. Committee,* pp. 119–142.
Nghiem, L.X., Fong, D.K. and Aziz, K., 1981. Compositional modeling with equation of state. *Soc. Pet. Eng. J.*, 8(12): 687–698.
Nolan, W.E. and Locker, G.R., 1954. How fractured limestone responds to gas injection. In: *Secondary Recovery. Oil Gas J., Repr. Ser.*, 54: 59–63.
Nolen, J.S., 1973. Numerical simulation of compositional phenomena in petroleum reservoirs. In: *Numerical Simulation. Soc. Pet. Eng., Repr. Ser.*, 11: 269–284.
Peaceman, D.W., 1966. Improved treatment of dispersion in numerical calculation of multidimensional miscible displacement. *Soc. Pet. Eng. J.*, 8(11): 213–216.
Peaceman, D.W. and Rachford, H.H., Jr., 1962. Numerical calculation of multidimensional miscible displacement. *Soc. Pet. Eng. J.*, 8(12): 327–346.
Pozzi, A.L. and Blackwell, R.J., 1963. Design of laboratory models for study of miscible displacement. *Soc. Pet. Eng. J.*, 8(3): 28–42.
Roebuck, I.F., Jr., Henderson, G.E., Douglas, J., Jr. and Ford. W.T., 1969. The compositional reservoir simulator: case 1. The linear model. *Soc. Pet. Eng. J.*, 8(3): 115–130.
Rowe, A.M., Jr. and Silberberg, I.H., 1965. Prediction of the phase behavior generated by the enriched-gas-drive process. *Soc. Pet. Eng. J.*, 5(2): 160–166.
Scott, E.Z. and Read, D.L., 1959. A study of variables in linear miscible displacement. *Soc. Pet. Eng. Reg. Meet., Pasadena, Calif., October 22–23,* SPE 1377-G.
Sessions, R.E., 1960. How Atlantic operates the Slaughter Flood. *Oil Gas J.*, July 4, pp. 91–98.
Settari, A., Price, H.S. and Dupont, T., 1977. Development and application of variational methods for simulation of miscible displacement in porous media. *Soc. Pet. Eng. J.*, 17 (3): 228–246.
Slobod, R.L. and Koch, H.A., 1953. High pressure gas injection mechanism of recovery increase. *Drill. Prod. Pract., API,* pp. 82–94.
Smith, C.R., 1966. *Mechanics of Secondary Recovery.* Reinhold, New York, N.Y.
Stalkup, F.I., Jr., 1984. *Miscible Displacement. SPE Monogr.*, 8: 204 pp.
Sturdivant, W.C., 1959. Pilot propane project completed in West Texas reef. *J. Pet. Technol.*, 11(5): 27–30.
Tarner, J., 1944. How different size gas caps and pressure maintenance programs affect the amount of recoverable oil. *Oil Weekly,* June 12, pp. 32–34.
Tittle, R.M. and From, K.T., 1967. Success of flue gas program at Neale Field. *Soc. Pet. Eng. 42nd Annu. Fall Meet., Houston, Texas, October 1–4,* SPE Preprint 1907.
Todd, M.R. and Longstaff, W.J., 1972. The development, testing, and application of a numerical simulator for predicting miscible flood performance. *J. Pet. Technol.*, 24(6): 874–882.
Welge, H.J., 1952. Simplified method for computing oil recovery by gas or water drive. *Trans. AIME,* 195: 91–98.
Welge, H.J., Johnson, E.F., Ewing, S.P. and Brinkman, F.H., 1961. The linear displacement of oil from porous media by enriched gas. *J. Pet. Technol.*, 13(8): 707–796.
Willman, G.J., 1966. Vertical miscible flood to hike recovery by 70 million barrels. *World Oil,* January, pp. 75–78.
Wilson, J.F., 1960. Miscible diplacement—flow behavior and phase relationships for a partially depleted reservoir. *Trans. AIME,* 219: 223–228.

Chapter 6

CARBON DIOXIDE FLOODING

C.S. MATTHEWS

INTRODUCTION

Carbon dioxide flooding as a tertiary recovery method entered the commercial era in a major way in 1985 with the completion of three large CO_2 pipelines to the West Texas area. The pipelines connect CO_2 sources at Sheep Mountain (Colorado), Bravo Dome (N.M.) and McElmo Dome (Colorado) to a large market in West Texas. The National Petroleum Council has estimated that use of CO_2 to recover residual oil left after waterflooding can add between 5.5 and 8.5 billion barrels to U.S. oil reserves. Thus, petroleum industry is entering an important new era in oil production.

In this chapter, the writer first discusses how CO_2 flooding arose, and then covers laboratory studies, early field tests, and what was learned from these. Next, the writer discusses later field tests and how these led to decisions to build large-scale CO_2 pipelines.

The CO_2 sources are described, along with some planned applications of CO_2 flooding. Sections on designing a carbon dioxide flood and on operating problems and reuse of produced CO_2 are included. Other problem areas in CO_2 flooding are also discussed.

The emphasis in this chapter is on the use of CO_2 to recover oil left as residual after a waterflood. A few secondary CO_2 floods, however, are also discussed.

HISTORY

Early enhanced oil recovery research

Research which led to currently applied enhanced oil recovery processes began largely in the U.S.A. in the late 1940's and 1950's. During this period, researchers developed an understanding of how waterflooding leaves residual oil dispersed as microscopic, disconnected globules in a porous rock matrix. This understanding led to a number of approaches aimed at eliminating the capillary forces which caused this disconnection.

One of the first techniques to be studied was miscible flooding. Propane and LPG were cheap in the early 1950's. Laboratory research showed that a slug of these followed by methane gave very high recovery of oil left as residual after waterflooding. Field tests were quickly carried out. These field tests in light-oil reservoirs using propane and LPG slugs showed that this miscible displacement process was very unstable. The solvent slug was quick to break through to the producers. In addition, only a small slug size (about 10% hydrocarbon pore volume) could be used economically. Because of the small slug size, the slug would often be injected mainly into the more permeable layers, or would be dispersed in the reservoir before sweepout could be achieved.

Early CO_2 research

These miscible flooding tests led to the realization that large slugs were necessary for success in the field. This eventually led to carbon dioxide flooding.

Laboratory research on use of CO_2 began in the 1950's and continued into the 1960's (Saxon et al., 1951; Beeson and Ortloff, 1959; Holm, 1959). Early research showed that carbon dioxide is not directly miscible with most crude oils at readily attainable reservoir pressures. The CO_2, however, is able to extract hydrocarbon in the gasoline boiling range as it flows past the residual oil trapped in the reservoir (Rathmell et al., 1971; Holm and Josendal, 1974). Under proper conditions of pressure, temperature, and oil composition, the CO_2 becomes enriched to the point where it is miscible with the oil. This has been called "dynamic miscibility". The pressure required for miscibility with CO_2 is normally much lower than that required for miscibility with methane.

PROCESS DESCRIPTION

Miscibility

As discussed above, CO_2 is a very powerful vaporizer of hydrocarbons. Hydrocarbon fractions as heavy as those in the gasoline and gas–oil range are vaporized into the injected CO_2 (Rathmell et al., 1971; Holm and Josendal, 1974; Metcalfe and Yarborough, 1979). This enables CO_2 to develop miscibility even though there may be very little of the ethane through hexane components in the crude oil (Holm and Josendal, 1974). The mechanism by which CO_2 miscibly displaces oil is through a multiple contact extraction of hydrocarbons in the C_5–C_{30} range. Suitable oils are usually in the 25–45° API range and are present in reservoirs deep enough so that displacement can take place above the minimum miscibility pressure. For 40° API oils, the minimum miscibility pressure ranges from 1200 psi at 100°F to 3000 psi at 250°F.

The multiple contact extraction required to achieve miscibility requires the CO_2 to move some distance through the reservoir. As the bank forms and moves, it tends to be dispersed both transversely and longitudinally. Additional extraction by CO_2 re-establishes miscibility. Two wells drilled in the Mead-Strawn

Field give some idea of the distance required to achieve miscibility in a reservoir. About four years after the CO_2 slug was injected, two wells were drilled and cored. In the well drilled 60 ft from the injector, the average oil saturation was 10% PV. In the second well drilled 400 ft from another injector, the average oil saturation was less than 5% PV. This was interpreted to mean that at a distance of 60 ft from the injector the slug was partially effective, but that it was still being formed and became more effective at greater distances. The oil saturation prior to CO_2 flooding was estimated to be about 25% PV.

At temperatures lower than about 120°F, the phase behavior is more complicated than that just discussed. Two liquid phases can coexist. These two liquid phases can then be in equilibrium with a vapor phase (Shelton and Yarborough, 1977; Gardner et al., 1981). Dynamic miscibility, however, can still be achieved under these conditions, as shown by slim-tube laboratory displacement experiments.

Measurement of miscibility

Flow experiments usually constitute the best technique for determining miscibility. This is especially true with CO_2, because miscibility is only obtained dynamically by mass transfer between the CO_2 and the oil.

Because of the adverse mobility ratio between CO_2 and oil, the injected CO_2 will tend to "finger" through the oil and give an unstable displacement. This fingering can be minimized in laboratory experiments by designing the flow experiment so that mixing by transverse dispersion minimizes fingering before injected CO_2 has traveled a significant distance. A very small-diameter tube and a low flow rate are required to achieve this.

A typical apparatus for conducting miscibility experiments is shown in Fig. 6-1 (Yellig and Metcalfe, 1980). The slim-tube consists of 40 ft of $\frac{1}{4}$-inch-diameter coiled stainless steel tubing packed with 160–200 mesh Ottawa Sand. For each test, the sand pack is saturated with oil and brought to the desired temperature and pressure. The CO_2 is injected at a velocity no greater than 40 ft/day until 70% of the pore volume has been displaced. After this the velocity is doubled.

The effluent from the slim-tube flows through a high-pressure sight glass for observation of whether one phase or two phases are exiting from the tube. Yellig and Metcalfe (1980) found that dynamic miscibility is established within a sufficiently small fraction of the total tube length so that this does not confuse interpretation of the data.

To determine whether miscibility has been achieved, a curve such as shown in Fig. 6-2 is plotted. The oil recovery plotted is that obtained after injection of 1.2 pore volumes of CO_2 into the slim-tube. As shown in Fig. 6-2, a sharp break in the recovery curve shows that immiscible displacement has changed to miscible displacement as the pressure is increased. Recoveries are usually 90–95% in the miscible region. Orr et al. (1981) have concluded that in some cases the breakover point may not indicate dynamic miscibility. At this point, however, the phase

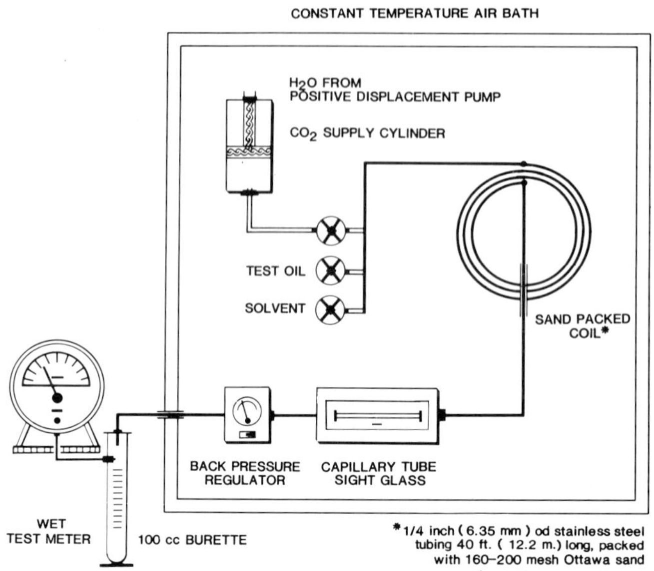

Fig. 6-1. Apparatus for miscibility determination. (After Yellig and Metcalfe, 1980; courtesy of the Society of Petroleum Engineers.)

behavior is favorable for an efficient displacement even if miscibility is strictly not achieved.

Observations in the visual cell are useful in judging whether or not miscibility has been achieved. Immiscible CO_2 displacements show sharp interfaces in the visual cell after CO_2 breakthrough. For CO_2 displacements that reach dynamic miscibility, the color of the oil changes gradually from dark to red, and then

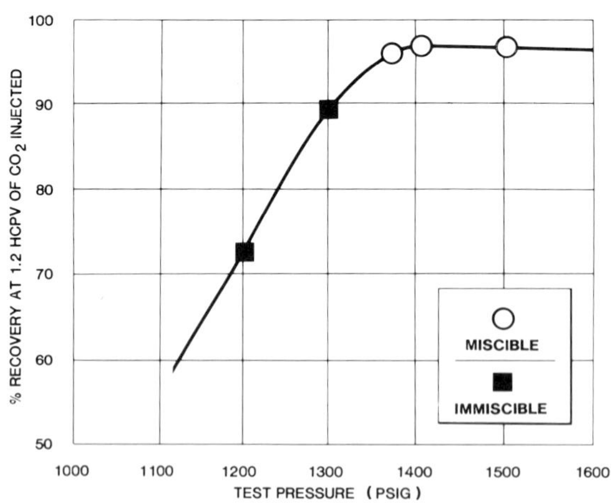

Fig. 6-2. Miscibility test results. (After Yellig and Metcalfe, 1980; courtesy of the Society of Petroleum Engineers.)

orange to yellow as the miscibility zone is approached. Two liquid phases may form and be visible along with a vapor phase. Finally, a single phase is reached as the CO_2 content increases.

In summary, the slim-tube method gives reproducible results (± 100 psi) for minimum miscibility pressure. It is convenient as well as accurate for this purpose.

Miscibility correlations

The simplest correlation is that of Yellig and Metcalfe (1980) as shown in Fig. 6-3. Oil composition is considered to have only a minor effect in this correlation. When there is little data on oil properties, this correlation can be used as a rough screening guide.

Holm and Josendal (1982) presented a correlation where oil composition is a parameter. Their chart as extended by Mungan is shown in Fig. 6-4. Their correlation indicates a small effect of oil composition below 100°F but an increasing one above that temperature.

As a CO_2 flood progresses, the CO_2-produced stream will contain a wide variety of components, including methane, nitrogen, ethane, propane, butane, and H_2S. Sebastian et al. (1985) have presented a useful correlation of minimum miscibility pressure for impure CO_2 streams with oils. This should be useful for modeling purification and reinjection alternatives. Their correlation is shown in Fig. 6-5. They showed that the amount of the gasoline fractions (C_5–C_{12}) has a much larger effect on the minimum miscibility pressure than the amount of the C_{30+} hydrocarbons or the type of hydrocarbons (paraffins, aromatics or naphthenics).

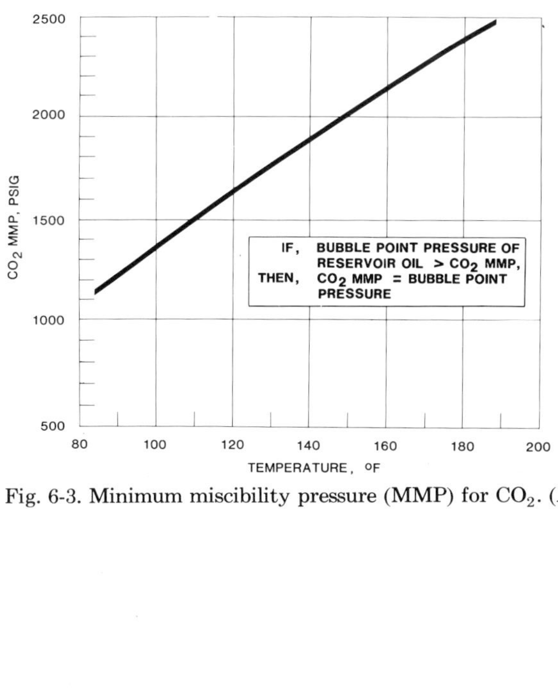

Fig. 6-3. Minimum miscibility pressure (MMP) for CO_2. (After Yellig and Metcalfe, 1980.)

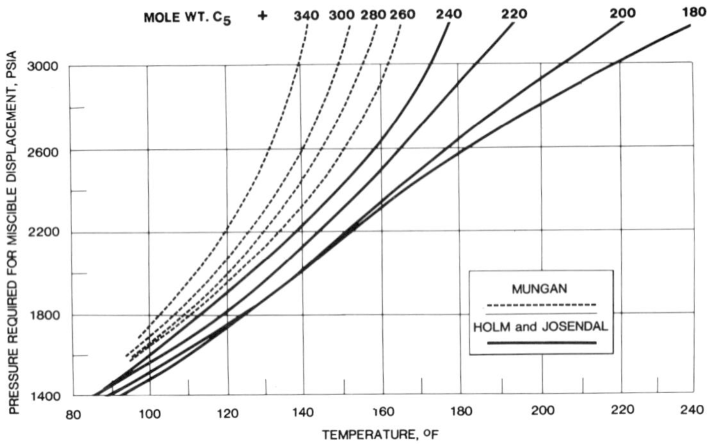

Fig. 6-4. Minimum miscibility pressure. (After Holm and Josendal, 1982.)

The pseudo-critical temperature for the impure CO_2 is calculated from the following equation:

$$\overline{T}_{Cm} = \sum_i x_i T_{Ci} \tag{6-1}$$

where x_i is the mole fraction of component i and T_{Ci} is the critical temperature of component i as given in Table 6-I. The critical temperature of H_2S has been adjusted in Table 6-I from 373 to 325 K to give a better fit to the data.

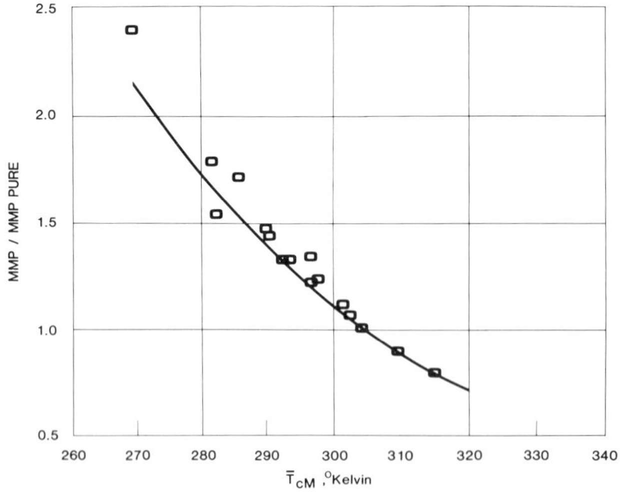

Fig. 6-5. Minimum miscibility pressure (MMP) for impure CO_2 streams. (After Sebastian et al., 1985.)

TABLE 6-I

Critical temperature of gases

Gas	Critical temperature (K)
Hydrogen	33.2
Nitrogen	126.2
Oxygen	154.6
Carbon monoxide	132.9
Methane	190.6
Carbon dioxide	304.2
Ethane	305.2
Hydrogen sulfide	325.0 [a]
Propane	369.8
Butane	425.2
i-Butane	408.1
n-Pentane	469.6
i-Pentane	433.8
n-Heptane	507.4

[a] Reduced from actual.

To use this correlation, it is necessary to first calculate the minimum miscibility pressure (MMP) for pure CO_2 and the oil from either Fig. 6-3 or Fig. 6-4. This pressure is called MMP Pure. Fig. 6-5 can then be used to calculate the minimum miscibility pressure (MMP) for the impure CO_2 stream.

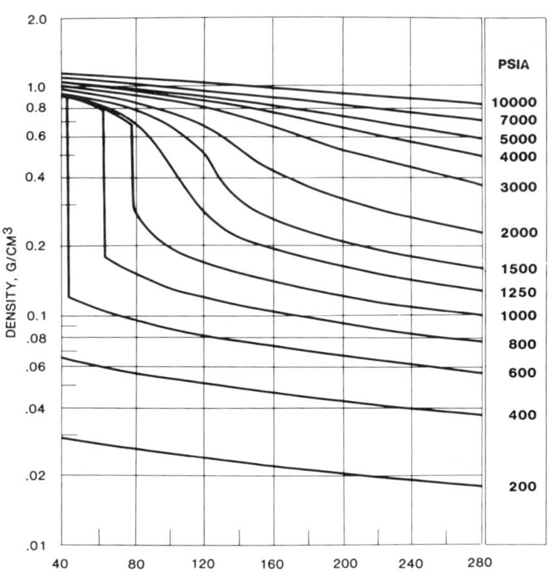

Fig. 6-6. Density of CO_2. (Redrawn from Sage and Lacey, 1955.)

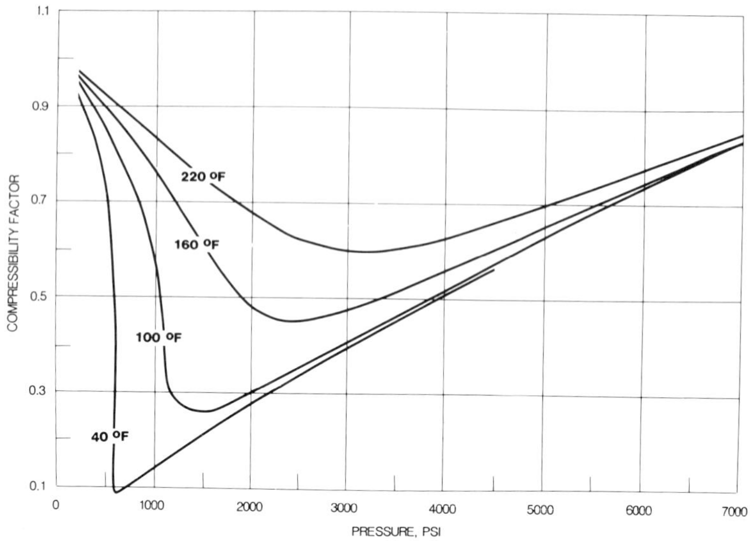

FIg. 6-7. Compressibility factor of CO_2. (Redrawn from Sage and Lacey, 1955.)

All of these correlations should be used only for screening analysis and preliminary design. Slim-tube experiments should be used to provide data for final design.

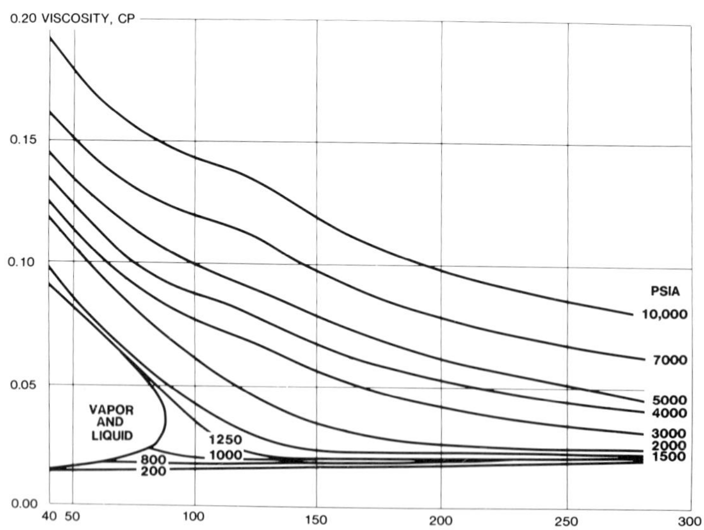

Fig. 6-8. Viscosity of CO_2. (Redrawn from Kennedy and Thodos, 1961.)

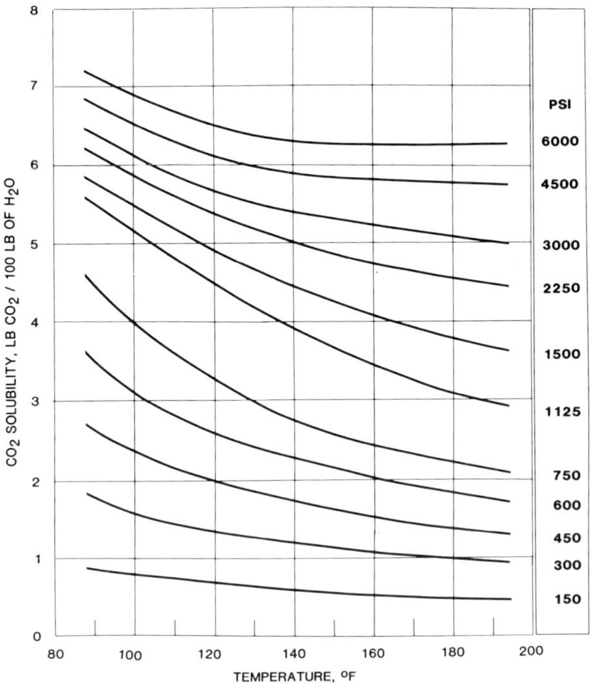

Fig. 6-9. Solubility of CO_2 in water. (Redrawn from Dodds et al., 1956.)

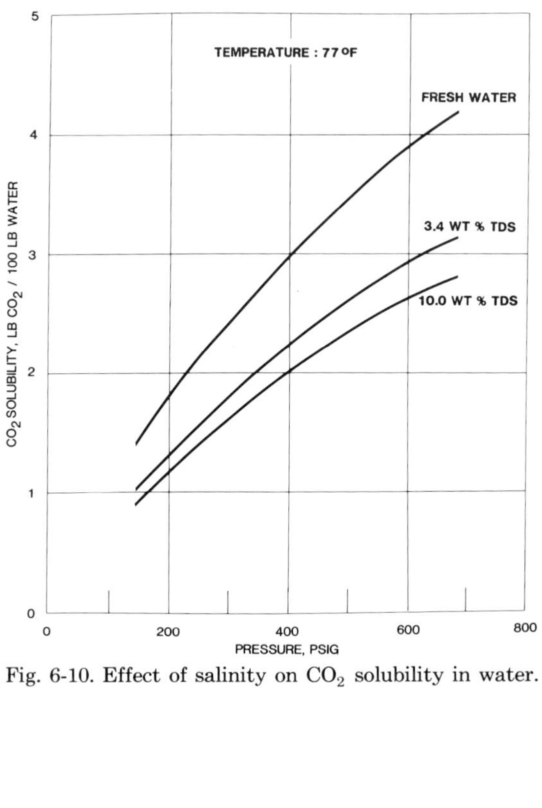

Fig. 6-10. Effect of salinity on CO_2 solubility in water. (Redrawn from Stewart and Munjal, 1970.)

Properties of CO_2

In most miscible floods, the reservoir temperature is above the critical temperature of CO_2 (88°F). Thus, CO_2 will be a single-phase fluid. The density will often be similar to that of oil, as shown in Fig. 6-6. For example, at 120°F and 3000 psi, the density is about 0.8 g/cm^3. The compressibility factor is shown in Fig. 6-7. At temperatures and pressures of interest in miscible flooding, the compressibility factor is in the neighborhood of 0.5.

The viscosity of CO_2 is shown in Fig. 6-8. At temperatures and pressures where miscible flooding is applicable, the viscosity will usually be in the range of 0.05–0.1 cP. This is unfavorable enough to cause viscous fingering through most oils.

Some of the injected CO_2 will dissolve in water which is already in the reservoir, or in water which may be injected alternately with the CO_2. The amount may be estimated from Figs. 6-9 and 6-10.

FIELD PROJECTS

Number and type

About 80 field trials of CO_2 flooding have been undertaken (Leonard, 1986). About 30 of these were of the non-miscible type. In these, use of CO_2 may improve recovery, though usually not to the extent of the miscible type.

The two earliest projects were not pilots but were commercial projects. These projects, started in 1972 at SACROC and Crossett, were secondary recovery projects. At Crossett, waterflooding had not been undertaken because of low formation permeability and, consequent, low injectivity to water.

In 1974, a large project began at Twofreds in West Texas in the previously waterflooded Delaware sands. That same year, a CO_2 pilot was undertaken in the waterflooded portion of the Tuscaloosa Sand in the Little Creek Field in Mississippi.

One of the first carbonate pilots was started in the Willard Wasson Unit in 1972. This test was aborted prematurely in 1975 because of an accident involving escape of the injected CO_2–H_2S mixture. Another test in the San Andres carbonates began in 1974 in the Wasson Denver Unit. This was a small (1-acre) mini-pilot test. The success of this pilot was a key factor in advancing commercial application of CO_2 flooding in the San Andres Formation in West Texas. Later tests in the San Andres Formation were undertaken at Slaughter (in 1976), Levelland (in 1979), and Maljamar (in 1981).

Larger projects were begun in 1981 at Ford Geraldine (Texas) in the Ramsay zone, in 1982 at Purdy (Oklahoma) in the Springer Sand, and in 1983 at Means (Texas) in the San Andres zone. Many more commercial San Andres projects are now underway, including those at North Cowden, Seminole, GMK South,

Wasson Denver Unit, Wasson ODC, Wasson Willard Unit, Wasson Cornell Unit, and Wasson Bennett Ranch.

A number of commercial CO_2 projects have also been undertaken recently in sandstone reservoirs. One of the largest is the Rangely (Colorado) flood in the Weber Sand. The Little Creek Field (Mississippi) is being flooded in the Tuscaloosa Sand.

Several gravity-stable CO_2 injection projects have been undertaken. The test at Weeks Island in the "S" Sand began in 1979. Another test at Bay St. Elaine in the 8000-ft-thick sand began in 1981. Both methane and butane were added to CO_2 in this latter flood to produce a gravity stable system that was still miscible with the oil.

The Lick Creek (Arkansas) CO_2 flood was an interesting non-miscible application in a reservoir containing a 17° API, 160-cP oil. The operator reported the project to be successful and profitable.

A summary of a few of these field applications are given below.

Slaughter Estate Unit, West Texas

A CO_2 miscible test was conducted at Slaughter Estate Unit in the Slaughter Field in West Texas (Adams and Rowe, 1981; Rowe et al., 1981). The test was in the San Andres carbonate. Drilling of the 12-acre double five-spot pilot began in 1972 in an area which had not been waterflooded. After waterflooding, injection of the miscible solvent (72% CO_2, 28% H_2S) began in 1976.

Production of incremental oil began in 1977 when solvent injection was 10% of the hydrocarbon pore volume. A total of 26% hydrocarbon pore volume solvent was injected. Chase gas * was then injected at a water to gas ratio (water alternated with gas) of 1 reservoir barrel of chase gas per reservoir barrel of water. This ratio was later increased to 1.33 to reduce the quantity of chase gas being recycled. After 1982 only water was injected.

As of July 1981, the enhanced oil recovery attributable to CO_2 flooding was 14.9% of the oil originally in-place. The ultimate incremental recovery was projected to be 20% of the original oil-in-place.

Wasson Denver Unit, Texas

The test was conducted in the Wasson Denver Unit of the San Andres Formation at a depth of about 5100 ft (Bremer, 1982). Residual oil saturation after waterflooding was estimated to be about 40%, based on pressure coring of one of the logging observation wells.

* Residue gas from the Slaughter gasoline plant and nitrogen were used at different times as chase gas.

The CO_2 pilot test consisted of five wells within a 1-acre area. There was one injection well, three logging observation wells, and a fluid sampling well. The logging observation wells were completed with fiberglass casing so that open-hole logging tools could be used.

A preflush of 65,000 ppm chloride brine was injected to establish a baseline salinity for logging. CO_2 was then injected at rates of 1.7–2.8 Mcf/day. This was followed by injection of an additional slug of 65,000-ppm chloride brine.

Pilot evaluation was based on measurements in the logging observers of oil saturation changes with time, and CO_2 saturation changes with time. These "snapshots in time" were confirmed by six pressure coreholes drilled at pilot completion. The minimum oil saturation found in the most thoroughly swept layers was 8% pore volume. This residual was a heavy tar-like fraction of the original 34° API oil.

The pressure core results showed that oil displacement was a function of CO_2 throughput in a particular layer. This result was used in scaling the 1-acre pilot to a 70-acre pattern used in the full-scale project. Recovery was estimated to be about 17% of original oil-in-place for the full-scale project.

Little Creek Field, Mississippi

The Little Creek Field in southern Mississippi had been depleted by waterflooding the 30-ft-thick Tuscaloosa Sand at 10,750-ft depth. CO_2 injection was started in 1974 in a 31-acre pilot near the eastern edge of the field (Hansen, 1977). Reservoir parameters were: porosity, 23.4%; permeability, 33 mD; initial oil saturation, 44%, residual oil saturation, 21%; reservoir temperature, 248°F; oil gravity, 39° API; GOR, 555 scf/bbl; and oil viscosity at bubble point, 0.3 cP. There was one injector and three producers spaced so as to model one-fourth of a nine-spot. Water injection continued in five wells around the pilot to confine it.

Total pilot oil production was projected to be 30–35% of the original oil-in-place. Gross CO_2 utilization including purchased and recycled gas was 27 Mcf/bbl.

Twofreds Field

The CO_2 is being injected continuously into the 4800-ft Delaware sandstone through 14 injectors (Thrash, 1979). The reservoir contains about 4400 acres with 25 ft of pay, 19% porosity, and permeability of 33 mD. Oil gravity is 36° API in this field which originally contained 53 million barrels of oil. The CO_2 source is the CO_2 separation plant in a nearby gas field. As CO_2 was injected, oil production rose steadily in the 800-acre injection pattern rising from 27 to 520 bbl/day. No recent data are available on this project.

Willard Wasson Unit mini-pilot

The San Andres carbonate is highly stratified with continuous dense intervals correlative over several well distances. These zones tend to prevent gravity overlay of CO_2 from occurring. Average porosity and permeability at the reservoir depth of 5200 ft are 8.5% and 1.5 mD, respectively. The 32° API oil has a viscosity of 1.3 cP at reservoir conditions.

A small three-well mini-pilot was run at Willard Wasson Unit (Charlson et al., 1978). Saturation changes were measured by periodically logging an observation well located 100 ft from the injector. Production tests were made in another well located 25 ft from the logging observation well.

Water was injected first to thoroughly waterflood the pilot area. When this was completed, CO_2 was injected alternately with water on a 1:1 reservoir volume ratio. A pressure core was taken at the end of the test at a location between the injector and the observation wells. The oil saturation in the permeable layers in the core varied between 3% and 30% pore volume.

These mini-pilot results were used with a miscible reservoir simulator to project oil recovery in a full-scale flood. Recovery was estimated to be about 12% of the original oil-in-place for a 20% hydrocarbon pore volume slug.

Rock Creek CO_2 pilot

Rock Creek pilot furnishes an example of an unsuccessful CO_2 pilot. Perhaps this should have been anticipated because three separate pilot waterfloods in the 1950's and 1960's were unable to move enough oil to make waterflooding economically feasible. This field in West Virginia produces from a 24.6-ft-thick sandstone at about 2000 ft depth, having 21.7% porosity and 21.5 mD permeability.

In spite of the waterflood failures, CO_2 flooding seemed possible because of high oil gravity (43° API), high oil saturation (34.4%) and low minimum miscibility pressure (1000 psi from slim-tube experiments). The pilot was a large one. It consisted of two adjoining 10-acre five-spots containing six injectors and two producers. This area was backed up by 13 surrounding water-injection wells.

The pilot area was pressured up initially with water and a surface pressure of 1000 psi was maintained. A small oil response of 11,300 bbl occurred during this time. CO_2 was then injected in four different periods over a 3-year interval. Field problems and CO_2 availability prevented continuous injection. Between 30% and 40% hydrocarbon pore volume of CO_2 was injected in each pattern. When CO_2 was not available, water was injected in the six injectors. Oil production from the pilot totaled about 16,000 bbl out of an estimated 380,000 bbl in place in the pilot area or about 4% of the oil-in-place at start of flooding. The CO_2 utilization was 16,000 cf/bbl oil.

During the latter part of the pilot, two additional injection wells were drilled in one of the 10-acre patterns and an additional producer was brought into production between these two wells and a corner injector. This producer recovered a substantial amount of oil, demonstrating that areal sweep in the original 10-acre five-spot waterflood was very poor.

A well was drilled and pressure cored about 75 ft from an injector in one of the two five-spots. Measured oil saturations in the upper 10 ft showed that CO_2 had reduced the oil saturation to about 4%. In the lower 20 ft, the oil saturation had been reduced from an estimated 33% after waterflood to about 19%.

These results indicate that CO_2 flooding can attain high recovery in the region that it sweeps. Poor sandstone continuity may be responsible for the failure of both waterflooding and CO_2 flooding.

West Sussex Shannon pilot

West Sussex Shannon pilot furnishes an example of a successful CO_2 pilot in a watered-out sandstone reservoir in the Rocky Mountain region. This reservoir occurs at a depth of about 3000 ft. Net pay averages 22 ft, porosity is 19.5%, permeability is 28.5 mD, and connate water saturation is 27%. Oil gravity was 39° API and reservoir temperature was 104°F. Residual oil saturation at the end of water injection was estimated to be 28%.

The reservoir showed favorable response to waterflooding in spite of its highly faulted nature. The pattern area selected had three producers and one injector all situated between two parallel faults. Pressure transient testing was used to verify continuity.

Slim-tube tests using 95% purity CO_2 showed the minimum miscibility pressure to be 1500–1600 psi, well below the reservoir pressure of 2000 psi at the time of the pilot. The CO_2 source was the McCallum Field in Jackson County, Colorado. Liquified CO_2 was transported 175 miles by tank truck over a 10-month period. The CO_2 injection was followed by water injection.

Increased oil cuts were noticed in one producer after one month and in another, in six months. The third producer did not respond. This producer had been used as a water injector at one time. This may have contributed to the lack of response.

A tritiated water slug was injected prior to CO_2 injection. This provided data to calculate the swept volume, which was consistent with the recovery of 16,000 bbl of oil, an oil saturation of 28% at the start of flood, a residual oil saturation for CO_2 of 10%, and a sweep efficiency of 52%. This sweep efficiency is very good for CO_2 injection alone and compares very well with the waterflood sweep. The CO_2 utilization was 12.9 Mcf/bbl of produced oil.

CARBON DIOXIDE SOURCES

Natural and industrial sources

In early applications, gas processing plants were used as a major source of CO_2 (see National Petroleum Council, 1984). Many West Texas natural gas deposits contain significant quantities of CO_2, which must be removed before this gas can

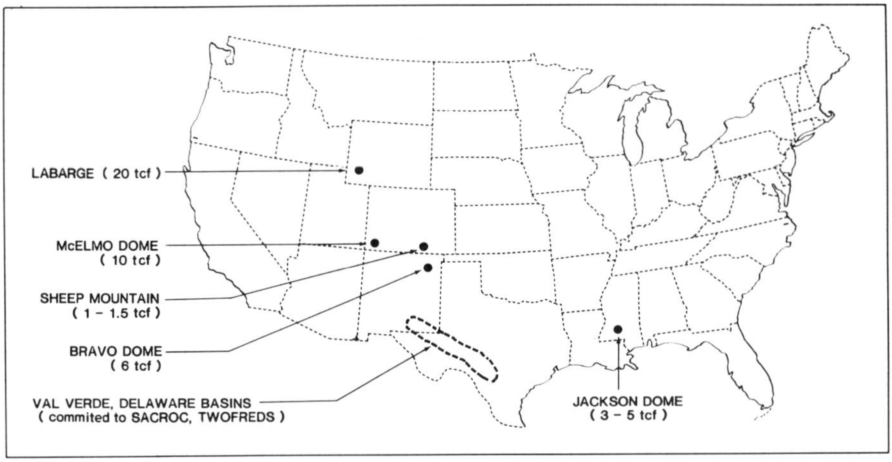

Fig. 6-11. Locations of several natural CO_2 deposits.

be sold. The SACROC project and the North Cross project were supplied in this manner from a large CO_2 pipeline beginning at the Pikes Peak processing plant in West Texas. The Twofreds field was supplied in a similar manner using CO_2 from the Mi Vida plant.

In Oklahoma, a fertilizer (ammonia) plant near Enid supplied 33 MMcf/day to two fields (Golden Trend and East Velma) through a 140-mile pipeline. Fertilizer plants were also used as the source of CO_2 for several pilot projects in southern Louisiana.

Most of the CO_2 sold today comes from natural sources as shown in Fig. 6-11. Sheep Mountain reserves are estimated to be about 1 Tcf of 97% pure CO_2. The pipeline from Sheep Mountain to the Seminole Field in West Texas was completed and delivery started in 1983. Capacity is 330 MMcf/day at Sheep Mountain and 500 MMcf/day past Bravo Dome.

The Bravo Dome Unit in New Mexico is estimated to have reserves of about 6 Tcf. A portion of this is allocated to the Sheep Mountain pipeline. An additional pipeline has now been built from Bravo Dome to Levelland, Texas, with a capacity of 400 MMcf/day.

In southwestern Colorado, the McElmo Dome and Doe Canyon Fields have reserves greater than 10 Tcf of 97% CO_2. A 500-mile pipeline has now been built to carry this CO_2 to West Texas. Initial design capacity is 650 MMcf/day.

In Mississippi, the Jackson Dome complex has 3–5 Tcf of CO_2 reserves. A 90-mile pipeline to southern Mississippi has been completed and an extension to southern Louisiana is under construction.

The largest CO_2 reserve in the U.S. is in the LaBarge–Big Piney area in southwestern Wyoming. The CO_2 reserves are in excess of 20 Tcf. Gas from the Madison consists of 70% CO_2 and that from the Big Horn, of 90% CO_2. Much of the remainder in both cases is methane, which can also be sold.

Carbon dioxide recycling

Nearly all projects will require that produced CO_2 be separated and reinjected. In most projects this will account for about half the required CO_2. A number of proven separation processes are available. Typically a processing scheme will involve first an acid gas removal process, which will remove both CO_2 and H_2S from the produced gas. The conventional removal method consists of using an amine to remove the acid gases by chemical reaction. An improved amine process using methyl diethanol amine has recently been introduced. Another class of processes developed to handle gas streams having high H_2S content involves use of solvents such as Solexol and propylene carbonate. Next, a selective H_2S removal process will be used. In many cases, a unit to convert the H_2S to solid sulfur will be added. Most sulfur conversion plants use the "Claus" process. This process converts the H_2S to sulfur by "burning" the acid gas stream with air in several catalytic beds. The purified carbon dioxide must then be recompressed for reinjection.

In addition to the conventional proven processes, several new approaches have been developed. One of these is a cryogenic extractive distillation process. The Ryan-Holmes technology is of this type. A hydrocarbon cut is added to alter the vapor–liquid equilibrium such that CO_2 emerges from the top of the column and ethane from the bottom. Semi-permeable membranes are also being developed to separate CO_2 from hydrocarbon gas streams. Spiral wound and hollow fiber type membranes have been developed successfully. At SACROC, 95% CO_2 comes through the membrane. The remaining gas containing about 30% CO_2 goes to a traditional sweetening plant using hot carbonate and MEA (monoethanol amine). Both extractive distillation and membrane separation have been installed in commercial facilities at this time.

It is not always necessary to produce pure CO_2 for reinjection. The correlation of Sebastian et al. (1985) may be used to estimate the minimum miscibility pressure for impure CO_2 streams. This can aid in modeling purification and reinjection alternatives. Direct determination of miscibility using a slim-tube apparatus, however, is mandatory before final design of a partial purification scheme.

PROBLEM AREAS

Operating problems

Among the operating problems which have been found in CO_2 flooding are corrosion, scale deposition, and precipitation of heavy ends from the crude. To prevent corrosion, dehydration of the CO_2 is necessary. This will usually be sufficient for injection systems where CO_2 is to be injected continuously. Injection wells at Crossett, Little Creek (Hansen, 1977), and Bay St. Elaine (Palmer et al., 1981) had conventional wellheads with no special provisions for corrosion protection. No water was injected with CO_2 in these projects. Stainless steel was used for a few items such as tubing hangers, master valves, and gravel packing assemblies.

When both CO_2 and water are injected alternately, a standard procedure in injection wells is to use separate CO_2 and water distribution lines. Some operators use a combination of stainless steel parts and plastic or metallic coatings to protect meter runs, wellheads, tubing, and packers. Plastic and metallic coatings are susceptible to mechanical damage. When this occurs, corrosion can be severe at the point where the coating is broken. Aluminum bronze and 316 stainless steel has been used in valves. Gates have been constructed of ceramic materials in some cases.

Plastic-coated tubing has been used in injection wells where both CO_2 and water are injected. The tubing must be handled carefully to maintain coating integrity. Tubing coupling leaks have been a problem in some cases. Special Teflon® seals were used at Crossett and Little Creek.

In production wells, the production of CO_2 along with brine creates a corrosive situation. Coatings and stainless steel have been used as in injection wells (Hansen, 1977). Inhibitors have also been used, usually as a monthly batch-type treatment. Epoxy coatings on tubing and on pumps have given satisfactory protection. Continuous inhibition down the tubing–casing annulus is thought to give more complete control of corrosion in some cases (Hansen, 1977).

As carbon dioxide is released from solution, carbonate scale tends to precipitate. Scale inhibitor treatments both downhole and in surface facilities are necessary to prevent this (Newton and McClay, 1977).

Reservoir problems

CO_2 breakthrough

Early CO_2 breakthrough has occurred in many CO_2 projects, usually after injection of 0.05–0.2 hydrocarbon pore volume of injectants. To correct this problem, alternate water injection is usually resorted to (Rowe et al., 1981). In some cases it has been found possible to isolate the thief zone, which is responsible for the early breakthrough (Kane, 1979). Reducing the injection pressure below the fracturing level has also been successful in some cases (Johnston, 1977).

Gravity overlay of CO_2 is a problem in reservoirs with good vertical permeability. Use of CO_2 foam has been proposed to solve this problem (Bernard et al., 1980). Gravity overlay is particularly important in relatively flat reservoirs with high permeability and good vertical continuity.

Unstable fronts

The viscosity ratio at the CO_2–oil bank front is unfavorable. This causes viscous fingering and an unstable front (see Fig. 6-12). Such viscous fingering,

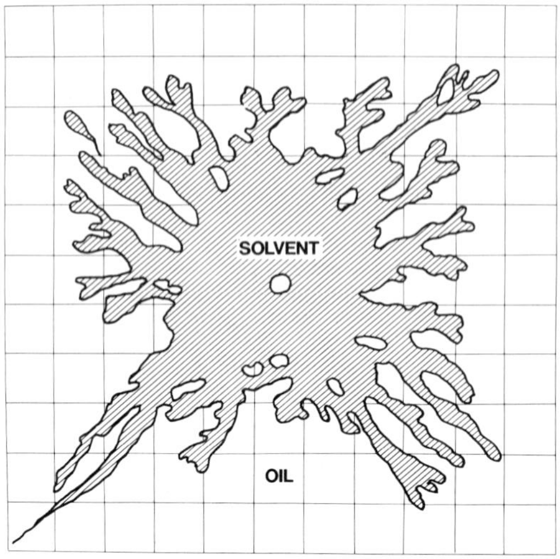

Fig. 6-12. Swept zone at solvent breakthrough for a five-spot. (Courtesy of the Society of Petroleum Engineers.)

which occurs during the displacement of a more viscous fluid by a less viscous one, has been studied extensively. Viscous fingering occurs because, at unfavorable viscosity ratios, the growth of the fingers dissipates less energy than does the maintenance and movement of a smooth front. The resultant poor sweep efficiency was one of the principal reasons for the failure of early propane and LPG floods in the 1950's and 1960's.

The early breakthrough of displacing fluid causes a continually increasing fraction of the injected fluid to be circulated through the reservoir without contacting or displacing oil. Viscous fingering also acts to destroy the zone of dynamic miscibility, which was achieved by extraction into the CO_2 of intermediate hydrocarbon components in the oil. This zone tends to be dissipated by transverse mixing of oil with CO_2, which has not been enriched by extraction.

One method of reducing the injectant mobility is to inject alternate slugs of water with the CO_2 (Rowe et al., 1981). The water decreases the relative permeability of the porous medium to CO_2 and, thus, lowers its mobility. Alternate water and CO_2 injection seems to work best when the porous medium is not strongly water-wet. In strongly water-wet formations, CO_2 may have difficulty in contacting the residual oil left after the slug of injected water. Contact may come only by diffusion through the water surrounding the residual oil globules.

Although water alternated with gas (WAG) seems to be successful in some field operations, it is not an ideal method. The fronts formed around the injection well after injected water is followed by CO_2 are unstable. The CO_2 will tend to finger through the high water saturation zones and may bypass regions where oil saturation is still high. In addition, the dynamically developed miscibility zones are destroyed by the water injection.

One method of achieving control of CO_2 viscous fingering is to produce a CO_2-foam. The foam will have a much higher effective viscosity than the CO_2 and thus would tend to inhibit viscous fingering. Bernard et al. (1980) have described laboratory investigations with foam. Heller et al. (1985) have described a recent field test with foam at Rock Creek, West Virginia.

Injectivity

When CO_2 contacts some reservoir oils, a heavy liquid or solid hydrocarbon phase may form (Pontious and Tham, 1978). This may reduce CO_2 injectivity. Alternate injection of CO_2 and water also reduces the injectivity to each injectant, especially at the time of changeover (Rowe et al., 1981). Injectivity to each injectant usually improves with time as the saturation of the injectant increases in the neighborhood of the wellbore. The saturation increase occurs both by displacement of the previous injectant and by solution of the previous injectant into the current injectant. Thus, residual saturations of CO_2 in the formation will dissolve gradually in the injected water slug. This will cause the injectivity to increase.

DESIGNING A CO_2 FLOOD

Geology

Because of the high cost of CO_2 when compared with water, a much higher degree of reservoir description is required for CO_2 flooding than for waterflooding. This makes the preparation of a geologic picture of the greatest importance. Using cores, logs, tracer tests and well tests, the best possible description should be made of the formation, including: (1) cross-sections showing pay zones and tight sections; (2) formation rock descriptions; (3) structure maps, and (4) isopachs.

A knowledge of the depositional environment is helpful in constructing such a model. The use of simulators to match prior performance is also helpful. The latter may point out zones of anomalous performance which can be correlated with the geologic picture.

Petrophysics

One of the most important items to determine is the oil saturation at the start of CO_2 flooding. Estimation from logs is not very accurate. Perhaps the best technique is pressure coring. A careful laboratory analysis is then required to determine how the residual oil saturation varies as a function of rock type and rock properties.

In addition to oil saturation at start of flooding, the petrophysicist must arrive at a description of porosity and permeability throughout the reservoir.

This must be integrated with the geologic picture. Rock wettability is important, as discussed previously. To determine rock wettability it is usually necessary to core with a mud containing no surface-active agents and to test for wettability immediately on site. Wettability can change rapidly as evaporation and oxidation take place.

Wettability

Determination of formation wettability is both important and difficult. Trapped oil is apparently more easily contacted by CO_2 after alternate water injection when the formation is partially oil-wet than when it is strongly water-wet. In strongly water-wet formations, CO_2 can only contact residual oil by diffusion through surrounding water.

The best method for determination of formation wettability is by use of native-state cores. Pressure coring is the most reliable method for retrieval of such cores. It allows their retrieval at near in-situ conditions. A proper mud system must be used, of course, to prevent changes in wettability during coring. The mud should be carefully designed to minimize flushing. An optimal penetration and mud circulation rate should be used.

Recommended coring fluids are: (1) synthetic formation brine, (2) unoxidized lease crude oil, and (3) water-base mud with minimum additives. Additives such as surfactants, caustic, and mud thinners should be avoided. De-aerated brine or formation water with an oxygen scavenger should be used in a water-base mud system. Some guidelines have been given by Goodlett et al. (1986).

The core must now be protected against loss of light components of the crude and against oxidation. It should be wrapped with a polyethylene film, then with aluminum foil, and then sealed in wax or plastic. In pressure coring, the cores are frozen before the pressure is bled off. They are then stored frozen. Samples from these cores may now be used for wettability studies.

Samples from these cores may also be used for flow and displacement studies. In choosing samples for analysis, X-ray tomography is helpful in identifying heterogeneities. In the case of sandstones, only small samples parallel to reservoir bedding planes should be chosen for flow experiments. These may be butted together with a capillary bridge such as facial tissue between them. The capillary discontinuity introduced in this manner should not be a problem in miscible flooding, if CO_2 enriched with enough intermediate hydrocarbons to be miscible with the residual oil is used for the displacement experiments.

Miscibility conditions

The correlations discussed earlier in this chapter (Yellig and Metcalfe, 1980; Holm and Josendal, 1982; Sebastian et al., 1985) are useful for early screening work. Before any application is made, however, the miscibility pressure should be determined by an apparatus such as the slim-tube device discussed previously.

Recovery prediction

With the above basic data at hand, the engineer must now arrive at a process design and recovery prediction. He must choose the following: injection pressure, injection slug size, continuous CO_2 injection or alternate injection of water, well pattern, and zonal isolation (if required or feasible.)

In a gravity-stable flood, the CO_2 is injected above the oil zone. Drive gas is then injected above the slug. In this case some additional design factors are: (1) critical rate for stable displacement, (2) initial placement of the CO_2, and (3) estimation of coning into the producing wells.

The critical rate, u_c, in ft/day may be calculated from the equation:

$$u_c = \frac{0.0439(\rho_o - \rho_c) \sin \alpha}{\mu_o/k_o - \mu_c/k_c} \qquad (6\text{-}2)$$

where ρ_o and ρ_c are oil and CO_2 densities in lb/ft^3, μ_o and μ_c are oil and CO_2 viscosities in cP, k_o and k_c are oil and CO_2 permeabilities in darcies, and α = dip angle, relative to horizontal.

Pressure transient data should be obtained and analyzed to determine directional permeability and barriers to flow. Analysis of a previous waterflood is critical to the success of a CO_2 flood. Production history matching with a reservoir simulator will be of enormous aid in guiding performance prediction using CO_2.

Reservoir simulation of the CO_2 flood is the next step in recovery prediction. Numerical finite difference simulators are the commonly used tools for this purpose.

Although reservoir simulators can treat nearly all aspects of CO_2 flooding, they cannot treat them all simultaneously. It is here that engineering judgment becomes important. Is vertical permeability so limited that the reservoir can be treated as a stack of non-connected parallel layers? * Can several representative areas and patterns be chosen? How can the complete reservoir behavior be modeled from those areas and patterns? Has a pilot been run? Can this be simulated and the results then scaled up to field scale?

Simulators for CO_2 flooding are basically of two types, black oil simulators and compositional simulators. In both, the difference equations which model the flow and other mechanisms are approximated by finite differences. The resulting finite difference equations are then solved on a high-speed computer. The reservoir is represented by a grid such as shown in Fig. 6-13. The pressure and saturation changes are then calculated for a series of small time increments.

---

* In many carbonate reservoirs, vertical permeability is equal to horizontal permeability and, sometimes, exceeds it. (Editorial comment.)

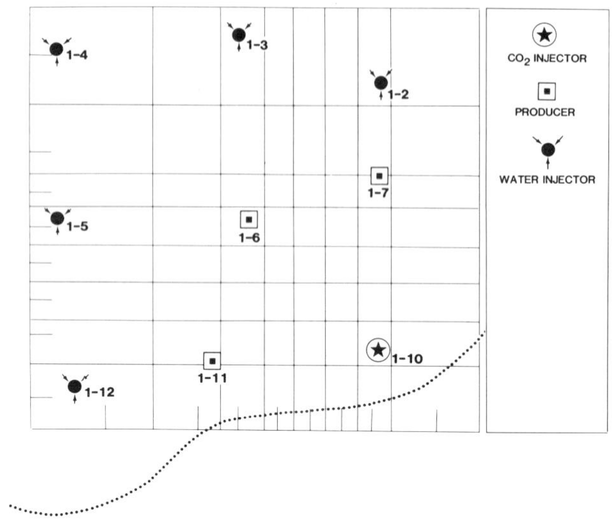

Fig. 6-13. Areal grid for a simulator. (From Youngren and Charlson, 1980; courtesy of the Society of Petroleum Engineers.)

In black oil simulators, reservoir oil is represented by two components, stocktank (black) oil and gas. It was noted by Lantz (1970) that black oil simulators could also be used for miscible flooding by modifying the expressions for relative permeability and capillary pressure.

In compositional simulators, multicomponent mixtures are used to represent the oil and gas phases (Nolen, 1973). Phase compositions are usually calculated using K-values. Phase properties are calculated either from correlations or from equations of state (Rowe, 1978). Calibration of the K-values and of the equations for phase behavior with experimental P-V-T data is recommended. Slim-tube experimental results are another calibration point. Viscous fingering is often represented in compositional simulators by a mixing parameter.

Todd (1979) has described the formulation of a simulator which takes advantage of the best features of both black oil and compositional simulators. Considerable experimental data is required to calibrate this system.

It is common practice with simulators to choose representative field patterns and to make the calculations for those patterns. This can be done for the different areas of the field, which have different physical properties and/or different patterns. Engineering judgment is required in choosing representative areas and patterns.

Example simulation

A miscible flood performance prediction was presented for CO_2 flooding in the San Andres Formation at Wasson by Bilhartz et al. (1978). In this reservoir, there are a number of well-defined non-permeable anhydrite units which effectively limit vertical permeability. Thus, the reservoir simulation was limited to the flow in non-connected vertical layers.

The authors used a black oil simulator with a mixing parameter. They used 16 vertical cells and 20 horizontal cells. The model was calibrated by first matching previous waterflood performance for selected areas of the field. Permeability of the various layers was adjusted during this match.

The CO_2 flood predictions were then made for a 20% hydrocarbon pore volume slug injected alternately with water. The mixing parameter was determined by matching the saturation changes observed in the logging observation well in a small pilot test. The pilot test indicated that a mixing parameter of about 2/3 best represented CO_2 flooding in this reservoir.

Oil recovery

Field tests conducted thus far showed that CO_2 can recover a significant fraction of the oil left as residual after a waterflood. Some of the highest recoveries have been reported for the San Andres carbonate reservoir in West Texas. The Slaughter Estate tertiary recovery pilot test (Rowe et al., 1981; Ader and Stein, 1982) in that formation was projected to recover about 20% of original oil-in-place for a 26% hydrocarbon pore volume slug. The mini-pilot in the Denver (San Andres) unit led to a projected recovery of about 17% of original oil-in-place for a 40% hydrocarbon pore volume slug.

The Little Creek pilot in the Tuscaloosa Sand in Mississippi reported a very high recovery of 30–35% of initial oil-in-place (Hansen, 1977). The slug size in this case was very large (160% hydrocarbon pore volume). The Tuscaloosa Sand has a very high initial water saturation (56%) due to a network of very fine pores. The larger pores which contain oil are apparently very accessible to injected CO_2.

Lower oil recoveries than these have been reported at SACROC (3.5% of original oil-in-place) and Levelland (6% of original oil-in-place). This makes it imperative to conduct both laboratory flooding tests and pilot tests in a formation different than those already tested.

Slug size

A 40% hydrocarbon pore volume CO_2 slug was chosen by the National Petroleum Council (1984) in their estimation of U.S. oil recovery by CO_2 injection. They assumed alternate injection of CO_2 and water in a ratio of 1.5 reservoir pore volumes of water per pore volume of CO_2. This was followed by water injection to the economic limit. Half the injected CO_2 was assumed to be purchased and half recycled. This reflects current practice in relatively homogeneous reservoirs such as the San Andres in West Texas.

Although the 40% slug size may be used for screening purposes, final selection should await a geological, petrophysical, and reservoir simulation study. More heterogeneous formations will require more CO_2 (possibly through recycling) and more water. This again points up the necessity of a thorough reservoir study prior to undertaking a field-scale flood.

Production of a significant fraction of the CO_2 slug should be anticipated. Again, reservoir simulation should give a good idea of when this should occur and how it will increase with time. In typical cases, 20–60% of the slug is expected to be produced.

Summary of design steps

The following list of design steps is from the recent paper by Holm and O'Brien (1986):

(1) Measure gravity, molecular weight, C_5–C_{30} content and type, and asphaltene content using oil sample from candidate reservoir.

(2) If the reservoir flow paths are not primarily fractures and if the gravity of the oil is lighter than 12° API, it is likely to be a candidate for CO_2 flooding, either miscible or immiscible.

(3) Estimate minimum miscibility pressure (MMP) using correlation. Conduct slim-tube experiments to establish MMP more accurately and observe phase behavior of oil sample in visual cell as CO_2 is added. Measure volume and viscosity of oil as a function of the amount of CO_2 dissolved in the oil.

(4) Based upon the above data and taking into account the reservoir pressure and depth, decide whether oil displacement will be miscible or immiscible.

(5) Process core and well log data to establish a geological model of the reservoir. Conduct pressure transient and oil saturation measurements in the field. Determine rock wettability.

(6) Construct a computer model based upon the reservoir data and determine the well pattern, injection and production rates, and slug size that will maximize sweep efficiency. The model will also provide oil and CO_2 production predictions.

(7) For thick reservoirs, consider zonal injection techniques.

(8) Plan to inject 20–40% PV of CO_2, and be prepared to inject surfactant solution or water in alternate slugs in the event CO_2 production becomes excessive.

(9) Consider installation of a CO_2 recovery plant, because it could be a sound economic investment, particularly if hydrocarbon gas sales are a prominent factor in the operation of the field.

GUIDELINES FOR SELECTION OF MISCIBLE CO_2 PROJECTS

Miscibility with CO_2 generally cannot usually be attained in reservoirs shallower than 2500 ft without fracturing the reservoir. Thus, this depth may be used as a guide for miscible projects. Oils with gravities lower than 27° API are generally not suitable because of high miscibility pressure requirements. As a check on this, the minimum miscibility may be estimated from Fig. 6-4.

Oils of high viscosity are poor candidates for CO_2 miscible flooding *. There is no single viscosity cutoff above which performance suddenly becomes unacceptable. Rather, performance continually worsens, with CO_2 requirements steadily increasing. This worsening is due to the unfavorable mobility ratio which lowers sweep, enhances viscous fingering, and hastens slug destruction.

Because of the unfavorable mobility ratio in miscible flooding, severe reservoir heterogeneity can cause excessive channeling and bypassing of injected CO_2. In this regard, experience with the previous waterflood is of very great value. It is advisable to match the waterflood performance with a reservoir simulator to obtain a composite picture of how reservoir heterogeneity affects performance. The matched model of waterflood performance provides a good starting point for CO_2 flood prediction.

How high must the oil saturation be for a successful miscible flood? In most reservoirs, about 20% pore volume is a lower limit. However, for reservoirs having a high proportion of very fine pores which contain only water, such as the Tuscaloosa Sand in Mississippi, the oil saturation may be lower for a successful CO_2 flood.

IMMISCIBLE CO_2 FLOODING

In a miscible application, over 95% of the oil contacted can be displaced. In broad terms, the minimum miscibility pressure is proportional to the specific gravity of the oil. It is usually not possible to achieve miscibility with heavy oils containing large amounts of C_{30+} hydrocarbons. The CO_2, however, is highly soluble in these heavy oils. As CO_2 dissolves, it swells the oil and reduces its viscosity. The oil can then be more readily displaced.

Several field projects have been conducted where CO_2 was not miscible with the oil, but where recovery was good. Projects at Lick Creek and Wilmington are two of the more notable ones. At Lick Creek in Arkansas, the 160-cP oil was successfully flooded with CO_2. The 55% initial oil saturation was estimated to be lowered to 46% at the end of the 1640-acre, 63-well project. The zone flooded was a 2500-ft-thick shallow sandstone of high permeability and porosity.

At Wilmington in California (Saner and Patton, 1983), the initial Champlin pilot was later expanded to the complete fault block. The Tar zone flooded at a depth of 2500 ft contained a 283-cP oil. Recovery is estimated to be 10–15% of original oil-in-place at a CO_2 requirement of only 6 Mcf/bbl oil.

---

* Two examples of CO_2 immiscible flooding discussed in this chapter indicate that *immiscible* CO_2 flooding of oils of 100–300 cP may be successful if the initial oil saturation is high.

Both these projects indicate that immiscible CO_2 flooding may have a much higher potential than has been thought in the past.

CONCLUSIONS

Recovery of residual oil left after waterflooding by CO_2 injection is now a proven commercial process. Major pipelines have been built to serve areas in West Texas, Colorado, Oklahoma and Mississippi. These serve to bring low-cost CO_2 from natural and industrial sources to the area of application.

A much higher degree of reservoir description is needed for a CO_2 flood than for a waterflood. A thief zone that would only be troublesome for a waterflood may be disastrous for injection of more expensive CO_2.

CO_2 flooding is applicable in both carbonate and sandstone reservoirs. At this stage of development, application of CO_2 flooding to new formations will require both laboratory and field testing to develop sufficient information for planning a full-scale project.

REFERENCES

Adams, G.H. and Rowe, H.G., 1981. Slaughter Estate Unit CO_2 pilot—surface and downhole equipment construction and operation in the presence of hydrogen sulfide gas. *J. Pet. Technol.*, 33(6): 1065–1074.

Ader, J.C. and Stein, M.H., 1982. Slaughter Estate Unit CO_2 pilot reservoir description via a black oil model waterflood history match. *SPE/DOE Symp. on EOR, Tulsa, Okla., April 4–7*, SPE 10727.

Beeson, D.M. and Ortloff, G.D., 1959. Laboratory investigation of the water-driven carbon dioxide process for oil recovery. *Trans. AIME*, 216: 388–391.

Bernard, G.G., Holm, L.W. and Harvey, C.P., 1980. Use of surfactant to reduce CO_2 mobility in oil displacement. *Soc. Pet. Eng. J.*, 20(8): 281–292.

Bilhartz, H.L., Charlson, G.S., Stalkup, F.I. and Miller, C.C., 1978. A method for projecting full-scale performance of CO_2 flooding in the Willard Unit. *Soc. Pet. Eng. Symp. on Improved Methods for Oil Recovery, Tulsa, Okla., April 16–19*, SPE 7051.

Bremer, C., 1982. *Report on the Denver Unit CO_2 Flood, Wasson Field, Texas*. Texas Petroleum Research Council, Austin, Texas.

Charlson, G.S., Bilhartz, H.L. and Stalkup, F.L., 1978. Use of time-lapse logging techniques in evaluating the Willard Unit CO_2 flood mini-test. *Soc. Pet. Eng. Symp. on Improved Methods for Oil Recovery, Tulsa, Okla., April 16–18*, SPE 7049.

Dodds, W.S., Stutzman, L.F. and Sollami, B.J., 1956. Carbon dioxide solubility in water. *I&EC Chem. Eng. Data Ser.*, 1.

Gardner, J.W., Orr, F.M., Jr. and Patel, P.D., 1981. The effect of phase behavior on CO_2 flood displacement efficiency. *J. Pet. Technol.*, 33(11): 2067–2081.

Goodlett, G.O., Honarpour, M.M., Chung, F.T. and Sarathi, P.S., 1986. The role of screening and laboratory flow studies in EOR process evaluation. *Soc. Pet. Eng. Rocky Mountain Reg. Meet., Billings, Mont., May 19–21*, SPE 15172.

Hansen, P.W., 1977. A CO_2 tertiary recovery pilot, Little Creek Field, Mississippi. *Soc. Pet. Eng. Annu. Meet., Denver, Colo., October 9–12*, SPE 6747.

Heller, J.P., Boone, D.A. and Watts, R.J., 1985. Testing CO_2 foam for mobility control at Rock Creek. *Soc. Pet. Eng. Eastern Rec. Meet., Morgantown, W. Va., November 6–8*, SPE 14519.

Holm, L.W., 1959. Carbon dioxide solvent flooding for increased oil recovery. *Trans. AIME*, 216: 225–231.

Holm, L.W. and Josendal, V.A., 1974. Mechanisms of oil displacement by carbon dioxide. *Trans. AIME*, 257: 1427.

Holm, L.W. and Josendal, V.A., 1982. Effect of oil composition on miscible-type displacement by carbon dioxide. *Soc. Pet. Eng. J.*, 22(2): 87–98.

Holm, L.W. and O'Brien, L.J., 1986. Factors to consider when designing a CO_2 flood. *Soc. Pet. Eng. Meet., Beijing, March 17–20*, SPE 14105.

Johnston, J.W., 1977. A review of the Willard (San Andres) Unit CO_2 injection project. *Soc. Pet. Eng. Conf. on Permian Basin Oil and Gas Recovery, Midland, Texas, March 10–11*, SPE 6388.

Kane, A.V., 1979. Performance review of a large scale CO_2–WAG enhanced recovery project. SACROC Unit—Kelly Snyder Field. *J. Pet. Technol..*, 31(2): 217–231.

Kenney, J.T. and Thodos, G., 1961. The transport properties of carbon dioxide. *AIChE J.*, 7(12): 625.

Lantz, R.B., 1970. Rigorous calculation of miscible displacement using immiscible reservoir simulators. *Soc. Pet. Eng. J.*, 10(6): 192–202.

Leonard, J., 1986. Increased rate of EOR brightens outlook. *Oil Gas J.*, 84 (April 14): 71–101.

Metcalfe, R.S. and Yarborough, L., 1979. Effect of phase equilibria on the CO_2 displacement mechanism. *Soc. Pet. Eng. J.*, 19(8): 242–252.

National Petroleum Council, 1984. *Enhanced Oil Recovery*. June, pp. E7–E17.

Newton, L.E., Jr. and McClay, R.A., 1977. Corrosion and operation problems. CO_2 project, SACROC Unit. *Soc. Pet. Eng. Conf. on Permian Basin Oil and Gas Recovery, Midland, Texas, March 10–11*, SPE 6391.

Nolen, J.S., 1973. Numerical simulation of compositional phenomena in petroleum reservoirs. In: *Numerical Simulation. Soc. Pet. Eng., Repr. Ser.*, 11: 269–284.

Orr, F.M. Jr., Yu, A.D. Jr., and Lein, C.L., 1981. Phase behavior of CO_2 and crude oil in low temperature reservoirs. *Soc. Pet. Eng. J.*, 21(8): 480–492.

Palmer, F.S., Nute, A.J. and Peterson, R.L., 1981. Implementation of a gravity stable miscible CO_2 flood in the 8000-foot Sand, Bay St. Elaine Field. *Soc. Pet. Eng. Annu. Meet., San Antonio, Texas, October 5–7*, SPE 10160.

Pontious, S.B. and Tham, M.J., 1978. North Cross (Devonian) Unit CO_2 flood—review of flood performance and numerical simulation model. *J. Pet. Technol.*, 30(12): 1706–1716.

Rathmell, J.J., Stalkup, F.I. and Hassinger, R.C., 1971. A laboratory investigation of miscible displacement by carbon dioxide. *Soc. Pet. Eng. Annu. Meet., New Orleans, La., October 3–6*, SPE 3483.

Rowe, A.M., Jr., 1978. Internally consistent correlations for predicting phase compositions for use in reservoir composition simulators. *Soc. Pet. Eng. Annu. Meet., Houston, Texas, October 2–4*, SPE 7545.

Rowe, H.G., York, S.D. and Ader, J.C., 1981. Slaughter Estate Unit tertiary pilot performance. *Soc. Pet. Eng. EOR Symp., Tulsa, Okla., April 5–8*, SPE 9796.

Sage, B.H. and Lacey, W.N., 1955. *Some Properties of the Lighter Hydrocarbons, Hydrogen Sulfide and Carbon Dioxide. Monograph on API Research Project 37*. API, Dallas, Texas.

Saner, W.B. and Patton, J.T., 1983. CO_2 recovery of heavy oil: the Wilmington Field test. *Soc. Pet. Eng. Annu. Meet., San Francisco, Calif., October 5–8*, SPE 12082.

Saxon, J., Jr., Breston, J.N. and MacFarlane, R.M., 1951. Laboratory tests with carbon dioxide and carbonated water as flooding mediums. *Prod. Mon.*, November 16, p. 8.

Sebastian, H.M., Wenger, R.S. and Renner, T.A., 1985. Correlation of minimum miscibility pressure for impure CO_2 streams. *J. Pet. Technol.*, 37(11): 2076–2082.

Shelton, J.L. and Yarborough, L., 1977. Multiple phase behavior in porous media during CO_2 or rich gas flooding. *J. Pet. Technol.*, 29(9): 1171–1178.

Stewart, P.B. and Munjal, P., 1970. Solubility of carbon dioxide in pure water, synthetic sea water, and synthetic sea water concentrates at $-5°C$ to $25°C$ and 10–45 atm pressure. *J. Chem. Eng. Data*, 15(1): 67.

Thrash, J.C., 1979. Twofreds Field—a tertiary oil recovery project. *Soc. Pet. Eng. Meet., Las Vegas, Nev., September 23–26*, SPE 8382.

Todd, M.R., 1979. Modeling requirements for numerical simulation of CO_2 recovery processes. *Soc. Pet. Eng. Calif. Reg. Meet., Ventura, Calif., April 18–20*, SPE 7988.

Todd, M.R. and Longstaff, W.J., 1972. The development, testing and application of a numerical simulator for predicting miscible flood performance. *J. Pet. Technol.*, 24(7): 874–882.

Yellig, W.F. and Metcalfe, R.S., 1980. Determination and prediction of CO_2 minimum miscibility pressures. *J. Pet. Technol.*, 32(1): 160–165.

Youngren, G.K. and Charlson, G.S., 1980. History match analysis of the Little Creek CO_2 pilot test. *J. Pet. Technol.*, 32(11): 2042–2052.

Chapter 7

POLYMER FLOODING

RYEN CAENN, DAVID B. BURNETT and GEORGE V. CHILINGARIAN

INTRODUCTION

One of the promising types of enhanced oil recovery is the use of water-soluble polymers. Their use to improve the sweep efficiency of standard waterfloods has been practiced since the early sixties. Many in the industry, however, still consider that the use of a polymer to augment a waterflood is just an extension of that secondary recovery process. Others insist that it should more properly be labeled as a tertiary recovery process, being considered a step away from simply pumping water through the formation. Regardless of the lable put on the process, the use of a polymer is basically to improve the relative mobility ratios of the oil–water front.

During the past 20–25 years many new processes and new polymer forms have been developed to assist the flooding operation. Polymers and other chemical flooding processes have become more sophisticated in recent years. Many organizations have ongoing research and development projects dealing with polymers and EOR processes. Thus, one can be assured that as the economical situation permits these newly developed methods will be made available.

This chapter covers the currently available polymer products and the field techniques in common use at this time. The chapter is organized into two sections as follows:

(1) Polymer products and theory of use
- (a) description of water soluble polymers
- (b) properties of polymer solutions
- (c) mobility control processes
- (d) wellbore profile control processes

(2) Planning polymer flood projects
- (a) screening tests
- (b) process design
- (c) equipment and operations
- (d) the economics of polymer projects

The first section covers a general description of polymers including some polymer chemistry. A few definitions of commonly-used polymer terminology are also given and the basic properties of the polymers are described. The data presented on these properties are taken from published sources and in general represent laboratory-derived numbers. The first section also covers the theory involved in mobility control with commonly used polymers.

The second section presents general guidelines on setting up and running field projects involving the use of polymers. The various stages of design and construction are covered and several examples of published projects are included. A brief discussion is also given on the economics of polymer flooding.

POLYMER PRODUCTS AND THEORY OF USE

The word polymer is derived from the two Greek words, "polys" and "meros", which literally mean "many parts" (Chatterji and Burchardt, 1981). To a chemist, "part" is an individual molecule that makes up a chemical species. A polymer is, therefore, a chemical that is composed of a number of individual molecules that are attached in some manner. These units are usually associated in a pattern that repeats itself throughout the length of each polymer. The repeating units are called monomers and the polymer can be either a homopolymer (one monomer), a dimer (two monomers), etc. Another common term used in polymer chemistry to represent the joining of two different monomers is copolymer. One published definition to differentiate a large molecule from a polymer is that a polymer should have a molecular weight greater than 200 and at least 8 or more repeating units (Clark, 1984).

Polymer flooding operations use high-molecular-weight organic chemicals to alter the flow of water in the formation. Molecular weights of flooding polymers may be as high as several millions. These large molecules are soluble in water because of hydrogen bonding between water molecules and the polymer's polar side chains. Strictly speaking, a water-soluble polymer is in actuality water dispersable, because it does not form a true solution as does, for example, table salt.

Some water-soluble polymers, such as the starches, have such large sizes that they do not readily flow through formation pore throats. The size of the hydrated polymer molecule, then, is important in designing a polymer flood for a specific formation.

TABLE 7-I

Polymers commonly used in EOR operations

Polyacrylamides	synthetic, usually anionic dry powder, water solution, emulsion, gel
Xanthan gum	natural biopolymer, anionic water solution, dry powder
Cellulosics	semi-synthetic, non-ionic or anionic dry powder
Polyacrylate copolymers	synthetic, anionic dry powder, water solution

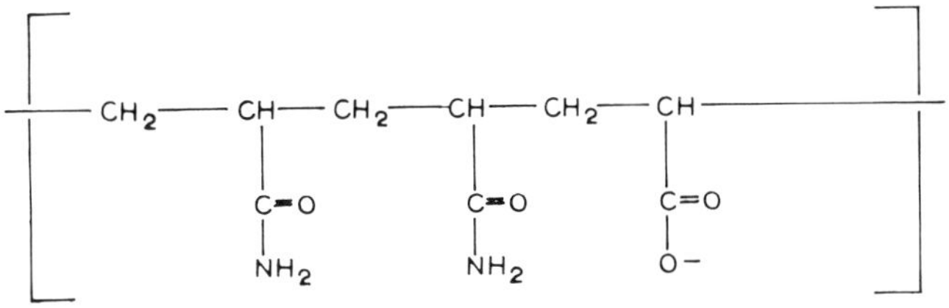

Fig. 7-1. Schematic diagram of polyacrylamide structure.

In theory, almost any water-soluble, non-toxic chemical can be used to increase the viscosity of water and improve the effectiveness of a waterflood process. Davidson (1980) devoted 23 chapters to the description of the most commonly used water-soluble polymers. A search of the patent literature will uncover many more chemicals claimed to be effective. In practice, very few polymer chemicals meet the requirements necessary for use in enhanced oil recovery processes. Four polymer types, listed in Table 7-I, are discussed in this section. Of the four, only two types (polyacrylamides and xanthan gums) are commonly used in the field, and the majority of the discussion is concerned with those materials. Schematic diagrams of the generally accepted structures of polyacrylamide and xanthan gum are shown in Figs. 7-1 and 7-2.

Description of water-soluble polymers

The following section gives general descriptions of four polymer types that have been used in the field. There are many different manufacturers of these polymers that may result in variations in final products even though the products have the same generic name. Also care must be taken that the manufacturer has adequate quality control, so that a consistent material is supplied to the field and for testing.

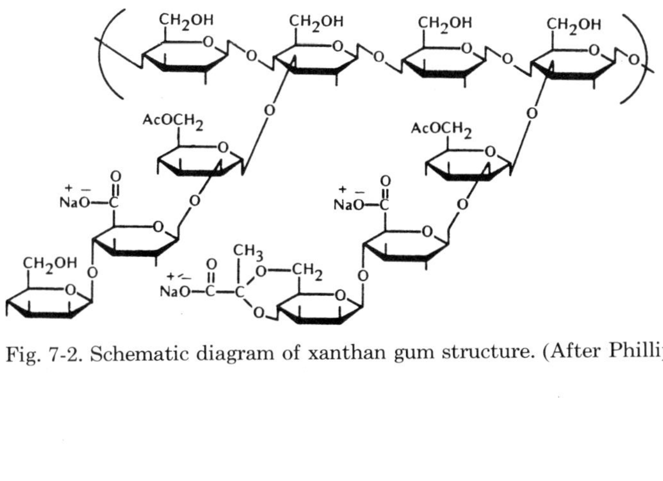

Fig. 7-2. Schematic diagram of xanthan gum structure. (After Phillips, et al., 1985, p. 595, fig. 1.)

Cellulosics

Cellulosics are modified natural materials derived from water-insoluble cellulose waste products. Two basic types of cellulosic polymers have been used in flooding. They are hydroxyethylcellulose (HEC) and carboxymethylcellulose (CMC). Side chains of either carboxymethyl or hydroxyethyl are added to the cellulose backbone, thereby rendering the polymer water-soluble. Cellulosics can be manufactured with a variety of molecular weights which, in general, will determine the viscosity imparted to water by the polymer. Most grades for flooding use have higher molecular weights, which give higher viscosities in water. The CMC's are anionic materials and HEC's are non-ionic.

Most HEC polymers are completely soluble in cold and hot water and in high-salinty brines. Because the molecule is non-ionic, it is compatible with practically all oilfield brines and is often used in drilling and completion fluids. Their use in polymer floods is mainly due to this high salinity tolerance.

Both CMC and HEC solutions are non-Newtonian, i.e., the solution viscosity decreases with increasing shear rates (see section on rheology). The molecules are resistant to mechanical shear degradation so that their viscosity is recovered when the shear rate is reduced. Cellulosics have temperature stabilities in the 250°F range, but lose viscosity rapidly as the temperature increases.

Cellulosic solutions are readily degraded by cellulose enzymes found in oilfield waters. Degradation results in a rapid viscosity loss and loss in EOR functionality. Different types of CMC and HEC have differing degrees of resistance to biological degradation, but biocides are necessary in all oilfield applications using cellulosics as well as with most other polymers.

Normally, cellulosics are used at concentrations ranging from 1000 to 5000 ppm. At the higher concentrations, the polymer solutions tend to lose injectivity and can cause plugging in injection wellbores. (See "Hydratability" later in this section.) Polymer retention within the wellbore is higher because of the relatively high polymer concentrations required for adequate viscosity, in comparison to xanthan gum and polyacrylamides. Cellulosics are easily crosslinked with multivalent cations.

In addition to HEC's and CMC's, there have been proposals to use carboxymethylhydroxyethylcellulose (CMHEC) for flooding operations. As the name implies, this polymer is a result of reacting both carboxyethyl and hydroxyethyl ethers with the cellulose backbone. To date, no particular advantage to use CMHEC over either CMC or HEC has been reported.

Polyacrylates (PAC)

Water-soluble acrylate polymers and copolymers are synthetic materials that have found limited usage in EOR applications. These polymers can be manufactured to have a wide range of physical and chemical properties that make them candidates for many applications in the oilfield. The PAC products offered for polymer flooding have shown resistance to mechanical shear degradation, have adequate injectivity, are compatible with some brines, and demonstrate much higher thermal stability than the other EOR polymers (Ryles, 1983).

Products which have been marketed for oilfield use have not had much success. The side chains on most PAC's are carboxyl groups (usually supplied as a salt), which may be sensitive to some high-hardness oilfield brines. Also, although they are non-Newtonian in solution, they are not nearly as highly non-Newtonian as other polymers.

Polyacrylamides (PAM)

Polyacrylamide polymers are synthetic chemicals which can be tailored to fit a broad range of applications. The molecule is unique in its strong hydrogen bonding, linearity, very high molecular weight, and a high degree of non-Newtonian viscosity. For enhanced oil recovery, the polyacrylamide molecule can be modified by copolymerization with ionic substituents or by partial hydrolysis of the amide side chain to the carboxylic acid group (see Fig. 7-1).

The hydrolysis of the side chain amide to a carboxyl group renders that part of the molecule to be a strongly water-soluble moiety. Different grades of polyacrylamides are hydrolyzed to different degrees. Generally, the highest degree of hydrolysis is approximately 35%. As the degree of hydrolysis increases, the water solubility increases and the viscosity of the polymer solution at a given concentration decreases.

An increasing number of copolymers of polyacrylamide are becoming available to the industry and more will be available in the future. Some copolymers are manufactured to have positively charged substitute groups. These cationic polymers have utility as industrial flocculants and as a component in wellbore profile processes to be discussed later. Also commonly used in these vertical conformance processes are crosslinked or 'gelled' polyacrylamide systems. The hydrolyzed polyacrylamide interacts with polyvalent metals, such as iron, aluminium, and chromium, to form three-dimensional gel networks.

Polyacrylamides are supplied in several forms. Historically, many field polymer projects used solid powder, which was then hydrated to form the polymer solution. Because of convenience, however, most field polymer projects today use polyacrylamides supplied in emulsion form as liquids. These materials are available in concentrations ranging from 30 to 50 wt.% solids and offer a fairly efficient and rapid way to get the polyacrylamide into solution. In high-salinity systems, the use of a surfactant to promote hydration may be necessary. Hydration procedures are discussed in a later section covering field techniques (p. 182 ff).

Xanthan gum (XG)

Another type of water-soluble polymer, which has found great acceptance in the oilfield, is xanthan gum. The XG is a high-molecular-weight, natural carbohydrate. It is a polysaccharide manufactured by a bacterial fermentation process. Commercial xanthan gum is obtained from mutant bacterial strains chosen for yield (amount of product produced during fermentation) and polymer functionality (injectivity and high non-Newtonian viscosity).

Xanthan gum is a polysaccharide (i.e., it contains differing sugar molecules) with unique intramolecular hydrogen bonding by side chains. The structure, as shown in Fig. 7-2, contains side chains which hold the molecule in a rigid, helical structure. Many of the unusual properties of the polymer, stem from its rodlike structure in solution. The side chains shield the backbone of the polymer molecule and protect against enzymatic attack and backbone cleavage. Thermal stability is also improved by this complex structure. Sensitivity to salt is also greatly reduced even though the molecule is anionic in character.

Unlike many polysaccharide solutions, the viscosity of which decreases with an increase in temperature, xanthan gum solutions show only a slight decrease in viscosity. This and other characteristics make it a preferred material in EOR applications despite its unusual origin and rather high cost (partially caused by the requirement for 'batch' manufacturing techniques).

The synthetic polymer, polyacrylamide, and the biopolymer, xanthan gum, offer the best performance/price ratio of the many types of polymers that have been proposed for EOR operations. The discussions in this chapter, therefore, are in general, limited to these two types of polymers.

Properties of polymer solutions

Both polyacrylamide (PAM) and xanthan gum (XG) polymer solutions give high viscosities at relatively low concentrations. Table 7-II shows a comparison of the concentrations of material needed to give 10-cP solutions of polyacrylamide and xanthan gum. The data shown in this table are for polymer mixtures in tap water.

Viscosity is also affected by salinity and divalent ions, such as calcium and magnesium, which cause difficulty in polymer hydration. Hard waters will also cause a decrease in the viscosity of solutions of polyacrylamide polymers to a much greater extent than those with xanthan gum. The following paragraphs give general descriptions of the physical and chemical properties that can be expected from polyacrylamides and xanthan gums.

TABLE 7-II

Comparison of wt.% concentrations of polyacrylamides and xanthan gum required to give a 10-cP viscosity at 12.5 s^{-1}

	Polyacrylamide (wt.%)	Xanthan gum (wt.%)
0.1% NaCl	0.05	0.055
2.0% NaCl	0.115	0.060

Hydratability

Dry powders are in general difficult to mix efficiently in the field. Care must be taken to slowly and completely wet the polymers to prevent the formation of "fish-eyes" (dry powder surrounded by wetted polymer) or microgels. Both xanthan gum and polyacrylamides can be supplied as either liquid, water-based products or as dry materials. Polyacrylamide can also be obtained as an emulsion or as a hydrated, gelled log. The emulsion material can also cause difficulty because it contains an oil-wet dispersion (Kampen and Haynes, 1987). Surfactants and surface-coating agents have been used to assist in dispersing dry powders, thereby allowing better hydration of each individual polymer particle. Care must be taken, however, to assure that any particular treatment has not impaired the functionality of the polymer. Anyone of these forms can be utilized in the field, but each has a different effect on the economics of an EOR project. This is discussed in more detail later in the chapter.

Rheology

Because of the basic nature of the polymers in solution and their high molecular weights, both types of polymers give non-Newtonian flow characteristics. Both polyacrylamide and xanthan gum solutions are shear thinning, i.e., as the shear rate increases, the relative viscosity decreases. Plots of typical viscosity versus shear rate development for both PAM and XG are shown in Figs. 7-3 and 7-4. Viscosity at a given shear rate can be calculated by the proper mathematical model. Further discussion of rheology is beyond the scope of this chapter. Viscosity terms as applied to mobility and flow through porous media is, however, discussed in later sections. Further information on the rheology of EOR polymer solutions can be found in the references (e.g., see Baijal, 1982; Liauh and Liu, 1984; Vossoughi et al., 1984; Aurbach, 1985; Sorbie et al., 1985).

The viscosity of polymer solutions decrease with increasing temperature. The loss in viscosity is reversible unless thermal degradation occurs. At moderate

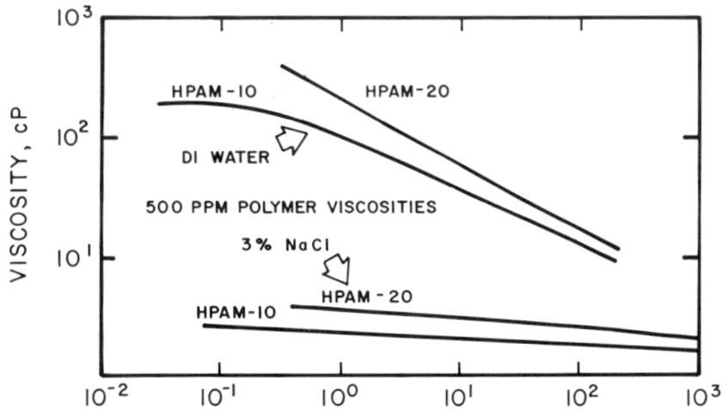

Fig. 7-3. Viscosity development of two 500-ppm polyacrylamide solutions in both distilled water and 3% NaCl water.

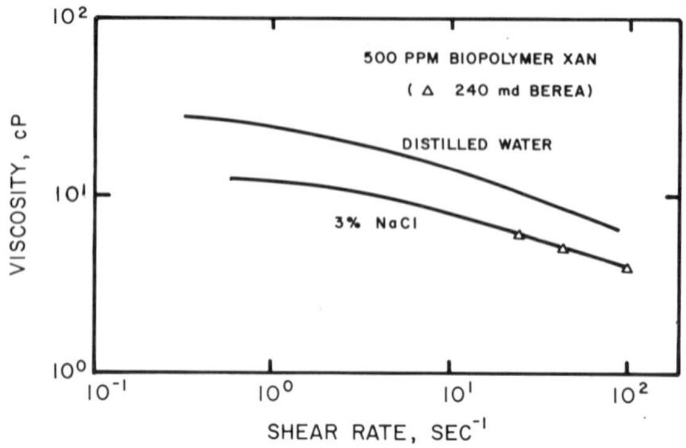

Fig. 7-4. Viscosity development of 500-ppm xanthan gum solution in both distilled water and 3% NaCl water.

temperatures, this change in polymer viscosity with temperature is less than the drop in brine water viscosity caused by the increase in temperature.

Stability

Stability refers to the maintenance of the integrity of the polymer molecule during oilfield operations. Polymers must have (1) mechanical strength, (2) thermal stability, (3) bacteriological stability, and (4) chemical stability.

Mechanical strength. Under most circumstances, xanthan gum solutions are shear stable, but polyacrylamides can be degraded by mechanical shear, such as in mixing equipment. Mixing shears and flow rates within the formation, therefore, can be of extreme importance to the economics of the EOR process. Equipment and field handling of these polymers are discussed later in this chapter.

Thermal stability. PAM and XG solutions are relatively thermally stable to above 200°F if protected against catalytic degradation by oxygen. Shupe (1981), Ash et al. (1983), Ryles (1983), and Wellington (1983) have discussed thermal degradation mechanisms of EOR polymers and have described protective chemical systems to retard their degradation. The following have all been found to be effective and have been recommended as possible temperature stabilizers: thiourea, isopropyl alcohol, nitrogen purges, sulfates, propylene glycol, formaldehyde, and acetone. Generally, it has been found that xanthan gum solutions are not as stable thermally as are polyacrylamides at temperatures above 225°F.

Both polymers can be protected to extend their usefullness at elevated temperatures for many months or even years. In addition, some special polyacrylamide copolymers have extended temperature stability. Most laboratory

TABLE 7-III

Types of bactericides used in treating oilfield waters

Chemical type	Physical properties	Recommended dosage (ppm)
Amine	water dispersable, oil soluble	5– 15
Polychlorophenols	relatively insoluble in oil	50–100
Quaternary ammonium	water dispersable, oil soluble	25– 70
Formaldehyde	water soluble, relatively oil soluble	30– 60

tests, however, have not shown long-term temperature stability much higher than 250°F for either polymer.

Bacteriological stability. Xanthan gum solutions are more susceptible to microbial attack than are polyacrylamides. Degradation of field solutions of xanthan gum have been reported by Exxon (Bragg et al., 1983). Bactericides are always recommended for use with xanthan gum. It should be kept in mind, however, that bacterial protection may well be needed in conjunction with any waterflooding project, whether or not polymers are being used. Aerobic bacteria can grow in surface systems causing polymer degradation and injectivity problems. Anaerobic bacteria can flourish within the formation causing plugging problems, polymer degradation, and reportedly can change a sweet crude to sour. Table 7-III shows some general types of bactericides that can be used in oilfield waters. The use of any bactericide, of course, must be approved by the appropriate environmental regulatory agency.

Chemical stability. Both polyacrylamides and xanthan gums are compatible with modest concentrations of the most commonly-occurring dissolved ions found in oilfield brines. Polyacrylamides, however, are considerably less tolerant to high concentrations of divalent cations, such as calcium and magnesium. As mentioned before, trivalent cations such as iron, chromium, and aluminum react with the polymers to form gels. Quaternary amine chemicals, often used as corrosion inhibitors, also have a deleterious effect on polymer performance, i.e., loss of viscosity or precipitation. Increased formation temperature will tend to accelerate any chemical incompatibilities that may be present.

The viscosity of polyacrylamide solutions is sensitive to pH. At low values of pH, the carboxylic side group is nonionic. As pH increases, the side group is converted to the anionic carboxylate ion. The anion exerts repulsive forces along the polymer chain, causing some uncoiling and a stiffening of the molecule with an attendent viscosity increase. At pH values above 10.5, hydrolysis of polyacrylamide side chains occurs even at room temperature. Xanthan gum solutions, however, are generally stable at the pH values found in oilfield systems.

The properties of a polymer solution are generally tested in field makeup and with formation waters as part of the EOR design. Both the type of polymer and the polymer-flooding technique to be used can then be tailored to the water in the candidate reservoir. This is critical to achieving a successfull polymer flood and is discussed in more detail in the next section.

Mobility control and profile control processes

Polymer flooding can be used in place of waterflooding, or to augment an existing flood. In mobility control processes, the polymer is used to alter the fractional flow characteristics of the water phase which is displacing the oil. In profile control processes, polymer gels are used to block water channels and divert flow to portions of the reservoir which have not been properly swept. Both concepts are discussed in detail in the following sections.

Mobility control

The correct polymer type to use in an enhanced oil recovery process depends upon the field conditions and upon the type of polymer process needed to improve oil production. Polymer solutions must be well understood before a mobility control flood can be properly designed. In addition to the previously mentioned viscosity effects and viscosity stability, the injectivity and residual resistance factor must be taken into account. The following sections describe the mobility control process in relation to these factors.

Definition of mobility. The function of any type of water-soluble polymer in EOR operations is to increase the viscosity of a brine and to reduce the relative permeability to water in the formation. If the flow of brine in a porous medium is decreased in relation to the flow of oil, then the oil's relative flow is improved. Mobility is a measure of the flow of a fluid through a permeable formation. It is the ratio of the relative permeability of the fluid divided by the apparent viscosity of the fluid. Relative mobility is the ratio of the displaced fluid's mobility to the mobility of the displacing fluid.

Mobility, M, and mobility ratio, M_r, are defined and can be calculated by the following mathematical relationships:

$$M = k_r/\mu_a$$

$$M_r = M_o/M_w$$

where k_r = relative permeability, mD; μ_a = apparent viscosity, cP; M_o = mobility to oil ($= k_{ro}/\mu_{ao}$); and M_w = mobility to water ($= k_{rw}/\mu_{aw}$).

A schematic of displacement process is shown in Fig. 7-5. The first diagram, an areal view of an inverted five-spot with water injection into the center well,

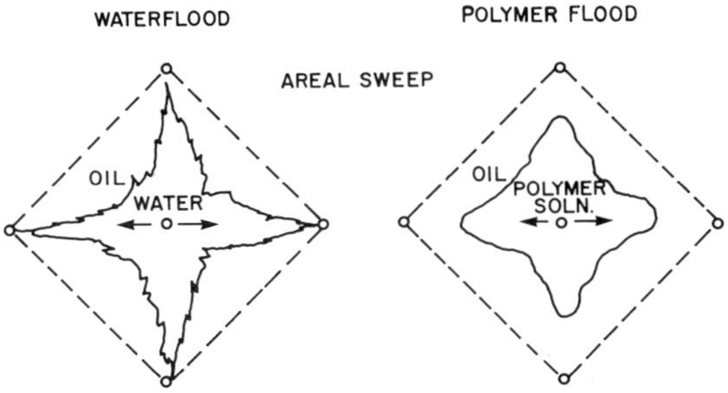

Fig. 7-5. Areal sweep schematic diagram showing mobility control with a polymer solution.

has created channels to the producing wells, causing early breakthrough of injection brine to those wells and increasing the producing water/oil ratios, perhaps to uneconomic values. Most of the oil in the pattern is left unproduced.

The second diagram in Fig. 7-5 shows the same pattern response to injection of fluid with reduced mobility in relation to the oil-in-place. In this case, the brine–polymer bank sweeps evenly out from the injection well resulting in increased sweep efficiency and more oil production at the producing wells with less produced brine.

The same phenomenon occurs in the vertical plane of the well, as shown in Fig. 7-6. In fact, the poor sweep efficiency of water injection is made more pronounced by the vertical permeability variation encountered in most reservoirs. As shown in the figure, the majority of the flow is traveling through a relatively small percentage of the total reservoir height.

Through mobility control, polymers improve oil recovery by changing the fractional flow characteristics of the oil and water flowing through the porous media. Volumetric sweep efficiency is improved because of the decreased mobility ratio. A decrease in mobility ratio means that more oil is produced for a given volume of water injected. Large mobility ratios denote inefficient displacements.

For displacement processes where two phases are flowing simultaneously through the porous media, Gogarty et al. (1970) described the use of total mobility to better represent a description of the fluid flow. Total mobility is the

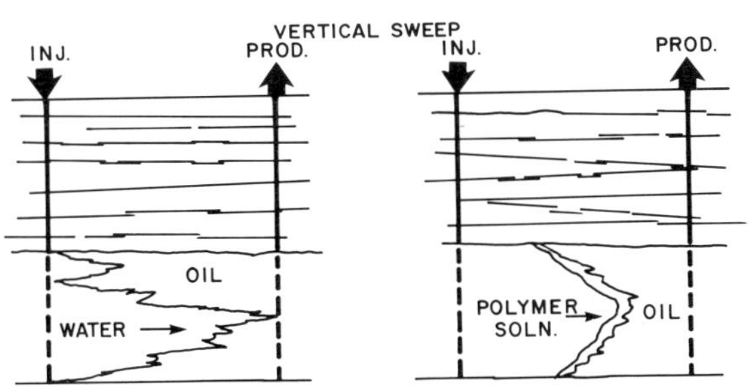

Fig. 7-6. Vertical sweep schematic diagram showing mobility control with a polymer solution.

TABLE 7-IV

Listing of data to obtain total mobility (based on Gogarty et al., 1970, p. 141, equation 1)

Water saturation (% pore space)	k_{ro}	k_{ro}/μ_o [a]	k_{rw}	k_{rw}/μ_w [b]	Total mobility Σ [c]
34	0.4	1.38	0.0014	0.0028	1.38
40	0.25	0.86	0.0060	0.0120	0.87
50	0.1	0.34	0.0340	0.0680	0.41
60	0.04	0.14	0.0900	0.1800	0.32
70	0.0065	0.02	0.2200	0.4400	0.46
72	0.0015	0.01	0.2500	0.500	0.60

[a] $\mu_o = 0.29$
[b] $\mu_w = 0.5$
[c] Total mobility $= cP^{-1}$.

sum of the oil and water mobilities at a particular water saturation. Properly, total mobility must be defined at a given saturation of one of the phases present in the flowing bank. Table 7-IV and Fig. 7-7 represent a study to determine total mobility. At a certain saturation value, the total mobility is at a minimum. According to Gogarty et al. (1970), the design point of a mobility control flood should be this minimum value. In many cases, the total mobility is significantly different than the mobility ratio.

Resistance factor

Resistance factor is a term referring to the relative pressure drop caused by polymer solutions in porous media, and normally only applies to polyacrylamide solutions. It would also apply to other polymers that adsorb onto the formation. Inasmuch as xanthan gums do not adsorb onto the formation, the following discussion is confined to polyacrylamides.

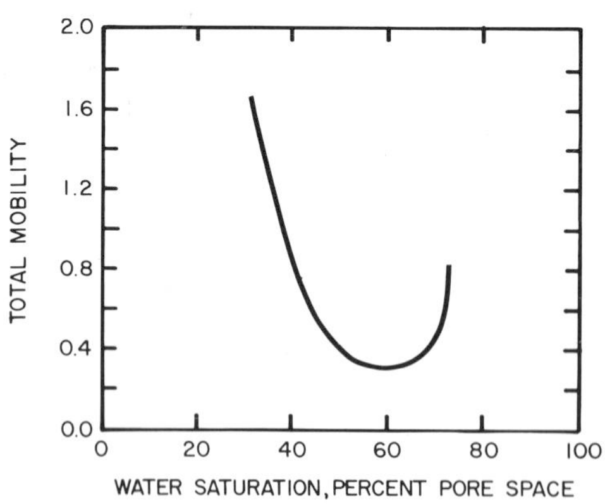

Fig. 7-7. Plot showing total mobility versus water saturation. Curve minimum indicates optimum mobility. (After Gogarty et al., 1970, based on Table 7-IV above.)

If a solution of polyacrylamide is passed through a reservoir core at a constant rate, the pressure drop of the system is much greater than would be calculated by the viscosity change using Darcy's law. This effect is caused by the polyacrylamide adsorbing onto the walls of the rock pore spaces and restricting flow. Resistance factor, R, is defined by the following equation:

$$R = M_w/M_p = (k_{rw}/\mu_w)/(k_{rp}/\mu_p)$$

where M_w = mobility to water; M_p = mobility to polymer; k_{rw} = water relative permeability, mD; k_{rp} = polymer relative permeability, mD; μ_w = water viscosity, cP; and μ_p = polymer viscosity, cP.

Under many conditions, the reduction in permeability to water caused by the polymer persists after the solution itself has apparently been displaced from the core matrix. When brine is injected after the polymer, this excess mobility reduction is termed residual resistance factor, R_r, and is defined by the following equation:

$$R_r = M_w/M_{wp}$$

where M_w = initial water mobility; and M_{wp} = water mobility after polymer injection.

Injectivity behavior

The injectivity behavior of a polymer solution is one of its most important characteristics. Polymer solutions with good injectivity behavior are *necessary* for successful field use. Systems with poor injectivity behavior must be physically or chemically treated before field injection. This treatment can be very expensive and severely affect the overall economics of a polymer flooding operation.

Poor injectivity behavior may be caused by:

(1) Incompatibility between the polymer and its makeup brine.

(2) Incompatibility between the polymer solution and the formation rock.

(3) Contamination.

Incompatibility among the brine, polymer and the porous media may be evaluated in the laboratory before beginning extensive design work. Care should be taken to determine if the incompatibility stems from the mixing and hydration of the polymer or from a reaction between dissolved ions or formation clays and the polymer itself.

Contamination can result from the polymer manufacturing process, makeup water, or careless field procedures. Contamination usually results in the formation of particles or microgels that can cause poor injectivity behavior.

Wellbore profile control

The second important function of polymers is the improvement of vertical sweep efficiency around the vicinity of injection wellbores. Water diverting

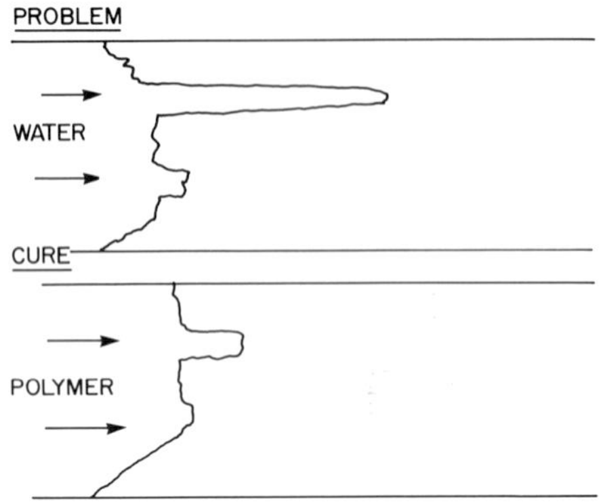

Fig. 7-8. Diagrams showing water breakthrough in high-permeability streaks being controlled by crosslinked polymers.

techniques (also called profile correction, conformation improvement, etc.) have been performed for many years.

Through laboratory and field testing, it has been determined that the materials which work successfully for this process were those that could be placed some distance into the formation.

Fig. 7-8 shows schematically the process of profile control. The top diagram shows the floodwater fingering through a high-permeability streak. The bottom picture represents the same formation after being treated by a gelled solution, which lowered the relative permeability of the streak allowing more of the water to sweep the rest of the formation. The polymer treatment is designed to place the material a significant distance into the formation so that water will not flow around the plugged zone back into the permeable zone. Well profile control techniques, then, affect the flow of water rather than improve the fractional flow characteristics of the oil–water displacement.

In profile control treatments, solutions contain higher concentrations of polymer than do the solutions designed for mobility control. Additionally, the polymer can be gelled or crosslinked to increase molecular weight and further restrict the flow of water. Several techniques have been developed as described in the following references: Green et al. (1984), Prud'homme and Uhl (1984), Barnes and Inker (1985), and Burkholder (1985). The following discussion describes in general terms some of the gelling techniques used for profile control.

Surface gels

The viscosity of polymer solutions can be increased by reacting the polymer with a gelling agent, usually a complex metal ion, in a mixing tank on the surface. The agent reacts with the polymer to create a three-dimensional polymer gel network, which is then injected into the formation. In some treatments, the gel is sheared before injection to break-up microgels, insuring that it will pass into the rock matrix.

In-situ gels

Viscosity enhancement by gelling can also be achieved by crosslinking the polymer within the formation. The theory is that a low-viscosity solution can be injected easily into the formation, and then permitted to chemically react, creating a gel in place to divert the flow of water. Triggering mechanisms for the reaction are salinity, temperature, or sequestrants. It is also possible to inject a monomer instead of the polymer and then to polymerize it, creating the gel structure within the formation.

Sequential injection

The "Phillips Process" is a technique patented by Phillips Petroleum Company (Gall, 1973; Clampett and Hessert, 1974) and licensed to various groups which perform the treatment. It involves a sequential injection of polymer and crosslinking agent to create a chemical which adsorbs on the walls of the formation pores. Reactants such as aluminium citrate are often employed with polyacrylamide solutions, such as those used in polymer mobility control processes. The technique can increase the resistance factor of reservoir rocks significantly (often tenfold).

PLANNING POLYMER FLOOD PROJECTS

The chemistry of polymers must be considered from the very beginning of the polymer flood design. The salinity of the injection and formation waters dictate both the type and the concentration of polymer to be used. Water must be of high quality to avoid injection-well plugging. The amount of water needed for the project depends on the response of the producing zones to more favorable water/oil mobility ratios.

At the outset of the polymer flood design, the reservoir must be screened to determine if it is suitable for this type of EOR process. The various options for polymer treatment must also be considered, e.g., use of a large slug of a solution having a relatively low concentration of polymer, versus a vertical conformance treatment. Design criteria for mobility control projects differ from wellbore profile control projects. If the project is to be a mobility control project, then a rather large slug of polymer is injected, beginning early in the life of the waterflood. On the other hand, remedial wellbore profile control treatments use small slugs of crosslinked polymer.

Fig. 7-9 shows the development (various stages) of a project in simplified form. The initial studies (Stage 1) identify the most promising reservoir candidate for polymer flooding. Screening studies include both laboratory tests and field reviews. At this stage the major reservoir characteristics are determined.

Stage 2 involves the initial process design work. Laboratory and engineering studies provide data in order to consider various design strategies. Performance predictions are made and a simplified study is made of the process economics.

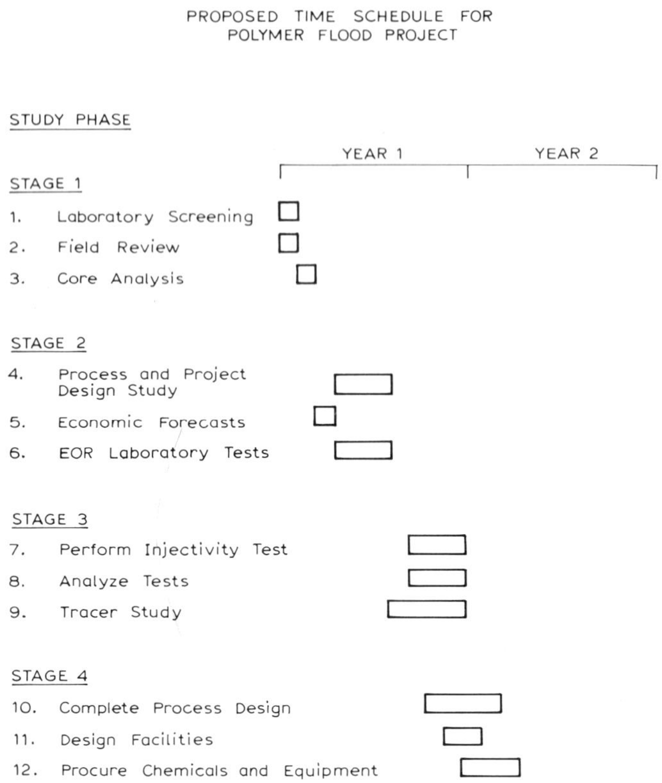

Fig. 7-9. Proposed time schedule for a polymer flood project.

Stage 3 consists of a field test of the EOR process. When working with chemicals such as polymers, injectivity is determined. Transient tests are run to determine (1) reservoir characteristics, and (2) in-situ fluid characteristics. If necessary, process design is changed to accommodate reservoir characteristics.

Stage 4 is field development. The EOR injection facility is planned during this stage. A field-wide simulation study is performed.

Stage 1—initial screening

When a candidate reservoir is being considered for an enhanced recovery polymer project, certain guidelines can be used to identify the best choice of techniques. Field characteristics such as reservoir temperature, oil/water mobility ratio, and oil properties have been shown to be good "identifiers" that have pointed out reservoirs that are amenable to EOR. Taber and Martin (1983) have compiled generally accepted principles into guidelines for polymer flooding design (as well as other EOR processes). Their guidelines for polymer flooding are shown in Table 7-V. These guidelines are known as binary screens. That is, a

TABLE 7-V

Guidelines for polymer-flooding design (after Taber and Martin, 1983, table 7)

Description

The objective of polymer flooding is to provide better displacement and volumetric sweep efficiencies during a waterflood. Polymer augmented waterflooding consists of adding water-soluble polymers to the water before it is injected into the reservoir. Low concentrations (often 250–2000 mg/l) of certain synthetic polymers or biopolymers are used; properly sized treatments may require 15–25% reservoir PV.

Mechanisms

Polymers improve recovery by:
- —increasing the viscosity of water
- —decreasing the mobility of water
- —contacting a larger volume of the reservoir

Technical screening guides

Crude oil:	
Gravity	> 25°API
Viscosity	< 150 cP (preferably < 100 cP)
Composition	not critical
Reservoir:	
Oil saturation	> 10% PV mobile oil
Type of formation	sandstones preferred but can be used in carbonates
Net thickness	not critical
Average permeability	> 10 mD (as low as 3 mD in some cases)
Depth	< about 9000 ft (see temperature)
Temperature	< 200°F to minimize degradation

Limitations

If oil viscosities are high, a higher polymer concentration is needed to achieve the desired mobility control.

Results are normally better if the polymer flood is started before the water/oil ratio becomes excessively high.

Clays increase polymer adsorption.

Some heterogeneities are acceptable but, for conventional polymer flooding, reservoirs with extensive fractures should be avoided. If fractures are present, the crosslinked or gelled polymer techniques may be applicable.

Lower injectivity than with water can adversely affect oil production rate in the early stages of the polymer flood.

Acrylamide-type polymers lose viscosity due to shear degradation, or increases in salinity and divalent ions.

Xanthan gum polymers cost more, are subject to microbial degradation, and have a greater potential for wellbore plugging.

candidate reservoir either meets the criteria or it does not. The advantage of such a screening technique is its speed and simplicity. The disadvantage is that it can be misleading, i.e., a poor process or an inappropriate reservoir can be chosen because the selection protocol was too simplistic.

TABLE 7-VI

Criteria for crude oil testing

BS & W (basic sediment and water)
Viscosity of dead oil
API gravity
Corrosion inhibitors
Acid number
Asphaltene content
Average molecular weight
Paraffin content

Initial screening criteria are general. Successful polymer projects are frequently performed that violate one or more of the suggested guidelines. It is up to the engineer in charge to make the final approval. For this reason, laboratory screening test data should be collected for the specific candidate reservoir so that accurate information is available for decision making. The screening program must determine the following variables: (1) the best process, (2) the best location, (3) the proper operating conditions, and (4) the project's performance estimate. Polymer screening programs consist of: (1) laboratory screening tests, (2) field reviews, and (3) core analysis studies.

Laboratory screening

Using screening guidelines and preliminary laboratory studies, the engineer should be able to determine if the candidate reservoir is suitable for polymer injection (Goodlett et al., 1986). The first data to be obtained from laboratory testing should be the characterization of the crude oil. The criteria for crude oil testing are shown in Table 7-VI. Oil viscosity should be low enough so that the polymer augmented flood can provide low mobility ratios. If the oil viscosity is too low, however, then water/oil mobility ratios may be favorable without polymer additions. The BS&W (basic sediment and water) determinations are made of the stocktank oil to be sure the sample is suitable for testing.

Also of crucial importance is the analysis of the injection brine. The total salinity and ionic composition of the brine to be injected must be known, because of the effect of salts upon polymer performance. If polyacrylamides are eventually chosen for the project, compatibility studies must be performed to insure that the brine hardness does not exceed the tolerance of the polymer. Generally, brines of greater than about 50,000 ppm total dissolved solids cause problems with polyacrylamide performance.

If different brines are to be mixed, then water compatibility must be considered. Polymers magnify the problem of injectivity impairment caused by scale buildup. Water softening can be used in some cases where brine salinity affects polymer performance. Conventional water softening techniques can be used for low-salinity brines. For high-salinity brines, weak acid ion exchange softening systems are used. Such systems have been used in large-scale field projects in California, U.S.A. (Morin and Gaucher, 1986).

Bacterial contamination is one of the greatest impediments to successful polymer flooding. The typical water injection system is not designed to prevent bacterial contamination. Even though most of the water passes quickly through a system, residual liquid trapped in dead spots in piping or in tanks allows accumulation of bacterial colonies. The problem of bacterial control in the water-injection system is further complicated by their diverse characteristics and rapid reproduction rate of bacteria.

Bacteria reproduce logarithmically, and under favorable conditions, one cell can produce millions of daughter cells within 24 hours. Bacterial growth is favored by neutral pH, moderately elevated temperature, the presence of nutrients, and, depending upon the species, available oxygen. A great deal of research has been performed to find the most effective biocide to use for a given set of field conditions and type of polymer. Most polymer suppliers will recommend preferred biocides for their specific type of polymer. Those recommendations should be considered as well as those of oilfield production chemical companies.

Stage 2—field reviews

The next step in the screening process is the preliminary field review. This consists of a study of the EOR potential for the candidate reservoir. The site of the proposed project should be visited by an enhanced oil recovery specialist. The specialist should inspect the field site, visit with key field personnel, and gather oil and water samples. The data needed for the field review are shown in Table 7-VII.

TABLE 7-VII

Listing of data needed for a complete field review

Facilities:	type and quality age monitoring facilities quality control
Process:	well spacing source and type of water current water/oil ratio current pressure
Oil-in-place:	reserves
Reservoir transmittance:	permeability net pay injectivity indices
Reservoir permeability variation:	logs core analysis

TABLE 7-VIII

Types of core analysis data generated for planning polymer flood projects

Description	Objectives	Data
Conventional tests	to define reservoir and its boundaries	—porosity, permeability, —oil and water saturations
	to obtain core for special tests	—grain density —lithology
Special core analysis	to determine petrography, wettability and saturation	—mineral contents —amounts and types of clays
	electrical properties for log analysis	—formation factor —resistivity index
	fractional flow characteristics	—relative permeability
Enhanced recovery tests	determine compatibility of reservoir injected fluids	—injectivity resistance —factor vs. concentration —polymer retention —shear degradation
	measure oil recovery characteristics of EOR process	—secondary or tertiary oil production
	define permeability alteration tendencies of polymer chemicals	

Data from the field review will help to characterize the reservoir. Four key parameters should be identified at this time: (1) volume of oil-in-place, (2) reservoir transmittance, (3) reservoir permeability variation, and (4) rock wettability. This will require some expenditure for well testing and analysis. If cores are not available, they will have to be obtained at this time to fully determine the acceptability of the reservoir for a polymer flood.

Core analysis

Characteristics of the reservoir in large measure determine the performance of a potential polymer project. For this reason, it is of the utmost importance to have core analysis data available for the candidate project. Table 7-VIII defines routine and special core analysis tests. There are only three stages of core analysis, with EOR testing being the final and most sophisticated stage. All these tests are expensive and time consuming.

Special core analysis

Special core analysis refers to the specialized tests, which are used to characterize core material. Special core analysis will elucidate the flow behavior of fluids in the porous media. This data will be used to determine how fast a fluid may flow through the rock. Relative permeability curves are needed for water–oil displacements to model a waterflood process. The curves are altered to represent polymer–oil displacements. Capillary pressure tests are performed to determine irreducible fluid saturations. Determination of the wettability of reservoir rocks is also of the utmost importance.

Injection waters should be tested for compatibility in the reservoir cores. The types and amounts of clay minerals, which might react with injection brines and chemicals, must be determined by X-ray analysis and/or petrographic tests. Additionally, cation exchange capacity data are used to support other evidence of the presence of clay minerals in the formation matrix. The above tests can only be used as guidance to possible clay–polymer incompatibilities. Actual core tests and field experience are the only definitive measures for incompatibilities.

It is essential to determine the types and amounts of various clays in the reservoir rocks. In addition to the deleterious effects of clays on chemical injectants, these minerals have significant effects on other reservoir properties, such as porosity, permeability, and well log responses.

Process selection

The type of polymer and the process to be used in the field depends upon the results of the above screening tests and the preliminary field review. Information from these sources are evaluated by the use of analytical models to determine the predicted performance of a polymer flood.

Design criteria for polymer mobility control and for polymer vertical conformance treatments are different. For instance, if the field has not been under waterflood, then low concentrations of polymer for mobility control are generally used. If the polymer treatment is to correct vertical conformance problems, then crosslinked, small-volume concentrated polymer injections are generally used. Data compiled in the laboratory and field screening stage are used by the engineer for determination of the most effective design possibilities.

Laboratory evaluations

Oil recovery projections are performed using basic reservoir engineering calculations to establish whether or not the process is feasible. It is helpful to perform laboratory enhanced recovery tests on field cores to get an idea of actual performance using the process.

Burnett and Shurz (1983) listed laboratory testing techniques which can be used to select a polymer. Laboratory testing includes (1) compatibility tests with the field oil and brine, and (2) core testing with polymer solutions to get more information on injectivity, polymer retention, and oil recovery efficiency. Goodlett et al. (1986) listed aspects of polymer core floods for polymer flood evaluations (Table 7-IX).

TABLE 7-IX

Listing of data obtained from laboratory polymer core floods that aids in polymer flood planning (after Goodlett et al., 1986, p. 82, table 1)

Test	Critical parameters	Information obtained	Limitations
Injectivity	flow rate (shear rate); rock pore geometry	polymer rheology in porous media (mobility vs. flow rate); plugging potential; shear degradation.	screening test only because of small sample size.
Plugging	flow rate (shear rate); rock pore geometry; core size large enough for multiple pressure taps.	magnitude and depth of plugging vs. time.	
Mobility control [a]	use of reservoir rock and fluids at reservoir temperature and pressure; no external shear degradation introduced; pressure taps between inlet and outlet of core.	fluid mobilities before and after flooding; polymer resistance and residual resistance factors.	baseline data; difficult to simulate effects of reservoir heterogeneity
Retention [a] and inaccessible pore volume	use reservoir rock and fluids at reservoir conditions including residual oil saturation.	sum effect of adsorption, entrapment, and inaccessible pore volume on transport of polymer.	used to determine minimum slug size.
Relative permeability w/polymer	use reservoir rock and crude at reservoir condition	relative permeability curves for oil and polymer; can be used to aid in recovery predictions.	no consideration of sweep efficiencies.
Oil saturation vs. recovery	accurate oil saturation determination.	optimum oil saturation at which polymer flooding should be initiated	one-dimensional model will not completely predict reservoir behavior.

[a] Mobility control and retention tests can be combined into a single series of floods.

Injectivity tests are performed to select the type of polymer to be used. Testing should be done on field core plugs using the brine planned for field injection. The laboratory tests give an indication of the best molecular weight range for polyacrylamides and the approximate concentrations for xanthan gum. Fig. 7-10 shows a plot of a typical injectivity test.

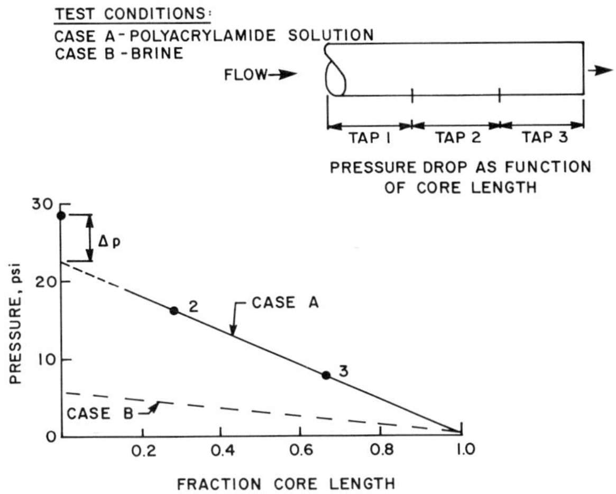

Fig. 7-10. Example of polymer injectivity tests.

Retention tests provide information on the adsorption of the polymer on reservoir rock. There are two methods of determining polymer retention:

(1) Polymer adsorption on crushed core may be determined by equilibrating a bulk solution of polymer stored over the core material. To correlate bulk polymer adsorption with polymer adsorption in consolidated media, specific surface areas of cores and crushed material must be known.

(2) Polymer retention in consolidated porous media may be performed by measuring and comparing the injected polymer concentrations to the effluent concentrations. This gives the relative adsorption and desorption characteristics of the polymer.

Inasmuch as xanthan gums do not normally degrade with mechanical shear, polymer shear degradation tests are normally performed only on polyacrylamides (Kochelek, 1984). The purpose of shear degradation tests is to determine

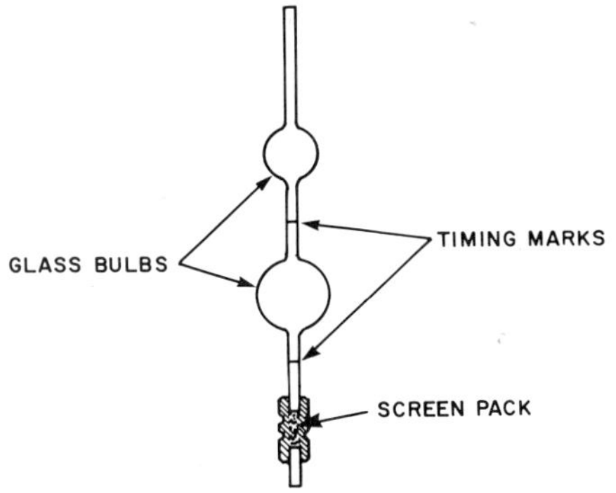

Fig. 7-11. Schematic diagram of a screen "viscometer".

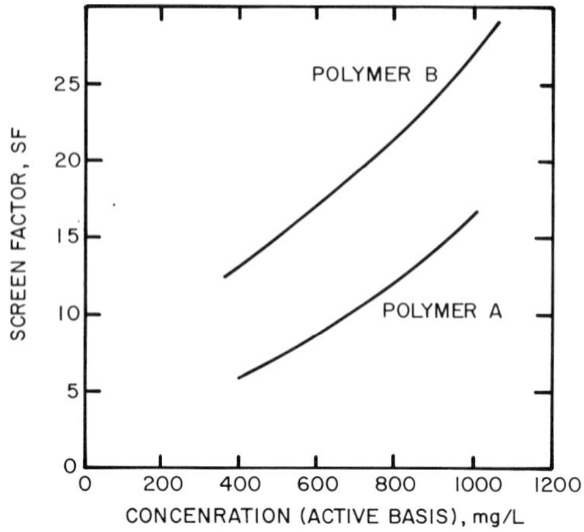

Fig. 7-12. Typical screen factors for polyacrylamide solutions.

the effect of polymer solution flux on polymer resistance factor and retention. The test is performed on a field core at varying injection rates. The higher the injection rate, the greater the loss in polymer performance if it shear degrades. The relative amount of shear degradation is defined by viscosity differences, differences in core test pressure drops, or by the use of a test called the screen factor.

Screen factor is a laboratory-designed test using a typical glass capillary viscometer in which the polymer fluid is also passed through five-layered stainless steel screens (Fig. 7-11). This test is specific for polyacrylamide, because xanthan gum passes through these screens relatively unhindered. Some typical screen factors for polyacrylamide solutions are shown in Fig. 7-12.

This type of behavior has been studied by Maerker (1975) and later by Seright (1983). Their correlations can be used to predict injection rates for particular polymer solutions and injection-well configurations. Laboratory testing is recommended in order to obtain more precise data.

Reservoir modeling

The mathematical modeling of the performance of a polymer flood can be simple or quite complex (Weiss and Baldwin, 1984). The degree of sophistication used in a simulation study depends upon the amount of information known about the reservoir. If the reservoir data is sparse, it is not wise to perform an expensive computer simulation of the process using assumed or "default" values for key process variables. If not many data are available, it is better to use simpler modeling techniques and to perform a number of "what if" studies to determine the impact of key field properties on the process efficiency. These simple models can often be done on hand-held calculators or microcomputers.

If a more sophisticated numerical simulation is to be performed, then the study should begin with a field history. If the candidate reservoir has been waterflooded, then several years of production–injection data are necessary, including a pressure history. It is often found that reservoir pressure data are not available. The engineer then must determine how detailed a waterflood study should be performed based on the overall economic factors that are known, i.e., estimated oil-in-place versus projected selling price.

Modeling the past history of a field permits the engineer to develop accurate estimates of the reserves and the fluid transmissivities. Upon their determination, accurate estimates of the effect of polymer injection can be made.

Screening models

Polymer flooding does not increase the displacement efficiency of the waterflood process; instead it increases the volumetric sweep efficiency or "coverage". Increased sweep efficiency occurs at the expense of injection rate. Screening models often use correlations of volumetric sweep as a function of mobility ratio to represent a displacement with a reduced mobility displacing phase. Screening models often calculate polymer injection rates based on polymer solution viscosity. If this practice is used, the polymer solution viscosity must be determined at a shear rate found within the wellbore.

Reservoir simulation

Reservoir simulators require much more information about the polymer process than does a screening model. If the candidate reservoir is well characterized, then a numerical simulator is recommended. There are several commercial simulators available. The preferred models offer the following options:

(1) Non-Newtonian polymer rheology can be modeled.
(2) Polymer retention is considered.
(3) Permeability reduction is considered.
(4) Salinity effects can be modeled.

Stage 3—pilot testing

To increase the accuracy of reservoir simulators, pilot field trials are often run. The more novel the technique, the more important it is to test the process in the field. Pilot tests also provide an opportunity to get additional data on the candidate reservoir. The information obtained from the pilot test provides: (1) more complete definition of the project area, and (2) testing all phases of the flooding process in an economical manner before scaling up to full size.

Defining the project area

There are a number of techniques used to define the project area. Tracer studies are very popular and effective in reservoirs where injection is already underway. Tracer tests identify the reservoir stratification and the presence of fractures. Whereas many companies use highly complex tracer studies, often a simpler analysis is satisfactory. Brigham (1970) has described a technique for determining fractures or directional permeability in EOR patterns. This technique can be used to design a tracer project and determine the amount of tracer to be injected. It can also be used to identify significant layering. In reservoirs where waterflooding has not been performed, pulse testing may be beneficial (Brigham, 1970).

Testing the process

Pilot injectivity tests can provide information on the following:
(1) Injection rates and pressures.
(2) Fluid properties in the reservoir.
(3) Reservoir characteristics.

Stage 4—field development

If all studies show that the polymer process is a viable option, then the project proceeds to Stage 4, i.e., field project development. The following sections concentrate on the facilities design aspects of field development. Some examples are given from current field practices.

Equipment and operations

Facilities design represents a significant part of the cost of a polymer project. The type of facilities differs for each process. Water injection facilities, polymer injection facilities, field laboratories, and some economic considerations are discussed in the following paragraphs.

Water injection facilities. Surface water is not the preferred choice for injection water, because it usually contains a high oxygen content. High levels of oxygen in the water not only threaten the stability of the polymer but also increase the potential for corrosion in the lines and storage vessels. Aerobic bacteria and fungal growth can foul injection wells and reduce injectivity. Standards should be for water quality. These standards should be based on the results of the special core testing performed on the candidate reservoir and upon the results of the field review. In many cases, a separate evaluation will have to be made for bacterial control. This evaluation in all likelihood will result in the development of special water-handling and water-treating facilities, such as shown in Fig. 7-13. (Also see Volume I, Chapter 6.)

Polymer injection facilities. An example of polymer injection facilities is shown in Fig. 7-14 (Bell Creek Field project). This project utilized a mixing plant designed to handle polyacrylamide emulsion. The diagram shows the location of the equipment for chemical mixing and chemical injection and the configuration of the injection and producing wells (see Hartshorne and Nikonchik, 1984).

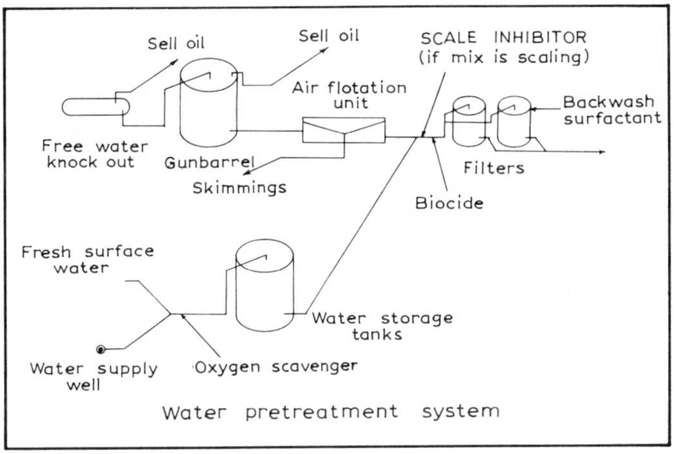

Fig. 7-13. Schematic diagram of a typical water pretreatment system. (After Kochelek, 1984, p. 3, fig. 4.)

Because of the difficulty encountered with most dry polymer mixing plants, emulsion systems are the most common choice for polyacrylamide operations. The problems encountered in dry powder mixing are microgel formation, fish-eyes, dusting, and inaccurate measurements. It is necessary to realize, however, that the polymers in these emulsions are not hydrated, but are only dispersed in the organic liquid phase of the emulsion. It still must be allowed to hydrate before injection (Kampen and Haynes, 1987). Typical field setups for mixing polyacrylamide solutions are shown in Fig. 7-15.

An example of a xanthan gum project is shown in Fig. 7-16 (Ware, 1983). Xanthan gum biopolymers can be supplied as either a dry powder or as a liquid slurry concentrate. The polymer in the slurry concentrate is fully hydrated so the material only needs to be diluted as it is pumped into the injection well. The slurry, or broth as it is sometimes called, can be supplied in 2–3% active content or as a 10–12% concentrate.

Field laboratories. Field laboratories are usually required on polymer flooding projects. Tests performed in such laboratories include water analysis, polymer injectivity, polymer viscosity and bacteriological tests (see Burnett and Shurz, 1983).

Economic considerations for field development

The oil recovery effectiveness of a polymer project must be considered as soon as possible. The use of simple models to provide estimates of oil recovery as early in the project as possible is recommended. The cost of designing a polymer project is directly proportional to the size and sophistication of the project.

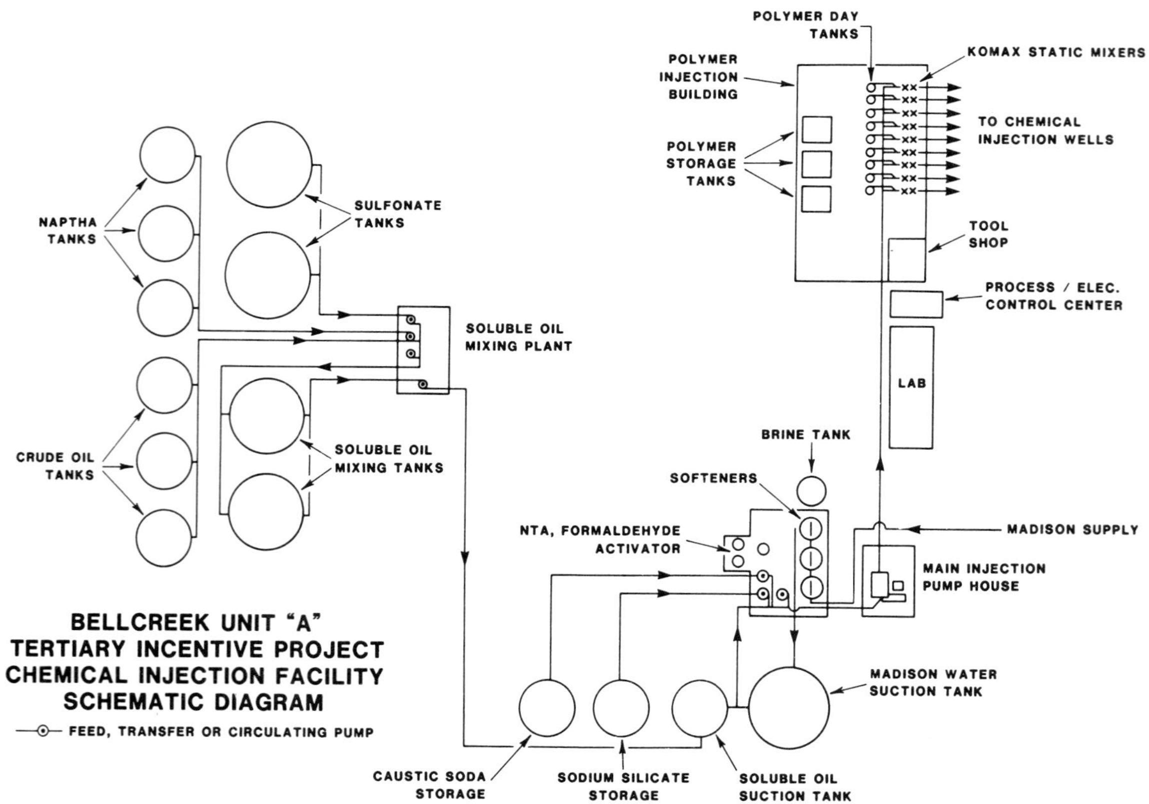

Fig. 7-14. Schematic diagram of the Bellcreek Unit "A" injection facility. (After Hartshorne and Nikonchik, 1984. fig. 6.)

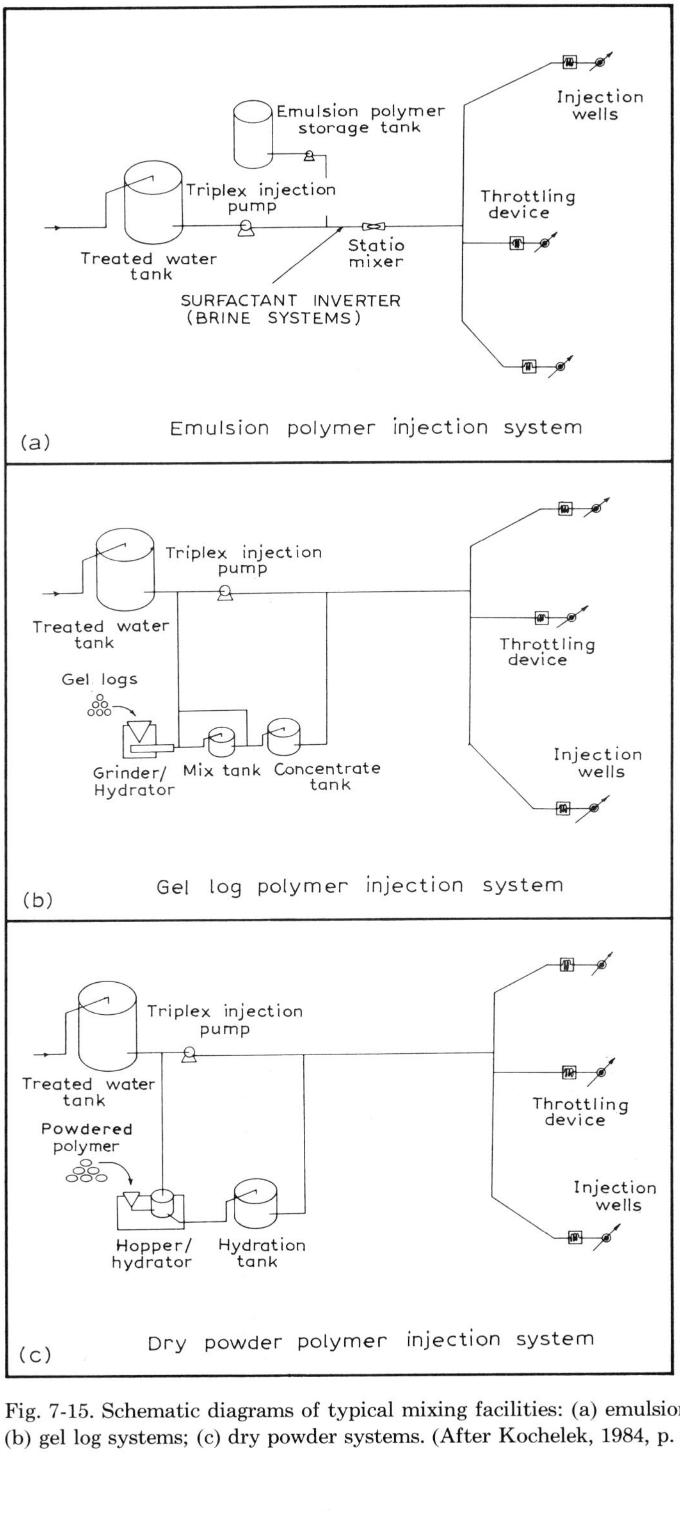

Fig. 7-15. Schematic diagrams of typical mixing facilities: (a) emulsion and liquid polymer systems; (b) gel log systems; (c) dry powder systems. (After Kochelek, 1984, p. 2, figs. 1, 2, and 3.)

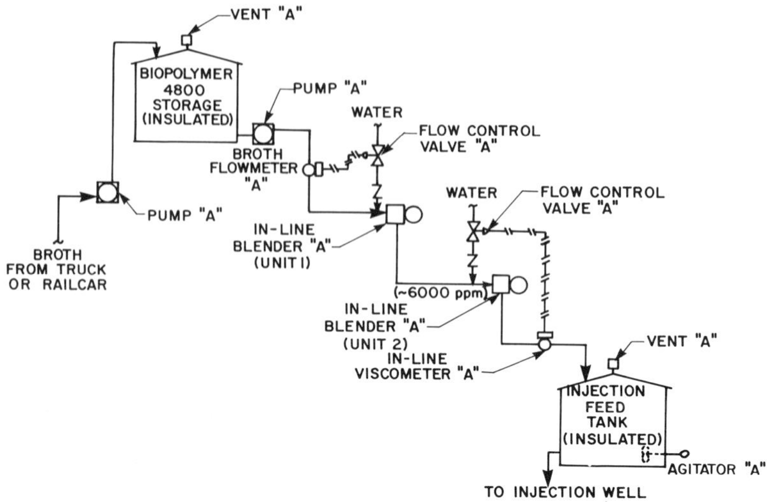

Fig. 7-16. Schematic diagram of a typical liquid xanthan gum injection system. (After Pfizer Oilfield Chemicals Division, 1982, fig. 25.)

The overall success of a polymer project can be directly related to the economic projections made on the amount of oil to be recovered. It is always wise to look *carefully* at the simple economic models for initial screening, but to be *careful* when rejecting projects strictly based on these simple models. The costs of polymers are constantly changing and need to be continuously evaluated along with the process involved in using a particular polymer. Polymer augmented flooding has proven itself to be an economically viable EOR process and other polymer EOR schemes are currently being developed that show much promise for the future.

REFERENCES

Ash, S.G., Clarke-Sturman, A.J., Calvert, R. and Nisbet, T.M., 1983. Chemical stability of biopolymer solutions. *Soc. Pet. Eng. Meet., Houston, Texas, October*, SPE 12085.

Aurbach, M.H., 1985. Prediction of viscosity for xanthan solutions in brines. *Soc. Pet. Eng. Chem. Symp.*, SPE 13591.

Baijal, S.K., 1982. *Flow Behavior of Polymers in Porous Media*. PennWell, Tulsa, Okla.

Barnes, P.F. and Inker, G.E., 1985. Production technology experience in Michigan waterflood. *J. Pet. Technol.*, 37(8): 1446–1458.

Bragg, J.R., Maruca, S.D., Gale, W.W., Gale, L.S., Wernau, W.C., Beck, D., Goldman, I.M., Laskin, A.I. and Naslund, L.A., 1983. Control of xanthan degrading organisms in the Loudon pilot. *Soc. Pet. Eng. Meet., Houston, Texas, October*, SPE 11989.

Brigham, W.E., 1970. Planning and analysis of pulse-tests. *J. Pet. Technol.*, 22(5): 618–624.

Burkholder, L.A., 1985. Xanthan gel system effective for profile modification. *Oil Gas J.*, 83(19): 68–69.

Burnett, D.B. and Shurz, G., 1983. *Polymer Flooding Seminar Manual*. Burnett Research Association, private publication.

Chatterji, J. and Burchardt, J.K., 1981. Application of water-soluble polymers on the oil field. *J. Pet. Technol.*, 33(11): 2042–2056.

Clampett, R.L. and Hessert, J.E., 1974. Method for controlling formation permeability. *U.S. Patent*, No. 3, 785, 437.

Clark, R.K., 1984. Applications of water-soluble polymers as shale stabilizers in drilling fluids. *Am. Chem. Soc. 188th Natl. Meet., Philadelphia, Pa.*

Davidson, R.L., 1980. *Handbook of Water Soluble Gums and Resins*. McGraw-Hill, Newark, N.J.

Gall, J.W., 1973. Subteranean formation permeability correction. *U.S. Patent*, No. 3, 762, 476.

Gogarty, W.B., Meabon, H.P. and Milton, H.W., 1970. Mobility control design for miscible-type waterfloods using micellar solutions. *J. Pet. Technol.*, 22(2): 141–147.

Goodlett, G.O., Honarpour, M.M., Carroll, H.B. and Sarathi, P.S., 1986. Lab data aid design and evaluation of polymer floods. *Oil Gas J.*, 84(26): 82–90.

Green, D.W., Willhite, G.P. and Huang, C.G., 1984. An experimental study of the in-situ gelation of chromium (+3) polyacrylamide polymer in porous media. *Soc. Pet. Eng. EOR Symp., Tulsa, Okla.*, SPE 12638.

Hartshorne, J.M. and Nikonchik, J.S., 1984. Micellar polymer flood shows success in Bell Creek Field. *Soc. Pet. Eng., Fall Meet., Houston, Texas*, SPE 13122.

Holm, L.W., 1983. Evaluation of additional data from Bell Creek micellar pilot indicates success. *Oil Gas J.*, 81(27): 140–142.

Kampen, D.T. and Haynes, H.H., 1987. Proper chemical use improves offshore water clarification. *Ocean Ind.*, 22(2): 38–42.

Kochelek, J.T., 1984. Operational problems associated with polymer applications in oil field EOR projects. *Southwestern Petrol. Short Course, Lubbock, Texas, April.*

Liauh, W.W. and Liu, T.W., 1984. A capillary viscometer for the study of EOR polymers. *Soc. Pet. Eng. EOR Symp., Tulsa, Okla.*, SPE 12649.

Maerker, J.M., 1975. Shear degradation of polyacrylamide solutions. *Soc. Pet. Eng. J.*, 15(8): 311–322.

Morin, R.J. and Gaucher, D.E., 1986. In-line continuous monitoring system improves water plant EOR performance. *Oil Gas J.*, 84(28): 104–105.

Pfizer Oilfield Chemicals Division, 1982. *FLOCON 4800 Brochure*.

Phillips, J.C., Miller, J.W., Wernau, W.C., Tate, B.E. and Aurbach, M.H., 1985. A high-pyruvate xanthan for EOR. *Soc. Pet. Eng. J.*, 25(4): 594–602.

Prud'homme, R.K. and Uhl, J.T., 1984. Kinetics of polymer metal ion gelation. *Soc. Pet. Eng. EOR Symp., Tulsa, Okla.*, SPE 12640.

Ryles, R.G., 1983. Chemical stability of water-soluble polymers used in enhanced oil recovery processes. *Soc. Pet. Eng. Chem. Symp.*, SPE 13585.

Savins, J.G., 1969. Non-Newtonian flow through porous media. *Ind. Eng. Chem.*, 61: 18–47.

Seright, R.S., 1983. The effects of mechanical degradation and viscoelastic behavior on injectivity. *Soc. Pet. Eng. J.*, 23(6): 475–485.

Shupe, R.D., 1981. Chemical stability of polyacrylamide polymers. *J. Pet. Technol.*, 33(8): 1513–1529.

Slaton, M., 1983. Micellar–polymer flood in Illinois showing increased oil cut. *Drilling*, 44(2): 13.

Sorbie, K.S., Parker, A. and Clifford, P.J., 1985. Experimental and theoretical study of polymer flow in porous media. *Soc. Pet. Eng., Fall Meet., September 22–25*, SPE 14231.

Taber, J.J. and Martin, F.D., 1983. Technical screening guides for the enhanced recovery of oil. *Soc. Pet. Eng., Fall Meet., San Francisco, Calif., October 5–8*, SPE 12069.

Vossoughi, S., Aslam, S. and Willhite, G.P., 1984. Viscometric measurement of chromium III polyacrylamide gels by Weissenberg rheogoniometer. *Soc. Pet. Eng. EOR Symp., Tulsa, Okla.*, SPE 12639.

Ware, J.W., 1983. Salem Unit Micellar/Polymer Project. *Soc. Pet. Eng., Fall Meet., San Francisco, Calif., October 5–8*, SPE 11985.

Weiss, W.W. and Baldwin, R., 1984. Planning and implementing a large-scale polymer flood. *Soc. Pet. Eng. EOR Symp., Tulsa, Okla.*, SPE 12637.

Wellington, S.L., 1983. Biopolymer viscosity stabilization–polymer degradation and antioxidant use. *Soc. Pet. Eng. J.*, 23(12): 901–912.

Chapter 8

POLYACRYLAMIDES

WEI S. YEN, ANTHONY T. COSCIA and SAMUEL I. KOHEN

INTRODUCTION

Among the various polymers that have been studied for chemical enhanced oil recovery (see Chapter 7), polyacrylamide (PAM) is the most important. PAM is used frequently for mobility control as well as permeability control. PAM (and its modified forms) is the principal polymer used in significant amounts on a commercial basis. Usually, anionic versions of PAM are used rather than the nonionic homopolymer itself. By far, poly(acrylamide–co-acrylic acid) is the most common, and the nomenclature PAM/AA is used here to refer to this group of polymers. PAM is made by the polymerization of acrylamide monomer, whereas PAM/AA is manufactured either by copolymerizing acrylic acid or via hydrolysis of the homopolymer. The structures for acrylamide monomer (AMD), PAM, and PAM/AA are shown in Fig. 8-1.

Acrylamide monomers

Acrylamide is a white, crystalline, water-soluble compound derived from natural gas via acrylonitrile. It was originally manufactured by the reaction of

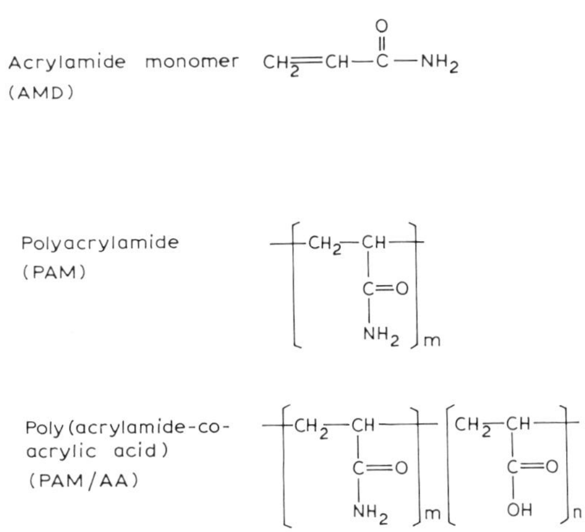

Fig. 8-1. Structures of acrylamide monomer (AMD), polyacrylamide (PAM) and poly(acrylamide–co-acrylic acid) (PAM/AA). The subscripts *m* and *n* represent the number of repeating units.

acrylonitrile with water and sulfuric acid at high temperatures. Current commercial processes for making this monomer involve catalytic hydration of the acrylonitrile over copper-containing catalysts.

The physical properties of acrylamide and its solubilities have been outlined in a number of articles; most recently, by Thomas and Wang (1984). Whereas acrylamide is chemically reactive both at the vinyl double bond at the carboxamide group, its greatest importance is as a vinyl monomer. Free radical polymerization of acrylamide leads to very high-molecular-weight water-soluble polymers, which are used widely in various applications.

Polymerization of acrylamide

Acrylamide may be polymerized via free radical initiation or by an anionic mechanism. The former pathway leads to vinyl polymers and provides the high-molecular-weight homopolymers and copolymers which are of interest in EOR. The most important free radical initiators employed commercially are persulfates, peroxides, redox systems involving a reducing component along with the latter two types of catalysts, azo compounds, UV light, and ionizing radiation. So long as chain transfer is kept under careful control, e.g., by avoiding high levels of reducing agent in a redox couple, and the temperature is maintained within certain limited ranges, these catalyst systems can provide equivalent results in terms of overall kinetics and molecular weights. Molecular weight distributions, however, may vary depending on the catalyst system used.

Acrylamide is unique in that its overall rate of polymerization in water is very high. The rate of polymerization is proportional to the propagation rate over the half power of the transfer rate $k_p/(k_t)^{1/2}$. Reported measurement of this parameter is about 5 (Suen et al., 1958). It is one of the highest such ratios for vinyl monomers. Thus, kinetic chain lengths are very large and extraordinarily high-molecular-weight polymers can be obtained. This is mostly due to the fact that chain transfer to monomer, polymer, and solvent (water) are low. Molecular weights of $5–15 \times 10^6$ daltons are readily achievable even in commercial production, which are usually driven to high conversions (see Scanly, 1968).

POLYACRYLAMIDE CHEMISTRY

General properties

Polyacrylamide is a water-soluble polymer which undergoes reactions typical of the amide group (Morris and Penzenstadler, 1978). Its importance lies in the solution properties, rheology, and ability to hydrogen bond to various surfaces. For the latter reason, it is widely used commercially as a flocculant. The polymer in the solid state is brittle and not a good film former because of its relatively high glass transition temperature ($> 150\,°C$). It has not found many industrial applications as a bulk material. A thorough review of PAM properties in solution has been reported by Kulicke et al. (1982).

The stability of PAM in solutions is affected by pH. At very low or high pH, some of the amide groups are converted to carboxylates via hydrolysis. The resulting polyelectrolyte expands in water, increasing the solution viscosity. Even without this hydrolysis, some viscosity changes have been observed in PAM solutions freshly prepared in water and aged afterwards. Kulicke and Klein (1978) presented evidence to show that initial solution viscosity losses result from changes in intramolecular conformations. These rearrangements, however, are not observed by all workers (see Machtle (1982), for example). It may also be related to the manner in which PAM solutions are prepared, either from freshly synthesized polymer or from commercial products which have different process histories.

Polyacrylamides are soluble in a few very polar organic solvents such as ethylene glycol, formamide, and certain acids. These are all inferior to water as solvents. Viscosity losses in water on aging, as discussed above, are not observed in ethylene glycol or formamide (Kulicke and Klein, 1978), presumably because of the weaker hydrogen bonding of polyacrylamide with these solvents.

Methods of manufacture

While acrylamide may be polymerized or copolymerized in water, in organic solvents, or in the solid state, commercial processes for making PAMs for EOR utilize aqueous solutions or dispersions. In water, the catalysts mentioned are used for solutions ranging in concentration from 5 to 60% by weight of monomer. The product may be left as an aqueous gel and sold as such, e.g., gel logs, if the polymer solids levels are high enough (> 20%). More commonly, the polymers are dried in drums, belts, or fluidized beds. Afterwards, the polymers are milled into powder to lower the mean particle size. The smaller the particle, the quicker it is to achieve complete dissolution.

The polymerization of acrylamide in inverse emulsions (water-in-oil) is also practiced commercially. In this method, acrylamide is initially dissolved in a high-solids aqueous solution and polymerized in a two-phase system with an inert high-boiling-point hydrocarbon as the continuous phase. Relatively low-HLB (hydrophilic lipophilic balance) surfactants are often used to stabilize the micron-size particles. After polymerization, a higher-HLB surfactant is usually added as a breaker, so that the emulsion may readily invert and the polymer particles dissolve when subsequently mixed with a larger quantity of water (Anderson and Frisque, 1971). The average particle size of an emulsion droplet usually ranges from 1 to 10 μm, which is considerably smaller than that of a comparable dry powder.

Another type of inverse emulsion, microemulsion of PAM/AA, has also been proposed for use in EOR (Candau et al., 1985, 1986). These emulsions have a droplet size of about 50 nm. Because of the much larger quantities of emulsifiers needed, however, microemulsion is not a practical system.

Suspension and precipitation polymerizations mentioned in the patent and technical literature are convenient laboratory techniques but are of lesser commercial importance.

Often in producing a high-molecular-weight PAM, very high conversion of the monomer (> 99%) is targeted in order to keep the residual monomer in the system to a minimum. This can lead to a broad molecular weight distribution. Also, some crosslinking of the polymers can occur during the drying process, resulting in the formation of larger insoluble structures sometimes referred to as microgels. Microgels are undesirable in EOR because they tend to plug the formation.

Molecular weight distribution

Inasmuch as polyacrylamide is used mainly to reduce the mobility of injection water during an enhanced recovery process, one must consider those polymer characteristics that are rheologically important. Of the many polymer attributes that one considers, rheology is most sensitive to the average molecular weight (MW) and molecular weight distribution (MWD). The MW is an estimate of the mass of the polymer, usually expressed in units of dalton, whereas MWD represents the mass variations found in a large sample. A typical MWD for PAM obtained using gel permeation chromatography (GPC), and showing effects of solution aging, is shown in Fig. 8-2 (Kulicke et al., 1982). The abscissa of this MWD is V_e, the elution volume, which can be converted into daltons using a proper calibration (see Klein and Westercamp, 1981).

In good solvents, the molecular dimensions of high-molecular-weight PAM/AAs can reach fractions of a micrometer, which is close to the average pore size in porous media. The passage of a small number of these very large polymers, therefore, can have a disproportionate effect on flow.

There are many existing methods for molecular weight determination, but for PAM/AA most of them are tedious and difficult to implement on a routine

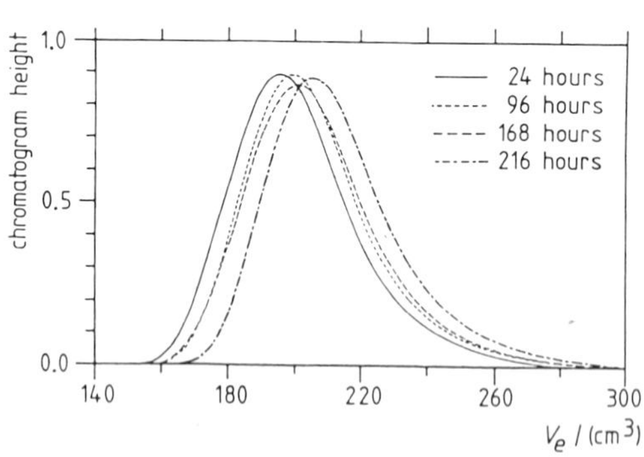

Fig. 8-2. GPC chromatogram showing elution profiles of PAM as a function of aging. (After Kulicke et al., 1982, p. 420; courtesy of *Progress in Polymer Science,* Pergamon Press.)

basis. The average molecular weight of a high-molecular-weight polymer can be estimated via light scattering, sedimentation, and viscometry, providing that certain power law relationships have been previously established (Klein and Conrad, 1978). The corresponding molecular weight distribution is much more difficult to determine, especially for molecular weight above 5×10^6 daltons. Aqueous size exclusion chromatography (SEC), dynamic light scattering, sedimentation, and hydrodynamic chromatography have all been attempted. Some of the techniques require labelling polymers with a fluorescent dye to enhance detection. Overall, the results appear mixed and seem to have limited value in routine quality control operations.

Despite a good deal of industry and university effort, there have been few successful measurements of MWD of high-molecular-weight PAM/AAs reported. Seright et al. (1981) used sedimentation on labeled PAM/AA to study the effect of polymer degradation. Aqueous SEC using controlled pore glass, together with specially synthesized standards, allows for some separation of homopolymer PAM, but only with molecular weight of up to 5×10^6 daltons.

Most commercial products are post-treated with additives for longer shelf life or dried at elevated temperatures. These treatments are known to modify the original distribution.

Argabright et al. (1982) studied the MWD of PAM/AA using sedimentation. The polymers they used were prepared by a commercial polymerization procedure under conditions reported to provide a very narrow molecular weight distribution. Photon correlation spectroscopy (PCS) is a relatively rapid measurement technique. This method, however, requires a great deal of care in sample preparation, and lengthy mathematical manipulation of the original data, to arrive at MWD (DiNapoli et al., 1982). The use of this method has not been widespread.

Hydrodynamic chromatography (HDC), originally designed to classify latex particles, has been adopted to obtain MWD for PAM/AA. The combination of HDC and low-angle light scattering has made it possible to determine MWD of polymers having average MW of up to 9×10^9 daltons (Martin et al., 1986).

Copolymers of acrylamide

Early reports by Sandiford and Keller (1958) and McKennon (1962) revealed that anionic PAMs were preferred as EOR polymers because of the improved mobility reduction and lower adsorption onto reservoir rocks. Martin and Sherwood (1975) developed data on the correlation between resistance factors in brines with degree of hydrolysis of PAMs, particularly in the 5–35 mole% carboxyl range. They also discussed the pros and cons of different carboxyl levels. According to them, resistance factors in unconsolidated sandpacks were improved with hydrolysis. Electrolyte effects were also investigated.

The polymers used in these studies were generally prepared by hydrolysis of homopolymers with alkaline reagents. Current commercial processes for making such polymers still involve post-treatments with caustic, although simultaneous polymerization–hydrolysis procedures have been reported (Proffitt, 1962). Carboxyl level heterogeneity in such anionic PAM/AA can occur in two ways: (1) by neighboring group effects which affect single chains, or (2) by higher degrees of hydrolysis of some chains than in others, possibly due to localized over-concentrations of base. Both effects have been observed. Nuclear magnetic resonance (NMR) is an analytical tool which can be used to determine sequences of carboxyl and amide groups in PAM/AA. Some triad distributions have been reported (Halverson et al., 1985). To date, no correlations have been made between these structural effects and performances in oil recovery experiments.

PAM/AA is also conveniently made by copolymerization of acrylamide and acrylic acid. These two monomers have similar electron densities. Therefore, reasonably homogeneous alternating structures can be made. Also, essentially the same average molecular weight can be achieved by copolymerization as in the homopolymerization of acrylamide. Care must be taken to adjust the pH properly, because the acrylate ion is much less reactive with acrylamide than it is with acrylic acid. At pH 3.3, the two monomers are equally reactive. One can obtain a 1 : 1 mole ratio random copolymer from a 1 : 1 mole ratio monomer feed. Bolton et al. (1980) reported some advantages in using anionic PAM/AA made by copolymerization in tests carried out on sandpacks and native cores from the Richfield East Dome Unit (REDU) in California. Lower retention and slightly improved oil recoveries were cited. Martin et al. (1983) have studied the viscosity retention in brine for hydrolyzed PAM/AA versus copolymerized PAM/AA and found little difference in the two. They also claimed shear resistance of the hydrolyzed PAM/AA was superior to the copolymers.

Other monomers have been used to copolymerize with acrylamide. Szabo (1979a, b) and McCormick et al. (1982) focused on sulfonated monomers such as 2-acrylamido-2-methylpropane sulfonic acid (AMPS). Molecular weights obtained for such copolymers were not as high as the acrylic acid counterparts. The AMPS copolymers did provide a somewhat improved calcium ion tolerance.

Modified polyacrylamides

A number of substituted and modified polyacrylamides (e.g., graft structures) have been reported. Norton and Falk (1973) and Martin (1976a) disclosed reactions of PAM/AA with formaldehyde to produce methylol derivatives, which have higher viscosities than the parent polymers. Martin (1976b) took this reaction a step further by aminomethylation using the well-known Mannich reaction sequence. An anionic version can be produced by a second reaction with sulfites after the methylolation step. All these products, however, are in equilibrium with their reactants in aqueous solution. The dilute conditions in a reservoir would lead to polymer instability. Probably as a result of this, such PAM derivatives have not found commercial application in EOR.

Platt (1981) prepared poly(alpha-alkoxy)acrylamides and claimed improved stability in brine solutions. Similar claims were made by Hunter (1981) who prepared N-substituted PAM/AA via ethoxylation. Martin et al. (1982) studied the synthesis of polyacrylamide containing imide rings and concluded that they are less susceptible to alkaline hydrolysis than PAM.

Cottrell et al. (1982) have made copolymers of AMD and N,N-dimethylacrylamide, followed by hydrolysis. These polymers appear to have improved brine and calcium tolerance. Murduchowitz (1982) disclosed terpolymers of acrylamide, acrylonitrile and acrylic acid, the latter optionally alkoxylated with ethylene oxide.

Graft copolymers of acrylamide, using polysaccharides as backbones, have been studied by a number of workers. McCormick (1981) synthesized and carefully characterized dextrans grafted with acrylamide. More recently, McCormick (1985) copolymerized acrylamide with diacetone acrylamide, and claimed an increase in viscosity in the presence of added salt. Starch has also been used as a substrate for AMD grafting by Pledger et al. (1981) and Meister (1981). Because very little performance data is available on these systems, it is very difficult to assess their relative merits in EOR.

To summarize, substituted PAMs appear to provide only marginal improvements over PAM/AA. Since a penalty is involved in terms of higher raw material costs or reduced stability, these polymer systems have yet to challenge PAM/AA commercially.

Thermal and oxidative stability

In discussing thermal stabilities of PAMs, distinctions must be drawn between solution and bulk properties, aerobic and anaerobic conditions, temperatures of storage, and polymer compositions. In the dilute state and in simulated anaerobic test conditions, Ryles (1983, 1985) studied the thermal stability of various polyacrylamides. His results indicated that high temperature ($> 50°C$) leads to increased hydrolysis of these polymers, with viscosity increasing in most cases. Ryles also discussed the data of previous workers in this area and showed the need to rigorously exclude oxygen during laboratory evaluation.

When oxygen is present, even at ppb levels, degradation of homo- and copolymer of acrylamide occurs via a radical abstraction mechanism followed by disproportionation. This oxygen sensitivity was recognized early in the history of polymer flooding by Pye (1966, 1968). Later, the use of stabilizing additives was reported by Knight (1973). Shupe (1981) elaborated by considering test procedures, material, and construction in the field, as well as the effects of biocides on polymer degradation.

The need to protect acrylamide polymers aboveground is well known. Procedures for protecting the products from the deleterious effects of heavy metals and oxygen are usually provided by chemical suppliers and service companies.

Resistance to biodegradation

Two main concerns with PAM/AA in EOR are possible biodegradation during application in the field, and the fate of residual polymer after injection. Often, biocides are used to retard bacterial action. Some typical oilfield biocides include aldehydes such as formaldehydes, paraformaldehydes, glutaraldehydes, and acrolein. Acrolein and glutaraldehyde may cause substantial polymer degradation in solutions containing oxygen. Formaldehyde may serve as a stabilizer for PAM solutions when oxygen is present. Other biocides include chlorinated phenols such as pentachlorophenoate and trichlorophenoate, both of which are compatible with PAM/AA. A good review in this area is provided by Shupe (1981).

Stability of polymer underground, where only anaerobic organisms are expected to flourish, is more difficult to establish because of the difficulty in establishing material balance. Grula and Sewell (1982) suggested that PAM/AA stimulated growth of *Desulfovibrio* sulfate-reducing bacteria in reservoirs. He showed that these bacteria could utilize nitrogen in laboratory experiments. Martin et al. (1982) also studied microbial growth in PAM/AA solutions, but the mechanism of growth was not clear. It is doubtful any enzymatic attack on the carbon–carbon backbone in PAM/AA is a source of concern, because no such carbon chain scission by bacteria is known. Oxygen would have to be involved in some manner in order to degrade the vinyl backbone.

APPLICATION OF PAM/AA IN ENHANCED OIL RECOVERY

PAM/AA is used in two principal areas in EOR: mobility control and profile modification. In this chapter, the emphasis is on mobility control. In this application, a small amount (>250 mg/l) of high-molecular-weight PAM/AA (MW $> 2\times 10^6$ daltons) is added to the injection water to increase the viscosity and to improve the mobility ratio, which is defined as follows:

$$M = \frac{\lambda_w}{\lambda_o} = \frac{k_w/\mu_w}{k_o/\mu_o} \tag{8-1}$$

where λ is the mobility of a fluid; k is the permeability of the medium calculated by using Darcy's law (equation 8-2); μ is the fluid viscosity; and subscripts "w" and "o" represent the injection fluid and the oil phase, respectively.

$$k = \frac{14.7 q\mu\, \Delta x}{A\, \Delta p} \tag{8-2}$$

where q = flow rate (ml/s); μ = viscosity (cP); Δx = length (cm); A = cross-sectional area (cm^2); Δp = pressure drop (psi); and k = permeability (darcy).

One can improve the overall sweep efficiency by making M closer to 1. In practice, the objective is to reduce the mobility of the injection water such that it does not bypass the oil phase during flooding.

When a polymer solution is injected into an oil-bearing formation, large polymers such as PAM/AA have a tendency to adsorb onto the rock surfaces. This is one mechanism of polymer retention. Adsorption effectively narrows passages in the porous media. Thus, the end result is a reduction in the permeability to the displacement fluid. According to equation (8-1) this serves to improve the mobility ratio. The retention property is distinctive of PAM/AA. It represents an advantage of PAM/AA over the other major EOR polymer, i.e., xanthan gum.

Polymer flooding is ineffective in reservoirs having severe permeability heterogeneities. Fractures or thief zones act as conduits for the injection fluid. In these situations, permeability adjustments aiming to alleviate gross heterogeneities are often performed prior to polymer flooding. PAM is often used to accomplish this task. In more severe cases where blockage of certain zones is required, gelling of the fluid after injection by means of in-situ gelation techniques can be used.

In designing an EOR process, the following should be obtained beforehand in the laboratory: (1) porous media flow properties, (2) interactions between the polymer and the candidate core sample obtained on-site, (3) influence of dissolved ions, (4) hydrodynamic degradation, and (5) aging characteristics of the solution under reservoir conditions.

These are all important parameters to be considered when assessing the overall soundness of a project. Their relative importance and the extent to which they influence each other, however, depends on the specific locale and the polymers available.

Solution preparation

The rheological properties of a PAM/AA solution is very sensitive to the way it is prepared. Different preparation schemes can often result in solutions having widely different flow properties, especially when shear is introduced in the dissolution process. Shearing a polymer solution usually lowers the mean molecular weight and shifts the molecular weight distribution.

Whether in the laboratory or in the field, the preferred dissolution method will depend on the particular product chosen (e.g., gel log, liquid, or dry powder) and the prevailing conditions. Polymer concentration, average particular size, temperature, brine composition, and brine concentration are all important factors. The rate of dissolution of PAM and any high-molecular-weight polymer is usually slow, especially in poor solvents or brines. Liquid polymers, other than emulsions, are easier to dissolve. The transportation cost, however, tends to be high because of the low polymer concentration (5% or less). Polymer gels are similar to liquid polymers except the polymer contents are usually higher, and they are not readily miscible with water. They are available usually in the form of logs and are extremely viscous. Special equipment is needed to properly dissolve them on-site. Usually this includes some device to chop the gel log into smaller granules before dilution water is introduced. This type of dissolution equipment is particularly necessary if one is to polymerize on site (Witheford et al., 1986).

The most common liquid emulsion for EOR is the water-in-oil inverse emulsion. It consists of micron-size polymer particles suspended in an oil phase, and stabilized by surfactants. Upon mixing the emulsion with dilution water, an inversion takes place whereby the polymer solution becomes the continuous phase and the original continuous phase is emulsified as oil droplets.

The advantage of an emulsion is its relatively high polymer solids level (up to 50%), ease of handling, and small particle size. The latter facilitates rapid dissolution. The disadvantage obviously is the higher cost of raw material and transportation. Microemulsions were proposed by some investigators (Candau et al., 1986). Because of the larger surface area, however, a much larger quantity of emulsifiers is needed to stabilize the system.

The dissolution of PAM/AA can be viewed as including at least the following stages:
(1) initial hydration of the individual granules from the diffusion of solvent into the polymer networks, which swells the particles; (2) fragmentation of the swollen particles by shear; and (3) diffusion of the outer polymer layer into the solvent.

Vigorous mechanical agitation produces large hydrodynamic stresses that cut down the average particle size, but at the same time introduces more surface area for better solvent–polymer contact. It may result, however, in severe polymer degradation. For dry PAM/AA, containing a limited amount of insoluble crosslinked species, a standing time of several hours with agitation is necessary for the solution to become homogeneous. Again, the optimum dissolution time depends on the polymer concentration, molecular weight, brine concentration, and temperature.

The existence of microgels as aggregates of insoluble crosslinked polymers have been postulated by Burcik (1969) and Burcik and Thakur (1974). Since microgels are prime suspects in formation plugging, their removal from the polymer solution before injection is essential. Techniques for the detection and removal of microgels have been developed, but mainly for biopolymers such as xanthan gum. These techniques are both tedious and difficult, often requiring the slow filtration of polymer solutions through membranes (Yost and Stokke, 1975; Kohler and Chauveteau, 1981); hence, they are not suitable for use on a large-scale basis in the oilfield. Solutions having a large amount of insolubles are undesirable as they often exhibit a progressive plugging behavior that severely reduces the rate of polymer injection.

Solution properties

In polymer flooding, the rheological properties of the injection fluid are greatly influenced by the equilibrium properties of the dissolved polymer. The viscosity of a solution is directly related to the average size of the polymer. It depends on the degree of polymerization, level of substitution (e.g., AA content), dissolved ion concentration, pH, and temperature. The larger the molecule, the more effective it is in increasing the solution viscosity.

Effect of polymer concentration

One measure of the average polymer size in a given solvent is the intrinsic viscosity or limiting viscosity number, $[\eta]$, which is usually expressed as a ratio of volume over weight with units in dl/g. The intrinsic viscosity is defined as follows:

$$[\eta] \equiv \lim_{C \to 0} \frac{\mu_{sp}}{C} \tag{8-3}$$

where $\mu_{sp} = (\mu_{solution} - \mu_{solvent})/\mu_{solvent}$, and C = polymer concentration. At equilibrium and polymer concentration below the overlap concentration (C^*), i.e., $C < C^*$ for $C^*\ [\eta] \sim 1$, polymers exist in solution as individual coils, and the intermolecular interactions between them is minimal. The viscosity enhancement in the dilute region originates from the friction between the individual polymer chains and the surrounding solvent. The solution viscosity can be approximated by the following relation:

$$\mu \approx \mu_{solvent}(1 + C[\eta]) \tag{8-4}$$

For polymer concentrations greater than the overlap concentration ($C > C^*$), the above expression becomes more complex and terms of higher order in C must be included. In the so-called semi-dilute region, polymer–polymer interaction begins to dominate and entanglement effects appear. In this regime, solution viscosity is non-linear with respect to polymer concentration. As a result, solution viscosity can increase dramatically with increasing polymer concentration. For most PAM/AA used in EOR, the overlap concentration, C^*, ranges from 300 to 800 ppm, depending on the molecular weight and water composition. Because of the extraordinary increases in solution viscosity from polymer–polymer interactions, polymer concentrations in mobility control applications are usually above C^*. In practice, the engineer first has to decide upon a targeted solution viscosity based on a desired mobility ratio. He or she then obtains a sample of the dilution water, either fresh or produced, from a candidate reservoir and measures solution viscosities for a series of polymer concentrations. The concentration that yields the most desirable viscosity is chosen.

Polyelectrolyte effect

PAM/AA exists as a polyelectrolyte in aqueous solutions. The negatively charged AAs on the PAM/AA backbone repel each other because of Coulombic interactions. This intra-chain repulsion causes the polymer chain to expand. In a solution where free counter-ions are available, the polyelectrolyte expansion is curtailed to a degree by the mobile counter-ions. These ions gather around the negatively-charged groups, and shield them from each other. The net result is a lowering of the effective charge of the polymer and a lower viscosity. In general, the higher the salt concentration, the lower the solution viscosity.

Divalent counter-ions such as Ca^{2+} and Mg^{2+} are more effective in charge neutralization than monovalent species such as Na^+ and K^+ (Ward and Martin, 1981). This interaction is also stronger for polymers with higher charge densities (Kowblansky and Zema, 1981a, b). The typical reservoir brine consists of a mixture of ions, such as Na^+, Ca^+, Mg^{2+}, Cl^-, SO_4^{2-} and CO_3^{2-}. The solution viscosity of PAM/AA in brine is dominated by the divalent ion fraction, particularly Ca^{2+} (Ward and Martin, 1981). Precipitation of PAM/AA by Ca^{2+} can occur if the degree of hydrolysis is high. The critical degree of hydrolysis for PAM/AA in the presence of Ca^{2+} has been reported by Kaczmar (1980) and Schwartz and François (1981) to be between 30 and 40%. The critical degree of hydrolysis, regardless of Ca^{2+} concentration, has been estimated by Zaitoun and Potie (1983) to be 35% at 30°C and 33% at 80°C. Redissolution of the precipitated polymer may occur if additional $CaCl_2$ (around 1 M $CaCl_2$) is introduced.

FACTORS AFFECTING FLOW IN POROUS MEDIA

Rheology *

Reservoir rock is a complex mixture of rock and mineral particles cemented together into a three-dimensional network. The oil to be recovered is situated in the interstitial space. The flow behavior of a high-molecular-weight polymer solution through porous media is necessarily complex. This is further complicated by interfering factors such as polymer retention and degradation. Retention reduces the permeability of the medium, whereas degradation lowers the solution viscosity. In addition, the permeability variation within an oil-bearing formation can be rather pronounced. This factor alone can make the design of a polymer flood difficult.

Although the porous media flow behavior of Newtonian fluids can be described adequately by Darcy's law (equation 8-2) this approach (i.e., the use of constant apparent viscosity and permeability in the analysis) is not suitable for non-Newtonian solutions, where viscosity is normally a function of shear rate. For dilute PAM solutions, the viscosity is independent of shear rate only at very low shear rates (Fig. 8-3). When a polymer solution is injected into the formation, it flows in a radial pattern. The exact velocity profile and hydrodynamic stresses at each point of the flow are difficult to obtain because of the complexity

* For a better understanding of rheology, the reader is referred to the book by G.V. Chilingarian and P. Vorabutr (1981).

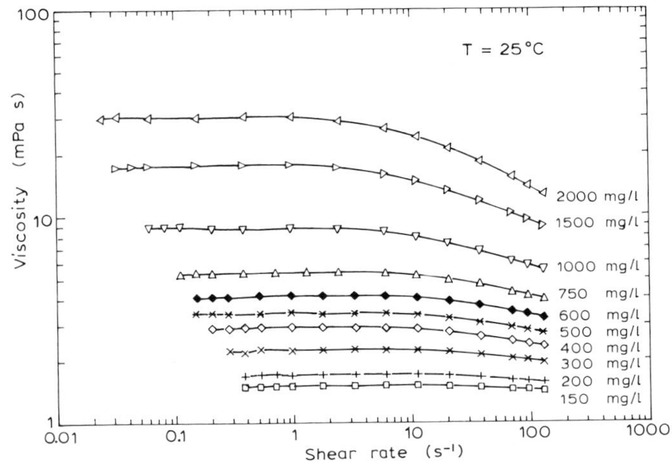

Fig. 8-3. Viscosity of PAM/AA in 3% NaCl as a function of concentration and shear rate. (After Martin et al., 1982, p. 71; courtesy of the U.S. Department of Energy.)

of the porous structure. In order to characterize the flow, simplifying assumptions on the pore geometry are often introduced. For example, a porous medium may be characterized by permeability (k), porosity (ϕ), and a constant (α) dependent on the type of medium (Chauveteau, 1982). Also, the flow can be assumed to have a frontal velocity, which is a function of $1/r$, where r is the radial distance away from the injection well.

The maximum hydrodynamic stress occurs near the injection well, where the flow rate is highest, but diminishes rapidly as the fluid propagates further into the formation. In this chapter, the term shear rate is used in an elementary manner to refer to the complex flow field experienced by the fluid. Shear rates corresponding to actual conditions can be estimated using the capillary bundle model (Chauveteau, 1982). It is of the order of 10^4 s^{-1} at the wellhead, and 1 s^{-1} when the fluid is further away.

Polyacrylamide is a flexible molecule that exhibits a coil-like conformation in solution. In porous media flow, the fluid experiences a combination of both shear and elongational flow fields. The resultant flow properties depend on the exact pore geometry. The individual contributions of these flow fields, however, are difficult to assess. In steady shear experiments using a rotational viscometer, the PAM/AA solutions exhibit pseudoplastic (shear thinning) behavior, i.e., the apparent viscosity decreases as a function of increasing shear rate. A large number of empirical relationships have been developed to describe the viscosity and shear rate relationships for this type of flow (Bird et al., 1977a). There are many manufacturers that produce viscometers suitable for EOR work. Contraves™, Haake™, or Brookfield™ viscometers can be found in most EOR laboratories, with the latter being the easiest to operate. It does not operate in the lowest shear rate range ($1\ s^{-1}$), however; thus it is difficult to obtain the low shear viscosities necessary for the most demanding polymer characterization.

The relationship between fluid velocity and pressure gradient in porous media is more complicated than can be described by Darcy's equation. Attempts to satisfactorily describe polymer flow behavior often involve the coupling of a structural model for the porous media and a rheological model for the polymer solution. One such approach involves the use of the Blake-Kozeny-Carman equation for the porous media and a power law model for the fluid (Wang et al., 1979). However, this does not explain the additional pressure drop usually observed for PAM. This dilatant (shear thickening) behavior in porous media causes the apparent viscosity to increase rapidly as a function of fluid velocity. The anomalous frictional phenomenon was first discovered by Pye (1964). He called it the "resistance factor". The resistance factor (RF) of a polymer solution is defined as the ratio of the mobility of the water to the mobility of the polymer solution:

$$RF = \frac{\lambda_w}{\lambda_p} = \frac{k_w/\mu_w}{k_p/\mu_p} \tag{8-5}$$

where λ is the mobility of the fluid, k is the permeability of the medium, and μ is the apparent viscosity. Resistance factor is usually both a function of polymer concentration and flow rate (shear rate).

Resistance factors for a solution can be obtained only by conducting laboratory core flow studies at several flow rates, which is a rather tedious and costly procedure. It is possible, however, to obtain preliminary information on viscoelasticity by means of a screen viscometer. The screen viscometer was first described by Jennings et al. (1971). A diagram of this very useful device is shown in Fig. 8-4. The device is used to measure the screen factor (SF) defined as follows:

$$SF = \frac{t_p}{t_w} \tag{8-6}$$

where t_p and t_w are the respective flow times in seconds for a fixed volume of a polymer solution and water to pass through the screen viscometer. The screen factor of a particular polymer solution has been shown to correlate well with its resistance factor when tests were carried out under similar conditions. The screen viscometer is one of the major polymer characterization tools used. It is convenient to operate either in the laboratory or on-site. A screen factor measurement is important in assessing the quality of a polymer solution. If the screen factor of a solution is high, there may be problems with actual injection. Ideally, one would want a solution to have high viscosity and low screen factor for good mobility control and ease of injection. Recently, the correlation of screen factor and polymer viscoelastic transitions has been established (Lim et al., 1986). The writers found that conventional screen viscometers operate in the

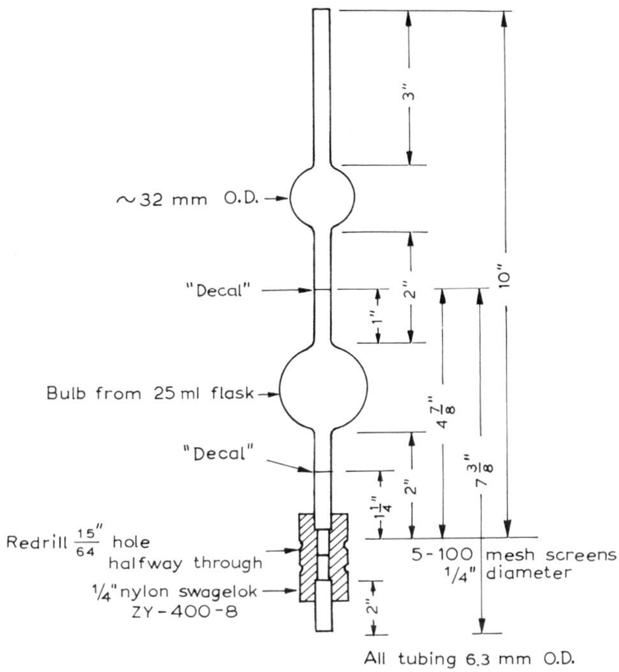

Fig. 8-4. Diagram of a typical screen viscometer. (After Foshee et al., 1976.)

regime where the polymers are in the highly stretched state; therefore, they mimic injection behavior.

The shear rate in porous media can be estimated by the equivalent capillary bundle model (Jennings et al., 1971; Chauveteau, 1982):

$$\dot{\gamma} = \alpha \frac{\overline{V}}{[(k/\phi)/2]^{1/2}} \tag{8-7}$$

where $\overline{V}$ is the average frontal advance velocity, k is the permeability, ϕ is the porosity and α is a constant factor relating to the type of media. The capillary bundle model is constructed from a collection of parallel thin capillaries which collectively mimic the resistance to flow of a porous medium. The model is valid only for Newtonian fluids. A major limitation of this model is the neglect of elongational effects associated with porous media flow (Duda et al., 1983).

Additional increases in flow resistance may be attributed to: (1) polymer adsorption onto the rock surface, which reduces the average pore size and the permeability; and (2) blocking of pore throats by large molecules (retention). The former is thermodynamic in origin, whereas the latter is a result of flow. The subject of polymer retention is discussed in more detail in a later section.

Viscoelastic effects

The flow behavior of dilute PAM solutions in porous media has been studied extensively (Jennings et al., 1971; Durst et al., 1981; Haas et al., 1981). Depending on the type of porous media used, there appears to be a corresponding critical fluid velocity above which the apparent viscosity of the polymer solution increases sharply. This critical onset velocity is inversely related to the fluid relaxation time, τ, which is a function of polymer molecular weight, solvent composition, and polymer concentration. Similar to the treatment of shear flow, one can introduce the term elongational rate, $\dot{\epsilon}$ (Maerker, 1975):

$$\dot{\epsilon} = \beta \frac{\overline{V}}{d} \tag{8-8}$$

where $\overline{V}$ is the average frontal advance velocity, β is a proportionality constant for a given porous media, and d is the average grain diameter of the media. The diameter d is related to k and ϕ by the Blake-Kozeny-Carman equation (Bird et al., 1977a):

$$d = \frac{1-\phi}{\phi}(180k/\phi)^{1/2} \tag{8-9}$$

The elongational rate $\dot{\epsilon}$ is used to describe the rate of deformation in the direction of the fluid velocity gradient. It is related to the diagonal terms in the rate of deformation tensor. The fluid relaxation time, τ, is an indication of how fast a polymer molecule responds in a given flow field. It can be estimated from several theoretical models available (Bird et al., 1977a).

Work on flow of dilute PAM solutions through glass bead packs suggests that the "Finite Extensible Non-linear Elastic" (FENE) dumbell model provides good correlations to experimental data (Durst et al., 1981). In this model, τ is defined as:

$$\tau = \frac{[\eta]_0 \mu_s MW}{RT} \tag{8-10}$$

where $[\eta]_0$ is the intrinsic viscosity extrapolated to zero shear rate, μ_s is the solvent viscosity, MW is the molecular weight of the polymer, R is the gas constant, and T is the absolute temperature. Results showing effects of polymer concentration in Durst et al.'s apparatus is shown in Fig. 8-5. The onset of dilatant behavior was found to occur when the ratio of fluid relaxation time and the characteristic flow time is of the order of unity. This ratio is termed the Deborah number (D_e) (Bird et al., 1977a):

$$D_e = \tau \dot{\epsilon} \tag{8-11}$$

A Deborah number of 0.5 for the onset of dilatancy provides the best fit to data obtained from flow through glass bead packs (Durst et al., 1981), and also for Nuclepore filters (Chauveteau, 1981). The exact mechanism behind such dilatant

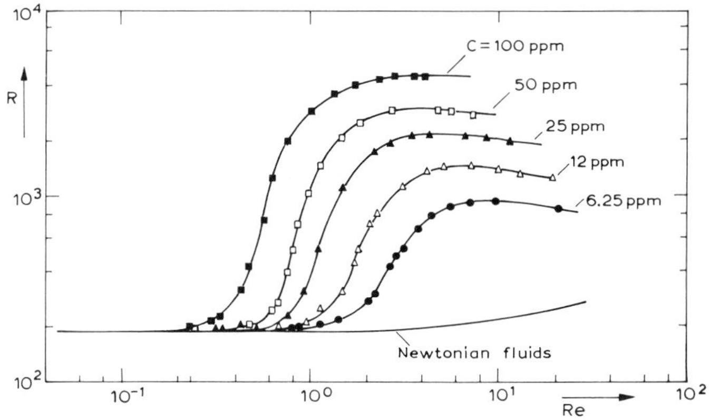

Fig. 8-5. Viscoelastic effects of PAM inflow through glass bead pack. R is the resistance factor and Re is the Reynolds number in the porous media. (After Durst et al., 1981, p. 3135; courtesy of the *Journal of Applied Polymer Science,* John Wiley & Sons.)

behavior is not clear. It was proposed earlier by DeGennes (1974) that increases in viscosity resulted from a polymer coil ↔ stretch transition under elongational flow. In this theory, a critical elongational rate is required to transform a flexible molecule from a random coil conformation to the one that is more stretched out. Viscosity enhancement results from dissipative effects of this transition, as well as the more complete exposure of polymer segments to flow.

Injectivity

In polymer flooding, the soundness of a project is directly related to the rate of incremental oil recovery. The relationship governing flow rate and pressure drop, therefore, warrants serious examination. A quantity that describes the throughput of an injection is called injectivity, I:

$$I = q/\Delta p \qquad (8\text{-}12)$$

where q is the volume flow rate in bbl/day, and Δp is the applied pressure in psi. In general, one would like to have as high a flow rate as possible for a given pressure drop. In practice, the applied pressure should never exceed the fracture pressure of the formation. For PAM/AA solutions, viscoelastic effects play an important role in determining injectivity.

The pressure drop profile of a polymer flood can be divided into two parts: (1) the near wellbore region where a discrete interface exists between the injection fluid and the porous media, and (2) the far-field region where the fluid velocity is significantly lower.

Because of the radial flow pattern, the far-field condition can be met when the distance is only several times the wellbore diameter (Seright, 1983; Sorbie and Roberts, 1983). In the near wellbore region, the sharp pressure drop is a direct result of the large viscoelastic contribution. At large radial distances, because of the low shear rate, the viscosity of the fluid approaches the low shear viscosity, provided that factors contributing to polymer loss such as polymer degradation and retention are insignificant.

Mechanical degradation

High-molecular-weight polymers tend to degrade mechanically (chain scission) when placed in the vicinity of large hydrodynamic stresses. This phenomenon has been well documented for a variety of polymer systems and flow configurations (Abdel-Alim and Hamielec, 1973; Nagashiro and Tsunoda, 1977). Clearly, polymer degradation results in lower fluid viscosity and poorer EOR performance.

In laboratory experiments where the molecular distribution of a polymer was determined before and after degradation, it was observed that: (1) high-molecular-weight fractions of the distribution degrade more readily, shifting the MWD lower; and (2) for a given molecular weight, there exists a critical stress rate below which little degradation is observed.

Similar critical conditions were observed for flow in porous media. In an attempt to quantify the changes, it was found that the percentage loss in screen factor was more severe than the loss in solution viscosity (Jennings et al., 1971). The mode of degradation was postulated to be the result of excessive stretching of the high-molecular-weight species (Seright et al., 1981). The assumption is that at Deborah numbers above some critical onset value, there is shear thickening of the fluid which leads to large shearing–elongational stress. It is this stress that degrades the polymer. More recently, Farinato and Yen (1987) studied polymer concentration effects on degradation, and found that polymer degradation is significantly more pronounced below the overlap concentration.

Using Berea Sandstone cores, different correlations have been established to relate the screen factor loss and functions containing parameters such as fluid velocity, average grain diameter, and length (Seright et al., 1981; Seright, 1983). In linear core flows, Seright (1983) discovered that PAM/AA degradation occurred at a critical flow rate well correlated with the entrance pressure drop. The degree of degradation was also more severe in solutions containing Ca^{2+}. Studies on the effect of degradation in porous media on molecular weight distribution suggested a trend similar to that obtained from other controlled shearing experiments.

Retention

Some polymers are inadvertently retained by the porous media during injection. This loss of polymer, often expressed in units of lb/acre-ft, affects the economics. However, in profile modification or permeability adjustment, where permeabilities of high permeability zones are reduced, polymer retention is often carried out intentionally. Profile modification improves the sweep efficiency of subsequent flooding. In cases where simple polymer adsorption is not enough, in-situ gelation may be considered.

In-situ gelation is a technique used when large thief zones exist. In this technique, polymer solution is first injected into the high-permeability zone and then gelled with a cross-linking agent. This amounts to plugging the cracks before beginning polymer flooding. This generates a barrier to the displacement fluid, thus diverting it to the oil-containing region. By varying the amount of polymer and crosslinkers, the desired gelation time can be achieved. Detailed description of profile modification is beyond the scope of this chapter.

The effectiveness of a profile modification process can be expressed by the residual resistance factor, *RRF*:

$$RRF = \frac{\lambda_{w,\text{before}}}{\lambda_{w,\text{after}}} \tag{8-13}$$

where $\lambda_{w,\text{before}}$ and $\lambda_{w,\text{after}}$ represent water mobilities for a given core before and after polymer treatment. The larger the *RRF*, without plugging, the better the treatment.

Whether or not a polymer molecule gets retained by the porous media depends on the relative dimensions of the polymer and the size of the pores. The average dimension of a PAM/AA polymer is governed by molecular weight and solvent conditions. There is also a finite distribution of pore sizes within any porous formation. When polymer solutions are subjected to flow through a core, the large molecules will be unable to enter the smaller pore throats. As a result, portions of the core will be inaccessible to the polymer. This is appropriately termed the effect of inaccessible pore volume (IPV) (Dawson and Lantz, 1972). In estimating the retention level of a polymer flood, IPV must be subtracted from the total reservoir volume to arrive at a realistic estimate of the volume being swept by the polymer solution.

If possible, polymer retention studies should be carried out on field cores obtained form the candidate reservoir in their native states. The level of retention can be estimated by performing material balance on the polymer before and after flow. If native cores are not available, alternative substrates such as Berea Sandstone could be used. The use of a model substrate such as Berea Sandstone in retention analysis does not yield values corresponding to actual field conditions, but rather provides a basis for comparison when different polymers are being considered, because polymer retention depends significantly on the type of formation (Meister et al., 1980).

Polymer retentions in porous media are believed to consist of three major mechanisms (Dominguez and Willhite, 1977). They are: (1) mechanical entrapment, (2) hydrodynamic retention, and (3) adsorption. The effect of mechanical entrapment is difficult to assess, but is usually considered to be independent of adsorption (Mungan, 1969). Static adsorption, as an approximation to the more desirable flow test, can be carried out on both synthetic and actual field substrates (Smith, 1970). The amount of PAM/AA adsorped on silica was seen to be a function of acrylic acid content. The minimum adsorption for this system occurs at 30% mole substitution (McKennon, 1962). Hydrodynamic retention describes polymers that are trapped behind crevices of the porous media due to an eddying effect, which occurs because of the irregular shapes of the porous passage. When the flow field is relaxed, the hydrodynamically trapped polymers will diffuse out from behind the crevices, and participate in flow when it is resumed.

Adsorped polymers form a depletion layer on the rock surface that effectively excludes additional polymer molecules from penetration. This is analogous to the inaccessible pore volume effect discussed earlier. The net result is a reduction in the average pore size and a simultaneous decrease in permeability. The thickness of this adsorped layer is a function of the radius of gyration of the adsorped polymer (Gramain and Myard, 1981). Depending on the original pore size, the depletion layer effect may or may not be significant.

FIELD CONSIDERATIONS

Injection tests

The injection test is one of the more important steps taken during the design and implementation of an EOR polymer flood. The objectives are:

(1) to develop field design parameters,

(2) to demonstrate that the proposed polymer concentrations can be injected at acceptable rates under field operating conditions, and

(3) to familiarize the production people with the handling, mixing, chemical testing, and injection techniques to be used.

It may be advisable to develop a Hall plot (Hall, 1963) to demonstrate the injectivity of the polymer solution.

Fig. 8-6. Emulsion polymer injection test skid. (Courtesy of the American Cyanamid Co.)

Equipment

Polymer suppliers have available equipment designed to handle their products. Injection tests are usually of a short duration of one to four weeks. Portable test units are desirable. Inasmuch as the tests are short term, the product(s) are generally not delivered in bulk or truckload.

A typical injection test skid for emulsion type PAM/AAs is shown in Fig. 8-6.

SITE FACTORS

Existing facilities

Before implementing polymer applications, it is good practice to examine the existing facilities. Should a waterflood be in place, it is feasible that the polymer injection facilities can be simply added to the pre-existing ones (an add-on). The number of injection wells involved, the mode for transporting water to each well, injection pressure, and volume limits are important factors in the design of equipment.

Injection limits

Pressure limits are usually set by the formation parting pressure. It is normal practice to operate waterfloods below this pressure level. With the introduction of polymer into the water, the rate will be lower at the same pressure. At this point, it is good practice to run a step-rate test if a recent one has not been conducted. This will furnish confirming data for the parting pressure.

Environmental and safety considerations

Environmental effects need to be considered, because the range of ambient temperatures and humidities affect the design of the polymer storage system. For dry products, high humidity conditions must be avoided to prevent the hygroscopic material from balling. The polymer manufacturers should be consulted in all cases.

Most PAM/AAs are considered to be safe to handle and generally fall into a class of compounds that are considered to be non-toxic and non-irritating. Provisions should be made, however, to handle spills should they occur. Containment measurements may be advisable. Usually, the products are scooped up prior to flushing with water. Each manufacturer will supply a material safety data sheet (MSDS) upon request. These will outline precautionary measures as well as acceptable disposable practices.

Another safety feature to consider in the design of equipment is the need for explosion-proof machinery. Offshore operations may require this by law or it may simply constitute a good practice. Should natural gas be used in the field, then explosion-proof equipment is required. The use of this special equipment can increase equipment costs several fold and the need should be clearly demonstrated.

Transportation

Transportation facilities available at the site can influence the design. Whenever possible, bulk type shipments should be used. This applies to either the dry or emulsion grade PAM/AA products. With the gel-log-type products, individual packages must be handled. The storage facilities must be designed for a capacity of about twice the expected delivery volume and provide for at least one week's operating needs. For more severe or remote situations (e.g., offshore and northern Canada) the on-site storage capacity should be extended to accommodate a supply interruption and an extended delivery time.

Well completion

The type of well completion (perforated or openhole) and the tubing coating can affect polymer performance. The type and condition of the piping, valves, and pumps in the system should be examined. These should be considered for replacement or modification to prevent side effects such as shear and/or chemical degradation effects upon PAM/AA solutions. Whenever possible, the use of iron, mild steel, or similar materials should be avoided if oxygen is present in the system. There are several acceptable pipe coatings available or plastic pipes may be used if permitted. Gate, choke, or similar valves should not be used with PAM/AA solutions because of shear effects.

Water quality and treatment

In addition to the mechanical aspects of waterflooding, the quality and treatment of the water is critical to success. If the injection water contains suspended solids or oil, they must be removed before the polymer is mixed. Often, the injection water does require chemical treatment. All chemical treatments should be checked for compatibility with the polymer.

It is important that a chemical oxygen scavenger be added to the system well before any polymer is injected. Typical scavengers such as ammonium bisulfite can be used. Average usage is about 10 ppm NH_4SO_3 to 1 ppm O_2. Care must be taken to prevent air from re-entering the solution after the oxygen scavenger has been introduced. If not, a severe loss of viscosity can result. Chemical treatments using cationic products should be avoided.

Emulsion products

A schematic diagram of an emulsion field mixing system is shown in Fig. 8-7. Typical polymer storage tanks and pumps are shown in Fig. 8-8. The storage tank should be constructed of acceptable materials (stainless steel, coated carbon, or mild steel). Galvanized metal, aluminum, bronze, and natural rubber should be avoided. Teflon^{TM} or other oil-resistant materials should be used for gaskets or seals. The tank should be vented using an inert gas or dry air to prevent moisture accumulation in the air space above the liquid level. Insulation is advisable to protect against temperature variation. Heat tracing is usually recommended, with the PAM/AA product maintained at 70°F.

PAM/AA emulsions are metered into the injection water system on the high-pressure side of the triplex injection pump. Positive displacement or gear pumps are satisfactory for transporting the emulsions. Materials of construction should be oil-resistant and non-corroding. Pumps and lines can be cleaned by flushing with a hydrocarbon solvent such as kerosene.

It is important that a check valve be installed in the polymer feedline near the point of entry into the waterline. Should water contact the emulsion product prematurely, plugging and line rupture could occur from the excessively high viscosity developed.

Immediately past the point of polymer entry, the polymer–water mixture should enter an in-line static mixer. The latter imparts a high shear to the mixture for a very short period of time. The high-energy input causes the water and polymer product to mix vigorously, and simultaneously emulsifies the oil into an internal phase. There are several manufacturers of these in-line mixers.

Sometimes it is necessary to add supplemental surfactants to assist the inversion of the emulsion. The quality of the water (total dissolved solids and hardness) will influence the need for extra surfactant. One method for minimiz-

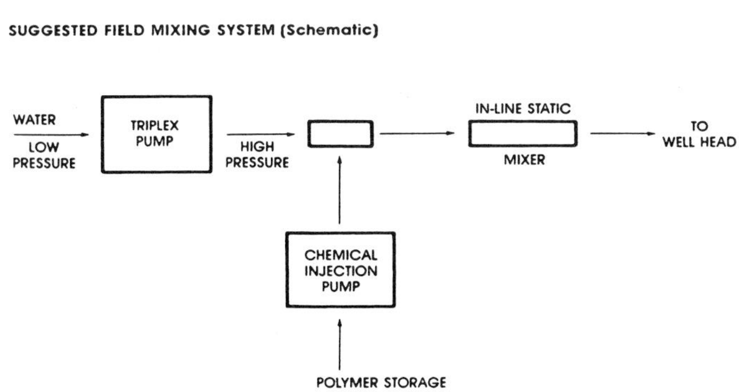

Fig. 8-7. Schematic diagram of an emulsion polymer field mixing system. (Courtesy of the American Cyanamid Co.)

Fig. 8-8. Photograph of an on-site emulsion storage tank and metering pumps. (Courtesy of the American Cyanamid Co.)

ing the usage of the extra surfactant is to split (80/20) the water flow so that a concentrated polymer–water mixture is prepared in one arm with only water in the other. The two flows are recombined and the correct polymer concentration is sent to the injection well.

Water flow to the injection wells can be controlled by:

(1) Adjusting this flow on the high-pressure water side before the emulsion product is introduced.

(2) Controlling the polymer solution flow at each wellhead with the use of coils to reduce the pressure drop without shear degradation of the polymer (Trantham, 1977).

(3) Controlling the polymer solution flow at each wellhead with the use of pinch valves designed to minimize shear effects.

Offshore operations can present a different set of constraints due to space and logistics. Boardman et al. (1982) described a polymer flood installation used in the bay water of southern Louisiana. This reference gives some details regarding the design of two polymer flood installations.

Gel-log polymers

Using the PAM/AA gel-log products in the field normally requires the preparation of a concentrated solution, about 5000 ppm, for metering into the injection water. This is accomplished by extruding the gel log into water where it is cut to form a slurry of particles which pass through 0.02-inch slots. This slurry is agitated and held for 30–60 minutes to assure complete dilution. The gel polymer is usually available in 40-lb logs, or in 850-lb slabs. An 8-inch grinder is capable of handling up to 10,000 lb/day of gel-log polymer. A 16-inch grinder is capable of handling 100 lb/min of slab polymer.

Dry polymer

When using a dry PAM/AA, it is important to use a mixing system that prevents, or reduces, "balling" of the product. Most field mixing systems used today employ a technique that wets the material almost particle by particle. When this is accomplished, retention time can be reduced, although not eliminated. It is also good practice to use a retention tank to assure complete dissolution of the polyacrylamide. Retention times can vary from 30 minutes up to several hours depending upon the nature of the product, water quality, design parameters, etc.

On-site liquid polymer

Another option for the use of polyacrylamide is to perform the polymerization at the field site (DOE, 1979). Specialized dissolution equipment may be required for on-site polymerization (Witheford et al., 1986). This is a relatively new concept, and has not received broad industry acceptance.

CASE HISTORIES

In 1982, chemically enhanced oil recovery was credited with 3769 bbl/day (Slaton, 1983) of production out of a total EOR production of 373,931 bbl/day. An estimate of success ratio on trials to date is clouded by several uncontrollable variables, such as oil prices, raw material costs, governmental actions, etc. Uneconomical trials conducted during the early 1970's have been expanded with commercial success expected. An overview of chemically enhanced oil recovery is given by Gogarty (1983).

On the average, approximately 1.5 lb of PAM/AA (dry basis) is required to recover one barrel of incremental oil. Assuming a price range of U.S.$2.00–2.25/lb of polymer, this infers a chemical cost of about U.S.$3.00/bbl of incremental oil recovered.

Commercial-scale polymer project—North Burbank, Oklahoma

The design of the North Burbank project (Trantham and Moffitt, 1982) is based upon a successful polymer pilot project started in 1970, with tertiary oil production still continuing. In this instance, the oil response to a change in the flooding pattern resulted in a 15,000 bbl oil production not credited to the pilot. PAM/AA consumption was 84,500 lb on a solids basis, credited with the 78,615 bbl of tertiary oil.

After considerable study, a commercial scale project is underway which covers 1440 acres. Estimated total PAM/AA consumption is 3,986,000 lb to recover a predicted 2,700,000 bbl of incremental oil.

The amount of polyacrylamide (solids basis) required for the pilot test was 1.07 lb PAM/AA per bbl of incremental oil. This has been increased to 1.48 lb PAM/AA per bbl of incremental oil in the design of the commercial project.

Alkaline–polymer field project—Isenhour Unit, Wyoming

The Isenhour alkaline–polymer project (Sloat and Zlomke, 1982) covers 173 producing acres with three injectors and six producers. The project was started in September, 1983, and is still continuing (1987). Projected performance shows a recovery of 62,000 bbl of incremental oil. Using the recovery expectation, the usage of PAM/AA is 1.25 lb/bbl of incremental oil.

Micellar–polymer pilot projects—Wilmington Field, California

The Wilmington Field, California, pilot project employed Marathon Oil Company's Maraflood^{TM} process and was designed by them. The project covered 596 acre-ft and included four injectors with six producers. During the design phase of this project, Marathon reported that chemical recovery was 3.68 bbl oil/bbl of slug. Using field performance data extrapolated to the economic limit, the recovery efficiency was predicted to be 2.38 bbl/bbl of slug. A dramatic decline in water-to-oil ratio (WOR) was noted shortly after the start of polymer injection. Polyacrylamide injection commenced in October, 1979, with WOR above 90 and oil production near 60 bbl/day of oil. By June, 1980, the WOR had dropped to 10 and oil production had risen to over 200 bbl/day of oil. By December, 1981, recovery was at 169,000 bbl of incremental oil with ultimate recovery expected to be 176,000 bbl of incremental oil at abandonment. This project recovered 43% of the oil-in-place after waterflooding (Gogarty, 1980). Total polymer usage was 1,134,250 lb as a 30% active emulsion product, which translates into 1.3 lb of polymer per bbl of incremental oil recovered.

REFERENCES

Abdel-Alim, A.H. and Hamielec, A.E., 1973. Shear degradation of water soluble polymers, I. Degradation of polyacrylamide in a high-shear Couette viscometer. *J. Appl. Polym. Sci.*, 17: 3769–3778.

Aguey, O.J., 1982. Fluid performance analyses of micellar polymer pilot flood. *Soc. Pet. Eng. of AIME, Pap.,* 10732.

Anderson, D.R. and Frisque, A.J., 1971. Process for rapidly dissolving water-soluble polymers. *U.S. Patent,* No. 3,624,019.

Argabright, P.A., Rhudy, J.S. and Phillips, B.L., 1982. Partially hydrolyzed polyacrylamide with superior flooding and injection properties. *Soc. Pet. Eng. of AIME, Pap.,* 11208.

Bird, R.B., Armstrong, R.C. and Hassager, O., 1977a. *Dynamics of Polymeric Liquids, Vol. 1. Fluid Mechanics.* John Wiley, New York, N.Y.

Bird, R.B., Hassager, R.C., Armstrong, R.C. and Curtis, C.F., 1977b. *Dynamics of Polymeric Fluids, Vol. 2. Kinetic Theory.* John Wiley, New York, N.Y.

Boardman, R.S., Moore, L.J., Julian, M.H., Bilbrey, D.G. and Morre, S.J., 1982. Design and implementation of four enhanced recovery projects in Bay Fields of South Louisiana, *SPE/DOE 3rd Jt. Symp. on Enhanced Oil Recovery, SPE/DOE Pap.,* 10697.

Bolton, H.P., Carter, W.H., Kamdar, R.S. and Nute, A.J., 1980. Selection of polymers for the control of mobility and permeability variation at Richfield East Dome Unit, Orange County, California. *Soc. Pet. Eng. of AIME, Pap.,* 8893.

Burcik, E.J., 1969. The mechanism of microgel formation in partially hydrolyzed polyacrylamide. *J. Pet. Technol.,* 21: 373–374.

Burcik, E.J. and Thakur, G.C., 1974. Some reactions of microgel in polyacrylamide solutions. *J. Pet. Technol.,* 26: 545–547.

Candau, F., Zekhnini, Z. and Durand, J-P., 1986. Copolymerization of water-soluble monomers in nonionic bicontinuous microemulsions. *J. Colloid Interface Sci.,* 114(2): 398–408.

Candau, F., Leong, Y-S., Kohler, N. and Dawans, F., 1985. Process for manufacturing a microlatex in a continuous oil phase by polymerization of a water-soluble monomer in a water-in-oil microemulsion. Resultant microlatices, and their use for enhanced oil recovery. *U.S. Patent,* No. 4,521,317.

Chauveteau, G., 1981. Molecular interpretation of several different properties of flow of coiled polymer solutions through porous media in enhanced oil recovery conditions. *Soc. Pet. Eng. of AIME, Pap.,* 10060.

Chauveteau, G., 1982. Rod-like polymer solutions flow through fine pores: influence of pore size on rheological behavior. *J. Rheol.,* 26(2): 111–142.

Chilingarian, G.V. and Vorabutr, P., 1983. *Drilling and Drilling Fluids. Developments in Petroleum Science, Vol. 11.* Elsevier, Amsterdam, 802 pp.

Cottrell, I.W., Baird, J.K. and Shim, J.L., 1982. Copolymers of N,N-dimethylacrylamide and acrylamide. *U.S. Patent,* No. 4,326,970.

Dawson, R. and Lantz, R.B., 1972. Inaccessible pore volume in polymer flooding. *Soc. Pet. Eng. J.,* 12: 448–452.

DeGennes, P.G., 1974. Coil-stretch transition of dilute flexible polymers under ultrahigh velocity gradients. *J. Chem. Phys.,* 60: 5030–5037.

Dominguez, J.G. and Wilhite, G.P., 1977. Retention and flow characteristics of polymer solutions in porous media. *Soc. Pet. Eng. J.,* 17: 111–121.

DiNapoli, A., Chu, B. and Cha, C., 1982. Molecular weight distribution of polyacrylamide by photon correlation spectroscopy. *Macromolecules,* 15: 1174–1180.

Duda, J.L., Hong, S-A., Seong, A. and Klaus, E.E., 1983. Flow of polymer solutions in porous media: inadequacy of the capillary model. *Ind. Eng. Chem. Fundam.,* 22(3): 299–305.

Durst, F., Haas, R. and Kaczmar, B.U., 1981. Flow of dilute hydrolyzed polyacrylamide solutions in porous media under various solvent conditions. *J. Appl. Polym. Sci.,* 26: 3125–3149.

Farinato, R.S. and Yen, W.S., 1987. Polymer degradation in porous media flow. *J. Appl. Polym. Sci.,* 33: 2353–2368.

Foshee, W.C., Jennings, R.R. and West, T.J., 1976. Preparation and testing of partially hydrolyzed polyacrylamide solutions. *Soc. Pet. Eng. Annu. Fall Conf., New Orleans, October 3–6,* SPE 6202.

Gogarty, W.B., 1983. Enhanced oil recovery through the use of chemicals—Part I. *J. Pet. Technol.,* 35: 1581–1590.

Gramain, P. and Myard, P., 1981. Elongational deformation by shear flow of flexible polymers absorbed in porous media. *Macromolecules,* 14: 180–184.
Grula, M. and Sewell, G., 1982. Polyacrylamide stimulation of *Desulfovibrio* growth and sulfate reduction. *Dev. Ind. Microbiol.,* 23: 419–424.
Haas, R., Durst, F. and Inerthal, W., 1981. Effects of dilute polymer solutions on porous media flow, II. Theory of molecule-flow interaction and its verification. In: A. Verruijt and F.B.J. Barends (Editors), *Proceedings of Euromech. 143, Delft.* A.A. Balkema, Rotterdam, pp. 163–168.
Hall, N.J., 1963. How to analyze waterflood injection well performance. *World Oil,* 10: 128–130.
Halverson, F., Lancaster, J.E. and O'Connor, M.N., 1985. Sequence distribution of carboxyl groups in hydrolyzed polyacrylamide. *Macromolecules,* 18: 1139–1144.
Harrington, R.E. and Zimm, B.H., 1965. Degradation of polymers by controlled hydrodynamic shear. *J. Phys. Chem.,* 69(1): 161–175.
Hunter, W.D., 1981. Process for secondary recovery, *U.S. Patent,* No. 4,297,226.
Jennings, R.R., Rogers, J.H. and West, T.J., 1971. Factors influencing mobility control by polymer solution. *J. Pet. Technol.,* 23: 391–401.
Kaczmar, B.U., 1980. On the effects of $CaCl_2$ upon hydrolyzed polyacrylamide in solution. *Univ. Karlsruhe, Rep.,* SFB 80/E/174.
Klein, J. and Conrad, K.D., 1978. Molecular weight determination of poly(acrylamide) and poly(acrylamide-co-sodium acrylate). *Macromol. Chem.,* 179: 1635–1638.
Klein, J. and Westerkamp, A., 1981. Peculiarities of polyacrylamide analysis by GPC. *J. Polym. Sci., Polym. Chem. Edit.,* 19: 707–718.
Knight, B.L., 1973. Reservoir stability of polymer solutions. *J. Pet. Technol.,* 5: 618–626.
Kohler, N. and Chauveteau, G., 1981. Xanthan polysaccharide behavior in porous media: preferential use of fermentation broth. *J. Pet. Technol.,* 23: 349–358.
Kowblansky, M. and Zema, P., 1981a. Interactions of sodium ions with the sodium salts of poly(acrylic acid/acrylamide) copolymers of varying charge density. *Macromolecules,* 14(1): 166–170.
Kowblansky, M. and Zema, P., 1981b. Effect of polyelectrolyte charge density on calcium ion activity coefficients and additivity in aqueous solutions of calcium acrylamide-acrylic acid copolymers. *Macromolecules,* 14(5): 1448–1551.
Kulicke, W.M. and Klein, J., 1978. Zur Frage der Instabilität von Polyacrylamidlösungen. *Angew. Makromol. Chem.,* 69: 189–210.
Kulicke, W.M., Kniewske, R. and Klein, J., 1982. Preparation, characterization, solution properties and rheological behavior of polyacrylamide. *Prog. Polym. Sci.,* 8: 373–468.
Lim, T., Uhl, J.T. and Prud'homme, R.K., 1986. The interpretation of screen-factor measurements. *Soc. Pet. Eng., Reservoir Eng.,* 1(3): 272–276.
Machtle, W., 1982. Zir Alterung von wässnigen Polyacrylamid-Lösungen. *Makromol. Chem.,* 183: 2515–2525.
Maerker, J.M., 1975. Shear degradation of partially hydrolyzed polyacrylamide in unconsolidated porous media. *Soc. Pet. Eng. J.,* 15: 311–322.
Maerker, J.M., 1976. Mechanical degradation of partially hydrolyzed polyacrylamide solutions in unconsolidated porous media. *Soc. Pet. Eng. J.,* 16: 172–174.
Martin, F.D., 1976a. Process for the secondary or tertiary recovery of petroleum. *U.S. Patent,* No. 3,945,929.
Martin, F.D., 1976b. Stabilization of polymer solutions. *U.S. Patent,* No. 3,953,341.
Martin, F.D. and Sherwood, N.S., 1975. The effect of hydrolysis of polyacrylamide on solution viscosity, polymer retention and flow resistance properties. *Soc. Pet. Eng. of AIME, Pap.,* 5339.
Martin, F.D., Donaruma, L.G. and Hatch, M.J., 1982. Development of improved mobility control agents for surfactant/polymer flooding. *Dep. Energy, Final Rep.,* DOE/BC/00047-19(DE82014918).
Martin, F.D., Hatch, M.J., Shepitka, J.S. and Ward, J.S., 1983. Improved water-soluble polymers for enhanced recovery of oil. *Int. Symp. on Oilfield Chemistry, Denver, Colo.,* SPE 11796.

Martin, A.L., Stanley, F.W., Jr., Cutie, S.S., Sugarman, J.H., Wilson, L.R., Hoagland, D.A. and Prud'homme, R.K., 1986. Determination of nonionic and partially hydrolyzed polyacrylamide molecular weight distributions using hydrodynamic chromatography. *Anal. Chem.*, 58: 2242–2247.

McCormick, C.L., 1985. Water soluble random and graft copolymers for utilization in enhanced oil recovery. *J. Macromol. Sci. Chem.*, A22(5/7): 955–982.

McCormick, C.L. and Lin, K.C., 1981. Water soluble copolymers, II. Synthesis and characterization of model dextran-g-acrylamides by Ce(IV)/HNO_3-induced initiation. *J. Macromol. Sci. Chem.*, A16(8): 1441–1462.

McCormick, C.L., Chen, G.-S. and Hutchinson, B.H., 1982. Water-soluble copolymers, V. Compositional determination of random copolymers of acrylamide with sulfonated comonomers by infrared spectroscopy and ^{13}C nuclear magnetic resonance. *J. Appl. Polym. Sci.*, 27: 3103–3120.

McKennon, K.R., 1962. Secondary recovery of petroleum. *U.S. Patent,* No. 3,039,529.

Meister, J.J., 1981. High radius-of-gyration, low-mass viscosifiers for water. *Am. Chem. Soc., Polym. Prepr.*, 22(2): 67–68.

Meister, J.J., Pledger, H.J., Hogen-Esch, T.E. and Butler, G.B., 1980. Retention of polyacrylamide by Berea sandstone, Baker dolomite and sodium kaolinite during polymer flooding. *5th Int. Symp. on Oilfield and Geothermal Chemistry, Stanford, Calif.*, SPE 8981.

Morris, J.D. and Penzenstadler, R.J., 1978. Acrylamide polymers. In: *Kirk Othmer Encyclopedia of Chemical Technology, Vol. 1.* John Wiley and Sons, New York, N.Y., 3rd ed., pp. 312–330.

Mungan, N., 1969. Rheology and adsorption of aqueous polymer solutions. *Can. J. Pet. Technol.*, 2: 45–50.

Murduchowitz, A., 1982. Secondary recovery process. *U.S. Patent,* No. 4,323,463.

Nagashiro, W. and Tsunoda, T., 1977. Degradation of polyacrylamide molecules in aqueous solutions by high-speed stirring, *J. Appl. Polym. Sci.*, 21: 1149–1153.

Norton, C.J. and Falk, D.O., 1973. Oil recovery with methylolated polyacrylamide thickener. *U.S. Patent,* No. 3,747,676.

Platt, J.L., Jr., 1981. Poly-(alpha-alkoxy) acrylamide and poly-(alpha-alkoxy) acrylamide complexes. *U.S. Patent,* No. 4,282,343.

Pledger, H., Meister, J.J., Hogan-Esch, T.E. and Butler, G.B., 1981. Starch-acrylamide graft copolymers for use in enhanced oil recovery. *Am. Chem. Soc., Polym. Prepr.*, 22(2): 72–73.

Proffitt, A.C., 1962. Process for making hydrolyzed polyacrylamide. *U.S. Patent,* No. 3,022,279.

Pye, D.J., 1964. Improved secondary recovery by control of water mobility. *J. Pet. Technol.*, 8: 911–916.

Pye, D.J., 1966. Water flooding process for the recovery of petroleum. *U.S. Patent,* No. 3,282,337.

Pye, D.J., 1968. Water flooding process for the recovery of petroleum and improved waterflooding process. *U.S. Patent,* No. 3,399,725.

Ryles, R.G., 1983. Elevated temperature testing of mobility control reagents. *Soc. Pet. Eng. of AIME, Pap.*, 12008.

Ryles, R.G., 1985. Chemical stability limits of water-soluble polymers used in oil recovery process. *Soc. Pet. Eng. of AIME, Pap.*, 13585.

Sandiford, B.B. and Keller, H.F., Jr., 1958. Secondary recovery of petroleum, *U.S. Patent,* No. 2,827,964.

Scanly, C.S., 1968. Simultaneous hydrolysis and polymerization of acrylamides. *U.S. Patent,* No. 3,414,552.

Schwartz, T. and François, J., 1981. Limités de solubilité des polyacrylamide partiellement hydrolyses en présence d'ions divalents. *Makromol. Chem.*, 182: 2775–2787.

Seright, R.S., 1983. The effects of mechanical degradation and viscoelastic behavior on injectivity of polyacrylamide solutions. *Soc. Pet. Eng. J.*, 23: 474–485.

Seright, R.S., Maerker, J.M. and Holzworth, G., 1981. Mechanical degradation of polyacrylamides induced by flow through porous media. *Am. Chem. Soc., Polym. Prepr.*, 22: 30–33.

Shupe, R.D., 1981. Chemical stability of polyacrylamide polymers. *J. Pet. Technol.*, 13: 1513–1529.

Slaton, M., 1983. EOR may account for 11.5 percent of production by 1993. *Drilling,* 9: 13.

Sloat, B. and Zlomke, D., 1982. The Isenhour Unit – a unique polymer augmented alkaline flood. *SPE/DOE 3rd Jt. Symp. on Enhanced Oil Recovery, SPE/DOE Pap.,* 19719.

Smith, F.W., 1970. The behavior of partially hydrolyzed polyacrylamide solutions in porous media. *J. Pet. Technol.,* 22: 148–156.

Sorbie, K.S. and Roberts, L.J., 1983. A model for calculating polymer injectivity including the effects of shear degradation. *At. Energy Auth., Rep.,* AEEW-R1626 (Winfrith, England).

Suen, T.J., Jen, Y. and Lockwood, J.V., 1958. The polymerization of acrylamide with chlorate-sulfite initiator. *J. Polym. Sci.,* 21: 481–497.

Szabo, M.T., 1979a. An evaluation of water soluble polymers for secondary oil recovery, Part I. *J. Pet. Technol.,* 11: 553–560.

Szabo, M.T., 1979b. An evaluation of water soluble polymers for secondary oil recovery, Part II, *J. Pet. Technol.,* 11: 561–570.

Thomas, W.M. and Wang, D.W., 1984. Acrylamide Polymers, In: *Encyclopedia of Polymer Science and Engineering, Vol. 1.* John Wiley and Sons, New York, N.Y., pp. 169–211.

Trantham, J.C., 1977. North Burbank Unit Tertiary Recovery Pilot. Second Annual Report, *U.S. Dep. Energy. Rep.,* BERC/TPR-77/5.

Trantham, J.C. and Moffitt, P.D., 1982. North Burbank Unit 1440 acre polymer flood project design. *SPE/DOE 3rd Jt. Symp. on Enhanced Oil Recovery, SPE/DOE Pap.,* 10717.

U.S. Department of Energy, 1979. *Commercial Scale Demonstration, Enhanced Oil Recovery by Micellar–Polymer Flooding and M-1 Project.* DOE/ET/13077.

Wang, F.H.L., Duda, J.L. and Klaus, E.E., 1979. Influence of polymer solution properties on flow in porous media. *Soc. Pet. Eng. of AIME, Pap.,* 8418.

Ward, J.S. and Martin, F.D., 1981. Prediction of viscosity for partially hydrolyzed polyacrylamide solutions in the presence of calcium and magnesium ions. *Soc. Pet. Eng. J.,* 21(5): 623–631.

Witheford, J.M., Pellon, J.J. and Colman, W.P., 1986. Preparation of aqueous acrylamide polymer solutions from polymer gels for enhanced oil recovery. *U.S. Patent,* No. 4,605,689.

Willhite, G.P. and Dominguez, 1977. In: D.O. Shah and R.S. Schechter, (Editors), *Improved Recovery by Surfactant and Polymer Flooding.* Academic Press, New York, N.Y., pp. 511–556.

Yost, M.E. and Stokke, O.M., 1975. Filtration of polymer solutions. *J. Pet. Technol.,* 27: 1271–1272.

Zaitoun, A. and Potie, B., 1983. Limiting conditions for the use of hydrolyzed polyacrylamides in brines containing divalent ions. *Soc. Pet. Eng. of AIME, Pap.,* 11785.

Chapter 9

ALKALINE FLOODING

SANJAY KUMAR, T.F. YEN, GEORGE V. CHILINGARIAN and ERLE C. DONALDSON

INTRODUCTION

Alkaline waterflooding is an enhanced oil recovery process where the pH of the injection water is increased by the addition of relatively inexpensive alkaline agents such as sodium carbonate, sodium silicate, sodium hydroxide and potassium hydroxide in an effort towards improving oil recovery. It was recognized as early as 1917 by F. Squires that the displacement of oil may be improved by introducing alkali into water. In spite of this early recognition, field applications are not so numerous and most of them have been unsuccessful. Presently, however, there seems to be a renewal of interest in alkaline flooding, the reason being the low cost and relative simplicity of the process and the ready availability of suitable alkaline chemicals (see Noran, 1978).

In the alkaline flooding process, the alkali reacts with the acidic constituents in the crude leading to lower water–oil interfacial tension, emulsification of oil and water, and solubilization of rigid, interfacial films (Mungan, 1981a). Also, the alkali may react with the reservoir rock, leading to wettability alteration. All of these mechanisms will potentially increase oil recovery.

The alkaline flooding process is a relatively simple process as compared to other chemical floods, but is still sufficiently complex to warrant careful laboratory investigation and field trials before application.

TYPES OF CAUSTIC USED

The alkaline materials used most commonly in recent floodings are sodium hydroxide and sodium orthosilicate (silicate buffer system of sodium hydroxide). The reactions of the latter in water have been described in detail by Campbell (1980) who has also claimed that the orthosilicate form is slightly more effective in cases where calcium ions are present in the injection fluids (see Noran, 1978).

Other alkali materials proposed and studied include sodium carbonate, ammonium hydroxide, polyphosphate, and hydroxyl amine. These do not seem to offer any improvement over sodium hydroxide. Chemicals such as calcium hydroxide and magnesium hydroxide do not even cause significant lowering of interfacial tension (IFT) (Jennings, 1975).

TABLE 9-I

Comparison of alkalis (after Mayer et al., 1983, table 5; courtesy of the Society of Petroleum Engineers of AIME)

	Formula	Molecular weight	1 wt.% alkaline solution in		Na_2O (%)	Solubility		Relative price range [a] (U.S.$/ dry ton)
			0% NaCl (pH)	1% NaCl (pH)		cold water (g/100 cm^3)	hot water (g/100 cm^3)	
Sodium hydroxide	$NaOH$	40	13.15	12.5	0.775	42	347	285–335
Sodium orthosilicate	Na_4SiO_4	184	12.92	12.4	0.674	15	56	300–385 [b]
Sodium metasilicate	Na_2SiO_3	122	12.60	12.4	0.508	19	91	310–415 [b]
Ammonia	NH_3	17	11.45	11.37	–	89.9	7.4	190–205
Sodium carbonate	Na_2CO_3	106	11.37	11.25	0.585	7.1	45.5	90– 95

[a] Mid-1982 U.S. prices not including any freight or bulk discounts.
[b] Based on price range of U.S.$6.80/cwt to U.S.$9.35/cwt of $Na_2O(SiO_2)_{16}$ at 42.55% solids.

A mixture of sodium hydroxide and sodium meta-silicate is also being used in field operations because the reaction between these two yields sodium orthosilicate. Table 9-I lists the properties and costs of the major alkalis used.

ENTRAPMENT OF RESIDUAL OIL

As pointed out by Stegemeier (1977), the factors which determine the entrapment and release of oil from a rock matrix are: (1) the geometry of the pore network and the macroscopic heterogeneities of the porous medium, (2) fluid–fluid properties, such as interfacial tension, phase behavior, and (bulk) viscosities, (3) fluid–rock properties, such as wettability, adsorption ion-exchange, and dissolution reactions, and (4) applied pressure gradients and gravity.

During primary depletion, crude oil is produced by the natural energies of the reservoir and the contained fluids, which manifest themselves in the form of gas-cap drive, water drive, compaction drive, and solution-gas drive. Below the bubble point, solution gas is released from the oil within the reservoir. Thus, gas eventually becomes mobile when the total gas saturation in the reservoir exceeds the critical gas saturation. This critical gas saturation is related directly to the relative permeability behavior of the gas in the reservoir, which depends on the microscopic structure of the porous matrix. The gas phase thus created and rendered mobile is much less viscous than the oil and bypasses (fingering, channelling, override, etc.) the oil, particularly in the high-permeability zones, resulting in isolated oil ganglia in the low-permeability zones.

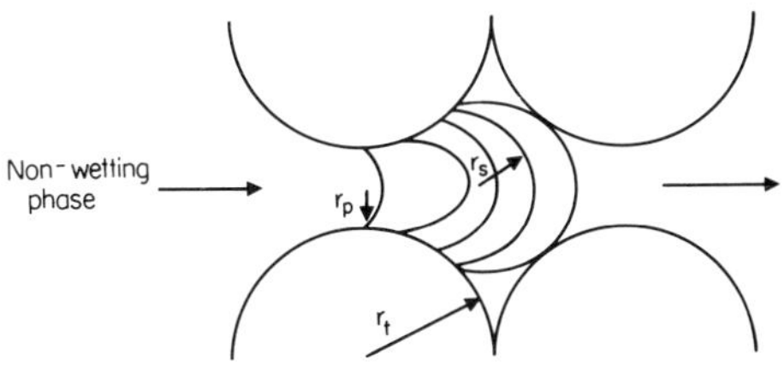

Fig. 9-1. A toroidal capillary model. Non-wetting phase displacing the wetting phase. r_p = pore radius, r_s = fluid radius, and r_t = grain radius.

Another mechanism for isolation of oil during primary depletion is interfacial instability. This is the dominant mechanism during the non-expansive, i.e., the decreasing gas/oil ratio (GOR), portion of production.

Haines (1930) first studied interfacial instabilities and postulated that desaturation occurs by a series of reversible saturation changes called "isons" and sudden irreversible pressure changes called "rheons" or "Haines jump" (Melrose, 1970; Morrow, 1970).

Fig. 9-1 shows a converging–diverging pore in an idealized porous medium consisting of spherical rock particles. The capillary pressure as a function of the non-wetting phase penetration is shown in Fig. 9-2. The non-wetting phase enters the pore when its pressure exceeds the pore entry pressure (A). It then encounters decreasing capillary pressures (because the wetting phase saturation is higher) and enters in completely until it reaches the pore wall at B. The non-wetting phase then enters narrower parts of the pore and the capillary pressure increases. On reaching the critical pore saturation, the capillary pres-

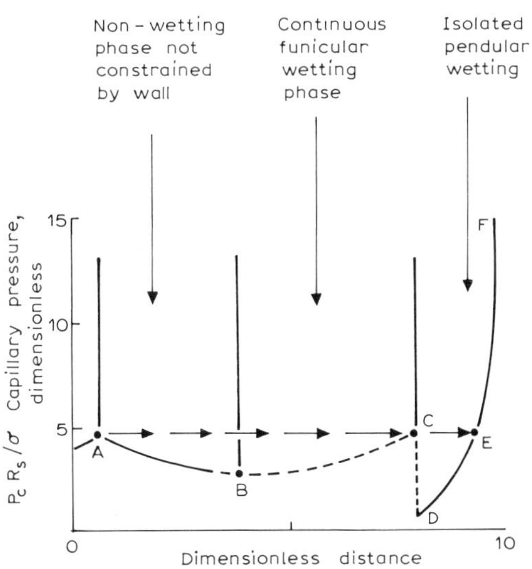

Fig. 9-2. Change in capillary pressure as a function of pore penetration.

sure drops discontinuously from C to D. The passage of the interface from pore entry to the pendular ring shape takes place in one discontinuous movement termed "Haines jump".

Waterflooding represents the secondary stage of production. Waterflooding improves recovery primarily through two mechanisms: (1) it provides the pressure energy to overcome the capillary forces that entrap the oil, and (2) it results in an improved mobility ratio as compared to the solution-gas drive the reservoir was subject to during primary depletion.

Some isolated oil ganglia are forced to move out from the low-permeability zones and coalesce with other ganglia to form localized regions of high oil saturation, leading to higher relative permeability to the oil. At the lowered oil saturations behind the flood front, the oil ganglia are left further apart than after primary depletion and are thus entrapped in the continous wetting phase. Extension of the advancing interface can result in separation of the oil phase at a pore constriction if the capillary pressure at the neck exceeds the capillary pressure at the interface front (Melrose and Bradner, 1974; Stegemeier, 1974).

The displacement of oil from the pores by water also occurs through a series of rheons and isons. Melrose and Bradner (1974) have suggested that oil ganglia, present as a discontinuous, non-wetting phase, are displaced simultaneously by oil drainage rheons at the downstream (low-pressure) end and the imbibition rheons at the upstream (high-pressure) end. The non-wetting phase, oil, is left behind as discontinuous blobs and filaments adhering to the pore surface.

This mechanism accounts for only a part of the residual oil-in-place after a waterflood. Poor areal and vertical sweep efficiencies resulting from reservoir heterogeneities and gravity override effects may reduce recovery efficiencies from 55–70% to 30–35%.

DISPLACEMENT MECHANISMS IN ALKALINE FLOODING

Alkaline flooding enhances the recovery of acidic oils by two-stage processes (Castor, 1979). The first stage involves the mobilization of the residual oil by configurational changes like emulsification and wettability alteration. Surface-active salts are formed in situ by the acid-base reaction between the alkali and the organic acids in the residual oil. The surfactants thus generated may: (1) adsorb at the oil–water interface to lower the interfacial tension, and in some cases cause spontaneous emulsification and phase swelling; and (2) react with or adsorb at the rock surface, changing the wettability characteristics of the rock and hence the configuration of the residual ganglia of the crude oil (Castor, 1979).

The second stage involves the modification of the macroscopic production characteristics of the mobilized oil phase. The overall recovery efficiency may be increased in this stage by improvements in the displacement efficiency through mobility control, i.e., by reduction in the floodwater mobility.

Johnson (1976) reviewed the mechanisms by which alkaline flooding may improve recovery of acidic crudes from partially depleted reservoirs: (1) emulsification and entrainment, (2) emulsification and entrapment, (3) wettability reversal from water-wet to oil-wet, and (4) wettability reversal from oil-wet to water-wet. Johnson has also discussed variations and combinations of these four basic mechanisms. These include: (a) cyclic emulsification and entrapment process (Sarem, 1974); (b) alternate emulsification–entrapment and emulsification–entrainment process (Wade and Lechtenberg, 1978); and (c) concurrent entrainment and entrapment after emulsification, wettability alteration, and chromatographic wettability reversal (Radke and Somerton, 1977, 1978; Castor et al., 1978).

Two additional very significant mechanisms are: (5) lowering of interfacial tension, and (6) solubilization of rigid, interfacial films. The extent to which each of these mechanisms may contribute to increasing recovery depends on the specific properties of the crude oil, injection water, and the reservoir (Mungan, 1981). The phenomena are discussed individually below in sequential order.

Lowering of interfacial tension

Early work by Reisberg and Doscher (1956) indicated that the interfacial tension between water and certain California crudes was extremely low at certain ranges of water pH. Their results are shown in Fig. 9-3. This suggests that increasing the concentration of sodium hydroxide will not reduce interfacial tension continuously or indefinitely. There exists an optimum concentration for which the lowest interfacial tension is obtained. De Ferrer (1977) has published similar results for interfacial tensions between caustic solutions and some heavy Venezuelan oils.

Donaldson and Crocker (1980) discuss the microscopic observation of residual oil displacement by 1000-ppm solutions of sodium hydroxide, a non-ionic surfac-

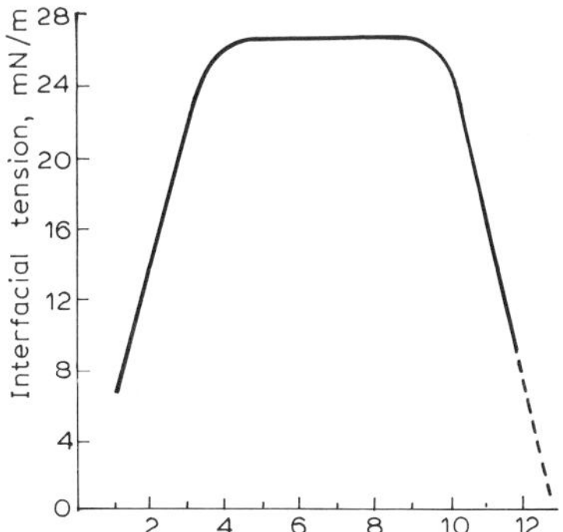

Fig. 9-3. Interfacial tension versus pH for a Ventura (California) crude. (After Reisberg and Doscher, 1956; courtesy of *Producers' Monthly*.)

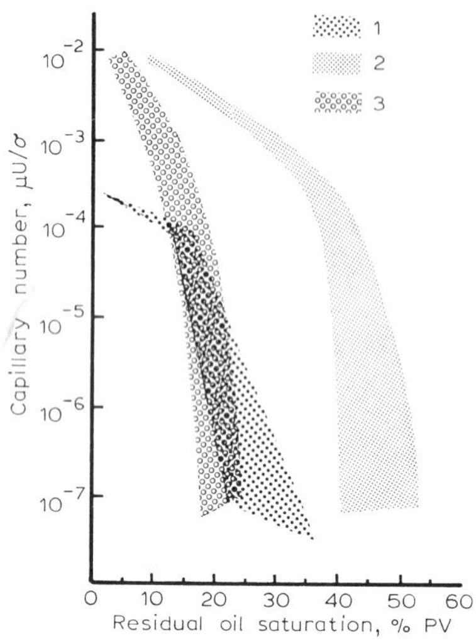

Fig. 9-4. Capillary number versus residual oil saturation. Data from *1*: Bardon and Longeron (1980); *2*: Foster (1973); and *3*: Lefebvre du Prey (1973). (After Mungan, 1981a, p. 210, fig 15; courtesy of *World Oil*.)

tant and a mixture of the alkaline and surfactant solutions. The displacements were made from sand packed Hele-Shaw cells using crude oils. The alkaline and surfactant solutions used alone only produced 1–3% additional oil, but the mixture removed virtually all of the residual oil.

Mungan (1981) has concluded from a review of all published literature that the interfacial tension minimum occurs, for the most part, in the NaOH concentration range of about 0.05–0.2% by weight and at a pH value of about 12.5.

The concept of capillary number, N_c, defined as:

$$N_c = v\mu/\sigma \tag{9-1}$$

(where v = darcy velocity of the displacement front; μ = viscosity of the flowing fluid; and σ = interfacial tension) is valuable in explaining and quantifying the effect of interfacial tension on oil recovery. Conventional waterfloods operate generally at or near a capillary number of 10^{-6}. The effect of capillary number on oil recovery has been discussed by Taber (1969), Lefebvre du Prey (1973), Melrose and Bradner (1974), and Bardon and Longeron (1980) among others. Fig. 9-4 shows the general relationship between capillary number and residual oil saturation.

It is evident from Fig. 9-4 that to improve oil recovery (i.e., to decrease residual oil saturation), the capillary number must be increased to at least 10^{-4}. Inasmuch as v and μ cannot be varied by an order of 10^2 or more, interfacial tension, which is amenable to being reduced by such orders of magnitude, is the only parameter that can be modified. Values as low as 0.002 dyne/cm can sometimes be achieved in alkaline flooding of certain crudes. Fig. 9-5 represents

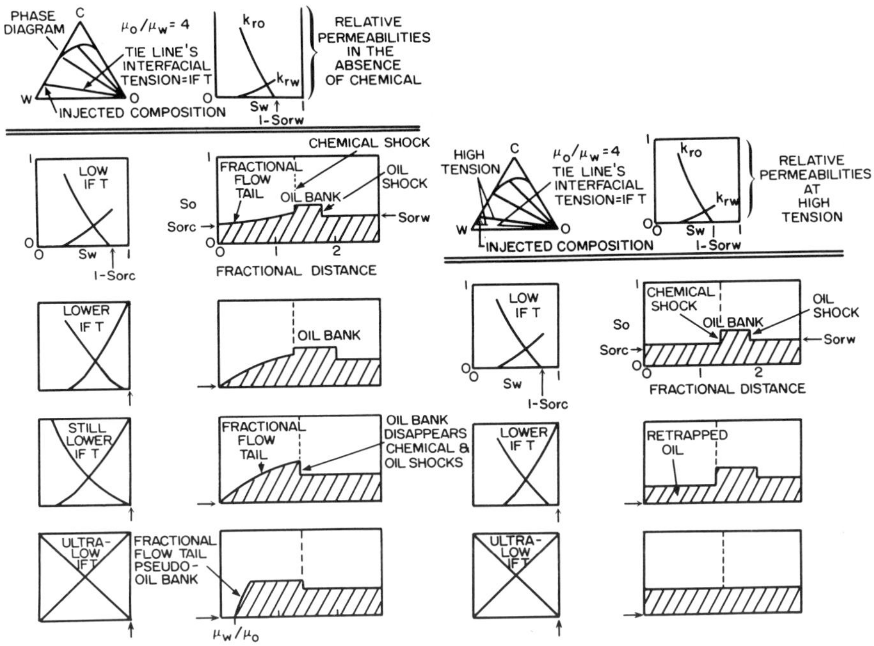

Fig. 9-5. Oil recovery by reduction of interfacial tension. (After Larson et al., 1982, p. 248; courtesy of the Society of Petroleum Engineers of AIME.)

the mechanism of oil recovery by reduction of interfacial tension (Larson et al., 1982). The idealized phase diagram shows the injection caustic solution at the end of a high-tension tieline. There are, however, low-tension tielines (one or more) at lower caustic concentrations (the reader is referred to the relationship between caustic concentration and interfacial tension that passes through a minimum). The assumed relative permeability behavior is shown alongside. The tendency of the relative permeability curves to approach the crossed lines of 45° inclination (lowermost in Fig. 9-5 at extremely low IFT) has been shown in the literature (Leverett, 1939; Lefebvre du Prey, 1973; Talash, 1976). It is assumed that there is no viscosity alteration. The interfacial tension is high behind the caustic front as well as ahead of it. Thus, no fractional flow tail can exist behind the chemical shock; instead there is a zone of retrapped oil of uniform saturation. If the tension associated with the low-tension lines is too low, the oil although mobilized, will move more slowly than the aqueous phase. Thus, instead of getting ahead of the shock to form an oil bank, the oil falls behind and is retrapped by the high interfacial tensions where it encounters the injected fluid (see Larson et al., 1982).

In the case of water-wet porous media, the relative permeabilities for which are shown in Fig. 9-5, the relative permeability to water increases more than that to oil on lowering IFT. Thus, the viscosity ratio being constant, the mobility ratio ($k_{rw}\mu_w/k_{ro}\mu_w$) increases with decreasing interfacial tension, when the relative permeabilities versus saturation curves become diagonal straight lines. The oil recovery is complete if viscosity ratio μ_o/μ_w approaches $1/(1 - S_{orw})$ and remains so at higher values of the viscosity ratio. This underscores the importance of controlling the mobility ratio (i.e., the viscosity ratio) in addition to lowering the IFT for increasing the capillary number.

Sarem (1974a) has proposed and patented (1974b) a two-step recovery process utlizing the emulsification and entrapment mechanism that is discussed in detail in the following sections. Sarem's process involves the alternate injection of (1) alkaline (sodium) silicate solution to emulsify and reduce the mobility of the advancing front, and (2) dilute calcium chloride solution to precipitate out the silicates and further reduce the permeability and divert the injected aqueous phase.

This cyclic process sequence, according to Sarem (1974a), leads to the formation of an oil bank and reduces viscous fingering of the alkaline phase through the oil bank.

Currently research is being carried out to implement a field pilot mobility controlled caustic flooding in the Huntington Beach field in California (see Weinbrandt, 1978).

Emulsification and flow

There is general agreement that the lowering of interfacial tension due to the reaction between the caustic and the organic acids in the oil is also responsible for the formation of emulsion in situ. There are disagreements, however, as to the type of emulsion (water-in-oil or oil-in-water) and also as to the exact role of the emulsion formation in the recovery of additional oil (McAuliffe, 1973a,b; de Ferrer, 1977; Bansal et al., 1978; Cooke et al., 1974; Jennings et al., 1974; Farouq Ali et al., 1979). The formation of emulsion can be inferred from laboratory studies in alkaline flooding where the production of very fine oil droplets from the outflow end of the core flood system has been reported by many researchers. Among others, Dranchuk et al. (1974) carried out laboratory displacement experiments on Lloydminster crude with various aqueous alkaline solutions. They noted that oil-in-water emulsions were produced in the core and that there was a reduction in water mobility during displacement tests.

Larson et al. (1982) provided a mathematical treatment of the emulsification mechanism. Fig. 9-6 is an example of an oil-in-water emulsion. The relative permeability curves are the same as in Fig. 9-5. The pseudophase diagram shown assumes that neither caustic nor the water are soluble in the oil. Hence all tielines pass through the oil corner of the ternary diagram. Also, it is assumed that neither relative permeabilities nor viscosities are affected by the presence of the alkali.

The dispersed-oil lag factor, r_o^w, is defined as the ratio of the volumetric rate at which the oil component in the water phase (or pseudophase) moves to the

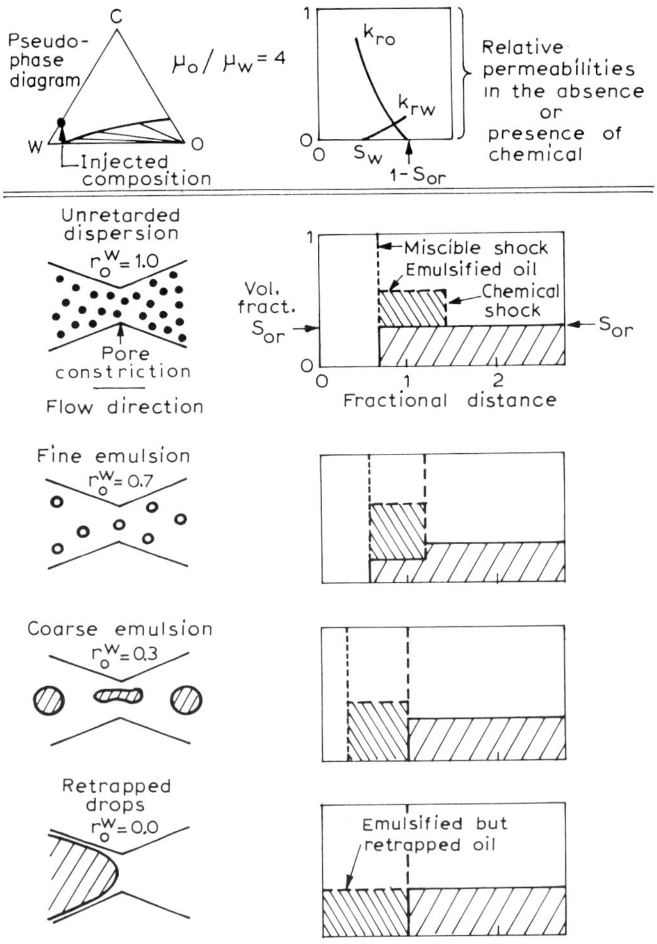

Fig. 9-6. Oil recovery by oil-in-water emulsification. (After Larson et al., 1982, p. 252; courtesy of the Society of Petroleum Engineers of AIME.)

volumetric rate of movement of water (see Larson et al., 1982). Fig. 9-6 shows three cases: (a) $r_o^w = 1$ corresponds to the limiting case where the oil droplets behave like a true dissolved component and travel unimpeded through even the tightest pore constrictions; (b) $r_o^w = 0.7$ and $r_o^w = 0.3$ represent a fine and coarse emulsion, respectively; (c) $r_o^w = 0$ corresponds to the limiting case where the droplets, like the trapped blobs initially in place, are too large and the interfacial tension is too high for them to move at all.

In the case of water-in-oil emulsion, the entrained water droplets may lag the average pseudophase velocity (i.e., the oil velocity). This emulsion phase, with $r_o^o > 1$ (because oil leads water), will probably have a relatively high effective viscosity. Thus, it is not permissible to assume the fractional flow of oil to be zero behind the chemical shock, and the analysis becomes quite complex.

There are essentially two mechanisms in the emulsification process: (1) emulsification and entraiment, and (2) emulsification and entrapment.

(1) *Emulsification and entrainment*: In this mechanism, first proposed by Subkow (1942), residual oil undergoing mobilization is emulsified and entrained into the flowing alkaline solution and is, therefore, produced as a fine emulsion (Mungan, 1981a). Subkow pointed out that the alkali concentration and pH should be such that stable oil-in-water emulsions are formed in the emulsification and entrainment process. Reisberg and Doscher (1956) concluded similarly but added that the incremental production rate by this process would be uneconomical. This is so because the adsorption and reaction of the alkali with the rock causes the alkali front to lag behind the oil–water displacement front by several pore volumes. The general consensus seems to indicate that this mechanism is the more applicable one for low caustic content field application. Castor (1979) has pointed out that some reduction in interfacial tension is required for the emulsification step. The ultra-low tension mechanism, however, may not be the dominant mechanism. The actual movement of the oil may be described better by coalescence behavior than by entrainment (Castor, 1979; Castor et al., 1979).

(2) *Emulsification and entrapment*: In this process, the emulsified oil is trapped again downstream in the porous medium at pore throats that are too small for the emulsion droplets to penetrate, forcing the injection water into pores that have not been previously displaced. This results in reduction in the mobility of the aqueous phase with a subsequent increase in the displacement efficiency and reduction in the amount of viscous fingering (Jennings et al., 1974).

This process requires very stable emulsions and has generally been postulated to occur with saltless alkaline solutions. De Ferrer (1977) described corefloods on heavy Venezuelan crudes that show a considerable increase in recovery on using sodium hydroxide without salt. Wasan (1979) and Wasan et al. (1979) noted a difference in emulsion stability due to the presence of salt.

Inasmuch as the emulsion is quickly entrapped, there is no reduction in residual oil saturation beyond a conventional waterflood. The rate of recovery, however, is improved significantly. The improvement in oil recovery by such a mechanism is only due to improved areal and volumetric sweep efficiencies. The viscosity of water-in-oil emulsions being higher than that of oil-in-water emulsions, the formation of water-in-oil emulsions may be more important in the entrapment mechanism (Mungan, 1981).

De Ferrer (1977) and Farouq Ali et al. (1979) observed that additional recovery from caustic flooding is less than that obtained by injecting laboratory prepared emulsions of comparable slug sizes. This indicates that the emulsion formation in situ is probably not as complete due to insufficient contact between injected caustic and the residual oil. Cooke et al. (1974) have shown, however, that an emulsion bank can be generated in a reservoir just as well as in the laboratory for a proper alkaline water and acidic oil flowing simultaneously under proper conditions of salinity, pH, and temperature.

Wettability alteration

The success of a tertiary caustic flood depends to a large extent on the initial distribution of residual oil in the pore spaces that is controlled by the reservoir wettability, and also on the ability to alter wettability.

Wettability has been studied at great length by many researchers. Denekas et al. (1959) found that the heavier components in the crude oil have a greater effect on the rock wettability, causing it to become oil-wet.

The main idea is to reduce the capillary forces holding the oil in the porous matrix. Wagner and Leach (1959) showed that a change in pH can result in wettability change through reaction with the surface-active materials in the crude adsorbed on the rock surface. An alkaline solution can alter pH very easily and inexpensively and is superior to surfactants in this respect. The role of reverse wetting agents on oil displacement efficiency was studied by Michaels and Timmins (1960). Fig. 9-7a shows an oil droplet trapped at a pore throat under water-wet conditions. The pore becomes gradually oil-wet (Fig. 9-7b) on arrival of the wettability reversal agent (caustic). This results in a decrease in the capillary resistance to the oil drop and it moves into tighter places in the pore and spreads out across the rock surface. Interfacial tension also is reduced. When the caustic concentration decreases and water-wetness returns, the oil clinging to the rock under oil-wet conditions will tend to be displaced (Fig. 9-7c,d).

Studies done by Ehrlich et al. (1974) on wettability alteration using a West Texas crude oil and a sandstone core are very illuminating. Fig. 9-8, which displays their results, indicates that alkaline flooding recovered more oil than

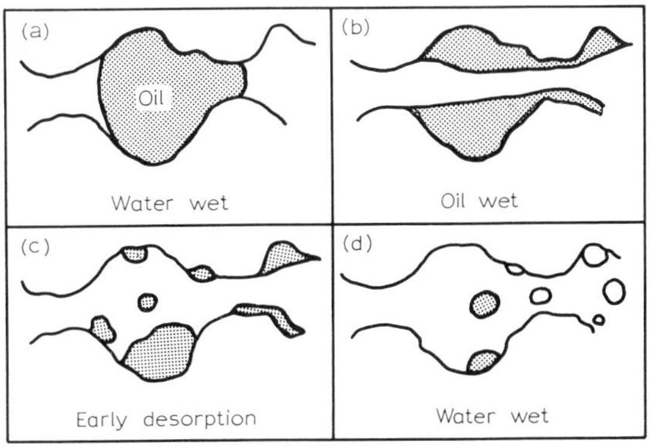

Fig. 9-7. Mechanism of trapped oil removal by wettability reversal. (After Michaels and Timmins, 1960; courtesy of the Society of Petroleum Engineers of AIME.)

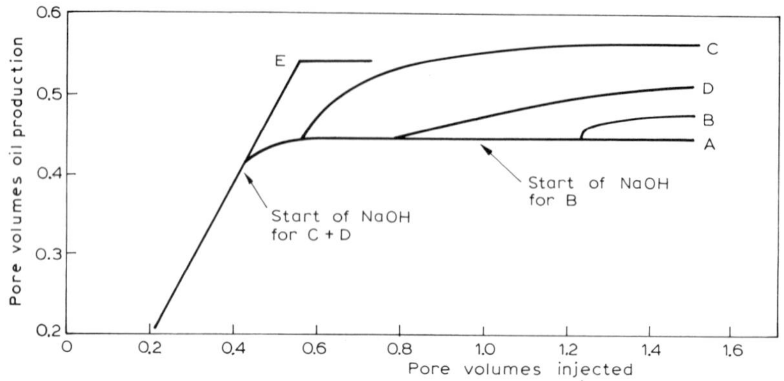

Fig. 9-8. Comparison of oil recovery by waterflooding and alkaline flooding. *A*: 2% NaCl brine; *B*: 2% NaCl brine followed by 3.5% NaOH at high WOR; *C*: 2% NaCl brine followed by 3.5% NaOH at water breakthrough; *D*: 2% NaCl brine followed by 2% NaOH at water breakthrough; *E*: 2% NaCl brine–mineral oil saturated pack. (After Ehrlich et al., 1974, fig. 4, p. 1340; courtesy of the Society of Petroleum Engineers of AIME.)

plain waterflooding. The wettability aspect was emphasized through controlling certain crude components. Weinbrandt (1979) found improved recovery in Huntington Beach corefloods at higher orthosilicate levels which may be attributable to wettability alterations.

(1) *Wettability reversal from oil-wet to water-wet*: The wettability is changed by the reaction between the caustic and acidic polar compounds attached to the oil-wet reservoir rock. This wettability change results in a favorable increase in the oil/water relative permeability ratio and a subsequent increase in displacement efficiency (see Leach et al., 1962).

Wagner and Leach (1966) have shown that wettability alteration from oil-wet to water-wet improves oil recovery. Raimondi et al. (1976) have also reported favorable results from a field test of this process in the Queen Sand lens of the North Ward-Estes Field in West Texas.

(2) *Wettability reversal from water-wet to oil-wet*: Cooke et al. (1974) suggested this mechanism where the alteration of wettability from water-wet to oil-wet causes the discontinuous non-wetting residual oil phase to spread out into a continuous wetting phase. Simultaneously, the low interfacial tension induces the formation of an alkali-in-oil emulsion that plugs flow paths and induces locally high pressure gradients. These high gradients overcome the already reduced capillary forces and reduce the residual oil saturation further.

Cooke et al. (1974) conducted some experiments. One interesting result reported by them was the rate dependence of alkaline waterflood effectiveness. They showed that the wettability reversal and emulsification process depended on the flow rate, and the amount of tertiary oil recovered decreased beyond some minimum value of the injection rate. The results from a field trial conducted by them, however, were not very conclusive.

According to Campbell and Krumrine (1979), wettability alteration from oil-wet to water-wet may be more favorable with sodium orthosilicate. They postulated that polysilicate anion species that alter the surface properties of both the aqueous and solid phases are responsible for this.

Solubilization of rigid interfacial films

Some crudes form insoluble films at the crude oil–water interface; it has been suggested that resins and asphaltenes in the crude are responsible for the formation of these films. These interfacial films may be redissolved in the oil, but this rate is very slow. During transport through the pores, these films will be lodged at the pore constrictions, resulting in a rigid, insoluble, immobile film and affecting the displacement efficiency (see Harvey, 1960).

Hasiba (1966) and Kimbler et al. (1966) have studied the film-forming constituents in crude oils. Among others, aldehydes, ketones, porphyrin-metal complexes, carboxylic acids, esters and organic nitrogen have film-forming tendencies (Hasiba, 1966).

Microvisual studies, linear core displacements, and imbibition experiments carried out by Bourgoyne et al. (1972) have demonstrated the formation of such films. Apparently, these films decrease displacement efficiency and hence the oil recovery in heterogeneous porous systems, but have no effect on the displacement efficiency in homogeneous porous media (Bourgoyne et al., 1972).

CRUDE OIL PROPERTIES

The lowering of oil–water interfacial tension is a function of the properties of the crude oil. It is generally accepted that the reaction of alkali with components of the crude oil leads to the generation of low interfacial tension through the formation of surface-active agents. A multitude of crude components directly responsible have been identified. These include (1) carboxylic acids, (2) carboxyphenols, (3) porphyrins, and (4) asphaltenes. Attempts have been made to characterize the crudes and correlate their properties to recovery efficiency. Some of the properties suggested by investigators are discussed here. A minimum acid number (the number of milligrams of KOH required to neutralize one gram of the crude oil to $pH = 7$) needed for successful alkaline flooding has been suggested by many researchers: a value of 0.5 or greater is generally considered very favorable. It is important to recognize that only organic acidity, determined by subtracting the inorganic acidity due to mineral acids or acidic gas from the total acidity, is correlatable. Most oils have acid numbers less than 1 (Jennings, 1975; Minnssieux, 1977).

Whereas the presence of acids in the crude appears to be a necessary requirement, no direct correlation has been observed between the acid number of the crude and the interfacial activity or the magnitude of the enhanced oil recovery (Ehrlich and Wygal, 1976; Campbell and Krumrine, 1979). The absence of such a correlation may be due to the existence of several acids in the crude oil, only a few of which can form good soaps on encountering alkali. The formation of these soaps is related to the reduction of interfacial tension.

Bansal et al. (1978) correlated minimum IFT with the maximum electrophoretic mobility related to the increase in charge density at the oil–caustic interface by acid ionization. They showed that the minimum IFT was achieved at the caustic concentration range which gives the maximum electrophoretic mobility. The ionization of carboxylic acids was credited for increasing the charge density at the interface. The pK_a values of these acids were not determined, however, due to the pH difference between the bulk solution and the elusive interface.

Farmanian et al. (1978), Pasquarelli and Wasan (1979), and Yen and Peterson (1982) have attempted to identify the interfacially active components of crude oils. Farmanian et al. determined that rather than being a single type of functional group, both aromatic as well as polar components of crude oils are interfacially active as judged by the interfacial tension response in alkaline solution. The active acids in the crude oil were shown to be a mixture of many species which ionize at different bulk solution pH.

Pasquarelli and Wasan (1979) investigated the effects of asphaltenes and acidic resins upon IFT behavior. They concluded that the heavy asphaltic components in crude oils are primarily responsible for the high interfacial activity and the resulting dynamic interfacial properties. Asphaltene adsorption at the oil–water interface creates a film which inhibits further reaction. The formation of such interfacial films that prevent water–oil emulsion coalescence has also been noted by Castor et al. (1979).

The concept of the equivalent alkane carbon number (EACN) introduced by Cash et al. (1976) has been used by Morgan et al. (1976) to correlate IFT behavior of crude oil–surfactant systems. The crude oil is replaced by that number of the homologous *n*-alkane series which exhibits interfacial tension behavior identical to that of the crude oil.

No single definitive criteria for crude properties needed for alkaline flooding exist (Mayer et al., 1983). The caustic flooding project at Long Beach, California, has demonstrated this and also provided evidence to the effect that crude oil handling affects the recovery enhancement in core floods.

ALKALI CONSUMPTION

Alkali consumption is one of the most critical design parameters in alkaline flooding. The injected alkali reacts with reservoir rock, oil, and the connate water (see Table 9-II). The loss to the rock is by far the most significant and considerable research work has concentrated on the loss of alkali to the rock-associated clays and minerals. Ehrlich and Wygal (1976) have correlated the loss of

TABLE 9-II

Caustic consumption reactions

Category	Reaction	Reference
Alkali–water reactions	$Ca^{2+} + 2OH^- \rightarrow Ca(OH)_2$ $Mg^{2+} + 2OH^- \rightarrow Mg(OH)_2$ $Ca^{2+} + CO_3^{2-} \rightarrow CaCO_3$ $Mg^{2+} + CO_3^{2-} \rightarrow MgCO_3$ $Ca^{2+} + H_2SiO_4^{2-} \rightarrow CaH_2SiO_4$	
Alkali–oil reactions	$HA_o \rightleftarrows HA_w$ $HA_w \rightleftarrows H_3O^+ + A^-$ $Na^+ + A^- \rightleftarrows NaA_w$ $NaA_w \rightleftarrows NaA_o$	Ramakrishna and Wasan (1982)
Alkali–rock reactions	Ion exchange: (a) $MH + Na^+ \rightleftarrows MNa + H^+$ (b) $M_2Ca + 2Na^+ \rightleftarrows 2MNa + Ca^{2+}$ Congruent dissolution: $SiO_2(s) + H_2O \rightleftarrows Si(OH)_4$ Incongruent dissolution: $CaSO_4(s) + 2NaOH \rightleftarrows Ca(OH)_2(s) + Na_2SO_4$	Bunge and Radke (1981)

alkalinity to the reservoir rock clay mineralogy. The consumption was found to increase with time, and then level off with no further consumption even at infinite time (which in this case was only one week). Other workers have reported continued consumption even after 60 days (Lieu, 1979).

Caustic reaction with the rock may be responsible for wettability changes (Goyal, 1982). Rock reactivity, however, has a major role. Too high a reactivity can affect the process in two ways: (1) mechanisms like emulsification and entrainment and wettability reversal from water-wet to oil-wet are vulnerable to high rock reactivity, and (2) the economics of caustic flooding may be severely impaired due to higher alkali consumption for a significantly lower oil production.

The alkali–rock reaction has been described as being approximately Langmuirian (Radke and Somerton, 1979a) and the reaction rate has been observed, as expected, to increase with temperature and decrease with increasing salt content. The alkali consumption comprises two types—reversible and irreversible.

Reversible reactions or the ion-exchange-type reactions seem to dominate at high flow rates and include several mineral reactions. One such reaction, that is particularly detrimental, is the one with calcium sulfate (Mayer et al., 1983). Calcium sulfate (as anhydrite or gypsum) reacts with the injected alkali to yield calcium hydroxide that dissolves in the water behind the alkaline slug. This redissolved calcium hydroxide does not contribute to any additional oil recovery, because calcium surfactants are generally ineffective in reducing surface and interfacial tensions. Thus, significant quantities of alkali are consumed (for every 0.1 wt.% gypsum in the rock, 1.5 meq alkali/100 g rock is approximately consumed) with no improvement in oil recovery. This has led to the use of softened water in all laboratory and field caustic flooding operations.

The most important irreversible reaction is the dissolution of the siliceous material in the rock. In caustic core floods, the effluent analysis can sometimes yield valuable information essential to the understanding of the mechanisms of rock dissolution. In the field, the well can be flowed back and the produced fluid analyzed. The dissolution process is affected primarily by the flow rate, surface area/unit volume ratio and mineralogical composition of the rock, and alkali concentration and its composition.

A convenient parameter to quantify dissolution is the Damkohler number (Radke and Somerton, 1979), which is a direct function of the dissolution rate constant, concentration gradient, and length of the flowpath. It is inversely proportional to the linear velocity. That the reduction in alkaline concentration due to dissolution is an exponential function of the Damkohler number has been confirmed by core flood experiments.

Also important in the dissolution process are the contents of clays and fines because of their relatively large surface area (Somerton and Radke, 1979).

An important caustic–rock interaction is the reversible sodium–hydrogen base exchange:

$$MH + Na^{+}OH^{-} \rightarrow MNa + H_2O \tag{9-2}$$

where M denotes a mineral base exchange site. The exchange isotherm has been measured for many sands and, as previously mentioned, found to be Langmuirian in shape. Although this is a reversible reaction, implying that the caustic is not permanently consumed, it has drastic consequences in terms of the pore volumes of caustic to be injected for breakthrough. Fig. 9-9 (after Radke and Somerton, 1979) shows the time τ_B expressed in terms of pore volumes for caustic to elute from a Wilmington Sand core at various injected caustic concentrations.

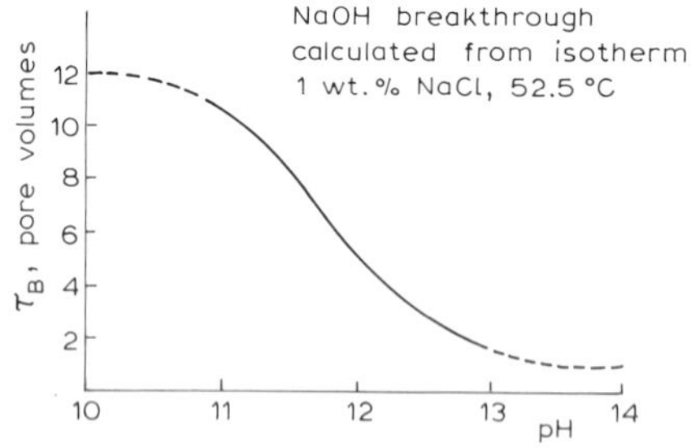

Fig. 9-9. Pore volume caustic injected to breakthrough as a function of injected caustic pH for Wilmington Sand, California. (After Radke and Somerton, 1979b; courtesy of the U.S. Department of Energy.)

This shows that the lower the injected caustic concentration, the longer is the time (or larger the number of pore volumes of caustic that need to be injected) for breakthrough. It has been demonstrated (de Zabala et al., 1980) that the chromatographic movement of caustic controls the rate of oil production in an alkaline flood and, therefore, a knowledge of the exchange or adsorption isotherm is critical.

The irreversible dissolution reaction has been subdivided into two types: congruent and incongruent. The congruent dissolution reaction type implies the mineral reaction which generates soluble aqueous species in the stoichiometric ratio of the mineral lattice. The incongruent reaction is the reaction where one mineral species dissolves and generates a second different mineral (such as in the case of the gypsum–alkali reaction).

Permeability damage or core plugging may sometimes occur due to reactions with alkali. This may happen due to: (1) clay swelling, and (2) migration of fines. Although this is an important problem in a conventional waterflood, laboratory and field studies suggest that permeability damage may not be a serious problem in a caustic flood in general (Goyal, 1982).

pH OF INJECTED CAUSTIC

Fig. 9-10 shows the influence of pH level on secondary recovery and Fig. 9-11 shows the effect in the case of a tertiary flood. These figures are from the work of de Zabala et al. (1980). At a lower pH, less surfactant is produced from the oil and consequently a lesser amount of oil is mobilized. Also, at the lower pH, the ion-exchange is high enough to drastically reduce the rate of oil production. Thus, alkaline flooding should be done at a high pH to overcome the chromato-

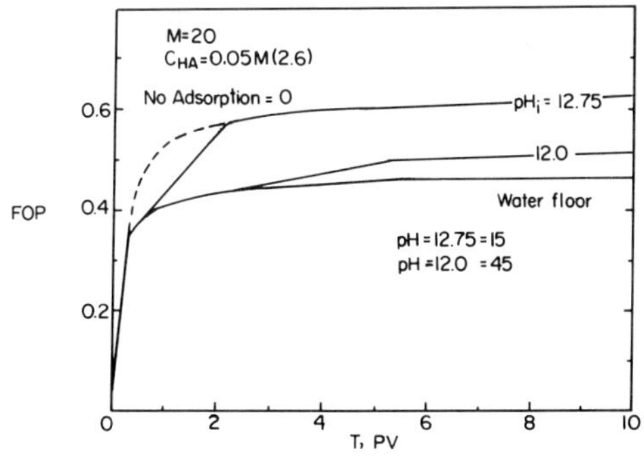

Fig. 9-10. The effect of injected pH on oil production from a secondary alkaline flood. (After de Zabala et al., 1980, fig. 8; courtesy of the Society of Petroleum Engineers of AIME.)

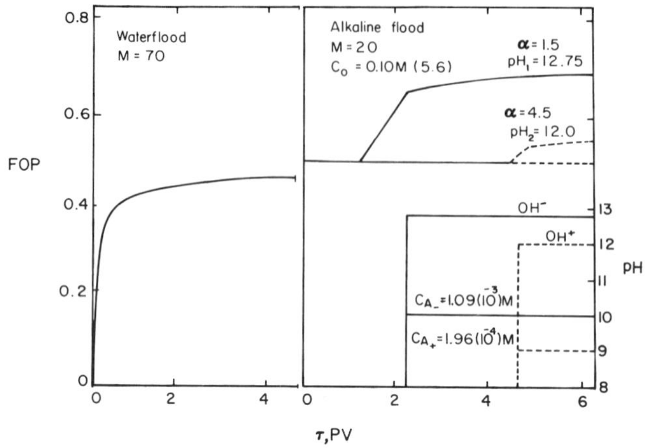

Fig. 9-11. The effect of injected pH on oil production from a tertiary alkaline flood. (After de Zabala et al., 1980, fig. 12; courtesy of the Society of Petroleum Engineers of AIME.)

graphic ion-exchange retardation of the alkali. This seems to be in contradiction with the requirement that pH be lower for achieving low interfacial tension.

EFFECT OF SODIUM IONS AND SODIUM CHLORIDE

The presence of sodium chloride has been found to help in enhanced oil recovery. Jennings et al. (1974) have shown that the presence of some sodium chloride reduces the bulk alkaline concentration required to achieve low IFT.

An ionic double layer is believed to exist on the surface of the oil in which the negatively charged acid species generate a negative potential on the surface which, in turn, creates another layer of counter ions, namely the positive sodium ions (Chan and Yen, 1981). Such a negatively-charged surface film will cause an electrostatic exclusion of hydroxide ions from the interface. The concentration of sodium hydroxide at the interface will, however, approach the bulk concentration. Inasmuch as the molecular sizes and activity coefficients of the chloride and hydroxide ions are very close, the addition of sodium chloride will also lead to the approach of the hydroxide concentration or pH of the interface to that of the bulk phase. Interfacial tension is reduced when the interfacial pH reaches the pK_a value of the native acids in the crude, which are thus ionized to form interfacially-active, charged acid species. Addition of salt lowers the bulk alkali concentration or the bulk pH needed to achieve the onset of the IFT drop, i.e., the point where the interfacial pH is equal to the pK_a values of the acids in the crude (Chan et al., 1982).

Thus, sodium chloride can be used to promote ultra-low interfacial tension of crude oils at a lower onset bulk aqueous pH. This is very useful in cases where

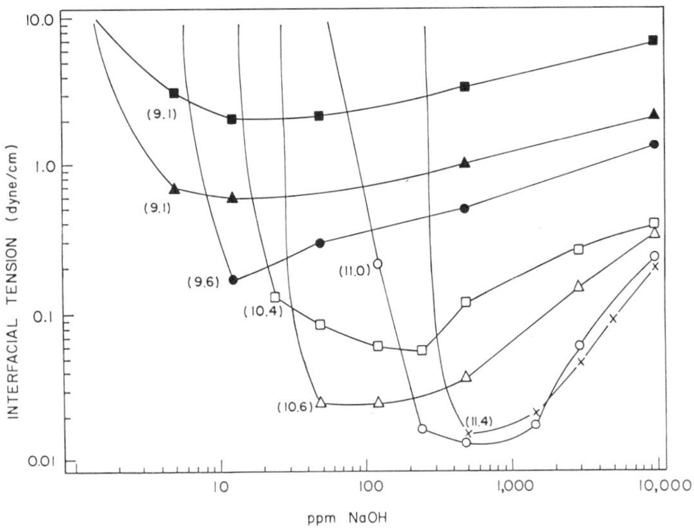

Fig. 9-12. Interfacial tension versus caustic concentration in the presence of different NaCl concentrations. The numbers in brackets indicate the onset bulk solution pH value. Wt% NaCl: × = 0.0; ○ = 0.1; △ = 0.75; □ = 2.0; ● = 5.0; ▲ = 10.0; ■ = 30.0. (After Chan and Yen, 1981, fig. 1; courtesy of IPC Science & Technology Press Ltd.)

too high an alkali concentration is undesirable due to economic, environmental, or technical reasons. There is, however, a disadvantage in doing this: the sodium ion concentration buildup accompanying the salt addition deactivates the soap formed and, therefore, increases the IFT value.

An optimum sodium chloride concentration may then be identified (Fig. 9-12) where a reasonably broad alkali response range is achieved without sacrificing too much in IFT (Chan and Yen, 1981).

EFFECT OF DIVALENT IONS

The presence of multivalent cations, such as magnesium or calcium, in reservoir brines can profoundly affect the oil recovery efficiency of an alkaline slug. The multivalent ions interact with the anionic surfactants generated by the caustic–reservoir oil reaction, rendering them water-insoluble. This precipitation of Ca^{2+} and Mg^{2+} (hardness) ions delays hydroxide transport and reduces the oil recovery efficiency. It can be avoided by using a soft brine preflush. Jennings et al. (1974) have shown the detrimental effects of divalent ions on interfacial activity. Inasmuch as calcium and magnesium hydroxides cannot be used as alkaline flooding chemicals and have been found to be ineffective in bringing about low IFT underscores the harmful effects of divalent ions. Mayer et al. (1983) pointed out that these precipitates can, if properly located, divert flow within the reservoir, leading to improved sweep efficiencies and higher recoveries. Falcone et al. (1982) described in detail alkali–surfactant reactions. From their studies, it can be inferred that the reduction of activity of the cations in the reservoir brine will lead to more surfactant activity, resulting in lower IFT values.

The divalent ion effect depends upon: (1) the ion exchange isotherm, (2) the divalent ion concentration, (3) injection pH and salinity, and (4) the solubility product of the precipitated salt(s). Bunge et al. (1980) have presented a linear chromatographic theory to quantitatively predict the hydroxide and divalent cation concentration profiles and effluent histories.

In laboratory tests, where the value of the dimensionless Damkohler number is very low, very little of the precipitated material redissolves. In the field, where flow rates are low and flow lengths are extremely large, resulting in a high Damkohler number, precipitate redissolution is probably complete (Bunge et al., 1980).

RESERVOIR SELECTION

It must be recognized that only a limited number of reservoirs may respond to alkaline flooding. After establishing such suitability, a comparative study must be done with other possible recovery techniques. General caustic flooding experience indicates that the basic criteria presented below may be useful in identifying reservoirs suitable for alkaline flooding.

Oil

Amount of residual oil-in-place. The amount of residual oil is fundamental to the design and evaluation of any recovery mechanism. Mungan (1981) suggested that the residual oil-in-place should at least be 400 bbl/acre-ft with a residual oil saturation of 40%. Higher oil saturations are obviously better. Lower oil saturations may not be displaceable by this immiscible displacement process. The caustic flooding process generally gives better results if applied to a field before a too advanced state of depletion has been reached.

Viscosity–gravity. In general, chemical floods are applicable to oils the viscosities of which are in the range of applicability of waterfloods (Poettmann, 1974). The upper limit is somewhat extended over that of a waterflood to about 100 cP. The gravity goes hand-in-hand and should be in the range 13–35° API for best results. Very heavy oils or very volatile oils make poor prospects. Crude oils with viscosities in the 100–200 cP range may be more amenable to thermal methods, but in certain cases may deserve additional investigation.

Chemical character of the oil. Inasmuch as the chemical character of the oil influences the IFT, it is important. In caustic flooding, the oil must have acidic components that can react with the caustic to form surface-active agents. As discussed before, an acid number of 0.5 is desirable. Ehrlich and Wygal (1976) suggested a minimum acid number of 0.1–0.2 mg KOH/g oil. The oil should preferably have no dissolved acid gases like H_2S and CO_2 that consume a lot of the injected alkali (Wilson, 1976).

Compressibility. Highly compressible oils are more suitable candidates for miscible flooding and should be generally avoided (Mungan, 1981).

Formation water

Connate water saturation. A high value of the connate water saturation usually implies a tight reservoir, a clayey–shaley formation, or both. In view of this, the initial connate water saturation should be less that 50% (Mungan, 1981).

Formation water salinity. Interfacial tension is sensitive to NaCl. Ideally, the salinity should be such as to produce the minimum interfacial tension (optimum NaCl concentration was discussed before). In an average situation, a formation water salinity less than 200,000 ppm is desirable.

Formation water chemistry. Besides NaCl, there are several other chemical species in the water that can strongly influence the behavior of all chemical floods. Divalent ions such as Ca^{2+} and Mg^{2+} are particularly detrimental, because they can deplete the injected alkaline slug by precipitation of insoluble hydroxides. In addition, the presence of Ca^{2+} results in a marked increase in interfacial tension. The acidic components, such as HCO_3^-, SO_4^{2-}, Cl^-, and H_2S, in the water that consume alkali also must be considered.

Rock mineralogy

Clays having a large surface area and high ion-exchange capacity can rapidly deplete the injected alkaline slug by transferring hydrogen, calcium and/or magnesium ions to the alkaline aqueous phase in exchange for sodium. Presence of anhydrite ($CaSO_4$) or gypsum ($CaSO_4 \cdot 2H_2O$) in greater than trace quantities is highly undesirable because they cause precipitation of insoluble $Ca(OH)_2$ and decrease in alkalinity of the injected slug. Thus, the reservoir rock should have a minimal amount of these minerals.

For these reasons carbonate reservoirs enriched in Ca^{2+} and Mg^{2+} are unsuited for alkaline flooding and only sandstone reservoirs are candidates for this process. The only reported case of alkaline flooding in a carbonate reservoir is in Hungary (Nagylengyel Field) where ammonium hydroxide has been used as the alkaline chemical.

Clays, especially of the expansible montmorillonite type, are also undesirable, because they may swell on encountering an injected water of a different salinity (and ionic composition) and reduce the permeability drastically. For example, if the Na/Ca ratio of injection water is higher than that of formation water, Ca-base clays can be converted to Na-base clays which swell more.

Reservoir characteristics

In addition, reservoirs that are not responsive to waterflooding are also not suitable for alkaline flooding. The major requirements and considerations are:

(1) The reservoir should not be very heterogeneous; otherwise some zones would be inaccessible.

(2) There should be no extensive fractures or thief zones causing the alkali to bypass the bulk of the rock with little or no response to the injection. Highly faulted and/or fractured reservoirs are unsuitable.

(3) Large degree of layering and shale barriers of large areal extent are detrimental to alkaline flooding. The net pay thickness should not be a small fraction of the gross formation thickness.

(4) The pay zone must be of a sizeable areal extent and sufficiently thick.

(5) An aquifer-connected water-drive reservoir is not suitable for alkaline flooding, particularly where a thin pay zone is underlain by a thick aquifer.

(6) The reservoir permeability and porosity should be well developed. An adequate injectivity must prevail, in view of the fact that the emulsions formed during alkaline flooding have lower mobility and some plugging always occurs due to formation damage and/or scale formation.

(7) The reservoir should not have a gas cap, because a sizeable amount of the mobilized oil could move up to resaturate the gas cap.

(8) The reservoir temperature is a critical parameter. Mungan (1966) found the alkaline flooding process to be temperature dependent. It has been shown (Cooke et al., 1974) that the alkali consumption increases rapidly at higher temperatures, and a reservoir temperature of less than 200°F is recommended. For the same reason, a reservoir depth of 6000 ft or less is preferable.

Other considerations may include an evaluation of the well completion data and equipment capability to ensure smooth operation.

Laboratory studies

Laboratory core flooding helps in the selection of suitable reservoirs and provides data essential to the design of the caustic flooding operation. It includes the measurements of the various parameters discussed before for oil, formation water, and rock. Subsequently, displacement tests are run.

Screening tests

(1) *Acid number*. The oil samples must be free from emulsion breakers, inhibitors and other oilfield chemicals. The two most popular techniques of measuring the acid number are:

(a) Color indicator titration (IP 139/69, ASTM D974-64). This method is used for crudes and crude products soluble or nearly soluble in mixtures of toluene and isopropyl alcohol. It is applicable for cases where the dissociation constants of the acids in water are larger than 10^{-9}. The method involves titration against KOH using *p*-naphtholbenzenium as indicator.

(b) Potentiometric titration. In this method, the constituents are resolved into groups having weak acid, strong acid, weak base, and strong base. A potentiometric titration is done against the KOH solution in a potentiometric titration cell.

(2) *Interfacial tension measurements*. These measurements can be made with the spinning drop instrument at ambient pressure–temperature conditions with dead-oil, or with the pendant drop apparatus at reservoir pressure–temperature conditions with the live-oil.

The interfacial tension may be measured as a function of caustic concentration, sodium chloride concentration, and the content of divalent ions. Thus, the "total" IFT behavior should be determined.

(3) *Wettability alterations*. Imbibition studies or contact angle measurements can be undertaken to determine the wettability of the rock and its change on addition of caustic. The samples have to be aged for a sufficient length of time (200 hours). Imbibition studies are generally preferred.

(4) *Emulsion formation*. The emulsion formation characteristics are studied using standard API testing techniques. In simplistic terms, this involves shaking together measured volumes of known concentration of NaOH solutions and crude oil. The type of emulsions formed and their viscosity are determined. Then emulsions are allowed to age to determine their stability.

(5) *Brine–crude oil interaction (formation of rigid films)*. The formation of rigid films on contact of the crude oil with brines can be studied using contact angle and interfacial tension measurements.

(6) *Caustic consumption (static test)*. Crushed core is cleaned by toluene extraction and dried. Then it is shaken in a jar or beaker with a known volume of caustic solution of a known concentration. The alkali loss is determined after a desirable period of time (for example, 200 hours or more).

This simplistic procedure may be valuable in view of the limitations of the dynamic consumption experiments where retention time is generally very low. The batch experiment can simulate the reservoir retention time in the laboratory. Reactions of the reservoir rock comprising silica, quartz, and clays with alkali are very complex. For clays, a fast, reversible sodium–hydrogen ion exchange occurs, whereas fine silica and quartz exhibit irreversible dissolution. Slower dissolving clay minerals also may be present. The batch experiment may be useful when reactions are irreversible and very slow, and the alkali consumption, consequently, increases steadily with time.

Laboratory displacement tests

After applying the screening criteria, positive results lead to the excruciating laboratory tests on reservoir core samples. First, the optimum injection rate and caustic concentration are established by running flooding experiments on synthetic cores. Then, reservoir core samples are used. Smaller core plugs are used at this stage to save core material. These plugs may be butted end to end to obtain the desired length. Unaltered reservoir crude oil and connate water and actual injection water should be used, and the experiments are generally conducted at reservoir conditions of temperature and overburden pressure.

These tests provide a first estimate of several properties. Reservoir permeability and oil recovery as a function of pore volumes of injected caustic are determined. The experiments can be conducted on water-flooded cores to determine the incremental recovery by using caustic. The emulsions formed are

TABLE 9-III

A. Completed alkaline flooding field tests (after Mayer et al., 1983, table 1; courtesy of the Society of Petroleum Engineers of AIME)

Project	Field	Operator	API gravity	Acid number (mg KOH /g)	Oil viscosity at reservoir temperature (cP)	Reservoir temperature (°F)	Oil-saturation at project start (%)
1	Bradford, Pa.	several tests by several operators	–	–	–	–	60
2	Southeast Texas, Texas	Exxon	low	2.4	75	112	40
3	Harrisburg, Nev.	Amoco	high	low	1.5	200	watered out, 30–40
4	North Ward-Estes, Texas	Gulf	34°	0.22	2.28	115	64
5	Singleton, Nev.	Sinclair	~ 40°	low	1.5	–	watered out, 40 [c]
6	Whittier, Calif.	Chevron	20°	–	40	120	51
7	Brea-Olinda, Calif.	Union (MCCF)	16°	–	90	135	watered out, 50–60
8	Orcutt Hill, Calif.	Union (MCCF)	22°	0.6	17–60	140	50
9	Dominguez, Calif.	Shell	–	–	–	–	30
10	Van, Texas	Union	–	–	–	–	25–35
11	Kern River, Calif. [a]	Getty	13.5°	1.6–1.8	1000 +	90	52.4
12	Epping, Sask., Canada [b]	Canadian Reserve (Getty)	14°	–	150	72	high

studied under the microscope. Their flow properties and stability are also studied.

Caustic consumption under dynamic conditions is determined by analyzing the effluent from the coreflood. This analysis provides other valuable data, such as the reaction sensitivity to injected caustic concentrations, salt concentration, and precipitation. Slim-tube displacement experiments are also desirable because they provide a long residence time for the injected alkali.

The laboratory evaluation program also includes thin-section studies on: (1)

Porosity (%)	Net thickness (ft)	Permeability/ variation (mD/–)	Project PV (MMbbl)	Spacing, pattern type	Injection rate (bbl/day per well)	Time to inject 1 PV (years)	Injection water salinity (ppm)	Depth (ft)
16–20	30	–	–	180-ft line drive	–	–	–	–
33–35	–	–	–	–	–	–	saline (?)	1250
15	10	119/–	0.47 [c]	40 acre/well irregular	800–1000	8.5	300	5900
20.6	36	20/0.6–0.7	0.283	5 acre/well five-spot	100	2.2	850	3140
16	–	280/–	0.756	12 acre/well peripheral	1000	2	–	–
30	137	320–495 0.66–0.74	7 (field total, 20)	1–2 acre/well line drive	1000	3.8	fresh hardness < 1	1500–2100
–	–	–	–	two injectors 800 ft apart, seven producers	–	–	–	–
22.5	155	71/0.7	113	11 acre/well irregular five-spot	1100	18	15,000	2200
–	–	–	–	–	–	–	–	–
–	–	–	–	–	–	–	–	–
35	88	2000	–	nine $2\frac{1}{2}$ acre/well five-spot	–	–	–	1000
32	20	1000	2	20 acre/well five-spot	300	18	fresh	1900

untreated reservoir samples, (2) cores that have been waterflooded, (3) cores that have been caustic-flooded, (4) cores that have been waterflooded–caustic flooded–waterflooded, sequentially, and (5) sand-packed Hele-Shaw models. Comparisons between these studies at the grain-level provide valuable insight into (1) behavior of fines—their generation and migration under the influence of various injected waters; (2) in-situ emulsification, and its advance through the sample; (3) adsorption of various materials on grains; (4) behavior and response of various clays to the various injected waters; and (5) formation of various precipitates in situ (Goyal et al., 1983).

PILOT TESTING

Before applying any enhanced oil recovery project on a field-wide scale, it is of utmost importance to pilot test it. This enables the operator to gain experience

TABLE 9-III (continued)

B. Operating data of completed tests (after Mayer et al., 1983, table 2; courtesy of the Society of Petroleum Engineers of AIME)

Project	Field	Operator	Type of alkaline material	Alkali concentration injection (wt.%)	Slug size (%PV)
1	Bradford, Pa.	several tests by several operators	Na_2CO_3	–	–
2	Southeast Texas, Texas	Exxon	Na_2CO_3	3.2	–
3	Harrisburg, Nev.	Amoco	NaOH	2.0	9
4	North Ward-Estes, Texas	Gulf	NaOH	5.0	15
5	Singleton, Nev.	Sinclair	NaOH	2.0	8
6	Whittier, Calif.	Chevron	NaOH	0.2	20
7	Brea-Olinda, Calif.	Union (MCCF)	orthosilicate	0.12	1.2
8	Orcutt Hill, Calif.	Union (MCCF)	orthosilicate	0.42	2
9	Dominguez, Calif.	Shell	NaOH	0.40	5.0
10	Van, Texas	Union	orthosilicate	0.14	12.0
11	Kern River, Calif. [a]	Getty	NaOH	0.15	48.0
12	Epping, Sask., Canada [b]	Canadian Reserve (Getty)	NaOH	1.0	planned 25

[a] Caustic and steam.
[b] Project suspended.
[c] Estimated.
[d] Based on NaOH at U.S.$0.15/lbm.
[e] See Mayer et al. (1983) for original references.

in applying the process in the field and acquaints him with the problems that may not have been anticipated before. There are many reasons for conducting a pilot test and it is almost always desirable, except when: (1) the reserves are insufficient or the field is undergoing rapid depletion; (2) the company has adequate experience and expertise with the process and sufficient successful field applications are known in similar formations; and (3) laboratory and other studies are overwhelmingly favorable. In addition to providing experience in applying the process in the field, a properly designed pilot flooding project can yield the following information:

(1) Define chemical requirements.

(2) Provide acquaintance with operating problems.

(3) Determine injectivity, feasibility of converting producers to injectors, equipment sizing, etc.

Chemical injected (10^6 lbm)	Pounds alkali barrel-PV	Incremental recovery (%PV)	Barrels incremental oil/ pounds chemical	Chemical cost of incremental oil recovery [d] (U.S.$/bbl)	Reference [e]
0.03 per acre	–	–	–	–	2–5
–	–	–	–	–	6
0.28	0.6	2	0.03	4.80	10
0.72	2.6	6 to 8	0.024–0.031	4.70–6.35	12
0.42	0.55	2	0.042	3.60	13
1.1	0.16	5 to 7	0.32–0.43	0.36–0.48	14
0.3	0.005	< 1	low	high	16
3.2	0.028	< 1	0.025–0.141	1.67–6.00	17
0.001	0.07	< 1	low	high	7
0.9	0.06	< 1	low	high	7
0.083	0.24	0	0	∞	9
planned 1.8	planned 0.9	–	–	–	–

(4) Determine incremental oil recovery and residual oil and the producing rates.

(5) Assess the overall process feasibility and economics.

Although it may not yield conclusive results, pilot testing nevertheless provides a direct indication of the performance that would be expected of a field-scale flood. It is not normally possible to design a pilot to yield all the desired information. For example, caustic consumption can be determined by injecting caustic and producing it back in a single-well test fairly easily. But such a test cannot provide any useful data on incremental oil recovery or oil-producing rates. Pilots, therefore, are generally designed to yield only the most critical and yet adequate information for meaningful extension of the system to a field-wide operation.

TABLE 9-IV

A. Reservoir characteristics of alkaline flooding field tests in progress (after Mayer et al., 1983, table 3; courtesy of the Society of Petroleum Engineers of AIME)

Project	Field	Operator	API gravity	Acid number (mg KOH /g)	Oil viscosity at reservoir temperature (cP)	Reservoir temperature (°F)	Oil saturation at project start (%)
1	Bison Basin, Wyo.	Gulf	16	2.5	220	85	62
2	San Miguelito, Calif.	Conoco	31	–	0.7	205	56
3	Huntington Beach, Calif.	Aminoil	22	0.7	15	165	30
4	North Ward-Estes, Texas, No. 2	Gulf	32	0.22	1.39	86	45
5	North Ward-Estes, Texas, No. 1	Gulf	32	0.22	1.39	86	32
6	Toborg, Texas	Gulf Oil Co.	21.9	3.7	82	76	50
7	Wilmington, Calif.	THUMS/ City of Long Beach/ U.S. DOE	17	2.05	23	125	51
8	Smackover, Ark.	Phillips	20	–	75	110	50
9	Quarantine Bay, La.	Gulf	33.2	–	4.85	77	47
10	Isenhour Unit, Wyo.	Belco	43	–	2.8	97	35

The important rules to follow in planning a pilot test are (after Smith, 1966):

(1) The location of the pilot test should be within the area of the final full-scale flood.

(2) The area selected for the pilot test should be representative of the entire reservoir. If there is no one such area, then several areas should be selected. The areas selected should have wells that are in good mechanical condition.

(3) The information required and the time period allowable must be specified before starting the test.

The first step in pilot flood project design is to choose the well patterns to perform the task adequately and economically. It should be emphasized that the primary purpose of a pilot test should not be to determine the recovery factor or percentage of oil recovery from the reservoir: recovery indicated by pilot floods

Porosity (%)	Net thickness (ft)	Permeability/ variation (mD/–)	Project PV (1000 bbl)	Spacing, pattern type	Injection rate (bbl/day per well)	Time to inject 1 PV (years)	Injection water salinity (ppm TDS)	Depth (ft)
22	16	144/–	5.25	10 acres line	600	3.9	brackish	1300
15.1	–	36/–	–	316 acres 18 producers 14 injectors	–	–	–	8000
24	267	150/0.7	32	7 acres/well five-spot	2500	4	7500	3500
19.8	39	9.1/–	2.7	10 acres inverted nine-spot	500	14.8	62,800	2450
21.2	25	65/–	1.75	10 acres inverted nine-spot	555	8.63	62,800	2600
20	21.6	216/–	0.642	2.5 acres five-spot	200	2.2	1678	400
25	305	240/0.7	100	8.4 acres/producer staggered line	4000	10	fresh < 1 hardness	2700
35	–	2000/–	–	14 acres six producers	–	–	–	2000
27.4	–	50/500	–	200 acres five producers three injectors	–	–	–	8230
15.5	15.1	21/–	–	173 acres six producers three injectors	–	7.5	–	3570

varies from a few percent to more than 150% of the actual field recovery.

The pilot caustic flooding process generally consists of a sequential injection of: (1) a preflush slug; (2) alkaline slug plus a tracer; (3) post-treatment (if mobility control slug is used, it is called MCCF, "mobility controlled caustic flooding") slug with tracer; and (4) drive (or chase) water plus a tracer.

(1) *Preflush slug.* The preflush slug is used to remove the divalent ions and any highly saline brines in the reservoir and thereby increase the alkaline bank efficiency. The preflush is usually a relatively fresh water, or a sodium orthosilicate solution. These can remove the divalent hardness ions by base exchange and also effectively screen the alkaline slug from the reservoir brine. The rear portion of the preflush slug may contain sacrificial inorganic salts (sodium tripolyphosphate, sodium carbonate, etc.) to improve the water-wetness of the rock and also aid in divalent ion removal. The effects of the preflush on injectivity must be pre-evaluated in the laboratory.

(2) *Alkaline slug.* The alkaline slug is the second slug to be injected. The primary purpose is to reduce the IFT to 10^{-3} dynes/cm or lower. The concentra-

TABLE 9-IV (continued)

B. Operating data of tests in progress (after Mayer et al., 1983, table 4; courtesy of the Society of Petroleum Engineers of AIME)

Project	Field	Operator	Type of alkaline material	Alkali concentration injection (wt.%)	Slug size (%PV)
1	Bison Basin, Wyo.	Gulf Oil Co.	NaOH	0.5	50
2	San Miguelito, Calif.	Conoco Inc.	NaOH	–	–
3	Huntington Beach, Calif.	Aminoil U.S.A.	orthosilicate	1.0	40
4	North Ward-Estes, Texas, No. 2	Gulf Oil Co.	NaOH	5.0	15
5	North Ward-Estes, Texas, No. 1	Gulf Oil Co.	NaOH	5.0	15
6	Toborg, Texas	Gulf Oil Co.	NaOH	0.2	50
7	Wilmington, Calif.	THUMS Long Beach Co./ City of Long Beach/ U.S. DOE	orthosilicate	0.4	60
8	Smackover, Ark.	Phillips Petroleum Co.	Na_2CO_3	–	–
9	Quarantine Bay, La.	Gulf Oil Co.	NaOH	5.0	15
10	Isenhour Unit, Wyo.	Belco Petroleum Corp.	Na_2CO_3 + polymer	–	10

[a] Alkaline slug 200 bbl/acre-ft Na_2CO_3 + 30 bbl/acre-ft polymer.
[b] Based on NaOH at U.S.$0.15/lbm.
[c] Estimated.
[d] See Mayer et al. (1983) for original references.

tion of the alkali must be sufficient to satisfy the consumption reactions in the reservoir. The salinity is designed to yield optimum conditions in the reservoir (the lowest IFT is obtained at an optimum sodium chloride concentration). Inasmuch as the alkali itself is consumed in the reservoir, it is not reliable or adequate as the indicator of breakthrough and sweep efficiencies. A tracer, therefore, is included in this bank. Tritiated water is a good tracer (Emery et al., 1970; Mungan, 1981a, b), because it is not adsorbed in the reservoir and can be detected readily even in low concentrations.

(3) *Post-treatment mobility control slug.* Dilute mineral acids can be used to liberate additional new organic acids from the nonreactive crude components, such as esters, amides, and acid-base complexes (Chan et al., 1982), resulting in an enhanced interfacial tension lowering of the oil. A slug of dilute acid can thus capture and mobilize some of the oil residual to the alkaline slug by regenerating interfacial activity in the oil. The slug must be designed to provide for the acid consumption by the reservoir rock. Sandstone reservoirs with low clay and carbonate content are suitable for such treatment.

Chemical injected (10^6 lbm)	Pounds alkali barrel-PV	Date started	Barrels incremental oil/ pounds chemical	Chemical cost of incremental oil recovery [b] (U.S.$/bbl)	Reference [d]
4.6	88	June 1980	–	–	19, 20
–	–	1981	–	–	20
45	1.4	March 1978	–	–	21, 22
7.0	2.6	July 1977	–	–	20
4.6	2.6	July 1977	–	–	20
0.22	0.34	April 1980	–	–	19, 20
84	0.84	Jan. 1979	–	–	23–28
–	–	–	–	–	20
21.2	–	Nov. 1981	–	–	29
[a]	–	Nov. 1979	0.13 [c]	4.40 [c]	30

Mobility control is another important feature that can be provided in this slug. Various polymers have been suggested. The salinity is kept lower than those of the other slugs. Sufficient polymer concentration is provided to yield a locally stable mobility condition at the rear of the oil bank. The viscosity of the slug may be graded in decreasing manner with distance from the rear of the oil bank.

(4) *Drive water*. The drive fluid, almost always water, is injected in the end and serves to push the three slugs ahead. A tracer is included to detect the arrival of the trailing edge of the slugs ahead and also to indicate the effectiveness of the alkaline bank in improving the volumetric sweep efficiency.

The field equipment during the pilot flooding includes storage for alkaline solutions of various concentrations, tracer mixing and storage tanks, injection pumps, surface lines and metering and sampling devices for the injected as well as produced fluids.

DOCUMENTED ALKALINE FLOODING FIELD TESTS

A complete documentation of all field tests on alkaline flooding is not available. A partial list of documented known field tests compiled by Mayer et al. (1983) is shown in Tables 9-III and 9-IV. Results of these tests vary greatly from one field to another, because of the differences in the injection chemical, injection rate and quantity of the chemical, nature of the reservoir, nature of the crude oil, etc. In many cases, detailed results are not available. Some of these projects have been successful, although recovery has generally been lower than that indicated in the laboratory (see Graue and Johnson, 1974).

Several important observations can be made from these published results of alkaline flooding field applications. Some of these are:

(1) Oil recovery in the field has been lower than that predicted on the basis of laboratory studies. Apparently, reservoir heterogeneities are responsible for this. A complete geological and reservoir engineering study is generally recommended before the implementation of any enhanced recovery process.

(2) Caustic consumption in the field has been higher than that measured in the laboratory. Laboratory investigations have to be made under more realistic conditions, simulating the reservoir conditions more closely. Even then, single-well tests may be necessary to define the actual chemical requirements (Mungan, 1981).

(3) At the injection end, laboratory studies have shown injectivity problems and permeability damage that has not been seen in the field. Such problems in the field have been almost absent.

(4) At the producing end in the field, scaling and plugging problems have been encountered; no such problems occurred in the laboratory.

SUMMARY

In conclusion, alkaline flooding has the following features:

(1) It is the cheapest enhanced oil recovery process.

(2) It is relatively simple.

(3) It has shown success in some field attempts and has a potential for giving greater incremental oil recovery in the future.

(4) The primary problem in caustic flooding is caustic consumption. It is difficult to predict and model, and is the most crucial design parameter.

(5) The mechanisms are fairly well understood only in a qualitative sense.

(6) Application is restricted to certain reservoirs and crude oils and project implementation requires careful studies.

A great variety of combinations of alkaline flooding with other chemicals have been suggested. Inasmuch as these processes (for example, caustic with steam, surfactant, etc.) have yet to be evaluated thoroughly, they are not discussed here.

REFERENCES

Ban, A., Balint, V. and Pach, F, 1977. Ammonia enhances production from Hungarian field. *Oil Gas J.*, 75(28): 149–152.

Bansal, V.K., Chan, K.S., McCallough, R. and Shah, D.O., 1978. The effect of caustic concentration on interfacial charge, interfacial tension and droplet size: a simple test for optimum caustic concentration for crude oils. *J. Can. Pet. Technol.*, 17(1): 56–67.

Bardon, C. and Longeron, D.G., 1980. Influence of very low interfacial tensions on relative permeability. *Soc. Pet. Eng. J.*, 20(5): 391–401.

Bourgoyne, A.T., Jr., Caudle, B.H. and Kimbler, O.K., 1972. The effect of interfacial films on the displacement of oil by water in porous media. *Soc. Pet. Eng. J.*, 12(1): 60–68.

Bunge, A.C. and Radke, C.J., 1981. Migration of alkaline pulses in reservoir sands. *56th SPE Annu. Fall Tech. Conf. and Exhibition, San Antonio, Texas, October 5–7*, SPE 10288: 11 pp.

Bunge, A.C., Klein, G. and Radke, C.J., 1980. Divalent ion exchange with alkali. *5th SPE Int. Symp. on Oilfield and Geothermal Chemistry, Stanford, Calif., May 28–30*, SPE 8995: 24 pp.

Campbell, T.C., 1980. The role of alkaline chemicals in the recovery of low gravity crude oils. *50th SPE Annu. Calif. Reg. Meet., Los Angeles, Calif., April 9–11*, SPE 8894: 6 pp.

Campbell, T.C. and Krumrine, R.H., 1979. Laboratory Studies on Alkaline Waterflooding. *54th SPE Annu. Fall Meet., Las Vegas, Nev., September 23–26*, SPE 8328.

Cash, R.L., Cayias, J.L., Fournier, G., Jacobson, J.K., Schares, T., Schechter, R.S. and Wade, W.H., 1976. Modelling crude oils for low interfacial tension. *SPE-AIME Symp. on Improved Oil Recovery, Tulsa, Okla., March 22–24*, SPE 5813.

Castor, T.P., 1979. *Enhanced Recovery of Acidic Oils by Alkaline Agents.* Ph.D. Thesis, Department of Mechanical and Chemical Engineering, University of California, Berkeley, Calif., 398 pp.

Castor, T.P., Kelly, J.F. and Somerton, W.H., 1978. Physical mechanisms of alkaline flooding. *48th Annu. SPE Calif. Reg. Meet., San Francisco, Calif., April 12–14*, SPE 7143.

Castor, T.P., Somerton, W.H. and Kelly, J.F., 1979. Recovery mechanisms of alkaline flooding. *3rd Int. Conf. on Surface and Colloid Science, Surface Phenomena in EOR Sect., Stockholm, August 20–25.*

Chan, M. and Yen, T.F., 1981. Role of sodium chloride in the lowering of interfacial tension between crude oil and alkaline aqueous solution. *Fuel*, 60(6): 552.

Chan, M., Sharma, M.M. and Yen, T.F., 1982. Generation of surface-active acids in crude oil for caustic flooding enhanced oil recovery. *Ind. Eng. Chem. Process Des. Dev.*, 21: 580–583.

City of Long Beach and THUMS Long Beach Co., 1977–1979. *Caustic Waterflooding Demonstration Project, Ranger Zone, Long Beach Unit, Wilmington Field, California*. First (1976-77), Second (1977-78), and Third (1978-79) Annual Reports, U.S. Department of Energy, DOE SAN/1396.

Cooke, C.E., Jr., Williams, R.E. and Kolodzie, P.A., 1974. Oil recovery by alkaline waterflooding. *J. Pet. Technol.*, 26(12): 1365–1374.

de Ferrer, M., 1977. Injection of caustic solutions in some Venezuelan crude oils. In: *The Oil Sands of Canada and Venezuela. CIM Spec. Vol.*, 17: 696–704.

de Zabala, E.F., Vislocky, J.M., Rubin, E. and Radke, C.J., 1980. A chemical theory for linear alkaline flooding. *5th Annu. SPE Int. Symp. on Oilfield and Geothermal Chemistry, Stanford, Calif., May 28–30*, SPE 8997: 199–211.

Denekas, M.O., Mattax, C.C. and Davis, G.T., 1959. Effect of crude oil components on rock wettability. *J. Pet. Technol.*, 11(11): 330–333.

Donaldson E.C. and Crocker, M.E., 1980. Characterization of the crude oil polar compound extract. *DOE Rep.*, DOE/BETC/RI-80/5 (October), 27 pp.

Dranchuk, P.M., Scott, J.D. and Flock, D.L., 1974. Effect of the addition of certain chemicals on oil recovery during waterflooding. *J. Can. Pet. Technol.*, 13(3): 27–36.

Ehrlich, R. and Wygal, R.J., Jr., 1976. Interrelation of crude oil and rock properties with the recovery of oil by caustic waterflooding. *SPE-AIME Symp. on Improved Oil Recovery, Tulsa, Okla., March 22–24*, SPE 5830.

Ehrlich, R., Hasiba, H.H. and Raimondi, P., 1974. Alkaline waterflood for wettability alteration—evaluating a potential field application. *J. Pet. Technol.*, 26(12): 1335–1343.

Emery, L.W., Mungan, N. and Nicholson, R.W., 1970. Caustic slug injection in the Singleton Field. *J. Pet. Technol.*, 22(12): 1569–1576.

Falcone, J.S., Krumrine, P.H. and Schweiker, G.C., 1982. The use of sacrificial agents in combination with surfactants in enhanced oil recovery. *J. Am. Oil Chem. Soc.*, 59 (October): 1826A–1832A.

Farmanian, P.A., Davis, N., Kwan, J.T., Yen, T.F. and Weinbrandt, R.M., 1978. Isolation of native petroleum fractions for lowering interfacial tensions in aqueous alkaline system. *Symp. on Chemistry of Oil Recovery, Div. Pet. Chem., Am. Chem. Soc., Anaheim, Calif., March 12–17.*

Farouq Ali, S.M., Figueroa, J.M., Azuaje, E.A. and Farquharson, R.G., 1979. Recovery of Lloydminster and Morichal crudes by caustic, acid and emulsion floods. *J. Can. Pet. Technol.*, 18(1): 53–59.

Foster, W.R., 1973. A low-tension waterflooding process. *J. Pet. Technol.*, 25(2): 205–210.

Goyal, K.L., 1982. *Improvement in Oil Recovery by Caustic Flooding*. Ph.D. Thesis, Indian School of Mines, Dhanbad.

Goyal, K.L. and Arora, P.D., 1981. Enhanced recovery of viscous oil by caustic flooding. *Soc. Pet. Eng. Annu. Fall Tech. Conf., Dallas, Texas*, SPE 7902.

Goyal, K.L., Tanwar, N.S., Chilingar, G.V. and Yen, T.F., 1983. Study of the effects of various waters on reservoir rocks during waterflooding using scanning electron microscope. *Energy Sour.*, 7(2): 95–120.

Graue, D.J. and Johnson, C.E., Jr., 1974. Field trial of caustic flooding process. *J. Pet. Technol.*, 26(12): 1353–1358.

Haines, W.B., 1930. Studies in physical properties of soil. *J. Agric. Sci.*, 20: 97.

Harvey, R.R., 1960. Theoretical approach to the investigation of films occurring at crude oil-water interfaces. *Trans. Soc. Pet. Eng. AIME*, 219: 350–353.

Hasiba, H.H., 1966. *Interface Active Compounds from Crude Oils*. Ph.D. Thesis, University of Texas at Austin, Austin, Texas.

Jennings, H.Y., Jr., 1975. A study of caustic solution-crude oil interfacial tensions. *Soc. Pet. Eng. J.*, 15(3): 197–202.

Jennings, H.Y., Jr., Johnson, C.E., Jr. and McAuliffe, C.D., 1974. A caustic waterflooding process for heavy oils. *J. Pet. Technol.*, 26(12): 1344–1352.

Johnson, C.E., Jr., 1976. Status of caustic and emulsion methods. *J. Pet. Technol.*, 28(1): 85–91.

Kimbler, O.K., Reed, R.L. and Silberberg, I.H., 1966. Physical characteristics of natural films formed at crude oil-water interfaces. *Soc. Pet. Eng. J.*, 6(2): 153–165.

Larson, R.G., Davis, H.T. and Scriven, L.E., 1982. Elementary mechanisms of oil recovery by chemical methods. *J. Pet. Technol.*, 34(2): 243–258.

Leach, R.O., Wagner, O.R., Wood, H.W. and Harpke, C.F., 1962. A laboratory and field study of wettability adjustment in water flooding. *J. Pet. Technol.*, 14(2): 206–212.

Lefebvre du Prey, E.J., 1973. Factors affecting liquid-liquid relative permeabilities of a consolidated porous medium. *Soc. Pet. Eng. J.*, 13(1): 39–47.

Leverett, M.C., 1939. Flow of oil-water mixtures through unconsolidated sands. *Trans. Soc. Pet. Eng. AIME*, 132: 149–171.

Lieu, V.T., 1979. Long term alkaline consumption in reservoir sand. *Prog. Rep./Minutes Meet., December 14, City of Long Beach, THUMS Long Beach Co.*

Mayer, E.H., Berg, R.L., Carmichael, J.D. and Weinbrandt, R.M., 1983. Alkaline injection for enhanced oil recovery—a status report. *J. Pet. Technol.*, 35(1): 209–221.

McAuley, R.G., 1977. Sodium hydroxide enhanced waterflood in the Wainwright Field. *28th Annu. Technol. Meet., Pet. Soc. CIM, Edmonton, Alta., May 30 – June 3.*

McAuliffe, C.D., 1973a. Crude oil-in-water emulsions to improve fluid flow in an oil reservoir. *J. Pet. Technol.*, 25(6): 721–726.

McAuliffe, C.D., 1973b. Oil-in-water emulsions and their flow properties in porous media. *J. Pet. Technol.*, 25(6): 727–733.

Melrose, J.C., 1970. Interfacial phenomena as related to oil recovery mechanisms. *Can. J. Chem. Eng.*, 48(6): 638–644.

Melrose, J.C. and Bradner, C.F., 1974. Role of capillary forces in determining microscopic displacement efficiency for oil recovery by waterflooding. *J. Can. Pet. Technol.*, 13(4): 54.

Michaels, A.S. and Timmins, R.S., 1960. Chromatographic transport of reverse-wetting agents and its effect on oil displacement in porous media. *Trans. Soc. Pet. Eng. AIME*, 219: 150–157.

Minnssieux, L., 1977. Waterflood improvement by means of alkaline water. *Symp. on Enhanced Oil Recovery by Displacement with Saline Solutions, British Petroleum, London, May 20.*

Morgan, J.C., Schechter, R.S. and Wade, W.H., 1976. Recent advances in the study of low interfacial tension. *AIChE Symp. on Improved Oil Recovery by Surfactant and Polymer Flooding, Kansas City, Kans.*

Morrow, N.R., 1970. Physics and thermodynamics of capillary action in porous media. *Ind. Eng. Chem.*, 62(6): 32.

Mungan, N., 1966. Certain wettability effects in laboratory waterfloods. *J. Pet. Technol.*, 18(2): 247–252.

Mungan N., 1981a. Enhanced oil recovery using water as a driving fluid, 4. Fundamentals of alkaline flooding. *World Oil*, 192 (6, June): 209–220.

Mungan, N., 1981b. Enhanced oil recovery using water as a driving fluid, 5. Alkaline flooding field applications. *World Oil*, 193 (1, July): 181–190.

Noran, D., 1978. Growth marks enhanced oil recovery. *Oil Gas J.*, 76(13): 113–140.

Nutting, P.G., 1927. Soda process for petroleum recovery. *Oil Gas J.*, 24(45): 76 and 150.

Nutting, P.G., 1928. Petroleum recovery by soda process. *Oil Gas J.*, 27(22): 146 and 238.

Pasquarelli, C.H. and Wasan, D.T., 1979. The effect of film forming materials on the dynamic interfacial properties in crude oil–aqueous systems. *3rd Int. Conf. on Surface and Colloid Science, Surface Phenomena in EOR Sect., Stockholm, August 20 – 25.*

Phillips Petroleum Co., 1979. Drilling-production U.S. briefs. *Oil Gas J.*, 77(15): 92.

Poettmann, F.H., 1974. *Microemulsion Flooding. Secondary and Tertiary Oil Recovery Processes.* Interstate Oil Compact Commission, Oklahoma City, Okla.

Radke, C.J. and Somerton, W.H., 1977. Enhanced recovery with mobility and reactive tension agents. *3rd ERDA Symp. on Enhanced Oil and Gas Recovery and Improved Drilling Methods, Tulsa, Okla.*, Paper B-5.

Radke, C.J. and Somerton, W.H., 1978. Enhanced recovery with mobility and reactive tension agents. *4th DOE Symp. on Enhanced Oil and Gas Recovery and Improved Drilling Methods, Tulsa, Okla., August 29 – 31*, Paper B-2.

Radke, C.J. and Somerton, W.H., 1979a. Enhanced recovery with mobility and reactive tension agents. *DOE Q. Prog. Rep.*, DOE BETC-Eng. 48-2 (April 1).
Radke, C.J. and Somerton, W.H., 1979b. Enhanced recovery with mobility and reactive tension agents. *5th Annu. DOE Symp., Tulsa, Okla., August 22–24*, Paper C-5.
Raimondi, P., Gallagher, B.J., Bennett, G.S., Ehrlich, R. and Messmer, J.H., 1976. Alkaline waterflooding: design and implementation of a field pilot. *SPE-AIME Symp. on Improved Oil Recovery, Tulsa, Okla., March 22–24*, SPE 5831.
Ramakrishna, T.S. and Wasan, D.T., 1982. A model for interfacial activity of acidic crude oil–caustic systems for alkaline flooding. *SPE/DOE Enhanced Oil Recovery Symp., Tulsa, Okla., April 4–7*, SPE 10716.
Reisberg, J. and Doscher, T.M., 1956. Interfacial phenomena in crude-oil-water systems. *Prod. Mon.*, November, p. 43.
Sarem, A.M., 1974a. Secondary and tertiary recovery of oil by MCCF (mobility-controlled caustic flooding) process. *44th Annu. Calif. Reg. Meet., San Francisco, Calif., April 4–5*, SPE 4901: 6 pp.
Sarem, A.M., 1974b. Mobility-controlled caustic flood. *U.S. Patent*, No. 3,805,893.
Smith, R.C., 1966. *Mechanics of Secondary Recovery*. Reinhold, New York, N.Y., Chapter 4.
Somerton, W.H. and Radke, C.J., 1979. Role of clays in enhanced recovery of petroleum. *5th Annu. DOE Symp., Tulsa, Okla., August 22–24*, Paper D-7.
Stegemeier, G.L., 1974. Relationship of trapped oil saturation to petrophysical properties of porous media. *SPE-AIME Symp. on Improved Oil Recovery, Tulsa, Okla., April*, SPE 4754.
Stegemeier, G.L., 1977. Mechanisms of entrapment and mobilization of oil in porous media. In: D.O. Shah and R.S. Schechter (Editors), *Improved Oil Recovery by Surfactant and Polymer Flooding*. Academic Press, New York, N.Y., pp. 55–92.
Subkow, P., 1942. Process for removal of bitumen from bituminous deposits. *U.S. Patent*, No. 2,288,857.
Taber, J.J., 1969. Dynamic and static forces required to remove a discontinuous oil phase from porous media containing both oil and water. *Soc. Pet. Eng. J.*, 9(1): 3–12.
Talash, A.W., 1976. Experimental and calculated relative permeability data for systems containing tension additives. *SPE-AIME Symp. on Improved Oil Recovery, Tulsa, Okla., March 22–24*, SPE 5801.
Wade, J.E. and Lechtenberg, H.J., 1978. *Caustic Waterflooding Demonstration Project, Ranger Zone, Long Beach Unit, Wilmington Field, California*. First Annual Report (1976-77), U.S. Department of Energy, DOE SAN/1396-1.
Wagner, O.R. and Leach, R.O., 1959. Improving oil displacement efficiency by wettability adjustment. *J. Pet. Technol.*, 11(4): 65–69.
Wagner, O.R. and Leach, R.O., 1966. Effect of interfacial tension on displacement efficiency. *Soc. Pet. Eng. J.*, 6(4): 335–344.
Wasan, D.T., 1979. The mechanism of oil bank formation, coalescence in porous media and emulsion stability. *DOE Annu. Rep.*, DOE/BC/110069-6 (August).
Wasan, D.T. et al., 1979. The mechanism of oil bank formation, coalescence in porous media and emulsion stability. *5th Annu. DOE Symp., Tulsa, Okla., August 22–24*, Paper C-2.
Weinbrandt, R.M., 1978. Improved oil recovery by alkaline flooding in the Huntington Beach Field. *4th Annu. DOE Symp. on Enhanced Oil and Gas Recovery and Improved Drilling Methods, Tulsa, Okla., August*, Paper B-70.
Weinbrandt, R.M., 1979. Improved oil recovery by alkaline flooding in the Huntington Beach Field. *5th Annu. DOE Symp., Tulsa, Okla., August 23–24*, Paper C-4.
Wilson, L.A., Jr., 1976. Physico-chemical environment of petroleum reservoirs in relation to oil recovery systems. *AIChE Symp. on Improved Oil Recovery by Surfactant and Polymer Flooding, Kansas City, Kans.*
Wotring, D.M., 1975. Orcutt Hill field. *SPE, Improved Oil Recovery Field Reports, June*, pp. 113–123.
Yen, T.F. and Peterson, G.D., 1982. Correlation of petroleum component properties for improved waterflooding. *DOE Final Rep.*, DOE/ET/12382-11 (August).

Chapter 10

USE OF SURFACTANTS IN OIL RECOVERY

M.K. SHARMA and D.O. SHAH

INTRODUCTION

Surfactant molecules are composed of two portions called lipophilic and lipophobic groups. Lipophilic group has sufficient solubility in the solvent and always tends to bring the entire molecule into solution, whereas lipophobic group is rejected by the solvent, because it has less affinity for the solvent molecules, and tends to expel entire molecules from the solution. This lipophilic–lipophobic structure is known as amphipathic structure. If the force due to the lipophobic group predominates, the surfactant molecules tend to concentrate at an interface and at least a part of the lipophobic group is not in contact with solvent molecules. The lipophilic group usually consists of a long hydrocarbon chain. A surfactant molecule is schematically represented in Fig. 10-1. A polymer can also act as a surface active molecule if it has two functional groups, one lipophilic (solvent loving) and the other lipophobic (solvent hating) as shown in Fig. 10-1. The surface properties of a surface-active compound are controlled by the balance between its lipophilic and lipophobic characteristics. If the surfactant contains a hydrocarbon chain with less than 12 carbon atoms, it is called water soluble because the polar head group drags the entire molecule in water. However, when the hydrocarbon chain length is greater than 14 carbon atoms,

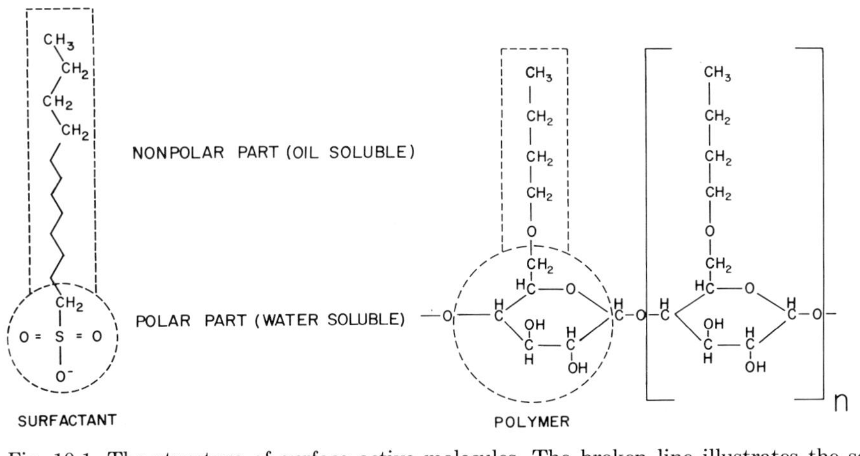

Fig. 10-1. The structure of surface active molecules. The broken line illustrates the separation of polar and non-polar parts of the molecule.

the compounds are called water-insoluble surfactants, because they do not dissolve in water due to the long hydrocarbon chains.

The molecular structure of surfactants causes not only the concentration of the surfactants at the surface and reduction in the surface tension of the solvent, but structure also affects orientation of the molecules at the surface with their lipophilic group in the solvent and a part of their lipophobic group oriented away from the solvent. The lipophilic group is ionic or highly polar. Based on the nature of polar group, surfactants can be classified as anionic, cationic, amphoteric or non-ionic. Among these types, anionic surfactants are widely used in enhanced oil recovery due to their lower adsorption on reservoir rocks as compared to other types of surfactants. Therefore, it is very important to describe briefly the anionic surfactants used in EOR processes.

CLASSIFICATION OF EOR SURFACTANTS

Among the various anionic surfactants, sulfonates have been widely used in EOR processes during the past 50 years. In this section, therefore, several sulfonates used for oil recovery are discussed.

Petroleum sulfonates

Most of the surfactant formulations for EOR contain petroleum sulfonates. These sulfonates were manufactured from certain refinery products. In refinery operations, the furfural extract from solvent lube operations is considered an attractive feedstock for preparation of oil recovery sulfonates. Historically, petroleum sulfonates used in EOR processes came as a byproduct of white oil manufacture. Lube oil stocks were sulfonated to remove aromatic components. In view of the limited demand for lube oil stocks, the petroleum sulfonate was in short supply. In order to increase the supply of petroleum sulfonates, a low-cost method was required.

In the early 1970's, the petroleum sulfonates were prepared by sulfonating gas-oil fractions with sulfuric acid. This process causes a sludge disposal problem. To eliminate the sludge disposal problem, the gas-oil fractions were sulfonated with sulfur trioxide. The process was expensive. Finally, crude oil was sulfonated with sulfur trioxide to provide a cost-effective process. In the mid 1970's, Marathon Oil Company prepared petroleum sulfonates by a crude oil sulfonation process. These sulfonates have been used for enhanced oil recovery in various Pennsylvania and Illinois projects.

Synthetic sulfonates

The synthetic sulfonates may lead to improvements in EOR surfactants. These sulfonates are more expensive, but are reported to be more effective in displacing oil. In addition, synthetic sulfonates have better resistance to metal ions in solution (particularly Ca^{2+} and Mg^{2+}). In terms of a general history, synthetic sulfonates were introduced during World War II, but they did not become viable commercial products until the late 1940's. The first commercially important synthetic surfactants were alkyl benzene sulfonates. The olefin feedstocks and aromatic feedstocks have been considered the major sources for preparation of alkylates. These alkylates were sulfonated with either sulfur trioxide or sulfuric acid. Suntech Inc. developed a procedure of neutralizing the product with caustic, which minimized the amount of water used in the process. The required amount of caustic was determined on an aliquot amount of the reaction mixture, and the final neutralization done with a proportionate amount of concentrated music.

Several companies (e.g., Exxon and Shell) used synthetic surfactants for enhanced oil recovery. The more expensive synthetic surfactants also are being used as additives. In this case, a part of the petroleum sulfonate prepared by sulfonating gas-oil or crude oil is replaced with synthetic sulfonates. The mixed petroleum sulfonate and synthetic sulfonate formulations have resulted in improved oil recovery.

Ethoxylates sulfonates

The results of laboratory and field tests indicate that the mixed surfactant formulation consisting of a petroleum sulfonate and ethoxylated sulfonate or alcohol can improve oil recovery as well as increase salinity tolerance. Ethoxylated sulfonates also lead to a greater flexibility for fluid system design because the degree of ethoxylation and/or the chain length may be varied to adjust slug viscosity. The condensation of fatty alcohols and ethylene oxide yields nonionic surfactants, which are good emulsifiers and low foamers. Sulfonated alcohols are excellent foamers as are the sulfonated versions of non-ionic surfactants. The ethoxylated sulfonates are also known as alcohol ether sulfonates.

Recently, Texaco used an ethoxylated alcohol as cosurfactant in EOR formulations. These formulations have resulted in improved oil recovery and they allow slugs to be tailored for higher salinity and temperature. Previous investigators (Bansal and Shah, 1978c) have shown that by replacing petroleum sulfonates with the ethoxylated sulfonate EOR-200, the optimal salinity can be increased to as high as 32% NaCl brine. This shows that the mixed petroleum sulfonates and ethoxylated sulfonate or alcohol formulations are promising candidates for reservoirs having formation waters of high salinity.

MECHANISM OF OIL DISPLACEMENT BY SURFACTANT FLOODING

Capillary forces and capillary number

At the end of waterflooding, the residual oil is believed to be in the form of immobile globules distributed through the pores of the rock in the petroleum

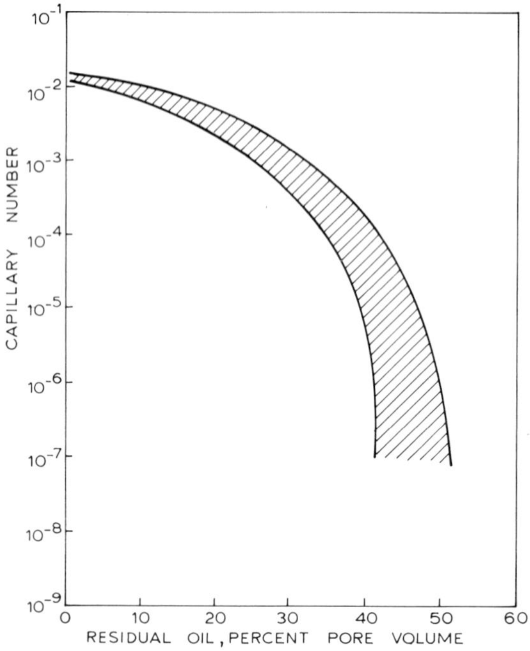

Fig. 10-2. A correlation between capillary number and residual oil saturation. (After Foster, 1973, p. 206; courtesy of the *Journal of Petroleum Technology*.)

reservoir. The two major forces acting on these immobile globules are capillary forces and viscous forces. A number of researchers have correlated the ratio of viscous to capillary forces with the oil recovery efficiency in porous media (Fairbrother and Stubbs, 1935; Leverett, 1939; Brownell and Katz, 1947; Ojeda et al., 1953; Moore and Slobod, 1956; Saffman and Taylor, 1958; Taber, 1969; Foster, 1973; Lefebvre du Prey, 1973; Ehrlich et al., 1974; Melrose and Brandner, 1974; Abrams, 1975; Reed and Healy, 1977; Stegemeier, 1977; Morrow, 1978; MacDonald et al., 1979; Oh and Slattery, 1979). Fig. 10-2 shows such a correlation between the capillary number and the residual oil saturation in a porous medium (Foster, 1973).

The capillary number is defined as the ratio of viscous to capillary forces and can be represented as follows:

$$N_c = \frac{\mu_w \phi q}{\gamma_{o/w}} \tag{10-1}$$

where N_c is the capillary number; μ_w and q are the viscosity and flow rate per unit cross-sectional area of water; $\gamma_{o/w}$ is the interfacial tension between oil and water; and ϕ is the porosity of the petroleum reservoir. It is generally recognized that at the end of waterflooding, the capillary number is in the range 10^{-6} to

10^{-7}. As the capillary number increases, the oil displacement efficiency also increases. It was reported that the increase in capillary number by 3 orders of magnitude would result in displacing 50% of the oil from waterflooded reservoirs. To displace 100% oil from the reservoir, it is necessary to increase the capillary number by 4 orders of magnitude. The maximum critical value of the capillary number for 100% displacement was reported to be on the order of 10^{-2} to 10^{-1} (Fig. 10-2). The capillary number can be increased either by increasing the aqueous phase viscosity (μ_w) and flow rate or by decreasing the interfacial tension between crude oil and water. Among these parameters, a change in the capillary number of 3–4 orders of magnitude can be achieved by reducing the interfacial tension. The interfacial tension between crude oil and water is generally in the range of 20–30 dynes/cm. Using an appropriate surfactant formulation, this interfacial tension can be reduced to 10^{-3} or 10^{-4} dynes/cm.

Dullien et al. (1972) proposed a correlation between the pore structure of sandstone and the residual oil saturation by surfactant flooding. Extended work from this group (Batra and Dullien, 1973) shows that the tertiary oil recovery efficiency increases with decreasing structural index of the porous media as defined by Batra and Dullien. Taber et al. (1973) also correlated the oil displacement efficiency with the ratio of $\Delta p/\gamma L$, where Δp is the pressure drop across the length (L) of the porous media and γ is the interfacial tension at oil–water interface. They proposed a critical value of the ratio $\Delta p/\gamma L$ for oil displacement below which no residual oil can be displaced from any porous media. The critical value of $\Delta p/\gamma L$ depends on the viscosity of oil and water as well as on the permeability of the porous media. It has been shown (Taber et al., 1973) that the value of the critical ratio increases either with decreasing permeability of the porous media or with increasing viscosity of the liquid phases flowing in the porous media. A quantitative correlation between the capillary pressure and various properties of fluid and rock has been proposed (Stegemeier, 1974, 1977). It was concluded that all of the non-wetting phases can be displaced from pores possessing capillary pressures less than that of the calculated value for a given combination of applied pressure drop and properties of the fluid.

A theoretical model for the critical pressure gradient in an ideal pore system, where the pore radius was assumed to be a sinusoidal function of axial position, has been proposed (Oh and Slattery, 1976). The results indicate that, for the most efficient displacement of residual oil, the porous media should be water-wet. Moreover, in the completely water-wet porous media oil is displaced more efficiently than in the intermediate-wettable reservoirs. In addition, some work has been carried out on the gravity and viscous forces needed for the mobilization of oil trapped in porous media. It was shown (Morrow, 1978) that gravity alone may dislodge oil when the interfacial tension is low enough to bring the system into the critical capillary number region at reservoir rates. It was reported (Sayyouh et al., 1978) that at very slow rates ($\ll 1$ ft/day), the recovery appears to increase slightly with further decrease of flow rates in the porous media.

Interfacial phenomena

It is well accepted that oil recovery efficiency can be influenced by producing ultra-low interfacial tension between crude oil and brine. The interfacial tension depends on other interfacial parameters such as interfacial charge, interfacial viscosity and wettability. It is worthwhile, therefore, to discuss the role of various interfacial parameters on oil displacement efficiency in porous media.

Interfacial tension

The flow of trapped oil droplets or oil ganglia through the narrow necks of pores is represented schematically in Fig. 10-3. An ultra-low interfacial tension between crude oil and brine phases should be required for easy flow of trapped oil drops, because it reduces the work of deformation needed for oil ganglia to move through the narrow necks of pore channels. It has been reported (Wilson and Brandner, 1973; Cayias et al., 1977; Doe et al., 1977; Franses et al., 1977; Bourrel et al., 1980; Chan and Shah, 1980) that in the presence of a suitable surfactant, an ultra-low interfacial tension can be achieved at an oil–brine interface.

It was calculated that, for the movement of oil through Berea Sandstone core, a pressure of 47,300 dynes/cm^2 is required for a system having an oil–brine interfacial tension of 25 dynes/cm (Taber, 1969). Many investigators, using similar types of calculations, have shown that the interfacial tension must be lowered in the range of 10^{-2} to 10^{-1} in order to displace residual oil from the reservoir. Previous investigators (Foster, 1973; Hill et al., 1973) have observed that relatively small concentrations of petroleum sulfonates can produce such low interfacial tensions between oil and brine. Extended work by Wilson et al. (1976) has shown that a petroleum sulfonate with a relatively narrow range equivalent weight distribution is the most effective in reducing interfacial tension. The effect of the equivalent weight of petroleum sulfonates on an ultra-low interfacial tension was also studied as a function of various alkanes. The main conclusion was that the preferred value of R corresponding to a minimum in interfacial tension increases with increasing equivalent weight of the surfactant.

It is evident that there are two regions of ultra-low interfacial tension in the TRS 10-410–IBA–dodecane–brine system (Shah, 1981). In the low surfactant concentration (around 0.1%) region, the system forms two phases, namely oil and brine in equilibrium with each other, whereas in the high concentration (around

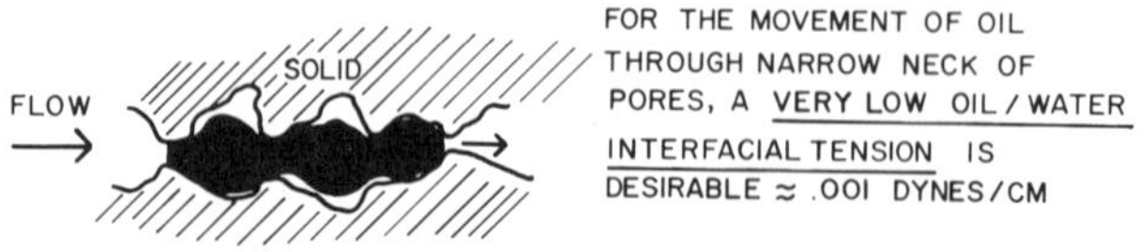

Fig. 10-3. Schematic diagram of the role of interfacial tension in surfactant flooding.

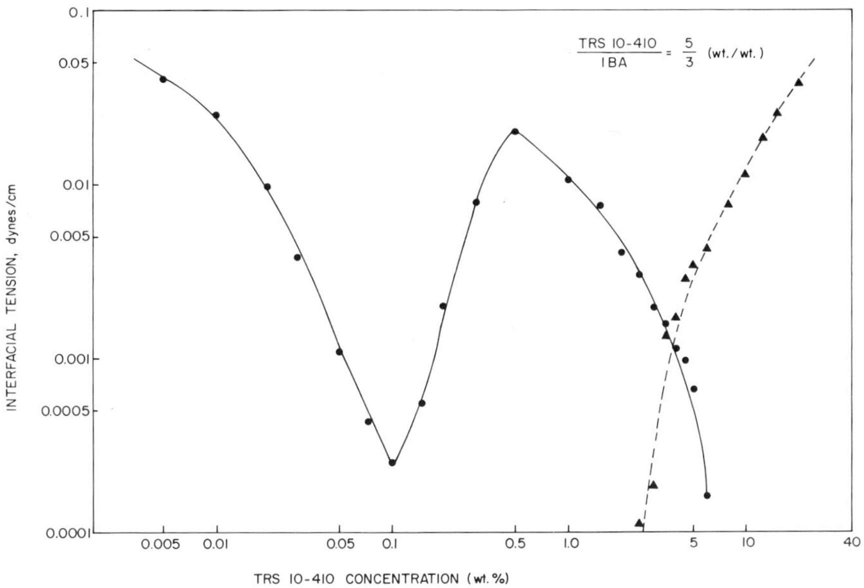

Fig. 10-4. Effect of surfactant concentration on the interfacial tension of TRS 10-410+IBA in 1.5% NaCl with dodecane.

4.0–8.0%) region, a middle phase microemulsion is in equilibrium with excess oil and brine (Fig. 10-4). It has been reported (Cayias et al., 1977) that the formation of a low interfacial tension requires an optimum electrolyte concentration and salinity, and the value of alkyl chain length (R) corresponding to the minimum interfacial tension is the same for alkylbenzene sulfonates and alkylsulfonates. The magnitude of the ultra-low interfacial tension, however, is lower for alkylbenzene sulfonates than that for alkylsulfonates.

Fig. 10-5 delineates the effect of salt concentration, oil chain length and surfactant concentration on the interfacial tension of pure and impure alkylbenzene sulfonates. It is evident that the ultra-low interfacial tension occurs at a specific surfactant concentration. Using various techniques such as osmotic pressure, dye solubilization, surface tension, and light scattering, it has been shown that a minimum in ultra-low interfacial tension for the TRS 10-410–IBA–dodecane–brine system correlates with the onset of micellization in the aqueous phase and a partition coefficient of the surfactant near unity. Similar results were also reported (Baviere, 1976) using different systems. Previous researchers (Morgan et al., 1977) have listed various parameters affecting ultra-low interfacial tension. This list includes the surfactant average molecular weight and molecular weight distribution; surfactant molecular structure; surfactant concentration; type and concentration of electrolyte; average molecular weight and structure of oil phase; elapsed time; and temperature of the system, which may be expected to vary during production for a reservoir or from reservoir to reservoir.

Interfacial charge

The role of interfacial charge in the displacement of oil from the reservoir rocks has not yet been emphasized in the literature. Fig. 10-6 schematically summarizes the relevance of surface charge in the oil displacement process from the reservoir channels. The writers strongly believe that a better understanding of the magnitude and nature of interfacial charge in the oil-displacement process can contribute significantly to the design of surfactant formulations for optimum performance under given reservoir conditions. The surface charge on minerals and clays present in the reservoir rocks would also play an important role in the oil displacement process. The sign and magnitude of the charge will influence the adsorption of the surfactant on minerals and clays of the reservoir.

A low surface charge density causes high interfacial tension and high interfacial viscosity as well as low electrical repulsion between the oil globules and sand particles (Fig. 10-6). The addition of a suitable surfactant would increase the interfacial charge density. This high charge density in the presence of surfactant would result in a decrease in interfacial tension and interfacial viscosity, and would also result in an increase in electrical repulsion between oil globules and sand particles, which in turn would facilitate the movement of oil ganglia through the pore channels if the oil does not adhere to the sand particles (Fig.

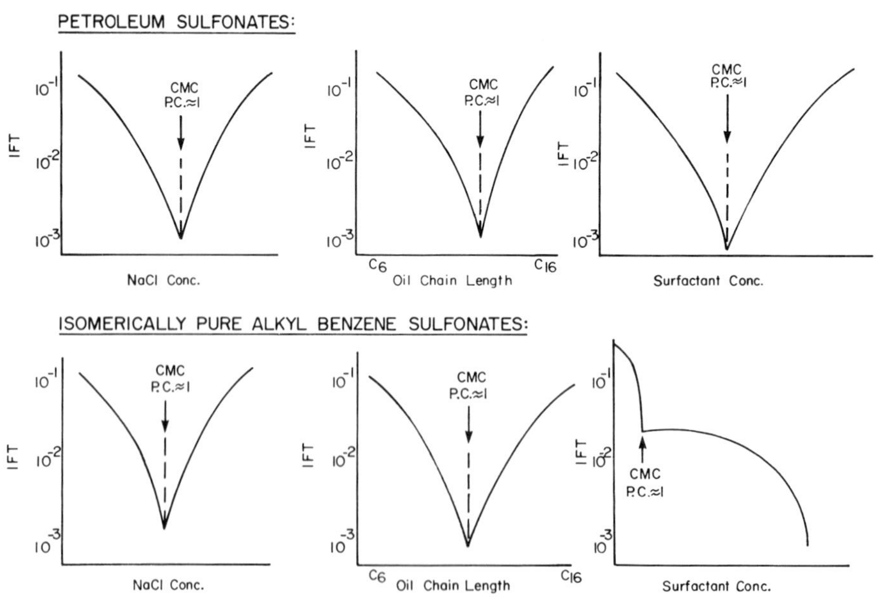

Fig. 10-5. Schematic diagram of the effect of salt concentration, oil chain length, and surfactant concentration on the interfacial tension of pure and impure alkyl benzene sulfonates.

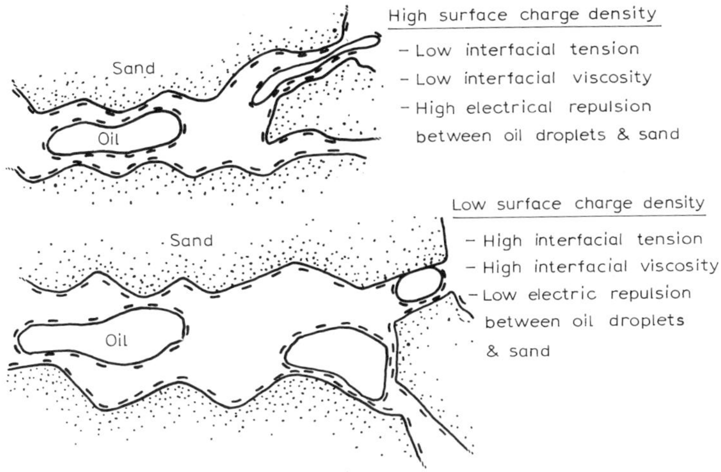

Fig. 10-6. Schematic diagram of the role of surface charge in the oil displacement process.

10-6). It was reported (Chan and Shah, 1977a, b) that a maximum in electrophoretic mobility (e.g., interfacial charge density) and a minimum in interfacial tension were observed at the same concentration of TRS 10-80–octane–brine system.

Interfacial viscosity

For efficient oil recovery, the displaced oil ganglia must coalesce and form an oil bank (Fig. 10-7). A very low interfacial viscosity is required for the coalescence of the oil ganglia to form an oil bank. Fig. 10-8 shows the movement during surfactant injection of an oil bank which coalesces with more oil ganglia resulting in further displacement of residual oil towards the producing sites.

It was concluded that the high interfacial viscosity in crude oil–brine systems containing a mixture of natural surfactants in the oil phase can be attributed to the presence of polar asphaltenes and resins (Berridge et al., 1968; Strassner,

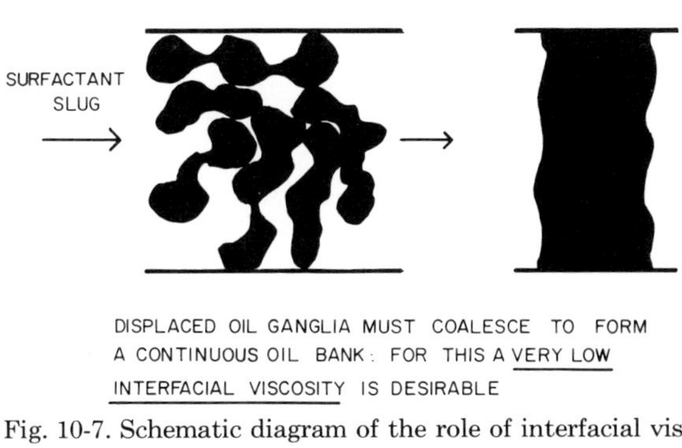

Fig. 10-7. Schematic diagram of the role of interfacial viscosity in the oil recovery process.

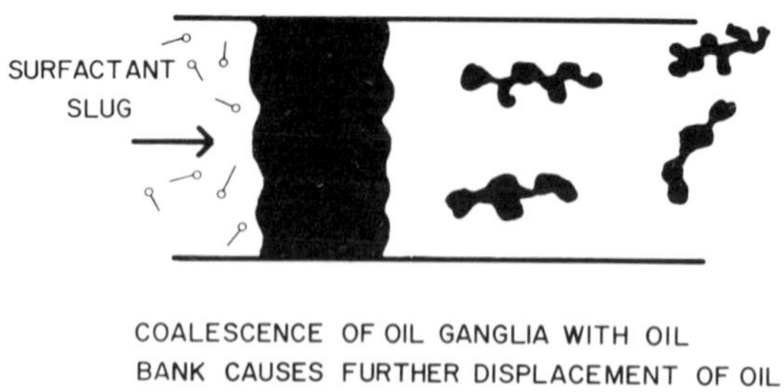

Fig. 10-8. Schematic diagram of the role of coalescence of oil ganglia in oil displacement processes.

1968; Petrov et al., 1969; Levchenko, 1970). The addition of commercial surfactants to a brine solution in contact with an oil can remove some of the natural surfactants from the crude oil–brine interface and reducing the interfacial viscosity. Further increase in concentration of sulfonates reduces the interfacial viscosity, because more natural surfactants are displaced from the interface. Similar results were also reported by Gupta and Wasan (1974). A number of papers on the role of coalescence behavior and interfacial shear viscosity in relation to oil bank formation have appeared (Wasan and Mohan, 1977; Wasan et al., 1979).

Fig 10-9 summarizes the results of coalescence time as a function of sodium chloride concentration in aqueous surfactant solution. The coalescence time decreases with increasing sodium chloride concentration and reaches a minimum at 1.5% NaCl concentration. At 1.5% NaCl concentration, the coalescence time is about 25 times lower compared to that in the absence of NaCl. The coalescence

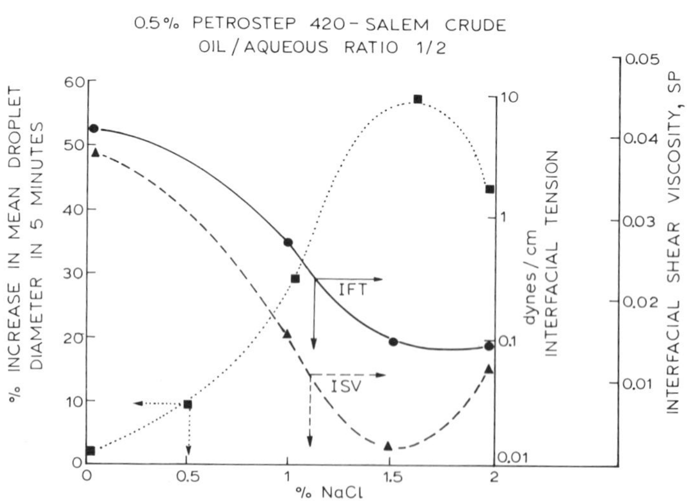

Fig. 10-9. Interdroplet coalescence and interfacial properties for varying brine concentrations. (After Wasan et al., 1978, p. 415, fig. 14; courtesy of the *Society of Petroleum Engineering Journal*.)

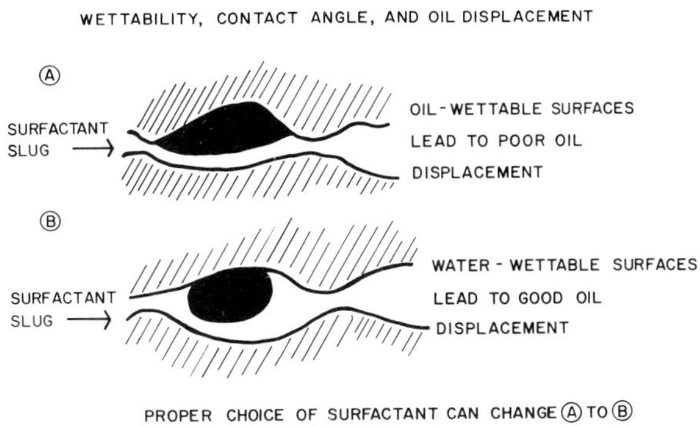

Fig. 10-10. Schematic diagram of the role of wettability and contact angle on oil displacement.

time is inversely related to the interfacial shear viscosity (Fig. 10-9). The highest coalescence rate, therefore, would result the maximum increase in the droplet diameter and would promote oil bank formation.

Wettability

A series of research papers has been published on the role of rock wettability in relation to oil displacement (Reisberg and Doscher, 1956; Wagner and Leach, 1959; Melrose, 1965; Raza et al., 1968; McCaffery and Mungan, 1970; Owens and Archer, 1971; Treiber et al., 1972; Morrow et al., 1973; Salathiel, 1973; Graue and Johnson, 1974; Kremesec and Treiber, 1976). Most of these studies suggested that the wettability of rock surfaces can be changed by adding a simple salt, acid, or base to the flood water. Leach et al. (1962) concluded that sodium hydroxide can effectively change rock surfaces from oil-wet to water-wet.

Fig 10-10 illustrates the role of the wettability of the solid surface on oil ganglia. The oil-wettable surfaces lead to poor oil displacement, whereas the water-wettable surfaces lead to efficient oil displacement. The proper choice of surfactant can selectively alter the rock wettability to oil and brine and can create favorable conditions for efficient oil displacement. It was observed (Moore and Slobod, 1956) that oil recovery at water breakthrough in water-wet systems is much higher than in oil-wet systems. Melrose and Brandner (1974) and Morrow (1978) suggested that the mobilization of trapped oil is most difficult for reservoir rocks having intermediate wettability. It has been reported (Morrow et al., 1973) that polar constituents in crude oil play an important role in determining the reservoir wettability due to their adsorption onto the rock surfaces. The effect of wettability on the efficiency of oil displacement during surfactant flooding has been reported (Kremesec and Treiber, 1976). Under given reservoir conditions, a lower apparent viscosity is required for mobility control in water-wet systems. Recent results on surfactants of different types, having the same surface activity, revealed the fact that even their effect on the wettability of the rock surface and on lowering the interfacial tension between oil and brine are strongly interdependent (Pinter and Wolfram, 1981). The use of additives to alter the rock wettability is a promising approach for enhanced oil recovery.

ULTRA-LOW INTERFACIAL TENSION IN RELATION TO OIL DISPLACEMENT BY SURFACTANT FLOODING

The importance of interfacial tension (IFT) in the oil displacement process has been discussed in the previous section. The effect of various factors on interfacial tension of surfactant formulations for efficient oil displacement is discussed in this section.

Effect of salt concentration on interfacial tension

It has been shown (Foster, 1973; Hill et al., 1973; Chan and Shah, 1977b) that the interfacial tension is strongly dependent on the salinity of the aqueous phase, and at a critical salt concentration the interfacial tension approaches its minimum value. By changing the salinity of the aqueous phase, the relative solubilities of the surfactant in oil and water can be varied significantly. It was reported (Foster, 1973) that for a particular petroleum sulfonate the minimum interfacial tension between a salt solution containing the petroleum sulfonate and an intermediate paraffinic crude would occur at lower surfactant concentrations and higher sodium chloride concentrations than for a typical naphthenic crude. Cash et al. (1976) extended work reported by Foster (1973) for different types of oils. Fig. 10-11 represents an interfacial tension contour map for a petroleum sulfonate–sodium chloride–water system against an intermediate-paraffinic crude (Cash et al., 1976). By replacing paraffinic crude with naphthenic crude, the region of minimum interfacial tension would move down and to the right as indicated by the arrow in Fig. 10-11.

It is obvious from the interfacial tension control map (Fig. 10-11) that a specific surfactant concentration and salinity is required for the formation of ultra-low interfacial tension. As the salt concentration is varied in the aqueous phase, the partition coefficient of the surfactant between oil and water is altered which seems to be responsible for achieving ultra-low interfacial tension. In order to delineate the effect of salt on interfacial tension and partition coefficient, a molecular mechanism (Fig. 10-12) has been proposed (Chan and Shah, 1977a, b). The surfactant concentration in the oil phase increases with increasing salt concentration in the aqueous phase. At low salt concentrations, most of the surfactant stays in the aqueous phase and, therefore, a very little of it partitions into the interface or oil phase. At high salt concentrations, the surfactant preferentially dissolves into the oil phase and, hence, very little of it partitions into the interface or aqueous phase. At intermediate concentration, however, the surfactant concentration is the same in both oil and water. This intermediate salt concentration is termed the optimal salinity. At optimal salinity, the

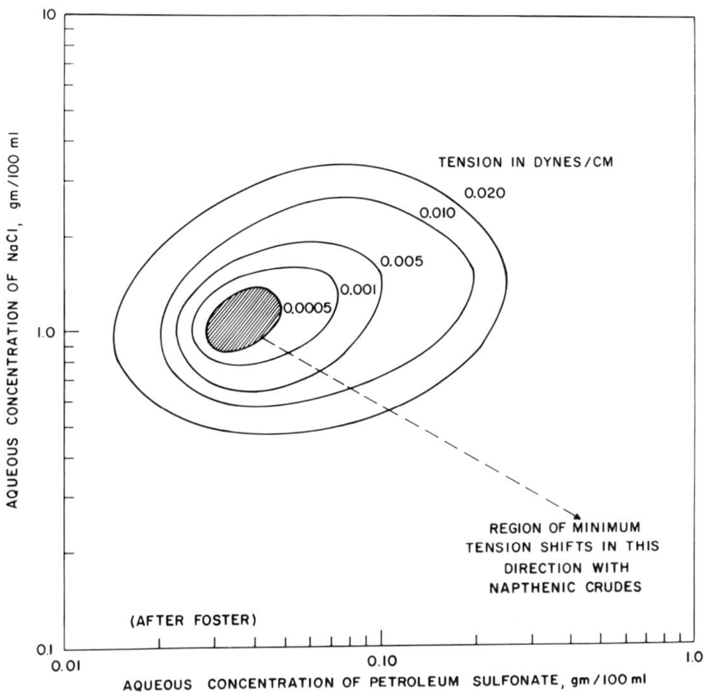

Fig. 10-11. Interfacial tension contour map for petroleum sulfonate–NaCl–water system against intermediate paraffinic crude. (After Wilson, 1977, p. 5, fig. 2; courtesy of the Academic Press, New York, N.Y.)

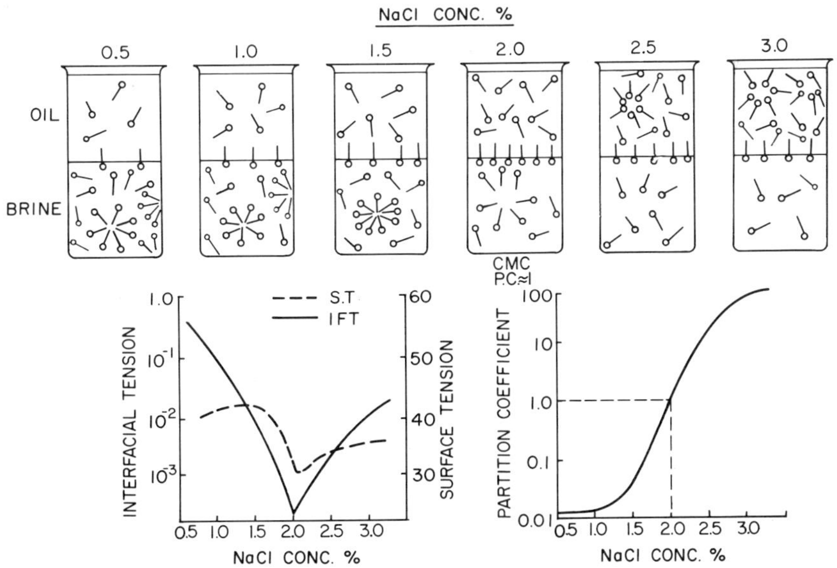

Fig. 10-12. The molecular mechanism for the effect of salt concentration on interfacial tension and surface tension.

surfactant concentration is highest at the oil–water interface which presumably produces the lowest interfacial tension. The partition coefficient at optimal salinity was also found to be unity. If one measures the surface tension of equilibrated brine phase, a minimum in surface tension also occurs at the optimal salinity (broken line in Fig. 10-12).

Effect of oil chain length on interfacial tension

The nature of the oil plays an important role in producing the ultra-low interfacial tension. The concentrations of both salt and surfactant required to produce a minimum in interfacial tension vary significantly with the properties and chain length of oil. It was shown that crude oils with high aromatic hydrogen content produced lower interfacial tension compared to crude oils with lower aromatic hydrogen content (Gale and Sandvik, 1973). Fig. 10-13 shows the interfacial tension for a petroleum sulfonate (TRS 10-80) in 1.5% NaCl as a function of oil chain length (Chan and Shah, 1980). The partition coefficient of surfactant corresponding to the minimum in interfacial tension is found to be unity (Fig. 10-13). Continued research showed that the minimum interfacial tension and surface tension of the equilibrated aqueous phase occur at the same

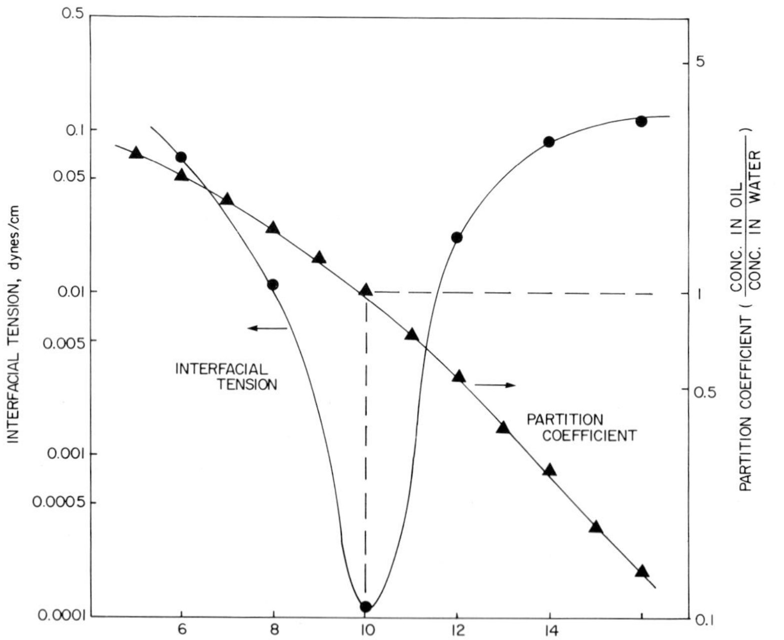

Fig. 10-13. Interfacial tension and partition coefficient for 0.1% TRS 10-80 in 1.5% NaCl with oil of various chain length. The concentration in oil was determined by UV and that in the brine by methylene blue absorption measurements.

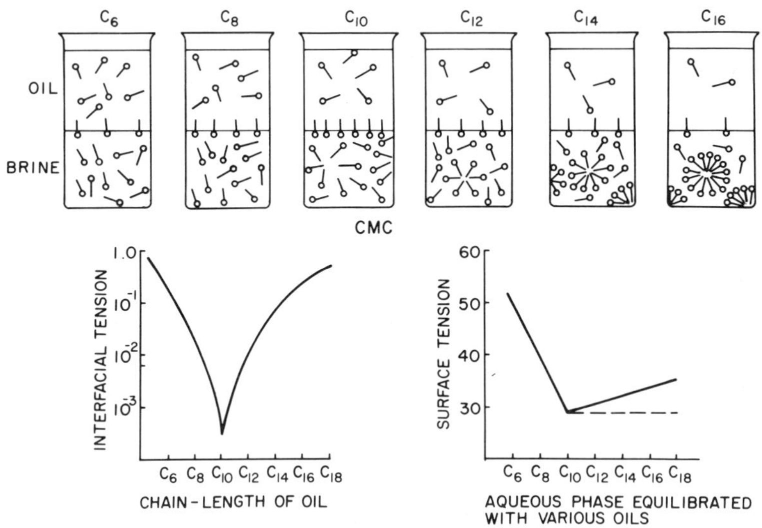

Fig. 10-14. The molecular mechanism for the effect of chain length of oil on interfacial and surface tensions.

oil chain length. One can conclude from the minimum in surface tension that the surfactant concentration in the aqueous phase equilibrated with decane must be at the CMC.

Fig. 10-14 illustrates schematically the proposed molecular mechanism for the effect of oil chain length on interfacial tension (Chan and Shah, 1980). It is evident from Fig. 10-14 that the partitioning of the surfactant in the oil phase decreases with increasing oil chain length. It was established that as the oil chain length increases, the monomer concentration of surfactant in the oil phase decreases (Fig. 10-14). The concentration of surfactant in the brine phase, therefore, must increase as the oil chain length is changed from C_6 to C_{16}. It was confirmed from surface tension and interfacial tension measurements that the concentrations of surfactant in the aqueous phase equilibrated with C_6 to C_8 oils below the CMC, whereas the surfactant concentration in the aqueous phase would be smaller as the chain length of the dissolved oil is increased. This explains why the interfacial tension should increase with increasing oil chain length from C_{10} to C_{16}. The molecular mechanism proposed in Fig. 10-14 explains clearly the effect of oil chain length reported by previous investigators.

The concept of an equivalent alkane-carbon number (EACN), which allows the substitution of an alkane or an alkane mixture for a crude oil for phase volume or interfacial tension studies, has been proposed by Cayias et al. (1976a, b). This concept arises from the observation that the interfacial properties of a given crude oil with surfactant can be modeled by the behavior of alkanes. In addition, the EACN of a mixture of hydrocarbons follows the simple mixing rule:

$$(\mathrm{EACN})_{\mathrm{mixture}} = \sum_i x_i (\mathrm{EACN})_i \tag{10-2}$$

where x_i is the mole fraction of component i. Cayias et al. (1976) found that the EACN of an oil (e.g., hydrocarbons, crude oil or pseudocrude oil) is independent of the surfactant formulation or cosurfactants used in the formulation, and this equivalence always holds. Since crude oil is dark in color and usually is quite viscous, it can attain equilibrium very slowly and phase observations are difficult. Replacing crude oil with alkane facilitates screening of surfactant formulation. A three-parameter representation to describe equivalent carbon number has been proposed by Puerto and Reed (1982). The three parameters used were the optimal salinity, solubilization parameter, and oil molar volume. It was concluded that this three-parameter representation provided a more nearly unique description of the phase behavior of surfactant formulations than that available in the past.

Effect of surfactant concentration on interfacial tension

One of the major variables in producing as well as in broadening the interfacial tension minimum is the concentration of the surfactant. Cayias et al. (1977) reported that the interfacial tension decreases with increasing surfactant concentration, and at a critical concentration the interfacial tension approaches its minimum value. Beyond this critical concentration, the interfacial tension increases with an increase in surfactant concentration. A.K. Sharma et al. (1983) showed that the addition of a non-ionic surfactant containing a phosphated ester (e.g., Klearfac AA-270) can broaden the interfacial tension minimum and increase the salt tolerance of the petroleum sulfonate system. The interfacial tension data for different sets of experiments suggest that the aqueous phase is predominantly responsible for the ultra-low interfacial tension. From surfactant partition measurements, it was shown that the number of surfactant monomers in the oil and brine phase increases with increasing concentration of surfactant. When the monomer concentration in the aqueous phase approaches the critical micelle concentration, both the interfacial tension and the surface tension show minimum values. Chan and Shah (1977b) proposed that the surfactant concentration, at which a minimum in interfacial tension and a minimum in surface tension occur, is the true critical micelle concentration of surfactant.

The molecular mechanism for the effect of surfactant concentration on interfacial tension minimum was proposed by Chan and Shah (1977b) for the TRS 10-80–*n*-octane–brine system (Fig. 10-15). These molecular mechanisms for the effects of salt concentration (Fig. 10-12), oil chain length (Fig. 10-14), and surfactant concentration (Fig. 10-15) explain all experimental results reported in previous investigations of ultra-low interfacial tension. This unified understand-

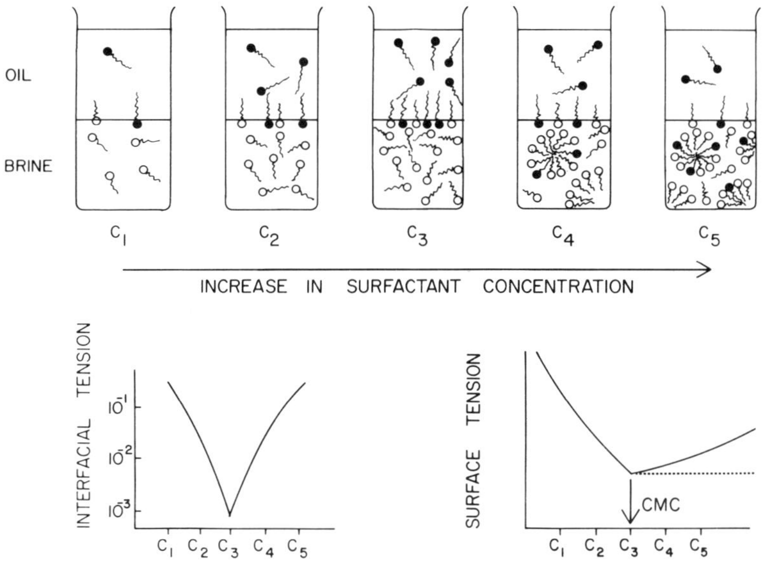

Fig. 10-15. The proposed molecular mechanism for the effect of surfactant concentration on interfacial and surface tensions.

ing of the molecular mechanisms for producing ultra-low interfacial tension could be utilized in designing surfactant formulations for enhanced oil recovery under particular reservoir conditions.

Effect of interfacial charge on interfacial tension

It has been shown (Chan and Shah, 1977b; Bansal and Shah, 1978a) that the interfacial charge strikingly influences the interfacial tension. For several systems, it was observed that a minimum in interfacial tension and a maximum in electrophoretic mobility, which is an indirect measure of interfacial charge density, has a definite correlation with the interfacial tension for oil–surfactant systems. Fig. 10-16 shows the electrophoretic mobility and interfacial tension as a function of surfactant concentration for the TRS 10-80–*n*-octane–brine system. This system exhibits a minimum in interfacial tension at 0.05% TRS 10-80 concentration, whereas the electrophoretic mobility exhibits a striking maximum. These results also demonstrate that the electrophoretic mobility and interfacial tension curves are almost a mirror image of each other.

It was also reported (Jennings et al., 1974; McCaffery, 1976; Bansal and Shah, 1978a) that at a particular concentration of NaOH, the interfacial tension between crude oil and caustic solution exhibits a minimum. In order to understand the mechanisms of ultra-low interfacial tension in crude oil–caustic systems, electrophoretic measurements were also carried out for these systems. Fig.

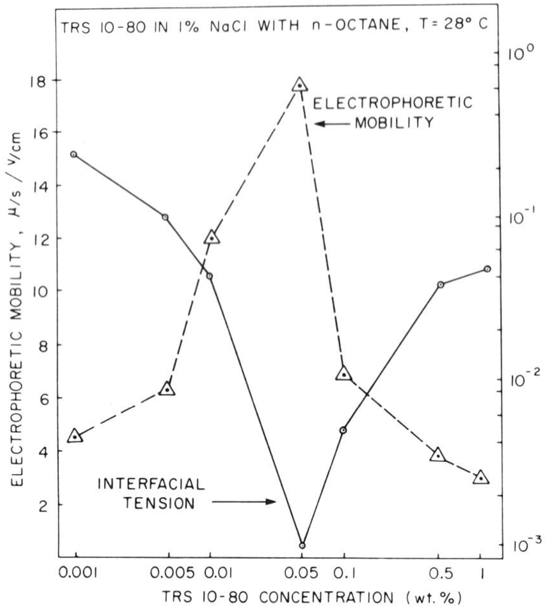

Fig. 10-16. A correlation between interfacial tension and electrophoretic mobility for TRS 10-80–n-octane–brine system.

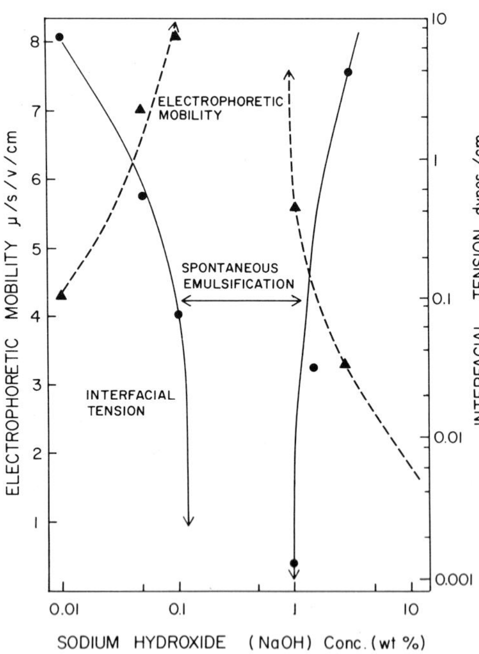

Fig. 10-17. A correlation between interfacial tension and electrophoretic mobility for crude oil–NaOH system.

10-17 represents the interfacial tension and electrophoretic mobility of a crude oil at several concentrations of NaOH. It is evident that the region of minimum interfacial tension coincides with that of the maximum electrophoretic mobility. These results can be explained on the basis of the maximum interfacial charge density at crude oil–caustic interface. The natural surfactants containing carboxylic groups present in the crude oil react with NaOH and determine the magnitude of the charge at crude oil–caustic interface. At low caustic concentration, carboxylic groups present in crude oil are in the unionized form (RCOOH). Charge at the crude oil–caustic interface, therefore, is less, which in turn is responsible for smaller values of the electrophoretic mobility. The carboxylic groups ($RCOO^-$) ionize with increasing caustic concentration, which results in an increase in the charge at crude oil–caustic interface and, therefore, in an increase in the electrophoretic mobility. If the caustic concentration is further increased, soap of carboxylic acid (RCOONa) is formed, which decreases surface charge at the crude oil–caustic interface and, consequently, decreases the electrophoretic mobility. These results suggest that the higher interfacial charge density is responsible for the minimum in interfacial tension in crude oil–caustic systems.

FACTORS INFLUENCING OIL RECOVERY

The oil displacement efficiency depends upon several variables, such as optimal salinity, mobility control, microstructures of surfactant formulations, and interactions between surfactant and polymer. In this section, the writers discuss how various parameters can influence the oil recovery. Some experimental results where oil displacement efficiency is influenced by these parameters are also presented.

Optimal salinity

The relative solubilities of petroleum sulfonates in oil and water phases are strongly dependent on the salinity of the aqueous phase. In a certain salinity range, the systems consisting of surfactant, cosurfactant, oil, and brine form three phases: lower-phase microemulsion, middle-phase microemulsion, and upper-phase microemulsion. The salinity at which an equal volume of oil and water is solubilized in middle-phase microemulsion is termed optimal salinity. It was also reported (Healy and Reed, 1974; Healy et al., 1975) that the interfacial tension at middle-phase microemulsion and oil (γ_{mo}) and middle-phase microemulsion and water (γ_{mw}) are equal at optimal salinity. As discussed previously at optimal salinity the value of the interfacial tension is minimum. If one considers only the interfacial tension, the oil displacement efficiency should be maximum at the optimal salinity. It has been shown (Chiang and Shah, 1977) that the tertiary oil recovery efficiency is a maximum at optimal salinity for 0.05 PV slug

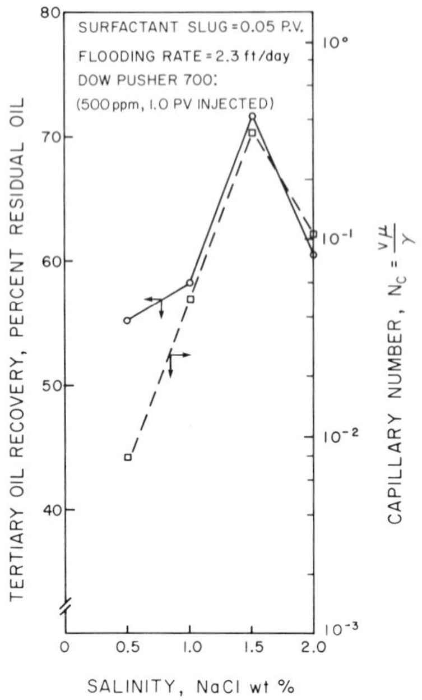

Fig. 10-18. Effect of salinity on the tertiary oil recovery and capillary number.

of 5% TRS 10-410 + 3% IBA + dodecane (Fig. 10-18). A correlation between oil displacement efficiency and capillary number is also illustrated in Fig. 10-18. At optimal salinity (1.5% NaCl), the maximum amount of oil (72%) was recovered. The oil recovery was further improved from 72% to 82% by increasing the length of the sand pack from 6.63 to 15 inches.

Table 10-I summarizes the results of tertiary oil recovery as a function of the salinity in sand pack and Berea core at 80°C (A.K. Sharma et al., 1983). It was observed that at optimal salinity the displacement efficiency of crude oil and pure alkane mixture in sand packs and Berea cores was a maximum. Fig. 10-19 shows that the surfactant concentration in the effluent is a function of the number of pore volumes of fluid injected. The surfactant breakthrough was earliest when the injected surfactant slug contained brine of higher than optimal salinity. The surfactant recovery or the area under the curve at this salinity was the lowest. This indicates a significant surfactant retention in porous media due to the precipitation of surfactant at higher salinities. This can decrease the stability or present the generation of in-situ microemulsions and, thus, can cause unstable heterogeneous phases to occur. These changes are frequently associated with extremely pronounced differences in the mobility of the phases flowing through the porous medium (Kraft and Pusch, 1982), which results in poor oil recovery.

Fig. 10-20 schematically illustrates several phenomena occurring at the optimal salinity in relation to enhanced oil recovery by surfactant flooding. It is

TABLE 10-I

Summary of flooding results [a]: effect of salinity

Salinity of surfactant slug (% NaCl)	Type of porous medium	Type of oil used	Residual oil after water-flooding (% PV)	Tertiary oil recovery efficiency (%)	Surfactant recovery (%)	PV at which surfactant concentration reached a maximum
2.0	sandpacks	a mixture of nonane + decane	18	31	68.7	0.86
2.5	sandpacks	a mixture of nonane + decane	16	28	0.59	0.86
3.0 (optimum salinity)	sandpacks	a mixture of nonane + decane	10	92	0.64	1.10
4.0	sandpacks	a mixture of nonane + decane	18	10	0.23	0.65
2.8 (optimum salinity)	Berea core	Ankleshwar crude oil	43	79	n.a.	n.a.

[a] 0.2 PV of surfactant slug, containing 5% (w/v) TRS 10-80, 2% (w/v) KF AA-270, 3% TAA in resident brine with 1% Na_2CO_3 and 0.3% sodium tripolyphosphate was injected in porous media (sandpack dimensions: 2.5 cm diameter, 30.0 cm length; permeability: 2–3 D. (Berea core dimensions: 30.5 cm by 2.5 cm; permeability: 0.415 D). Back pressure = 40 psi; temperature = 80°C; flow = 1 ft/day. 1500 ppm Dow Pusher 1000 in resident brine was used when flooding was carried out with a mixture of nonane and decane; 2000 ppm Dow Pusher 1000 in resident brine was used when flooding was carried out with Ankleshwar crude oil.

evident that the maximum oil recovery correlates well with dynamic and equilibrium properties of surfactant–oil–brine systems. Inasmuch as optimal salinity leads to favorable conditions for optimum oil recovery, one would like to design approaches to alter the optimal salinity of a given surfactant formulation (Bansal and Shah, 1978a, b). Fig. 10-21 shows the optimal salinity of a mixed surfactant formulation containing petroleum sulfonate and ethoxylated sulfonate (EOR-2000). By replacing petroleum sulfonate with ethoxylated sulfonate, the optimal salinity increases and can reach as high as 32% NaCl brine. Interestingly, when these formulations were equilibrated with oil, middle-phase microemulsions having very low interfacial tension were produced. Thus, the mixed surfactant formulations consisting of petroleum sulfonates and ethoxylated sulfonates or alcohols are promising candidates for high-salinity formulations (Bansal and Shah, 1978c, d).

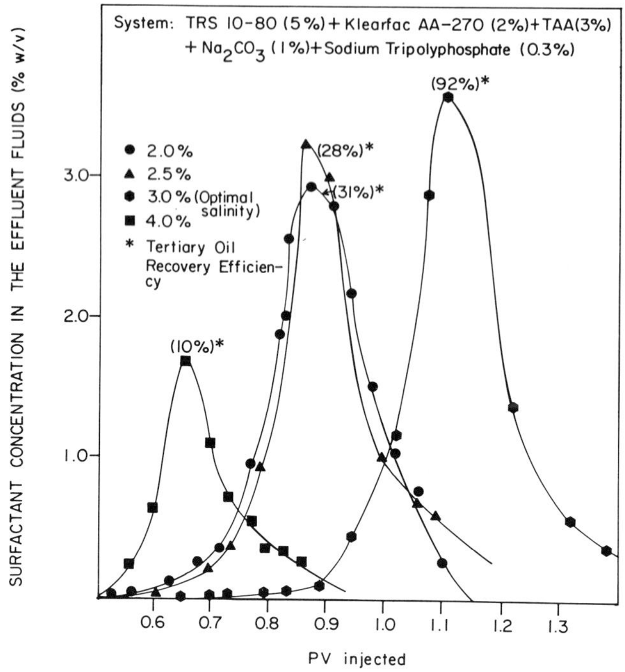

Fig. 10-19. Effect of salinity on surfactant concentration in the effluent fluids.

Mobility control

For an efficient oil displacement by surfactant flooding, the mobility-controlling polymer should be less mobile than the surfactant slug and oil bank. Loss of mobility control in the fluid sequence causes fingering which, in turn, reduces oil displacement efficiency. It has been stated (Gogarty, 1967) that mobility control is one of the most important considerations in designing a surfactant formulation. Many polymers such as polyacrylamides and polysaccharides have been used as effective mobility control agents (Gogarty, 1967; Trushenski et al., 1974). For polyacrylamides, the mobility control was achieved through a reduction in both the viscosity and the permeability. The effectiveness of the polymer solutions as mobility control agents depends on the polymer concentration, shear degradation, adsorption, gel formation, pseudodilatant flow, salt concentration, and temperature.

The effect of temperature on the stability of polyacrylamide solutions has been reported by Mungan et al. (1966) and Knight (1973). In the temperature range of 120–150°C, thermal degradation of polymer causes permanent loss of solution viscosity. Mechanical, chemical and thermal stability of polymers is desirable for oil displacement process. The effect of salt on the rheological properties of various polymers has been investigated in relation to oil recovery (Smith, 1970; MacWilliams et al., 1973; White et al., 1973; Sparlin, 1975; Castor

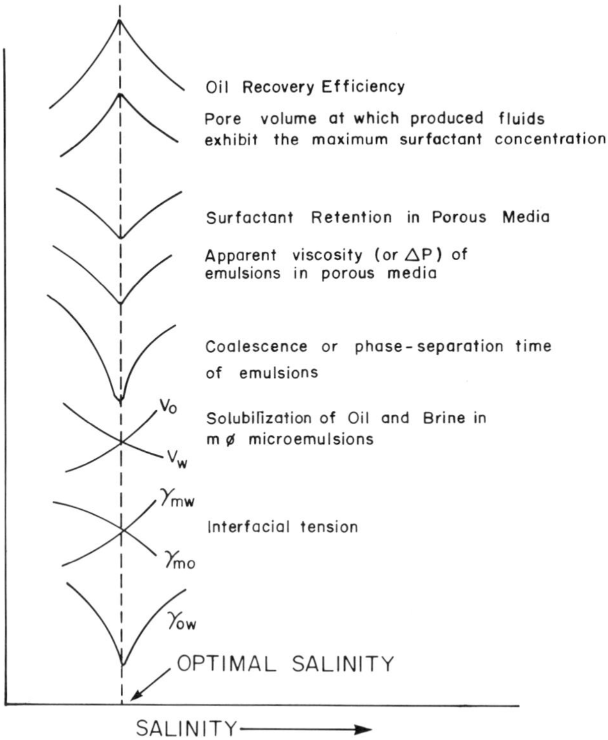

Fig. 10-20. A summary of various phenomena occurring at the optimal salinity in relation to enhanced oil recovery by surfactant flooding.

et al., 1981). It was observed (Castor et al., 1981) that the polyacrylamide copolymers are much more viscous than the polysaccharide biopolymers at equivalent concentrations in fresh water, whereas in saline water the viscosities of these copolymers are strikingly decreased as compared to biopolymers. The viscosity of the copolymers decreases by an order of magnitude in the presence of 1.0% NaCl. Results of mobility control and rotational shear experiments suggest that some permanent loss of viscosity would occur for the polyacrylamide and not for polysaccharide. The salt effect was found to be most dominant in the measurements of capillary viscosity as compared to screen viscosity. This indicates that salt has less effect on viscoelastic properties or screen factor as compared to shear viscosity.

The polymers such as acrylamides and partially hydrolyzed acrylamides form a gel-type structure in the presence of certain organic cations, as well as multivalent inorganic cations. The important parameters to be considered in the application of gels for plugging high-permeability zones are the gelation time and the yield strength of the formed gel. The former controls the distance that a mixture of polymer and crosslinking agent can travel inside the reservoir,

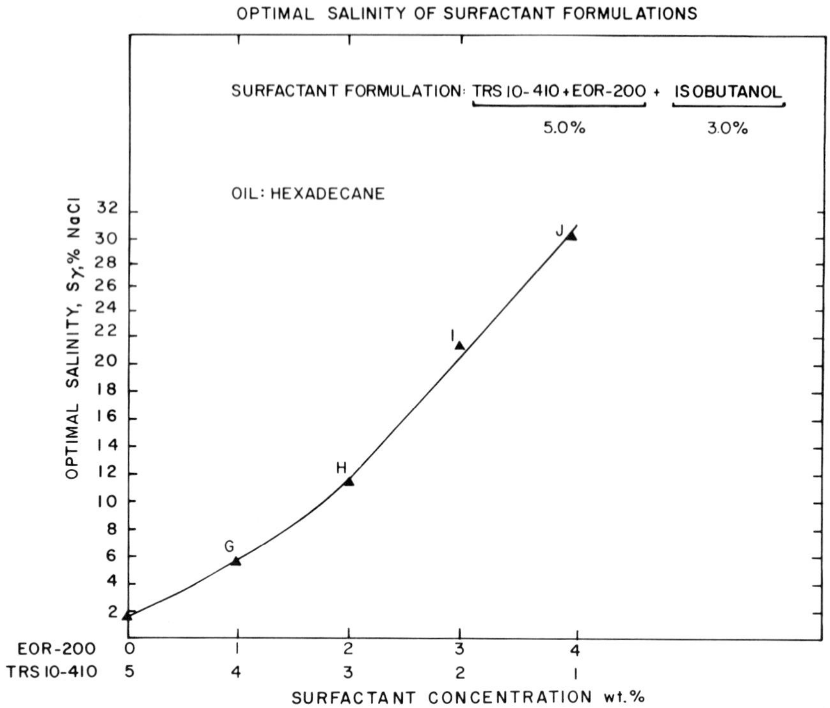

Fig. 10-21. Increase in the optimal salinity of surfactant formulation by addition of EOR-200.

whereas the latter determines the effectiveness of the treatment in the contacted zone. The effects of temperature, polymer concentrations, and crosslinking agents on gelation time have been reported in the literature (Terry et al., 1979; Jordon et al., 1981). The degree of permeability reduction by a polymer depends upon its concentration, and beyond certain concentration, the permeability reduction becomes independent of polymer concentration (Jennings et al., 1971). It was also reported that at low flow rates, the effect of polymer concentration on permeability reduction is negligible.

It has also been reported (Pye, 1964; Hill et al., 1974; Castor et al., 1981) that polyacrylamide solutions are highly sensitive to shear degradation. Several practical problems in the injection, mixing, and delivery of polymer solutions in the field occur due to shear degradation of polyacrylamides. Despite all precautions, polymer solutions delivered to the wellbore may lose typically 15–40% of the non-sheared factor. It was observed (Lakatos et al., 1981) that, in general, the unhydrolyzed and partially hydrolyzed polyacrylamides behave differently. The dynamic adsorption of hydrolyzed polyacrylamides is more sensitive than the unhydrolyzed ones. The adsorption of polyacrylamides strongly influences the flow through porous media of polymer solutions and carrier phases in the porous systems. It seems to be accepted that polyacrylamides can withstand moderate shear and adsorption and that they can effectively reduce permeability to improve mobility control.

Surfactant–polymer interactions

It has been shown that surfactant–polymer interactions (SPI) can lead to a considerable reduction in efficiency of the oil recovery process (Trushenski, 1977). The SPI not only occur in the surfactant–polymer mixing zone, but also throughout the entire surfactant slug due to invasion of the micellar slug by polymer molecules, which move more rapidly than the carrier water. The surfactant–polymer mixing takes place due to both dispersion effects as well as excluded volume effects during the flow of polymer molecules in porous media. The mixing and invasion of the surfactant slug by polymer molecules generally forms two phases of different viscosities. The entrapment of the high-viscosity phase can effectively remove some components from the surfactant formulation, which in turn reduces the efficiency of the flooding process.

The effect of salinity on surfactant–polymer mixing has been investigated by Shah (1981). It was observed that there are two regions of phase separation, one at low salinity and another at high salinity separated by a metastable colloidal dispersion. The separation of micelles at low salinity is promoted by the presence of polymer molecules. The surfactant–polymer interactions strikingly influence the phase formed at low salinity (Desai, 1983). Fig. 10-22 shows the effect of polymer on an oil–brine–surfactant–alcohol system. It is evident that the formation of middle-phase microemulsion is promoted by the presence of polymer (Desai, 1983). The transition from middle-phase to upper-phase microemulsion, however, is not at all influenced by the presence of polymer, whereas the formation of middle-phase microemulsion shifts to lower salinities with increasing polymer concentration. It was also observed that the optimal salinity is not influenced significantly by polymer concentration in the oil–brine–alcohol–surfactant system.

Fig. 10-23 illustrates schematically the explanation of the surfactant–polymer interactions and concomitant phase separation (Desai, 1983). It was proposed that around each micelle there is a region of solvent from which polymer molecules are excluded. As micelles approach each other, however, there is overlapping of excluded regions. If all micelles separate out, therefore, then the excluded region diminishes due to the overlap of the shells, and more solvent becomes available for the polymer molecules. This effect is similar to the polymer depletion stabilization (Cash et al., 1976). Furthermore, this effect is similar to the osmotic effect where polymer molecules tend to maximize the solvent for all possible configurations.

Recently, various investigators (e.g., Trushenski, 1977; Shah, 1981) have studied the effect of several variables on surfactant–polymer interactions and oil displacement efficiency. Their observations can be summarized as follows:

(1) Surfactant–polymer interactions are minimized when salinity of the mobility buffer bank (polymer) is lower than the salinity of the surfactant slug.

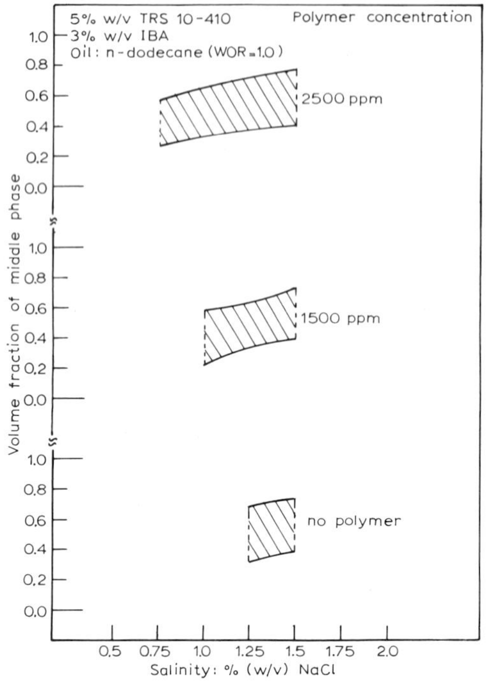

Fig. 10-22. Effect of polymer concentration on the salinity range for formation of middle-phase microemulsion.

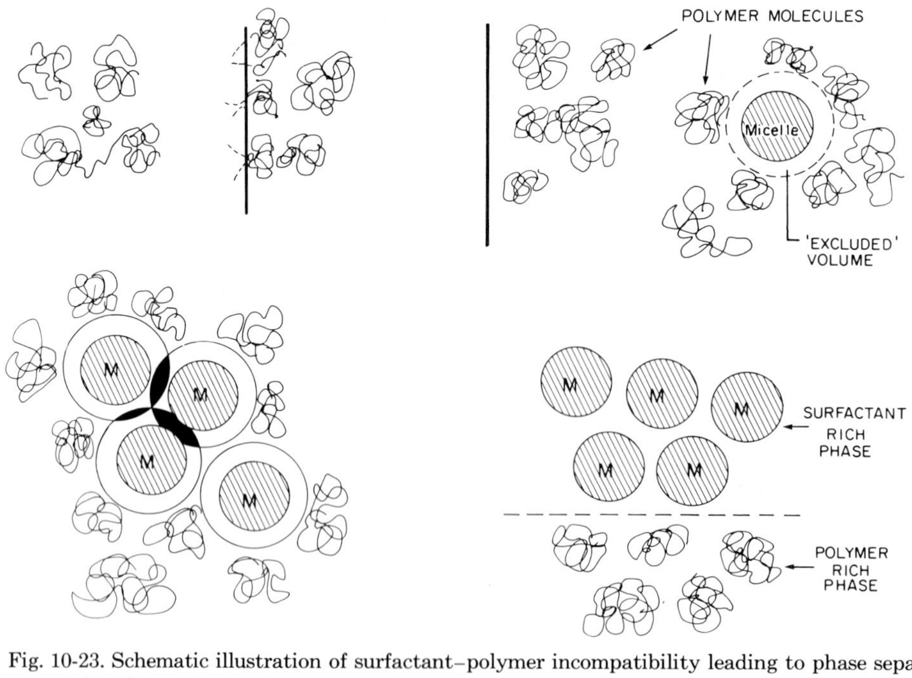

Fig. 10-23. Schematic illustration of surfactant–polymer incompatibility leading to phase separation in mixed surfactant–polymer systems.

(2) Increasing temperature facilitates surfactant–polymer interactions.

(3) Polymer invasion of the micellar bank (surfactant slug) and surfactant–polymer interactions can be eliminated if the polymer is insoluble in the micellar fluid.

(4) Surfactant–polymer interactions can also be eliminated by addition of surfactant solubilizers (e.g., cosurfactants or cosolvents) in the surfactant slug or polymer.

(5) The salinity shock design for the polymer solution employs two slugs of polymer solution in which the first polymer slug is at the optimal salinity. This provides ultra-low IFT and maintains mobility control, whereas the second polymer slug at a much lower salinity is capable of reducing the surfactant loss. Oil recovery in Berea cores was as high as 86% even in the presence of 3% NaCl + 1% $CaCl_2$ in connate water.

(6) If the salinity of the polymer solution was at the optimal salinity of the preceding surfactant formulation, oil recovery in porous media was favorable over a wide range of connate water salinities for both aqueous and oleic surfactant formulations. Oil recovery drastically decreased when the salinity of

Fig. 10-24. Electron-micrograph of 5% TRS 10-410 + 3% isobutanol in 2% NaCl. 36,000×, bar = 1 μm.

the polymer solution was shifted from the optimal salinity, even when the connate water was at the optimal salinity. These results indicate that the processes occurring at the surfactant–polymer mixing zone dominate the oil displacement efficiency.

Microstructures of surfactant formulations

The surfactant formulations used for tertiary oil recovery can be in the form of several structures: micellar solutions, microemulsions, and liquid crystalline dispersions. The formation of these microstructures depends upon the composition of surfactant formulation. It is very likely that the oil displacement efficiency may be influenced by the structure of surfactant formulations. Surfactant formulations used for tertiary oil recovery consist of petroleum sulfonates as surfactants and various short chain alcohols as cosurfactants. The microstructure of these formulations can be studied by NMR, electrical, birefringence and viscosity measurements, as well as by freeze-etching microscopy.

Fig. 10-24 shows the electronmicrograph of a TRS 10-410 (5%) + IBA (3%) formulation in 2% NaCl aqueous solution at 36,000 magnification. It is evident

Fig. 10-25. Electron-micrograph of 5% TRS 10-410 + 3% isobutanol in 2% NaCl. 60,000×, bar = 0.5 μm.

that this formulation consists of a foamlike structure in which the surfactant and alcohol molecules are presumably present in membranes, whereas the aqueous phase is trapped inside the bubbles. When placed between polarized plates, this formulation exhibited a strong birefringence. The electronmicrograph clearly shows (Fig. 10-24) a continuous membrane structure and each bubble is about 2–3 μm in diameter. Fig. 10-25 shows the electronmicrograph of the same formulation fractured at a different angle at 60,000 magnification. It is clear that the membranes are made up of several bilayers of surfactant molecules (right-hand side of the electronmicrograph). It was observed (Shah et al., 1976) that this formulation with foamlike structure exhibits an extremely low viscosity. It was also found that the interfacial properties (e.g., interfacial tension and surface viscosity) are strikingly influenced by the microstructure of the surfactant formulation.

It has been reported (Gogarty and Tosch, 1968; Gogarty, 1977; Holm and Robertson, 1981; Shah, 1981) that the middle-phase microemulsion at optimal salinity plays an important role in enhanced oil recovery process. The elucidation of the structure of the middle-phase, therefore, is one of the major objec-

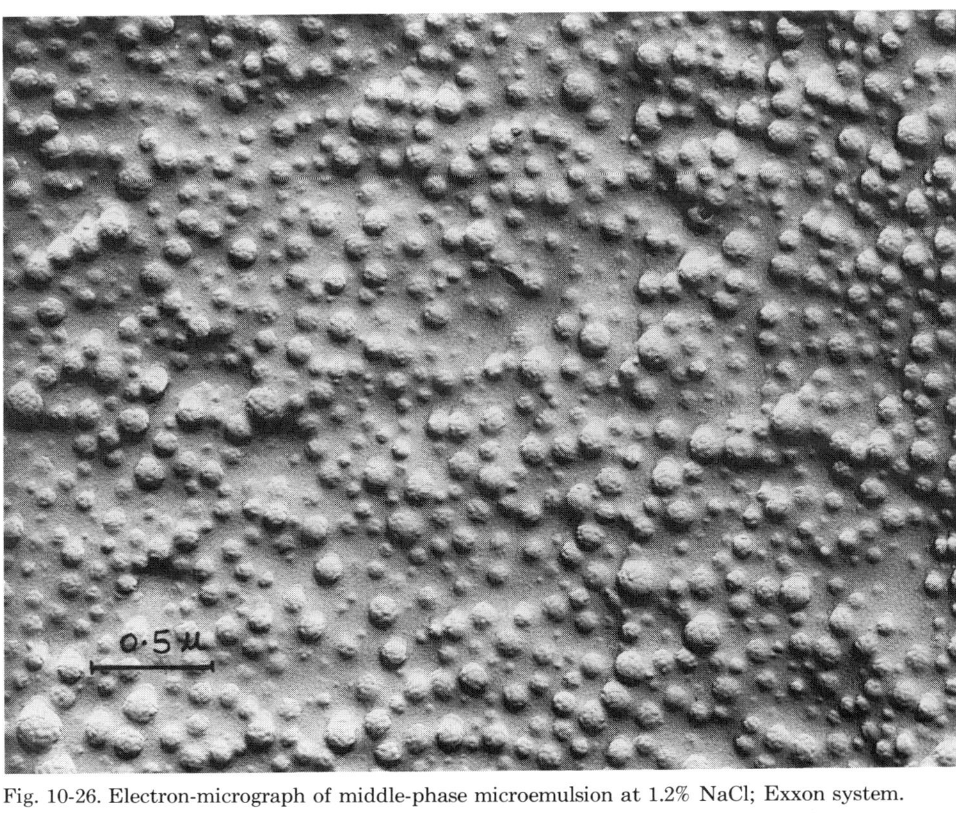

Fig. 10-26. Electron-micrograph of middle-phase microemulsion at 1.2% NaCl; Exxon system.

tives of the research in the field of oil recovery. A series of research papers has been published on the structure of middle-phase microemulsions (Scriven, 1976, 1977; Miller et al., 1977; Hatfield et al., 1978). It was suggested (Friberg et al., 1976) that the middle-phase microemulsion may be a mixture of spherical aggregates and planar surfactant layers. The structure of middle-phase microemulsion has also been proposed (Shah and Hsieh, 1981) to be a water-external microemulsion with monodispersed oil droplets suspended in a continuous aqueous phase. In support of this view that the middle-phase microemulsion is a water-external microemulsion, electron micrographs of the middle-phase of various salinities were obtained using freeze-fracture electronmicrography at 60,000 magnification (Healy and Reed, 1976). The electronmicrograph of the middle-phase at optimal salinity (1.2% NaCl) is shown in Fig. 10-26. This electronmicrograph clearly indicates that the middle phase is composed of swollen spherical micelles dispersed in a continuous aqueous phase. The swollen micelles are almost arranged in a regular fashion and are of nearly uniform size. It was also found that the size of microemulsion droplets decreases with increasing salt concentration. This decrease in the size of oil droplets to almost micellar size appears to facilitate the phase inversion from water-external to oil-external microemulsion with increasing salt concentration. Several experimental observations and theoretical considerations have been discussed (Shah and Hsieh, 1981) in support of this proposed structure of the middle-phase microemulsion. A.K. Sharma et al. (1983) and Chiang (1979) have studied oil recovery efficiency by injecting slug of middle-phase microemulsion in porous media. It was observed that at optimal salinity, the middle-phase microemulsion slug of various surfactant formulations displaced maximum amount of oil.

SURFACTANT–GAS (FOAM) FLOODING FOR OIL RECOVERY

Problems related to gas–steam flooding

Steam can be injected in petroleum reservoir for heavy oil recovery, whereas gases such as carbon dioxide and nitrogen can be used for light oil displacement. The common problems encountered in gas–steam flooding processes are illustrated in Fig. 10-27. In these processes, the lighter gas phase tends to flow on top and ride over the liquid phases in the reservoir because of the density differences between gas and liquid phases. This preferential movement of gas through the upper part of the reservoir is called gravity override (Fig. 10-27A). At reservoir conditions, the viscosity of the injected gases is often 10–100 times lower than oil viscosity. At these unfavorable viscosity (mobility) ratios, gases have the great potential to channel (finger) through the oil (Fig. 10-27B). In addition to this, if a petroleum reservoir consists of rocks with variable permeability, steam or gases have the inherent tendency to flow preferentially through

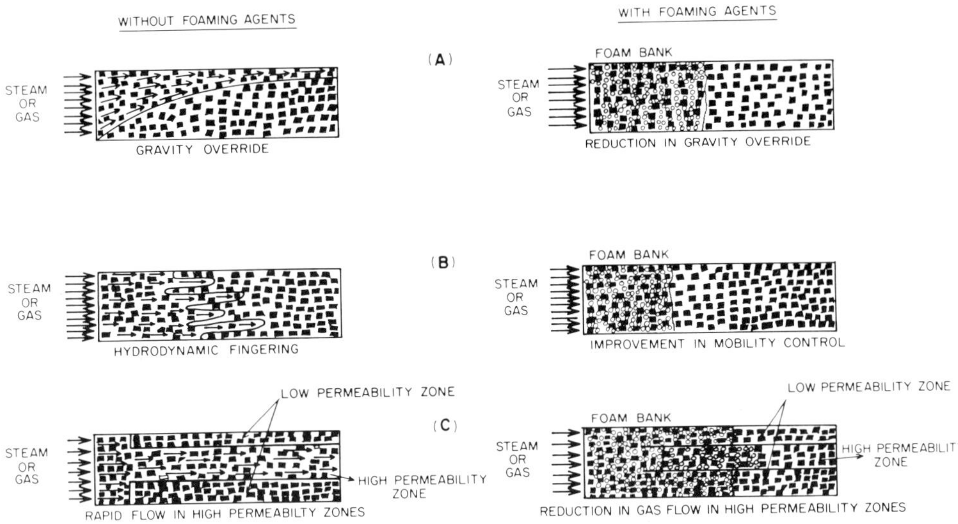

Fig. 10-27. Possible effects of foam on transport of steam or gas in porous media.

the more permeable zones compared with low-permeability zones (Fig. 10-27C).

As a consequence of channeling and gravity override, the heat distribution through the reservoir is not uniform during steamflooding. During gas–steam flooding, early breakthrough of the gas phase at the production wells is common, with a considerable volume of the petroleum reservoir bypassed.The efficiency of these processes, therefore, is significantly reduced and they often are uneconomical. To minimize these problems in gas–steam flooding processes for efficient oil displacement, one possible mechanisms is to have steam and/or gas present as the gas phase of a foam.

Role of surfactants in gasflooding processes

Surfactants with foaming ability can be used to reduce the mobility of the gas phase, which in turn decreases the channeling and gravity override (Fig. 10-27). To generate in-situ foam, a water-soluble surfactant is injected into the reservoir as an aqueous slug, which is followed by steam or gas. It has been shown (Sharma et al., 1982a, b; Sharma and Shah, 1983) that foam increases trapped gas saturation in porous media. As gas saturation increases, oil saturation decreases. In addition, a high trapped gas saturation would result in a higher pressure gradient and usually would reduce gas mobility.

As shown by Owete et al. (1982), Sharma and Shah (1982) and Sharma et al. (1982a, b), foam has a high potential for reducing gas mobility. The important advantage of foam over water for controlling mobility is that a foam is 80–95% gas, with a relatively small amount of water required to decrease gas mobility. In addition, foam possesses other properties that are considerably different from those of its components, (i.e., the gas or the aqueous surfactant solution) and that are favorable for oil displacement efficiency by gas–steam flooding. Foam is a relatively low-density material which can easily overcome gravitational effects and can pass through most of a heterogeneous petroleum reservoir. Moreover, foam viscosity is greater than that of its components. Both of these factors are favorable for efficient oil displacement, because increased viscosity is reflected in an improved mobility ratio. Futhermore, foam has other beneficial effects for oil displacement. The surface activity of foam reduces oil–water interfacial tension and facilitates wetting of oil-wet surfaces. It was also observed that foam generation capacity was high at low oil saturation (i.e., high water saturation) of the reservoir, which indicates that foam tends to have the greatest effect in waterflooded zones of the reservoir.

Surfactant–steam flooding for heavy oil recovery

Inasmuch as the importance of the foam in steam-driven heavy oil recovery has been realized, the examination of various properties of surfactant solutions is one of the major objectives of research in the field of heavy oil recovery. The ability of both pure and commercial surfactants to function at elevated temperatures and over extended periods of time has been investigated (Elson and Marsden, 1978; Sharma et al., 1982a). It was observed that some surfactants lose their ability to function after several days at 100°C, whereas others are still fully effective for a period of one week. Most non-ionic surfactants can produce foam at room temperature but not at 100°C (Elson and Marsden, 1978). The ability of both pure and commercial surfactants to generate foam increases with increasing temperatures, whereas the foam stability decreases with increasing temperatures (Sharma et al., 1984).

The thermal stability of several petroleum sulfonates and non-ionic surfactants has been investigated in relation to heavy-oil recovery (Handy et al., 1979; Handy, 1981; Isaacs et al., 1981). Stability of the surfactants is reported in terms of the time required for half of the surfactant to decompose. Most of the surfactants investigated decomposed by first-order kinetics (Handy, 1981). At a minimum steamflood temperature (177°C), the half-lives for the most stable surfactants evaluated were about 11 days. In order to determine the half-lives for petroleum sulfonates at different temperatures, the activation energies of the surfactants were measured. It was observed that petrostep 465 showed a half-life of 33 days at 93°C (Handy et al., 1979). The effect of surfactant concentration and salt on the stability of TRS 10-80 has been investigated (Isaacs et al., 1981) at various temperatures. The presence of salt promotes the decomposition process of the surfactants at high temperatures. The stabilities, however, appear to increase for surfactant solutions with initial concentrations in excess of the critical micelle concentration.

It has been reported (Celik et al., 1979; Isaacs et al., 1981; Ziegler and Handy, 1981) that precipitation of sulfonates occurs at the higher temperatures in natural sandstones. This precipitation appears to occur as a result of an interaction with minerals in the porous media and not due to the direct effects of temperature on the solubility of the surfactants. A significant increase in the solubility of rock minerals at steamflooding temperatures has been observed (Reed, 1980). The sulfonate ions from surfactants react with divalent cations present in rocks and form precipitates, the solubility of which decreases with increasing temperature. In general, the presence of salt in solutions decreases the solubility of the petroleum sulfonates. The solubility of non-ionic surfactants, however, is not as sensitive to salt concentration as that of the ionic surfactants. At specific temperatures, termed as cloud points, the ethoxy groups in non-ionic surfactants lose associated water and the solubility decreases abruptly to form precipitates. The effect of temperature on surfactant adsorption has been reported by several investigators (Handy, 1981; Isaacs et al., 1981). The results suggest that adsorption decreases for both sodium dodecylbenzene sulfonate and for Igepal CO-850, but the effect is not as significant as one might have expected. Additional work is required to understand the mechanism of surfactant adsorption at various temperatures in porous media. It was also shown (Handy, 1981; Sharma et al., 1984) that the bubble size in foams and interfacial tension between aqueous surfactant solution and oil decrease with increasing temperature.

A comparison of the results of steamflooding alone and steam–surfactant (foam) flooding shows that the latter has the potential to recover 25% more bitumen from oil sand than steamflooding alone (Isaacs et al., 1981). It was also reported (Sharma and Shah, 1983) that steam foams are more effective than air foams in displacing oil. Fig. 10-28 shows the effect of oil viscosity on residual oil saturation after air and steam foam flooding. Compared to steam foam flooding, the effectiveness of air foams in recovering additional oil decreases with increasing oil viscosity. The higher oil recovery by steam foam flooding is presumably due to reduction in the oil viscosity and higher foaminess of the surfactants during steam injection compared with air injection. Further studies are needed to delineate the separate contribution of each factor to oil recovery.

Surfactant–CO_2 flooding for light oil recovery

The mechanism and various basic principles involved in CO_2 flooding have been investigated for the last 50 years by several researchers (e.g., Holm, 1959, 1963; Holm and Josendal, 1974). It has been reported that at elevated pressure CO_2 flooding can recover additional oil which is not recoverable by conventional methods. In addition, super-critical CO_2 develops multicontact miscibility with various crude oils, resulting in a very efficient oil displacement in porous media. In general, oil displacement efficiency with CO_2 is much higher in the laboratory than in the field because field conditions are more complex for all recovery processes. Carbon dioxide is costly compared with water. Several methods have been tried, therefore, to minimize the amount of CO_2 required for optimum oil recovery. First, a CO_2 slug was injected rather than continuous injection of CO_2.

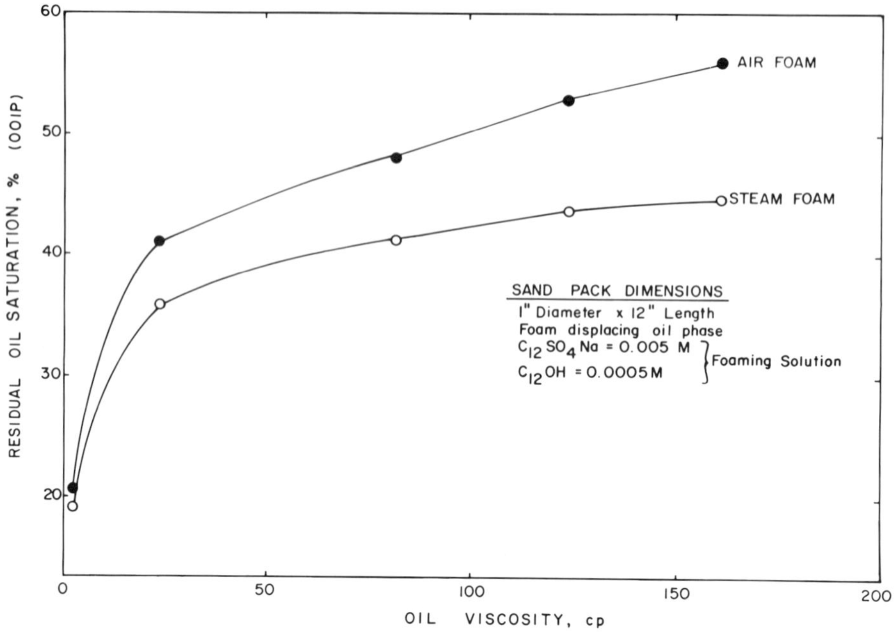

Fig. 10-28. Effect of oil viscosity on residual oil saturation after air and steam foamflooding.

This slug was followed by water. In the second method, CO_2 slugs were injected while the producing sites were shut in to avoid severe fingering and pressure sinks (Holm and O'Brien, 1971). In the third method, alternate slugs of CO_2 and water were injected in porous media (Caudle and Dyes, 1958; Holm and O'Brien, 1971). The main objective was to reduce CO_2 channeling by reducing its mobility. This method has both negative and positive effects. Although water reduces CO_2 mobility to a certain extent, it also traps oil which, in turn, increases the flow of water and decreases extraction of hydrocarbons from oil by CO_2. To reduce the amount of water required for CO_2 mobility improvement and efficient CO_2 oil recovery, surfactant–CO_2 systems have been used rather than water in recent years (Bernard et al., 1980).

A number of surfactants have been investigated for CO_2 mobility reduction in porous media (Bernard et al., 1980). The primary requirement for CO_2 mobility control is a surfactant with proper chemical stability, adsorption characteristics, and foaming or emulsifying capability. An anionic surfactant (Alipal CD-128) was recommended for the improvement of the CO_2 process (Bernard et al., 1980). Alipal CD-128 performs well in a CO_2–oil–water system, remains chemically stable in a CO_2–rock system for long period of time and shows minimum adsorption on rocks. It was also reported that Alipal CD-128 was effective in reducing CO_2 mobility in an oil–water system. The mobility reduction, however, was slightly less than in the oil-free system. This difference in mobility reduction is favorable for selective mobility reduction. For waterflooded reservoirs, it is highly desirable to maintain the greatest mobility reductions in oil-free reservoir zones and lesser reductions in oil-saturated zones. It appears, therefore, that this process has a high potential to achieve selective mobility reduction of CO_2.

It was observed experimentally (Bernard et al., 1980) that in the presence of 0.1% Alipal CD-128 the oil recovery was 9% of the residual oil after 4% PV CO_2 injection. If the experiment was repeated without Alipal CD-128, about 120% PV of CO_2 was required to displace the same amount of oil. This indicates that surfactant improves both oil displacement efficiency and conformance in a CO_2 injection. It was also shown that the primary mechanism of residual oil production was CO_2 extraction of light and middle range hydrocarbons from crude oil. More work on the influence of surfactant on interfacial properties of foaming solutions, oil recovery by surfactant–CO_2 flooding, and reduction of CO_2 mobility needs to be done.

INTERFACIAL PHENOMENA IN SURFACTANT–GAS (FOAM) FLOODING

The writers believe that the efficiency of gas- or steamflooding depends on the interfacial properties of surfactant formulations. Fig. 10-29 schematically shows the interrelationship of molecular properties of surfactants with EOR efficiency of steam- or gas-driven processes. It is evident that the molecular properties of foaming agents can influence the microscopic characteristics of foam which, in turn, can influence its flow behavior in porous media and ultimately the oil recovery efficiency. To prove the above hypothesis (Fig. 10-29), various pure systems ($C_{12}H_{25}SO_4Na + C_nH_{2n+1}OH$, where $n = 8$, 10, 12, 14 and 16) were extensively studied (Sharma et al., 1982, 1983; Sharma and Shah, 1983), knowing that these pure surfactants will not be employed in the petroleum reservoir for oil displacement. In this section, the various surface properties of surfactant solutions in relation to the foamflooding process for oil recovery are also discussed.

Surface properties of surfactant solutions

Surface tension

It was observed (Holcomb et al., 1981) that low surface tension foaming solutions tend to produce stable foams at lower concentrations. Surface tension determines the surface activity which, in turn, can influence the properties of

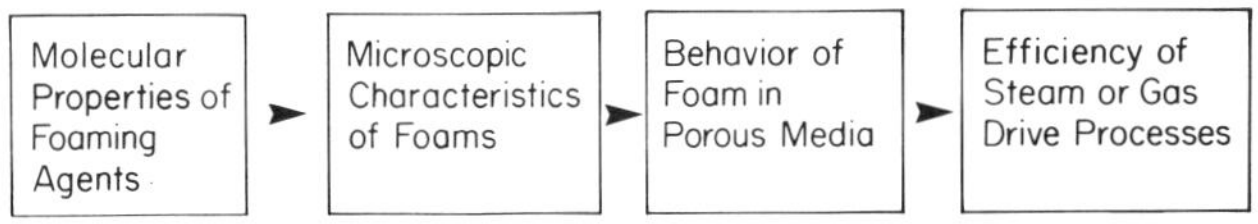

Fig. 10-29. A correlation of molecular properties of surfactants with the efficiency of steam- or gas-drive processes.

TABLE 10-II

The effect of chain length compatibility of foaming agents on surface tension, surface viscosity, bubble size, breakthrough time, fluid displacement efficiency and effective air mobility

System	$C_{12}H_{25}SO_4Na + C_8OH$	$C_{12}H_{25}SO_4Na + C_{10}OH$	$C_{12}H_{25}SO_4Na + C_{12}OH$	$C_{12}H_{25}SO_4Na + C_{14}OH$	$C_{12}H_{25}SO_4Na + C_{16}OH$
Surface tension (dynes/cm)	25.88	24.26	22.97	25.61	30.32
Surface viscosity (s.p.)	0.98×10^{-2}	1.45×10^{-2}	3.20×10^{-2}	2.62×10^{-2}	2.10×10^{-2}
Bubble size, radius (cm)	10.60×10^{-2}	6.50×10^{-2}	2.50×10^{-3}	8.00×10^{-2}	8.70×10^{-2}
Breakthrough time (min):					
in sandpacks at 5 psi	25.1	41.3	46.8	30.6	27.2
in Berea cores at 40 psi	18.5	23.5	29.5	18.0	17.0
Fluid displacement efficiency (%):					
in sandpacks at 5 psi	77.8	78.1	83.0	69.6	68.9
in Berea cores at 40 psi	54.3	58.6	62.9	47.1	45.7
Effective air mobility (cm^3/s psi):					
in sandpacks at 5 psi	11.91×10^{-3}	7.27×10^{-3}	6.86×10^{-3}	8.76×10^{-3}	9.63×10^{-3}
in Berea cores at 40 psi	1.81×10^{-4}	1.54×10^{-4}	1.32×10^{-4}	1.62×10^{-4}	1.66×10^{-4}

foams. In order to correlate the surface tension with the flow behavior of foam through porous media, the surface tensions of various pure and commercial surfactants have been measured (Sharma et al., 1982a). For mixed surfactant systems, a minimum in surface tension was observed when both components of the system had equal chain length (Table 10-II). An increase in surface tension was observed with increasing difference in chain length of mixed surfactants. Further studies showed that a decrease in surface tension was observed with increasing temperature. As the temperature is raised, the kinetic energy of the molecules increases, resulting in a decrease in attractive forces between the molecules which, in turn, reduces the surface tension of the surfactant solutions. A similar trend in the variation of surface tension with temperature was observed in the presence of NaCl.

The behavior of a mixed surfactant ($C_{12}H_{25}SO_4Na + C_{12}H_{25}OH$) system is more complicated in the presence of brine. The addition of brine to the surfactant solutions results in an increase in surface tension of the system. This increase in surface tension is probably caused by a decrease in dodecanol in sodium dodecyl sulfate solution in the presence of brine. Similar results were also reported by Klevens (1950) with different systems. The concentration of the mixed surfactant solutions for these observations was well above the critical micelle concentration. It has been reported (Corrin and Harkins, 1947) that the addition of long-chain alcohols or electrolytes lowers the CMC of ionic surfactants. If the effect of temperature is considered, the situation becomes much more complicated. In the absence of additives, the CMC of ionic surfactants generally increases with increasing temperature (Ginn et al., 1960). To solubilize dodecanol in surfactant solutions, the surfactant concentration used for foam formation was always above the CMC at all temperatures.

The effect of surfactant concentration on the surface tension of commercial surfactant solutions (Suntech IV) was determined (Sharma and Shah, 1984). The CMC of various samples of Suntech IV varies in the surfactant concentration range from 0.1% to 0.3%. It was concluded that, as the surface tension decreases, the breakthrough time and fluid displacement efficiency in porous media increases. Furthermore, the abrupt change in breakthrough time and fluid recovery corresponds to the break in surface tension at the same concentration of surfactant solution. One can determine the CMC of surfactant solutions, therefore, by measuring the breakthrough time and fluid displacement efficiency in porous media.

Surface viscosity

It is recognized that the surface viscosity can influence the stability of foams (Shah and Dysleski, 1969; Shah et al., 1978). A high surface viscosity reduces the rate of thinning of liquid films resulting in high foam stability (Brown et al., 1953). The surface viscosity of various pure and commercial surfactant systems was reported by Sharma et al. (1982a), in order to delineate the relationship with the flow through porous media behavior of foams. For mixed surfactant systems, a maximum in surface viscosity was observed when both components of the mixed surfactant systems had the same chain length (Table 10-II). This indicates that chain length compatibility strikingly affects the surface viscosity of surfactant solutions. Surfactant molecules of similar chain length are tightly packed at the air–liquid interface compared with the molecules of dissimilar chain length (Sharma et al., 1982a). Addition of long chain alcohols in surfactant solutions enhances the surface viscosity of the system which, in turn, increases foam stability. At pH 9.0, foams of decanoic acid and a 1 : 3 mixture of decanoic acid–decanol had maximum stability (Shah, 1970a, b).

Recently, the effect of surfactant concentration on surface viscosity has been studied (Sharma and Shah, 1982) using various commercial surfactants. It was observed that the surface viscosity increases with increasing surfactant concentration, and beyond the CMC the surface viscosity remains constant. These results were also correlated with other surface properties of surfactant solutions, microscopic characteristics of foams, and flow behavior of foam in porous media. To understand the role of the surface viscosity in the foam flooding process for enhanced oil recovery, more research work on the influence of salt concentration, temperature, pressure, and surfactant concentration needs to be done.

Foaminess of surfactants

Bennett (1963) reported that among other properties of surfactant formulations for foam flooding processes, the foaminess of a surfactant solution is more important than the foam stability. The presence of an in-situ foam bank between the displacing gas phase and displaced liquid improved both breakthrough time and ultimate recovery. The ability of several surfactants to generate foam has been studied in relation to enhanced oil recovery by the foamflooding process. The foam producing capacity of the surfactant solution increases with increasing surfactant concentration and temperature. Beyond the CMC, the foaminess of surfactant solutions almost remains constant. The static foaminess of surfactant solutions is strikingly increased by varying temperature from 20° to 40°C (Sharma et al., 1984). With further increases in temperature from 40° to 80°C, a gradual increase in the foaminess of the surfactant solutions was observed.

For mixed surfactant systems, the addition of brine to surfactant solutions results in a slight decrease in static foaminess (Sharma et al., 1984). The trend in the variation of the foaminess with temperature was found to be the same in the presence or the absence of brine. The foaminess of the surfactant solutions as a function of temperature can be correlated by the following equation:

$$F = F_{20}\, e^{A(T-20)} \qquad (10\text{-}3)$$

where A is a constant (1.6×10^{-3}), F is the foaminess of surfactant solution at temperature T (°C) and F_{20} is the foaminess at 20°C. Using above equation, the foaminess of surfactant solutions can be evaluated at any desired temperature between 20° and 80°C for the mixed sodium dodecyl sulfate and dodecanol system. The authors have shown that, for mixed surfactant systems, a maximum in foaminess of the surfactant solutions occurs when both components of the mixed surfactants have an equal chain length. These results were also correlated with the flow behavior of foams through porous media and bubble size (Sharma et al., 1984).

It was reported (Belkin et al., 1979) that, in general, anionic surfactants appear to be superior in the static foam test. Sulfate esters of ethoxylated linear alcohols were slightly better than alkylated aromatic sulfonates.

Foam stability

Surfactants play an important role in the foam stability and the formation of foams. Several investigators (Belkin et al., 1979; Sharma and Shah, 1984; Sharma et al., 1984) have determined foam stability by measuring half-life of foams. Half-life is the time required to reduce foam volume to half of its initial value. Belkin et al. (1979) studied ninety-three samples of different surfactants, both alone and as mixtures, in a static screening test in order to identify the suitable surfactant for efficient foamflooding. They observed that the ethoxylated sulfonates are the most effective compared with other classes of ionic surfactants.

The effect of temperature on foam stability of both pure and commercial surfactants has been studied in relation to the foamflooding process by Sharma and Shah (1983). It was observed that the foam stability (e.g., half-life) decreased with increasing temperature. The foam half-life decreased strikingly in the temperature range of 20–40°C, whereas the half-life exhibited a gradual decrease between 40° and 80°C for mixed surfactant systems. Among all surface properties of the surfactant solutions measured, the effect of temperature on foam stability was most pronounced. Similar conclusions can also be drawn from the relationship between bubble size and temperature. A similar trend in the variation of foam stability with temperature was also observed with commercial surfactants.

Comparison of the results of static foam stability with results for the dynamic behavior of foam in porous media reveals that the foam stability is not required for efficient fluid displacement or a decrease in effective air mobility in a porous medium. Moreover, the ability of the surfactants to produce in-situ foam is one of the important factors in the displacement of fluid in a porous medium. Similar results were also reported by Bernard and Holm (1964).

In summary, only a small amount of work has been done on the interfacial properties of surfactant solutions in relation to the foamflooding process for oil recovery. To elucidate the correlation between interfacial properties of surfactant solutions and flow through porous media behavior of foams, more research on various interfacial parameters using several commercial and pure surfactants is needed.

Microscopic characteristics of foams

Average bubble size

Sharma et al. (1982b) showed that the bubble size can influence the efficiency of the foamflooding process. Fig. 10-30 shows photomicrographs of various foams containing sodium dodecyl sulfate and dodecanol. Mixed surfactants of equal chain length ($C_{12}H_{25}SO_4Na + C_{12}H_{25}OH$) produced smaller foam bubbles as compared to mixed surfactants of dissimilar chain length ($C_{12}H_{25}SO_4Na + C_nH_{2n+1}OH$, where $n = 8$, 10, 14 and 16). The average size of foams as a function of chain length compatibility of mixed surfactants is shown in Table 10-II. It has been shown (Shah and Shiao, 1975) from monolayers studies that the surfactant molecules are tightly packed in mixed films when both components have the same chain length. It is very likely, therefore, that the thermal motion and area per molecule increase with increasing difference in chain length of mixed surfactants. One can conclude from these results that the molecular packing at the interface is strongly influenced by the chain length compatibility of the mixed

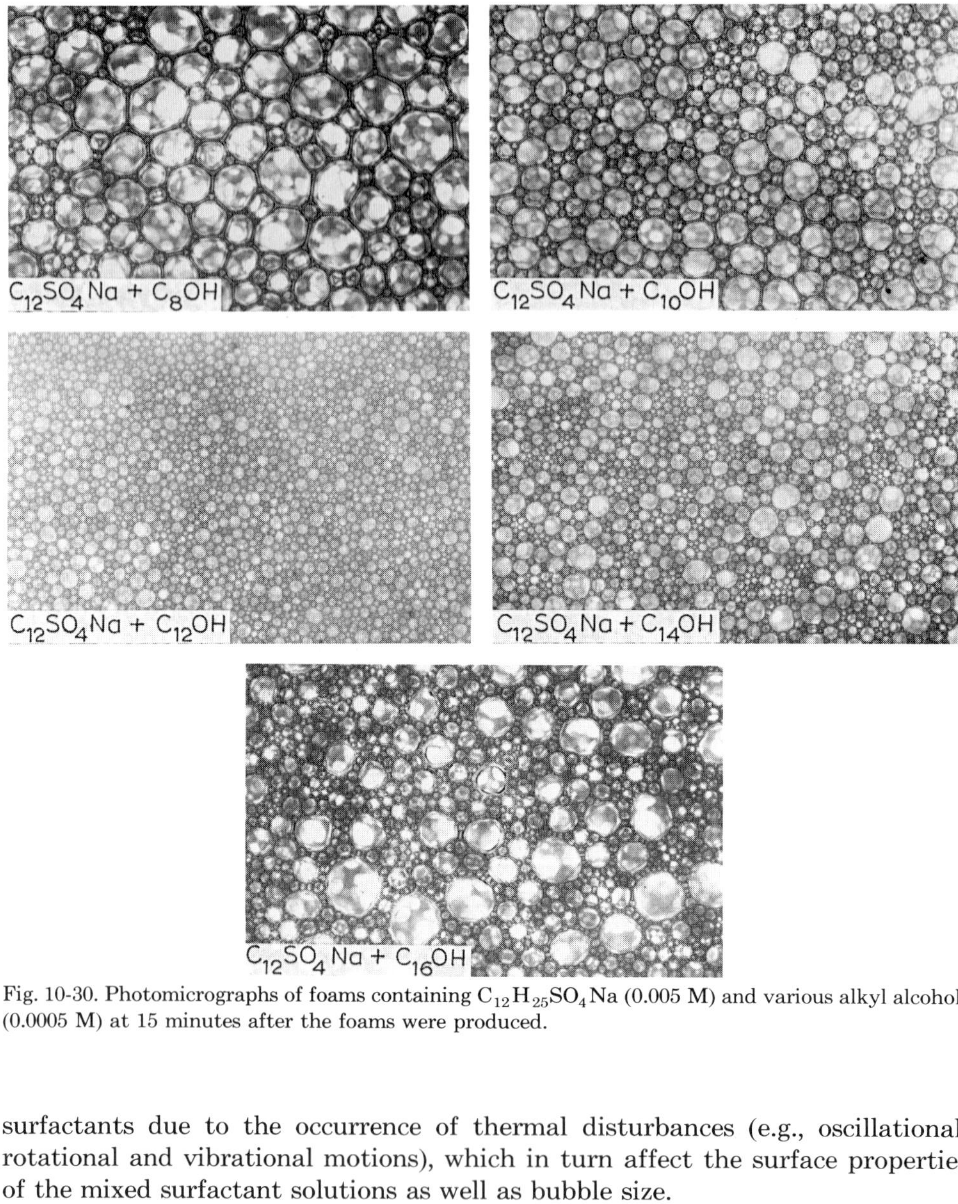

Fig. 10-30. Photomicrographs of foams containing $C_{12}H_{25}SO_4Na$ (0.005 M) and various alkyl alcohols (0.0005 M) at 15 minutes after the foams were produced.

surfactants due to the occurrence of thermal disturbances (e.g., oscillational, rotational and vibrational motions), which in turn affect the surface properties of the mixed surfactant solutions as well as bubble size.

The bubble size of several commercial surfactants was also measured in relation to the behavior of foam flow in porous media (Sharma and Shah, 1984). It was observed that the smaller bubbles displaced more fluid as compared to the large size foams. The rate of increase of bubble size was found to be smaller for surfactants containing more surface-active material compared with surfactants having less surface-active material. The bubble size decreases with increasing surfactant concentration; however, beyond the CMC, the bubble size remains almost constant for both pure and commercial surface-active agents. These results are consistent with foam stability and surface-tension data.

Effect of temperature on bubble size

The temperature of most petroleum reservoirs is higher than the surface temperature. It is important, therefore, to determine the effect of temperature on bubble size in relation to the improved oil recovery by foamflooding. Fig. 10-31 illustrates the photomicrographs of foams containing sodium dodecyl sulfate and dodecanol at various temperatures (Sharma et al., 1982b). It is obvious that the surfactants can produce initially smaller size bubbles at higher temperatures compared with lower temperatures. At a particular temperature, the bubble size increases slightly in the presence of brine.

The variation in bubble size as a function of time at various temperatures is shown in Fig. 10-32 (Sharma et al., 1984). The bubble size increases with elapsed time due to coalescence. It increases rapidly with time at higher temperatures. Moreover, it is evident that the surfactant solutions produced smaller bubbles

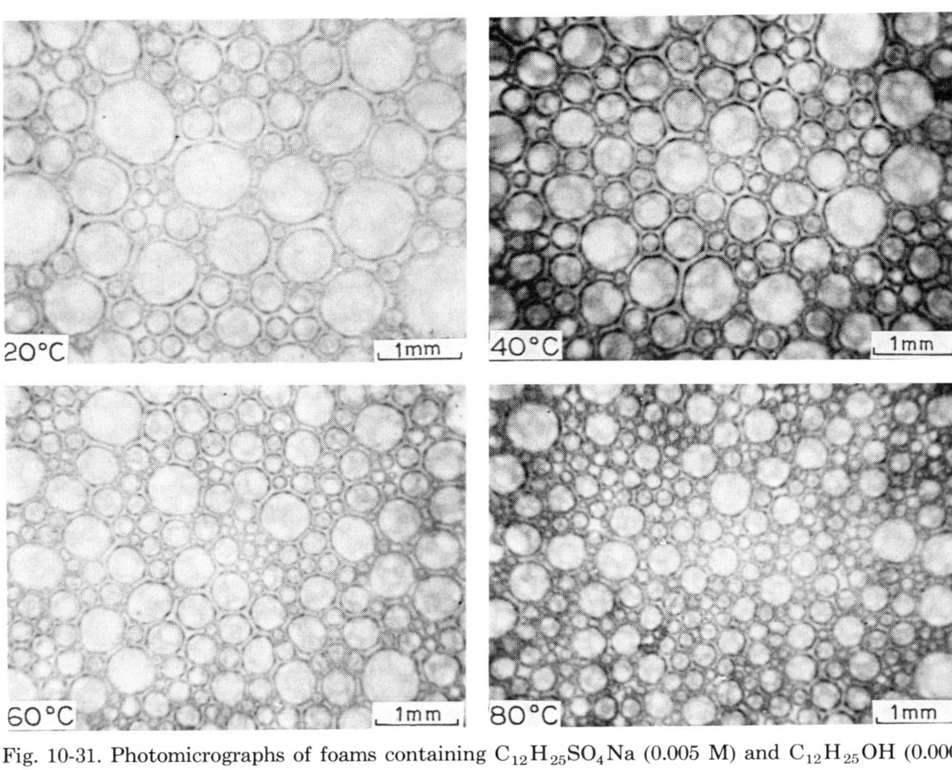

Fig. 10-31. Photomicrographs of foams containing $C_{12}H_{25}SO_4Na$ (0.005 M) and $C_{12}H_{25}OH$ (0.0005 M) at various temperatures.

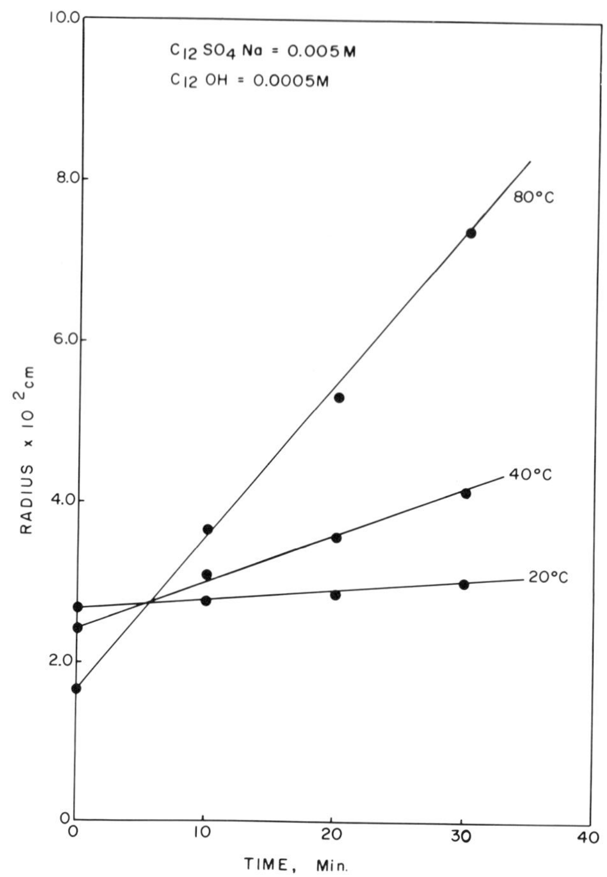

Fig. 10-32. Bubble size in foams as a function of time at different temperatures.

initially at higher temperatures. In Fig. 10-32, the curves intersect each other at a particular elapsed time which indicates that, at a specific time, the surfactant solutions produce the same size foam at various temperatures. To calculate the size of foam bubbles, a quantitative correlation among bubble size, temperature and time has been established (Sharma et al., 1984):

$$r = (aT + b)t + (cT + d) \tag{10-4}$$

where r is the radius of foam bubbles, T is the temperature, t is the elapsed time, and a, b, c and d are the constants. The values of these constants for the system studied are: $a = 2.96 \times 10^{-4}$ cm/min °C; $b = -5.70 \times 10^{-4}$ cm/min; $c = -1.82 \times 10^{-4}$ cm/°C; and $d = 3.14 \times 10^{-2}$ cm.

Using these values of the constants, one can evaluate the bubble size at any time at a particular temperature. The above correlation is in good agreement (within $\pm 4\%$) with the experimental values of bubble size.

It is evident from these observations that the foam stability decreases with rising temperature as well as with elapsed time. These findings support previously discussed results for surface tension and foam stability. To broaden the understanding of this subject, it is suggested that several classes of surfactants be examined and that bubble size be correlated with foam behavior in porous media.

Effect of pressure on bubble size

It is known that with increasing pressure the concentration of surfactants at an interface increases and surface tension decreases. If one considers only the surface tension parameter, the change in pressure should influence the bubble size. Using video-micrographic techniques, the structure of foams was studied at high pressures (Holcomb et al., 1981). It was observed that the bubble size of flowing foam in a capillary tube decreases with increasing pressure. Moreover, the static bubbles were found to be larger in diameter as compared to those of the flowing foam at the same pressure. It has been shown (Sharma et al., 1982b) that the smaller size foams are more effective in displacing fluid and in reducing gas mobility in porous media. In view of this fact, it appears that the foamflooding process under pressure can efficiently displace fluid and reduce gas mobility in porous media. No group has attempted to correlate the bubble size with foamflooding at reservoir pressures. To delineate the relationship between bubble size with foamflooding efficiency for oil recovery, the microscopic characteristics of foams and the surface properties of surfactant solutions should be studied at reservoir temperatures and pressures.

Flow through porous media behavior of foams

Fluid displacement efficiency

Laboratory studies have shown (Sharma et al., 1982b) that the presence of foam in a porous medium considerably increases the fluid recovery both in sand packs and Berea cores. The volume of fluid collected at gas breakthrough at a producing site is termed as fluid displacement efficiency. The fluid displacement efficiency in sand packs was higher as compared to Berea cores. This indicates that foams are more effective in high-permeability zones. Similar conclusions were also drawn (Bernard and Holm, 1964) by measuring the permeability of porous media to gas in the presence of foam. For sand packs, the fluid displacement efficiency in the absence of foam continuously decreased with increasing pressure drop, whereas in the presence of foam, a maximum occurred at 4 psi. In the case of Berea cores with foams, however, a continuous decrease in fluid displacement efficiency was observed for both pure and commercial surfactants (Sharma et al., 1982b). Furthermore, an increase in the length of porous media improved fluid displacement efficiency.

The effects of temperature and brine on fluid displacement efficiency has been studied by Sharma et al. (1984). The fluid displacement efficiency increased with temperature. A steep increase in fluid recovery was observed in the temperature range of 20–40°C. The fluid displacement efficiency increased gradually beyond 40°C. The presence of 0.5% and 1.0% NaCl in aqueous solutions slightly reduced fluid recovery at various temperatures. The writers have proposed a quantitative correlation between the fluid displacement efficiency at temperature T in °C (E) and the displacement efficiency at 20°C (E_{20}):

$$E = E_{20}\, e^{B(T-20)}, \qquad (10\text{-}5)$$

where B is a constant (4.0×10^{-3}). A correlation of the static foaminess of surfactant solutions (equation (10-3)) with fluid displacement efficiency or dynamic behavior of in-situ foam (equation (10-5)) can be expressed as follows:

$$\ln E = 2.5 \ln F + C \qquad (10\text{-}6)$$

where:

$$C = \ln E_{20} - 2.5 \ln F_{20} \qquad (10\text{-}7)$$

Constant C decreases with increasing concentration of sodium chloride in the system. The value of the constant C is about -10.25 for this system. Using equation (10-6), one can evaluate the fluid displacement efficiency in porous media from a knowledge of the static foaminess of the surfactant solutions. For mixed systems of various chain lengths, a maximum in fluid displacement efficiency was observed (Sharma et al., 1982a) when both components of the system had a similar chain length (Table 10-II). The mixed surfactant system of similar chain length (e.g., $C_{12}H_{25}SO_4Na + C_{12}H_{25}OH$) improved fluid recovery by approximately 15% as compared to that by mixed surfactant systems of dissimilar chain length (e.g., $C_{12}H_{25}SO_4Na + C_{16}H_{33}OH$). Hence, the molecular properties of the surfactants can influence the flow through porous media behavior of foam. A detailed discussion of this subject is given in later sections.

Breakthrough time

The breakthrough time is defined as the time required by gas to travel from injection site to production site. Sharma et al. (1982b) observed that the breakthrough time strikingly increased using foam as compared to that without foam. As the gas travels in porous media in the presence of surfactant solution, the in-situ foam generation increases trapped gas saturation which, in turn, increases the residence time of the gas in porous media. As the gas saturation increases, the fluid saturation decreases. The breakthrough time decreased sharply up to a pressure drop of 4 psi. Beyond this pressure, a gradual decrease in breakthrough time was observed in sand packs. A similar trend in the variation of breakthrough time as a function of pressure drop was observed in Berea cores (Sharma et al., 1982b).

The effect of temperature on breakthrough time was studied both in sand packs and in Berea cores (Sharma and Shah, 1983). The breakthrough time increased with increasing temperature. A steep increase in breakthrough time

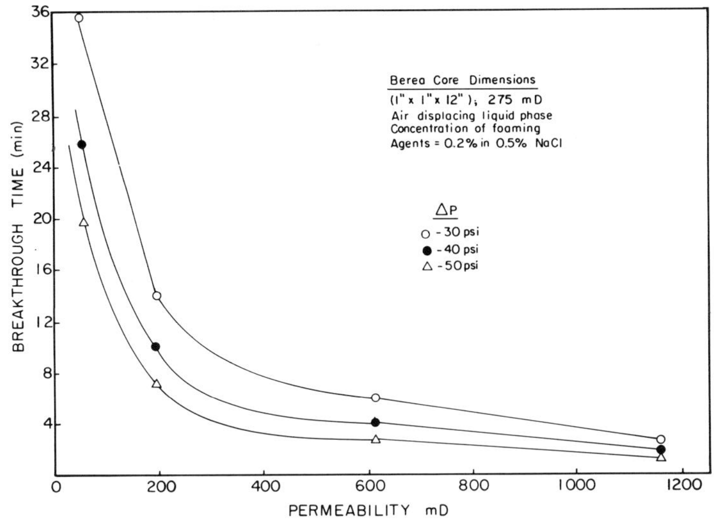

Fig. 10-33. Effect of the permeability of porous media on breakthrough time with Suntech (IV) at various pressure drops (Δp).

was observed up to 40°C. Beyond this temperature the breakthrough time increased gradually. Using a commercial surfactant (Suntech IV), the effect of permeability of porous media on breakthrough time was studied at different pressure drops (Sharma et al., 1982b). The decrease in breakthrough time was observed with increasing permeability of the sandstones (Fig. 10-33). The breakthrough time decreased sharply in the permeability range of 50–200 mD. A gradual decrease in breakthrough time was observed beyond 200 mD.

The effect of the length of porous media on breakthrough time was studied in the presence and the absence of surfactant solutions (Sharma et al., 1982b). In the presence of surfactant solutions, the breakthrough time per unit length decreased with increase in the length of the porous media. The trend in the variation of breakthrough time was found to be the same in the absence of surfactant solution. To delineate the effect of chain length compatibility on breakthrough time, the mixed anionic and non-ionic surfactant systems of varying chain lengths were employed for flow through porous media studies. A maximum in breakthrough time was observed for mixed surfactant systems of the same chain length (Table 10-II). This shows that the molecular properties of surfactants can influence the breakthrough time and gas saturation in porous media.

Gas mobility

The effective and relative mobilities of a gas in porous media can be evaluated from various expressions proposed by different researchers (Sharma et al., 1983). Fig. 10-34 illustrates the effect of the permeability of porous media on relative

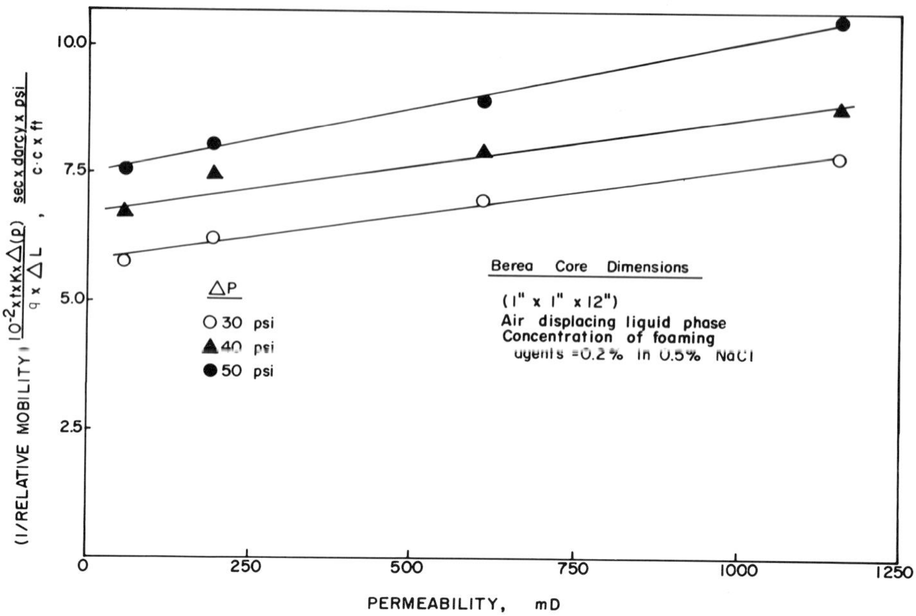

Fig. 10-34. Effect of the permeability of porous media on relative air mobility at different pressure drops (Δp).

air mobility at different pressure drops. A linear relationship between the inverse of the relative mobility and absolute permeability of the porous media was observed (Fig. 10-34). As the pressure drop and absolute permeability of porous media are increased, the relative mobility of air decreases. From the slopes of the curves in Fig. 10-34, it is observed that the effective air mobility decreases with increasing pressure drop at a given absolute permeability of the porous media. These results show that the foams are more effective in reducing gas mobility in high-permeability as compared to low-permeability sandstones. Using a different approach, similar conclusions were also drawn by Bernard and Holm (1964).

The effect of temperature and length of porous media on effective air mobility was studied in relation to oil recovery (Sharma and Shah, 1983; Sharma et al., 1984). As the temperature and pressure drop increases, the effective air mobility in porous media decreases (Fig. 10-35). A sharp decrease in effective air mobility was observed up to 40°C, and beyond this temperature a gradual decrease in effective air mobility was observed. It has been reported that the foaming tendency of surfactants increases with increasing temperature and pressure drop (Bikerman, 1973; Sharma et al., 1982b). The effectiveness of foam to recover additional fluid as well as to reduce the gas mobility in porous media, therefore, increases with increasing temperature and pressure drop. These results also indicate that high foam stability is not required for reducing gas mobility or for higher fluid recovery. As the length of the porous medium was increased, the

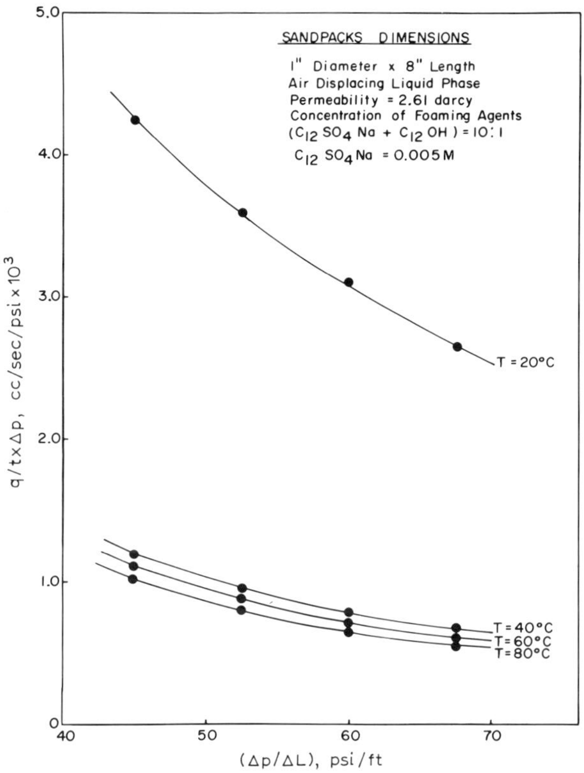

Fig. 10-35. Effect of pressure gradient ($\Delta p/\Delta L$) on effective air mobility at different temperatures.

effective air mobility also decreased in the presence of pure and commercial surfactants.

Dilgren et al. (1982) developed a surfactant formulation to reduce steam mobility in the steam-driven process. The commercially available dodecyl-benzene sodium sulfonates, both straight and branched chain, as well as hexadecyl alpha-olefin sodium sulfonates, have been shown to be effective in reducing steam mobility in the presence of brine. It was also concluded that the field results were consistent with laboratory findings. Inasmuch as a lower permeability reduction factor is observed in the laboratory, a higher bottomhole injection pressure was obtained in the field in the presence of surfactant solution in the steamflooding process.

The effect of chain length compatibility of mixed surfactants on air mobility in porous media was studied by Sharma et al. (1982a). For mixed surfactant systems, a minimum in effective air mobility was observed for the mixed system of the same chain length (Table 10-II). The mixed systems of the same chain length are more effective in reducing air mobility in porous media.

In summary, the results of studies of chain length compatibility of mixed surfactants show that a minimum in surface tension, a maximum in surface viscosity, a minimum in bubble size, a maximum in breakthrough time, a minimum in effective air mobility and a maximum in fluid displacement efficiency were observed when both components of the system had the same chain length (Table 10-II). A detailed explanation of these findings is given in a later section. It is evident that the molecular properties of a surfactant formulation for the gas-driven process can influence surface properties of the surfactant solution, microscopic characteristics of the foam, and flow behavior of the foam in porous media. These results can be utilized in developing surfactant formulations for efficient gasflooding processes under reservoir conditions.

Correlation between chain length compatibility of mixed surfactants and heavy oil recovery

As shown previously, in the absence of oil, the molecular properties of surfactant formulations can improve the efficiency of the foamflooding process. In this section, the effect of chain length compatibility of mixed surfactants on oil displacement efficiency in porous media is briefly discussed. To delineate the correlation between the chain length compatibility of mixed surfactant systems and oil recovery, the heavy-oil displacement efficiency in porous media was measured by surfactant solution flooding, air foamflooding and steam foamflooding processes (Sharma and Shah, 1983). Fig. 10-36 shows a correlation of chain length compatibility with interfacial properties of surfactant solutions, microscopic characteristics of foam, and heavy oil recovery by various processes. An air injection at the stage of surfactant solution breakthrough recovered nearly 2% additional oil. Replacing air by steam injection recovered about 12% additional oil. The steam injection presumably decreases the viscosity of heavy oil as well as generates more foam in porous media as compared to air injection, resulting in an increase in heavy oil recovery.

For mixed surfactant systems, a minimum in surface tension, a maximum in surface viscosity, a minimum in bubble size, and a maximum in heavy oil recovery by surfactant solution flooding as well as foamflooding were observed when both components of the system had the same chain length (Sharma and Shah, 1983). The explanation for these chain length compatibility effects is illustrated schematically in Fig. 10-37, based upon the concept of thermal motion of hydrocarbon chains at an air–liquid interface (Adam, 1938). The mixed surfactant molecules of equal chain length form a tightly packed mixed monolayer at the air–liquid interface. When the chain lengths of surfactants are not equal, the portion of the hydrocarbon chain above the height of the adjacent hydrocarbon chain will exhibit thermal motion (vibrational, rotational and oscillational). If thermal motions were limited to the portion above the height of the adjacent hydrocarbon chain, it would not change the packing of the molecules or the average area per molecule. The thermal disturbances, however,

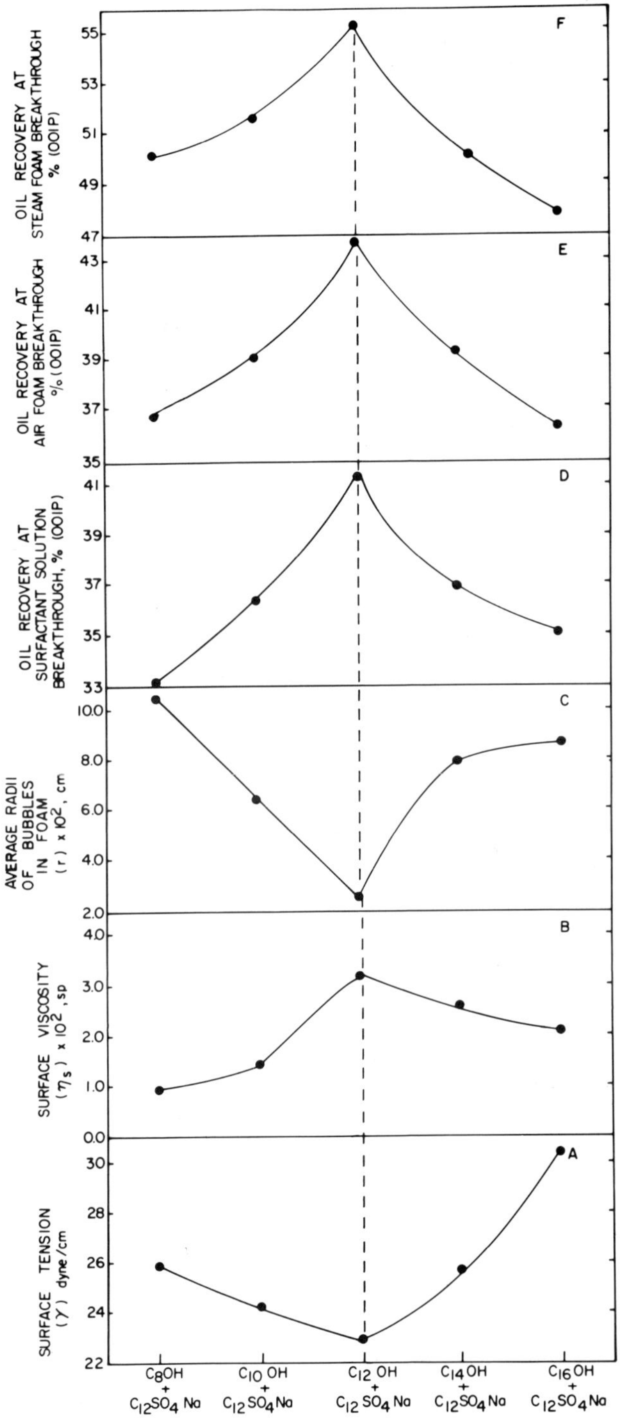

Fig. 10-36. Correlation of interfacial properties of surfactant solutions and microscopic characteristics of foams with heavy oil recovery efficiency in porous media.

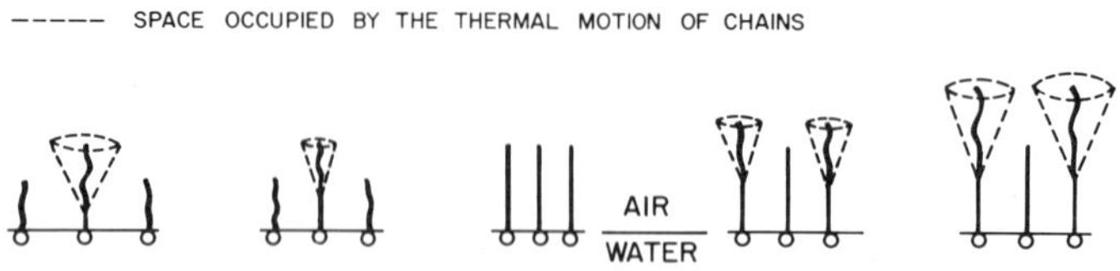

Fig. 10-37. Schematic diagram of the thermal motion of the upper segments of hydrocarbon chains at the air–liquid interface. The cones shown by dashed lines represent the space occupied by the thermal motion of the segments.

probably extend along the alkyl chain towards the polar part of the surfactant molecules which, in turn, exhibit a greater area per molecule in mixed monolayers at the air–liquid interface (Shah and Shiao, 1975). Moreover, as the difference in hydrocarbon chains in foaming agents increases, the expansion of the mixed monolayer also increases because the length of the segment undergoing rotation, vibration, and oscillation increases. The cones with dashed line (Fig. 10-37) show the time-average space occupied by segments of hydrocarbon chains of surfactants. A similar concept was also used by Phillips et al. (1970) and Shah and Shiao (1975) to explain results on mixed monolayers of various ionic and nonionic surface-active agents. It was observed (Shah and Shiao, 1975) that in the mixed monolayers of various alcohols, the average area per molecule is a minimum when both components possess the same chain length. The results indicate that chain length compatibility also influences the bubble size and stability (Shiao, 1976). These results suggest that the molecular packing at air–liquid interface and the surface properties of surfactant solutions exhibit a striking correlation with bubble size and oil displacement efficiency in porous media.

MECHANISM OF SURFACTANT LOSS IN POROUS MEDIA

Success or failure of various oil recovery processes which use surfactant injection depends on the degree of retention of surfactants during the course of the flooding. The retention of surfactants by reservoir minerals causes decrease in their concentration in surfactant slugs injected for oil displacement from waterflooded reservoirs. The following is a brief description of several mechanisms responsible for the loss of surfactants in porous media.

Surfactant adsorption

In the surfactant flooding process, as the surfactant slug comes into contact with the reservoir rock and brine, there would be a loss of surfactant due to adsorption at solid–liquid interface. Only part of the total surfactant injected in the reservoir, therefore, is available for oil displacement. Several investigators

(Fuerstenau et al., 1963; Somasundaran et al., 1964; Ginn, 1970; White, 1970; Hill et al., 1973; Trushenski et al., 1974; Somasundaran and Grieves, 1975; Bae and Petrick, 1976) have investigated the adsorption of various petroleum sulfonates on solid–liquid interfaces. It was observed that a maximum in adsorption occurs near the critical micelle concentration, followed by a minimum around the concentration region where fluid appears to be changed from turbid to transparent (Trushenski et al., 1974). According to Bae et al. (1974), the results of adsorption of petroleum sulfonates on sandstones from aqueous solutions also indicate the presence of an adsorption maximum, which is generally in agreement with findings of Trushenski et al. (1974).

It has been reported (Hill et al., 1973) that the adsorption of petroleum sulfonate can be reduced in the presence of a Dow pusher polymer and tripolyphosphate. The adsorption of sulfonates on oil-wet cores was found to be greater than on the water-wet cores (Boneau and Clampitt, 1977). The effect of flow rate on the adsorption of surfactants in sandstones has been studied (Bae and Petrick, 1976). For sulfonates of broad equivalent weight distribution, the amount of maximum adsorption and surfactant concentration at which maxima occur are dependent on flow rate. The adsorption of petroleum sulfonates increased with increasing salt concentration (Foster, 1973; Gale and Sandvik, 1973; Hurd, 1976). Some inorganic salts were found to decrease the adsorption of petroleum sulfonates (Foster 1973; Hill et al., 1973). The adsorption of petroleum sulfonates strikingly decreased in the presence of sodium tripolyphosphate. Moreover, some inorganic salts can also be used as sacrificial agents to reduce the adsorption of petroleum sulfonates (Foster, 1973). The effect of equivalent weight of surfactant on its adsorption was reported in the literature (Hurd, 1976). A sharp increase in adsorption was observed at a petroleum sulfonate equivalent weight of nearly 450. It was also observed (Hanna and Somasundaran, 1977) that the adsorption of mahogany sulfonate AA on Berea Sandstone increased with increasing temperature.

Recently, the adsorption of surfactant has been studied in the presence of short-chain alcohols (Novosad, 1981). Surfactant adsorption can be reduced by the addition of low-molecular-weight alcohols (e.g., secondary butyl alcohol, *n*-propanol). Moreover, the adsorption of pure and commercial surfactants on Berea core can easily be compared. The effect of microemulsion composition on the adsorption of petroleum sulfonates on Berea Sandstone has been investigated (Derderian et al., 1981). It was observed that the petroleum sulfonates which are preferentially oil-soluble tend to be adsorbed to a lesser extent from oil-external microemulsions than from brine-external microemulsions under similar conditions.

Surfactant precipitation

Waterflooded reservoirs contain irreducible oil, connate water, and injected water. In various reservoirs, the ionic content of the injected water during the waterflooding process may differ considerably from the connate water in the reservoir. During surfactant flooding, the surfactant slug may contact both the connate water and injected water. The presence of divalent cations in connate water may result in precipitation of the surfactant. Their presence may also alter the oil–surfactant solution interfacial tension due to the difference in ionic content between injected water and connate water. The precipitation of surfactants caused by the high concentrations of divalent ions decreases the efficiency of oil displacement.

The results reported by Bansal and Shah (1978b) indicate that the presence of ethoxylated sulfonate in petroleum sulfonate formulations can increase the tolerance to divalent cations (e.g., $CaCl_2$ and $MgCl_2$) as well as produce ultra-low interfacial tension at the optimal salinity. It was also observed (Dauben and Froning, 1971; Maddox et al., 1975; Gale et al., 1976) that the ethoxylated alcohols and ethoxylated sulfonates can improve the salt tolerance of surfactant formulations for oil recovery.

Surfactant degradation

In several oil recovery processes, the surfactants have been employed to improve the efficiency of the processes. In these processes, the thermal stability of surfactants at reservoir temperatures as well as during steam injection is required for efficient oil displacement in porous media. Several investigators (Handy et al., 1979; Handy, 1981; Isaacs et al., 1981) have delineated the effect of temperature on the stability of surfactants. The anionic surfactants were observed to be more stable than non-ionic surfactants. Most of the surfactants tested are sufficiently stable at normal reservoir temperatures, whereas the stability of the best sulfonate was marginally acceptable at a temperature of 180°C (Handy, 1981). Thermal degradation of a petroleum sulfonate in aqueous media can be explained by equilibrium hydrolytic desulfonation reactions (Isaacs et al., 1981). The surfactants were found to be effective even after heating at 200°C for 24 hours. Suntech IV and DOWFAX 2AI exhibited high thermal stability with a slight decrease in thermal stability observed at 205°C (Wang et al., 1982).

Surfactant–polymer mixing

The loss of surfactants is the major factor which determines the optimum volume of the expensive micellar fluid required to displace residual oil. Extensive oil displacement tests in consolidated cores indicate that several micellar fluids are not compatible with the polymer from the mobility buffer bank. When polymer mixes and invades the micellar fluid, multiple phases often occur. If the interfacial tension between these phases is sufficient, phase trapping occurs in porous media. This surfactant loss reduces efficiency of oil displacement as well as increases chemical requirements.

The polymer inaccessible pore volume determines the magnitude of surfactant–polymer mixing in a porous medium (Dawson and Lantz, 1972). This inaccessible pore volume phenomenon occurs due to the greater molecular volume of the polymer. Polymer molecules, therefore, are excluded from the small pores and can move only through larger pores in the porous media, whereas water molecules can flow through both the small and large pores. The pore volume available to the polymer molecules is, therefore, less than that available to water. This phenomenon leads to polymer molecules moving faster than the carrier water and results in interaction between micellar slug and polymer. This surfactant–polymer interaction can be reduced if the salinity of the polymer slug is lower than the salinity of the surfactant slug, and phase separation can also be eliminated by the addition of cosurfactants to the polymer or surfactant slug (Trushenski, 1977). Desai (1983) reported that the addition of polymer to surfactant formulations can appreciably lower the interfacial tension between these formulations and oil in certain ranges of composition of the system.

Surfactant partitioning in residual oil phase

When the surfactant slug propagates through a reservoir, it comes into contact with the residual oil. The partitioning of surfactant with the oil phase has been reported by Baviere (1976) and Chan and Shah (1980). It has also been reported (Chan and Shah, 1977b) that one of the mechanisms of producing ultra-low interfacial tension between brine and oil phase is the equal partitioning of surfactant with the brine and oil phase. If the interfacial tension is sufficiently high, oil-phase trapping occurs in porous media, which also causes surfactant loss due to its partitioning in oil.

PRESENT STATUS OF THE USE OF SURFACTANTS IN OIL RECOVERY

The various oil recovery processes involving surfactant injection account for approximately 1% of the enhanced oil recovered in the United States. Surfactant- as well as foamflooding processes, however, have a high potential for recovering additional oil from waterflooded reservoirs. These processes have developed slowly so far because of associated high risks, high costs, and complicated technology. In the United States, active projects involving surfactant injection from 1977 to 1980 have increased from approximately 5 to 30.

Many research laboratories are actively involved in investigating the use of surfactants in light and heavy oil recovery. Surfactant- and foamflooding may become more widely accepted as more is learned about the mechanism of oil displacement and as these processes are improved and tested in the field. At present, all of these processes are complicated and are not well understood. The contact of surfactants with residual oil, connate water, and reservoir rock decreases their efficiency in displacing oil.

The state of the art in the use of surfactants in light and heavy oil recovery is at the research stage so far. New developments and improvements are required for these processes to become economically feasible. These improvements will benefit not only the micellar flooding process but also foamflooding processes for heavy oil recovery by steam injection and light oil recovery by CO_2 or N_2 injection because of their use as steam or gas mobility control agents.

ACKNOWLEDGEMENTS

It is a pleasure to express our gratitude to Dr. G. Westermann-Clark, Assistant Professor of Chemical Engineering and Dr. B.M. Moudgil, Director of Center for Research in Mining and Mineral Resources, University of Florida, Gainesville, Florida, for their valuable suggestions. The help extended by Drs. T.F. Yen and George V. Chilingar was also invaluable.

REFERENCES

Abrams, A., 1975. The influence of fluid viscosity, interfacial tension and flow velocity on residual oil saturation left by waterflood. *Soc. Pet. Eng. J.*, 15(5): 437–447.

Adam, N.K., 1938. *The Physics and Chemistry of Surfaces*. Oxford University Press, London, pp. 67–69.

Bae, J.H. and Petrick, C.B., 1976. Adsorption of petroleum sulfonates in Berea cores. *Soc. Pet. Eng. Symp. on Improved Oil Recovery, Tulsa, Okla., March 22–24*, pp. 303–312, SPE 5819.

Bae, J.H., Petrick, C.B. and Ehrlich, R., 1974. A comparative evaluation of microemulsions and aqueous systems. *Soc. Pet. Eng. Symp. on Improved Oil Recovery, Tulsa, Okla.*, pp. 165–180, SPE 4749.

Bansal, V.K. and Shah, D.O., 1978a. The effect of caustic concentration on interfacial charge, interfacial tension and droplet size: a sample test for optimum caustic concentration for crude oil. *J. Can. Pet. Technol.*, 17(1): 1–4.

Bansal, V.K. and Shah, D.O., 1978b. The effect of divalent cations (Ca^{++} and Mg^{++}) on the optimal salinity and salt tolerance of petroleum sulfonate and ethoxylated sulfonate mixtures in relation to improved oil recovery. *J. Am. Oil Chem. Soc.*, 10: 437–442.

Bansal, V.K. and Shah, D.O., 1978c. The effect of addition of ethoxylated sulfonates on salt tolerance, optimal salinity and impedance characteristics of petroleum sulfonate solutions. *J. Colloid Interface Sci.*, 65(3):451–459.

Bansal, V.K. and Shah, D.O., 1978d. The effect of ethoxylated sulfonates on salt tolerance and optimal salinity of surfactant formulations for tertiary oil recovery. *Soc. Pet. Eng. J.*, 18(3): 167–172.

Batra, V.K. and Dullien, F.A.L., 1973. Correlation between pore structure of sandstones and tertiary oil recovery. *Soc. Pet. Eng. J.*, 13(5): 256–258.

Baviere, M., 1976. Phase diagram optimization in micellar systems. *51st Annu. Fall Tech. Conf., Soc. Pet. Eng. of AIME, New Orleans, La., October 3–6*, SPE 6000.

Belkin, H., Holbook, S. and Kuntamukkula, M., 1979. *Enhanced Oil Recovery: CO_2 Foam Flooding, 1st Annual Report*, New Mexico State University, Las Cruces, N.M., pp. 1–66.

Benham, A.L. and Poettmann, F.H., 1958. The thermal recovery process—an analysis of laboratory combustion data. *Trans. AIME*, 213: 406–408.

Bennett, G.S., 1963. *A Study of the Foam Drive Process for Removal of Brine from a Consolidated Sandstone Core*. M.S. Dissertation, Pennsylvania State University, University Park, Pa.

Bernard, G.G. and Holm, L.W., 1964. Effect of foam on permeability of porous media to gas. *Soc. Pet. Eng. J.*, 4(3): 267–274.

Bernard, G.G., Holm, L.W. and Harvey, C.P., 1980. Use of surfactant to reduce CO_2 mobility in oil displacement. *Soc. Pet. Eng. J.*, 20(4): 281–292.

Berridge, S.A., Thew, M.T. and Loriston-Clarke, A.G., 1968. The formation and stability of emulsions of water in crude petroleum and similar stocks. *J. Inst. Pet.*, 54(539): 333 pp.

Bikerman, J.J., 1973. *Foams*, Springer-Verlag, Berlin, pp. 138–140.

Boneau, D.F. and Clampitt, R.L., 1977. A surfactant system for the oil-wet sandstone of the North Burbank Unit. *J. Pet. Technol.*, 29(5): 501–506.

Bourrel, M., Koukounis, Ch., Schechter, R.S. and Wade, W.H., 1980. Phase and interfacial tension behavior of nonionic surfactants. *J. Dispersion Sci. Technol.*, 1(1): 13–35.

Brown, A.G., Thuman, W.C. and McBain, J.W., 1953. The surface viscosity of detergent solutions as a factor in foam stability. *J. Colloid Sci.*, 8: 491–507.

Brownell, L.E. and Katz, D.L., 1947. Flow of fluids through porous media, II. *Chem. Eng. Progr.*, 42: 601.

Burger, J.G. and Sahuquet, B.C., 1973. Laboratory research on wet combustion. *J. Pet. Technol.*, 25(10): 1137–1146.

Cash, R.L., Cayias, J.L., Fournier, G.R., Jacobson, J.K., Schares, T., Schechter, R.S. and Wade, W.H., 1976. Modeling crude oils for low interfacial tension. *Soc. Pet. Eng. Symp. on Improved Oil Recovery, Tulsa, Okla., March 22–24*, pp. 217–226, SPE 5813.

Castor, T.P., Edwards, J.B. and Passman, F.J., 1981. Response of mobility control agents to shear, electrochemical and biological stress. In: D.O. Shah (Editor), *Surface Phenomena in Enhanced Oil Recovery*. Plenum Press, New York, N.Y., pp. 773–820.

Caudle, B.H. and Dyes, A.B., 1958. Improving miscible displacement by gas–water injection. *Trans. AIME*, 213: 281–284.

Cayias, J.L., Hayes, M.E., Schechter, R.S. and Wade, W.H., 1976a. Surfactant aging: a possible detriment to tertiary oil recovery. *J. Pet. Technol.*, 28(9): 985–988.

Cayias, J.L., Schechter, R.S. and Wade, W.H., 1976b. Modeling crude oils for low interfacial tension. *Soc. Pet. Eng. J.*, 16(6): 351–357.

Cayias, J.L., Schechter, R.S. and Wade, W.H., 1977. The utilization of petroleum sulfonates for producing low interfacial tension between hydrocarbon and water. *J. Colloid Interface Sci.*, 59(1): 31–38.

Celik, M., Goyal, A., Manev, E. and Somasundaram, P., 1979. The role of surfactant precipitation and redissolution in the adsorption of sulfonate on minerals. *54th Annu. Tech. Conf., Soc. Pet. Eng. of AIME, Las Vegas, Nev., September 23–26*, SPE 8263.

Chan, K.S. and Shah, D.O., 1977a. *Improved Oil Recovery Research Program. Semi-annual Report, A.* University of Florida, Gainesville, Fla., 22 pp.

Chan, K.S. and Shah, D.O., 1977b. The molecular mechanism for achieving ultra-low interfacial tension. *51st Symp. on Surface and Colloid Science, Grand Island, N.Y., June 19–22*, Paper No. 510.

Chan, K.S. and Shah, D.O., 1980. The molecular mechanism for achieving ultra-low interfacial tension minimum in a petroleum sulfonate/oil/brine system. *J. Dispersion Sci. Technol.*, 1(1): 55–95.

Chiang, M., 1979. *Role of Surfactant Mass Transfer and the Formation of an Oil Bank in Displacement of Oil Through Porous Media*. Ph.D. Dissertation, University of Florida, Gainesville, Fla.

Chiang, M. and Shah, D.O., 1977. *Improved Oil Recovery Research Program. Semi-annual Report, B.* University of Florida, Gainesville, Fla.

Corrin, M.L. and Harkins, W.D., 1947. The effect of salts on the critical concentrations for formation of micelles in colloid electrolytes. *J. Am. Chem. Soc.*, 69: 683–688.

Dauben, D.L. and Froning, H.R., 1971. Development and evaluation of micellar solutions to improve water injectivity. *J. Pet. Technol.*, 23(5): 614–620.

Dauben, D.L. and Menzie, D.E., 1967. Flow of polymer solutions through porous media. *J. Pet. Technol.*, 19(8): 1065–1073.

Dawson, R. and Lantz, R.B., 1972. Inaccessible pore volume in polymer flooding. *Soc. Pet. Eng. J.*, 12(5): 448–452.

Derderian, E.J., Glass, J.E. and Bryant, G.M., 1981. The effect of microemulsion composition on the adsorption of petroleum sulfonates on Berea Sand/montmorillonite clay adsorbants. In: D.O. Shah (Editor), *Surface Phenomena in Enhanced Oil Recovery*. Plenum Press, New York, N.Y.,pp. 649–674.

Desai, N.N., 1983. *Surfactant–Polymer Interactions in Oil Recovery Systems*. Ph.D. Dissertation, University of Florida, Gainesville, Fla., 70 pp.

Dietz, D.N., 1970. Wet underground combustion state of the art. *J. Pet. Technol.*, 22(5): 605–617.

Dilgren, R.E., Deemer, A.R. and Owens, K.B., 1982. The laboratory development and field testing of steam/non-condensible gas foams for mobility control in heavy oil recovery. *Calif. Reg. Meet., Soc. Pet. Eng. of AIME, San Francisco, Calif., March 24–26*, pp. 591–601, SPE 10774.

Doe, I.H., Wade, W.H. and Schechter, R.S., 1977. Alkyl benzene sulfonates for producing low interfacial tensions between hydrocarbons and water. *J. Colloid Interface Sci.*, 59(3): 525–531.

Dullien, F.A.L., Dhawan, G.K., Gurak, N. and Babjak, L., 1972. A relationship between pore structure and residual oil saturation in tertiary surfactant floods. *Soc. Pet. Eng. J.*, 12(4): 289–296.

Ehrlich, R., Hasiba, H.H. and Raimondi, P., 1974. Alkaline waterflooding for wettability alteration—evaluating a potential field application. *J. Pet. Technol.*, 26(12): 1335–1343.

Elson, T.D. and Marsden, S.S., 1978. The effectiveness of foaming agents at elevated temperatures over extended period of time. *Calif. Reg. Meet., Soc. Pet. Eng. of AIME, San Francisco, Calif., April 12–14*, pp. 1–7, SPE 7116.

Fairbrother, F. and Stubbs, A.E., 1935. Studies in electrode-endosmosis, VI. The (bubble-tube) method measurement. *J. Chem. Soc.*, pp. 527–529.

Foley, C., 1976. *The Energy Question*. Penguin Books, Harmondsworth, 210 pp.

Foster, W.R., 1973. A low-tension waterflooding process. *J. Pet. Technol.*, 25(2): 205–210.

Franses, E., Bidner, M.S. and Scriven, L.E., 1977. Interfacial tension minima in two-phase micellar systems. In: K.L. Mittal (Editor), *Micellization, Solubilization and Microemulsions, Vol. 2*. Plenum Press, New York, N.Y., pp. 855–876.

Friberg, S.E., Lapexynsra, I. and Gilliberg, G., 1976. Microemulsions containing non-ionic surfactants—the importance of the PIT value. *J. Colloid Interface Sci.*, 56(1): 19–32.

Fuerstenau, M.C., Martin, C.C. and Bappu, R.B., 1963. The role of hydrolysis in sulfonate flotation on quartz. *Trans. AIME*, 226: 449–454.

Gale, W.W. and Sandvik, E.J., 1973. Tertiary surfactant flooding: petroleum sulfonate composition-efficiency studies. *Soc. Pet. Eng. J.*, 13(4): 191–199.

Gale, W.W., Saunder, R.K. and Ashcraft, T.L., Jr., 1976. Oil recovery method using a surfactant. *U.S. Patent*, No. 3,977,471.

Ginn, M.E., 1970. Adsorption of cationic surfactants on minerals and miscellaneous solid substrates. In: E. Jungermann (Editor), *Cationic Surfactants*. Marcel Dekker, New York, N.Y., pp. 341–385.

Ginn, M.E., Kinney, F.B. and Harris, J.C., 1960. Effect of temperature on CMC. *J. Am. Oil Chem. Soc.*, 37: 183–187.

Gogarty, W.B., 1967. Mobility control with polymer solutions. *J. Pet. Technol.*, 7(2): 161–173.

Gogarty, W.B., 1977. Oil recovery with surfactants: history and current appraisal. In: D.O. Shah and R.S. Schechter (Editors), *Improved Oil Recovery by Surfactant and Polymer Flooding*. Academic Press, New York, N.Y., pp. 27–54.

Gogarty, W.B. and Tosch, W.C., 1968. Miscible type water flooding: oil recovery with micellar solutions. *J. Pet. Technol.*, 20(12): 1407–1414.

Grant, B.F. and Szasz, S.E., 1954. Development of an underground heat wave for oil recovery. *Trans. AIME*, 201: 108–118.

Graue, D.J. and Johnson, C.E., Jr., 1974. Field trial of caustic flooding process. *J. Pet. Technol.*, 26(12): 1353–1358.

Gupta, L. and Wasan, D.T., 1974. Surface shear viscosity and related properties of adsorbed surfactant film. *Ind. Eng. Chem., Fundamental*, 13(1): 26–33.

Handy, L.L., 1981. Behavior of surfactants in EOR applications at high temperatures. In: *Proceedings, European Symposium on Enhanced Oil Recovery, Bournemouth, England, September 21–23.* Elsevier, Lausanne, pp. 149–160.

Handy, L.L., Amaefule, J.O., Ziegler, V.M. and Ershaghi, I., 1979. Thermal stability of surfactants for reservoir applications. *Int. Symp. on Oil Field and Geothermal Chemistry, Houston, Texas, January 22–24*, pp. 61–70, SPE 7867.

Hanna, H.S. and Somasundaran, P., 1977. Physico-chemical aspects of adsorption at solid-liquid interfaces, II. Mahogany sulfate, Berea sandstone, kaolinite. In: D.O. Shah and R.S. Schechter (Editors), *Improved Oil Recovery by Surfactant and Polymer Flooding*. Academic Press, New York, N.Y., pp. 253–274.

Hatfield, J.C., Davis, H.T. and Scriven, L.E., 1978. Freeze-fracture electron microscopy of microemulsions. *85th Natl. Meet., Am. Inst. Chem. Eng., New York, N.Y.*, Paper No. 45.

Healy, R.N. and Reed, R.L., 1974. Physicochemical aspects of microemulsion flooding. *Soc. Pet. Eng. J.*, 14(5): 491–512.

Healy, R.N. and Reed, R.L., 1976. Immiscible microemulsion flooding. *Soc. Pet. Eng. Symp. on Improved Oil Recovery, Tulsa, Okla., March 22–24*, pp. 265–280, SPE 5817.

Healy, R.N., Reed, R.L. and Carpenter, C.W., 1975. A laboratory study of microemulsion flooding. *Soc. Pet. Eng. J.*, 15(2): 87–100.

Hill, H.J., Reisberg, J. and Stegemeier, G.L., 1973. Aqueous surfactant system for oil recovery. *J. Pet. Technol.*, 25(2): 186–194.

Hill, H.J., Brew, J.R., Claridge, E.L., Hite, J.R. and Pope, G.A., 1974. The behavior of polymers in porous media. *Soc. Pet. Eng. Symp. on Improved Oil Recovery, Tulsa, Okla., April 22–24*, pp. 149–164, SPE 4748.

Holcomb, D.L., Callaway, E. and Curry, L.L., 1981. Chemistry, physical nature and rheology of aqueous stimulation foams. *Soc. Pet. Eng. J.*, 21(4): 410–414.

Holm, L.W., 1959. Carbon dioxide solvent flooding for increased oil recovery. *Trans. AIME*, 216: 225–231.

Holm, L.W., 1963. Carbon dioxide slug and carbonated water-oil recovery process. *Prod. Mon.*, 27(9): 6–8.

Holm, L.W. and Josendal, V.A., 1974. Mechanism of oil displacement by carbon dioxide. *J. Pet. Technol.*, 26(12): 1427–1436.

Holm, L.W. and O'Brien, L.J., 1971. Carbon dioxide test at the Mead-Strawn field. *J. Pet. Technol.*, 23(4): 431–442.

Holm, L.W. and Robertson, S.D., 1981. Improved micellar/polymer flooding with pH chemicals. *J. Pet. Technol.*, 33(1): 161–172.

Hurd, B.G., 1976. Adsorption and transport of chemical species in laboratory experiments. *Soc. Pet. Eng. Symp. on Improved Oil Recovery, Tulsa, Okla., March 22–24*, pp. 281–302, SPE 5818.

Isaacs, E.E., Prowse, D.R. and Rankin, J.P., 1981. The role of surfactant additives in the in-situ recovery of bitumen from oil sands. *32nd Annu. Tech. Meet., Pet. Soc. CIM, Calgary, Alta., May 3–6*, pp. 1–20.

Jennings, H.Y., Jr., Johnson, C.E., Jr. and McAuliffe, C.D., 1974. A caustic waterflooding process for heavy oils. *J. Pet. Technol.*, 26(12): 1344–1352.

Jennings, R.R., Rogers, J.H. and West, T.J., 1971. Factors influencing mobility control by polymer solutions. *J. Pet. Technol.*, 23(3): 391–410.

Jordon, D.S., Green, D.W., Terry, R.E. and Willhite, G.P., 1981. The effect of temperature on gellation time for polyacrylamide-chromium (III) systems. *56th Annu. Fall Tech. Conf., Soc. Pet. Eng. of AIME, San Antonio, Texas, October 5–7*, SPE 10059.

Klevens, H.B., 1950. Effect of electrolytes upon the solubilization of hydrocarbons and polar compounds. *J. Am. Chem. Soc.*, 72: 3780–3785.

Knight, B.L., 1973. Reservoir stability of polymer solutions. *AIChE Symp. Ser.*, 69(127): 40–42.

Kraft, H.R. and Pusch, G., 1982. Mobilization and banking of residual oil in high salinity reservoir systems with the use of aqueous surfactant solutions of ethoxylated carboxymethylates. *SPE/DOE Improved Oil Recovery Symp., Tulsa, Okla., April 4–7,* pp. 587–610, SPE/DOE 10714.

Kremesec, V.J. and Treiber, L.E., 1976. Effect of system wettability on oil displacement by micellar flooding. *51st Annu. Fall Meet., Soc. Pet. Eng. of AIME, New Orleans, La., October 3–6*, SPE 6001.

Kuhn, C.S. and Koch, R.L., 1953. In-situ combustion, newest method of increasing oil recovery. *Oil Gas J.*, 52(14): 92–96, 113–114.

Lakatos, I., Lakatos-Szabo, J. and Toth, J., 1981. Factors influencing polyacrylamide adsorption in porous media and their effect on flow behavior. In: D.O. Shah (Editor), *Surface Phenomena in Enhanced Oil Recovery*. Plenum Press, New York, N.Y., pp. 821–842.

Leach, R.O., Wagner, O.R., Wood, H.W. and Harpke, C.F., 1962. A laboratory and field study of wettability adjustment in water flooding. *Trans. AIME*, 225: 206–212.

Lefebvre du Prey, E., 1973. Factors affecting liquid-liquid relative permeabilities of a consolidated porous medium. *Soc. Pet. Eng. J.*, 13(1): 39–47.

Levchenko, D.N., Khydyakova, A.D. and Ratich, L.I., 1970. *Khim. Tekhnol. Topl. Massel*, October, p. 21.

Leverett, M.C., 1939. Flow of oil-water mixtures through unconsolidated sands. *Trans. AIME*, 132: 149–171.

MacDonald, I.F., El-Sayed, M.S., Mow, K. and Dullien, F.A.L., 1979. Flow through porous media—the Ergun equation revisited. *Ind. Eng. Chem. Fundam.*, 18(3): 199–208.

MacWilliams, D.C., Rogers, J.H. and West, T.J., 1973. Water soluble polymers in petroleum recovery. In: N.M. Bikales (Editor), *Polymer Science and Technology, Vol. 2. Water Soluble Polymers*. Plenum Press, New York, N.Y.

Maddox, J., Jr., Tale, J.F. and Shupe, R.D., 1975. Secondary recovery method. *U.S. Patent,* No. 3,916,996.

Martin, W.L., Alexander, J.D. and Dew, J.N., 1958. Process variables of in-situ combustion. *Trans. AIME*, 213: 28–35.

McCaffery, F.G., 1976. Interfacial tension and aging behavior of some crude oils against caustic solutions. *J. Can. Pet. Technol.*, 15(3): 71–74.

McCaffery, F.C. and Mungan, N., 1970. Contact angle and interfacial tension studies of some hydrocarbon-water-solid systems. *J. Can. Pet. Technol.*, 9(3): 185–196.

McCain, W.D., Jr., 1973. *The Properties of Petroleum Fluids*. Petroleum Publishing Company, Tulsa, Okla., Chapter 6.

McNeil, J.S., Jr., and Moss, J.T., 1958. In-situ combustion. *Oil Gas J.*, 56(37): 232, 235–241.

McNeil, J.S. and Nelson, T.W., 1959. Thermal methods provide three ways to improve oil recovery. *Oil Gas J.*, 57(3): 86–98.

Melrose, J.C., 1965. Wettability as related to capillary action in porous media. *Soc. Pet. Eng. J.*, 5(3): 259–271.

Melrose, J.C. and Brandner, C.F., 1974. Role of capillary forces in determining microscopic displacement efficiency for oil recovery by water flooding. *J. Can. Pet. Technol.*, 16(4): 54.

Miller, C.A., Hwan, R.N., Benton, W.J. and Fort, T., Jr., 1977. Ultra-low interfacial tensions and their relation to phase separation in micellar solutions. *J. Colloid Interface Sci.*, 61(3): 554–568.

Moore, T.F. and Slobod, R.L., 1956. The effect of viscosity and capillary pressure on the displacement of oil by water. *Prod. Mon.*, August, pp. 20–30.

Morgan, J.C., Schechter, R.S. and Wade, W.H., 1977. Recent advances in the study of low interfacial tensions, In: D.O. Shah and R.S. Schechter (Editors), *Improved Oil Recovery by Surfactant and Polymer Flooding*. Academic Press, New York, N.Y., pp. 101–118.

Morrow, N.R., 1978. *29th Meet. Pet. Soc. CIM, Calgary, Alta., June 13*, Paper No. 78-29-24.

Morrow, N.R., Cram, P.J. and McCaffery, F.G., 1973. Displacement studies in dolomite with wettability control by octanoic acid. *Soc. Pet. Eng. J.*, 13(4): 221–232.

Mungan, N., Smith, F.W. and Thompson, J.L., 1966. Some aspects of polymer floods. *J. Pet. Technol.*, 18(9): 1143–1150.

Nelson, T.W. and McNeil, J.S., Jr., 1961. In-situ combustion project. *Oil Gas J.*, 59(23): 59–65.

Novosad, J., 1981. Adsorption of pure surfactant and petroleum sulfonate at the solid-liquid interface. In: D.O. Shah (Editor), *Surface Phenomena in Enhanced Oil Recovery*. Plenum Press, New York, N.Y., pp. 675–694.

Oh, S.G. and Slattery, J.C., 1976. Interfacial tension required for significant displacement of residual oil. In: *ERDA Symposium on Enhanced Oil–Gas Recovery, Tulsa, Okla*. U.S. Government Printing Office, Washington, D.C.

Oh, S.G. and Slattery, J.C., 1979. Interfacial tension required for significant displacement of residual oil. *Soc. Pet. Eng. J.*, 19(1): 83–96.

Ojeda, E., Preston, F. and Calhoun, J.C., 1953. Correlation of oil residuals following surfactant floods. *Prod. Mon.*, 18(2): 20–29.

Owens, W.W. and Archer, D.L., 1971. The effect of rock wettability on oil–water relative permeability relationship. *J. Pet. Technol.*, 23(7): 873–878.

Owete, O.S., Mahmood, M.R., Castanier, L.M. and Brigham, W.E., 1982. Foam for gas drive improvement. *Annu. Convent. DOE-Contractor, Enhanced Oil Recovery Projects, San Francisco, Calif., July 27–29*, pp. 227–234.

Petrov, A.A., Pozdnyzhev, G.N. and Borizov, A.A., 1969. *Khim. Tekhnol. Topl. Masel*, March 11.

Phillips, M.C., Ladbrooke, B.D. and Chapman, D., 1970. Molecular interactions in mixed lecithin systems. *Biochem. Biophys. Acta*, 196(1): 35–44.

Pinter, J. and Wolfram, E., 1981. Wettability of and oil film stability on glass capillaries as basic processes in tertiary oil recovery. In: D.O. Shah (Editor), *Surface Phenomena in Enhanced Oil Recovery*. Plenum Press, New York, N.Y., pp. 479–494.

Puerto, M.C. and Reed, R.L., 1982. A three-parameter representation of surfactant–oil–brine interaction. *SPE/DOE Improved Oil Recovery Symp., Tulsa, Okla., April 4–7*, pp. 51–76, SPE/DOE 10678.

Pye, D.J., 1964. Improved secondary recovery by control of water mobility. *J. Pet. Technol.*, 16(8), 911–916.

Ramey, H.J., Jr., and Nabor, G.W., 1954. A blotter-type electrolyte model determination of aereal sweeps in oil recovery by in-situ combustion. *Trans. AIME*, 201: 119–123.

Raza, S.H., Treiber, L.E. and Archer, D.L., 1968. Wettability of reservoir rocks and its evaluation. *Prod. Mon.*, 32(1): 2–8.

Reed, M.G., 1980. Gravel pack and formation sandstone dissolution during steam injection. *J. Pet. Technol.*, 32(6): 941–949.

Reed, R.L. and Healy, R.N., 1977. Some physicochemical aspects of microemulsion flooding: a review. In: D.O. Shah and R.S. Schechter (Editors), *Improved Oil Recovery by Surfactant and Polymer Flooding*. Academic Press, New York, N.Y., pp. 383–438.

Reisberg, J. and Doscher, T.M., 1956. Interfacial phenomena in crude oil–water systems. *Prod. Mon.*, 21(1): 43–50.

Saffman, P.G. and Taylor, G., 1958. The penetration of a fluid into a porous medium of Hele-Shaw cell containing a more viscous liquid. *Proc. R. Soc. London, Ser. A*, 245: 312 pp.

Salathiel, R.A., 1973. Oil recovery by surface film drainage in mixed-wettability rocks. *J. Pet. Technol.*, 25(10): 1216–1224.

Sandiford, B.B., 1964. Laboratory and field studies of waterfloods using polymer solutions to increase oil recoveries. *J. Pet. Technol.*, 16(8): 917–922.

Sayyouh, M.H.M., Farouq Ali, S.M. and Stahl, C.D., 1978. Rate effects in tertiary micellar flooding of Bradford crude oil. *Eastern Reg. Soc. Pet. Eng. of AIME, Washington, D.C., November 1–3.*

Scriven, L.E., 1976. Equilibrium bi-continuous structure. *Nature (London)*, 236: 123.

Scriven, L.E., 1977. Equilibrium bi-continuous structure. In: K.L. Mittal (Editor), *Micellization, Solubilization and Microemulsion, Vol. 2.* Plenum Press, New York, N.Y., 877 pp.

Shah, D.O., 1970a. The effect of buffer ions on stearic acid monolayers in relation to foam stability. *J. Colloid Interface Sci.*, 32(4): 570–576.

Shah, D.O., 1970b. Significance of the 1:3 molecular ratio in mixed surfactant systems. *J. Colloid Interface Sci.*, 37(4): 744–752.

Shah, D.O., 1981. Fundamental aspects of surfactant-polymer flooding process. In: *Proceedings, European Symposium on Enhanced Oil Recovery, Bournemouth, England, September 21–23.* Elsevier, Lausanne, pp. 1–40.

Shah, D.O. and Dysleski, C.A., 1969. Molecular interactions in monolayers: molecular association and foam stability of fatty acids and alcohols. *J. Am. Oil. Chem. Soc.*, 46(12): 645–648.

Shah, D.O. and Hsieh, W.C., 1981. Microemulsions, liquid crystals and enhanced oil recovery. In: J.A. Fitzpatrick and M.P. Freeman (Editors), *Theory, Practice and Process Principles for Physical Separations.* Engineering Foundation, New York, N.Y., pp. 137–154.

Shah, D.O. and Shiao, S.Y., 1975. The chain length compatibility and molecular area in mixed monolayers. *Adv. Chem. Ser.*, 144: 153–164.

Shah, D.O., Hsieh, W.C. and Deamer, D.N., 1976. The structure and interfacial properties of birefringent cellular fluids. *Am. Inst. Chem. Eng. Meet., Chicago, Ill., November 28–December 2.*

Shah, D.O., Djabbarah, N.F. and Wasan, D.T., 1978. A correlation of foam stability with surface shear viscosity and area per molecule in mixed surfactant systems. *Kolloid Z. Z. Polym.*, 256: 1002–1008.

Sharma, A.K., Pithapurwala, Y.K. and Shah, D.O., 1983. Selection criteria and formation of a surfactant slug for high-temperature, moderate salinity reservoir conditions. *Int. Symp. on Oilfield and Geothermal Chemistry, Denver, Colo., June 1–3*, pp. 33–44, SPE 11772.

Sharma, M.K. and Shah, D.O., 1982. Surface properties of foaming agents in relation to foam flow in porous media. *184th Natl. Meet., Am. Chem. Soc., Washington, D.C., August 21–26*, GEOC 039.

Sharma, M.K. and Shah, D.O., 1983. Interfacial phenomena in foam flooding process for heavy oil recovery. In: *Proceedings, 18th Intersociety Energy Conversion Engineering Conference, Los Angeles, Calif., August 21–26*, pp. 527–534.

Sharma, M.K. and Shah, D.O., 1984. Effect of oil viscosity on recovery processes in relation to foam flooding. *J. Am. Oil Chem. Soc.*, 61(3): 585–590.

Sharma, M.K., Shah, D.O. and Brigham, W.E., 1982a. The chain length compatibility and surface properties of foaming agents in relation to fluid displacement efficiency in porous media. *Int. Symp. on Oilfield and Geothermal Chemistry, Dallas, Texas, January 25–27*, pp. 171–180, SPE 10612.

Sharma, M.K., Shah, D.O. and Brigham, W.E., 1982b. Surface properties of foaming agents and behavior of foam in porous media. *Annu. Convent. DOE-Contractor, Enhanced Oil Recovery Projects, San Francisco, Calif., July 27–29*, pp. 74–84.

Sharma, M.K., Shah, D.O. and Brigham, W.E., 1984. The influence of temperature on surface and microscopic properties of foaming solutions in relation to fluid displacement efficiency in porous media. *Am. Inst. Chem. Eng. J.*, 31(2): 222–228.

Shiao, S.Y., 1976. *The Effect of Chain-Length Compatibility on the Properties of the Mixed Surfactant Systems.* Ph. D. Dissertation, University of Florida, Gainesville, Fla., 238 pp.

Smith, F.W., 1970. The behavior of partially hydrolyzed polyacrylamide solutions in porous media. *J. Pet. Technol.*, 22(2): 148–156.

Somasundaran, P. and Grieves, R.B., 1975. Advances in interfacial phenomena of particulate solution/gas systems. *Am. Inst. Chem. Eng. Symp. Ser.*, 71(150): 189.

Somasundaran, P., Healy, T.W. and Fuerstenau, D.W., 1964. Surfactant adsorption at solid/liquid interface: dependence of mechanism on chain length. *J. Phys. Chem.*, 68(12): 3562–3566.

Sparlin, D., 1975. An evaluation of polyacrylamides for reducing water production. *Soc. Pet. Eng. Meet., Dallas, Texas, September 28–October 1,* SPE 5610.

Stegemeier, G.L., 1974. Relationship of trapped oil saturation to petrophysical properties of porous media. *Soc. Pet. Eng. Symp. on Improved Oil Recovery, Tulsa, Okla.*, pp. 225–238, SPE 4754.

Stegemeier, G.L., 1977. Mechanism of entrapment and mobilization of oil in porous media. In: D.O. Shah and R.S. Schechter (Editors), *Improved Oil Recovery by Surfactant and Polymer Flooding.* Academic Press, New York, N.Y., pp. 55–92.

Strassner, J.E., 1968. Effect of pH on interfacial films and stability of crude oil–water emulsions. *J. Pet. Technol.*, 20(3): 303–312.

Taber, J.J., 1969. Dynamic and static forces required to remove a discontinuous oil phase from porous media containing both oil and water. *Soc. Pet. Eng. J.*, 9(1): 3–12.

Taber, J.J., Kirby, J.C. and Schroeder, F.U., 1973. Studies on the displacement of residual oil: viscosity and permeability effect. *Am. Inst. Chem. Eng., Symp. Ser.*, 69(127): 53–56.

Terry, R.E., Huany, C., Green, D.W., Michinik, M.J. and Willhite, G.P., 1979. Correlation of gellation time for polymer solutions used as mobility control agents. *54th Annu. Fall Tech. Conf., Soc. Pet. Eng. of AIME, Las Vegas, Nev., September 23–26*, SPE 8419.

Treiber, L.E., Archer, D.L. and Owens, W.W., 1972. A laboratory evaluation of the wettability of fifty oil producing reservoirs. *Soc. Pet. Eng. J.*, 12(6): 531–540.

Trushenski, S.P., 1977. Micellar flooding: sulfonate-polymer interaction. In: D.O. Shah and R.S. Schechter (Editors), *Improved Oil Recovery by Surfactant and Polymer Flooding*. Academic Press, New York, N.Y., pp. 555–575.

Trushenski, S.P., Dauben, D.L. and Parrish, D.R., 1974. Micellar flooding, fluid propagation, interaction and mobility. *Soc. Pet. Eng. J.*, 14(6): 633–645.

Uren, L.C., 1939. *Petroleum Production Engineering, Vol. 2.* McGraw Hill, New York, N.Y., p. 32.

Wagner, O.R. and Leach, R.O., 1959. Improving oil displacement by wettability adjustments. *Trans. AIME*, 216: 65–72.

Wang, P.F., Al-Khafaji, A., Castanier, L.M. and Brigham, W.E., 1982. Steam-surfactant systems at reservoir conditions. *Annu. Convent. DOE-Contractor, Enhanced Oil Recovery Projects, San Francisco, Calif., July 27–29*, pp. 219–226.

Wasan, D.T. and Mohan, V., 1977. Interfacial rheological properties of fluid interfaces containing surfactants. In: D.O. Shah and R.S. Schechter (Editors), *Improved Oil Recovery by Surfactant and Polymer Flooding*. Academic Press, New York, N.Y., pp. 161–204.

Wasan, D.T., Shah, S.M., Shah, R., Chan, M. and McNamara, J.J., 1977. Interrelationship of emulsion stability and interfacial viscosity in improved oil recovery. In: *Proceedings, Engineering Foundation Conference on Theory, Practice and Process Principles for Physical Separation, Asilomar, October.*

Wasan, D.T., Shah, S.M., Aderangi, N., Chan, M.S. and McNamara, J.J., 1978. Observations on the coalescence behavior of oil droplets and emulsion stability in enhanced oil recovery. *Soc. Pet. Eng. J.*, 18(6): 407–417.

Wasan, D.T., Shah, S.M., Aderangi, N., Chan, M.S. and McNamara, J.J., 1979. The mechanism of oil bank formation and coalescence in porous media. In: *Proceedings, Third Annual ERDA Symposium on Enhanced Oil Recovery and Gas Recovery and Improved Drilling Methods, Vol. 1.* U.S. Government Printing Office, Washington, D.C., p. 70, Paper No. B-4.

White, H.J., Jr., 1970. Adsorption of cationic surfactants by cellulosic substrates. In: E. Jungermann (Editor), *Cationic Surfactants*. Marcel Dekker, New York, N.Y., pp. 311–341.

White, J.L., Phillips, H.M., Goddard, J.E. and Bakey, B.D., 1973. Use of polymers in control of water production in oil wells. *J. Pet. Technol.*, 25: 143–150.

Wilson, L.A., Jr., 1977. Physico-chemical environment of petroleum reservoirs in relation to oil recovery systems. In: D.O. Shah and R.S. Schechter (Editors), *Improved Oil Recovery by Surfactant and Polymer Flooding*. Academic Press, New York, N.Y., pp. 1–26.

Wilson, P.M. and Brandner, C.F., 1973. Aqueous surfactant solutions which exhibit ultra-low tensions at the oil–water interface. *165th Natl. Meet., Am. Chem. Soc., Dallas, Texas, April 8–13.*

Wilson, P.M., Murphy, G.L. and Foster, W.R., 1976. The effects of sulfonate molecular weight and salt concentration on the interfacial tension of oil–brine–surfactant systems. *Soc. Pet. Eng. Symp. on Improved Oil Recovery, Tulsa, Okla., March 22–24*, pp. 201–216, SPE 5812.

Ziegler, V.M. and Handy, L.L., 1981. Effect of temperature on surfactant adsorption in porous media. *Soc. Pet. Eng. J.*, 21(2): 218–226.

Chapter 11

STEAMFLOODING FOR ENHANCED OIL RECOVERY

KUNDAN L. GOYAL and SANJAY KUMAR

INTRODUCTION

In recent years steamflooding has acquired a major role in the tertiary recovery of crude oils, especially heavy, viscous oils. Steam injection is the most widely used and profitable enhanced recovery technique available today. The process involves the injection of steam generated at the surface or downhole (to reduce heat losses) continuously, or in cycles. Continuous injection, called steam drive or steamflooding, provides a higher ultimate recovery and is, therefore, used more often. Cyclic steam injection, also known as "steam soak" or "huff and puff", is a single well operation. Steam is injected into a (producing) well for some time, is allowed to "soak" for a period of time, and the well is subsequently returned to production. Steam heats up the areas close to the wellbore, and with a close well spacing, this process can generate a very good production rate at a relatively low cost. Recoveries from steamflooding are typically in the range of 50–60% (sometimes up to 75%) of the oil-in-place. The ultimate recovery from cyclic steam injection is considerably lower, typically in the range of 10–25% of the oil-in-place.

Steam is believed to increase recovery by: (1) reducing the oil viscosity in situ, allowing the oil to flow more readily; (2) reducing the residual oil saturation, and improving relative permeability to oil; (3) improving the sweep efficiency; (4) increasing the formation volume factor; (5) vaporizing and distilling condensable hydrocarbons from the crude; and (6) providing a gas drive mechanism. Gates and Ramey (1958) illustrated very well the benefits of thermal methods for increasing oil recovery. An example of how thermal recovery can increase production markedly is shown in Fig. 11-1, which gives data for California from 1964 to 1978. According to these estimates by the Conservation Committee of California Oil Producers (1979), most of the thermal oil produced in California is of 10–20° API gravity and the incremental production rate in 1979 was 199,000 bbl/day, which amounted to 20% of the state's total production. Currently, the process accounts for the production of more than 500,000 bbl/day of heavy oils worldwide, of which about 400,000 bbl/day are in the U.S.A.

THEORY

It is quite clearly understood, at least in the macroscopic sense, that the process of steamflooding depends upon the following parameters:

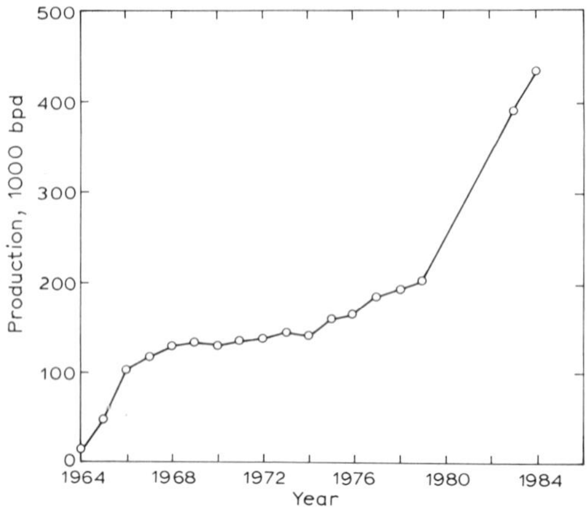

Fig. 11-1. Total production increase in California due to thermal recovery processes. (Data from *Annual Review of California Oil and Gas Production,* 1979–1984.)

(1) Alterations in the fluid properties in situ. These comprise changes in phase behavior, densities, viscosities, composition, compressibilities, and *P-V-T* relationships.

(2) Rock properties such as absolute permeability, porosity, rock compressibility, and the attendant changes in these properties on the injection of steam.

(3) Properties related to fluid–rock interaction. These include residual saturations (related in turn to wettability, interfacial tension, etc.), relative permeabilities, capillary pressures, and their dependence upon temperature.

(4) Thermal properties of the formation and the contained fluids, such as specific heats, thermal conductivities, thermal expansion coefficient, and the changes induced in these.

(5) The reservoir environment: net/gross ratio (presence of shale barriers, etc.), heterogeneity, properties of the overburden and underburden, the initial oil saturation, temperature, and pressure.

(6) Flood geometry: pattern shape and spacing, producing–injecting interval (well completion) location and thickness.

(7) Parameters within the operator's control, such as steam injection rate, steam quality, injection pressure (temperature), cumulative amount of injection, etc.

Clearly, the system is quite complicated. Some of these parameters are amenable to measurement, whereas others are not quantified so easily. Further, the functional dependence upon temperature (and pressure) is only approximately empirically known. Of major concern are the fluid–rock properties such as residual saturations, relative permeabilities, and capillary pressure, that have not been fully understood in a quantitative sense even today.

Fortunately, the mechanisms involved in the enhanced recovery of oil by injection of steam are known enough to model the process at least macroscopically, unlike chemical flooding where the system is far more complex. In the absence of fundamental theoretical analyses in this area, a working hypothesis is to include the dominant temperature effects on the system in any study designed to evaluate or predict process performance. Thermal processes must produce mechanisms different but analogous to chemically induced mechanisms in the porous medium. Chemical diffusion and dispersion is replaced by thermal diffusion. One can also visualize thermally induced changes in rock wettability and fluid interfacial tensions (affecting residual saturations, relative permeabilities, and capillary pressures), similar to chemically induced changes in a chemical flooding process. The displacement, or drive, behavior is similar to other drive mechanisms. The process, however, is quite complex and the present discussion precludes phenomena such as the interaction of steam with the formation (clay swellling, modifications in the pore geometry configurations due to the imposed thermal stresses), multidirectional diffusion of steam, and dispersion in the hot water flood-front resulting from the condensation of steam.

The mechanism of oil recovery by steam injection can be well understood by considering steam in a long, porous medium initially containing oil and connate water. The oil in the immediate vicinity of the injection end is vaporized and pushed ahead. A fraction of the oil is not vaporized and is left behind, but at a higher temperature due to the heat provided by steam. The advancing steam eventually condenses, due to heat losses into the overburden and underburden, into water at the steam temperature thereby generating a hot condensate bank. This hot water bank drives oil ahead as it moves, cooling down to reservoir temperature. From this point onwards, the displacement process continues as it would be in a conventional waterflood. It is clear, therefore, that there are three distinct flow regimes: (1) the steam zone, (2) hot condensate zone, and (3) the cold water zone.

Steam zone

In the steam zone, the predominant effect is steam distillation. High temperature and the presence of a gas phase lead to the vaporization of the light ends, which are carried forward by the advancing steam until they condense in the cooler portion of the reservoir. The relatively heavier components of the oil, characterized by a high vapor pressure, are left behind. The actual oil recovery by steam distillation is determined by the composition of the oil involved. Willman et al. (1961) estimated that the additional recovery by steam distillation may account for as much as 20% of the oil-in-place for a 50% distillable crude at 340°F. The oil ahead of the steam zone gets richer in lighter ends serving to extract and displace the original oil, thus enhancing oil recovery.

Hot-water zone

The oil recovery for the hot-water zone is largely governed by the thermal characteristics of the oil involved. If the viscosity of the oil exhibits a sharp decrease with an increase in temperature, considerable amount of oil will be recovered by the hot waterflood. Thermal expansion of oil accounts for the recovery of 3–5% of the oil-in-place. The displacement efficiency is related to the reduction in residual oil saturation with temperature. Based upon residual saturation and relative permeability models by Kumar et al. (1985), this can lead to the recovery of as much as 10–20% of the oil in the underlying zones of the reservoir not swept by steam. Injection of chemicals, proposed to improve recoveries from the hot- (and cold-) water zones, is currently an area of active research.

Cold-water zone

The oil recovery from the cold-water zone is approximately equal to that for an equivalent waterflood: a residual oil saturation of about 20–35% is achieved.

SCREENING CRITERIA FOR STEAMFLOOD PROSPECTS

Several authors have presented screening guides for steamflooding (Geffen, 1973; Farouq Ali, 1974; Lewin and Assocs., 1976; Iyoho, 1978; Chu, 1985). Chu (1985) provided an interesting review of past screening guides, and presented the most recent analysis of reservoir data from 28 steamflood projects worldwide. Described below are some of the important results of these investigations.

Depth

In designing steam injection projects, the depth of the reservoir is a significant parameter. A shallow depth is preferable because: (1) wellbore heat losses are minimal in shallow reservoirs; (2) deeper formations already have a high temperature, and incremental benefits from steam injection are low; and (3) higher formation pressures at greater depths require higher temperature operations, entailing greater chances of well failure during steaming. For a steam injection project to be successful, Chu (1985) assigned a lower depth limit at 400 ft. The upper limit for depth has been suggested to be 3000 ft by Farouq Ali (1974), 4000 ft by Geffen (1973), and 5000 ft by Lewin and Assocs. (1976). Iyoho (1978) proposed a depth in the range of 2500–5000 ft.

Reservoir pressure and temperature are related to its depth. Reservoir pressures in excess of about 700 psia require a steam temperature (for saturated steam) of 500°F or greater. The latent heat of vaporization of steam is also decreased at higher steam pressures. Thus, low formation pressures are desirable. Usually, a steam temperature of 300–400°F is considered optimum.

Formation thickness

Formation thickness primarily affects heat loss to the overburden and underburden. Thus, thicker the formation, better is the steamflood process performance. Farouq Ali (1974) has suggested an h greater than 30 ft, whereas Iyoho (1978) reported h in the range of 30 – 400 ft. Less restrictive criteria have been found by Geffen (1973) and Lewin and Assocs. (1976), who reported h greater than 20 ft, and Chu (1985) who reported that an h greater than 10 ft yields satisfactory performance.

Residual oil-in-place

The residual oil-in-place should be at least 40% (Chu, 1985). Lewin and Assocs. (1976) and Iyoho (1978) suggested an in-place oil saturation S_o of greater than 50%. The porosity–oil saturation product, ϕS_o, is another measure of the oil-in-place. A ϕS_o greater than 0.065 has been suggested by Lewin and Assocs. (1976) and Iyoho (1978), greater than 0.08 by Chu (1985), and greater than 0.10 by Geffen (1973). Farouq Ali (1974) proposed a ϕS_o in the range of 0.15–0.22. Iyoho (1978) has also suggested a ϕ greater than 0.30, whereas Chu (1985) reported a value greater than 0.2 as a less restrictive criteria. Generally, an oil content of 1200 bbl/acre-ft or more is considered economically viable for steamflooding.

Oil viscosity

The optimum oil viscosity is placed in the range of 1000–4000 cP. Farouq Ali (1974) suggested an oil viscosity less than 1000 cP, whereas Chu (1985) reported a viscosity range of 200–1000 cP for successful steamflood projects.

Oil gravity

Oil gravity, indicative of the composition of the oil, also affects recovery. Iyoho (1978) suggested a gravity range of 10–20° API. Geffen (1973) and Lewin and Assocs. (1976) suggested an oil gravity greater than 10° API, whereas Chu (1985) suggested an oil gravity less than 36° API. Thus, in the broadest terms, oil gravity should be in the range of 10–36° API.

Formation permeability

Permeability should be high enough to permit steam injection at adequate rates and to ensure rapid oil flow into the wellbore. A permeability range of 100–4000 mD is considered very satisfactory. Iyoho (1978) suggested a formation permeability of at least 1000 mD.

Formation permeability (k), thickness (h) and oil viscosity (μ) are frequently lumped together to state a composite kh/μ criteria. Geffen (1973) suggested a kh/μ greater than 20, whereas Iyoho (1978) and Lewin and Assocs. (1976) suggested a kh/μ greater than 50 and greater than 100, respectively.

An understanding of many of the governing physical relationships is only beginning to evolve now. Described below are some of the studies that have been reported in the literature.

Residual saturations

Wettability changes with temperature: the oil–water–rock contact angle decreases as temperature increases, i.e, the rock becomes more water-wet at higher temperatures. The thermodynamically unstable polar–non-polar adsorbed layers on the rock (from such components in the crude oil in situ) may be desorbed at elevated temperatures, resulting in the rock becoming more water-wet. Interfacial tension between oil and water is also temperature-dependent; usually an increase in temperature decreases the interfacial tension (see, for example, Torabzadeh and Handy, 1984). Residual saturations are, therefore, strongly temperature-dependent.

This is corroborated by laboratory studies undertaken by several workers who have studied the effect of temperature on residual saturations and relative permeabilities (Edmondson, 1965; Davidson, 1969; Ehrlich, 1970; Poston et al., 1970; Sinnokrot et al., 1971; Lo and Mungan, 1973; Weinbrandt et al., 1975; Nakornthap and Evans, 1982; Maini and Batycky, 1983; Miller and Ramey, 1983; Torabzadeh and Handy, 1984). The following generalizations can be inferred from their work (Kumar et al., 1985):

(1) There is no effect of interfacial tension (IFT) on residual saturations and relative permeabilities for high-tension ($\sigma > 1$ dynes/cm) systems.

(2) With increasing temperature, residual oil saturation decreases, whereas residual water saturation increases.

(3) With increasing temperature, the relative permeability curves shift to the right, i.e., relative permeability to oil increases, whereas the relative permeability to water decreases at any given saturation.

Kumar et al. (1985) described two approaches for correlating residual oil and water saturations. The first neglects viscous forces, which for many systems are low relative to surface tension at low reservoir flow rates, and correlates the residuals as a function of temperature directly. Kumar et al. have also shown a more general approach, applicable to thermal and chemical enhanced recovery methods, that uses capillary number, N_c, as the correlating parameter.

Neglecting rate effects

Temperature primarily affects two parameters: (1) interfacial tension, and (2) wettability (or contact angle). Interfacial tension governs the capillary forces tending to keep the oil globules trapped. Contact angle measurements on reservoir rocks at elevated temperatures are almost nonexistent, though contact angle measurements at high temperatures on clean, smooth quartz and other polished surfaces show a decrease in contact angle with increasing temperature. Assuming that contact angle is primarily a function of temperature, Kumar et al. found the following relationships for residual oil saturation, S_{or}, and residual water saturation, S_{wr}, as a function of temperature, T:

$$S_{or} = a_1 - b_1 T \tag{11-1}$$

$$S_{wr} = a_2 T^{b_2} \tag{11-2}$$

where a_1, b_1, a_2 and b_2 are regression constants.

The capillary number approach

The capillary number, N_c, represents the ratio of viscous to capillary forces:

$$N_c = \mu V / \sigma \tag{11-3}$$

where μ is the fluid viscosity, V is the fluid velocity, and σ is the interfacial tension between the displacing and displaced fluids. The fluid viscosity to use in the capillary number has been a matter of much disagreement among workers in this area. Some propose using the oil viscosity, others water viscosity, and some others like Abrams (1975) suggest using the water viscosity along with the viscosity ratio (μ_w/μ_o). Kumar et al. (1985) pointed out that the rate and viscosity of the displacing fluid are the logical parameters that need be considered. The wettability–contact angle changes must also be included in any parameter used to correlate residual saturation. Using temperature itself as a proxy for contact angle, the following capillary number has been proposed (Kumar et al., 1985):

$$N_c = (V\mu_w/\sigma)(T/T_b) \text{ against } S_{or}, \tag{11-4}$$

and

$$N_c = (V\mu_o/\sigma)(T/T_b) \text{ against } S_{wr}. \tag{11-5}$$

where T_b is a reference temperature, usually room temperature. The viscosity ratio can also be included in the above capillary number. The residual oil and water saturation models presented are as follows:

$$S_{or} = a_1 + b_1[\tanh(N_c)] + c_1[\tanh(N_c)]^2 + d_1[\tanh(N_c)]^3 \tag{11-6}$$

$$S_{wr} = a_2 + b_2[\tanh(N_c)] + c_2[\tanh(N_c)]^2 + d_2[\tanh(N_c)]^3 \tag{11-7}$$

where a_1, b_1, c_1, d_1, a_2, b_2, c_2 and d_2 are regression constants, and the hyperbolic tangent is defined for any quantity x as:

$$\tanh(x) = (e^x - e^{-x})/(e^x + e^{-x})$$

Relative permeabilities

From an experimental study, Casse and Ramey (1979) concluded that absolute permeability depends upon the nature of the saturating fluid and upon the overburden (confining) pressure. Temperature was found to have no effect on absolute permeability if the fluid in the pores was oil or gas. With water as the interstitial fluid, absolute permeability was found to decrease with increasing temperature. The effect, however, is not very significant, and can in any case be neglected for an oil recovery process.

Relative permeability functions can be generated using the residual saturations. The general relationship has been found by almost all investigators to be exponential in a "normalized" saturation which, however, has been defined in several different ways. The important idea in modelling relative permeabilities is to replicate the end-point permeabilities as accurately as possible, and to try to approximate the region in-between with standard mathematical functions.

Several investigations have been undertaken for determining the important effect of temperature on relative permeability, and some predictive models using some sort of "normalized" relative permeability versus a "normalized" fluid saturation have been proposed (see Davidson, 1969; Ehrlich, 1970; Sinnokrot et al., 1971; Weinbrandt and Ramey, 1972; Lo and Mungan, 1973). Kumar et al. (1985) have proposed general relative permeability models for steam and steam-additive flooding processes. The Kumar et al. method is quite general and is described below.

First, the end-point relative permeabilities are correlated with the end-point (i.e., residual) saturations. The residual saturations directly influence the relative permeability curves: they determine the movable saturation range and the shape of the relative permeability curves. Kumar et al. (1985) correlated end-point relative oil and water permeabilities against residual water saturation and the mobile water saturation range, S_{wm} $(=1-S_{\mathrm{or}}-S_{\mathrm{wr}})$ as follows:

$$k_{\mathrm{ro}}(S_{\mathrm{wr}}) = uS_{\mathrm{wr}}^{\mathrm{o}}S_{\mathrm{wm}}^{\mathrm{w}} \tag{11-8}$$

$$k_{\mathrm{ro}}(S_{\mathrm{wr}})/k_{\mathrm{rw}}(S_{\mathrm{or}}) = xS_{\mathrm{wr}}^{y}S_{\mathrm{wm}}^{z} \tag{11-9}$$

To generate the intermediate relative permeability values and the curvature of the permeability versus saturation relationship, a model proposed by Chierici (1981) is used:

$$k_{\mathrm{rw}}^{*} = \exp(-mS_{\mathrm{w}}^{**-n}) \tag{11-10}$$

$$k_{\mathrm{ro}}^{*} = \exp(-pS_{\mathrm{w}}^{**q}) \tag{11-11}$$

where m, n, p, and q are regression constants, and:

$$k_{\mathrm{rw}}^{*} = k_{\mathrm{rw}}/k_{\mathrm{rw}}(S_{\mathrm{or}}) \tag{11-12}$$

$$k_{\mathrm{ro}}^{*} = k_{\mathrm{ro}}/k_{\mathrm{ro}}(S_{\mathrm{wr}}) \tag{11-13}$$

$$S_{\mathrm{w}}^{**} = (S_{\mathrm{w}} - S_{\mathrm{wr}})/(1 - S_{\mathrm{or}} - S_{\mathrm{w}}) \tag{11-14}$$

The S^{**}, unlike the "normalized" saturations used elsewhere, has a value ranging from 0 to ∞. This kind of relationship implicitly satisfies the end points:

At $S_{\mathrm{w}} = S_{\mathrm{wr}}$,

$S_{\mathrm{w}}^{**} = 0$, and $k_{\mathrm{rw}}^{*} = \mathrm{e}^{-\infty} = 0$

implying that:

$k_{rw} = 0,$

which is the condition desired.

At $S_w = 1 - S_{or}$,

$S_w^{**} = \infty$, and $k_{rw}^* = \exp(-m\infty^{-n}) = e^0 = 1$

$k_{rw}^* = k_{rw}(S_{or})$,

which is the condition desired.

Similar conclusions can easily be verified for oil relative permeability at the end points.

Gas relative permeability

Hydrocarbon gas–oil relative permeabilities also pose a difficult problem. The relationships presented below are typical, and can be used to calculate the gas relative permeability, k_{rg}, in an oil–gas system as a function of gas saturation, S_g, and temperature, T:

$$k_{rg} = a_1 \exp\left[a_2(1 - S_g) + a_3(1 - S_g)^2 + a_4(T - T_b)S_g(1 - S_g)\right] \tag{11-15}$$

where T_b is the base temperature, and a_1, a_2, a_3, and a_4 are regression constants.

Steam relative permeabilities are very difficult to measure, and no studies have been reported in the literature to date. Steam can also significantly affect absolute permeability of the formation through reactions with the clays and minerals in the reservoir rock. Further, hot alkaline condensate near the injectors can dissolve clays and silica that may be re-deposited at the lower temperatures near the producers. These are difficult to characterize, and extensive laboratory studies are recommended for evaluating the permeability behavior in the presence of steam.

Conversion to three-phase system

Three-phase oil–water–gas relative permeability measurements are scarce, and schemes for synthetically generating these from two-phase data are not well established. Published work includes that of Stone (1973) who has described a simple method using two sets of relative permeabilities: oil–water system and oil–gas system. The water permeability is assumed to be the same in the oil–water–gas system as in the oil–water system. The gas permeability is also assumed to remain the same in the three-phase system as in the two-phase oil–gas system. The relative permeability to oil is then derived as follows:

$$k_{ro} = (k_{row} + k_{rw})(k_{rog} + k_{rg}) - k_{rw} - k_{rg} \tag{11-16}$$

where k_{row} is the relative permeability to oil in an oil–water system, and k_{rog} is the relative permeability to oil in an oil–gas system.

Capillary pressures

The dependence of wettability as well as interfacial tension on temperature implies that capillary pressure is also strongly temperature-dependent. Laboratory studies corroborate this postulated behavior, though it is difficult to quantify capillary pressures as a function of temperature. Literature reports limited data (Sinnokrot et al., 1971; Samaroo and Guerrero, 1981), which indicates that with increasing temperature, capillary pressures increase at a given saturation for an oil–water system, but decrease for a gas–water system.

Kumar (1989) has proposed the following general analytic relationship for oil–water capillary pressures as a function of temperature and interfacial tension:

$$P_c(S_w) = a \exp\left[-\frac{S_w - S_{wr}}{S_w - S_a}\right] + b \ln\left[\frac{S_w(S_a - S_{wr})}{S_a(S_w - S_{wr})}\right] + c \qquad (11\text{-}17)$$

where a, b, and c are regression constants, and $S_a = S_a(\sigma, T)$ is the saturation at which the capillary pressure approaches c, a small valued constant. This may not be easily seen on a regular cartesian plot. A semilog plot of P_c versus S_w reveals this constant c readily. The capillary pressure remains almost constant for $S < S_a$.

The relationship above was chosen so as to satisfy the following criteria:

(1) As $S_w \to S_a$, $P_c \to c$,

$$P_c(S_w) = a \exp\left[-\frac{S_a - S_{wr}}{S_a - S_a}\right] + b \ln\left[\frac{S_a(S_a - S_{wr})}{S_a(S_a - S_{wr})}\right] + c$$

$$= a \exp(-\infty) + b \ln(1) + c = 0 + 0 + c = c$$

(2) As $S_w \to S_{wr}$, $P_c \to \infty$,

$$P_c(S_w) = a \exp\left[-\frac{S_{wr} - S_{wr}}{S_{wr} - S_a}\right] + b \ln\left[\frac{S_{wr}(S_a - S_{wr})}{S_a(S_{wr} - S_{wr})}\right] + c$$

$$= a \exp(0) + b \ln(\infty) + c = a + \infty + c = \infty$$

Thus, this combination of a logarithmic and exponential function satisfactorily replicates the general shape of a drainage capillary pressure curve.

Enhanced recovery modelling requires imbibition capillary pressure data which is relatively easier to correlate as compared to drainage. The two limiting points on the imbibition capillary pressures are governed by the residual fluid saturations, and there are no complications related to entry pressure as on drainage data. Kumar (1989) has proposed the following simplified model for imbibition capillary pressure:

$$P_c(S_w) = a \exp\left[-\frac{S_w - S_{wr}}{S_w - 1 + S_{or}}\right] + b \ln\left[\frac{S_w(1 - S_{or} - S_{wr})}{(1 - S_{or})(S_w - S_{wr})}\right] \qquad (11\text{-}18)$$

which is the same as for the drainage case, with S_a replaced by $1 - S_{or}$. The constant c is no longer needed, because capillary pressure is zero at the water saturation corresponding to the residual oil saturation. The advantage of using the flow residual saturations is that the correlations determined for residual saturations as a function of temperature (and IFT) can be used here to construct a capillary pressure curve as a function of temperature (and IFT).

Viscosity

Most heavy oils are highly viscous, with viscosities ranging from 100 cP to a few million centipoises at reservoir conditions. A simple, accurate and computationally convenient technique, such as the one described by Beggs and Robinson (1975) is desirable. In the Beggs-Robinson method, the temperature dependence of viscosity is accounted for as a function of oil gravity (which is representative of the oil composition). It yields viscosities correct to within 1–2%, compared to an average error of 200—300% for the classic Beal and Chew-Connally techniques (Lohrenz et al., 1964).

A much superior method employing the critical pressures and temperatures for viscosity determination by an iterative procedure has been described by Lohrenz et al. (1964). The functional dependence on composition and temperature is very satisfactorily taken into account. It is designed for use in reservoir simulation models and has proved quite reliable in even compositional simulators. Other methods are available which are generally quite satisfactory, but they do not account directly for oil composition.

A reasonable approach is to take oil viscosity, μ_o, as the product of the oil viscosity at composition X_i (mole fraction of component i in liquid), i.e., $\mu_o(X_i)$, and the oil viscosity at temperature T, i.e., $\mu_o(T)$:

$$\mu_o(X_i, T) = \mu_o(X_i) \cdot \mu_o(T) \tag{11-19}$$

The effect of temperature on hydrocarbon gas viscosity, μ_g, can be taken as follows:

$$\mu_g(p, T) = \mu_g(p)(a_1 T + a_2) \tag{11-20}$$

Unlike oil viscosity, gas viscosity is easily correlatable with composition (see Carr et al., 1954). The viscosity of the steam–gas mixture is determined from:

$$\mu_g(p, T, Y_{HC}) = \mu_g(p, T) Y_{HC} + \mu_s(T)(1 - Y_{HC}) \tag{11-21}$$

where Y_{HC} is the mole fraction of hydrocarbon in vapor and μ_s is the viscosity of steam.

Water viscosity can be readily estimated using the following equation by Gottfried (1965):

$$\mu_w = \frac{1776 - T}{26.5T - 89} \tag{11-22}$$

where T is the temperature in °F, and μ_w is the water viscosity in cP.

Fluid densities

Several correlations are available for estimating oil density. Oil density can be calculated as a function of temperature, pressure, and composition using the thermal expansion coefficient C_{To} and compressibility C_{o}:

$$\rho_{\mathrm{o}}(p, T, X_i) = \rho_{\mathrm{o}}(X_i)[1 - C_{\mathrm{To}}\, \Delta T + C_{\mathrm{o}}\, \Delta p] \tag{11-23}$$

where $\rho_{\mathrm{o}}(X_i)$ is the oil density for a composition X_i.

Gas densities can be calculated as a function of temperature, pressure, and composition using the gas compressibility factor, Z:

$$\rho_{\mathrm{g}}(p, T, Y_{\mathrm{HC}}) = \frac{5.6146\, pM_{\mathrm{g}}}{ZRT} \tag{11-24}$$

where M_{g} is the gas molecular weight. The Z-factor is calculated for the gas mixture of steam and hydrocarbon as follows:

$$Z = Y_{\mathrm{HC}} Z_{\mathrm{hg}} + (1 - Y_{\mathrm{HC}}) Z_{\mathrm{s}} \tag{11-25}$$

Water density as a function of temperature and pressure can be calculated using the water compressibility, C_{w}, as follows:

$$\rho_{\mathrm{w}}(p, T) = \rho_{\mathrm{w}}(T)[1 + C_{\mathrm{w}}\, \Delta p] \tag{11-26}$$

where $\rho_{\mathrm{w}}(T)$ is the water density at temperature T. Generally, water and steam densities are determined from steam tables.

Phase behavior parameters

The system consists of three phases: aqueous (water), oleic (oil), and gaseous. The components are: (1) *aqueous phase*: water; (2) *oleic*: hydrocarbon components; and (3) *gaseous*: hydrocarbon components, and water (steam). The phase behavior involves two different activities: (1) hydrocarbon partitioning between the liquid (oleic) and gaseous phases, and (2) steam partitioning between the liquid (aqueous) and gaseous phases. The aqueous and oleic phases are often assumed completely immiscible to simplify flash calculations. Steam table data is used for determining steam saturation pressure, quality, density, etc. Hydrocarbon phase relationships are established using equilibrium ratios for the various hydrocarbon components given by Katz and Hachmuth (1937) and others, or calculated using an equation of state (EOS). Among the several available EOS, the Redlich-Kwong and Peng-Robinson EOS are widely used. Use of an EOS enables accurate determination of other properties such as fluid densities also.

Thermal properties of the rock

Thermal conductivity and diffusivity of the adjacent formations is required for computing the heat loss to overburden and underburden. Thermal diffusivity,

α (ft^2/hr), is the ratio of thermal conductivity and volumetric heat capacity M. An excellent review of correlations and data available for estimating thermal properties of several types of rocks has been presented by Prats (1982). Generally, rock thermal conductivity increases with an increase in liquid saturation or a decrease in porosity or temperature.

For the fluid-saturated reservoir, the heat capacity M is the sum of the contributions by the rock grains and the contained fluids (water, oil and gas). Usually, about 2/3 of the injected heat is utilized in heating the rock matrix, with the result that enormous amounts of heat are left behind after the injection fluid has displaced most of the oil-in-place. Injection of cold water following steam injection has been successfully used in transporting this heat downstream. For the same reason, water is injected with air in the in-situ combustion processes.

Enthalpy

Hydrocarbons

Hydrocarbon gas enthalpies, H_g, are calculated from the specific heat:

$$H_g(p, T) = c_{pg}(p, T)\,\Delta T \qquad (11\text{-}27)$$

The specific heat c_{pg} of a gaseous hydrocarbon component is found as a function of pressure and temperature using correlation equations based upon the work of Edmondson (Green and Perry, 1984), Kesler and Lee (1976), and Touloukian et al. (1970). These correlations, presented graphically in the original work, may also be converted to suitable equations for computational purposes.

The enthalpy of any non-volatile hydrocarbon component, H_{nv}, in the oleic phase can be determined from its specific heat in the liquid state:

$$H_{nv}(p, T) = c_{pnv}(p, T)\,\Delta T \qquad (11\text{-}28)$$

For the volatile components in the oil, enthalpy, H_v, can be determined by subtracting the latent heat of vaporization, H_{vap}, from the vapor enthalpy for the given component:

$$H_v(p, T) = H_g(p, T) - H_{vap}(p, T) \qquad (11\text{-}29)$$

where the latent heat of vaporization, H_{vap}, is calculated from the work of Kesler and Lee (1976). The phase enthalpies for gas and oil are then calculated from:

$$H_g(p, T, Y_{HC}) = H_g(p, T)Y_{HC} + H_s(p, T)(1 - Y_{HC}) \qquad (11\text{-}30)$$

for gas, and:

$$H_o(p, T, X_i) = H_v(p, T)X_v + H_{nv}(p, T)(1 - X_v) \qquad (11\text{-}31)$$

for oil.

Steam and water

Water and steam enthalpies and internal energies can be determined from steam tables.

In a steamflooding process, steam is the heat transfer medium used for heating up the formation. Only a part of the heat supplied to the steam in the boiler goes into formation heating. From the steam boiler (or generator) to the heat front in the reservoir, the following are the sources of heat loss: (1) surface (steam generator and piping), (2) wellbore (tubing–casing), (3) heat loss to adjacent formations (over- and underburden).

Heat losses at the surface

About 20% of the heat supplied to a surface steam generator is lost to the flue gas that leaves the stack at approximately 400°F. Another 3–5% is lost even in a well-designed, insulated pipeline that carries the steam to the injection wellhead. The basic equation used for calculating the heat loss per unit length of pipe, Q_{ls} (Btu/hr-ft pipe) is:

$$Q_{ls} = (T_b - T_a)/R_h \tag{11-32}$$

where T_b is the bulk temperature of the fluid in the pipe (°F); T_a is the ambient (atmospheric) temperature (°F); and R_h is the resistance to heat transfer (Btu/hr-°F-ft pipe)$^{-1}$, commonly represented as $1/(2\pi rU)$, r being the radius (ft) of the surface across which the heat transfer is occurring and U, the overall heat transfer coefficient (Btu/hr-°F).

For a surface pipe, R_h can be written as follows:

$$R_h = (1/2\pi)\big[1/h_f r_i + 1/h_{pi} r_i + (1/\lambda_p)\ln(r_o/r_i) + 1/h_{po} r_o + (1/\lambda_{ins})\ln(r_{ins}/r_o) + 1/h_{fc} r_{ins}\big] \tag{11-33}$$

where h is the film coefficient of heat transfer, Btu/hr-ft^2-°F, subscript "f" indicating the h between fluid and inside pipe wall, subscript "pi", across any dirt or scale on the inside pipe wall, and subscript "po", across the contact between the outside pipe wall and the insulation. λ is the thermal conductivity, Btu/hr-ft-°F, subscript "p" indicating the pipe wall, and "ins" the insulation. This gives the steady-state heat transfer. Transients are usually neglected because they prevail for a very short time only (less than a day).

The reduction in steam quality due to heat loss in the surface pipe can be calculated as follows:

$$f_{sw} = f_{sg} - Q_{ls} L/(m L_v) \tag{11-34}$$

where f_{sw} is the wellhead steam quality; f_{sg} is the generator discharge steam quality; L is the pipe length (ft); m is the mass flow rate of steam (lb/hr); and L_v is the (average) enthalpy of vaporization of steam (Btu/lb).

Heat loss in the wellbore

As steam moves through the wellbore, there is a loss of heat to the surrounding formations due to the difference between fluid and geothermal temperatures. Wellbore heat losses can be high enough, especially in wells that are deep or where steam injection rates are low, to reduce steam to hot water at the formation rockface. Inasmuch as thermal oil recovery methods depend upon the ability to provide heat to the reservoir, heat losses between the surface and the injection interval must be minimized. For this reason, steam is usually injected through an insulated tubing, with the casing–tubing annulus packed-off and vented to the atmosphere. Other techniques to reduce heat transfer through the casing–tubing annulus include: (1) use of concentric tubing strings, with insulators between the strings; (2) crude oil in the annulus to prevent steam leaks (from packer, expansion joint, etc.) and prevent radiation; and (3) "gas pack"—displacing the annulus with high-pressure gas.

The typical elements offering resistance to heat loss from the tubing include: (1) fluid film on inside of tubing wall (subscript "f"), (2) scale or dirt on tubing pipe wall (subscript "pi"), (3) tubing pipe wall (subscript "p"), (4) contact between outside of tubing and the insulation (subscript "po"), (5) insulation (subscript "ins"), (6) radiation and convection from insulation (subscript "rins"), (7) casing pipe wall (subscript "pc"), (8) cement (subscript "cem"), (9) damaged zone (altered earth) around the cement (subscript "Ea"), and (10) variable (unsteady state) conduction into the earth (subscript "E"). R_h can, thus, be written as follows:

$$\begin{aligned} R_h = (1/2\pi)\big[&1/h_f r_i + 1/h_{pi} r_i + (1/\lambda_{pc})\ln(r_o/r_i) + 1/h_{po} r_o \\ &+ (1/\lambda_{ins})\ln(r_{ins}/r_o) + 1/h_{rins} r_{ins} + (1/\lambda_{pc})\ln(r_{co}/r_{ci}) \\ &+ (1/\lambda_{cem})\ln(r_w/r_{co}) + (1/\lambda_{Ea})\ln(r_{Ea}/r_w) + (1/\lambda_E) f(t_D)\big] \end{aligned} \tag{11-35}$$

where $f(t_D)$ is a time function that represents the thermal resistance of the earth as a function of time, given by Willhite (1967). Heat losses from wells never reach a steady state, but a quasi-steady state in which the rate of heat loss is a monotonically decreasing function of time, as shown by Ramey (1962) and Willhite (1967). The dimensionless time, t_D, is based upon the thermal properties of the earth:

$$t_D = \alpha_E t / r_{Ea}^2$$

where α_E is the thermal diffusivity of the earth, ft^2/hr, and t is the time from the initiation of heating, hours.

Assuming a constant steam pressure (or temperature) in the tubing is a poor assumption in some cases. Pacheco and Farouq Ali (1972) and Farouq Ali (1981) have derived heat loss procedures that account for the geothermal gradient, the pressure variations in the wellbore with depth due to friction losses and fluid hydrostatic head, etc., and provided an interesting study of various parameters on heat loss.

Once the heat loss is known, equation (11-34) can be used for computing steam quality at the rockface. Steam quality at the bottomhole (or rockface) is usually lower than that at the surface. A combination of high steam rate, small tubing size, and high injection steam quality, however, can result in lower bottomhole pressures and, consequently, higher bottomhole steam quality.

Growth of heated volume with heat loss to adjacent formations

Steam injection

Marx and Langenheim (1959) provided the basic model for reservoir heating. They considered the injection of hot fluid at both constant rate and constant temperature, assuming the temperature everywhere in the heated zone to be equal to the downhole injection fluid temperature, and a constant injection rate and temperature. The areal extent A (ft^2) of the equivalent heated zone is as follows (Marx and Langenheim, 1959):

$$A = \frac{Q_i M_R h_t}{4 \Delta T \alpha_S M_S^2} F(t_D) \qquad (11\text{-}36)$$

where the injection rate Q_i is given by:

$Q_i = 350[h_w + f_s L_v - c_w(T_R - 32)]$;

the time-dependent function $F(t_D)$ is given by:

$F(t_D) = e^{t_D} \operatorname{erfc} t_D^{1/2} + 2(t_D/\pi)^{1/2} - 1$;

the dimensionless time t_D is given by:

$$t_D = \frac{4 M_S^2 \alpha_S t}{M_R^2 h_t^2}; \qquad (11\text{-}37)$$

ΔT is the temperature difference (°F) between the surrounding shales (overburden and underburden), T_S, and the reservoir, T_R:

$\Delta T = T_S - T_R$;

M_R and M_S are the volumetric heat capacities of the reservoir rock and shales, respectively, Btu/ft^3-hr; α_S is the thermal diffusivity of the shales, ft^2/hr; h_t is the formation thickness, ft; and t is the time, hours.

The rate of growth of the equivalent heated zone is:

$$\frac{dA}{dt} = \frac{Q_i e^{t_D} \operatorname{erfc} t_D^{1/2}}{\Delta T M_R h_t} \qquad (11\text{-}38)$$

The heat remaining in the reservoir is given by:

$$Q = \frac{Q_i M_R^2 h_t^2}{4 \alpha_S M_S^2} F(t_D) \qquad (11\text{-}39)$$

The cumulative heat lost to the adjacent formations, Q_l, can be calculated using Q_i and Q:

$$Q_l = Q_i t - Q \tag{11-40}$$

The fraction of the injected heat utilized in heating the reservoir, called reservoir heat efficiency, E_h, is given by:

$$E_h = Q/Q_i = F(t_D)/t_D \tag{11-41}$$

For variable rates of injection, superposition techniques are used with the above procedure. For a given reservoir, heat loss depends on t_D only, increasing directly with time, and inversely as the square of the formation thickness. This is indicative of the fact that steam injection is uneconomical in thin formations.

For thin formations, or in cases where pressure is high or steam quality is low, steam zone volume can be better estimated from the work of Mandl and Volek (1969). They identified a critical time beyond which the latent heat content of the injected steam is unable to supply the heat losses to the adjacent formations and raise the temperature of newly-invaded formation to the steam temperature. Prior to this critical time, all the heat in the reservoir is within the steam zone, and the Mandl-Volek model reduces to the Marx-Langenheim model. Beyond the critical time, the zone downstream of the condensation bank is heated by hot water. It is assumed that the heat injection rate is constant, there is no gravity override of the steam zone, and all points on the condensation front advance at the same rate. Departure from these assumptions will reduce the critical time. Other authors, such as Willman et al. (1961), considered the presence of hot water beyond the condensation bank from the beginning of injection, though Myhill and Stegemeier (1978) ignored the contribution of the hot-water zone in oil recovery. Neuman (1975) considered the presence of a hot-water zone below steam zones affected by gravity override.

Myhill and Stegemeier (1978) have presented charts for estimating steam zone volume with the Mandl-Volek model. The expressions for the steam drive process described in these works are outlined below. The volume of the steam zone, V_s, is given by:

$$V_s = \frac{Q_i E_{hs}}{M_R \, \Delta T} \tag{11-42}$$

where the time-dependent function E_{hs} is given by:

$$E_{hs} = \left(\frac{1}{t_D}\right)\left\{F(t_D)(1 - f_{hv})U(t_D - t_{cD})/\pi^{1/2}\left[2t_D^{1/2} - 2(1 - f_{hv})(t_D - t_{cD})^{1/2} - \pi^{1/2}F(t_D) - \int_0^{t_{cD}} \frac{e^x \operatorname{erfc} x^{1/2}}{(t_D - x)^{1/2}} \mathrm{d}x\right]\right\} \tag{11-43}$$

The dimensionless critical time, t_{cD}, is defined as follows:

$$e^{t_{cD}} \operatorname{erfc} t_{cD}^{1/2} \equiv 1 - f_{hv} \tag{11-44}$$

where the fraction of the heat injected in latent form, f_{hv}, is calculated as follows:

$$f_{hv} = (1 + c_w \, \Delta T / f_{sdh} L_{vdh})^{-1}$$

where c_w is the heat of vaporization of water, and f_{sdh} and L_{vdh} are the downhole steam quality and latent heat, respectively.

Hot water injection

The temperature T at any point x in the reservoir at any time t for a hot water injection process is given, in dimensionless terms, by Lauwerier's (1955) equation:

For $t_D > x_D$,

$$T_D = \frac{T - T_R}{T_f - T_R} = 1 - \operatorname{erf}\left[\frac{x_D}{2(t_D - x_D)^{1/2}}\right] \tag{11-45}$$

And for $t_D \leqslant x_D$,

$$T = T_R \tag{11-46}$$

where subscript "f" indicates fluid property, and x_D is given by:

$$x_D = \frac{4 k_S M_S A}{M_S \rho_f c_f i_f h_t}$$

where i_f is the injection rate of the fluid, bbl/day; and ρ_f is the fluid density, lb/bbl.

Willman et al. (1961) proposed an approach where the reservoir is discretized into a number of zones. Within each of these constant temperature zones, isothermal Buckley-Leverett displacement equations are applicable, so that the rate of growth of the saturation front is constant. But the rate of growth of the area encompassed by the saturation fronts changes as each new temperature zone is entered, as follows:

$$\frac{dA}{dt} = \frac{1}{\phi h_n} q \frac{\partial f_w(S, T_i)}{\partial S} \tag{11-47}$$

where ϕ is the porosity, h_n is the net sand thickness, f_w is the fractional flow of water, q is the flow rate at (ambient) reservoir temperature and pressure, T_i is the temperature in zone i, and S is the water saturation.

OIL RECOVERY CALCULATIONS

There are several ways of predicting the performance of steam drive, cyclic steam injection, and hot water injection processes: (1) mathematical (computer) modelling; (2) scaled physical models (laboratory studies); (3) analytical methods, using estimates of the volume heated or steamed; and (4) correlations developed using any of the three earlier techniques. Scaled physical models, if designed properly, can offer valuable data, but are expensive and time-consuming. Analytical methods, such as the Marx-Langenheim and Mandl-Volek models discussed earlier, offer an over-simplified analysis that is usually not applicable to real reservoir systems. Correlations have limited applicability, and are useful only for the reservoir system for which they were designed. Computer modelling, therefore, offers the only reliable means of predicting reservoir performance, provided enough information is available.

This section describes some simple analytic methods and correlations. In a real reservoir the analysis is far more complex. Some of the phenomena that defy description are: (1) viscous fingering of the steam, (2) flow description for multiphase flow (fluid–rock properties), (3) gravity segregation of steam, and (4) reservoir heterogeneity.

Steamflooding

Computer models and laboratory and field tests show that the ultimate oil recovery depends, in many cases, upon the volume of the reservoir that is heated. At the end of a steamflood, the ultimate oil recovery, N_p, can be calculated as follows:

$$N_p = R\phi V_s(h_n/h_t)(S_{oi} - S_{ors}) \qquad (11\text{-}48)$$

where R is the ratio of actual to theoretical oil production, S_{oi} is the initial oil saturation, and S_{ors} is the residual oil saturation at the end of the steamflood. Calculation of the steam zone volume, V_s, has been discussed earlier. R has been reported by Myhill and Stegemeier (1978) to be in the range of 0.7–1.0 for most cases. But equation (11-48) gives very poor predictions of oil production rate versus time when compared with field data (Myhill and Stegemeier, 1978). One of the reasons could be that the initial rates in the field tend to be lower because of the high viscosity of the cold oil near the producers. Jones (1981) has presented estimates of R as a function of steam injection from field data that yield better results.

Gomaa (1980) developed correlations for predicting oil production rates as a function of time for thick low-pressure reservoirs in the Kern County, California, area, based upon computer simulation studies. This method gives good results for Kern County steamfloods, but is hardly applicable to reservoirs with conditions much different from these. The method also does not include the effect of steam zone pressure, a critical parameter in steamflood performance.

Cyclic steam injection

In cyclic steam injection, steam is injected for 2—4 weeks at a high rate. The well is returned immediately to production, or shut-in for a few days. Subsequently, the well produces oil for a few months, or up to one year, at an improved rate. These cycles are repeated several times till the economic limit is reached. Generally, oil production rate and amounts of oil produced fall off in the later cycles.

The response of a reservoir to cyclic steaming depends upon several factors, such as reservoir pressure, saturations, volume of steam injected, etc. Cyclic steaming works best in thick sands containing viscous oil, with close well spacings. The underlying concept in the mathematical treatment of this process is the decline of temperature with time in the zone heated by the steam. A simplified calculation procedure assumes single-phase flow. For a total drainage radius of r_e, heated zone radius of r_h, well radius of r_w, and oil viscosities in the cold and heated zones of μ_{oc} and μ_{oh}, respectively, the stimulation ratio, q_{oh}/q_{oc}, is given by:

$$\frac{q_{oh}}{q_{oc}} = \frac{(\mu_{oc}/\mu_{oh})\ln(r_e/r_w)}{\ln(r_h/r_w) + (\mu_{oc}/\mu_{oh})\ln(r_e/r_h)} \tag{11-49}$$

For a high cold-oil viscosity, μ_{oc}, equation (11-49) reduces to:

$$\frac{q_{oh}}{q_{oc}} = \frac{\ln(r_e/r_w)}{\ln(r_e/r_h)}$$

Thus, the stimulation ratio is essentially a function of the heated zone radius and it is advantageous to inject large steam slugs in very viscous oil formations.

AN OVERVIEW OF STEAMFLOOD MODELLING

Early efforts in mathematical modelling of thermal floods (Lauwerier, 1955; Marx and Langenheim, 1959; Willman et al., 1961; Mandl and Volek, 1969; Satter and Parrish, 1971) concentrated on simulating the heat flow and heat losses. Shutler described three-phase models for linear (1969) and two-dimensional (1970) steamflooding. Abdalla and Coats (1971) treated a two-dimensional steamflood model using the IMPES (Implicit Pressure, Explicit Saturation) method of solution. Approximate analytic methods have been presented by several investigators; notable among them are Shutler and Boberg (1972) and Neuman (1975). Neuman presented a theoretical analysis that accounts for the longitudinal as well as areal displacement by steam, and steam override.

One of the most widely used formulations in computer modelling of steamflood processes is that by Coats (1976b), an improvement over previous formulations by Coats and coworkers (Coats et al., 1974; Coats, 1976a). This three-phase model considers two hydrocarbon components in the oleic phase, one hydrocarbon component in the gaseous phase, representing a step towards a fully compositional thermal model without incurring the computational expense of the latter. Steamflooding processes are generally used for heavy oils, where

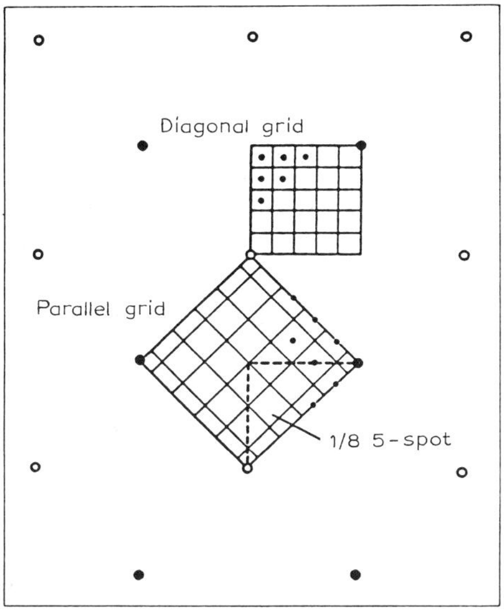

Fig. 11-2. Parallel and diagonal grids for five-spot pattern. (After Coats et al., 1974; courtesy of the Society of Petroleum Engineers of AIME.)

reducing the oil to two pseudocomponents (one volatile component and the other non-volatile) is sufficient to characterize it fairly well. Such an approach is very desirable.

The material and energy balances for the system are solved using finite difference methods. Interest has been shown in using finite element methods, but these are quite complex, and offer no particularly significant advantage (except perhaps the excellent treatment of reservoir boundaries). In a finite difference method of solution, the differencing scheme to use is an important issue. Explicit methods give faster, but inaccurate and sometimes unstable solutions. Implicit schemes are slower, but give more accuracy and unconditional stability. Results from any simulation model are subject to numerical dispersion. To control this, no universally applicable method has been developed yet, and it remains one of the most serious concerns. Kumar (1989) has developed some semi-analytic upstream and downstream weighting schemes for controlling dispersion of saturation and chemical profiles. These schemes give a greatly enhanced control on numerical dispersion.

Fluid and rock properties used in the model must include the effect of temperature in the proper manner. Otherwise, the results would hardly be meaningful. It is also more desirable to include effects of composition, wherever possible.

Grid orientation effects are also a major concern. Fig. 11-2 shows two grid types representing a five-spot well pattern. As shown by Coats et al. (1974), the parallel grid results in a far more pronounced steam finger progressing along the

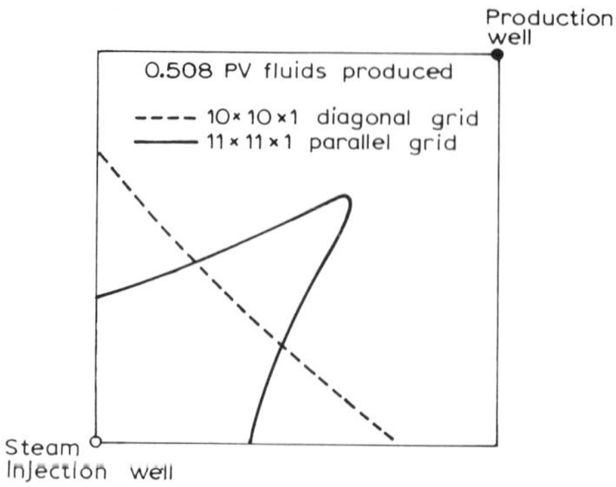

Fig. 11-3. Comparison of 20% steam saturation contours calculated using parallel and diagonal grids. (After Coats et al., 1974; courtesy of the Society of Petroleum Engineers of AIME.)

injector–producer diagonal (Fig. 11-3). The diagonal grid results in near-radial growth of the steam front, with fingering occurring only as the front approaches the producer. Computations using these two grid systems differed by a factor of three for steam breakthrough times, as demonstrated by Coats et al. (1974). The results converged when the grids were better defined (i.e., smaller grid sizes were used). Todd et al. (1972) pointed out that the diagonal grid gave more accurate results up to an adverse mobility ratio of ten. In steamflooding, the mobility ratio is typically of the order of 10^5. This would seemingly suggest using the parallel grid, but as pointed out by Baker (1973), because of the condensation of steam, a mobility ratio based upon kinematic viscosity might give a better indication of the effective mobility ratio. Further, two fronts are created in situ: the steam–trailing edge of the oil bank, and the leading edge of the oil bank–water (at almost irreducible saturation), the mobility ratio of the latter being highly favorable. This further complicates the resolution of the question of the choice of the grid system to use. Coats et al. (1974) also found that the choice is not very clear. They determined that the disparity in steam front shapes, calculated using the two grids, persists long after the leading edge of the oil bank has been swept from the reservoir. Inasmuch as most of the total mobility in the oil bank ahead of the steam front comprises near-irreducible water saturation (rather than oil), the use of oil viscosity or mobility in calculating mobility ratios is ill-justified.

Fortunately, grid orientation has little or no effect on oil recovery computations. To avoid unnecessary indecisiveness about the results, however, it is imperative to use small grid blocks, for example, 16×16 for the diagonal and 22×22 for the parallel, for a 1/4 of a five-spot.

The solution scheme used affects the existence and rate of convergence. Price and Coats (1974) have shown that the iterative reduced bandwidth direct solution method with D4 grid-ordering is faster than the SOR, ADI, and SIP methods. Currently, this is an extensively used solution method. Sparse matrix solution methods that exploit the sparseness of the matrix that results after the finite difference expansion, however, are becoming more common. Some char-actersitics of the SOR, ADI, and SIP procedures are described below:

SOR: Reguires only diagonal dominance of the matrix for convergence. Point SOR is advantageous when the grid system is essentially uniform. Line SOR is somewhat slower for this case, but is much faster for anisotropic systems.

ADI: Requires a positive definite matrix to guarantee convergence. Two categories: non-iterative (ADIP) and iterative (ADIPIT). ADIP is applicable primarily to parabolic-type problems having smooth, continuous solution surfaces and fairly constant coefficients, such as systems with large grid blocks in which the dependent variables change only slightly during a time step. ADIPIT is most useful for the solution of flow problems employing a horizontal, two-dimensional grid network, or a vertical cross-section. Both ADI and ADIPIT have been found to have a speed advantage over SOR in areal two-dimensional problems.

SIP: Stone (1968) demonstrated that SIP (Strongly Implicit Procedure) achieved greater convergence rates than ADI on all types of problems, except the simple model problem where the coefficients were constant and isotropic. The advantage of SIP over ADI increases as the complexity of the problem increases. The details are well described by Weinstein et al. (1969).

The choice of the particular solution technique depends upon the eigenvalues (maximum and minimum) of the matrix to be solved, homogeneity of the problem, and the grid orientation. For many simulation models, the direct solution method with D4 grid-ordering scheme shows good performance, reducing computation time by a factor of almost 4, and storage requirements by a factor of 3.

PARAMETRIC STUDIES IN STEAMFLOODING

The effect of various parameters on oil recovery can be investigated relatively easily using a simulation model. Described below are some of the results obtained by Gomaa (1980).

Porosity

High-porosity reservoirs produce more oil than low-porosity reservoirs per barrel of steam injected due to the larger heat loss in the rock matrix in the latter case. But when converted to fractional recovery basis (percentage oil-in-place), the recovery is, according to Gomaa (1980), almost independent of the porosity for the same steam injection rate per unit reservoir volume (see Fig. 11-4). Gomaa used a "tank model" approach, and these results are perhaps a manifestation of such an approach. A higher porosity suggests larger pore sizes and higher pore connectivity which will significantly alter the flow resistance offered to the oil bank. Even on a fractional recovery basis, the recovery should

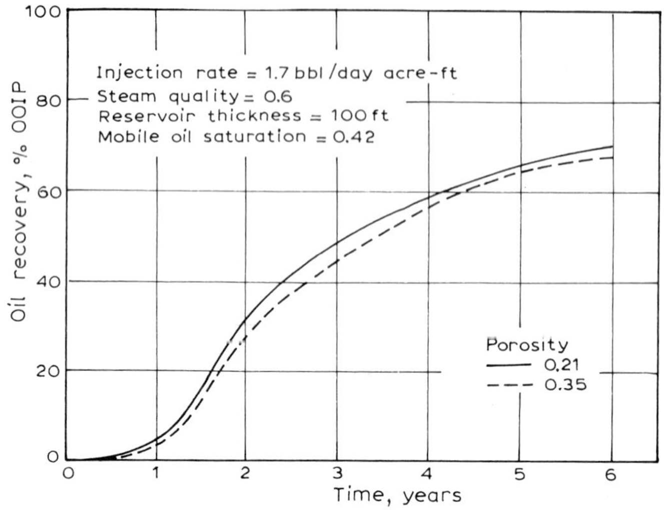

Fig. 11-4. Effect of porosity on steamflood oil recovery. (After Gomaa, 1980; courtesy of the Society of Petroleum Engineers of AIME.)

be higher for a higher porosity system, depending upon the extent of viscosity reduction, heat loss to the rock matrix, the permeabilities achieved, etc.

Reservoir thickness

Reservoir thickness has been traditionally conceptualized as having repercussions primarily on the heat loss to the overburden and underburden. Fig. 11-5 shows that recovery is higher for thicker reservoirs. Using this same approach, it may be concluded that after accounting for heat lost to the surroundings, oil recovery curves for reservoirs with different thicknesses should become nearly identical, as shown in Fig. 11-6 from Gomaa's (1980) work. Reservoir thickness does, however, play an important role and Gomaa's results are merely a simplification. Reservoir thickness determines the vertical temperature gradient, extent of steam override, the steam–oil saturation profile, and the logistics of the advance of this profile in the reservoir during the steamflood process. Capillary pressures also are sensitive to height in situ.

Net/gross ratio

The net/gross ratio is a measure of the presence of any discontinuous shale streaks that reduce the net productive thickness without diminishing vertical communication. This situation may be modelled using (1) an effective porosity and permeability equal to the product of the net/gross ratio and the producing sand porosity and permeability, respectively; and (2) the gross reservoir thick-

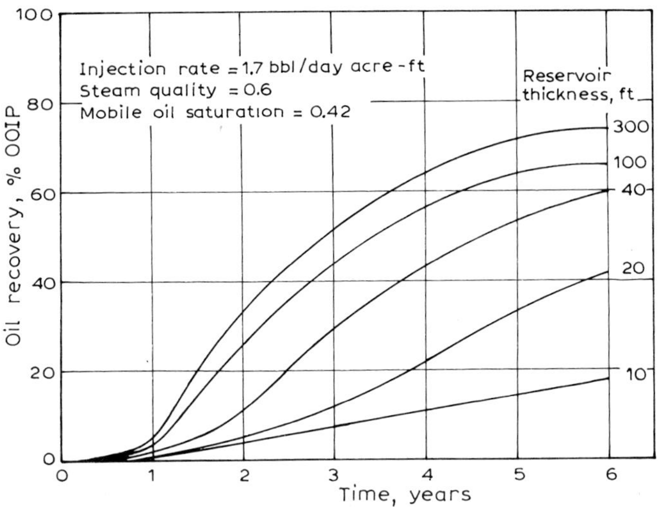

Fig. 11-5. Effect of reservoir thickness on steamflood oil recovery. (After Gomaa, 1980; courtesy of the Society of Petroleum Engineers of AIME.)

ness as the effective thickness. It is obvious that such a treatment would result in the shaly reservoir providing a better recovery than a clean one because of the greater thickness of the former, everything else remaining the same. There are, however, other factors that result in improved recoveries in shalier systems.

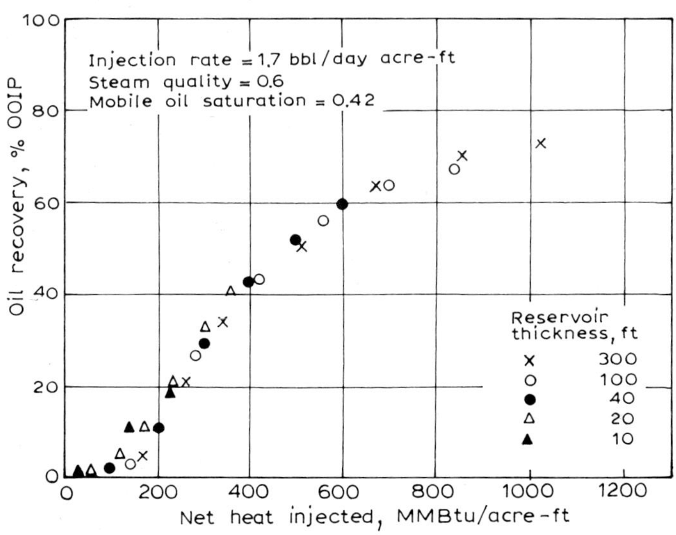

Fig. 11-6. Effect of reservoir thickness on steamflood oil recovery. (After Gomaa, 1980; courtesy of the Society of Petroleum Engineers of AIME.)

Gravity override of steam is impaired by these non-communicating barriers, leading to better sweep efficiency. Also, shales are generally water-saturated and, therefore, have a very high specific heat. In cyclic steam stimulation, these can act as storage areas for heat, reducing heat loss during injection and releasing heat to sustain a higher temperature during the latter part of the steam cycle. Shaly sands require higher steam/oil ratios than clean sands.

Mobile oil saturation

Higher the mobile oil saturation, defined as the difference between the oil saturation prior to the steamflood and the residual oil saturation after the steamflood, higher is the ultimate recovery as well as the rate of recovery. This is true for any enhanced recovery process. In simplistic terms, the higher rate of recovery can be attributed to the higher permeability to oil at the higher oil saturation level, whereas the increased ultimate recovery results simply from the greater amount of recoverable oil-in-place per unit reservoir volume.

Pattern geometry

In reservoir simulators, frequently there are no limitations imposed on well productivity. This will lead to a model prediction that neither the pattern shape nor the spacing influence oil recovery for a fixed injection rate per unit reservoir volume. This does not hold true for actual field projects where such limitations do exist and patterns with higher producer-to-injector ratios yield better recoveries. Also, in patterns with non-uniform locations of producers relative to the injectors (e.g., the nine-spot pattern), loss in ultimate recovery could be caused by lower sweep efficiencies, earlier steam breakthrough, etc.

Injection rate

To eliminate the effect of several geometrical factors, steam injection rate is commonly expressed on a unit reservoir volume basis. Gomaa (1980) found the injection rate in this form to have almost no effect on oil recovery. Several investigators have described the effect of steam injection rate as being significant. As pointed out by Willman et al. (1961) during a very early laboratory study, a higher steam injection rate gives a higher recovery because heat losses are minimized.

Steam quality

Higher steam quality results in higher and faster recovery (Fig. 11-7). Some authors have discovered an optimum steam quality for maximizing oil recovery on a net heat injected (total heat injected minus the losses) basis. Gomaa (1980) found that this optimum occurs at about 40% steam quality for the system

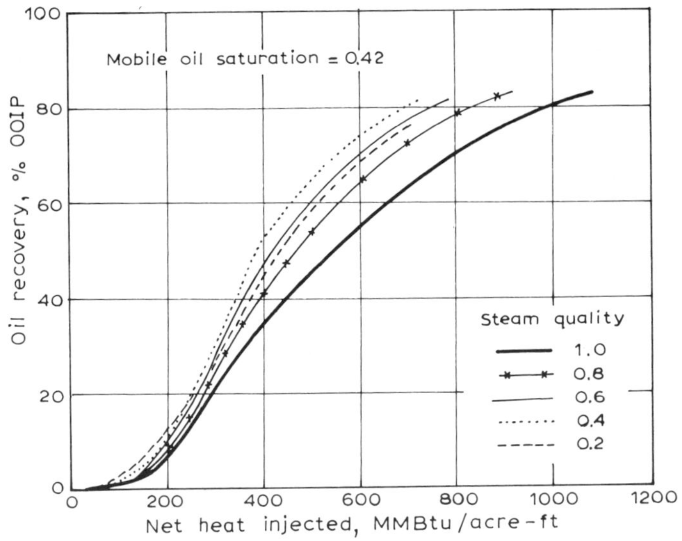

Fig. 11-7. Effect of steam quality on steamflood oil recovery. (After Gomaa, 1980; courtesy of the Society of Petroleum Engineers of AIME.)

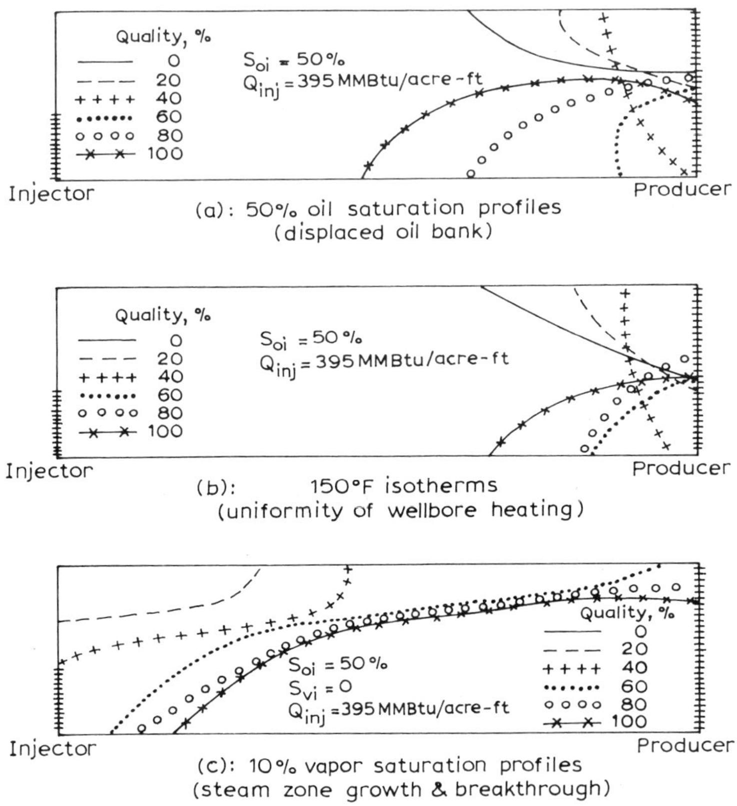

Fig. 11-8. Effect of steam quality on displacement parameters. (After Gomaa, 1980; courtesy of the Society of Petroleum Engineers of AIME.)

considered by him. Two factors may be responsible for this: the combined effects of steam volume and viscosity, and the vapor override and liquid underrunning in the reservoir. On increasing steam quality, steam volume increases, whereas steam viscosity decreases. Higher steam volume results in greater volumetric contact with the oil, whereas lower steam viscosity results in poorer displacement and sweep efficiencies due to a poorer mobility ratio. Thus, a countereffect is generated, leading to the existence of an optimum steam quality. The effect of vapor override and liquid underrunning is shown in Fig. 11-8. The oil bank and the temperature profiles (Fig. 11-8a, 11-8b, respectively) are very close to being vertical at the optimum steam quality of 40%. The steam saturation profile is also indicative of the recovery: higher as well as lower steam qualities result in greater bypassing.

Heat loss to overburden and underburden

Large reservoir thickness, high injection rate, and high steam quality result in low heat loss, thus favoring higher recovery efficiencies.

ECONOMICS OF THE STEAMFLOODING PROCESS

Apart from the technical criteria discussed earlier, certain economic criteria that must be considered are described in this section.

Profitability

Steamflood profitability depends mainly upon the steam/oil ratio (SOR) during the project. The ultimate SOR from a steamflood can roughly be predicted by the following equation given by Neuman (1975):

$$SOR = \frac{1}{2.7\phi(S_{oi} - S_{ors})f_{sdh}} \tag{11-50}$$

where ϕ is the formation porosity, fraction: S_{oi} is the initial oil saturation (when steamflooding is initiated), fraction; S_{ors} is the residual oil saturation in the steam zone, fraction; and f_{sdh} is the downhole steam quality, fraction. Empirical equations by Chu (1985) may also be used; for $SOR > 5.0$:

$$\begin{aligned}(SOR)^{-1} = &-0.011253 + 0.00002779D + 0.0001579h - 0.001357\theta \\ &+ 0.000007232\mu + 0.00001043(kh/\mu) + 0.5120\phi S_o\end{aligned} \tag{11-51}$$

and for $SOR \leqslant 5.0$:

$$\begin{aligned}SOR = &18.744 + 0.001453D - 0.05088h - 0.0008864k \\ &- 0.0005915\mu - 0.0002938(kh/\mu) - 14.79S_o\end{aligned} \tag{11-52}$$

where θ is the angle of dip in degrees.

The incremental cost per barrel of oil, ICBO, is approximately equal to:

$$\text{ICBO} = SOR(C_s I_w + N_w C_{lw})/I_w \tag{11-53}$$

where C_s is the steam cost, \$/bbl steam; I_w is the steam injection rate, bbl/day water equivalent; N_w is the number of producing wells; and C_{lw} is the combined lease and well cost, U.S.\$/producing well/day.

Economic life of the project

An economically successful flood must be profitable long enough to repay the investment costs and generate an acceptable return on investment. It is essential, therefore, to calculate the time taken by the steam to displace oil at the estimated rate. The project life time, t^* (days), can be calculated by using the following equation (Neuman, 1975):

$$t^* = \frac{\pi}{4}\left\{\frac{hL_v M_s}{(2Kc_w\,\Delta T/d_{th}^{1/2})\left[1 + (\pi^{1/2}/2)(M_s/M_w)(L_v/c_w\,\Delta T)f(T_c - T_o/\Delta T)\right]}\right\}^2 \tag{11-54}$$

where L_v is the heat of vaporization of water, Btu/lb; M_s is the volumetric heat capacity of the steam zone, Btu/ft^3-°F; M_w is the volumetric heat capacity of the water-saturated region, Btu/ft^3-°F; K is the thermal conductivity, Btu/ft°F-day; c_w is the heat capacity of water at steam temperature, Btu/lb-°F; ΔT is the temperature difference between steam and the unheated reservoir, °F; T_c is the critical temperature above which enough oil has been removed (the equivalence temperature, which gives the longest conceivable life time, is generally used), °F; d_{th} is the thermal diffusivity, ft^2/day; and $f(x)$ is the inverse complementary error function of the argument x.

Total cost of produced oil

The total cost of produced oil, TCPB, can be estimated as follows:

$$\text{TCPB} = \text{ICBO} + (SOR\ I_{ic})/(I_w t^*) \tag{11-55}$$

where I_{ic} is the investment cost, U.S.\$.

Optimum steam injection rate

It is not possible to derive a general equation for estimating the optimum steam injection rate, because there is no good way to predict what fraction of usable heat will be produced. Neuman (1975), however, has given an equation for estimating the optimum injection rate, $(I_w)_{opt}$, provided the rate of oil displacement, Q_o, is known:

$$(I_w)_{opt} = \frac{Q_o}{2.7\phi(S_{oi} - S_{ors})f_{sdh}} \tag{11-56}$$

The areal coverage of steam, A, can be calculated if the injection rate of steam is known:

$$A = \frac{6.1 I_w L_v t^{1/2}}{(2K/f_{sdh})\,\Delta T(L_v + c_w\,\Delta T)} \tag{11-57}$$

Equation (11-57) can also be used to estimate the injection rate that will give a complete areal coverage in a given time.

Well spacing

Well spacing is another factor that is critical to the performance as well as the economics of any enhanced recovery project. When a field is developed, the wells are drilled in such a way as to be suitable for secondary and tertiary recovery operations. This helps in avoiding drilling new wells when such projects are initiated.

The effect of well spacing on maximum recovery and net profits has been a disputed subject for many years. Advocates of closer spacing have shown that more wells drilled per acre have resulted in increased recovery from certain fields, whereas increased net profits and greater conservation of reservoir energy have been attributed by others to wider spacing of wells. A definite mathematical relationship exists between increased recovery and closer spacing; recovery is inversely proportional to the square root of the well density expressed in acres per well.

Arbitrary spacing regulations may impose definite handicaps on the adequate exploitation of small accumulations on sharp folds or against faults, and increase the ratio of dry holes to producers. In choosing well spacing, reservoir characteristics such as depth, permeability, and porosity, should be taken into consideration. If the reservoir is shallow and massive, closer spacing is generally adopted. If the spacing is closer, interference may occur. In secondary recovery operations, breakthrough of the injected fluid may occur too early in the life of the project, causing it to be an economic failure.

Another important consideration in selecting well spacing is drilling costs. It may not be economically feasible to drill many wells if the reservoir is at a greater depth. Thus, well spacing depends upon reservoir characteristics as well as economics. Generally, steam injection projects have a well spacing of 3–6 acres in the field, 2–2.5 acres for pilots, and 4–8 acres for full-scale patterns.

SUMMARY

Steamflooding is an important and economically efficient enhanced recovery process for heavy, viscous crudes, as proven in fields worldwide. The chances of success are very great under the following conditions:

(1) The crude in situ has a gravity of 10–36° API, with a viscosity around 1000 cP.

(2) The oil-in-place in the reservoir is at least 1200 bbl/acre-ft, with an average oil saturation in excess of 40%, and reservoir porosity greater than 20%.

(3) The reservoir depth is less than 5000 ft.

(4) Reservoir injectivity is high enough to permit steam injection.

(5) The formation pressure and reservoir productivity are sufficient to ensure at least moderate production rates.

(6) Reservoir pressure and temperature are such that an optimum steam temperature of 300–400°F can be maintained.

ACKNOWLEDGEMENTS

The authors would like to express their sincere gratitude to Dr. George V. Chilingarian, who initiated the idea of writing this chapter, and provided valuable suggestions and comments. His help has indeed been a major factor in the development of this work.

REFERENCES

Abdalla, A. and Coats, K.H., 1971. A three-phase, experimental and numerical simulation study of the steamflood process. *46th Annu. Fall Meet. Soc. Pet. Eng. of AIME, New Orleans, La., October 3–6*, SPE 3600.

Abrams, A., 1975. The influence of fluid viscosity, interfacial tension, and flow velocity on residual oil saturation left by waterflood. *Soc. Pet. Eng. J.*, 15(5): 437–447.

Baker, P.E., 1973. Effect of pressure and rate on steam zone development in steam flooding. *Soc. Pet. Eng. J.*, 13(5): 274–284.

Beggs, H.D. and Robinson, J.R., 1975. Estimating the viscosity of crude oil systems. JPT Forum, *J. Pet. Technol.*, 27(9): 1140–1141.

Carr, N.L., Kobayashi, R. and Burrows, D.B., 1954. Viscosity of hydrocarbon gases under pressure. *Trans. AIME*, 201: 264–272.

Casse, F.J. and Ramey, H.J., Jr., 1979. The effect of temperature and confining pressure on single-phase flow in consolidated rocks. *J. Pet. Technol.*, 31(8): 1051–1059.

Chierici, G.L., 1981. Novel relations for drainage and imbibition relative permeabilities. *56th SPE Annu. Tech. Conf. Exhib., Soc. Pet. Eng. of AIME, San Antonio, Texas, October 5–7*, SPE 10165.

Chu, C., 1985. State-of-the-art review of steamflood field projects. *J.Pet. Technol.*, 37(10): 1887–1902.

Coats, K.H., 1976a. Simulation of steamflooding with distillation and solution gas. *Soc. Pet. Eng. J.*, 16(5): 235–247.

Coats, K.H., 1976b. A fully implicit steamflood model. *51st Annu. Fall Meet., Soc. Pet. Eng. of AIME, New Orleans, La., October 3–6*, SPE 6105.

Coats, K.H., George, W.D. and Marcum, B.E., 1974. Three-dimensional simulation of steamflooding. *Soc. Pet. Eng. J.*, 14(6): 573–592.

Conservation Committee of California Oil Producers, 1979–84. *Annual Review of California Oil and Gas Production*. Conservation Committee of California Oil Producers, Los Angeles, Ca.

Davidson, L.B., 1969. The effect of temperature on the relative permeability ratio of different fluid pairs in two-phase systems. *J. Pet. Technol.* 21(8): 1037–1046.

Edmondson, T.A., 1965. Effect of temperature on waterflooding. *J. Can. Pet. Technol.,* 4(4): 236–242.

Ehrlich, R., 1970. The effect of temperature on water–oil imbibition relative permeability. *Soc. Pet. Eng. Eastern Reg. Meet., Pittsburgh, Pa., November 5–6,* SPE 3214.

Farouq Ali, S.M., 1974. Current status of steam injection as a heavy oil recovery method. *J. Can. Pet. Technol.,* 13(1): 1–15.

Farouq Ali, S.M., 1981. A comprehensive wellbore steam–water flow model for steam injection and geothermal applications. *Soc. Pet. Eng. J.,* 21(5): 527–534.

Gates, C.F. and Ramey, H.J., Jr., 1958. Field results of South Belridge thermal recovery experiment. *Trans. AIME,* 213: 236–244.

Geffen, T.M., 1973. Oil production to expect from known technology. *Oil Gas J.,* 71(19): 66–76.

Gomaa, E.E., 1980. Correlations for predicting oil recovery by steamflood. *J. Pet. Technol.,* 32(2): 325–332.

Gottfried, B.S., 1965. A mathematical model of thermal oil recovery in linear systems. *Soc. Pet. Eng. J.,* 5(3): 196–210.

Green, D.W. and Perry, R.H. (Editors), 1984. *Chemical Engineers' Handbook.* McGraw-Hill, New York, N.Y., 6th ed.

Iyoho, A.W., 1978. Selecting enhanced oil recovery processes. *World Oil,* 187(6): 61–64.

Jones, J., 1981. Steam drive model for hand-held programmable calculators. *J. Pet. Technol.,* 33(9): 1583–1598.

Katz, D.L. and Hachmuth, K.H., 1937. Vaporization equilibrium constants in a crude oil–natural gas system. *Ind. Eng. Chem.,* 29(9): 1072–1077.

Kesler, M.G. and Lee, B.I., 1976. Improve prediction of enthalpy of fractions. *Hydrocarbon Process.,* 55(3): 153–158.

Kumar, S., 1989. *Simulation of Steam and Surfactant Flooding.* Ph.D. Thesis, University of Southern California, Los Angeles, Calif. (submitted for approval).

Kumar, S., Torabzadeh, S.J. and Handy, L.L., 1985. Relative permeability functions for high and low tension systems at elevated temperatures. *55th Annu. Soc. Pet. Eng. Calif. Reg. Meet., Bakersfield, Calif., March 27–29,* SPE 13670.

Lauwerier, H.A., 1955. The transport of heat in an oil layer caused by the injection of hot fluid. *Appl. Sci. Res., Sect. A,* 5: 145–150.

Lewin and Assocs., Inc., 1976. The potentials and economics of enhanced oil recovery. *Fed. Energy Admin. Rep.,* B76/221 (April).

Lo, H.Y. and Mungan, N., 1973. Effect of temperature on water–oil relative permeabilities in oil-wet and water-wet systems. *48th Annu. Tech. Conf. Exhib., Soc. Pet. Eng. of AIME, Las Vegas, September 30–October 3,* SPE 4505.

Lohrenz, J., Bray, B.G. and Clark, C.R., 1964. Calculating viscosities of reservoir fluids from their compositions. *J. Pet. Technol.,* 16(10): 1171–1176.

Maini, B.B. and Batycky, J.P., 1983. Effect of temperature on heavy-oil/water relative permeabilities in horizontally and vertically drilled core plugs. *58th Annu. Tech. Conf. Exhib., Soc. Pet. Eng. of AIME, San Francisco, Calif., October 5–8,* SPE 12115.

Mandl, G. and Volek, C.W., 1969. Heat and mass transport in steam-drive processes. *Soc. Pet. Eng. J.,* 9(1): 59–79.

Marx, J.W. and Langenheim, R.H., 1959. Reservoir heating by hot fluid injection. *Trans. AIME,* 216: 312–315.

Miller, M.A. and Ramey, H.J., Jr., 1983. Effect of temperature on oil/water relative permeabilities of unconsolidated and consolidated sands. *58th Annu. Tech. Conf. Exhib., Soc. Pet. Eng. of AIME, San Francisco, Calif., October 5–8,* SPE 12116.

Myhill, N.A. and Stegemeier, G.L., 1978. Steam drive correlation and prediction. *J. Pet. Technol.,* 30(2): 173–182.

Nakornthap, K. and Evans, R.D., 1982. Temperature dependent relative permeability and its effect upon oil displacement by thermal methods. *57th Annu. Tech. Conf. Exhib., Soc. Pet. Eng. of AIME, New Orleans, La., September 26–29,* SPE 11217.

Neuman, C.H., 1975. A mathematical model of the steam drive process—applications. *45th Annu. Soc. Pet. Eng. Calif. Reg. Meet., Ventura, Calif., April 2–4,* SPE 4757.

Pacheco, E.F. and Farouq Ali, S.M., 1972. Wellbore heat losses and pressure drop in steam injection. *J. Pet. Technol.,* 24(2): 139–144.

Poston, S.W., Ysrael, S., Hossain, A.K.M.S., Montgomery III, E.F. and Ramey, H.J., Jr., 1970. The effect of temperature on irreducible water saturation and relative permeability of unconsolidated sands. *Soc. Pet. Eng. J.,* 10(2): 171–180.

Prats, M., 1982. *Thermal Recovery. Henry L. Doherty Series, Monograph No. 7.* Society of Petroleum Engineers of AIME, Richardson, Texas, 283 pp.

Price, H.S. and Coats, K.H., 1974. Direct methods in reservoir simulation. *Soc. Pet. Eng. J.,* 14(3): 295–308.

Ramey, H.J., Jr., 1962. Wellbore heat transmission. *J. Pet. Technol.,* 14(4): 427–440.

Samaroo, B.H. and Guerrero, E.T., 1981. The effect of temperature on drainage capillary pressure in rocks using a modified centrifuge. *56th Annu. Fall Tech. Conf. Exhib., Soc. Pet. Eng. of AIME, San Antonio, Texas, October 5–7,* SPE 10153.

Satter, A. and Parrish, D.R., 1971. A two-dimensional analysis of reservoir heating by steam injection. *Soc. Pet. Eng. J.,* 11(2): 185–197.

Shutler, N.D., 1969. Numerical three-phase simulation of the linear steamflood process. *Soc. Pet. Eng. J.,* 9(2): 232–246.

Shutler, N.D., 1970. Numerical three-phase model of the two-dimensional steamflood process. *Soc. Pet. Eng. J.,* 10(4): 405–417.

Shutler, N.D. and Boberg, T.C., 1972. A one-dimensional analytical technique for predicting oil recovery by steamflooding. *Soc. Pet. Eng. J.,* 12(6): 489–498.

Sinnokrot, A.A., Ramey, H.J., Jr. and Marsden, S.S., Jr., 1971. Effect of temperature level upon capillary pressure curves. *Soc. Pet. Eng. J.,* 11(1): 13–22.

Stone, H.L., 1968. Iterative solution of implicit approximations of multidimensional partial differential equations. *SIAM J. Num. Anal.,* 5(3): 530–558.

Stone, H.L., 1973. Estimation of three-phase relative permeability and residual oil data. *24th Annu. Tech. Meet., CIM, Edmonton, Alta, May 8–12.*

Todd, M.R., O'Dell, P.M and Hirasaki, G.J., 1972. Methods for increased accuracy in numerical reservoir simulators. *Soc. Pet. Eng. J.,* 12(6): 515–530.

Torabzadeh, S.J. and Handy, L.L., 1984. The effect of temperature and interfacial tension on water/oil relative permeabilities of consolidated sands. *4th SPE/DOE Symp. on Enhanced Oil Recovery, Tulsa, Okla., April 15–18,* SPE 12689.

Touloukian, Y.S., Kirby, R.K., Taylor, R.E. and Lee, T.Y.R., 1970. Specific heat—nonmetallic liquids and gases. In: *Thermophysical Properties of Matter, Vol. 6.* IFI/Plenum, New York, N.Y.

Weinbrandt, R.M. and Ramey, H.J., Jr., 1972. The effect of temperature on relative permeability of consolidated rocks. *47th Annu. Tech. Conf. Exhib., Soc . Pet. Eng. of AIME, San Antonio, Texas, October, 8–11,* SPE 4142.

Weinbrandt, R.M., Ramey, H.J., Jr. and Casse, F.J., 1975. The effect of temperature on relative and absolute permeability of sandstones. *Soc. Pet. Eng. J.,* 15(5): 376–384.

Weinstein, H.G., Stone, H.L. and Kwan, T.V., 1969. Iterative procedure for solution of systems of parabolic and elliptic equations in three dimensions. *I. & E.C. Fund.,* 8(2): 281–287.

Willhite, G.P., 1967. Over-all heat transfer coefficients in steam and hot water injection wells. *J. Pet. Technol.,* 19(5): 607–615.

Willman, B.T., Valleroy, V.V., Runberg, G.W., Cornelius, A.J. and Powers, L.W., 1961. Laboratory studies of oil recovery by steam injection. *J. Pet. Technol.,* 13(7): 681–690.

Chapter 12

OPERATIONAL ASPECTS OF STEAM INJECTION PROCESSES

MICHAEL PRATS

INTRODUCTION

This chapter discusses field operations in steam drives and cyclic steam injection projects. That is, it discusses the most common operational requirements and problems associated with these thermal recovery processes. It starts by considering the treatment of the water used to generate steam, and follows the steam as it enters the reservoir. The produced oil is followed until it is shipped out of the lease or field, and the produced water and gas are followed until they are properly disposed. A schematic diagram for the flow of the injected water and the produced fluids is shown in Fig. 12-1. In addition to the handling of the fluids, a number of areas related to the generation of steam and production of viscous oils using steam are also covered. Equipment related to thermal operations is also discussed, with emphasis on those items which are either specific or of special importance to thermal operations. Most of the equipment discussed is identified schematically in Fig. 12-1.

There generally is very little difference in either equipment or fluid handling between steamflood and cyclic steam injection operations. Two exceptions come to mind. From the nature of the cyclic steam injection process, which aims primarily at stimulating the oil production rate, it is understood that steam is injected into a reservoir through a well for a limited period of time, generally one to three weeks. The well is shut in to "soak" for perhaps as much as a few days, and then the well is placed on production until the oil production rate drops to a value such that a new stimulation treatment (i.e., another steam injection) appears worthwhile. This series of steps is repeated as warranted by the operator to maintain the oil production rate at adequate levels, sometimes over ten times, or cycles, from which the process takes its name. This heating and cooling of the wells is different from the conditions which prevail in a steamdrive. There, steam is injected continuously (barring mechanical problems) for a number of years, so that the wells are not intentionally subjected to as frequent significant temperature cycles. The repeated and wide temperature variations to which the wells are subjected during cyclic steam injection operations sometimes require that special attention be paid to their design and operation.

The other exception is the portability of the steam generation equipment preferred by some cyclic steam project operators. Permanent, centralized steam generator facilities are generally preferred in steamdrive operations where injec-

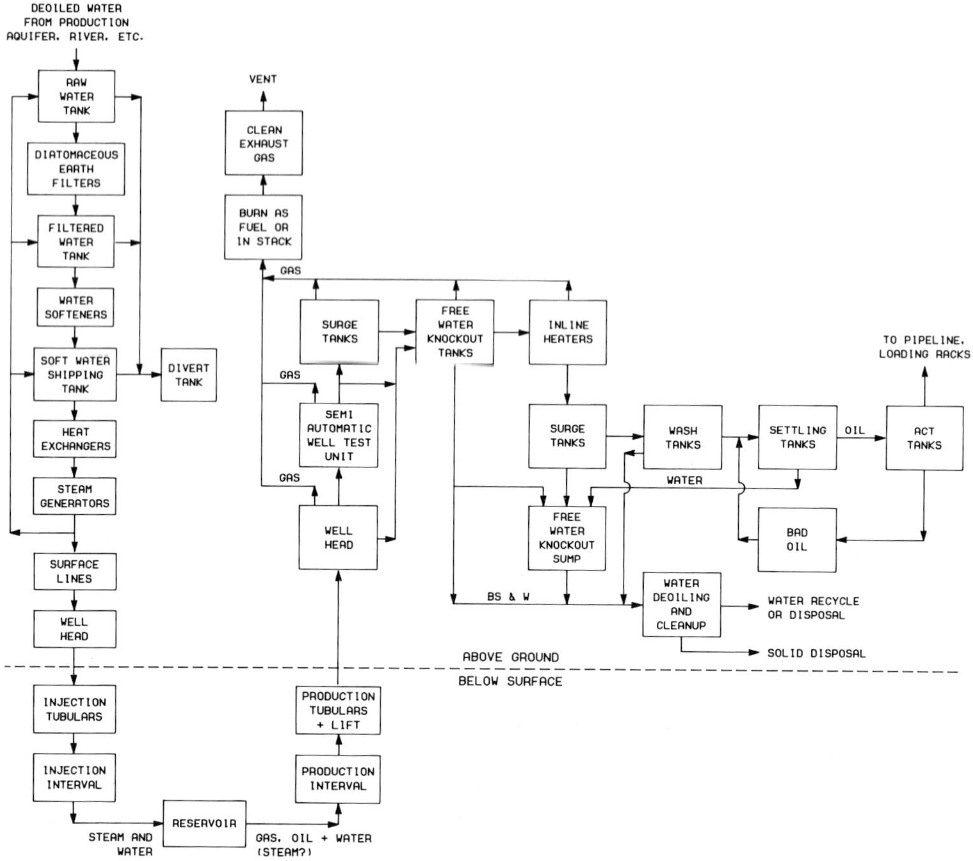

Fig. 12-1. Schematic flow diagram for water in steam injection operations. (After Prats, 1982, p. 138, fig. 11.1; courtesy of the Society of Petroleum Engineers of AIME.)

tion of steam is anticipated to continue for several years. In very large projects, a number of permanent steam-generating plants are distributed throughout the field to minimize the length of the steam distribution system, and thus reduce heat losses. These comments should be considered only as trends, however, as the economics of each project determines how it is designed and operated.

WATER TREATMENT FOR STEAM GENERATION

Steam generation requires clean water (Burns, 1965; McLennan, 1968; Roberts and Williams, 1970; Bradley and Gatzke, 1975; Prats, 1982): (1) it should not contain suspended solids, and (2) the dissolved minerals, liquids and gases must not be harmful to the steam generation equipment, the steam distribution system, and the wells.

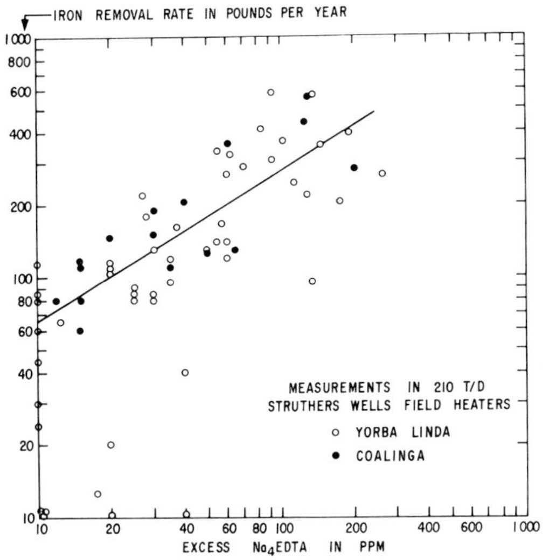

Fig. 12-2. Relationship between excess Na_4EDTA in feedwater and heater corrosion. (After Bradley and Gatzke, 1975, p. 171, fig. 1; courtesy of the Society of Petroleum Engineers of AIME.)

Basically, the water must not contain suspended or dissolved matter that would precipitate on or attack the tubes in the steam generator. Suspended solids tend to foul the water softeners and promote the formation of sludge. In addition, any solids passing through the steam generation system would have the tendency to accumulate at the face of the injection interval, thus reducing the injectivity of the well. Oil in the feedwater reduces resin life, and by promoting scale adherence to the tubes of the steam generators, tends to cause film boiling and subsequent coke formation.

The formation of scales that adhere to the tubes of the generators tends to reduce the heat transfer rates, thus promoting localized overheating and possible tube burnouts. Furthermore, any deposits forming on the tubes may promote and accelerate corrosion. The scales most commonly found contain calcium, magnesium, iron and silica. Chelating agents may be used to sequester any excess Ca^{2+} and Mg^{2+} cations passing through the water softeners. As shown in Fig. 12-2, however, the use of an excess of EDTA (ethylenediaminetetraacetic acid) may have a negative effect, as it appears to remove iron from the tubes in the steam generation system, which can lead to tube thinning and eventual leaks (Bradley and Gatzke, 1975.)

Because at least part of the feedwater entering a steam generator is converted to steam, the dissolved matter remaining in the liquid phase will tend to reach saturation point. Thus, minerals may precipitate even where solubilities are independent of temperature. The solubility of carbonates actually decreases with increasing temperature, thus aggravating the problem. Carbonates of limited solubility (those of calcium and magnesium) are present in all naturally occurring water sources. Dissolved gases, such as oxygen, carbon dioxide, and hydrogen sulfide can also affect the operation of the steam generators by fostering various types of corrosion or supporting bacterial life. The main problem of treating water is treating it as economically as possible while considering that insufficient or improper water treatment would likely result in increased maintenance costs of the steam generation equipment, and even in well workovers.

Water treatment is affected by the source and chemical composition of the water stream. The two are often related. Waters from rivers and lakes tend to have a high concentration of dissolved oxygen and suspended solids, and low concentration of dissolved solids. Subsurface waters tend to have the opposite characteristics.

Water produced from any oilfield often is more difficult and expensive to treat than water from other sources (Hull, 1967; Smith, 1971; Elias et al., 1980), because of its increased hardness, dissolved solids content, salinity, and oil content. It may, however, be the only water available in the required amounts. Also, its use may increase in those areas where it becomes increasingly difficult and expensive to meet any requirements for disposal of the water produced from the field operation.

It is generally appropriate to consider the following procedures even though not all of them are necessarily applied in every water-treating case:

(1) Use of bactericide. Usually liquid or gaseous chlorine, or formaldehyde, is used. Less frequently, methylene bisthiocyanate (MBT) and barium peroxide are used.

(2) Filtering out of the suspended solids and bacterial debris. The levels of suspended solids in the filtered water should be below 5 ppm, and preferably lower (API, 1983).

(3) Removing dissolved oxygen by adding catalyzed sodium sulfite or by stripping it with sweet natural gas or steam. Oxygen content should be less than 0.01 ppm (API, 1983). A slight excess of scavenging chemical is normal practice.

(4) Controlling scale-forming ions such as calcium, magnesium, and iron to less than 1 ppm of total hardness. Calcium and magnesium are removed by ion-exchange methods. Typically, a scale inhibitor is used as an additional precaution. Total dissolved solids should be maintained at levels where they will not precipitate (at the conditions prevailing in the steam generator) when 80% of the liquid water is converted to steam vapor. Silica can be maintained in solution by controlling the pH. In the absence of scaling ions, concentrations of the order of 150 ppm may be acceptable (API, 1983).

(5) Removing oil from produced waters by a combination of settling tanks or basins, flotation cells, and diatomaceous earth filters. Levels should be below 1 ppm (API, 1983).

(6) Ascertaining that the pH of the feedwater is in the range of 7–12. Control of the pH is often not required. A pH below 7 suggests the likelihood of acidic corrosion of the pipes, which can result in pickup of iron by the feedwater. Iron concentrations should be below 0.1 ppm, but lower limits may be required in the

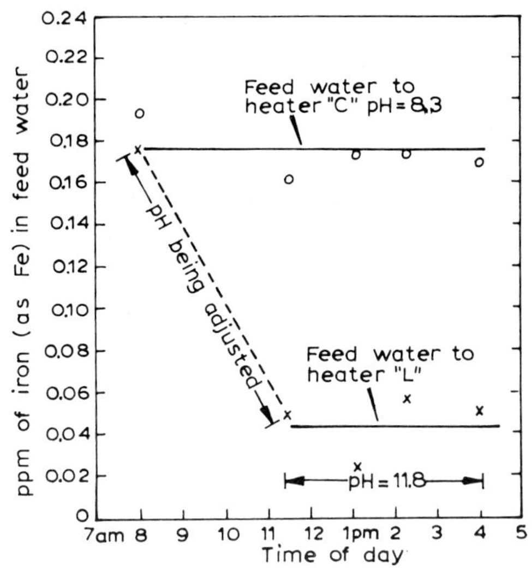

Fig. 12-3. Iron concentration in feedwater to heaters at different pH values. (After Bradley and Gatzke, 1975, p. 176, fig. 10; courtesy of the Society of Petroleum Engineers of AIME.)

presence of high concentrations of silica (API, 1983). Fig. 12-3 shows the effect of pH on the iron concentration in the feedwater in one California operation. On the other hand, a pH above 13 may lead to caustic corrosion or embrittlement.

The sulphate-reducing bacteria sometimes found in the subsurface waters can generate hydrogen sulfide. In the presence of iron, iron sulfide is formed. It tends, along with bacterial bodies, to plate out on the ion-exchange resin in the water softeners. This reduces the effectiveness of the ion-exchange resin. Suspended solids, including bacteria, and oil in the water can deposit in the ion-exchange resin beds and in the steam generators and reduce their effectiveness. Iron compounds can affect adversely the operation of ion-exchange resins (and are difficult to remove from them), and soluble iron compounds can collect in the generator pipes and impair their heat transfer. Removal of oxygen and acid gases (like carbon dioxide and hydrogen sulfide) reduces corrosion rates. Silica deposits and scale can reduce heat transfer and cause hot spots in steam generator pipes.

If the casing of a water well should prove to be a source of iron, cathodic protection may help. Non-metallic flowlines from the water source to the treatment plant (or plastic coatings) may be used where their strength is adequate at the maximum operating pressure and temperature. Any corrosion inhibitor used in the feedwater should be checked to determine that it does not foul the resins, as do some type of amines and diamines. If harmful to the system, such chemicals may be introduced as close to the steam generator as convenient, downstream of the water softeners or any other part of the equipment which they may affect. *

Inasmuch as not all of these problems necessarily exist in every operation, not all of these six steps listed may be required to maintain a high degree of efficiency. In certain circumstances, however, water-treatment facilities may be more complex than indicated.

The softened water usually is gravity-fed from a storage tank to a centrifugal pump to distribute the water to individual steam generators. Positive displacement pumps (the feedwater pumps), operating at adequate pressure to prevent cavitation and satisfy the required net positive suction head are then used to force the water through the steam generator. The storage tank preferably should be galvanized and have a nitrogen blanket to reduce iron pickup and dissolved oxygen, respectively.

STEAM GENERATORS

Types of steam generators

Steam generators used in oilfields differ significantly from conventional boilers. Conventional boilers are generally used to generate saturated and even superheated steam to drive impellers. Because of the high fluid velocities, it is necessary to separate the vapor from the liquid before directing the steam at the impellers, for otherwise the liquid drops would damage them. Alternatively, superheated steam can be used to avoid liquid–vapor separation. This separation is done by means of revolving drums, making use of the centrifugal and inertial forces resulting from their rotation. The condensed steam is collected downstream of the impellers for reuse, requiring little makeup water. Once the initial inventory of water is treated, the additional treating costs are limited to those associated with the makeup water. On the other hand, oilfield operations require large quantities of steam for continued and long-term injection into reservoirs. Inasmuch as there is essentially no steam condensate available for reuse, this requires that the cost of treating water be relatively low.

Steam generators of the once-through or single-pass type, also known as wet-steam generators, are used almost exclusively in the oilfields (Burns, 1965; Fanaritis and Kimmel, 1965; API, 1983). They were specifically developed for oilfield applications in the early 1960's, and differ from a true boiler in that they have no steam-separating drum and do not require recirculation and blowdown. Because the generators do not have a separating drum, the maximum steam quality must be limited to about 80% to prevent the precipitation and deposition of dissolved solids (scale) on the the coils, thus reducing the likelihood of localized film boiling and subsequent tube failure. There are indirectly-heated generators, however, that use unsoftened or produced water as feed (Hull, 1967). These types of steam generators have not gained wide acceptance.

---

* In addition, some corrosion inhibitors change the wettability of reservoir rocks. (Editorial comment)

Although essentially all field steam generators today are fired with either natural gas or oil, there is at least one which is coal fired (Jones and Davis, 1983), and others are at various stages of design and construction. The trend in the design is in the direction of using fluidized beds, which appears to offer the potential for providing uniform and low heat fluxes, minimum heat retention, suitable response to load changes, low tube corrosion rates and fouling tendencies, removal in solid form of the sulfur burned from the fuel (which if proven would eliminate the need for scrubbers), and low levels of nitrogen oxides. The incentive is the low cost of coal relative to the fuels normally used. It will be a few years before a definite trend in the construction of coal-fired steam generators for field use is established.

Another factor that will probably change the type of steam generation equipment used, at least in the United States, is a recent law affecting cogeneration. This law provides incentives to generate both electricity and steam for commercial use (PURPA, 1980). It has already resulted in the field use of true steam boilers to generate electricity and ultimately provide steam for oilfield enduse (Davis and Young, 1984). Many other systems are at the design or construction phase (Stambler, 1983). A basic configuration consists of a conventional boiler to generate electricity, which the local electric company must buy at close to their avoided cost, i.e., the cost they would have had to incur to increase their capacity by the amount provided. The spent steam and stack gas are used to generate about 80%-quality steam by means of heat exchangers. Thus, the steam for oilfield use is still generated in the equivalent of the ordinary once-through steam generators, the difference being that these are not fuel-fired. Then the effluent steam is transported conventionally and injected into reservoirs (or used elsewhere). But well over 90% of the steam currently used for steam drives and cyclic steam injection processes is still generated by means of oil-fired, single-pass oilfield steam generators.

Fig. 12-4 is a photograph of a typical unit, whereas Fig. 12-5 is a schematic representation of an oil-fired, once-through steam generator. Treated water is delivered to the generator at the pressure necessary to inject the steam into the wells by the feed pump, which usually operates at a constant rate. Water is bypassed to the upstream side of the pump if the wells take less water than is fed by the pump. In the schematic process diagram shown in Fig. 12-5, water under pressure is first forced through a feedwater preheater, raising the temperature to 280°F when it enters the top part of a heat exchanger (also known as an economizer) in the top part of the convection section, counter-current to the flow of gas, to improve the heat efficiency. Preheating of the water is not necessary where the heat exchanger in the convection section is designed to be contacted directly by cold water. In the convection section, heat present in the escaping flue gas is used to further increase the temperature of the feedwater. At a point in the heat exchanger where the water temperature is 477°F, the water is diverted to the water preheater, from which it returns at a temperature of 280°F for additional heat scavenging from the convection section.

Fig. 12-4. A typical steam generator. (Courtesy of Struthers Thermo-Flood Corporation.)

The hot flue gas, entering the convection section at 1600°F, first contacts the bare lowermost tubes of the heat exchanger. The flue gas, at a reduced temperature, then contacts a series of finned tubes. Finned tubes, which improve the heat exchange rate to the feed water, cannot be used effectively at the entrance to the convection section because of increased corrosion at the high temperature present there. Packing of the fins with fly-ash would also be aggravated at the conditions prevailing at the entrance to the convection section. The preheated water next enters the radiant section of the steam generator, in which the arrangement of the heat exchange tubes or coils and of the burner are designed to maximize the enthalpy of the steam and reduce transverse heat losses. In this case, the generator delivers steam at a pressure of 1000 psia and a saturated temperature of 544°F.

Steam generators can be equipped with various instruments and devices to ensure the safe and efficient delivery of steam. Sensor and control devices are used to automatically adjust the rates of fuel, air, and water injection to generate steam of given quality at predetermined rates and at sustained high efficiencies. Important elements of steam generators are automatic monitoring and safety shutdown features. These include instruments to control steam quality, excessive tube temperatures, hard water, power failure, blower failure, low feedwater rates, and high and low steam pressure or temperature. They protect the equipment by shutting the fuel supply if the feedwater stops flowing,

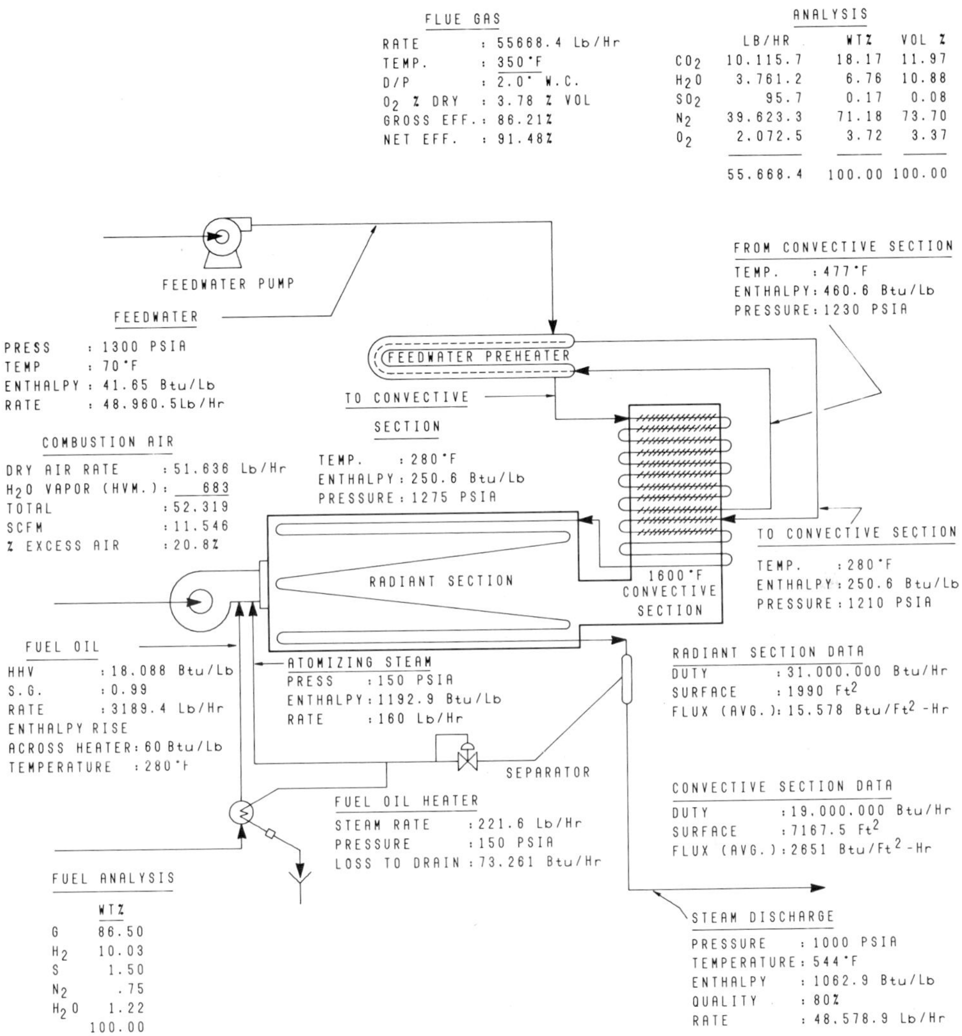

Fig. 12-5. Process flow diagram and thermodynamic heat balance for a 50×10^7-Btu/hr steam generator. (After Prats, 1982, p. 139, fig. 11.4; courtesy of the Society of Petroleum Engineers of AIME.)

if there is flame failure, or if there is some other malfunction or abnormal condition.

Typically, steam generators experience frequent shutdowns. These are due to the nature of the sophisticated automatic equipment used to protect them, which must fail on the safe side. An alarm then sounds to alert personnel to the fact that the unit is down and needs to be reactivated. Such frequent shutdowns and restarts result in cyclic thermal stresses on the refractory material and the coils. As a result, the refractory material may develop cracks, allowing the flue gas to contact the outer metal shell of the generator, which can cause skin failure due to corrosion. If holes should develop, then not all the flue gas would exit through the stack. Even if holes do not develop in the shell, its surface temperature would be higher than normal, and would result in an increase of heat losses from the steam generator. Also, expansion and contraction of the coils may force out any sealing packing from the openings in the steam generator shell, through which the coils exit and reenter, with analogous consequences.

Steam generator failure can be kept to a minimum with the proper use of fail-safe instruments, especially at the start of a project. As with any piece of equipment, the design affects the operating difficulty and the downtime required for repairs, as well as the frequency of repairs. Where possible, it is advisable to compare the refractory and tube repair records of competing equipment when considering the purchase or lease of steam generators. The API has prepared recommended practice for specifications, installation, commission, operation, and maintenance of wet steam generators (API, 1983).

Efficiency of steam generators

Typical efficiencies of steam generators are about 80–85%, and can exceed 90% with special equipment design. Most of the lost heat goes through the stack. The amount of heat lost via stack gases is directly proportional to the temperature and to the mass flow rate of the vented gas. Accordingly, the stack heat losses are reduced by using heat exchangers in the convective section to preheat the feedwater, as already discussed, and by using special burners that allow reducing the excess air to less than 5%. The amount of heat recoverable from the exhaust often is limited by the corrosive effects of sulfur oxides contained in the flue gas, which tends to condense and form acids on the stack as the temperatures decrease. Thus, it is normal to operate stacks at temperatures slightly higher than the acid dewpoint, about 350°F under typical stack conditions, even though it limits the efficiency of the generators when sulfur-containing fuels are used. Some manufacturers can provide special alloys for use in the stack section to withstand corrosion, thus allowing a higher thermal efficiency.

Efficient flame patterns in the radiant section of the generator are obtained easily with gaseous and liquid fuels. It is important, however, for liquid fuels to have constant properties (such as viscosity, volatility, heating value, and dissolved gas and water content) to reduce the frequency of automatic shutdowns. Coal-fired steam generators are not common because of intrinsic drawbacks related to burner design and to reduced heat transfer to the coils resulting from soot coatings. These problems are being resolved (Makansi and Schweiger, 1982; Plass and Daradimos, 1984), however, and one such generator has been in operation since 1980 (Jones and Davis, 1983).

Loss in the thermal efficiency of a unit is usually an indication that heat transfer to the coils is impaired. In gas-fired heaters, where soot should not be a

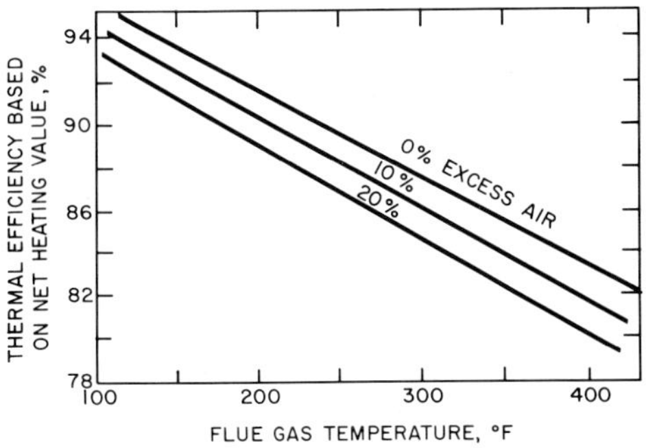

Fig. 12-6. Thermal efficiency versus flue gas temperature. Net heating value of fuel oil = 17,500 Btu/lb (9722 cal/g); gravity = 15° API. Thermal efficiency has been reduced by 2% to allow for steam generator setting radiation loss. (Based on information provided through the courtesy of Struthers Thermo-Flood Corporation.)

problem, this loss in thermal efficiency generally indicates that scale has formed inside the coils. Scale is usually removed by circulating dilute acid solutions through the unit until the effluent acid concentration indicates the reaction is complete. The unit is, of course, "cold" during such treatment. When the tubes are coiled vertically, as opposed to horizontally as shown in Fig. 12-5, high circulation rates are preferred to minimize accumulation of hydrogen against the top portions of the coils, and to assure even acidizing.

The recommended API (1983) practice is to base the rated thermal efficiency of steam generators "on the higher heating value of the fuel to be burned, the specified exhaust gas temperature, and the excess oxygen or excess air in the flue gas." Furthermore, "the higher heating value of a fuel is defined as the heat evolved, as measured by a constant pressure calorimeter, when a fuel is completely burned at stochiometric conditions and the products of combustion are cooled to 60°F and the water vapor produced is completely condensed to liquid at that temperature." The API, however, also allows the rated efficiencies to be based differently. But thermal efficiencies determined during the course of a project to rate the performance of the steam generators should be based on the same criteria.

Typical thermal efficiencies, based on the ratio of the total heat losses (the convective ones through the stack and the radiant ones from the body of the generator) to the lower (net) heating value of the fuel, are shown in Fig. 12-6 as a function of the excess air and the flue gas temperature, for the case where the fuel is a 15°API crude. As can be seen, the efficiency increases as the temperature of the stack gas decreases. This has led at least one manufacturer to make available a special add-on convection section with stainless steel or plastic-coated heat exchange surfaces, which can cool the exhaust gases to about 150°F even when burning sulfur-bearing crudes. This can increase the thermal efficiency by about 8–10%, while at the same time reduce the amount of cooling required in the scrubbers (where these are used).

Sizing steam generators

Steam generators are generally available at three nominal coil pressure ratings: 1000, 1500, and 2500 psig. For pressure ratings of 2700–2800 psig, molybdenum steels are used, and stainless steel is required for higher pressures. More commonly, the ratings of the positive displacement feedwater pumps limit the pressures that can be made available at the generator.

Typical capacities of steam generators range from 10×10^6 to 150×10^6 Btu/hr, corresponding to rates of about 650 and 10,000 BWPD (barrels of feedwater per day) converted to 80%-quality steam. Generators are not normally designed to operate at less than half their maximum rating. Units up to about 40×10^6 Btu/hr are available on trailer mounts, and essentially all field units are available on skid mounts. Major suppliers also provide complete portable water treating facilities, as well as equipment to meet any unusual operating and pollution control requirements.

It is relatively important to size properly steam generators when the amount of steam needed is small. There is no need to buy more than what is necessary. For example, if it is necessary to treat each well in a cyclic operation with 10,000 bbl of equivalent steam every year (on the average), and the injection period per well is between one and two weeks, then a small generator would be more than sufficient to treat as many as 30 wells. In this case only 300,000 bbl of equivalent steam are needed per year, and a 1000-BWPD steam generator would provide that with a little spare capacity, allowing for some downtime for maintenance. Such a steam generator is small enough to be trailer mounted and easily moved from well to well. The advantage of moving the generator is to save on surface distribution lines and heat losses from them; however, it results in increased maintenance because of a greater likelihood of developing cracks in the refractory material or leaks at flanged joints.

Where a large number of wells is involved, the size of individual generators loses significance, because one or more steam-generating plants are located throughout the oilfield, each plant consisting of several steam generators. The choice of generator sizes is determined then primarily on the basis of overall cost effectiveness. Even where central plants are available, portable steam generating units may be used to provide steam to wells in peripheral areas, where a few wells may be too far from the plants to receive from them the required steam quality, or to provide additional steam on a short-term basis. In steamdrive projects, the steam generator capacity is determined from the desired steam injection rate per injector and the number of injectors to be used at any one time. Here, it is important to recognize that steam requirements will decrease with time as the injected steam reaches the producers. This means that if the steam-generating capacity were to be sized to start the entire project at one time, there would be idle steam capacity (or an inefficient steamdrive operation) soon after steam breakthrough. For that reason steam generating capacity in steam drives is undersized relative to the total project potential, unless there were other uses for it in other projects as the steam capacity exceeds the efficient demand of the main project. Normally, the steam capacity that becomes available, as parts of a steam drive mature, is used to develop new areas of the field. Obviously, the poorer parts of the field are developed last.

In small projects, consideration should always be given to having two or more steam generators instead of one of equivalent capacity. This increases the likelihood that at least some steam is available at all times; for example, even when one generator would require unforeseen or unscheduled maintenance.

Scrubbers

Where oil is burned to fire the steam generators, the stack gases leaving the convection section contain particulate matter, and sulfur and nitrogen oxide gases. The rate at which these emissions may be vented to the atmosphere is regulated in many communities. Produced crude from the field and natural gas are the fuels most commonly used. Gas yields a cleaner stack gas, seldom requiring treatment. In the U.S.A., most steam generators are fired with produced crude. Inasmuch as the bulk of steam injection is for the recovery of heavy crudes and these always contain sulfur, sulfur dioxide is one of the major contaminants removed from most U.S. operations in which produced crude is burned.

There are three main sources of the particulate matter in the stack gas: (1) metals bound to the molecular structure of the crude; (2) small amounts of unfiltered solids in the crude; and (3) particles of the refractory material lining the radiation and convection sections of the steam generator, which are eroded and abraded by the high gas velocities. This particulate matter is known as fly-ash. Because of the contribution of the refractory material, even the use of gas as fuel would result in some fly-ash, but usually at acceptably low concentrations.

The requirements for emission control vary from one community to another, and change with time. As an example of the emission levels permitted in 1984 in Kern County, California, the emission of sulfur and particulates from a nominal 50×10^6 Btu/hr steam generator was limited to less than 1.88 and 3.6 lb/hr, respectively.

These low levels of sulfur emission rates are attained by passing the stack gas through scrubbers (Wendt, 1978; Sachtschale and Dydo, 1982). Components in scrubbing units may be of the dry or wet type, but most scrubbing units in current use in California are made solely of wet components. The dry components in these units are not scrubbers themselves. They are used solely for the removal of the fly-ash, and consist of a number of filter bags placed in a baghouse at the entrance of the scrubbing unit. In some scrubbing units, there are hundreds of such filter bags arranged in parallel and series configurations. These filters are messy to replace because of the dust, but are relatively efficient and present economical ways to trap particulates.

The wet (scrubbing) components are used both to entrain the particulates in the aqueous stream and to remove the sulfur dioxide from the stack gas. The particulates are entrained as a result of wetting (by contacting them with water). The sulfur dioxide is removed from the stack gas by absorbing and reacting it in an alkaline solution. To this end, different types of wet components are used in scrubbers: tray towers, packed towers, spray towers, and ejector venturi. Although a scrubber unit could in principle be made up of a single component, say a spray tower, most of the scrubbers in current use in oilfields are made of of two or more of these components. Brief descriptions of these components are presented below:

Tray tower. The flue gas from the generator convection section is forced to enter the bottom of the tray tower. The alkaline solution flows counter-current to the rising gas, dissolving and neutralizing the sulfur dioxide. As in any tray tower, the staggered trays provide a large surface for these mechanisms to occur, while at the same time increasing the path length and residence times of the gas in the tower. This increases the efficiency with which the neutralization takes place.

Packed tower. Instead of trays, the surface area and residence time required for the neutralization are obtained by filling the tower with industrial packings such as Raschig rings and Berl saddles.

Spray tower. Alkaline water is sprayed through nozzles to contact the rising gas in an otherwise empty tower.

Ejector venturi. An alkaline aqueous solution is injected into the top of a large venturi tube to draw the flue gas into contact with the solution.

These four techniques are of the wet-contact type, and are the most commonly used because the aqueous solution also is effective in removing the particulate matter. Sodium and ammonium hydroxides (NaOH and NH_3OH), anhydrous ammonia (NH_4), and sodium and calcium carbonates (Na_2CO_3 and $CaCO_3$) are some of the chemicals commonly used to make the alkaline solutions. The choice is governed by local costs, but the chemical is normally sodium hydroxide (caustic soda) or carbonate (soda ash), which have the advantage of being relatively easy to mix. The carbonates are relatively inexpensive to buy, but have slightly higher maintenance costs because of scaling inside the scrubber. All scrubbing solutions contribute some small amount of particles (mainly sodium salts) to the effluent stack stream.

A schematic of a typical scrubbing unit is shown in Fig. 12-7. The flue gas is forced through the scrubber unit by means of a centrifugal variable-speed blower. This fan is designed to operate at relatively high temperatures in a corrosive environment and with some buildup of particulates on its moving parts. Hot gases are cooled quickly upon entering the scrubber by contacting them with the recycle liquid. The trays provide an alkaline solution blanket through which the flue gas must then pass, as shown schematically in Fig. 12-8. The trays have self-adjusting valves which open on impingement of the flue gas stream. The high velocity of the flue gas through the valves and the overlying

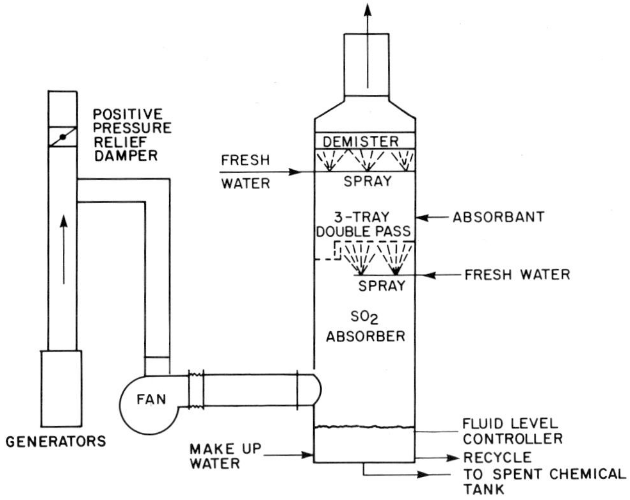

Fig. 12-7. Schematic diagram of typical scrubbing unit (based on information provided through the courtesy of AirPol, Inc.).

alkaline liquid blanket forms a froth, which provides efficient contact between the flue gas and the scrubbing solution and results in efficient removal of the SO_2. The clean flue gas then passes through chevron-type demisters to trap water droplets before being exhausted through the stack. The chevrons and the lower trays are sprayed with fresh water to reduce salt buildup and plugging. The absorbent solution is recycled from the bottom of the scrubbing unit to the trays, with fresh water added to make up for moisture losses in the effluent gas.

This particular unit makes use of soda ash (Na_2CO_3) as the base, which reacts with sulfur dioxide to form sodium sulfite and carbon dioxide. In the presence of water, sodium sulfite reacts with sulfur dioxide to form sodium bisulfite, which

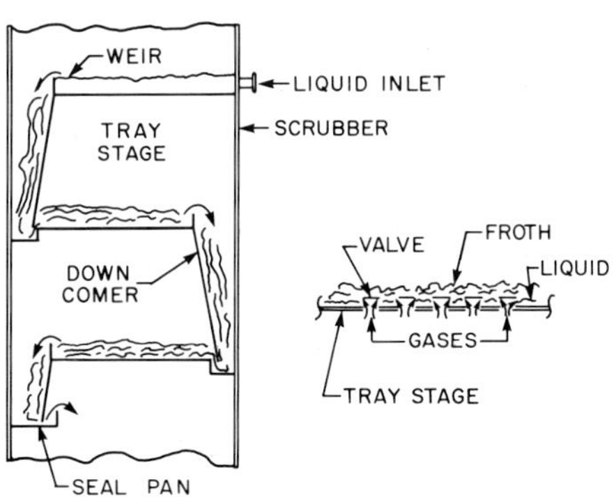

Fig. 12-8. Detail of flue gas passing through trays (based on information provided through the courtesy of AirPol, Inc.).

in turn yields sodium sulfite, carbon dioxide and water upon reaction with soda ash. The ratio of sodium sulfite and sodium bisulfite present depends upon the pH of the system, with the percentage of the absorbent (sodium bisulfite) increasing as the pH increases. In this system the sulfur is removed by bleeding the recycle stream to maintain its density at a predetermined level, while soda ash is added to the system to maintain the proper level of absorbent.

Another system, not as common, is the double-alkali scrubber. Here, a portion of the recycle solution is taken from the scrubber unit to a tank. The sodium bisulfite in the solution is reacted with hydrated lime (calcium hydroxide) to form the sulfites of calcium and sodium. The sodium sulfite solution and its suspended sulfite and sulfates of calcium then go to a thickener tank, from which the sodium sulfite solution is reintroduced into the recycle absorbent stream (Sachtschale and Dydo, 1982). The sludge containing calcium sulfite and sulfate precipitates and is then passed through a filter to reduce excess moisture, thus forming a thick paste which can be disposed as a solid waste to an approved landfill.

Typically, scrubbing units remove about 95% of the sulfur dioxide and from 30 to 60% of the particulates. Special designs can remove as much as 90% of the particulates. The temperature of the gas leaving the scrubbing unit is about 140°F, which is the dew point of the sulfur dioxide at the reduced concentration and conditions in the effluent stream. The humidity of the gas is about 18%. The level of SO_2 in the effluent gas stream is affected by the total sulfur concentration in the absorbent solution, and the pH and temperature of the solution. Fig.

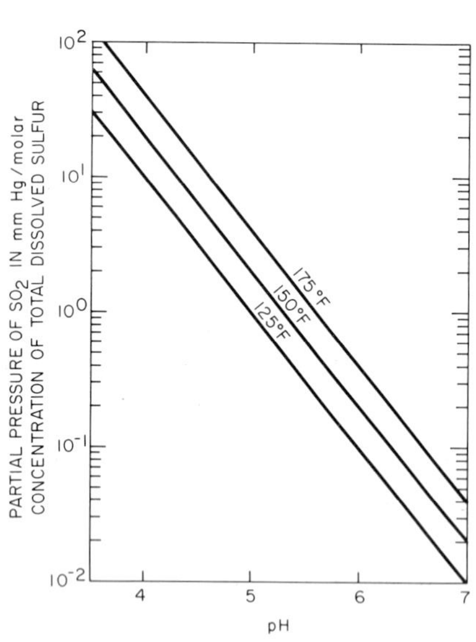

Fig. 12-9. Effect of pH, temperature, and total dissolved sulfur on the partial pressure of SO_2. (After Oestreich, 1976, p. 13, equation 17.)

Fig. 12-10. Installed scrubbers. A cloud of clean steam effluent is typical—there are no pollutants in the steam. (Courtesy of Struthers Thermo-Flood Corporation.)

12-9 shows the calculated relationship between these variables (Oestreich, 1976). At a fixed temperature, as can be seen, high pH favors reduced concentrations of SO_2 in the stack gas. Generally each scrubbing unit is designed for each specific application, taking into account the expected fuel and air injection rates, the characteristics of the steam generator, and the composition of the fuel. Some manufacturers are preparing the more popular units as off-the-shelf items.

Fig. 12-10 shows a bank of several vertical scrubbing units being evaluated under field use in California. Vertical units are used to conserve space in permanent installations. Horizontal units are easier to maintain and move from one location to another.

Oxides of nitrogen can usually be maintained at acceptable levels in low NO_x burners by monitoring temperatures and oxygen concentrations and controlling the air/fuel ratio in the initial combustion region in order to prevent excessively high temperatures. Additional air is introduced further downstream to complete the combustion, while at the same time reducing the excess air to a minimum. Reducing excess air helps by itself. Other methods include the recirculation of the flue gas, injection of ammonia, and catalytic conversion.

The efficiency of a generator is commonly determined from the enthalpy of the produced steam relative to the total energy used to generate it. The enthalpy of the produced steam is affected by its pressure and quality. Of these, the steam quality is the most difficult to determine accurately (Pryor, 1966; Spink, 1975; Wilson, 1976; White and Moss, 1983). Enthalpies could, in principle, be measured with a condensing calorimeter. But passing the entire steam stream through a calorimeter is seldom attempted in the field, as it requires a large and specially built unit which would be difficult to operate under field conditions. Sampling part of the stream and passing it through a calorimeter is done sometimes. It is notoriously difficult, however, to obtain a representative sample from a two-phase stream, thus affecting the significance of any quality determination based on sampling a wet-steam stream.

As long as there are no branches in the effluent line, the steam quality at any point downstream of a generator operating at stable conditions commonly is calculated as the increase in the chloride concentration in the liquid phase relative to that in the feedwater, divided by the chloride concentration in the liquid phase at the point of interest. The method is based on the conservation of the chloride ion. Any other ion that remains in solution and does not react or decompose also could be used. Sampling problems may be avoided by separating the liquid and vapor from the entire stream, and recombining them downstream of the point at which the sample is taken for the chloride content determination. Concentrations are best determined chemically. Inasmuch as for a particular water-treating method and steam-generation system the concentration of any ion is often a measure of the total amount of dissolved solids, steam qualities measured chemically are often used to provide a correlation with more easily determined quantities such as electrical conductivities and dissolved solids.

Orifice meters, which are familiar metering instruments in oilfields and do not require sampling or phase separation, are preferred by some operators. Then the steam quality is given by the square of the ratio of the apparent mass flow rate through the orifice meter (assuming the stream is 100% steam vapor) to the mass feedwater rate (Pryor, 1966; White and Moss, 1983). The steam qualities obtained with orifice meters also can be correlated with those obtained chemically.

Beyond the first split in the surface lines, there is no easy way to determine steam quality with any degree of confidence. This again is the result of not knowing how the phases split between the branches. The use of orifice meters, which is the only conventional technique that samples the entire stream, requires knowledge about the total mass flow rate in the portion of pipeline of interest, which is not available. Because of the lack of appropriate, easy-to-use, and relatively accurate procedures for determining the steam quality anywhere in the surface distribution system, most operators don't attempt such determinations. This still remains one of the unsolved areas of steam injection operations. Instead, estimates of the rates and qualities of steam delivered to individual wells are based on orifice meter readings coupled with calculated heat losses from the last point at which a valid steam quality determination has been made, and assuming that the quality of the steam is the same for both branches of every split in the surface distribution lines (White and Moss, 1983). The total steam injection rate and average quality injected is a reliable number, but the distribution among wells must be considered nominal when done in this fashion, with a relatively high degree of uncertainty. There also are no tools or methods for measuring the steam quality within an injection well. Reductions in the steam quality from the value at the wellhead are usually calculated from wellbore heat losses (Prats, 1982).

Surface lines

Steam distribution lines from the steam generators to the wellheads of the injectors are at appreciably higher pressures and temperatures than those from the production wells to the separation facilities. Schedule 80 pipeline is more than adequate for the lines from the production wells, and are generally used for the steam distribution lines as well where the saturated steam pressures are below about 1500 psi. Splits in the steam distribution lines should be designed so that both phases of the wet steam have the same tendency to go on either side of the legs. A minimum requirement is for the incoming steam stream to meet a "T" dead on, with the legs level (horizontal) to minimize gravity separation effects. Although this approach may split the stream adequately, it may result in accelerated abrasion of the "T" joint.

Even though heat losses from surface lines in steam injection operations may be a small fraction of the total heat injected, it is generally worthwhile to insulate them to reduce fuel costs, and as a safety precaution. Accordingly, surface lines, and especially those from steam generators, are usually insulated (Trujano et al., 1974). Lines may be buried, laid on the ground, or supported off the ground. This last approach is preferred for long-term installations because allowances must be made for expansion due to changes in temperature. This usually is done by providing various configurations of loops and anchors and installing sliding shoes at intermediate points (Owens and Bramley, 1966). The objective of these loops is to ensure that the stresses on the lines that could result from either cyclic steam injection operations or the unanticipated shut down of steam would be tolerable. Parted steam injection lines are not only potentially hazardous to personnel, but could also cause rapid depressuring of the reservoir near an injection well, which in unconsolidated reservoirs often results in sand-outs and expensive well workovers. When uninsulated lines are buried, it is often advisable to lay them in a sand-filled large-diameter conduit pipe to reduce excessive heat losses through wet soil.

Heat losses are estimated at steady-state conditions in most oilfield operations. The steady state rate of heat loss is directly proportional to the length of the pipe and the temperature difference, and is inversely proportional to the overall thermal resistance of the system. The temperature difference normally is known. The high temperature is that of the fluid being handled. The low temperature is the ambient one at the surface of the earth. Generally, the mean ambient temperature over the period of interest is used to account for diurnal and even seasonal variations.

For insulated pipes supported off the ground, the thermal resistance is primarily that of the insulation, which is inversely proportional to its thermal conductivity. For uninsulated pipes, the thermal resistance would be very low, and heat would be lost from the surface of the exposed pipe by radiation and convection. Typically, a few inches of weather-proofed calcium silicate insulation reduces the rate of heat losses by a factor of 20 to 100. Annual savings resulting from insulation typically amount from 1 to 20 bbl of fuel oil per foot of line, depending on the line diameter, the type and thickness of the insulation, and the temperature levels (Prats, 1982).

WELLS

Casing design considerations

Discussion of well designs in this section is limited to the effect of temperatures on thermal stresses in the casing and, briefly, to criteria for designing failure-free casing systems.

Leutwyler and Bigelow (1965) and Willhite and Dietrich (1967) have published excellent discussions on the design of steam-injection wells. Although they directed their studies toward wells used in cyclic steam injection operations, where the temperature cycles are most pronounced, the same considerations apply to wells in any thermal recovery operation.

The injection or production of hot fluids can be interrupted by (1) mechanical failures, such as burn-outs of tubings in steam generators or disruptions in fuel or feedstream supply; and (2) workovers to restore injectivity or productivity or to correct an unexpected situation, such as an undesirable injection or production profile.

If the shutdown of the hot fluid injection period is long, the wellbore temperatures will fall, leading to tensile casing stresses (to be discussed). In the case of workovers, where injection or circulation of extraneous fluids may be required to control the well, cooling may be fast and severe. Inasmuch as workovers almost certainly will be needed to correct impairment or mechanical failures, it is well to consider the effect of heating and cooling in designing wells for any thermal operation.

Casing tends to elongate when heated. For casing cemented on bottom only, buckling is almost inevitable in the unsupported sections of any significant length unless the casing is free to expand through a packing gland at the surface pipe. In all but very shallow wells, hole deviations and formation sloughing will

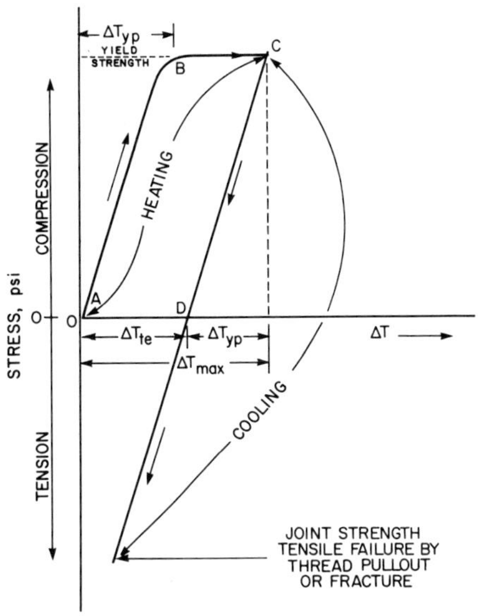

Fig. 12-11. Stress history of casing overstressed during steam injection. (After Willhite and Dietrich, 1967, p. 15, fig. 1; courtesy of the Society of Petroleum Engineers of AIME.)

cause enough loading at intermittent joints to build up compressive stresses and possibly buckle the heated casing.

The stresses that result when heated casing cools can be understood with the aid of Fig. 12-11. As long as the compressive stress of the casing has not exceeded the yield strength of the steel, or the casing does not get stuck in an expanded position, cooling will not cause difficulties; the stress is reversible and follows the line *AB*. If the yield strength of the steel is exceeded, however, the reduction in compression will follow a line such as *CD* when the casing cools. The decompression cycle parallels *AB*, but is displaced by the difference between the maximum temperature increase (ΔT_{max}) and the temperature increase at which the steel yields (ΔT_{yp}). If the temperature drops below this value (denoted by ΔT_{te}), the casing and joints actually will be in tension. Casing joints will separate if the tensile stress load exceeds the joint-pullout strength.

An obvious way to prevent joint pullout is to ensure that the compressive casing stress does not exceed its yield strength. In new completions, this can be done by selecting a suitable combination of materials, by controlling casing temperature, and by cementing the casing in tension. In existing wells, one can control only the casing temperature.

Table 12-I compares calculated minimum tensions resulting from heat–cool cycles in carbon steel with field observations regarding failure of wells completed with 7-inch, 20-lbm/ft, J-55 short-threaded casing. The joint pullout strength in tension of this casing (Willhite and Dietrich, 1967) is 32,000 psi (based on the

TABLE 12-I

Temperature changes and tensile forces resulting from cooled pipe (after Willhite and Dietrich, 1967, p. 17, table 2; courtesy of the Society of Petroleum Engineers of AIME)

Casing temperature (°F)	ΔT_{max} (°F)	ΔT_{yp} (°F)	ΔT_{te} (°F)	Tension when cool, (psi)	Remarks
500	400	275	125	25,000	no failure
550	450	275	125	35,000	no failure
600	500	275	225	45,000	failure
650	550	275	275	55,000	failure

cross-sectional area of the casing wall). The field data are consistent with the calculations, thus supporting the design theory.

If casing lacks lateral support, it can buckle. Buckling, in turn, can reduce the temperature increase at which the casing reaches its yield strength and, thus, increase the likelihood of joint pullout on subsequent cooling. For a 1/2-inch radial clearance in a rigid borehole, Table 12-II gives calculated allowable temperature increases for the buckling of a 7-inch, 23-lbm/ft, 30-ft length of casing the yield strength of which would not exceed that of carbon steel. The allowed temperature increase for a perfectly supported casing is also given. As can be seen, buckling significantly reduces the temperature increase at which the yield strength is reached.

The length and clearance of unsupported casing in old wells depend on the cementing practice followed, the number and distribution of hole washouts, the tendency of the formations to slough, and the hole straightness and deviation. In new wells, lateral deformation (and buckling) can be reduced significantly by drilling a relatively straight hole and by cementing to surface with a temperature-stable cement.

Another way to prevent buckling in new wells is to use materials with a higher yield strength than carbon steel. This, of course, is more expensive. One must weigh the possibility of failure of the existing wells against the cost of properly designing and completing new wells. If the choice is to drill new wells, decisions

TABLE 12-II

Allowable temperature changes for buckling and compressive loading (after Willhite and Dietrich, 1967, p. 16, table 1; courtesy of the Society of Petroleum Engineers of AIME)

Material grade	Minimum yield stress (psi)	Allowable temperature change (ΔT, °F)	
		buckling [a]	uniform loading
J	55,000	237	275
N	80,000	346	400
S	95,000	410	475

[a] 7-inch, 23-lbm/ft, 30-ft long joint with 1/2-inch radial clearance.

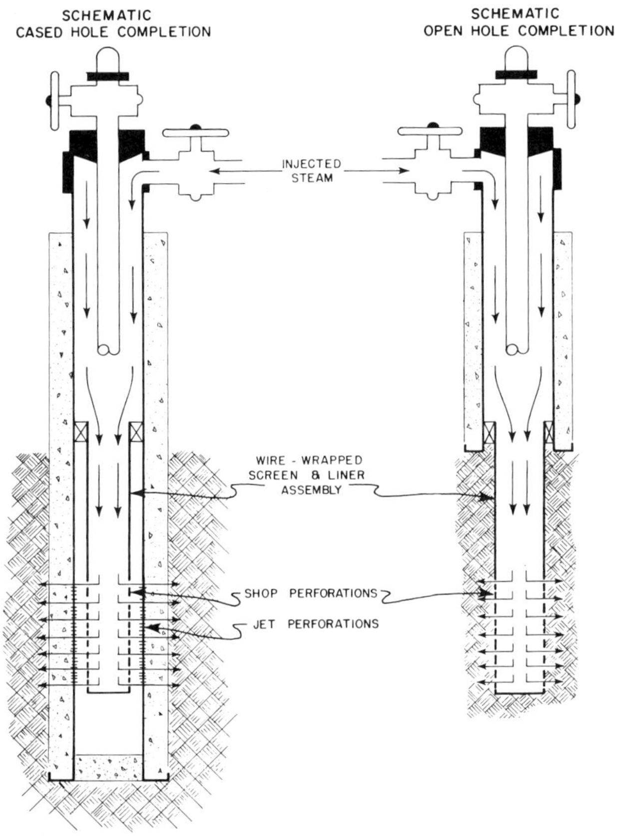

Fig. 12-12. Schematics of well completions in the Slocum Field. (After Hall and Bowman, 1972, p. 153, fig. 7; courtesy of the Society of Petroleum Engineers of AIME.)

must be made about insulation, the quality of the materials, and cementing techniques.

In some shallow reservoirs steam is injected directly down the casing. In the example shown schematically in Fig. 12-12, the wells are only 500 ft deep, so the steam temperatures are low. Either 7-inch, 22-lbm/ft K-55, or 5-inch, 14-lbm/ft K-55 casing are used, with Class G cement modified with Perlite, silica flour and calcium chloride. Silica flour and calcium chloride are used to reduce the loss of cement strength at high temperature, as discussed in more detail in the next section. All figures showing well completion designs, such as Figs. 12-13 to 12-17, have radial–horizontal/vertical scales generally in excess of 100.

Steam, however, is seldom injected directly down the casing in deep wells. Casing temperatures can be reduced below that of the injected steam by injecting the steam through tubing and by keeping the steam from the casing–tubing annulus. This serves not only to reduce the likelihood of well failure but also to reduce heat losses to the reservoir. Thermal packers are commonly used to keep the steam from the tubing–casing annulus. Another alternative is to inject inert gas down the annulus at low rates while steam is injected through the tubing.

Tubing strings

A well design using a small gas flow in the tubing–casing annulus to insulate the casing is shown in Fig. 12-13. In this case, no packer is used and the gas enters the formation along with the steam. This is usually the case, although packers can be used where the injected cooling gas is circulated to the bottom of the annulus through a separate string (Fraser et al., 1982). The casing completion shown in Fig. 12-13 is designed for a high-temperature (about 645°F) steamdrive operation at Peace River, Alberta, in which the process calls for large reductions in the temperature of the injected steam. It is important to note here that (1) the casing is designed for direct exposure to the high-temperature steam, (2) both the casing and its connections have been selected to withstand the loads resulting

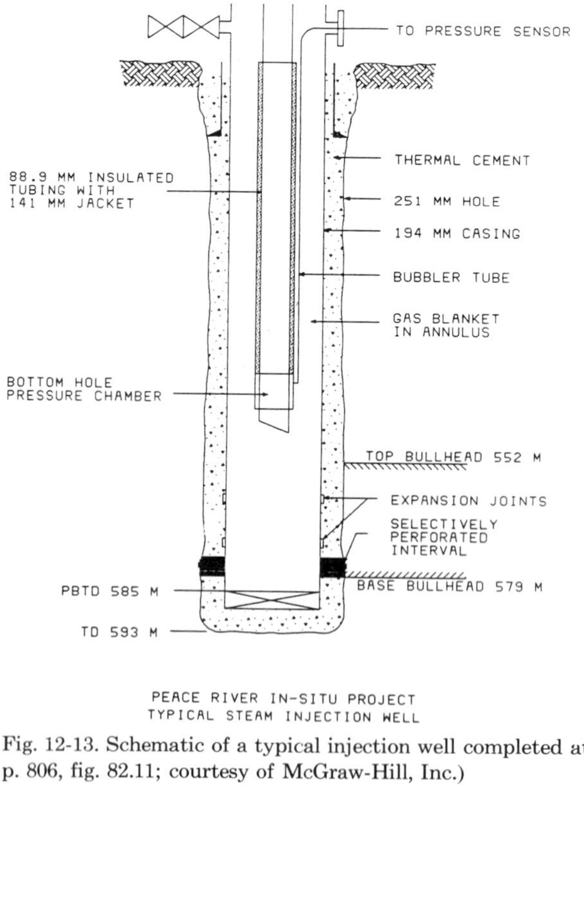

Fig. 12-13. Schematic of a typical injection well completed at Peace River. (After Fraser et al., 1982, p. 806, fig. 82.11; courtesy of McGraw-Hill, Inc.)

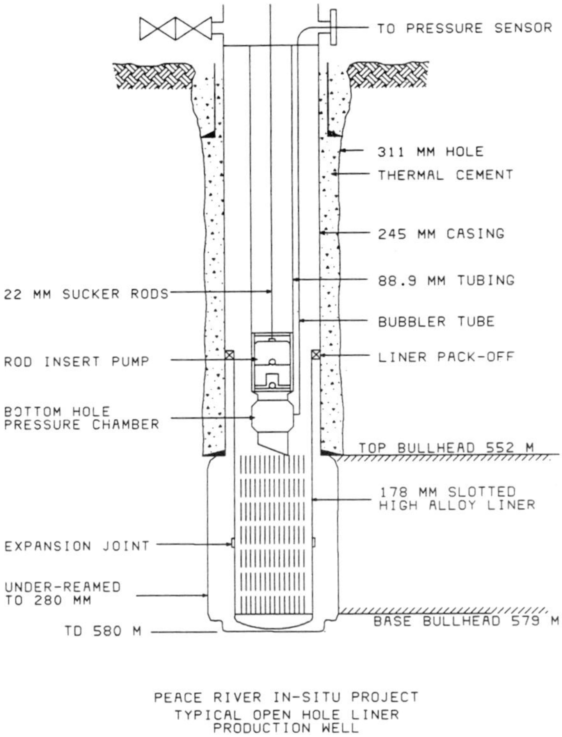

Fig. 12-14. Schematic of a typical openhole liner production well at Peace River. (After Fraser et al., 1982, p. 808, fig. 82.12; courtesy of McGraw-Hill, Inc.)

from the expected temperature variations, and (3) additional precautions have been taken to reduce the temperature on the casing over most of the injection interval by insulating the tubing and by intermittently injecting gas at a low rate down the annulus. Other unusual features of this well design include a bottomhole pressure sensor strapped to the tubing, a selectively perforated casing for steam injection into the base of the bitumen-bearing sand, and the use of two expansion joints within the 90-ft bitumen-bearing sand to allow for additional loads on the casing resulting from anticipated compaction of the reservoir.

The production wells at the Peace River project are designed for the same temperature levels, and for numerous temperature variations resulting from steam soaking, hot production, and remedial workovers. The three production well designs being tested are shown in Figs. 12-14 to 12-16. Fig. 12-14 is a schematic of an openhole completion with high alloy liner. Fig. 12-15 shows an openhole gravel pack completion with a high alloy liner, whereas Fig. 12-16 illustrates a cased and perforated completion in which a high-alloy liner will be installed as required. Produced fluids are corrosive and the liquid levels in the

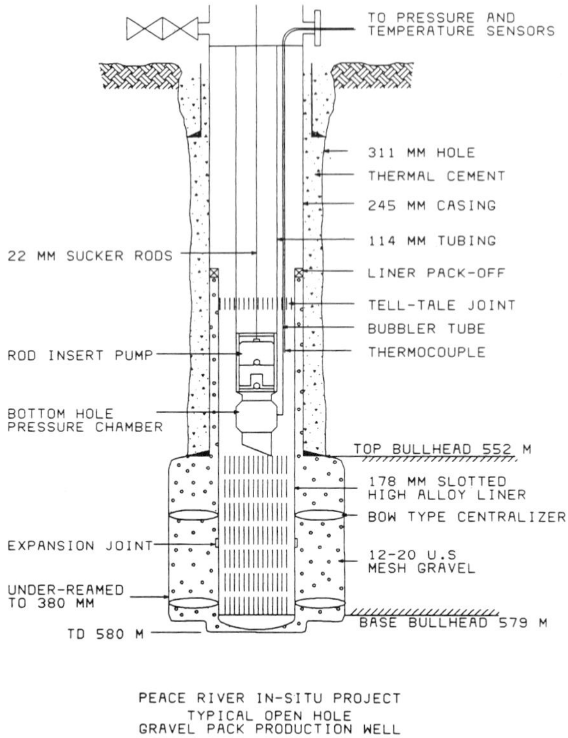

Fig. 12-15. Schematic of a typical openhole gravel pack production well at Peace River. (After Fraser et al., 1982, p. 809, fig. 82.14; courtesy of McGraw-Hill, Inc.)

wellbore are depressed by injecting gas down the annulus to a position below the top of the bitumen-bearing sand. The gas again will help to reduce the casing temperatures. The use of downhole pressure sensors strapped to the tubing, and of centralizers and expansion joints in the liners, are worth noting.

These Peace River completions are unusual because of the high temperatures present at both injectors and producers, the relatively large temperature variations expected during the life of the project, and the corrosive nature of the produced fluids at process conditions. Normally, much simpler completions are possible, such as the ones shown in Fig. 12-17.

Fig. 12-17 shows schematically some tubing completions that have been used in the Schoonebeek Field, in Holland (Schafer, 1974). In two of the three steam injection wells shown in the top row, the tubing was insulated with a calcium silicate cover protected (not sealed) by a thin-walled steel jacket. In these three completions the brines within the annulus were boiled off at atmospheric pressure when steam injection started. The reduction in casing temperature due to the insulation (compared to the uninsulated case) was estimated to be about

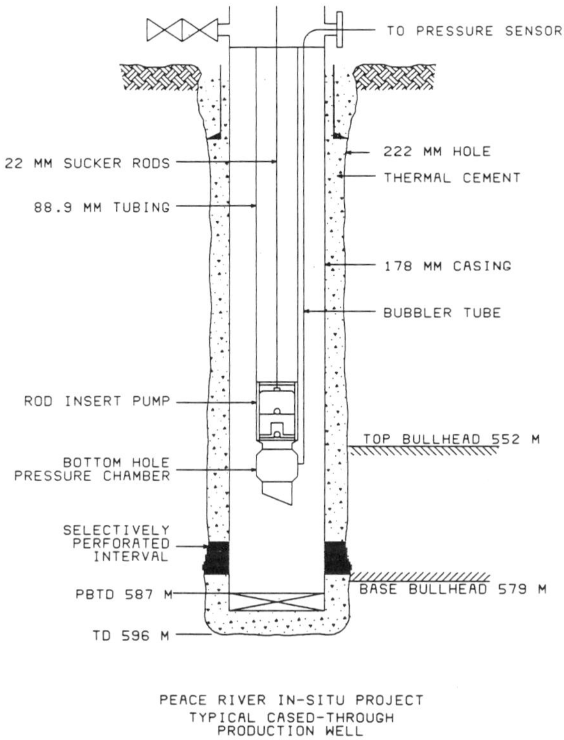

Fig. 12-16. Schematic of a typical cased-through production well at Peace River. (After Fraser et al., 1982, p. 810, fig. 82.15; courtesy of McGraw-Hill, Inc.)

400°F.

In general, any insulation has to be maintained dry in order to remain effective. Calcium silicate insulations have been foamed in the annulus to adhere to the tubing, but in the early days it was more common to band the insulation to the tubing as it was run in. Nowadays there are available pre-insulated tubing strings which can be coupled and run in routine fashion. These units allow for differential thermal expansion between the insulation-filled inner and outer pipes, and are sealed to keep the insulation dry (Traynor, 1980).

It is noted that all injection wells shown in Fig. 12-17 have the tubing connected rigidly to the casing at the thermal packer. Tubing expansion on heating is accommodated by allowing the tubing to slip through a stuffing box at the wellhead. At the present time, it is far more common to hang the injection tubing from the wellhead, allowing it to expand downward into the wellbore as it heats, as shown schematically in Figs. 12-12 to 12-16.

Bissoondatt (1976) discussed the effect of temperature cycling in cyclic steam injection operations and how to modify well designs to avoid failures, using examples from a commercial operation. Other casing completions allowing expansion have been discussed by Bleakley (1966).

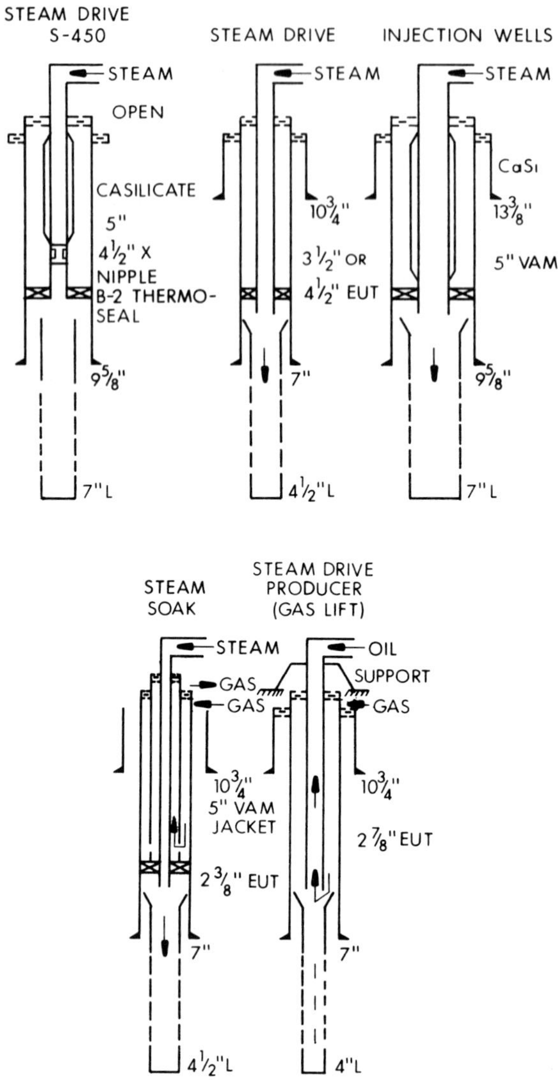

Fig. 12-17. Schematics of well completions in the Schoonebeek Field. (After Schafer, 1974, p. 376, figs. 9 and 10; courtesy of Urban-Verlag Hamburg/Wien GmbH.)

The successful development of a downhole steam generator (Johnson and Fox, 1980) involving injection of all fluids into the reservoir would avoid the need for insulation in steam injection wells. Insulation is still needed during the production phase of cyclic steam operations unless it is safe to expose the well to the producing temperatures, which are generally significantly lower than that of the injected steam.

One other general aspect of well design that must be considered is the need for materials and equipment that can survive and operate at high temperatures. These include cement (Gates and Holmes, 1967), thread dope (welding is an alternative), expansion sleeves or joints, elastomers and other sealants used in packers, and any equipment that must engage and disengage under varying amounts of thermally induced loads. In steam injection processes, such equipment may normally be subjected to temperatures of up to 550°F, and higher on special projects and in certain circumstances.

Cementing

Cementing thermal wells differs from that in conventional wells in the following aspects (Ludwig and Pence, 1956; Carter and Smith, 1958; Gates and Holmes, 1967):

(1) Cement should be circulated to the surface on every string of casing. The objective is to provide a complete fillup of cement in the casing hole annulus to anchor the casing firmly to the ground through the cement sheath, thus reducing buckling and casing creep. Also, the hardened cement sheath should protect the casing against possible corrosion and prevent the uncontrolled flow of hot fluids outside the casing.

(2) API Class G or H cement should be used with at least 30% silica flour. Retarders, friction reducers, and additives to control lost circulation and the density of the cement slurry may be added as required. Bentonite, perlite, pozzolan and very small glass beads are the normal additives used to reduce the slurry density. The use of this type of cement should be considered in all wells where sustained temperatures above 230°F are anticipated, as the compressive strength of hardened cements is significantly reduced when exposed to higher temperatures even for periods of weeks. Production wells in steam drives as well as all wells where steam is to be injected (and, of course, many conventional deep wells) fall in this category.

(3) Consistent and very careful exercise of the following usual cementing and completion practices: (a) estimating the amount of cement slurry required; (b) estimating the times to run the casing and circulate the cement to surface, while displacing the mud under conditions which would not lead to fracturing the formation; (c) deciding whether or not to reciprocate and/or rotate the casing while cementing; and (d) determining the need for mud thinners, preflushes, centralizers, scratchers, and floating equipment. These practices become more important at the higher temperatures.

The role of the silica flour in Class G cements subjected to high temperatures is not that of a filler. A small amount of the silica flour reacts with calcium hydroxide to form a type of calcium silicate which loses strength at high temperatures. When there is more silica flour than required to form this type of calcium silicate, the remainder reacts to form a cementing phase with excellent temperature resistant characteristics.

Ideally, cements should retain their design compressive stress, low permeability, and bonding to both formation and casing even after exposure to high-temperature cycles over a prolonged period of time. Normal neat cements used for shallow conventional wells have been shown (Ludwig and Pence, 1956; Gates and Holmes, 1967) to lose over 80% of their compressive strength when subjected to temperatures of around 400°F for one week. This loss in strength is generally accompanied by increases in the permeability of the cured cement, sometimes by as much as a factor of 200. This permeability increase is normally of little importance except in reservoirs of low permeability, of which not many are of interest in steam injection operations. A more important type of fluid communication behind casing can result from losing the bond between the cement and the casing, which can be easily broken under the thermal expansion of the metal. The expansion of a joint of casing, which normally indents into the cement at the collars, could also rupture the cement in tension. Other tension breaks could result where the cement bond does not give, or where there is sufficient friction between the cement and casing.

TABLE 12-III

Slurry properties of silica flour cements (after Cain et al., 1966, p. 432, table 1; courtesy of the Society of Petroleum Engineers of AIME)

Silica flour (%)	Gel content (%)	Perlite (ft^3)	Slurry volume (ft^3/sack of cement)	Weight of slurry (lb/ft^3)
API Class G cement				
30	0	0	1.50	117.0
40	0	0	1.62	116.5
30	0	1	2.14	102.0
30	2	1	2.35	99.0
40	0	1	2.26	102.5
40	2	1	2.45	100.0
30	0	2	2.78	94.5
30	2	2	2.97	93.0
40	0	2	2.90	95.0
40	2	2	3.09	93.5
30	0	3	3.42	89.5
30	2	3	3.61	88.5
40	0	3	3.54	90.5
40	2	3	3.73	89.5
50:50 Class G – Pozzolan cements				
30	0	0	2.86	104.0
30	2	0	2.86	100.0
40	0	0	3.04	105.0
40	2	0	3.33	101.0
Calcium aluminate cement				
40	0	0	1.69	114.0

TABLE 12-IV

Compressive strength of cementing compositions (after Cain et al., 1966, p. 432, table 2; courtesy of the Society of Petroleum Engineers of AIME)

Silica flour (%)	Gel content (%)	Perlite (ft^3)	Strength at 100 °F (psi)		Strength at 400 °F (psi)	
			8 hours	1 day	1 day	7 days
API Class G cement						
30	0	0	585	2515	4915	6640
40	0	0	330	1490	4890	6500
30	0	1	315	925	3915	3565
30	2	1	230	710	3075	1990
40	0	1	335	775	3825	3130
40	2	1	205	400	2560	1905
30	0	2	120	420	2145	1415
30	2	2	60	325	1595	1480
40	0	2	145	375	1850	2025
40	2	2	30	265	1440	1220
30	0	3	60	270	1080	1220
30	2	3	20	180	840	880
50:50 Class G – Pozzolan cements						
30	0	0	130	690	4050	4100
30	2	0	100	480	2925	3280
40	0	0	135	670	3930	3850
40	2	0	95	460	1505	2945
Calcium aluminate cement						
40	0	0	5	965	900	620

Cements in normal use in steam injection operations have been developed over the years to avoid these drawbacks. They consist of an API Class G or H cement typically containing from 30 to 60% of silica flour (by weight of cement), 1–3 ft^3 of expanded perlite, and about 2% bentonite gel, with 10% or more common salt (NaCl) by weight in the mixing water. Perlite reduces the density of the cement slurry and the thermal conductivity of the set cement. The bentonite gel improves the mixing of the perlite in the cement slurry. Salt reduces the slurry viscosity, improves bonding to shales and bentonitic sands, and may prevent or reduce hole sloughing. The salt also causes the cement to expand slightly (as much as 0.17% linearly), thus tending to maintain continued bonding (Slagle and Smith, 1963; Carter et al., 1966). Tables 12-III to 12-V provide compressive strengths, slurry properties, and thermal conductivities of several compositions of cement slurries suitable for use in wells exposed to steam temperatures.

Proper cementing is important in any well where steam is injected. Inasmuch as production wells in steamdrive operations are often stimulated with steam, it is common practice to cement all wells in a project the same way. Chemical preflushes containing surfactants and mud thinners, and the circulation of the preflush and the placement of the cement slurry at low-velocity turbulent

TABLE 12-V

Thermal conductivity of API Class A cements (after Halliburton Services Bulletin C-1142, revised)

	Slurry composition				
Water (gal/sack of cement)	5.20	6.80	12.50	24.30	14.60
Bentonite (%)	0	0	2.0	2.0	0
Silica flour (%)	0	40	30	30	30
Perlite (ft^3/sack)	0	0	1.0	3.0	0
Vermiculite (%)	0	0	0	0	30
Curing time (hours)	Thermal conductivity (Btu ft^{-1} hr^{-1} $°F^{-1}$)				
24	1.068	0.955	0.535	0.653	0.324
48	1.005	0.828	0.420	0.515	0.268
120	0.803	0.475	0.400	0.332	0.324
144	n.d.	n.d.	0.319	0.333	n.d.
329	n.d.	n.d.	n.d.	n.d.	0.320

n.d. = not determined.

conditions (if achievable) would promote removal of the drilling mud and good bonding to the formation and to the casing. Buckling and elongation of the casing would be minimized by using floating and guiding equipment, whereas centralizers would promote a uniform cement sheath.

Sand control

A large fraction of the reservoirs to which thermal recovery methods are applied are composed of unconsolidated or friable formations. Well completions are designed to provide low resistance to fluids and at the same time keep sand from entering the wellbore. Although production wells receive the most attention, it is also advisable to control sand in injection wells. When, because of equipment malfunction or for any other reason, injection pressure decreases, the well may backflow. Where there is no space below the injection interval to accommodate the sandfill, the injectivity may be decreased when injection operations resume. Well completions also are designed to promote or ensure the desired distribution of the injection fluid among multiple sands or sand stringers. Sometimes it is also desirable to separate producing intervals, as, for example, when two sands are at different pressures (referred to a common datum).

Fig. 12-12 shows a sand exclusion technique using a wire-wrapped liner assembly, hung either inside perforated casing or from casing inside open hole. The same technique was used in the production wells. Another practice is to gravel-pack a preslotted liner in an underreamed hole (see Figs. 12-15 and 12-18). Slot-opening and gravel sizes are important in effective sand control (Coberly and Wagner, 1938; Holke and Huebner, 1971; Saucier, 1974; Gulati and May, 1975). Saucier (1974) recommended that "the ratio of pack median grain size to formation median grain size should be between 5 and 6 where there is severe flow

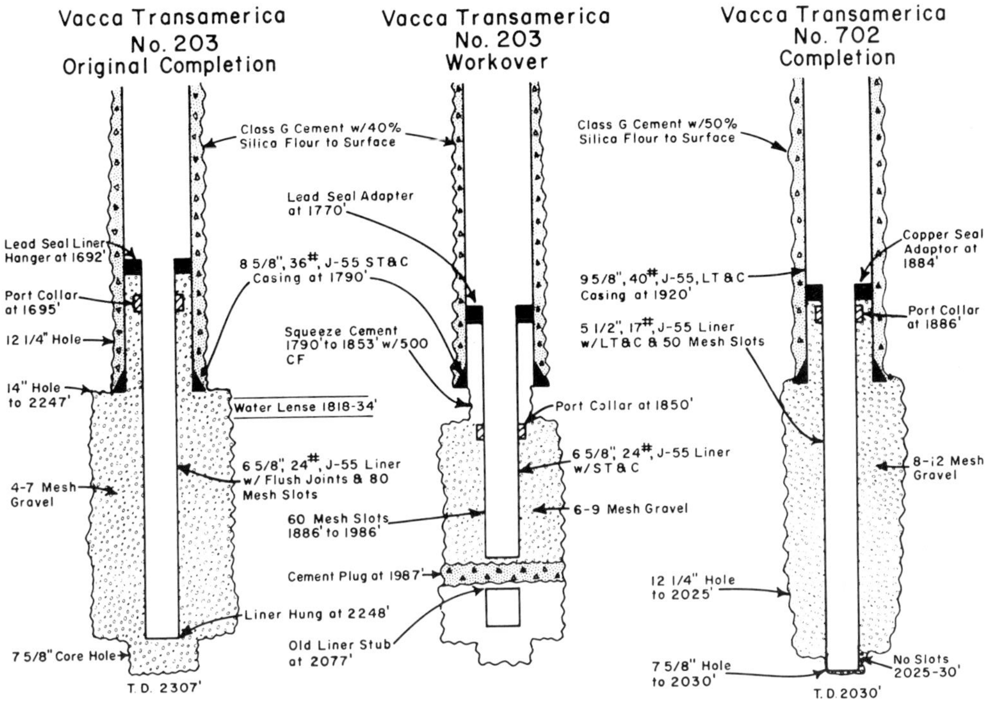

Fig. 12-18. Vacca Transamerica well completions. (After Bott, 1967, p. 588, fig. 4; courtesy of the Society of Petroleum Engineers of AIME.)

disturbance." The slots are then sized to retain the gravel. Because some gravel is relatively soluble at the steam-condensate conditions prevailing downhole in injection wells, voids can develop in the gravel pack (Reed, 1980). Local experience is an important factor in proper gravel-pack design. To reduce the likelihood of parting due to stresses, liner joints are sometimes welded (Gates and Brewer, 1975), especially in cyclic steam injection operations.

Other mechanical and chemical types of sand control have been discussed by Suman (1975). In wells where steam is injected, the hot, high-pH liquid water from the generator and that condensed as a result of heat losses will tend to dissolve the sand grains in the reservoir, the gravel in the gravel pack (Hall and Bowman, 1972), and even chemical agents used to consolidate the grains. Perhaps because of dissolution, methods for consolidating sand grains have not been very successful in hot-fluid injection wells where temperatures are high.

Profile control

Profile control is a much more difficult problem, especially for unconsolidated sands separated by thin shale stringers. As the "conventional" completion shown

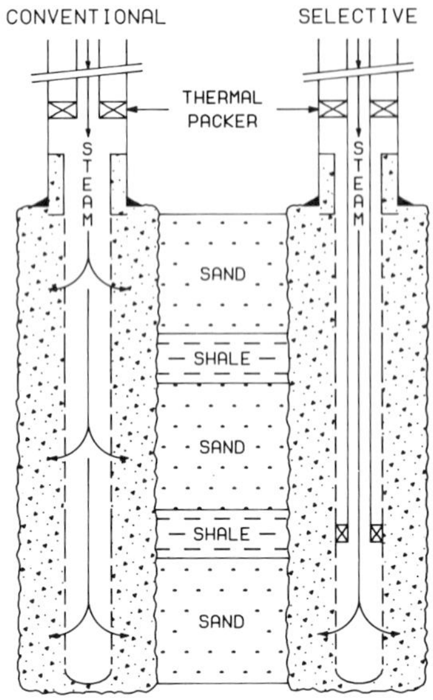

Fig. 12-19. Schematic diagram of conventional and selective completions in gravel-packed wells. (After Burkhill and Leal, 1974, p. 365, figure in Appendix 1.)

in Fig. 12-19 indicates, the injected steam enters each of the sands according to its relative flow resistance. There is no control of the relative amount of steam entering each sand. In fact, with the completion as shown, the lower sand can be expected to receive more wet steam than the upper sand even when the flow resistances are equal, because the liquid water will tend to move to the lower part of the wellbore. Depending on the quality of the steam entering the injection interval (and on other factors), the effective liquid–vapor level may be above the first shale stringer, effectively cutting off the lower sand from steam vapor injection. One way to remedy this downhole steam separation effect is to extend the tubing below the lowermost steam injection interval (or as close to it as feasible), so that the steam coming up the annulus can help distribute the liquid water and steam vapor more uniformly (Hall and Bowman, 1972). The "selective" steam injection completion shown in Fig. 12-19 aims at injecting steam into the lowest sand interval, which is usually the least depleted and the one having the highest flow resistance (all other variables being the same). Although the figure shows steam entering the formation, it must be recognized that part of the steam can move up through the gravel pack behind the packers and enter the upper sands. The effectiveness of such a selective completion obviously depends on the amount of steam leaking behind the packer through the gravel pack.

Profile correction falls in three different categories, and in each the completion and condition of the well must be considered.

(1) In the case of set-through completions where the well is perforated, selective injection is a straightforward matter of using straddle packers (and

perhaps bridge plugs). Essentially any mechanical means that delivers the fluid in the desired interval will be successful if there are a competent cement job and adequate separation of the sands within the reservoir.

(2) Gates and Brewer (1975) reported the successful use of limited-entry perforation techniques in a properly cemented casing to control the injection profile in steam injection wells in the South Belridge field. With this method it is necessary to ensure that each perforation is open; the technique works best in low-pressure reservoirs.

(3) In underreamed and gravel-packed wells, profile correction is much more difficult and two choices are available. One is to place an impermeable plug over a short interval of the gravel-packed region between the liner and the sandface. A packer, bridge plug, or other mechanical device can then be set inside the liner to divert the flow. Attempts have been made to seal the gravel pack across the shale intervals by injecting sealants such as polymers, cement, mud, and different chemicals. Burkhill and Leal (1974) reported continued routine success using a mixture of bentonite, silica flour, barite, cement and other chemicals in the Bolivar Coast of Venezuela. Furfuryl alcohol is an example of a low-viscosity chemical that can be catalyzed to rock hardness. It has been reported to be moderately successful in field applications (Hess et al., 1971; Anderson, 1978). Chemicals also have been reported to be successful to seal thief zones in cyclic steam injection wells (Fitch and Minter, 1976).

Instead of forming an impermeable plug within the gravel pack opposite a shale section, it is sometimes desirable to seal the rock matrix in the thief zone. Many chemicals are available to plug thief zones—either temporarily or permanently—to divert the injected steam into desired intervals; however, few are economically attractive and easy to place. If a permanent seal is desired, few, if any, chemicals can withstand the effect of high temperatures over a prolonged period of time. Lignosulfonate, for example, forms a gel that has been used to plug thief zones, but it loses effectiveness with prolonged exposure to high temperatures.

Wells are sometimes completed initially in such a manner that subsequent profile control is facilitated. Possible steps that would aid in later operations included the use of external casing packers (or equivalent), and cementing zones between the gravel-packed intervals (Marsden and Khan, 1967). The latter technique requires relatively large nonpay zones within the total production interval.

A problem related to injection well profile control is the control of the distribution of the injected steam within the reservoir. Because of the high permeability of most of the heavy oil reservoirs into which steam is injected, the steam tends to rise to the top of the sand intervals during the course of the operation, thus bypassing oil columns of significant thicknesses below the steam zones. There is an increasing effort to develop and field test steam foams to improve the vertical distribution within the reservoir of the injectant (Dilgren et al., 1982), with the intent of improving the amount of oil recovered.

Lifting

Although produced fluids can always be lifted to the surface, this must be done efficiently and in such a manner as to promote the desired reservoir response, which usually entails maintaining a low wellbore pressure to increase the pressure difference available from the reservoir. For heavy crudes, lifting problems normally fall in three categories: (1) the crude oil is very viscous, (2) the temperatures are very high, and (3) there is sand production. These are discussed primarily as they affect the performance of subsurface sucker rod pumps, which represent the most prevalent lift form for fluids in steam injection operations.

When the crude oil is viscous at downhole conditions, it generally means that the temperature is still low. Where the viscosity of the crude oil is very high, the fall of the rods is significantly delayed, requiring the pump speed to be reduced in order to keep the polished-rod clamp and hanger bar from separating on the pump downstroke. This may result in the well not being pumped-off. A common solution to this problem is to inject a diluent through hollow sucker rods or side strings to improve rod fall and pump performance. Load limitations, resulting from their smaller metal cross-section, limit the use of hollow sucker rods to probably no more than about 4000 ft in depth. Other uses of diluent for improving lifting performance were discussed by Franco (1974). The diluent generally is obtained from the light ends of the produced crude oil by means of a topping unit, but lighter crudes are often used where they are easily available and compatible with the heavier crude oil.

Rod fall also has been improved by lifting through the annulus, with the rods in the tubing being in contact with water under pressure. A downhole stuffing box just above the perforated tubing nipple isolates the tubing fluids. This system, reported by Ballard et al. (1977), is illustrated in Fig. 12-20. A similar system was discussed by Franco (1974).

Lifting produced fluids at elevated temperatures may present problems. Pumps may perform poorly, even where downhole gas separation is effective, because hot water may flash to steam on the intake pump stroke and cause vapor locking and reduced pump efficiency. The remedies are few: (1) shutting in the well and allowing it to cool so that it can be produced intermittently; (2) increasing the back pressure on the well to make flashing less likely; and (3) adding a liquid (through the annulus, hollow sucker rods, or a separate string) to reduce the downhole temperature. None of these methods is efficient. Allowing the well to cool and increasing its back pressure would probably result in a loss in the average production rate. Adding coolants, of which water is the most cost-effective, may not be too effective in reducing temperatures where the cause of the high temperatures is the inflow of steam. In addition, it requires ad-

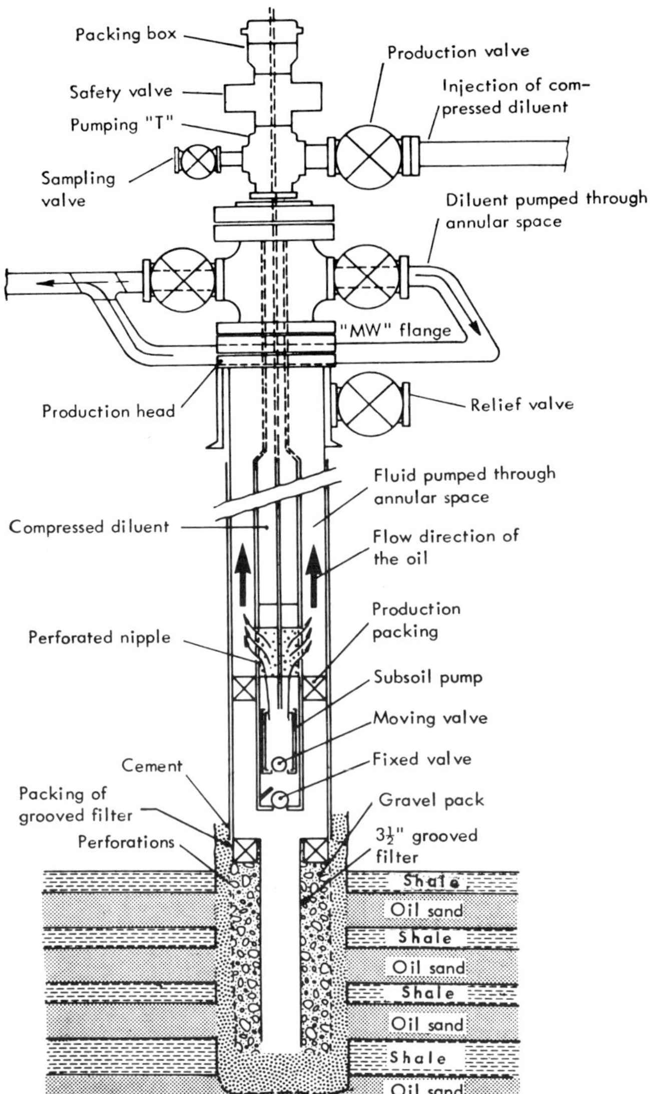

Fig. 12-20. Jobo Field—typical well completion. (After Ballard et al., 1977, p. 23, fig. 4; courtesy of the Canadian Institution of Mining.)

ditional facilities (lines, pumps, increased capacity of tanks and separators, etc.) and increased capacity of lifting equipment.

Steam or gas flowing through the pump would also cause gas locking and severely curtail pumping efficiency. This would be likely to occur only where the bottomhole pressures are sufficiently high to allow the gas to flow through the tubing.

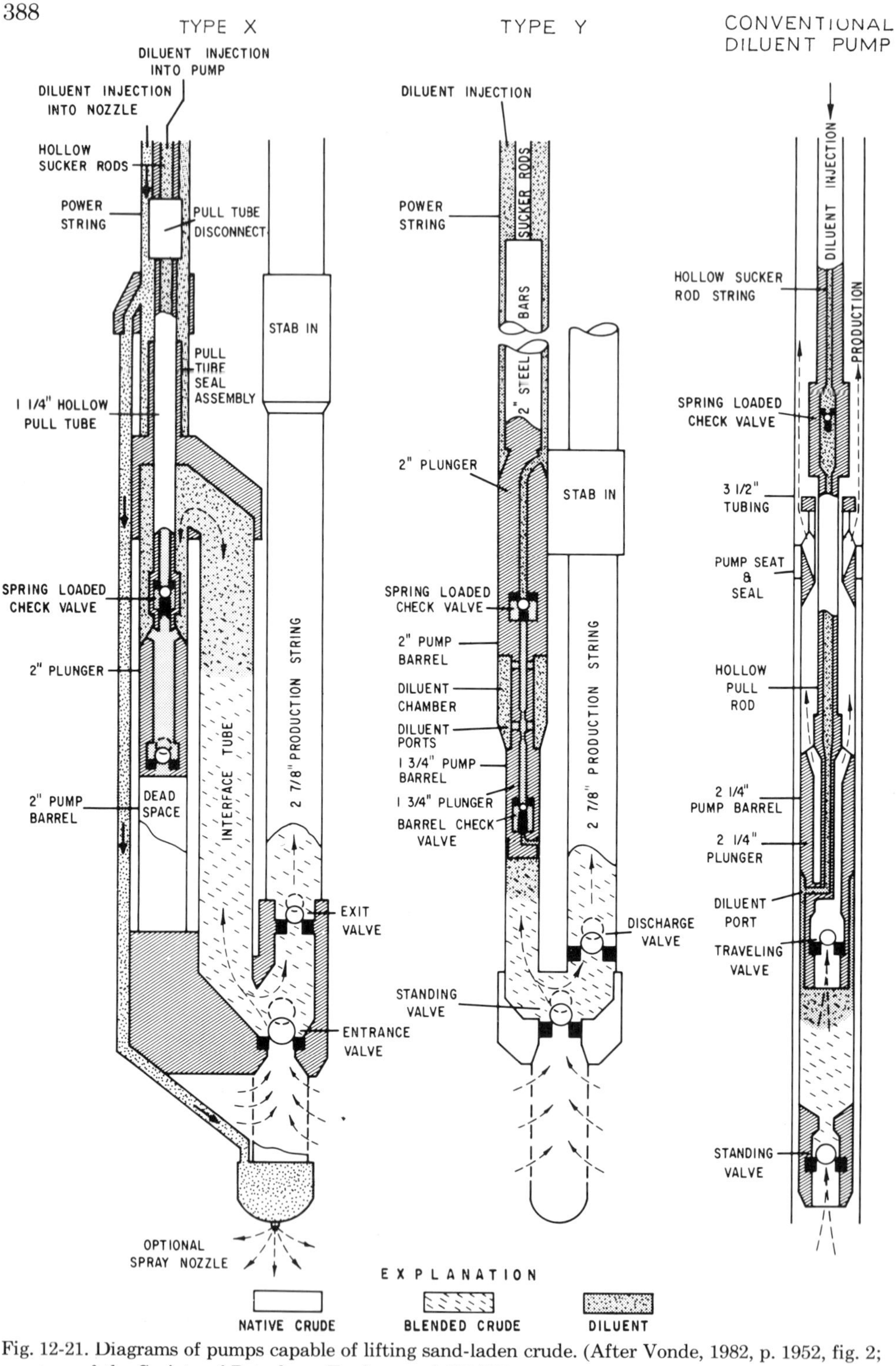

Fig. 12-21. Diagrams of pumps capable of lifting sand-laden crude. (After Vonde, 1982, p. 1952, fig. 2; courtesy of the Society of Petroleum Engineers of AIME.)

If gas lift is used, the pressure drop in the tubing is difficult to anticipate because of the changes in temperature that occur as the fluids move towards the surface. The changes in pressure and temperature affect the viscosity of the crude oil, the relative amounts of steam and liquid water (through phase changes), and the flow regime (mist annular, transition to slug flow, etc.) in the tubing. As pointed out by Boberg et al. (1973), the performance of gas-lifted or flowing well producing hot fluids is influenced by the total pressure drop in the tubing and the surface flowlines.

A large clearance between the plunger and the barrel of pumps is normal when lifting viscous crude oils, especially where there is sand entry. The greater pump clearance used to lift the more viscous crude oils allows sand to move through the pump somewhat freely, thus prolonging the life of the pump and reducing the frequency of its replacement. At the same time, the slippage losses are relatively small because of the viscous nature of the crude oil. In shallow wells, this is also the case because of the small pressure differential across the plunger. Clearances of the order of 0.01 inches have been reported by Juch and Watson (1968). Special pumps are also used. Fig. 12-21, for example, shows cross-sections of a conventional diluent pump and two modifications used successfully to lift as much as 150 bbl of 4° API crude per day, with up to 70% sand by volume.

Gas lift is not efficient for viscous crude oil. Steam has potential because it reduces the oil viscosity, but probably cannot be justified economically compared to other alternatives. Where steam is available, it also would be used to heat and stimulate production by cyclic steam injection. Production from cyclic steam injection usually flows at the outset and often can be maintained at acceptable levels by pumping or gas lifting. Where pumps are used in cyclic steam injection, they are generally installed from the start and left unseated so the injected and produced fluids can flow through them until they are required for lifting.

Other methods used to help lift heavy crude oils include the use of downhole heaters, when the crude oil viscosity can be reduced substantially, and emulsification of the crude to form low-viscosity readily pumpable oil-in-water emulsions (Simon and Poynter, 1968; Strassner, 1968; Dilgren et al., 1982).

Separation

From the wellhead, the produced stream usually flows first to an oil and gas separator. It is designed to separate the free water and gas from the crude oil, and is usually sized to handle the highest (rather than the average) rate of production anticipated. Wells on pump may not head too much, but flowing wells may produce fluids in slugs, and thus at highly variable rates. The rate of separation of the phases is approximately proportional to their density difference and to the retention time, and is inversely proportional to the viscosity of the crude. The viscosity of the crude may be reduced by heating, which increases the separation rate. This may require additional heating if the production stream is not sufficiently hot. Another way to reduce the viscosity of the oil phase is to

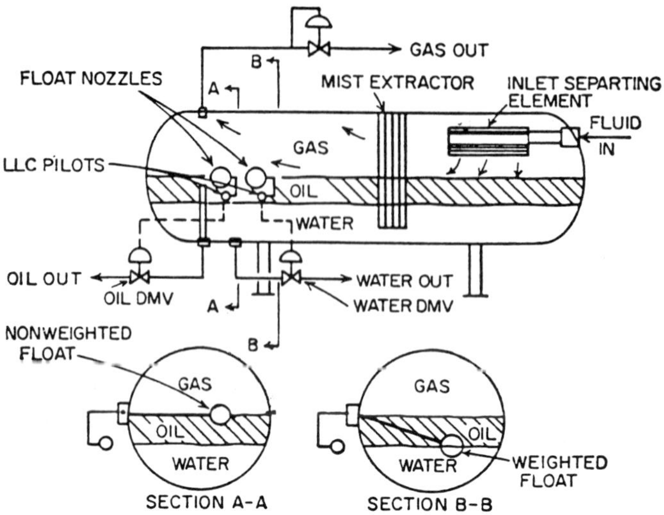

Fig. 12-22. Schematic diagram of typical horizontal three-phase oil–gas–water separator. (After Frick, 1962, p. 11.13, fig. 11.17; courtesy of the Society of Petroleum Engineers of AIME.)

add diluents, either at the separator or at the well. Reducing the viscosity of the oil phase also increases the separation rate of foam and free gas from the oil. For crude oils having gravity of about 10° API and above, density differences between water and crudes generally are increased by heating, so that temperatures have a dual beneficial effect on separation. The adverse effect of the diluents on the density differences is usually small unless large quantities are used. Another way to increase density differences between the water and oil phases is to add salt to the produced water (but not if the water is to be treated for use in the generation of steam). Because of large differences between the liquid and gas densities, gas can generally be separated quickly unless foams are formed. Foams may require the use of antifoamers as an alternative to longer retention times.

There is a large variety of separators. A schematic diagram of a typical horizontal oil, gas and water separator is shown in Fig. 12-22. In this unit, the inlet discharges above the oil-free surface through a separating element. The shallow depth of the oil column inherent in horizontal oil separators promotes the relatively quick removal of gas and water. Production rates are relatively stable in thermal operations, so that surges in production that would cause upsets in the pump control valves are not frequent. Because of these factors, horizontal separators are usually preferred to vertical ones where viscous crude oils are present. Vertical separators occupy less area, are less likely to suffer upsets, and may be somewhat more efficient than the horizontal ones.

The retention or residence time of the stream in the separator is the factor that affects the size or numbers of the separators to be used. Retention times are normally just a few minutes, but may range up to about 20 minutes where foaming oils are present. In any case, simple separation as described is fairly effective on free fluids and coarse emulsions.

In heavy oil steam projects, the producing pressure is normally low, well below the minimum rating of 125 psig for the separators. Then tanks instead of pressure vessels are used for separation. The separation principles are the same, of course. Because of the near atmospheric pressures at which the fluids are separated in most heavy oil projects, the vessels are much larger than conventional separators, which helps to reduce unit costs. Most separation in California thermal operations use low-pressure tanks.

Further processing may be necessary before the oil meets specified sales (or refinery) requirements. The crude stream from the primary separators is likely to contain water in emulsions in amounts too large to meet specifications. It then passes through a second series of vessels to remove the remaining water. The water present at this stage is more difficult to separate because the droplets are dispersed more finely. To coalesce these droplets and ease their removal, a number of approaches are used (Bansbach, 1965, 1970; Lucas, 1966):

(1) Chemicals (demulsifiers) are added to reduce the forces tending to stabilize the emulsion.

(2) Diluents, chemicals, and heat are added to increase the density contrast between phases and to increase droplet mobility by reducing the viscosity of the crude oil.

(3) The rate of contact between the droplets is increased to improve their rate of coalescence and their eventual removal. This is accomplished by agitating the fluid, by forcing the fluids through packed beds of materials promoting spreading of water on their surfaces, and by creating electrical fields to induce dipole attraction between droplets (Burris, 1977).

Separators, treatment vessels, heater treaters, or wash tanks usually combine most of these approaches (Fazio and Banderob, 1966; Lucas, 1966). These vessels can be designed so that a single type can handle the production from most fields.

The degree of difficulty of removing water from crude oils to acceptable levels is related to the stability of the emulsions. The latter, in turn, is controlled greatly by the properties of the crude oil, as well as by the operating environment.

No field project is known to have been abandoned because the produced emulsion could not be broken economically. Nevertheless there are emulsions that require more effort to break than others. Emulsions can be water-in-oil, oil-in-water, or both. The discontinuous phase itself may be an emulsion.

The character of the emulsion produced from a well changes over time. For example, Table 12-VI shows the amount of emulsion and free water produced as a function of time from cyclic steam injection wells in a California field, together with the voltage required to break the emulsions. The amount of the emulsion tends to decrease with time and the proportion of free water tends to increase, whereas the stability of the emulsion (as measured by the voltage required for

TABLE 12-VI

Stability of emulsions from several wells in a cyclic steam operation (after Bertness, 1965, p. 10, table 3; courtesy of the Society of Petroleum Engineers)

Property	Well 1	Well 2	Well 3	Well 4	Well 5	Production sump
Gravity (°API)	13.5	14.3	13.9	13.6	13.7	13.6
Days after steaming	7	21	36	55	197	n.d.
Wellhead temperature (°F)	225	190	175	150	110	n.d.
Sample temperature (°F)	175	138	138	130	110	85
Production rate						
gross (bbl/day)	160	160	180	300	20	n.a.
net (bbl/day)	60	102	140	78	9	n.a.
Free water (%)	0	0	0	34	25	n.d.
Emulsion cut (%)	60	36	22	44	30	32
Voltage required to break emulsion with 2 qt of chemicals per 100 bbl (volts):						
(a) without chemicals	480	520	540	440	400	⩾ 600
(b) with chemicals	200	120	120	135	90	n.d.

n.d. = not determined; n.a. = not applicable.

separation) decreases with time when a demulsifier is used, but goes through a maximum when no additive is used. This implies that the demulsification treatment should be altered during the production period to reflect the properties of the emulsion. This is indeed the procedure followed by Hall and Bowman (1972) in the steam drive at Slocum, Texas. But individual wells seldom receive special treatment except to avoid upsets in the treating facilities or where resultant cost savings are important.

The rate at which emulsions can be broken affects the size of the surface facilities required to do the job. Obviously, there are economic incentives to reduce the size of those facilities to a minimum. This can be done at the expense of using (1) high temperatures and solvents to reduce the crude oil viscosity and the separation time, (2) increased electric voltages, and (3) relatively expensive demulsifiers. The challenge is to achieve a compromise between cost of facilities and operating costs, with some intangible value given for a system that will work trouble-free with little attention after the initial design and test period.

Foams, or gas emulsions, generally do not affect the delivery requirements of the crude oil, but are difficult to dispose of and may hinder the operation of the dehydration units.

The clean oil is generally kept in storage tanks until the time of shipping it from the lease. All pipes through which the crude oil flows, and vessels in which the crude oil is treated or stored, are generally insulated to maintain the crude oil as hot and as fluid as possible. It is also common practice to have facilities for heating these vessels as necessary; for example, in the case of cold ambient temperatures. The heating is most commonly done by diverting part of the steam through heat exchangers within the vessels. Also, produced water used as boiler feedwater is transported through insulated pipe in order to take advantage of its high temperature, which is usually well above 120° in mature steam drives.

Well tests

Except in special field tests, the liquid production from several wells is commingled before it is measured. It is customary to measure individual well rates on a schedule ranging from a few days to a few weeks, and then allocating a fraction of the lease of block production to individual wells on the basis of their performance. Such a procedure saves production equipment and is adequate to define trends in the production rates of all wells. It can be appreciated, then, that cumulative total volumes are known much better than individual well rates.

A well is tested from a few hours to a few days by passing its production through a separator. Gauging the amount of oil and water produced is closely associated with the separation process. In some cases, the number of oil and water dumps are counted and multiplied by the calibrated associated volumes. In other cases, calibrated capacitance probes are used to measure the water cut. Oil and water rates are then calculated from the liquid densities and the total measured liquid production rate.

In operations having several hundred wells, scheduling these and other tests, as well as gathering data about the project, may become a formidable task. It is not surprising, then, to find that such tasks are sometimes automated (Otto, 1974).

MOVING THE CRUDE OIL FROM THE FIELD

The responsibilities of the operators of a steam injection project do not extend beyond the delivery of the crude oil to a suitable transportation system, even when the transportation system is operated by the same company. Arrangements normally are made so that the operator of the steam injection project has minimal responsibilities beyond those of producing oil. Crude oil is normally pipelined from the field to a common carrier, who in turn is responsible to deliver it to its destination. In fields where the production is more than just a few thousand barrels per day, the produced crude oil is almost always transferred to the carrier by means of surface pipe. Surface pipe is by far the most commonly used method of handling crude. The crude oil leaving the separators is still hot, and there are normally no problems in transferring the heated crude oil to nearby terminals from which the carriers take over. This transfer is often done automatically by means of approved Automatic Custody Transfer units.

Unusual situations arise when the crude oil is extremely viscous, the pour point of the oil is high, the ambient surface temperatures are low, or the distances to the terminal are long. Crude oils can be transported by pipeline even when they have high viscosities. Several methods are used to achieve this. In one, the crude is piped as a core surrounded by a water lining between it and the wall—the water acts as a low-viscosity lubricating film. In other cases, the carrier may require that the operator add a diluent, which normally would be a locally available light crude oil. In this case, the crudes are tested to ensure that no organic precipitates are formed on mixing. If no suitable diluent is available, the alternatives are for the carrier to heat its pipeline, truck the oil to the user, or emulsify the crude oil in an aqueous phase. In these cases, the transportation costs would be above normal.

DISPOSAL

As in any oilfield operation, solids, water, and gases must be disposed of properly. Typically, the waste water stream itself can be expected to contain some solids, oil, and gas. Solids, which usually are contaminated with oil, include sand and silt particles, inorganic scales, rust and metal-organic compounds, and sometimes organic matter (Hall and Bowman, 1972).

Even though viscous crude oils do not have large quantities of gas in solution, the low pressures and high temperatures at which they are produced give rise to substantial amounts of gas which must be properly collected and treated. If the oil is produced separately through the casing and the tubing, then casing gas stream should also be properly collected and treated. This is done by using casing vapor recovery systems, commonly referred to as CVR systems. If the flowline pressure is lower than the casing-head pressure, the two can be commingled and the fluids treated at the separators. One type of CVR system uses a small compressor driven by the beam pump to force the casing gas into the flowline when the pressure differential requires it. More commonly, separate gathering lines are used. Hydrocarbon condensates are sometimes obtained from these streams, but this is seldom enough to pay for the CVR system. Separate gathering lines place a back pressure on the casing, which often results in lower oil production rates. The problem lies in selecting a CVR system which reduces the loss of current revenues from oil sales to acceptable levels while meeting environmental demands.

The waste gas stream, whether it flows from the separator, CVR system, or test station, may contain hydrocarbons and sulfur and nitrogen compounds. The only unusual disposal problems related to thermal operations are those resulting from the prevalence of hydrogen sulfide and, sometimes, from the considerable amount of pollutants that must be handled. Within the U.S.A., the treatment of these effluents varies markedly from one location to another, reflecting the variability of the problem, legal requirements, and local practices.

For gases, condensation, incineration, and scrubbing are the usual procedures. Condensation may be cost effective where there are enough recoverable hydrocarbon liquids. Incineration is used to convert the hydrogen sulfide to sulfur dioxide, as well as to burn the hydrocarbon gases. Burning the gas itself in the steam generators or for some other use is ideal, but may not be possible or may not result in complete combustion of the components if the fuel value of the gas is too low, as could be the case if large concentrations of carbon dioxide are present. In such cases, additional fuel gas is added to the stream to attain the temperature necessary to achieve complete combustion. Finally, scrubbing before venting to the atmosphere can reduce sulfur dioxide in the gas stream to acceptable levels, as is done with the stack gas from the generators. High stacks may be necessary to ensure an acceptable low ground-level concentration of contaminants. Where the volume of the hydrocarbon gases is high enough, large processing plants may be built to recover them.

The method of waste water treatment depends on the type and concentration of the contaminants and on the ultimate disposition of the water, whether to the surface or the subsurface. Subsurface disposal of water may require prior filtering, the removal of oil, and use of chemicals such as bactericides to keep the injection wells from plugging. Furthermore, no effluent should be injected where there is the likelihood of contaminating sources of potable water. Oil and solids can be removed from a water stream by the use of filters (Anonymous, 1975; Evers, 1975), flotation cells, and retention basins. Hydrogen sulfide can be removed from water by stripping with natural gas, by reacting it with oxygen in the presence of a catalyst (Snavely and Blount, 1969), and by frothing (Martin and Levanas, 1962; Bheda and Wilson, 1969). Waste water sometimes is treated for use as feedwater to steam generators (Hull, 1967). An example of a typical water treating sequence, as used in the Kern River field, California, was reported by Smith (1971).

Solids are usually disposed of in pits or retaining basins. Depending on the nearness to habitation and sources of potable water and on the character of the solids, these retaining basins may be lined with materials to preclude or prevent drainage leakoff and rain runoff. Good practice is to cover these when they are full or out of use, and to restore the land at the completion of the project.

SPECIAL SITUATIONS

Offshore operations

One of the several problems posed by steam injection projects from platforms over water is that the tubing through which steam is to be injected must be better insulated than normal, as heat losses from the wellbore to the water are relatively large. Obviously, the importance of these heat losses is in proportion to the depth of the water. The main problem of offshore thermal operations is one of economics. The high cost of leases, platforms, and barge work (specially in salt water environments) make the economics of any offshore operation to be less attractive than on land, everything else being the same. So there are not many such projects. But steam has been injected offshore since as early as 1965 (Cuff, 1966).

The main problems appear to be related to the limited space on the platform in which the facilities must be located, while at the same time a few extra precautions resulting from the new environment have to be taken into account. Space for feedwater treating facilities and the steam generator itself must be provided. The water treating can be done onshore, with the clean water piped in. Transporting the steam generator may pose difficulties. At Platform Emmy (Cuff, 1966), the steam generation equipment was essentially broken down into four units for easy transportation and reassembly. Allowances for expansion of surface injection and production lines require special attention in platforms. In addition the wells must be designed so that the casing can expand without transferring any significant stress to the platform. All this requires extra care and planning. Also, the platform would probably be designed for a large number of pumping units, which is not normal.

Where the offshore production intervals targeted for steam injection are sufficiently deep and close to land, deviated wells have been drilled from onshore. Such steam operations from deviated onshore wells have been practiced since at least 1964 (Cuff, 1966). Unusual aspects of these types of operations, where the tubulars may have high curvatures in places, include difficulties in running and in seating and unseating subsurface tools of conventional length, and in gravel packing the wells. Short tools, and prepacked liners, may be used instead.

Compacting reservoirs

Compacting reservoirs provide additional energy to produce the oil heated by the steam. From the production point of view, compaction is an asset, for it may significantly extend the period of time over which cyclic steam injection remains attractive. In fact, this is the reason why over 80% of the thermal production from the Bolivar Coast of Lake Maracaibo, Venezuela, is still being produced by cyclic steam injection, after some 20 years of thermal operations. In addition to the obvious problems related to possible subsidence and changes in the surface drainage patterns, compacting reservoirs have a higher incidence of well failures. In the Bolivar Coast, well tubulars cannot withstand the loads resulting from compaction, and are replaced at least twice the normal rate.

OPERATIONS

The main responsibility of field and office operations personnel is to ensure that the field operation is running smoothly and on target. This entails a lot more than just making sure the surface equipment is performing adequately. Foremost, it requires that the production be maintained at the highest possible levels for the manpower and resources available, and in a manner consistent with prudent oilfield practices. A harmonious relationship should exist between the field and office personnel. Office staff should be aware at all times about the loss in potential production resulting from lack of manpower or equipment. There always will be work remaining to be done in any operation involving a large number of wells. This work will be prioritized to make sure that the production is maintained at near top levels. Resources are adjusted to meet the project goals.

The identification of problem wells and areas is everyone's responsibility. The engineers in the offices can spot changes in the production behavior of wells after examining trends in the data. Field personnel can sometimes observe these trends several weeks in advance, thus providing an early identification of possible remedial work. There is no set way of going about the identification of problem areas—it simply requires an intimate familiarity with the performance behavior of each well and the ability to observe and relate facts. Factors to be aware of include not only changes in the production rates of all the fluids, but any atypical change in the pressure or temperature of either the flowline or wellhead, in the amount of sand cleaned from the separators, in the ability to break emulsions, and in the performance of beam pumps during rod fall.

REFERENCES

Anderson, G.W., 1978. New methods make downhole liquid plugging practical. *World Oil*, 186(2): 37–43.

Anonymous, 1975. Continuous steam injection, heat scavenging spark Kern River. *Oil Gas J.*, 73(4): 127–142.

API, 1983. *Recommended Practice for Installation and Operation of Wet Steam Generators*. American Petroleum Institute, Washington, D.C., API RP 11T.

Ballard, J.R., Lanfranchi, E.E. and Vanags, P.A., 1977. Thermal recovery in the Venezuelan heavy oil belt. *J. Can. Pet. Technol.*, 16(2): 22–27.

Bansbach, P.L., 1965. Treating emulsions produced by thermal recovery operations. *36th Annu. Calif. Reg. Meet., Soc. Pet. Eng. of AIME, Bakersfield, Calif., November 4–5*, SPE 1328.

Bansbach, P.L., 1970. The how and why of emulsification. *Oil Gas J.*, 68(36): 87–93.

Bertness, T.A., 1965. Thermal recovery: principles and practices of oil treatment. *40th Annu. Meet., Soc. Pet. Eng. of AIME, Denver, Colo., October 3–8*, SPE 1266.

Bheda, M. and Wilson, D.B., 1969. A foam process for treatment of sour water. In: *Water—1969. Chem. Eng. Prog. Symp. Ser.*, 274.

Bissoondatt, J.C., 1976. Casing failure in steam injection wells—a case history of the Guapo Field. *Soc. Pet. Eng. Conf., Trinidad and Tobago Sect., April 2–3*, SPE 5951.

Bleakley, W.B., 1966. Steamed wells need good completions. *Oil Gas J.*, 74(14): 136–138.

Boberg, T.C., Penberthy, W.L., Jr. and Hagedorn, A.R., 1973. Calculating the steam-stimulated performance of gas-lifted and flowing heavy oil wells. *J. Pet. Technol.*, 25(10): 1207–1215.

Bott, R.C., 1967. Cyclic steam project in a virgin tar reservoir. *J. Pet. Technol.*, 19(5): 585–591.

Bradley, B.W. and Gatzke, L.K., 1975. Steam flood heater scale and corrosion. *J. Pet. Technol.*, 27(2): 171–178.

Burkhill, G.C.C. and Leal, A.J., 1974. Aspectos operacionales de la inyección selectiva de vapor en pozos de la C.S.V. de la Costa Bolívar, Estado Zulia. *Proc., Simposio Sobre Crudos Pesados, Inst. Invest. Pet., Univ. del Zulia, Maracaibo, July 1–3.*

Burns, J.A., 1969. A review of steam soak operations in California. *J. Pet. Technol.*, 21(4): 25–34.

Burns, W.C., 1965. Water treatment for once-through steam generators. *J. Pet. Technol.*, 17(4): 417–421.

Burris, D.R., 1977. Dual polarity oil dehydration. *Pet. Eng. Int.*, 49(9): 30, 32, 34, 36, 41.

Cain, J.E., Shryock, O.H. and Carter, G., 1966. Cementing steam injection wells in California. *J. Pet. Technol.*, 18(4): 431–436.

Carter, G. and Smith, D.K., 1958. Properties of cementing compositions at elevated temperatures and pressures. *J. Pet. Technol.*, 10(2): 20–28.

Carter, L.G., Waggoner, H.F. and George, C., 1966. Expanding cements for primary cementing. *J. Pet. Technol.*, 18(5): 551–558.

Coberly, C.J. and Wagner, E.M., 1938. Some considerations in selection and installation of gravel packs for oil wells. *J. Pet. Technol.*, 1(3): 1–20.

Cuff, F., 1966. Steaming from offshore ... mechanical aspects, results. *Oil Gas J.*, 64(28): 92–94.

Davis, J.S. and Young, W.W., 1984. Cogeneration could provide economic benefits in steamflood EOR operations. *Oil Gas J.*, 82(10): 81–86.

Dilgren, R.E., Deemer, A.R. and Owens, K.B., 1982. The laboratory development and field testing of steam/noncondensible gas foams for mobility control in heavy oil recovery. *Calif. Reg. Meet., Soc. Pet. Eng. of AIME, San Francisco, Calif., March 24–26*, SPE 10774.

Elias, R., Jr., Johnson, J.R., Krause, J.D., Scanlan, J.C. and Young, W.W., 1980. Steam generation with high TDS feedwater. *50th Annu. Calif. Reg. Meet., Soc. Pet. Eng. of AIME, Los Angeles, Calif., April 9–11*, SPE 8819.

Evers, R.H., 1975. Mixed media filtration of oily waters. *J. Pet. Technol.*, 27(2): 157–163.

Fanaritis, J.P. and Kimmel, J.D., 1965. Review of once-through steam generators. *J. Pet. Technol.*, 17(4): 409–416.

Fazio, P.J. and Banderob, L.D., 1966. Thermal well problems and completion techniques in the Midway-Sunset Field. *Prod. Mon.*, 30(11): 10–15.

Fitch, J.P. and Minter, R.B., 1976. Chemical diversion of heat will improve thermal recovery. *51st Annu. Tech. Conf. Exhib., Soc. Pet. Eng. of AIME, New Orleans, La., October 3–6*, SPE 6172.

Franco, A., 1974. How MGO handles heavy crude. *Oil Gas J.*, 72(25): 105–109.

Fraser, J.R., Henderson, I.G., Kitzan, P., Schmitz, R.V. and Myhill, N.A., 1982. *Reservoir Performance and Thermal Completions of the Peace River In-Situ Project.* Unitar/CF10/VIII/9, Caracas, pp. 798–814.

Frick, T.C., 1962. *Petroleum Production Handbook*. Society of Petroleum Engineers, Dallas, Texas.

Gates, C.F. and Brewer, S.W., 1975. Steam injection into the D and E Zones, Tulare Formation, South Belridge Field, Kern County, California. *J. Pet. Technol.*, 27(3): 343–348.

Gates, C.F. and Holmes, B.G., 1967. Thermal well completion and operation. In: *Proceedings, Seventh World Petroleum Congress, Mexico City.* pp. 419–429.

Gulati, M.S. and May, G.P., 1975. Thin-section and permeability studies call for smaller gravels in gravel packing. *J. Pet. Technol.*, 27(1): 107–112.

Hall, A.L. and Bowman, R.W., 1972. Operation and performance of Slocum Field thermal recovery project. In: *Thermal Recovery Techniques. Soc. Pet. Eng., Repr. Ser.*, 10: 144–157.

Hess, P.H., Clark, C.O., Haskin, C.A. and Hull, T.R., 1971. Chemical method for formation plugging. *J. Pet. Technol.*, 23(5): 559–564.

Holke, D.C. and Huebner, W.B., 1971. Thermal stimulation and mechanical techniques permit increased recovery from unconsolidated viscous oil reservoir. *42nd Annu. Calif. Reg. Meet., Soc. Pet. Eng. of AIME, Los Angeles, Calif., November 4–5*, SPE 3671.

Hull, R.L., 1967. The Thermosludge water treating and steam generation process. *J. Pet. Technol.*, 19(12): 1537–1540.

Johnson, D.R. and Fox, R.L., 1980. Examination of techniques for thermally efficient delivery of steam to deep reservoirs. *SPE/DOE Symp. on Enhanced Oil Recovery, Tulsa, Okla., April 20–23*, SPE 8820.

Jones, O. and Davis, J.S., 1983. Solid fuels proven for oil-field steam generation. *Oil Gas J.*, 81(8): 109–12.

Juch, A.H., and Watson, R.J., 1968. New concepts in sucker-rod pump design. *43rd Annu. Fall Meet., Soc. Pet. Eng. of AIME, Houston, Texas, September 29–October 2*, SPE 2172.

Leutwyler, K. and Bigelow, H.L., 1965. Temperature effects on subsurface equipment in steam injection wells. *J. Pet. Technol.*, 17(1): 93–101.

Lucas, R.N., 1966. Dehydration of heavy crudes by electrical means. *41st Annu. Meet., Soc. Pet. Eng. of AIME, Dallas, Texas, October 2–5*, SPE 1506.

Ludwig, N.C. and Pence, S.A., 1956. Properties of Portland cement pastes at elevated temperatures and pressure. *J. Am. Concr. Inst.*, 6(1): 27.

Makansi, J. and Schweiger, R., 1982. Fluidized bed boilers. *Power*, 126(8): 8.1–8.16.

Marsden, S.S. and Khan, S.A., 1967. Foam proves effective way to solve production problems. *Oil Gas J.*, 65(23): 126–129.

Martin, J.D. and Levanas, L.D., 1962. Air oxidation of sulfide in process water. *Oil Gas J.*, 60(24): 184–187.

McLennan, I.C., 1968. Raising of high pressure steam from brackish Lake Maracaibo waters. *J. Inst. Pet.*, 54(536): 213–231.

Oestreich, D.K., 1976. *Equilibrium Partial Pressure of Sulfur Dioxide in Alkaline Scrubbing Processes*, EPA-600/2-76-279, 33 pp.

Otto, S.H., 1972. SCAN-automation at Kern River. *Am. Pet. Inst. Pacific Coast Jnt. Chapter Meet., Bakersfield, Calif., October.*

Owens, M.E. and Bramley, B.G., 1966. Performance of equipment used in high-pressure steam floods. *J. Pet. Technol.*, 18(12): 1525–1531.

Plass, L. and Daradimos, G., 1984. Combustion of heavy oil and upgrading residues in the circulating fluid bed for steam production. In: *Second International Conference on the Future of Heavy Crude and Tar Sands*. McGraw-Hill, New York, N.Y., pp. 638–652.

Prats, M., 1982. *Thermal Recovery, Soc. Pet. Eng. Monogr.*, 7: 283 pp.

Pryor, J.A., 1966. Orifice meter measures steam quality. *Oil Gas J.*, 64(22): 86–88.

PURPA, 1980. *Public Utility Regulatory Policies Act 210.*

Reed, M.G., 1980. Gravel pack and formation sandstone dissolution during steam injection. *J. Pet. Technol.*, 32(6): 941–949.

Roberts, J.C. and Williams, J.W., 1970. Steam injection oil recovery, techniques and water quality. *J. Water Pollut. Control Fed.*, 42(8/1): 1437–1445.

Sachtschale, J.R. and Dydo, J.F., 1982. Operation and performance of a double-alkali scrubber. *J. Pet. Technol.*, 34(11): 2630–2636.

Saucier, R.J., 1974. Considerations in gravel pack design. *J. Pet. Technol.*, 27(2): 205–212.

Schafer, J.C., 1974. Thermal recovery in the Schoonebeek oil field—fifteen years of experience. *Erdoel, Erdgas Z.*, 90(10): 372–379 (English translation).

Simon, R. and Poynter, W.G., 1968. Downhole emulsification for improving viscous crude production. *J. Pet. Technol.*, 20(12): 1349–1353.

Slagle, K.A. and Smith, D.K., 1963. Salt cement for shale and bentonitic sands. *J. Pet. Technol.*, 15(2): 187–194.

Smith, M.L., 1971. Waste water reclamation for steam generator feed, Kern River Field, California. *42nd Calif. Reg. Meet., Soc. Pet. Eng. of AIME, Los Angeles, Calif., November 4–5*, SPE 3689.

Snavely, E.S. and Blount, F.E., 1969. Rates of dissolved oxygen with scavengers in sweet and sour brines. *Proc., NACE 25th Annu. Conf.*, pp. 397–404.

Spink, L.K., 1975. *Principles and Practice of Flowmeter Reading*. The Foxboro Company, Foxboro, Mass., 9th ed., 543 pp.

Stambler, I., 1983. Plans for 300 MW & 800 MW oil field cogeneration plants. *Gas Turbine World Cogenerat.*, 13(3): 38–41.

Strassner, J.E., 1968. Effects of pH on interfacial films and stability of crude oil–water emulsions. *J. Pet. Technol.*, 20(3): 303–312.

Struthers Thermo-Flood Corporation, 1979. *Enhanced Oil Recovery Steam Generators*. Bull. TF-100-79-7-7500.

Suman, G.O., 1975. *World Oil's Sand Control Handbook*, Gulf, Houston, Texas.

Traynor, B.V., 1980. New double-walled tubulars can aid thermal-recovery operations. *Oil Gas J.*, 78(7): 103–107.

Trujano, D.N., Garza, B.T. and Lecona, S.C., 1974. How ambient conditions affect steamline heat losses. *Oil Gas J.*, 72(3): 83–86.

Vonde, T.R., 1982. Specialized pumping techniques applied to a very low gravity, sand-laden crude: Cat Canyon Field, California. *J. Pet. Technol.*, 34(9): 1951–1957.

Wendt, R.E., 1978. Review of stack gas scrubber operation experience on an oil field steam generator. *49th Calif. Reg. Meet., Soc. Petr. Eng. of AIME, San Francisco, Calif., April 12–14*, SPE 7125.

White, P.D. and Moss, J.T., 1983. *Thermal Recovery Methods*. PennWell, Tulsa, Okla.

Willhite, G.P. and Dietrich, W.K., 1967. Design criteria for completion of steam injection wells. *J. Pet. Technol.*, 19(1): 15–21.

Wilson, T.M., 1976. Steam quality and metering. *J. Can. Pet. Technol.*, 15(2): 33–38.

Chapter 13

IN-SITU COMBUSTION TECHNOLOGY

MAHENDRA K. SHARMA

INTRODUCTION

This chapter describes a state-of-the-art review of the in-situ combustion process for heavy oil recovery. Among various processes of enhanced oil recovery (Holm, 1968; Minssieux, 1974; Wang and Locke, 1980; Fayers et al., 1981; Isaacs et al., 1981; Johnson et al., 1982; Sharma and Shah, 1983, 1984, 1985; Sharma et al., 1986), in-situ combustion has received an increased attention during the past two decades. Several excellent review articles and research papers have been published on this subject: Aldea et al. (1968), Bousaid and Ramey (1968), Byars (1970), Earlougher et al. (1970), Tadema and Weijdema (1970), Bleakley (1971), Farouq Ali (1972), Gadelle et al. (1981), and Chu (1982, 1983). In-situ combustion has been known since 1888: Mendeleev was the first scientist to suggest the in-situ conversion of coal into combustible gases. Based on the earlier laboratory results, Sheinman and Dubrovai (1934) proposed the process of oil displacement by means of a moving underground firefront. A number of field tests, were performed in various regions in the late 1940's and early 1950's. The results from these tests indicated that the heat losses were large; therefore, the injected hot gases reached the formation zone with zero thermal energy. These studies, however, were followed by laboratory research, field tests and development of mathematical models to simulate in-situ combustion, as a result of which this process has been recognized and can be used as a promising method of recovering heavy oil from petroleum reservoirs.

Several screening methods for in-situ combustion processes have been proposed in the literature (Geffen, 1973; Chu, 1977; Iyoho, 1978). Among these methods, two screening guides were developed by Chu (1982) based on the confidence limit approach and dry- and wet-combustion processes together. No statistical inference, however, can be obtained from these screening guides regarding permeability, porosity, dip angle, and net-to-gross pay ratio as well as other variables of the reservoirs.

RESERVOIR CHARACTERISTICS

The in-situ combustion process was applied to petroleum reservoirs with a wide range of characteristics in order to recover oil. Selected parameters of thirty

TABLE 13-I

Reservoir characteristics of fields where the in-situ combustion process was utilized (modified after Chu, 1982, 1985)

No.	Field and location	Reservoir formation	Operator	Year production started	Average reservoir permeability (mD)	Reservoir porosity (%)	Initial reservoir pressure (psi)	Reservoir temperature (°F)	Oil viscosity (cP)	Gravity (°API)	Oil saturation (%)	Depth (ft)
1	South Belridge, Calif.	Tulare Sand	General Petroleum Corp.	1956	8,000	36.0	–	87	2,700 (87°F)	12.9	59.0	700
2	Niitsu Field, Japan	Sand	–	1957	750	–	–	75	173 (77°F)	18.0	25.4	590–633
3	West Newport, Calif.	Miocene B Sand	General Crude Oil Co.	1958	1,070	37.0	195	105	700 (105°F)	15.2	67.0	800–2,100
4	North Tisdale, Wyo.	Curtis Sand	Continental Oil Co.	1959	1,034	24.5	290	73	175 (73°F)	21.0	64.6	770–1,090
5	Midway Sunset, Calif.	Moco Sand	Mobil Oil Corp.	1960	1,575	36.0	1,000	125	110 (125°F)	14.5	75.0	2,100–2,700
6	East Venezuela	Uncon. sand	Mene Grande	1960	3,500	35.0	1,300	–	400	9.2	94.0	4,500
7	Robinson, Fry, Ill.	Robinson Sand	Marathon Oil Co.	1961	320	19.7	20	65	40 (65°F)	28.7	68.0	880–936
8	Schoonebeek, The Netherlands	–	Shell Oil Co.	1962	3,000	–	–	–	175	–	82.0	2,720–2,850
9	Sloss, Nebr.	Muddy Jl Sandstone	Amoco Production Co.	1963	191	19.3	2,274	200	0.8 (200°F)	38.8	30.0	6,200
10	Bellevue, La.	Nacatoch Sand	Getty Oil Co.	1963	1,094	38.3	–	75	450 (75°F)	19.0	52.2	350

11	South Belridge, Calif.	Tulare Sand	Mobil Oil Co.	1964	3,000	34.0	180	95	1,600 (95° F)	13.0	67.5	1,080
12	Miga, Venezuela	P2-3 Sand and MG 517 Reservoir	Gulf Oil Co.	1964	5,500	22.6	–	146	335 (135° F)	13.5	75.0	4,000–4,350
13	Suplacu de Barcau Romania	Uncon. sand	IFP/ CPPG	1964	1,700	32.0	–	64.4	2,000 (64° F)	15.9	85.0	164–656
14	Pavlova, Gora, U.S.S.R.	Sand	–	1966	1,100	25.0	214	70	170 (70° F)	18.0	71.0	820
15	May-Libby, La.	Tuscaloosa Sandstone	Sun Oil Co.	1966	1,070	31.2	–	135	3 (135° F)	40.0	42.8	3,400
16	East Tia Juana, Venezuela	Lagunillas Sand	Shell Oil Co.	1966	5,000	39.2	–	104	6,000 (104° F)	12.5	78.0	1,540–1,600
17	Trix Lix, Texas	Wood-bine C Sandstone	Sun Oil Co.	1968	500	30.0	–	138	26 (138° F)	24.0	55.9	3,650
18	Glen Hummel, Texas	Poth A Sandstone	Sun Oil Co.	1968	1,200	36.0	800	114	52 (114° F)	20.8	63.1	2,432
19	Caddo Parish, La.	Nacatoch Sand	–	1969	606	35.2	–	80	280 (80° F)	21.1	52.0	1,031
20	Gloriana, Texas	Poth A Sandstone	Sun Oil Co.	1969	1,000	35.0	300	112	111 (112° F)	20.8	53.3	1,600
21	Fosterton Northwest, Sask.	Roseray Sand	Mobil Oil Corp.	1970	958	28.8	240	125	13.5 (125° F)	23.6	45.2	3,100
22	Heidelberg, Mich.	Cotton Valley Sand No. 5	Gulf Oil Corp.	1971	39	16.4	1,500	221	4.5 (221° F)	24.0	77.8	11,500

TABLE 13-I (continued)

No.	Field and location	Reservoir formation	Operator	Year production started	Average reservoir permeability (mD)	Reservoir porosity (%)	Initial reservoir pressure (psi)	Reservoir temperature (°F)	Oil viscosity (cP)	Gravity (°API)	Oil saturation (%)	Depth (ft)
23	Bellevue, La.	Nacatoch Sand	Cities Service Co.	1971	695	34.9	40	75	676 (77°F)	19.0	60.6	350
24	Midway Sunset, Calif.	Potter Sand	CWOD	1972	2,500	33.0	10	110–230	1,630 (120°F)	11.5	53.0	1,500
25	Brea-Olinda, Calif.	First Miocene Sand	Union Oil Co. of California	1972	300	29.0	150	135	20 (135°F)	22.0	54.0	2,550
26	Little Tom, Texas	San Miguel Sandstone	Hanover Petroleum Corp./ ERDA	1975	64	22.3	750	125	90 (125°F)	18.0	60.0	2,800
27	Balaria, Romania	Uncon. sand	IFP/CPPG	1975	500	30.0	–	113	120 (113°F)	19.0	–	2,625
28	Paris Valley, Calif.	Ansberry Sandstone	Husky Oil Co./ERDA	1976	3,266	32.3	250	85	140,000 (84°F)	10.0	62.6	840
29	Bellevue, Bodcau, La.	Nacatoch Sand	Cities Services Co./ US-DOE	1976	700	34.0	40	75	676 (75°F)	19.0	72.6	400
30	Silverdale, Alta.	Sparky Sand	General Crude Oil Co.	1977	3,500	31.0	100	70	804 (70°F)	15.6	86.0	255

reservoirs in various parts of the world are listed in Table 13-I. The process was applied in reservoirs with average permeability ranging from 40 to 8000 mD, whereas the oil saturation varied from 25 to 95%. It is evident from the results presented in Table 13-I that one cannot draw positive conclusion regarding the reservoir suitability for the in-situ combustion process. In addition, fuel content is one of the most important factors influencing the success of a fireflood process.

The fuel content (lbm/ft^3 of burned volume) of the reservoir is the amount of coke available for combustion that is deposited on the reservoir rock as a result of distillation and thermal cracking. If the fuel content is too low, the combustion process in the reservoir cannot be self-sustained. Moreover, a high fuel content requires a large amount of air and high power cost, which means low oil production. Gates and Ramey (1980) compared the estimated fuel content by various methods including laboratory results with that of the field project data. It has been shown that fuel content determined experimentally in the laboratory by tube-run method can provide a reasonably good estimation of the fuel content obtained in the field.

In a separate study, Showalter (1963) presented a correlation between the fuel content and API gravity of crude oil. In addition, the mathematical treatment proposed by Chu (1977) based on the data from several field projects can be used in order to evaluate the fuel content of the petroleum reservoir. For a specific reservoir, it has been shown from field projects as well as from laboratory experiments that the fuel content decreases as the ratio of water to air increases. An attempt was also made to correlate fuel content with water-to-oil ratio in the reservoir using data of several field projects listed in Table 13-I; however, the correlation coefficient was found to be 0.022. This low value suggests that there is no statistically significant relationship between fuel content and water-to-oil ratio in the presence of the widely varying properties of the petroleum reservoir.

Among the various firefloods listed in Table 13-I, twenty-eight field projects are reported to be technical successes. It is evident that the in-situ combustion process has been applied under a wide range of properties and conditions of the petroleum reservoirs, which indicate that the fireflood process is readily applicable. One should keep in mind, however, that many firefloods which have failed, were not reported. In the majority of cases, the failure or success of the process mainly depends on its design and application.

IGNITION

In this section, the pre-ignition operations and ignition process are described. In order to establish permeability to air in a selected petroleum reservoir, it is important to inject air at the designed rate prior to ignition. If the gas saturation in the reservoir formation is nearly zero, the injection process requires high compressor pressures. In the case of heavy oils, the slow oxidation of crude oil may occur during the process of air injection. In most of the cases, the injection gas for ignition is usually air; however, the fuel gas and recycle gas have been injected to lower the oxygen content in order to control the fireflood velocity (Grant and Szasz, 1954). The air fluxes at the injection site have been reported in the range of 0.5–2.9 scf/hr-ft ^{2} (Parrish et al., 1962; Earlougher et. al., 1970). The early breakthrough of injected gas may occur during the pre-injection period, which does not seem to prevent a successful combustion operation (Parrish et al., 1962).

During the early stages of the in-situ combustion, several alternate methods were recommended in order to ignite oil. In one of the methods, high-powered heaters were lowered in the air injection wells. This operation resulted in the formation of coke around the wells, which is much more difficult to ignite than crude oil. An attempt was also made to use a strong reactive and oxidizing agent such as nitric acid. Violent explosions occurred, however, which resulted in destruction of the tubular equipment in the well. In order to reduce the ignition time, a moderate heating of the injected air may be required. If the reservoir temperature is increased to 200°F, a one- or two-day time lapse is needed for ignition. The initial location of ignition can be controlled by preheating. Without preheating, ignition may occur at some distance beyond the face of the injector screen which may result in a reverse combustion. Moderate preheating will cause ignition near the wellbore and high combustion reversal temperatures can be avoided.

The ignition time for a fireflood can be evaluated by the following relation (Tadema and Weijdema, 1970):

$$t = \frac{\rho_1 c_1 T_o (1 + 2T_o/B) e^{B/T_o}}{864 \phi S_o \rho_o H A p_x^n B/T} \times 10^{-2} \tag{13-1}$$

where t = ignition time, days; ρ_o = density of oil, kg/m^3; ρ_1 = density of oil-bearing formation, kg/m^3; c_1 = specific heat of oil-bearing formation, kcal/kg°C; p_x = partial pressure of oxygen (atm abs) = 0.209 p, where p is the injection pressure in atm abs; ϕ = porosity, fraction; H = heat of reaction, kcal/kg O_2; S_o = oil saturation; T_o = initial temperature, °K; A = constant, s^{-1} atm^{-n}; B = constant, °K; and n = pressure exponent.

The specific heat of the formation ($\rho_1 c_1$) can be determined by using the following equation:

$$\rho_1 c_1 = (1-\phi)\rho_s c_s + \phi S_o \rho_o c_o + \phi S_w \rho_w c_w \tag{13-2}$$

where ρ_w = formation grain density, kg/cm^3; c_s = specific heat of the formation grain, kcal/kg °C; c_o = specific heat of oil, kcal/kg °C; ρ_w = water density, kg/m^3; c_w = specific heat of water, kcal/kg °C; and S_w = water saturation.

The values of constants A, B and n can be determined by measuring the oxidation rates of several crude oil–sand mixtures at different pressures and temperatures. The oxidation rate (K) is related to the oil temperature (T_o) and partial pressure of the oxygen (p_x) by the following equation:

$$K = A p_x^n \, e^{-B/T_o} \tag{13-3}$$

TABLE 13-II

Values of A, B and n for various crude oils (table rearranged from Tadema and Weijdema, 1970; courtesy of *Oil and Gas Journal*)

Crude oils	A	B	n
1	84,800	10,270	0.48
2	7,380	9,480	0.31
3	3,080	8,860	0.46
4	1,210	8,680	0.45
5	925	8,640	0.57
6	498	8,880	0.79

The values of the constants A, B and n for various crudes evaluated by using equation (13-3) are listed in Table 13-II (Tadema and Weijdema, 1970). From the knowledge of these constants and field data, one can easily calculate the ignition time from equation (13-1). For the Venezuelan field and a South Belridge field, the ignition times were obtained to be 49 and 99 days, respectively, which are in good agreement with the observed values of 35 and 106 days.

In the laboratory, the following procedure can be followed to determine the oxidation rate of crude oil. A cylindrical autoclave containing 250 mg of experimental sample (e.g., crude oil–sand mixture) can be used. The air is displaced by passing nitrogen. The contents of the autoclave are heated and pressurized to desired conditions, whereas the autoclave is kept in the rotating position. After achieving the desired temperature and pressure for an appropriate period, the autoclave is allowed to cool, and gas content is analyzed for CO_2, O_2 and CO contents. The liquid and solid materials of the sample are also analyzed for total hydrocarbons, bound oxygen, and water. Using these data, one can determine the uptake of oxygen. This procedure is repeated for the same sample at different temperatures and at a constant pressure in order to determine the oxidation rate of crude oil by using equation (13-3). From the slope of the log K versus $(1/T)$ curve, the constant B can be determined, and the value of Ap_x^n can also be evaluated. In order to determine the values of A and n, the experiments for the oxygen uptake should be performed at different pressures.

IGNITION METHODS

The oil characteristics and reservoir conditions play an important role in the combustion process. In order to achieve the combustion in the petroleum reservoir, mainly the following two ignition methods are widely used for heavy oil recovery.

Spontaneous ignition

If the reservoir temperature is very high or sufficient heat is released by the oxidation of oil, spontaneous ignition will occur a few days after air injection is started. The probability of the spontaneous ignition can be determined by

TABLE 13-III

Spontaneous ignition methods (modified after Chu, 1982)

Serial No.	Field location	Operator	Description
1	Brea-Olinda, Calif.	Union Oil Co. of California	ignition confirmed after 0, 9, 15 days
2	East Venezuela	Mene Grande	ignition occurred in 4 days
3	Fosterton Northwest, Sask.	Mobile Corp.	linseed oil added; ignition confirmed after 20 days
4	Heidelberg, Mich.	Gulf Oil Corp.	–
5	Midway Sunset, Calif.	CWOD	steam injected; ignition occurred in a week
6	Midway Sunset, Calif.	Mobil Oil Corp.	ignition occurred after 18 days
7	Miga, Venezuela	Gulf Oil Corp.	–
8	South Belridge, Calif.	Mobil Oil Corp.	ignition occurred shortly after injection
9	South Belridge, Calif.	General Petroleum Co.	ignition occurred after $3\frac{1}{2}$ months
10	Sloss, Nebr.	Amoco Production Co.	reactive crude added; ignition occurred in 1 week

measuring oxidation rate of the oil in the laboratory as described in the previous section or reported in the literature (Burger, 1976). It appears from previous studies that spontaneous ignition is likely to occur when the reservoir temperature is above 131–140°F. In the South Belridge Field operated by the General Petroleum Corporation, however, the ignition occurred at 87°F which is the lowest ignition temperature reported in the literature. This field represents the shallowest depth, nearly 700 ft.

In many projects, several additional means were provided to facilitate spontaneous ignition. In the Sloss project operated by the Amoco Company, a reactive crude oil was used, whereas linseed oil was added to facilitate the ignition in the petroleum reservoir operated by Mobil Oil Corporation in the Fosterton Northwest project. In addition, steam was injected in the Midway Sunset project of CWOD due to the following reasons: (a) to increase the temperature, (b) to shorten the required ignition time, (c) to clean the perforation, and (d) to lower the oil saturation near the wellbore so that the possibility of injector damage due to burning can be reduced. Table 13-III lists various successful fireflood projects in which spontaneous ignition occurred during the process. It has been shown (Gates and Holmes, 1967) that spontaneous ignition requires careful control to minimize well damage caused by burning process.

Artificial ignition

If spontaneous ignition does not occur, ignition can be achieved by artificial means such as a special downhole ignition equipment positioned opposite to the

TABLE 13-IV

Artificial ignition methods (modified after Chu, 1982)

Serial No.	Field location	Operator	Description
Gas burners			
1	May-Libby, La.	Sun Oil Co.	–
2	Bellevue, La.	Cities Service Co.	propane used as fuel
3	Glen Hummel, Texas	Sun Oil Co.	–
4	Sloss, Nebr.	Amoco Production Co.	operated 3 days; 30 MMBlu generated.
5	Shannon Pool, Wyo.	Pan American Petroleum Corp.-Caspar	operated 4 days; 15 MMBlu generated.
Electrical heaters			
6	Bellevue, La.	Getty Oil Co.	440-V, 20-kW heater
7	Robinson, Fry, Ill.	Marathon Oil Co.	40-kW heater
8	Bellevue, La.	Cities Service Co./US-DOE	440-V, 30-kW heater
9	West Newport, Calif.	General Crude Oil Co.	25-kW heater
10	Tisdale, Wyo.	Continental Oil Co.	40-kW heater

reservoir. In many fields, artificial ignition devices, which include gas burners, electrical heaters and catalytic ignition systems were used (Aldea and Petcovici, 1963, 1976). The burner or electric igniter raises the temperature of the area around the injection well to a desired value in order to achieve rapid ignition without damaging downhole equipment. The various ignition equipments and methods of their operation have been described previously (Strange, 1964).

Several artificial ignition methods used in different fireflood projects are listed in Table 13-IV. Electric heaters with a capacity ranging from 20 to 40 kW were used for in-situ combustion. In the Bellevue project operated by the Getty Oil Company 20-kW electric heaters were used. The Continental Oil Company for its North Tisdale project, and the Marathon Oil Company for its Robinson Fry project utilized 40-kW heaters. The Cities Service Company and United States Department of Energy (US-DOE) ignited the Bellevue project with a 30-kW electric heater; whereas a 25-kW electric heater was used by the General Crude Oil Company in the West Newport project.

In artificial ignition with gas burners, natural gas was used as fuel in most of the projects. Among several projects, the Cities Service Company used propane in the burners in the Bellevue fireflood project. One can avoid the chances of well damage due to burnback by controlling the ignition. It has been reported (Gates and Holmes, 1967) that little or no injector damage was experienced in reservoirs requiring artificial ignition.

PROCESSES OF IN-SITU COMBUSTION

This section includes the processes of in-situ combustion used for the displacement of heavy oil from reservoirs. Basically, there are two different processes of in-situ combustion: forward combustion and reverse combustion.

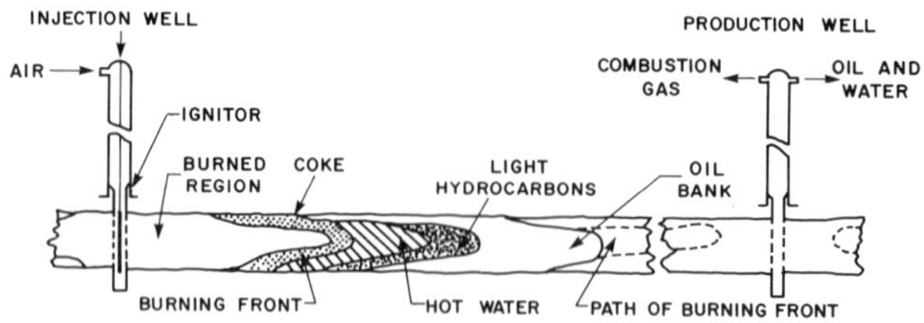

Fig. 13-1. Schematic illustration of the in-situ combustion process. (After Nelson and McNeil, 1961; courtesy of the *Oil and Gas Journal*.)

Forward combustion

The forward combustion process is schematically shown in Fig. 13-1. In this process, the reservoir is ignited in the vicinity of an air injection well. Moreover, the combustion front propagates away from the injection well. The continuous injection of air maintains and drives the combustion front through the reservoir; in general, towards the production wells or the direction of air flow. In addition, if the air injection is reversed (e.g., air injection at the injection well is stopped and then started at the producing well), the direction of flow of the oil bank will be reversed (e.g., in the direction of air injection). The oil bank is thus forced to move through the burned zone, whereas the flow pattern of combustion front will remain unchanged (Fig. 13-1).

The temperature distribution in a reservoir during the forward combustion process is shown in Fig. 13-2. It is evident that the temperature and heat content of the burned zone are quite higher than those of the rest of the reservoir. The oil, therefore, will be heated to temperatures ranging from 500° to 700°F, which results in the reduction of oil viscosity by several orders of magnitude. This reduction in oil viscosity, therefore, allows high-viscosity crude oils to flow freely through the reservoir rock towards the production well. The forward combustion is also known as dry combustion. Moreover, several variations of the forward in-situ combustion have been suggested to enhance the efficiency of this process. In several laboratory and field tests, mainly two variations, i.e., wet and partially quenched combustion were employed. The pilot test operated by the Cities Service Company demonstrated that dry combustion could recover the low-gravity crude oil economically, whereas laboratory studies showed that wet combustion would be superior to dry combustion in the Bellevue Field. Although wet

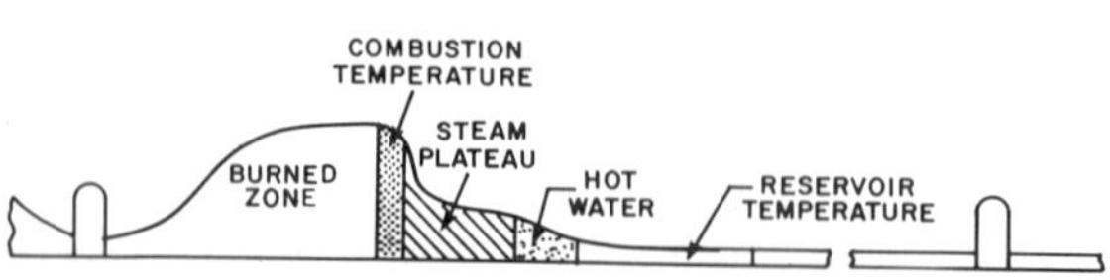

Fig. 13-2. Temperature distribution in the reservoir during a fireflood. (After Nelson and McNeil, 1961; courtesy of the *Oil and Gas Journal*.)

and partially quenched combustion methods have not extended beyond pilot field and laboratory tests, the general opinion is that these processes offer significantly higher potential than does the dry process.

Comparison of wet and dry combustion

Figs. 13-1 and 13-2 show that much heat is left in the burned zone behind the combustion front. During the dry combustion process, this heat is dissipated through the base rock and reservoir cap. Moreover, when water in a moderate amount is injected simultaneously with the air, it converts into superheated steam in the reservoir. The steam travels a short distance from the injection well, whereas the water evaporation front remains close to the injection well (Fig. 13-3a). As the mixture of superheated steam and air approaches the combustion front, only oxygen is utilized in the burning process. Upon passing the combustion front, the superheated steam mixes with nitrogen from air, and the flue gas phase consists mainly of the mixture of carbon monoxide and carbon dioxide.

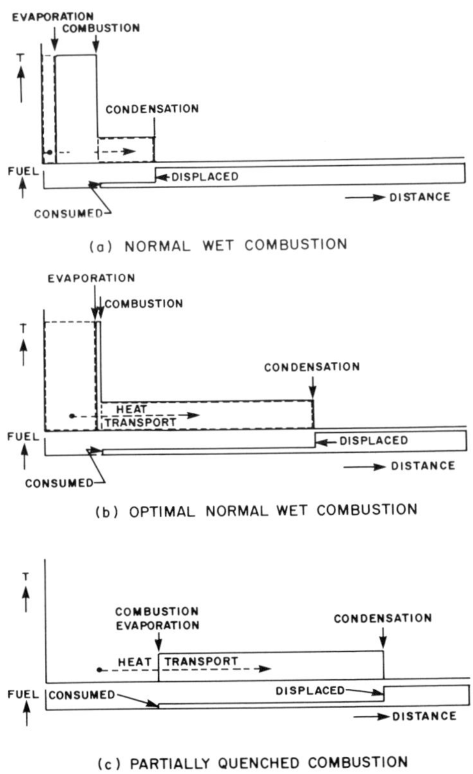

Fig. 13-3. Schematic illustration of (a) normal wet combustion, (b) optimal normal wet combustion, and (c) partially quenched combustion process. (After Dietz and Weijdema, 1968; courtesy of the *Producers Monthly*.)

The flue gas phase displaces the oil in front of the combustion zone and condenses as soon as its temperature drops to nearly 400°F. The size of the steam zone depends on the amount of heat recovered from the burned zone upstream. In order to bring the evaporation front as close as possible to the combustion zone, water in sufficient amount is required for injection (Fig. 13-3b), which, in turn, would result in transporting major part of excess heat of the burned zone to the steam zone. The size of the steam zone, therefore, would increase, and for a given well spacing the combustion front would travel a much shorter distance than that for dry combustion. This would result in displacing more oil towards the producing site. It has been shown in the laboratory that optional wet combustion requires about one-third of the air needed for dry combustion.

In order to compare the dry and wet combustion processes, a field test was performed by the Cities Service Company in 1974 (Joseph and Pusch, 1980). It was observed that the air volume required to process a specific volume of reservoir reduced to 63% with wet combustion as compared to dry combustion. As the major portion of the costs for a combustion project is attributable to maintenance and operation of the air compressor, wet combustion offers significant economic incentives. Moreover, the time required to process pattern volume will also be decreased due to less air needed per pattern in wet combustion. A realistic comparison of the two recovery processes on the basis of oil production, however, was very difficult due to measurements and allocation problems. In addition, a direct method of comparison can be developed from the study presented by Gates and Ramey (1980). Their technique showed that the oil production for a given field operation can be predicted from the relationship between the volume of oil recovered and that burned. If such a relationship is valid, then two combustion processes can be estimated by comparing the burned volumes generated by each process.

The selection between dry combustion and wet combustion is an important decision in conducting a field project. Several laboratory results indicated that the injection of water either simultaneously or alternatively with air in the ignition wells reduces the air/oil ratio. Based on data of various field projects, the correlation between air/oil ratio and water/air ratio was observed to be statistically insignificant in the presence of widely varying properties of the reservoirs (Chu, 1982). The field test of wet and dry combustion conducted by Cities Service Company (Joseph and Pusch, 1980) yielded the following conclusions:

(1) The volumetric sweep was improved to a great extent with wet combustion as compared to dry combustion which, in turn, indirectly implies an increase in oil recovery.

(2) The project life is significantly reduced, which means improving the economics.

(3) The reduced air requirement in wet combustion for a specific volume of reservoir decreases the operating cost and improves the economic future.

It should be noted that these encouraging results may not be entirely applicable to other reservoirs having different characteristics and conditions. The possible advantages of wet combustion over dry combustion, however, can be examined by numerical simulation, laboratory experiments and pilot field tests.

The partially quenched combustion is schematically shown in Fig. 13-3c. When water is introduced in insufficient amounts into the combustion zone, the partially quenched combustion would occur. In a laboratory test, it has been shown that oxygen is utilized within 0.4–3.3 ft at a temperature of 400°F. Thus, the introduction of water in the combustion zone would partially quench the combustion, allowing the oxygen to travel until it comes in contact with oil at 400°F. The combustion zone, therefore, travels at the speed of the cooling water. In this way, all the processes such as partial evaporation and heating of water, heat production by combustion, and heat recovery from the formation occur in one front, which travels fast (Fig. 13-3c).

Reverse combustion

The reverse combustion process is schematically shown in Fig. 13-4. In this process, air is injected through ignition wells that eventually become oil-producing wells. Initially, the reverse combustion process starts as a forward combustion process. After the burning zone moves to within a short distance from the ignition well, air injection is stopped in the ignition wells, and is started in adjacent wells. The air injection is continued in the adjacent wells in order to drive the oil towards the wells which previously were ignition wells. The combustion front moves in the opposite direction towards the adjacent wells (Fig. 13-4). The oxygen required for combustion is only supplied by the air, which is continuously injected in the adjacent wells. In addition, if the oil in the adjacent wells ignites spontaneously, the air injection for reverse combustion is stopped, and the process essentially converts to the forward combustion process.

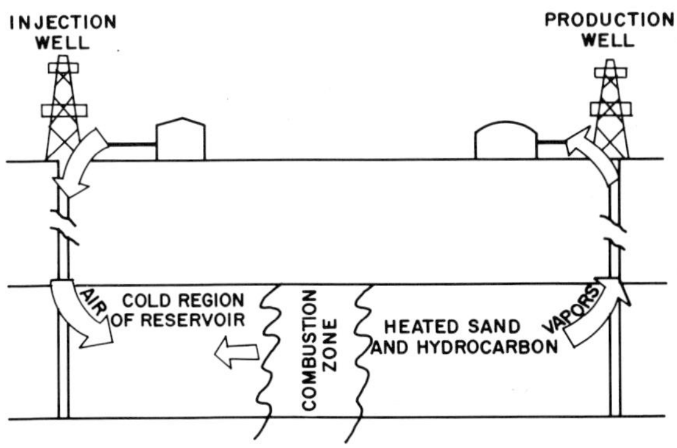

Fig. 13-4. Schematic illustration of the reverse combustion process. (After Berry and Parrish, 1960; courtesy of the Society of Petroleum Engineers of AIME.)

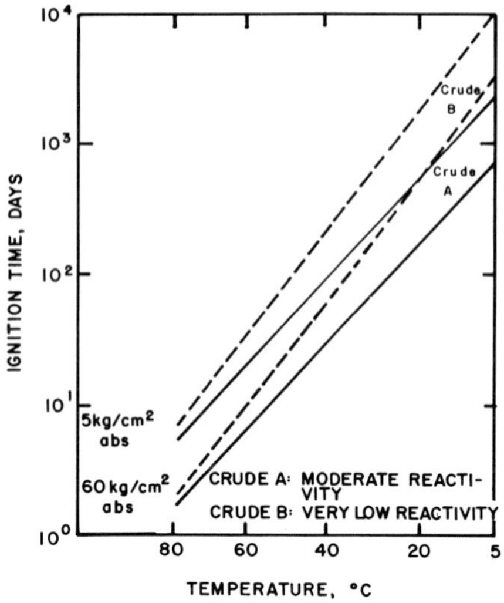

Fig. 13-5. Variation of ignition time as a function of temperature for several air-injection pressures. (After Dietz and Weijdema, 1968; courtesy of the *Producers Monthly*.)

It has been shown that the ignition occurs spontaneously within a period of a few days, except for low-reactivity crudes, i.e., oils at near arctic reservoir temperatures and low reservoir pressures (Dietz and Weijdema, 1968). Fig. 13-5 illustrates the results of ignition time as a function of reservoir temperature for crude oils having different reactivity. These results probably explain the reason why several pilot field tests of reverse combustion process have failed, even though the laboratory experiments have shown satisfactory results.

USE OF IN-SITU COMBUSTION

In this section, the use of in-situ combustion as primary, secondary, and tertiary recovery processes is described. Most of the in-situ combustion field projects have been performed for secondary oil recovery purposes, whereas in some cases, combustion process was also employed as primary and tertiary recovery processes. Several examples are given which explain why in-situ combustion was selected as a primary, secondary, or tertiary recovery process for recovering oil at various depletion stages of a reservoir.

Use as a primary recovery process

In a review article, Poettmann (1964) stated that the sooner the combustion process is applied to a reservoir, the better the results. The application of the in-situ combustion to a virgin reservoir is just one step ahead in the right

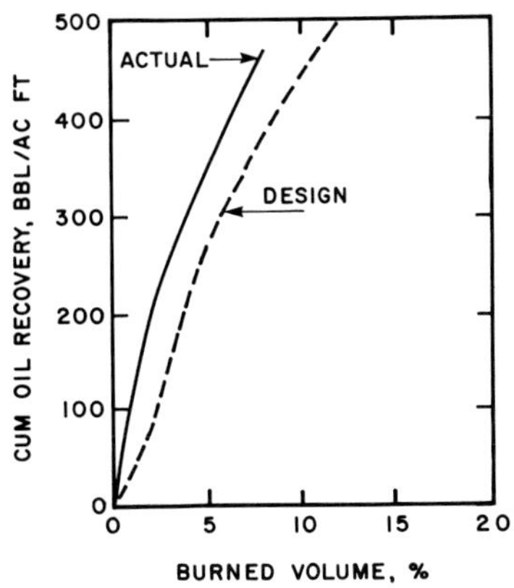

Fig. 13-6. Cumulative oil recovery as a function of burned volume for the in-situ combustion process in the Moco Zone, Midway Sunset Field, Kern County, California. (After Gates and Sklar, 1971; courtesy of the Society of Petroleum Engineers of AIME.)

direction. It should be emphasized, however, that economics is the only criterion which determines the optimal timing of starting not only an in-situ combustion but also other enhanced oil recovery processes.

In a Moco zone reservoir of the Midway Sunset Field, Kern County, California, evaluation of several operating methods indicated that the heavy crude oil would be recovered most economically if the combustion process was applied in the beginning of the project. Accordingly, the combustion was started in November 1959, whereas the air injection begun in January 1960, at which time the cumulative oil production from the reservoir was 0.4% of the oil-in-place. The reservoir pressure was only slightly lower than the initial value (1000 psi) of the virgin reservoir. In 1971, the total recovery of the heavy crude was 9 million bbl or 24% of the original oil-in-place and amount of air injected was 26 Bcf. The volume of the burned oil was estimated at 7.5% of the reservoir in 1971, and was mostly confined to the area at the top of the structure around the injectors and nearby producers. Based on the results of the South Belridge thermal recovery experiment (Gates and Ramey, 1958), a relationship between oil recovery and burned volume used in the design of the Moco combustion project is shown in Fig. 13-6 (dashed line). The actual performance using same assumptions is shown as a solid line (Fig. 13-6).

Use as a secondary recovery process

Several field projects were chosen to employ combustion as a secondary recovery process to provide a possibly more efficient alternative method to waterflooding under certain reservoir conditions. In the Brea-Olinda Field, California, the combustion project was started by the Union Oil Company, because two other secondary recovery methods (waterflooding and cyclic steam processes) had failed (Showalter and MacLean, 1974). The waterflooding was unsuccessful because of the extreme reservoir heterogeneity in which water tended to channel through the formation excessively. Although the cyclic steam injection did increase crude oil production to some extent, the added recovery was not economically beneficial compared with the costs.

A combustion project in a thin reservoir containing high-gravity oil in the May-Libby reservoir, Louisiana, was operated by the Sun Oil Company. A detailed analysis of waterflood project showed that nearly 53% of the original oil-in-place would remain in the reservoir at the end of waterflooding. An efficient recovery method was needed, therefore, to recover the rest of the oil. Based on the analysis of laboratory and reservoir modeling results, it was decided to initiate a fireflood in the western portion of the reservoir, while continuing waterflood in the eastern portion of the reservoir.

In the Sun Production Company's Glen Hummel Field, Texas (Buchwald et al., 1973), waterflood, polymer flood and in-situ combustion were evaluated. Among these three processes, combustion was finally selected based on the expected recovery, profits, and anticipated life of the project.

In the Amoco Production Company's Shannon pool, Salt Creek Field, Wyoming (Parrish et al., 1962), primary production method recovered only about 2% of the original oil-in-place in several decades. An analysis of the reservoir indicated that oil recovery by waterflooding may not exceed 30% of the original oil-in-place. In addition, several research results showed that this reservoir was favorable for in-situ combustion process. Finally, it was decided to start in-situ combustion to recover oil from this reservoir.

A research group of the Continental Oil Company (Martin et al., 1972) discussed in detail why in-situ combustion was selected as a secondary recovery method in the North Tisdale Field, Wyoming. Water cuts had increased to 50% by the time the original development was completed. The primary recovery production declined to nearly 60% of its peak rate by 1957, and it was estimated that economical recovery by primary method would be less than 5% of the original oil-in-place. It was determined that conventional waterflood could yield additional recovery; however, the presence of bottom water, the relatively high oil viscosity, and high water cuts were not favorable for such a process. Moreover, other reservoir characteristics appeared to be favorable for fireflood, including porosity, permeability, depth, and pay thickness. Among several variables, the main concern was the presence of the bottom-water zone. Thus, the combustion process was selected to be employed for oil displacement.

Use as a tertiary recovery process

In several reservoirs, a considerable amount of oil still remains after successful completion of primary and secondary recovery processes. A combustion pilot project was started when a successful waterflood was approaching its economic limit in the Fosterton Northwest Unit, Fosterton Field, Saskatchewan, Canada

(Marberry and Bhatia, 1974). When the combustion project was started, the cumulative oil recovery by primary and secondary methods was 23% of the original oil-in-place. The main aim of this combustion pilot study was to evaluate the feasibility of employing in-situ combustion as a tertiary oil recovery process. If this pilot project is successful, combustion process could be utilized in several reservoirs in southwestern Saskatchewan which are similar in characteristics to those of the Fosterton Northwest Unit. Inasmuch as the oil viscosity is as low as 14 cP at the reservoir temperature, other EOR methods were also considered. It was decided that in-situ combustion was only one of the possible methods to enhance productivity of the waterflooded reservoir.

Three in-situ combustion pilots were conducted in the Potter Sand, Midway Sunset Field, California (Counihan, 1977). One of the three projects was unique and combustion was used as a tertiary oil recovery method after steamflooding process. The main aim of this pilot project was to determine its feasibility and to compare the results obtained from steamflooding and in-situ combustion flooding processes under similar conditions.

The combustion process was also employed as a tertiary recovery method in the Sloss Field, Nebraska, by Amoco. The results of several studies, however, indicated that the reservoir was not an ideal candidate for in-situ combustion. Reservoir properties such as permeability and porosity are reasonably good, but not outstanding. The reservoir is thin and deep, containing light oil with low viscosity. In addition, the residual oil saturation at the end of waterflood was low. A pilot project, however, was started. Later it was expanded to a full-scale project to evaluate the economic success of employing combined forward combustion and waterflooding processes under these reservoir conditions (Parrish et al., 1974).

These examples illustrate that the in-situ combustion process can be applied to recover the residual oil after completion of waterflooding or other secondary methods.

CONCLUSIONS

It has been shown that the in-situ combustion process is suitable to displace oils of gravities greater than 10° API. The average oil recovery by employing in-situ combustion is nearly 50%. The major amount of oil is recovered before breakthrough of the combustion zone. For heavy oils, about 50% crude oil recovery occurs after breakthrough, whereas low-viscosity oil production declines very rapidly following breakthrough. The breakthrough of combustion zone can be recognized by an increase in gas production and its oxygen content. This is followed by a sharp increase ranging from 100° to 200°F in bottomhole temperature. In addition, the increase in water cut of the produced oil also indicates the breakthrough of the combustion zone. At the same time, the pH of the produced water decreases, which is usually due to increase in the content of ions such as iron and sulfate.

Economic evaluation

It is recognized that the success or failure of an enhanced oil recovery process depends on the economic evaluation. An economic study completed by Wilson and Root (1966), which is based on a modified form of two-dimensional model presented by Chu, compares the cost of heating a reservoir. The cost comparison was studied for a reservoir either in the presence of steam injection or forward combustion without considering oil production. The main consideration was to determine heating cost of the same dimensions of a reservoir by either steam injection or by forward combustion. The following conclusions were drawn from this study:

(1) Combustion is favored over steam injection as the sand thickness decreases and the pressure increases.

(2) As the coke deposition increases, steam injection is favored over the combustion process.

(3) As the heated distance in the reservoir increases, reservoir heating by combustion is more favorable as compared to steam injection.

(4) Decreased injection rates favor the cost of steam injection relative to air.

(5) Increased wellbore losses with increasing depth favors combustion.

Disadvantages

There are several disadvantages of in-situ combustion, which can be summarized as follows:

(1) The major portion of heat generated by in-situ combustion is not utilized in heating the oil. Instead, it heats the oil-bearing strata, base and cap rocks, and interbedded shales. In-situ combustion, therefore, would be economically feasible in those reservoirs which have high oil saturation and high porosity, as well as moderate sand thickness (with less rock formation to be heated).

(2) Several studies indicate that the reservoirs with viscous and low-gravity crudes are best suited for in-situ combustion, because they provide the needed fuel for combustion. In addition, the required air/oil ratio for viscous crudes is high, whereas their prices are, in general, lower than the high-gravity crudes. Some projects for higher-gravity and lower-viscosity crudes, however, have also been reported to be successful.

(3) The fireflood process has a tendency to displace oil only from top portion of the oil zone. The vertical sweep in very thick reservoir formation, therefore, is likely to be poor.

(4) There are several serious operational problems associated with in-situ combustion process. For example, formation of emulsions causes pumping problems and reduces well productivity. The production of acidic (low pH) hot water with high content of iron and sulfate ions causes environmental pollution and well corrosion problems, whereas increased sand production and cavings cause plugging of well liner. In addition, there is a possibility of (a) plugging of the

producing wellbore due to deposition of carbon and wax as a result of thermal cracking of the oil, and (b) liner and tubing failure due to excessive temperatures at the production well.

(5) There is a large investment required for the installation of in-situ combustion equipment. Surface installation, however, needs less fuel than hot-water or steam-generating equipments.

(6) There is a possibility of explosion in the compressor. In order to minimize explosion hazards in the air injection system, an explosion-proof lubricant should be employed.

CURRENT STATUS OF IN-SITU COMBUSTION

The in-situ combustion process is attractive economically, provided it is applied to petroleum reservoirs containing approximately 50% oil saturation. The fuel content is one of the important parameters for combustion support at a relatively low air/oil ratio. Laboratory studies on fuel content and air requirements can establish the possibility of starting a field project. Although laboratory experiments can provide some basic understanding of the process, the primary evaluation factor is a field application before the process is employed on a large scale.

The present status of oil production by in-situ combustion in the United States is nearly 11,000 bbl/day, whereas the world oil production is approximately 58,000 bbl/day. It is likely that very little laboratory research can be performed to improve the displacement efficiency of this process. With continued improvement of the in-situ combustion technology, it is almost certain that some form of this process, such as dry, wet, and partially quenched combustion, will find greater application in the coming years.

REFERENCES

Aldea, G. and Petcovici, V., 1963. Aprinzator de strat (gas). *SRS Patent*, No. 43685 (Rumanian Patent).

Aldea, G. and Petcovici, V., 1976. Aprinzator de strat (electricity). *SRS Patent*, No. 87379 (Rumanian Patent).

Aldea, G., Petcovici, V., Dumitrescu, H. and Palada, T., 1968. Aplicarea experimentala a metodei de exploatare prin combustie subterana in republica socialista Romania. *Pet. Gaze*, 19: 26–36.

Berry, V.J., Jr. and Parrish, D.R., 1960. A theoretical analysis of heat flow in reverse combustion. *J. Pet. Technol.*, 12(5): 15–16.

Bleakley, W.B., 1971. Fry unit fireflood surviving economic pressures. *Oil Gas J.*, 69(18): 92–97.

Bousaid, I.S. and Ramey, H.J., Jr., 1968. Oxidation of crude oil in porous media. *Soc. Pet. Eng. J.*, 4(2): 137–148.

Buchwald, R.W., Jr., Hardy, W.C. and Neinast, G.S., 1973. Case histories of three in-situ combustion projects. *J. Pet. Technol.*, 25(7): 784–792.

Burger, J.G., 1976. Spontaneous ignition in oil reservoirs. *Soc. Pet. Eng. J.*, 16(2): 73–81.

Byars, C., 1970. Firefloods blossom in North Louisiana. *Oil Gas J.*, 68(31): 58–62.

Chu, C., 1977. A study of fireflood field project. *J. Pet. Technol.*, 29(2): 171–179.

Chu, C., 1982. State-of-the-art review of fireflood field project. *J. Pet. Technol.*, 34(1): 19–36.

Chu, C., 1983. Current in-situ combustion technology. *J. Pet. Technol.*, 35(8): 1412–1418.

Chu, C., 1985. State of the art review of steam flood field projects. *J. Pet. Technol.*, 37(11): 1887–1902.

Counihan, T.M., 1977. A successful in-situ combustion pilot in the Midway Sunset Field, California. *Calif. Reg. Meet., Soc. Pet. Eng. of AIME, Bakersfield, Calif., April 13–15*, SPE 6525.

Dietz, D.N. and Weijdema, J., 1968. Reverse combustion seldom feasible. *Prod. Mon.*, 32: 19–23.

Earlougher, R.C., Jr., Galloway, J.R. and Parsons, R.W., 1970. Performance of the Fry in-situ combustion project. *J. Pet. Technol.*, 22(5): 551–557.

Farouq Ali, S.M., 1972. A current appraisal of in-situ combustion field tests. *J. Pet. Technol.*, 24(4): 477–486.

Fayers, F.J., Hawes, R.I. and Mathews, J.D., 1981. Some aspects of the potential application of surfactants or CO_2 as EOR processes in North Sea reservoirs. *J. Pet. Technol.*, 33(9): 1617–1627.

Gadelle, C.P., Burger, J.G., Bardon, C.P., Machedon, V., Carcoana, A. and Petcovici, V., 1981. Heavy-oil recovery by in-situ combustion—two field cases in Rumania. *J. Pet. Technol.*, 33(1): 2057–2066.

Gates, C.F. and Holmes, B.G., 1967. Thermal well completions and operation. *7th World Pet. Congr., Mexico City*, Paper PD11.

Gates, C.F. and Ramey, H.J., Jr., 1958. Field results of South Belridge thermal recovery experiments. *Trans. AIME*, 213: 236–244.

Gates, C.F. and Ramey, H.J., Jr., 1980. A method for engineering in-situ combustion oil-recovery projects. *J. Pet. Technol.*, 32(2): 285–294.

Gates, C.F. and Sklar, I., 1971. Combustion as a primary recovery process—Midway Sunset Field. *J. Pet. Technol.*, 23(8): 981–986.

Geffen, T.M., 1973. Oil production to expect from known technology. *Oil Gas J.*, 71(19): 66–76.

Grant, B.F. and Szasz, S.E., 1954. Development of an underground heat wave for oil recovery. *Trans. AIME*, 213: 236–244.

Holm, L.W., 1968. The mechanism of gas and liquid flow through porous media in the presence of foam. *Soc. Pet. Eng. J.*, 8(4): 359–369.

Isaacs, E.E., Prowse, D.R. and Rankin, F.P., 1981. The role of surfactant additives in the in-situ recovery of bitumen from oil sands. *32nd Annu. Tech. Meet., Pet. Soc. CIM/83rd Annu. Gen. Meet., CIM, Calgary, May 3–6*, pp. 1–18.

Iyoho, A.W., 1978. Selecting enhanced oil recovery processes. *World Oil*, 187(6): 61–64.

Johnson, L.A. Jr., Fahy, L.J., Romanowski, L.J. Jr., Thomas, K.P. and Hutchinson, H.L., 1982. An evaluation of a steamflood experiment in a Utah tar sand deposit. *J. Pet. Technol.*, 34(5): 1119–1126.

Joseph, C. and Pusch, W.H., 1980. A field comparison of wet and dry combustion. *J. Pet. Technol.*, 32(9): 1523–1527.

Lewin and Assocs., Inc., 1976. The potentials and economics of enhanced oil recovery. *U.S. Fed. Energy Admin. Rep.*, April B-76/221.

Marberry, J.E. and Bhatia, S.K., 1974. Fosterton Northwest—A tertiary combustion case history. *Soc. Pet. Eng. Symp. on Improved Oil Recovery, Tulsa, Okla., April 22–24*, SPE 4764.

Martin, W.L., Alexander, J.D., Dew, J.N. and Tynan, J.W., 1972. Thermal recovery at north Tisdale field, Wyoming. *J. Pet. Technol.*, 24(5): 606–616.

Minssieux, L., 1974. Oil displacement by foam in relation to their physical properties in porous media. *J. Pet. Technol.*, 26(1): 100–108.

Parrish, D.R., Rausch, R.W., Beaver, K.W. and Wood, H.W., 1962. Underground combustion in the Shannon Pool, Wyoming. *J. Pet. Technol.*, 14(2): 197–205.

Parrish, D.R., Pollock, C.B., Ness, N.L. and Craig, F.F., Jr., 1974. A tertiary Cofcaw pilot test in the Sloss field, Nebraska. *J. Pet. Technol.*, 26(6): 667–675.

Poettmann, F.H., 1964. In-situ combustion: a current appraisal, 2. *World Oil*, 158(5): 124–126, 128; 158(6): 95, 96, 98.

Sharma, M.K. and Shah, D.O., 1983. Interfacial phenomena in foam flooding process for heavy oil recovery. In: *Proceedings, 18th Intersociety Energy Conversion Engineering Conference*, pp. 527–534.

Sharma, M.K. and Shah, D.O., 1984. Effect of oil viscosity on recovery processes in relation to foam flooding. *J. Am. Oil Chem. Soc.*, 61: 585–590.

Sharma, M.K. and Shah, D.O., 1985. Macroemulsions and microemulsions in enhanced oil recovery. *Am. Chem. Soc. Ser.*, 272: 149–172.

Sharma, M.K., Shah, D.O. and Brigham, W.E., 1986. The effect of mixed chain length surfactants on fluid displacement in porous media by in-situ foaming process. *SPE Res. Eng.*, 1(3): 253–260.

Sheinman, A.B. and Dubrovai, K.K., 1934. *Underground Gasification of Oil Reservoirs and a Thermal Method of Oil Recovery*. Nedra Press, Moscow (in Russian).

Showalter, W.E., 1963. Combustion drive tests. *Soc. Pet. Eng. J.*, 3(1): 53–58.

Showalter, W.E. and MacLean, M.A., 1974. Fireflood at Brea-Olinda Field, Orange County, California. *Soc. Pet. Eng. on Improved Oil Recovery, Tulsa, Okla., April 22–24*, SPE 4763.

Strange, L.K., 1964. Ignition: key phase in combustion recovery. *Pet. Eng.*, November, pp. 105–109; December, pp. 97–106.

Tadema, H.J. and Weijdema, J., 1970. Spontaneous ignition of oil sands. *Oil Gas J.*, 68(50): 77–80.

Wang, G.C. and Locke, C.D., 1980. A laboratory study of the effects of CO_2 injection sequence on tertiary oil recovery. *Soc. Pet. Eng. J.*, 20(4): 278–280.

Wilson, L.A. and Root, P.J., 1966. Cost comparison of reservoir heating using steam or air. *J. Pet. Technol.*, 18(2): 233–239.

Chapter 14

MICROBIAL ENHANCED OIL RECOVERY

REBECCA S. BRYANT, ERLE C. DONALDSON, TEH FU YEN
and GEORGE V. CHILINGARIAN

MICROORGANISMS

Types and properties of microbes

The term microorganism encompasses five major groups of organisms: viruses, fungi, algae, protozoa, and bacteria. These are organisms that exist as individual cells or undifferentiated aggregates of cells (cells that are not differentiated into tissues that have distinct functions). The size of microbial cells is so small that a microscope is required for observation. The viruses are the smallest form of recognized microbial life. They are not cells because they differ in many respects from the multifunctional cells: the viruses are much simpler in structure containing only protein and nucleic acid surrounded by a lipid or protein membrane and they do not reproduce by growth followed by division as cells do. Viruses must use other living cells in order to reproduce.

A second division of microorganisms is the eucaryotic microbes which include the fungi, algae, and protozoa. Eucaryotic microbes are distinguished from viruses and bacteria by virtue of possession of a true nucleus, which is enclosed by a membrane that contains the genetic material of the cell (the deoxyribonucleic acid (DNA)), organized into structures known as chromosomes. Eucaryotic microorganisms also have specialized organelles in their cells, such as Golgi bodies that conduct specialized functions of transport of materials within the cell and secretion of materials to the exterior (Gaudy and Gaudy, 1980). Although eucaryotic microbes are probably responsible for some microbial plugging problems of injection wells, they are not important to enhanced oil recovery processes at this time; therefore, the reader is referred to the literature for additional information on them.

The third division of microorganisms that can be distinguished by its physical characteristics is the procaryotes. The procaryotes are in general about ten times smaller than the eucaryotes and the structural features within the cells are not distinguishable with an optical microscope. The intracellular features of some procaryotes can be observed by staining, but an electron microscope is required for detailed structural observation. The only food utilized by procaryotes (or bacteria) comprises soluble molecules that can be assimilated through the cell wall. On the other hand, the eucaryotic protozoa contain a flexible outer membrane that can surround particles of food to form a vacuole where digestion of the food takes place.

Bacteria have two distinctive structural features that are: a rigid cell wall that determines the shape of the organism, which may be either cylindrical or spherical, and those that are mobile possess flagella which are responsible for their movement. Differences in the cell walls of bacteria furnish the basis for classification into two broad groups: Gram-positive and Gram-negative bacteria. The cell wall of Gram-positive bacteria consists of multiple layers of peptidoglycan, cross-linked through amino acid bridges, and teichoic acids bonded to the peptidoglycans. The three-dimensional network of molecules provides a strong (rigid) structure. The wall of the Gram-negative bacteria also contains peptidoglycan; however, the wall is very thin and surrounded by a lipid layer of lipoprotein and lipopolysaccharide, sometimes referred to as the outer membrane.

The Gram stain test is conducted by applying crystal violet, which is adsorbed by the bacteria. Next, a solution of iodine is added, which forms an iodine-dye complex. Then the mixture is treated with ethanol which removes the iodine-dye complex from the Gram-negative cells where the complex is not chemically bound. Finally, a red stain is applied, which colors the Gram-negative bacteria (that could not retain the dye complex) red. The two types of bacteria are then easily distinguished under the microscope: Gram-positive bacteria appear purple and the Gram-negatives are pink in color. Retention of the iodine-dye complex is related to the structure of the cell wall; it is retained by the Gram-positive bacteria that contain the thick, multilayered, cross-linked cell walls described above (Nester et al., 1978; Gaudy and Gaudy, 1980).

The bacterial cytoplasm is a concentrated solution of organic and inorganic molecules which are prevented from leaving the cell by the cell membrane; however, water and other small molecules can move freely through the semipermeable membrane. This concentration gradient between the molecules inside the cell and the surroundings produces an osmotic pressure within the cell. If the cell did not have a rigid wall that can withstand the osmotic pressure, it would expand and burst. The osmotic pressure may be as much as 2.5 MPa (25 atm). If the concentration of low-molecular-weight compounds in the surroundings is high, the osmotic pressure in the cell decreases. Indeed, bacteria that occur in the Dead Sea water (containing approximately 30% salt) do not have a rigid cell wall and will burst if the salt concentration is decreased (Nester et al., 1978).

There are many species of bacteria having a variety of sizes and shapes. Some have flagella that are used for movement within an aquatic environment. Bacteria that are not mobile are transported only by motion of the fluid. Some bacteria are cylindrical (or rod-shaped), whereas others are spherical. Some exist as individual cells, whereas others grow in aggregates or chains of cells.They range in size from 0.2 to about 5 μm and are able to penetrate consolidated rocks that typically have pores as large as 100 μm.

Bacteria are the only microbes that have thus far been proposed for development of processes for enhancement of oil recovery, because they possess several important properties: small size, exponential growth rate when supplied with essential nutrients, and production of metabolic compounds, such as gases, acids, low-molecular-weight solvents, surfactants, and polymers. Various types of bacteria also tolerate harsh environments similar to those encountered in subsurface geological formations, such as high salinity, high pressure, and high temperature. Also, many bacteria are anaerobic (grow in the absence of oxygen). The ranges of physical and chemical properties of bacteria are so wide that in many cases it may be prudent to select a culture of mixed bacteria to live in a symbionic relationship in a petroleum reservoir to improve the recovery of oil. Thus, microorganisms have a large capacity for chemical synthesis.They produce a wide range of products, generally from relatively simple nutrient compound, and propagate vigorously under favorable conditions.

Microbial growth and metabolism

Populations of microbes are found everywhere in nature; in fact, they are even found in areas that will not support any other form of life. The actual species growing in a particular environment comprise those that have been able to successfully adapt to the prevailing environmental and nutritional conditions and the extremes of variations of those conditions. They also have been the most successful in competition with other microbes that may have entered the particular ecosystem. There may be several species living in an area, apparently as a homogeneous population. On closer inspection by division of part of the material supporting the growth (soil, decaying matter, etc.) into squares of approximately 100 μm^2, however, one will often find that only one species occupies this microzone where it has been able to competitively exclude all other microbes. The combined effects of all of the species in a given zone develops a symbiotic relationship that results in recycling of essential chemical compounds and elements required for maintenance of life. The population will usually remain stable as long as the environmental and nutrient conditions do not undergo drastic changes. Small changes of the environmental conditions, however, can result in rapid changes of the relative populations of the species living in a particular zone.

Utilization of nutrients in the environment to maintain metabolism and growth depends on the enzyme inventory of a given species of bacteria. Enzymes are protein molecules endowed with the specific characteristics of organic catalysts. They lower the chemical activation energy of metabolic materials (substrates) in the surroundings, allowing them to undergo various organic reactions at low temperatures which are compatible to the living organism. The enzymes are true catalysts in that they remain unchanged by the reaction which causes a rearrangement of substrate molecules or decompositions into smaller units. The enzymes increase the rates of reaction as well as facilitating low-temperature reactions. In the absence of the enzyme, the same reaction will only take place at an elevated temperature, which may be too high for living organisms to tolerate. The relationship among the rate of reaction, temperature, and the activation energy is expressed by the Arrhenius equation (Levenspiel, 1972):

$$\log(K) = \log(A) - E/(2.3RT) \tag{14-1}$$

where K = reaction rate constant; A = constant, mol/l; E = activation energy, cal; R = gas constant (1.987 cal/deg-mol); and T = temperature, ° K.

Equation (14-1) is used to determine the activation energy by conducting the reaction at various temperatures and measuring the rate of product formation. A plot of the rates of reaction is then made versus the reciprocal of the absolute temperature. The activation energy is obtained from the slope of the line and the constant, A, from the vertical intercept.

Differentiation, as to whether one or another type of bacteria will survive in a given environment containing food resources, depends on the types of enzymes associated with a given species. This is due to the fact that the enzymes have a high degree of specificity with regard to the substrates with which it will interact as a catalyst. Thus, one type of bacteria can assimilate paraffin hydrocarbons, whereas another cannot do so. In the majority of cases, the enzymes remain within the cell and the substrates must penetrate the cell wall, with the aid of a protein which is made specifically for transportation of substrates through the cell wall. This special protein is called permease. In some cases, enzymes are present on the outside of the cell wall, or may even be released as free molecules into the solution surrounding the microbe. Nevertheless, it is the enzyme inventory of the given species, and the permease, which control the type of substrate that can be utilized by the microbes and the rate of transport into the cell. Excretion of molecules from the cell also appears to be a controlled type of process (Nester et al., 1978).

The catalytic function of enzymes depends on the presence of special groups and the spatial configuration of other groups on the protein. Any change of these special functional groups results in a decrease, or complete termination, of catalytic activity with fatal consequences to the microorganism. A change of pH, above or below the optimum for the microbes, will change the functional groups enough to slow down their activity. With a more severe change of pH (+ or −, 2–4 pH units) the functional groups may be destroyed (denatured).

The temperature of the environment also has a profound influence on the microbe. As indicated by equation (14-1), with increasing temperature the rate of reaction also increases; but, an optimum temperature level exists for any given species. When the temperature is raised above the optimal level, the microbial rate of metabolism decreases and finally stops as the proteins making up the enzymes are denatured. Most of the enzymes cannot withstand temperatures greater than 70°C; however, a few enzymes which are possessed by thermophilic microbes remain active at temperatures up to 100°C (Moses and Springham, 1982).

Bubela (1983) found that an increase of pressure increased the optimum metabolic temperature and rate of growth of rod-shaped bacteria (6–8 μm long, 3–4 μm in diameter). At atmospheric pressure (101 kPa) the optimum growth temperature was 50°C with a mean generation time of 17 hours, but when the pressure was increased to 20 MPa the maximum growth rate occurred at 65°C with a mean generation time of 12 hours. The morphology of the microbe also changed from rod-shaped to coccoidal (about 5 μm in diameter). Cores from offshore wells in Brazil's Namorado oilfield at a depth of more than 3000 m (about 30 MPa pressure), contained large populations of *Desulfovibrio* and clusters of coccoidal (spherical) bacteria. When the coccoidal bacteria were grown at one atmosphere pressure in the laboratory, they reverted to rod-shaped morphology, indicating either *Clostridium* or *Bacillus* (Petzel and Williams, 1986). Similar results were discussed by Marquis (1983). Moses and Springham (1982) also reported that bacteria have been found to be catalytically active at 179 MPa and that other bacteria were found to raise their optimum growth temperature from 65° to 85°C, when the pressure of the growth medium was increased from 101 kPa to 60 MPa. Thus it is evident that the denaturization of some enzymes is inhibited by an increase of pressure.

Another environmental factor that is important to microbial enhancement of oil recovery is the salt concentration (NaCl and $CaCl_2$) of the surroundings. As discussed earlier, a difference of concentration between the surroundings and the cytoplasm within the cell produces a large osmotic pressure difference which can result in dehydration of the cell followed by growth inhibition or death of the cell. This is the basis of the high salt or sugar concentrations used for preservation of foods. Some bacteria, however, can tolerate a low concentration of salt within the cell walls, which diminishes the osmotic pressure. Others, called halophiles, may actually require high salt concentrations for growth. Grula et al. (1983) readily grew *Clostridium* in salt concentrations up to 75,000 ppm and 45°C, which are representative of conditions that are encountered in petroleum reservoirs that are generally less than 1000 m deep.

Three divisions of microbes are based on their ability to utilize oxygen. Aerobic bacteria contain enzymes that can decompose peroxides, which are formed as part of the metabolic processes involving oxygen, but the strict aerobes cannot grow in an oxygen-free environment. The facultative bacteria contain a group of enzymes that allow growth in both aerobic and anaerobic environments. The strict (obligate) anaerobes cannot utilize oxygen because they do not contain the appropriate complement of enzymes that are necessary for growth in an aerobic environment.

Many biological reactions in microbes will occur only if adenosine triphosphate (ATP) is present for interaction with the enzymes, because even though the activation energy of the reaction is lowered by the enzymes, ATP is required to furnish additional energy. This leads to another method of classification of microbes on the basis of their ability to use carbon dioxide as a source of carbon for synthesis. Autotrophic bacteria can use carbon dioxide as their source of carbon and they can make ATP from oxidation of inorganic compounds, deriving their energy from the sun (photosynthesis) or from metabolism of inorganic compounds. The other class of microbes, known as heterotrophic microbes, must have preformed organic compounds as both their source of carbon and for energy (Nester et al., 1978; Gaudy and Gaudy, 1980; Moses and Springham, 1982).

In assessing microbial nutrient requirements and their metabolic products from the standpoint of microbial enhancement of oil recovery, it is more appropriate to classify the microbes according to their ability to utilize oxygen, as discussed previously. Numerous microbial cultures (pure and mixed) are capable of synthesizing a variety of biochemical products using petroleum fractions as the substrates. The range of metabolic products from microbial consumption of petroleum is very broad, depending on environmental conditions (pressure, temperature, salinity, pH, and the presence or absence of oxygen), supporting nutrients available for cell metabolism (nitrogen, phosphorus, minerals, etc.), and the specific bacterial cell interaction with petroleum (Donaldson and Clark, 1982).

Obligate aerobes are organisms that must have oxygen for their metabolism and growth, but they can exist dormant in the absence of oxygen. The bacteria decompose sugars to carbon dioxide, water and, often, acids: $C_6H_{12}O_2 + 8O_2 = 6CO_2 + 6H_2O +$ energy. When the microbes are furnished hydrocarbons and an ample supply of oxygen, a wide variety of products will result which depend on the type of microbe, environmental conditions, and the specific type of hydrocarbon substrates. The products may be carbon dioxide, low-molecular-weight acids, phenols or biopolymers (proteins, polyanionic lipids, glycolipids, or polysaccharides). The compounds are waste products from the microorganism and some may even be toxic if accumulated in the surrounding fluid.

Obligate anaerobes cannot utilize free oxygen; in fact, small quantities of oxygen (10 ppm) are even toxic to some anaerobes. Spores produced by some anaerobes, however, can remain dormant for long periods in an aerobic environment, germinating when they enter anaerobic environments. The anaerobes use low-molecular-weight organic compounds, such as sugars, as a source of carbon and energy. In the process of metabolism, the microbes release various products. Sugars undergo anaerobic fermentation yielding acids, alcohols, ketones, aldehydes, carbon dioxide, and hydrogen. Some species of the anaerobic genus *Clostridium* have been found to produce all of these compounds (Grula et al., 1983). These anaerobes also may reduce sulfur occurring as inorganic sulfates, or as part of the molecular structure of organic compounds, to hydrogen sulfide. Petroleum reservoirs have been known to become sour (produce large quantities of H_2S with the hydrocarbons) when infected by *Desulfovibrio* bacteria from injected water used in secondary recovery (Crawford, 1983). Anaerobes also can produce chemicals, some of which are surface-active agents that lower the oil–water interfacial tension and promote emulsification of oil. The anaerobic bacteria also may produce biopolymers (primarily polysaccharides) that can be used as mobility control agents.

Facultative bacteria can change their metabolism for growth either in an oxidizing or reducing environment. Their products of metabolism are quite similar to those described above. They may produce hydrogen sulfide from organic sulfides and inorganic sulfates, and reduce low-molecular-weight compounds (sugars, aldehydes, etc.) to methane, hydrogen and carbon dioxide. In the proper environment, they produce biosurfactants and biopolymers.

The three classes of bacteria noted above are generally mesophilic (existing within a temperature range of 20–50 °C), but there are thermophilic and caldoactive bacteria that live quite well in the temperature range from 40° to 100 °C. These have not yet been studied in reference to MEOR, but they reduce sugars and other carbohydrate compounds to methane and carbon dioxide, while reduction of sulfur compounds yields hydrogen sulfide. They are used in the secondary (anaerobic) treatment of sewage.

There has been some concern regarding the injection of microorganisms into petroleum reservoirs in the past because of possible pathogenicity (disease-causing) of the microorganisms. To our knowledge, pathogenic microorganisms have never been used for any MEOR field test in the United States. There have been reports of raw sewage injection in Hungary in 1969 and 1970 (Hitzman, 1983). Raw sewage may contain pathogenic or potential pathogenic bacteria and viruses; however, microbial researchers in the United States who have injected petroleum reservoirs with bacteria are not using bacteria that can cause disease. The literature on bacteria genetics indicates that microorganisms do not easily mutate into pathogenic strains. Specific microorganisms can cause human and animal diseases, and these microbes are generally fastidious, i.e., they require specific growth factors and specialized environments in order to survive. The environment is not favorable for pathogenic microorganisms in a petroleum reservoir; however, any microorganism to be used in field experiments should be tested for pathogenicity before injection.

HISTORICAL DEVELOPMENT OF MICROBIAL ENHANCEMENT OF OIL RECOVERY

The earliest realization that microbes could live in petroleum reservoirs came from the discovery by oil producers that bacteria were causing plugging of wells and the generation of hydrogen sulfide in crude oil stored in tanks and in reservoirs. Beckman (1926) suggested that bacteria might be beneficial to the production of oil, because products of their metabolism would assist in the release and transport of the oil in the geologic structures. No other reports appeared in the literature until ZoBell (1946, 1947) published the results of his work that began late in the decade 1930–1940. ZoBell patented a process for the secondary recovery of petroleum using anaerobic, sulfate-reducing bacteria in situ, such as *Desulfovibrio hydrocarbonoclasticus*. ZoBell observed that a mixed culture of *Clostridium* and *Desulfovibrio* would produce hydrogenase in an aqueous nutrient solution. Hydrogenase was used by the bacteria to produce low-molecular-weight oxygenated compounds (acids, ketones, etc.) and carbon dioxide. The gas and solvents assisted in the release of oil from sand packs in laboratory tests. In later work, ZoBell (1953) used other fermenting types of bacteria to produce substantial amounts of organic acids and carbon dioxide to enhance oil recovery in laboratory experiments. He observed that several strains of bacteria would grow at temperatures up to 88°C, and would migrate through 1/25-cm unglazed porcelain in a few hours, and through tightly packed sand at a rate of more than 2.5 cm/day.

Updegraff and Wren (1954) repeated the work of ZoBell and investigated other microorganisms, recommending the injection of *Desulfovibrio* and molasses for enhancement of oil recovery. In 1954, Updegraff assisted Yarbrough and Coty (1983) with a field trial in Arkansas, U.S.A., using *Clostridium acetobutylicum* and molasses to recover additional oil from a waterflooded zone.

Davis (1967) reviewed the current knowledge of petroleum microbiology and emphasized the potential of microbes for enhancement of oil recovery. This book and the Proceedings of the 1975 Engineering Foundation Conference (Anonymous, 1976) formed the basis of the resurgence of interest in MEOR that was sponsored by the U.S. Department of Energy by a conference at San Diego, California, in 1979 (Anonymous, 1979) and research grants at the University of Oklahoma, Oklahoma State University, University of Georgia, and University of Southern California. The U.S. Department of Energy continues to support research on MEOR by funding field projects at Universities and the National Institute for Petroleum Research at Bartlesville, Oklahoma.

Hitzman (1962) patented a process for the injection of bacterial spores and nutrients into a petroleum reservoir. He tested his hypothesis in the laboratory by using an oil-saturated sand-packed column. An aqueous solution of spores of *Clostridium roseum* and molasses was passed through the column and an improved release of oil (about 30%) was obtained. Later patents by Hitzman concern the use of microorganisms that consume injected polymers and the byproducts of carbon dioxide floods (Hitzman, 1972). In polymer floods, the injected organisms consume polymers that are adsorbed on the reservoir rock surface, whereas in carbon dioxide floods, the microbes feed on soluble compounds of carbon, nitrogen, and sulfur left behind the carbon dioxide–crude oil slug. The process has mobilized residual oil in sandpack tests, but no core or field tests have been reported.

Kuznetsov et al. (1963) discovered that bacteria found in some oil and gas reservoirs in the Soviet Union produced 2 g of carbon dioxide per day, per ton of rock. Kuznetsov suggested that methane was probably formed at the surface of the rock from carbon dioxide and hydrogen by the bacteria. Attempts to employ microbes to aid the recovery of oil were later made by Senyukov et al. (1970). Ivanov and Belyaev (1983) reported on a study of aerobic and anaerobic populations of bacteria associated with injection wells.

LABORATORY EXPERIMENTS SHOW THE POTENTIAL OF MICROBIAL ENHANCEMENT OF OIL RECOVERY

Microbial cells always contain a large number of both inorganic and organic compounds in ratios that are characteristic of the particular microbe, but different from the chemical composition of their environment. The most complex molecules are the nucleic acids and the proteins, some of which function as enzymes. The metabolism of the organism is considered to be the overall result of the total number of reactions occurring within the cell. Nutrients are assimilated and chemically coupled with the formation of products that are excreted as metabolic wastes. Heat also is produced as a result of the enzyme-catalyzed reactions. The total energy available to the cell from its environment must be greater than the energy contained in its products and dissipated heat. This energy is obtained from the process of photosynthesis, fermentation, or respiration. Cell metabolism is the consequence of perhaps thousands of chemical reactions that are catalyzed by enzymes (Donaldson and Clark, 1982).

In MEOR research, the study of specific microorganisms, or types of microorganisms, and their metabolic products is done for one of three reasons: (1) for the surface production of various compounds which, when injected into a petroleum reservoir, will enhance oil production; (2) for the injection of cells into a reservoir for in-situ production of metabolic products, which will enhance oil recovery; and (3) for the study of the microbial ecology of the reservoir. The environmental parameters of the reservoirs limit the types of microorganisms that can be used for in-situ processes. Clark et al. (1981) made a survey of the environmental conditions that exist in ten petroleum reservoirs in the United States and found that (1) the average temperature ranged from 49° to 90°C, which includes the temperatures suitable for mesophilic and thermophilic bacteria; (2) the reservoir fluid pressure averages 3.1 kPa (0.445 psi) per ft of depth, but, as shown earlier, an increase in pressure is apparently beneficial because it raises the optimum growth temperature and salt tolerance of the microorganisms; (3) they found that 66% of the reservoirs had a pH range of 6–8, but varied between the extremes of 3 and 10; and (4) salt concentrations (principally sodium, potassium, and calcium chlorides) ranged from 1.3 to 15.6%. High concentrations of salts (greater than 6%) may be the most limiting growth factor due to dehydration of the microbes.

McInerney (1983) stated that thermophilic microbes will be required for deeper reservoirs, because the temperature gradient (25°C + 20°C/km) will probably restrict their use in reservoirs deeper than 3500 m. The lack of certain essential nutrients such as sulfur, phosphate, and nitrate may pose severe limitations to MEOR processes, unless these nutrients are added to the injection waters. A suitable carbon and energy source also will be required, although some microorganisms use various components of crude oil as a carbon and energy source. Petroleum reservoirs are anaerobic and it would be difficult to supply oxygen at the required amounts for hydrocarbon metabolism. This analysis leads to the conclusion that thermophilic anaerobic bacteria capable of growing in at least 5% salt solution would be most suitable for general in-situ MEOR processes. In-situ growth of the bacterial cultures could be maintained by supplying the required nutrients (nitrogen and phosphates, if necessary) and a carbon source such as molasses.

Bacteria grow at an exponential rate (Monod, 1950):

$$dX/dt = uX \tag{14-2}$$

where u = specific growth rate, s^{-1};

$$u = u_{max}[S/(K+S)]$$

where u_{max} = maximum value of u when the growth nutrients are present in excess; X = cell concentration, g/ml; t = time, s; S = substrate concentration; K = saturation rate constant (equal to the substrate concentration at which $u = (\frac{1}{2}U_{max})$.

When the substrate concentration is maintained constant, the exponential growth rate can occur at rates between 0 and u_{max}. This is important to in-situ MEOR processes, because the rate of microbial growth can be controlled by the rate at which substrate is provided. The amount and type of product may be controllable by limiting essential nutrients. For example, *Bacillus licheniformis* produces large amounts of extracellular polysaccharides when the substrate is limited in ammonia (McInerney, 1983).

A potentially useful group of microorganisms for in-situ MEOR is *Clostridia*, because petroleum reservoirs are anaerobic. Many species of the anaerobic genus *Clostridium* produce carbon dioxide, low-molecular-weight acids, alcohols, and ketones as metabolic products, when carbohydrates are present as the substrate. The nascent production of these metabolic products in a petroleum reservoir might serve as a means for release of residual oil if the bacteria and required nutrients are placed in the injection water. The bacteria also may be used to generate a high pressure in the pores of the rock in the immediate vicinity of the wellbore, which will serve as a clean-out process if the well is allowed to blow after a shut-in period for pressure buildup. Grula et al. (1983, 1985) conducted laboratory experiments to isolate salt-tolerant strains of *Clostridia* and then conducted field tests using them. These bacteria seem to be well suited for MEOR, because they were found to tolerate up to 7.5% salt; however, the rate of growth was slowed down as much as 7 times in the presence of high salt concentrations. The bacteria produce spores which also are favorable for in-situ applications, because spores were found by Jang et al. (1983, 1984) to travel through rocks without interaction with the rock surface.

Grula et al. (1983) also discovered that sandstone and limestone rocks used in their experiments provided the supply of phosphate and other inorganic ions that are necessary to obtain good growth of the bacteria.

Donaldson and Grula (1985) found that several species of *Clostridia* produce emulsifiers in salt concentrations up to 7.5%. The emulsifiers are synthesized from nonhydrocarbon precursors such as molasses. The biosurfactants were not chemically characterized, but some preliminary work indicated that the biosurfactant is a lipopeptide with a molecular weight greater than 1000 and that it is non-ionic (not precipitated by multivalent cations such as calcium and strontium). Production of the biosurfactants along with other solvents and carbon dioxide within a petroleum reservoir would tend to release and mobilize residual oil, which is the principal objective of MEOR. In addition to the exocellular emulsifiers, the cells of *Clostridia* have emulsifying ability due to the cell wall surface composition. The bacteria have electrical charges distributed between their cell surface and interior. Usually a new negative charge exists at the cell surface that can exhibit marked interfacial activity if the cells concentrate in an oil–water interface. One effect of this property is to lower the interfacial tension and enhance emulsification properties of other surfactants in the aqueous phase in a manner similar to a cosurfactant. A second effect that the cells have is to stabilize emulsions. Adsorption of the cells at the oil–water interface of the oil droplets dispersed in water increases the repulsive forces between the droplets, stabilizing the emulsion.

Potential field applications of microbes for displacement of oil are the in-situ production of gases, solvents, or polymers, and the reduction of high-permeability zones between the injection and production wells. Both of these processes require the deep emplacement of the microbes within the petroleum reservoir (Donaldson, 1982). Although live cells may in some cases be difficult to inject into rocks because of filtration and cake buildup at the surface, the endospores that are formed by many microbes are not adsorbed on the rock surfaces. The endospores are highly refractile dormant structures, which are formed within the parent cell, that contain all of the genetic information required for the formation of another cell under appropriate conditions. The spores of *Clostridium* possess inert rigid walls and are less than 1 μm in diameter. The injection of spores of gas–acid forming bacteria, such as *Clostridium*, was suggested in a patent by Hitzman (1962). Hitzman recommended that this would be a method for the widespread distribution of the spores within the petroleum reservoir. Upon germination by injection of nutrients, the bacteria would generate carbon dioxide in the reservoir to assist in production by addition of gas pressure and solution of the gas in the hydrocarbons.

Porous geological materials are composed of a network of interconnected pores with a wide range of pore sizes. The distribution of pore sizes has a significant effect on displacement of oil and all EOR processes. Ewall (1985) showed that in many rocks, 80% of the flow of fluid occurs in less than one third of the pores (the larger pores). Thus most of the injected fluids will flow through these pores leaving hydrocarbons behind as a residual oil saturation in the smaller pores of the rock. Selective plugging has been proposed to block the large pores and force the injected fluids through the smaller pores to displace the remaining oil. Many methods have been proposed for selective plugging of rocks containing hydrocarbons, some of which include the use of clays, gels, waxes, resins and cements

(Garland, 1966). The in-situ growth of microorganisms in sandstones results in significant reductions of permeability, because the organisms selectively grow in the more permeable regions of the rock (Jang et al., 1983, 1984; Jenneman et al., 1983, 1984).

Torbati et al. (1986) measured the pore-size distribution of Berea Sandstone samples before and after injection and growth of bacteria in the rock. Their results showed that the larger pores are plugged by the bacteria, which caused a reduction of permeability in one case from 282 to 24 mD and production of 10–35% additional oil, above the maximum recovery by waterflood.

Due to the unfavorable mobility ratio, the recovery of heavy oil (API gravity less than 20°) by primary and secondary methods is very low. Jack et al. (1983) estimated that the ultimate recovery from the Lloydminster oilfield in Canada is less than 8%. The principal problem is low sweep efficiency due to water channeling, because the oil viscosity ranges from 200 to 9000 cP at 25°C. Jack et al. suggested two approaches to the use of micro-organisms for enhancement of heavy oil recovery: (1) repressurization of the reservoir by in-situ production of gases, and (2) oil release due to anaerobic fermentation of molasses, which can cause in-situ production of gas, acids, solvents, and surfactants. Laboratory experiments conducted by Jack et al. showed that nascent production of gas in situ was the major mechanism causing release of additional oil as a result of decrease in oil viscosity accompanied by swelling from dissolved gas.

Jack et al. (1982) conducted a microbial culture screening search for anaerobic microbes that could grow and produce desirable metabolic products in 6% salt solutions (NaCl and $CaCl_2$) and the crude oil. Many cultures, including *Clostridium*, did not grow well in the artificial medium; however, isolates of *Enterobacter cloacae*, Gram-negative, facultative, rod-shaped bacteria were found that grew rapidly and released 1.6 moles of gas per mole of sucrose. During the screening tests, Jack et al. (1982) discovered 11 cultures that produced a biopolymer, which was effective in stopping water channeling in artificial porous media made of glass beads. The injection into a reservoir of bacteria that are actively producing biopolymers, will result in plugging at the face of the rock, rather than their transport into the rock. Control of the growth medium, however, can stop the microbial production of biopolymer. Thus, the microbes can be injected into the porous medium without development of a filter cake at the face of the rock and, once injected, the injection of a nutrient solution which promotes the production of biopolymers can result in the desired in-situ plugging of the high-permeability zones of the rock. In other experiments, the addition of sodium pyrophosphate to the nutrient medium facilitated the injection of the microbes by stopping the production of biopolymers (Chang and Yen, 1985; You-Im and Yen, 1985).

A different approach to laboratory research on microbial enhancement of heavy oil recovery was reported by Singer et al. (1983) and Singer (1985), and summarized by Donaldson and Clark (1982). An aerobic culture labeled H-13 was added to a basal salt medium and a Venezuelan heavy crude oil (API gravity of

8°) and incubated aerobically for 7 days. The oil phase was increased by 2.5 times due to emulsification with water and the extracellular products. This oil emulsion exhibited physical properties that were very different from the original crude oil. It did not adhere to glass and behaved as a thixotropic liquid, and the viscosity was 10 times less than the viscosity of the crude oil which was 6,500 cP. When the microbially treated oil was separated from all traces of water by thin-layer distillation, the resulting oil was found to be lower in density and boiling point than the crude oil. Chromatographic analysis showed that the microbes had removed most of the paraffin compounds from the crude oil, but the emulsified oil could be readily pumped and transported in pipelines because of its non-wetting properties and reduced viscosity.

Singer et al. (1983) isolated the surface-active extracellular compounds and discovered that they were a mixture of glycolipids. The glycolipids generally contained a disaccharide (trehalose) and an oligosaccharide with 3–5 glucose units. The molecule also contained glycerol, uronic acid, saturated and hydroxy fatty acids, fatty alcohols, and a peptide. The order and nature of the chemical bonding was not determined; however, the complexity and the high molecular weight (3000 daltons) of the molecule, suggests that it may be a glycolipopeptide or a polysaccharide–peptide–fatty acid polymer.

Comparisons of biosurfactants to synthetic surfactants show that the biosurfactants exhibit chemical and surface properties that are superior to the synthetic surfactant for lowering the interfacial tension, emulsion formation and stability, and tolerance of cations. The biosurfactants are applicable to EOR when used in combination with, or as an alternative to, synthetic surfactants. They might be used for micellar or surfactant flooding processes, or for reduction of heavy crude oil viscosity for pumping and transportation (Singer, 1985).

Pfiffner et al. (1985) isolated a strain of *Bacillus licheniformis* that grew anaerobically (strain JF-2) from produced water from an oil reservoir. This strain of bacteria produced an extracellular biosurfactant when grown in either aerobic or anaerobic environments. The properties of the biosurfactant resemble those of surfactin produced by *Bacillus subtilis*; however, strain JF-2 grows anaerobically and is not a strain of *B. subtilis*. The JF-2 surfactant is apparently a lipid that does not contain proteins or amino acids and, thus, differs from surfactin, a lipoprotein. The biosurfactant reduces the interfacial tension of 5% NaCl to 27 mN/m and has a critical micellar concentration of about 50 μg/ml. The lowest interfacial tension was obtained with the JF-2 surfactant and octane at 30°C, which was 0.05 mN/m. Thus, low interfacial tension values are obtained using low surfactant concentrations without the use of cosurfactants. A field pilot study has been initiated by the University of Oklahoma to test this microbe for enhanced oil recovery.

Other laboratory research involved the use of Hele-Shaw type cells for rapid analysis of the potential for oil displacement of various types of microbes. Zajic et al. (1985) used the classical Hele-Shaw model, which represents a simple single pore that can be used for comparisons of the oil displacement efficiency of synthetic surfactant to biosurfactant and growing cultures of biosurfactant-producing microorganisms. The technique reported by Zajic et al. (1985) allows rapid evaluation of the optimum concentration of surfactant for displacement of oil; for example, they found that the synthetic surfactants (pluronics) exhibited a peak oil displacement from the cells of 50% at an optimum concentration of 0.002%.

A modification of the Hele-Shaw cell was used by Kianipey and Donaldson (1986) to determine changes of wettability, capillary pressure, and ultimate production using Hele-Shaw cells that were packed with a thin layer of sand. Observations were made, and photographed, through a stereo microscope which allowed observation of incipient gas production, emulsification, and movement of oil–water–rock interfaces. Three genera of bacteria (*Bacillus, Pseudomonas*, and *Clostridium*) were injected into water-wet and oil-wet cells of unconsolidated sand, saturated with brine and crude oil. The flow cells were placed under a microscope for visual and photographic observations, and were connected to a water manometer to examine the complete capillary pressure hysteresis loop before and after introduction of bacteria. The cells were first saturated with brine and, then, water was displaced with crude oil to irreducible water saturation. Oil and brine displacements were then made to determine the complete capillary pressure relationship. Next, the bacteria and nutrients were introduced and incubated in the cell for 24–48 hours. During incubation, some emulsification of the oil at the oil–water–sand interfaces was observed. Biogenic gas production caused pressure increases in the cells. Final measurements of the capillary pressure hysteresis loop showed decreases of residual oil saturation from 9 to 24% and changes in wettability to more water-wet systems.

Bryant and Douglas (1987) used acid-etched glass micromodels to demonstrate crude oil displacement mechanisms by microorganisms. They correlated their micromodel studies with Berea Sandstone core experiments to evaluate microbes for in-situ oil recovery.

FIELD APPLICATIONS OF MICROBIAL ENHANCEMENT OF OIL RECOVERY

There are two general methods for use of microbes in enhanced oil recovery: (1) extracellular products of metabolism may be extracted from cultures grown at the surface and solutions of these materials may be injected into the petroleum reservoir to aid in oil recovery, (2) the bacteria and or spores may be injected into the reservoir along with required nutrient solutions where they are expected to grow and produce bioproducts that are in some way beneficial to oil production. The first procedure received attention before the renewed interest in MEOR was developed by the Department of Energy beginning with a small workshop in 1979 (Anonymous, 1979). The biopolymer, xanthan gum, was considered early in the 1970's, along with polyacrylamide, as an additive to flood waters to improve the water/oil mobility ratio.

Slodki and Cadmus (1983, 1985) discussed the preparation, structure and properties of xanthan gum. Xanthan gum solutions display a combination of useful properties that are not typical of other polysaccharides from algae, or synthetic water-soluble polymers. These properties include high viscosity of dilute aqueous solutions, resistance to shear degradation, shear thinning (pseudoplasticity), rheological yield, insensitivity of viscosity to mono- and divalent cations over a broad range of pH, and good thermal stability. In addition, the xanthan gum exhibits properties that are of particular interest to EOR: resistance to chemical degradation and low adsorption on rock surfaces (relative to polyacrylamide).

Pseudoplasticity and rheological yield are among the most important properties of xanthan gum dispersions. Pseudoplastic dispersions thin rapidly with shear, but do not begin to flow until sufficient shear stress is applied. Such dispersions are easy to pump and are good suspending agents. Shear thinning is thought to result from disruption of weak, but extensive, associations between rod-like molecules. Apparently, the tenuous, gel-like networks are re-established quickly when shear forces are removed, i.e., the molecular rods align in the direction of flow. Newtonian behavior can be observed at low rates of shear for solutions of sufficiently low concentration (less than 0.3%). Difficulties relevant to the use of commercial preparations of xanthan gum as a mobility control agent arise from (1) the presence of bound calcium ions which are particulates of cell origin, (2) the formation of microgels, (3) degradation of aqueous solutions by aerobic bacteria, and (4) filtration of particulates remaining after pasteurization (Slodki and Cadmus, 1983, 1985).

Filterability of xanthan gum solutions can be enhanced by treatment with cellulase and a protease. The cellulases remove microgels that arise from incomplete hydration of the powdered gum.

The U.S. Department of Energy sponsored a field project at Coalinga, California, with Shell Oil Company to test xanthan gum as a mobility control agent. Peterson (1980) reported that the overall production performance of the pilot field test was significantly lower than anticipated from the initial measurements of oil saturation and computer simulation studies. The pre-injection history was used to extrapolate the exponential decline rate for the project area for comparative analysis. Oil production after the initiation of full-scale injection was approximately equal to the pre-injection production even though the active well count was increased by four. Water injection was continued for one year after polymer injection was discontinued; however, there was no change in the oil production decline. The poor performance was attributed to considerably lower reservoir oil saturations that indicated by the initial measurements: the original estimate of oil saturation was 54%, but this was later revised to 39%. The waterflood residual oil saturation also was revised upward from 24% to 28%. If these later estimates of S_o and S_{or} were correct, this would indicate that the Coalinga Field was a poor choice for the pilot test. Nevertheless, the report by

Peterson is an excellent discussion of facility design, polymer slug design, and properties of the biopolymer.

Hitzman (1983) presented a review of field applications of MEOR and summarized them in tabular form with brief discussions. According to Hitzman, Kuznetsov in 1955 reported that bacteria were present in certain petroleum-bearing strata in the U.S.S.R. in large quantities and were apparently involved in the generation of a considerable quantity of carbon dioxide that was present in these reservoirs. Kuznetsov et al. (1963) injected a mixed culture of aerobic and anaerobic bacteria together with acid-hydrolyzed substances from peat and soils into a waterflooded field and shut the field in for 6 months. When the wells were produced again, oil production rose from 44 to 48 m^3/day (275 to 300 bbl/day) for four months.

Von Heningen et al. (1958) reported on two field tests performed in The Netherlands. The year and location of the tests are unknown. In one test they used *Betacoccus dextranicus* in a sucrose–molasses medium of 10% total sugar content and obtained a 30% increase in oil recovery. The second test used a mixed culture of slime-forming bacteria in 50% molasses. The water/oil production ratio was improved by the treatment from 50 to 20 (m^3 water per m^3 of oil produced).

Dostalek and Spurney (1957), in Czechoslovakia, conducted field tests by injecting a mixed culture of sulfate-reducing bacteria (*Desulfovibrio*) and paraffin-oxidizing bacteria (*Pseudomonas*) with a molasses-base nutrient solution. They stated that the daily average oil production increased by approximately 7% during the 6-month period of the experiment.

Jaranyi et al. (1963), working in Hungary, used a mixture of anaerobic, thermophilic bacteria that fermented molasses in a field containing naphthenic-type crude oil. In later field trials, they used raw sewage as the inoculum (1 liter of sewage mixed in 300 liters of 2–3% molasses). The deepest reservoir treated was 2500 m (8200 ft). They reported beneficial results to oil production in 7 out of 10 reservoirs that were treated, but the specific increases of production were not given.

Karaskiewicz (1962) conducted 18 field trials in Poland between 1961 and 1969 with microbial cultures obtained from soil and water samples taken in the vicinity of the oilfields and from sugar factory waste-waters. The mixed cultures contained bacteria of the following genera: *Arthrobacter, Clostridium, Mycobacterium, Peptococcus*, and *Pseudomonas*. The cultures were grown in 10-liter bottles containing formation water plus 4% molasses at 32°C. The wells that were treated ranged in depth from 500 to 1525 m (1650 to 5000 ft). According to Hitzman (1983), Karaskiewicz reported increases of oil production rates that ranged from 20 to 200% over the original rate of production.

Lazar and Constantinescu (1985) and Lazar (1986) reported on an extensive review of MEOR work that has been conducted in Romania during the last decade. Three areas of research were emphasized by Lazar: (1) examination of the bacterial population in the formation water of the reservoir, (2) adaptation of the microorganisms to field conditions prior to injection, and (3) field testing of the adapted microorganisms. Lazar inoculated seven reservoirs and stated that two of the reservoirs have shown a consistent 200% increase in production for five years. The other five reservoirs have not responded, but additional field testing is in progress. Yarbrough and Coty (1983) reported on a field test performed in Arkansas in 1954. Laboratory experiments with inoculation of cores from the Nacatoch Formation with *Clostridium acetobutylicum* showed that gases (and other metabolic products) produced within the reservoir rock prior to waterflooding led to better oil recoveries than could be obtained by either waterflooding or gas flooding. A field test was, therefore, initiated in which a 2% solution of molasses, dissolved in fresh water, was injected into the formation for a period of six months. Eighteen inoculations consisting of about 800 liters containing *Clostridium acetobutylicum* were injected at scheduled intervals. Fresh water began to appear at the production wells 70 days after the initial injection of bacteria, and fermentation products (carbon dioxide and low-molecular-weight oxygen compounds) appeared at the production well 80 days after the first inoculation and continued throughout the duration of the test. The principal fermentation products were seven short-chain fatty acids (formic through caproic, and caprylic acids) and carbon dioxide, including small amounts of ethyl and butyl alcohol and acetone. The total organic acid production during the test period of 286 days was 35,000 kg and the carbon dioxide was 5600 m^3 (200,000 ft^3) or 11,400 kg. An increase of oil production began soon after breakthrough of the fermentation products. Based on the normal decline curve for the single production well that was monitored, the average oil production rate from November 1954 to May 1955 increased from 0.10 to 0.33 m^3 per day (0.6 to 2.1 bbl/day), which represents a 250% increase of production.

Bond (1961) injected an agar medium containing *Desulfovibrio hydrocarbonoclasticus* into a sandstone reservoir at a depth of 1000 m, which was producing 2.4 m^3 (15 bbl) per day prior to inoculation. The well was shut-in for 3 months after injection of bacteria to allow time for fermentation and dispersion. When the well was placed on production once more the rate had increased to 4.0 m^3 (25 bbl) per day.

During the period 1977–1981, Johnson (1979) inoculated approximately 150 stripper wells in the United States (wells producing about 0.3 m^3 (2 bbl) per day) with a mixed culture of *Bacillus–Clostridium*. The carbonate reservoir depths ranged from 60 to 300 m (200 to 1000 ft) and porosity from 10 to 30%. Johnson injected the inoculum into production wells, shut the wells in for 10–14 days, and then opened them once more for production. Production increases of 20–30% were obtained when the API gravity of the oil ranged from 15° to 30° and the formation water contained less than 100,000 ppm salt.

Hitzman (1983) reported that Petrogen, Inc., tested the Johnson method in 24 wells ranging in depth from 90 to 1400 m (300 to 4600 ft). The production rates of four of the wells doubled for a period of 6 months, and 12 of the wells showed a 3-month production increase of 50%.

A joint microbial–waterflood project was recently initiated by the National Institute for Petroleum and Energy Research, Bartlesville, Oklahoma, two industrial participants (Microbial Systems Corp. and INJECTECH, Inc.), and the U.S. Department of Energy (Anonymous, 1987). The site of the field project is an ongoing freshwater flood in the Delaware-Childer oilfield, Nowata, Oklahoma, on a 24-hectare lease (60 acres). A major goal of this project is to fully document the field test and make the information available through the Department of Energy publications.

MICROBES ASSOCIATED WITH OILFIELD PROBLEMS

A great deal of the early work by microbiologists working with petroleum was oriented toward the control of the deleterious effects of microorganisms in oilfields. Numerous reports of the presence of microbes in reservoirs have been published (Crawford, 1983; Lazar and Constantinescu, 1985; Singer, 1985). These reports discuss the microbial population increases that occur with the application of secondary oil recovery methods where the injected water is exposed to the atmosphere in open ponds. Lazar and Constantinescu (1985) found abundant microbial flora indigenous in oilfield formation waters which included species of *Bacillus, Pseudomonas, Micrococcus, Mycobacterium, Clostridium*, and *Escherichia*. Spore-forming bacilli and some cocci were the usual bacteria found in deep reservoirs, whereas aerobic pseudomonads and facultative anaerobes were the dominant species in shallow reservoirs. Pseudomonads are one of the most difficult microbes to control in industrial water systems; they exhibit a high rate of growth and can produce a slime that apparently protects them from biocides to some extent and also results in plugging of the reservoir rock pores (Chakrabarty, 1982). *Escherichia* is reported to contain hydrogenase, an enzyme that utilizes molecular hydrogen and may be associated with cathodic hydrogen depolarization, causing corrosion of steel casings and pipes in the oilfield. Natural mixed populations of microorganisms growing in oil wells and in petroleum reservoirs, produce chemicals that corrode production equipment and plug reservoirs. Sulfate-reducing bacteria cause a sweet crude to turn sour (produce hydrogen sulfide), which has happened in the Wilmington Field in California (Gates and Parent, 1976).

Several patents deal with biocides that are recommended for injection into reservoirs to control indigenous microbial flora. Some examples of these biocides include: formaldehyde, benzene, toluene, and several quaternary ammonium compounds (Jack and Thompson, 1983). Other processes involve the use of physicochemical methods for bacterial control, including treatment of refinery waste waters and filtering of injection water to remove microorganisms.

Many corrosion problems in the oilfield are caused by hydrogen sulfide produced in situ by *Desulfovibrio*. The hydrogen sulfide also reacts with iron in the reservoir, producing ferrous sulfide which oxidizes to ferric sulfide in the produced water.

Bacteria that form slime (some form of polysaccharides), such as *Achromobacter* sp. and *Flavobacterium* sp., will adhere to each other forming a large mass. They also adhere to the walls of the pores, causing severe plugging problems at injection wells (McCoy and Costerton, 1982).

Ivanov and Belyaev (1983) showed that a population of aerobic bacteria around the injection well, which is receiving oxygen and nutrients dissolved in the injected water, produce low-molecular-weight oxygen compounds. The latter are utilized as essential food for growth by *Desulfovibrio* and other anaerobic bacteria in the oxygen-depleted region beyond the vicinity of the wellbore.

MICROBIAL INTERACTIONS WITH PRODUCED OIL

Microorganisms have been shown to be normal inhabitants of interstitial waters (Singer, 1985). Microbial metabolism on petroleum in the subsurface, particularly where surface waters carry oxygen and nutrients to the oil deposit, can reduce the value of the crude oil because the aerobic bacteria use the paraffins as a carbon source (Crawford, 1983). A staff report in *World Oil* (Anonymous, 1972) claims that 10% of the world's crude oil has been destroyed, and another 10% considerably reduced in value, by microorganisms.

Singer et al. (1983) reported that microbial oxidation of specific sulfur or nitrogen-containing components of crude oil is feasible. Microorganisms have been isolated that oxidize specific sulfur- and nitrogen-containing aromatic compounds in crude oils. These isolates exhibit a specificity for sulfur- and nitrogen-containing aromatic compounds. They are unable, however, to oxidize or grow on a variety of other hydrocarbons, including aliphatic alkanes, cycloalkanes, and mono- and polynuclear hydrocarbons. Isolates identified to date can metabolize only under strictly aerobic conditions. In a related study, Fedorak et al. (1983) found that several isolates of pure and mixed bacterial cultures could degrade the aromatic compounds in Prudhoe Bay oil.

It is estimated that 10 million tons of oil enter the marine environment each year from accidental spills associated with routine operations (Atlas, 1981). The chemical and biological decomposition of petroleum pollutants in the environments is complex and has not been elucidated, but the degradation of these compounds occurs at a rapid rate. Microorganisms are certainly responsible for a large part of the degradation as shown by the wreck of the "Amoco Cadiz" off the coast of France. It is estimated that microorganisms biodegraded 10 tons of the crude oil per day in the area (Atlas, 1981). The susceptibility of petroleum hydrocarbons to biodegradation is determined by the structure and molecular weight of the hydrocarbon molecules (Bailey et al., 1973; Davies and Westlake, 1979; Bryant, 1986). The *n*-alkanes of intermediate chain length (C_{10}–C_{24}) are degraded more rapidly than alkanes and branched hydrocarbons. Short-chain alkanes (less than C_9) are toxic to most microorganisms, but they generally evaporate from oil slicks. Branching reduces the rate of biodegradation, and aromatic compounds are degraded much more slowly. The mechanism of hydrocarbon uptake by microorganisms is not yet clearly understood. Two of the theories that have been proposed are: (1) emulsification of the hydrocarbon by secreted emulsifiers from the bacterial cells, and (2) adherence of the bacterial cell directly to the hydrocarbon for assimilation of the compound by the cells.

To clarify the relationship between biodegradation of crude oils and the geochemical implications of the process, Bailey et al. (1973) observed bacterial degradation of oil under controlled laboratory conditions and compared the results to a group of altered and unaltered crude oils from fields in Saskatchewan, Canada. Alkanes, cycloalkanes, and aromatics were all metabolized by bacteria and the lighter paraffin members were preferentially removed. The bacterial cultures consisted of a member of each of the following genera: *Pseudomonas, Flavobacterium, Achromobacter*, and *Bacillus*. The loss of the lighter saturated hydrocarbons and aromatics from these crude oils treated with the cultures resulted in creation of more viscous (and heavier) crude oils, which were comparable to the natural crude oils obtained in Saskatchewan.

The literature provides ample evidence that aerobic strains of microorganism will degrade crude oils; however, it has not been proven yet that crude oil will undergo biodegradation under anaerobic conditions, although Moses et al. (1983) indicated that some very slow anaerobic biodegradation of crude oils may occur.

POTENTIAL OF MICROBIAL ENHANCEMENT OF OIL RECOVERY

The production of oil in the United States is entering a tertiary phase because the discovery of new oil fields has decreased with time, whereas waterflooding of the existing oilfields is steadily becoming less productive as the water-to-oil production ratio approaches the economic limit of operation. The oil left in-place is trapped by capillary forces and chemical adhesion to the reservoir rock minerals. Thus, efforts to displace this residual oil require addition of considerable energy to the petroleum reservoir in the form of chemical reactants, heat, or solvents. A potential source of supplemental energy for petroleum reservoirs and other petroleum operations is from microorganisms (Donaldson, 1982).

Thousands of different species of microorganisms have been classified and examined to determine their nutrient requirements, products of their metabolism, and the limits of environmental conditions that they can tolerate. Three very broad divisions have emerged based on environmental tolerance: (1) aerobes, capable of life processes only in the presence of oxygen; (2) anaerobes, which exist in the absence of oxygen and derive their energy from degradation of oxygenated molecules; and (3) facultative, which are able to exist aerobically or anaerobically. These three categories can be further subdivided into mesophilic (capable of life below 45°C) and thermophilic (able to live in environments at a constant temperature greater than 45°C). Within these broad groupings there are microbes that metabolically produce gases (methane, hydrogen, nitrogen, and carbon dioxide), polymers (polysaccharides and proteins), surface-active compounds that are generally poly-anionic lipids, and many other compounds from simple alcohols to very complex macromolecules. The microbial cells always contain a large number of both inorganic and organic compounds in ratios that are characteristic of the particular microbe and the environment. The most complex molecules are the nucleic acids and the proteins, which function as enzymes to catalyze a host of organic reactions at ambient temperature. The metabolism of the organism is considered to be the result of all of the reactions occurring within the cell. Nutrients are assimilated and chemically coupled with the formation of products that are excreted as metabolic wastes. Heat is also produced as a result of the enzyme-catalyzed reactions.

There are three types of metabolism that are recognized as providing energy for non-photosynthetic organisms:

(1) Fermentation involves a series of oxidation–reduction reactions in which organic compounds are the final electron acceptors. Waste products are carbon dioxide, ammonia, acids, alcohols, and other low-molecular-weight solvents.

(2) Respiration occurs in an oxygen environment in which oxygen is the final electron acceptor. Incomplete oxidation (alcohol to acid plus water) is characteristic of this type of metabolism for some organisms.

(3) Respiration under anaerobic conditions involves the reduction of an inorganic compound containing oxygen (such as sulfate, carbonate, or nitrate). The waste products may be hydrogen sulfide, carbon dioxide, nitrogen, methane, and acids depending on the specific microbe, the nutrients, and the environmental conditions (Davis, 1967). Many of these properties of microbes are in common industrial use. Examples are the production of alcohols, vinegar, cheese, leavening of bread, acetone, citric and lactic acids, butyl alcohol, vitamins, and antibiotics. The type and yield of the metabolic products can be controlled to a large extent by modifications of environmental conditions and nutrients. Additionally, the microbes exhibit amazing properties for adaptation to new circumstances, both environmentally and nutritionally, through mutation. These properties give the engineer considerable flexibility for applications of microbial systems (Donaldson, 1982).

Although many distinct types of microbes have been isolated and thoroughly characterized, this by no means implies that all microbes have been found. There may be many undiscovered microbial systems that are either already existing in deep subsurface reservoirs or are capable of existence there if the required nutrients are supplied for their metabolism. The potential for their discovery and use for oil recovery and other operations certainly exists. The fundamental research effort launched by the U.S. Department of Energy at four major universities in the United States in 1979 was specifically directed toward the discovery and application of microbial systems to petroleum engineering (Donaldson and Clark, 1982; Zajic and Donaldson, 1985).

INJECTION OF CELLS AND SPORES

Two potential uses of microbes for displacement of oil are the in-situ production of gases, solvents, or polymers, and the reduction of high-permeability zones between the injection and production wells. Both of these processes require the deep emplacement of microbes within the petroleum reservoir. The endospores of microbes are highly refractile, dormant structures that are formed within the parent that contains all of the genetic information required for the formation of another cell under appropriate conditions. The spores of *Clostridium* possess inert rigid walls that are less than one micron in diameter. The injection of spores of gas–acid forming bacteria, such as *Clostridium*, was suggested in a patent by Hitzman (1962). Hitzman recommended that this would be a method for the widespread distribution of the spores within the petroleum reservoir. Upon germination by injection of nutrients, the bacteria would generate carbon dioxide in the reservoir to assist in production by addition of gas pressure and solution of the gas in the hydrocarbons.

Johnson (1979) used *Clostridium* to open the pores of the formation around production wells causing a two- to three-fold increase of production. The cells were injected with molasses, allowed to ferment for 12–14 days, and then the well was opened allowing the pressure of carbon dioxide to expel particles and residual hydrocarbons plugging the pores around the injection well. Donaldson (1985) discussed the changes of capillary pressure phenomena at the wellbore that result in the additional production after treatment with *Clostridium*.

Improvement of sweep efficiency

The work at the University of Oklahoma (Torbati et al., 1985) has shown the feasibility of using microbes to plug high-permeability zones in petroleum-saturated sandstones to improve sweep efficiency and displace bypassed oil. A field project with Tenneco, Inc. has been initiated to test the technique which will involve the injection of bacteria and nutrients along with secondary recovery injection water to improve the recovery from a field under waterflood.

Reduction of crude oil viscosity

The worldwide distribution of heavy crude oils and tar sands (viscous hydrocarbons that cannot be recovered by standard production method using wells), makes this resource a candidate for unconventional recovery methods. The factor that inhibits conventional recovery is the high viscosity of the hydrocarbons; therefore, techniques for recovery involve the application of heat to reduce the viscosity sufficiently to allow movement of the oil to a well for production. Retorting of mined tar is used on a very limited scale.

Bacteria certainly aided in the transformation of organic matter into petroleum and in the formation of heavy crude oils which are near the surface. Both evaporation of volatile hydrocarbons and destruction of paraffinic compounds by microorganisms led to deposits high in polynuclear aromatic compounds (Bailey et al., 1973). These are microbes which are able to degrade asphaltic material, especially in the soil exposed to the oil for a long time, such as near oil wells, storage tanks, and oil seeps. Oil spilled on the ground or in the ocean and which is not continuously replenished, will completely disappear in a few months due to microbial degradation and subsequent evaporation of low-molecular-weight compounds. Aerobic microbes that produce glycolipids, which emulsify crude oil resulting in an oil-in-water emulsion with a viscosity reduction of 90% over the original crude oil, were discovered by Singer et al. (1983). The glycolipids might be employed (1) to enhance oil recovery by addition to injection water, (2) in steam enhanced release of tar from sands instead of retorting, and (3) in transportation of heavy crude oils through pipelines.

In a different approach for separation of kerogen from oil shale, Findley et al. (1974) proposed that sulfur-oxidizing bacteria of the genus *Thiobacillus* could be used to attack and degrade the mineral matrix of oil shale through production of sulfuric acid. They concluded that dolomite in the matrix could be attacked and dissolved by the action of *Thiobacillus* on crushed oil shale, leaving a porous matrix from which the kerogen could be easily removed. The same process may be feasible for separation of tar from tar sands.

Surface treatment processes

The metabolism product of the bacterium *Xanthomonas* is a polysaccharide polymer which is used in a large number of commercial applications. The petroleum industry uses xanthan gum in drilling fluids. It also has been field tested for use as a mobility control agent in the Coalinga Field, California (Peterson, 1980). Other bacteria produce biosurfactants, such as lipoproteins with a strong capability for emulsification of crude oils. Other produced biosurfactants are equally efficient as oil–water emulsion breaking reagents. The oil emulsifiers are generally quite resistant to precipitation by multivalent cations, such as calcium and strontium, found in oilfield brines. Thus, a potential exists for their use as additives to injection water and for the preparation of micellar–polymer solutions for oil displacement.

Activation of indigenous microbes

Desulfovibrio has been isolated from numerous oil reservoirs and recommended for numerous oil recovery processes. It has been tested, as previously mentioned, but no commercial processes have been developed for its use as an oil recovery agent. There may be many other types of bacteria, however, that are native to petroleum reservoirs or have been introduced by oilfield operations and have managed to adapt to the subsurface environment. It is quite possible that some of these organisms may have properties useful for enhanced oil recovery if their growth is accelerated by introduction of a proper blend of nutrients. A careful study of petroleum reservoirs to isolate and characterize indigenous microbes may lead to discoveries of lasting importance.

An example of such a discovery was made from oil obtained at a depth of 1000 m from the Wilmington Field, California (Grula and Sewell, 1983). Samples of the oil and brine were studied to determine if they contained microbes capable of using a polyacrylamide as a nitrogen–carbon–oxygen source. Five strains of a pigmented bacterium were found associated with *Desulfovibrio*. These bacteria were facultative and produced a sticky polymer. The fact of their existence in a petroleum reservoir is proof that the reservoirs are not sterile environments and the potential exists for many other bacteria that may be useful in enhanced oil recovery if they can be properly activated.

SUMMARY

Microbial enhancement of oil recovery still remains only as a promising method. Significant advances have been made in a relatively short period of time. The technology has advanced from laboratory concepts to field applications and field pilot studies, but the commercialization of the technology has not yet been widespread. The first phases of laboratory research have shown that microorganisms can produce chemical compounds that can mobilize oil. Microbes can move through porous media and can adapt to a variety of environmental conditions. The field studies being sponsored by the U.S. Department of Energy should define the technical and economic feasibility of several methods for applications of microbes in petroleum reservoirs. A considerable amount of work in this direction is also currently conducted in the U.S.S.R. and Romania.

REFERENCES

Anonymous, 1972. Bacteria have destroyed 10 percent of the world's crude. *World Oil*, 174(2): 28–29.

Anonymous, 1976. *The Role of Microorganisms in the Recovery of Oil. Eng. Found. Rep.*, NSF/RA-770201, U.S. Government Printing Office, Washington, D.C., 140 pp.

Anonymous, 1979. *Conference on Microbiological Processes Useful in Enhanced Oil Recovery, San Diego, Calif., August 29–September 21.* NTIS, Springfield, Va., DOE Conf. 79087, 54 pp.

Anonymous, 1987. Microbial oil recovery test starts in Oklahoma. *Oil Gas J.*, 85 (4): 30.

Atlas, R.M., 1981. Microbial degradation of petroleum hydrocarbons: An environmental perspective. *Microbiol. Rev.*, 45: 180–209.

Bailey, N.J.L., Jobson, A.J. and Rogers, M.A., 1973. Bacterial degradation of crude oil: Comparison of field and experimental data. *Chem. Geol.*, 11: 203–221.

Beckman, J.W., 1926. *Ind. Eng. Chem.*, 4 (10 November): 3.

Bond, D.C., 1961. Bacteriological method of oil recovery. *U.S. Patent*, No. 2,975,835, March 21.

Bryant, R.S., 1986. Microbial transformation of hydrocarbons. *U.S. Dep. Energy, Rep.*, NIPER-213, 19 pp.

Bryant, R.S. and Douglas, J., 1987. Evaluation of microbial systems in porous media for enhanced oil recovery. *Symp. on Oil Chemistry, San Antonio, Texas, February 4–6*, SPE 16284.

Bubela, B., 1983. Combined effects of temperature and other environmental stresses on MEOR. In: E.C. Donaldson and J.B. Clark (Editors), *Proceedings, 1982 International Conference on Microbial Enhancement of Oil Recovery, Afton, Okla., May 16–21.* NTIS, Springfield, Va., pp. 118–123.

Chakrabarty, A.M., 1982. *Biodegradation of Detoxification of Environmental Pollutants.* CRC Press, Boca Raton, Fla., 160 pp.

Chang, Y. and Yen, T.F., 1985. Effects of sodium pyrophosphate additive on the "huff and puff"/nutrient flooding MEOR process. In: J.E. Zajic and E.C. Donaldson (Editors), *Microbes and Oil Recovery.* Bioresources Publications, El Paso, Texas, pp. 247–256.

Clark, J.B., Munnecke, D.M. and Jenneman, G.E., 1981. In situ microbial enhancement of oil recovery. *Dev. Ind. Microbiol.*, 22: 695–701.

Crawford, P.B., 1983. Possible reservoir damage from MEOR. In: E.C. Donaldson and J.B. Clark (Editors), *Proceedings, 1982 International Conference on Microbial Enhancement of Oil Recovery, Afton, Okla., May 16–21.* NTIS, Springfield, Va., pp. 76–79.

Davies, J.S. and Westlake, D.W.S., 1979. Crude oil utilization by fungi. *Can. J. Microbiol.*, 25: 146–156.

Davis, J.B., 1967. *Petroleum Microbiology.* Elsevier, Amsterdam, 572 pp.

Donaldson, E.C., 1982. The potential of petroleum microbiology. In: *Proceedings, 17th Intersociety Energy Conversion Engineering Conference, Los Angeles, Calif., August 8–12.* IEEE, American Chemical Society, New York, N.Y., pp. 1009–1014.

Donaldson, E.C., 1985. Effect of *Clostridium* on production well capillary pressure phenomena. In: J.E. Zajic and E.C. Donaldson (Editors), *Microbes and Oil Recovery.* Bioresources Publications, El Paso, Texas, pp. 181–189.

Donaldson, E.C. and Clark, J.B., 1982. Conference focuses on microbial enhancement of oil recovery. *Oil Gas J.*, 80(51): 47–52.

Donaldson, E.C. and Grula, E.A., 1985. There are bugs in my oil well. *CHEMTECH*, October, pp. 602–604.

Dostalek, M. and Spurney, M., 1957. Release of oil by microorganisms; pilot experiment in an oil deposit. *Cesk. Mikrobiol.*, 2(5): 300–306.

Ewall, N.R., 1985. *Relationship of Pore Size Distribution to Fluid Flow.* M.S. Thesis, Department of Petroleum Engineering, University of Oklahoma, Norman, Okla., 152 pp.

Fedorak, P.M., Foght, J.M. and Westlake, D.W.S., 1983. Comparative studies on microbial degradation of aromatics and saturates in crude oil. In: J.E. Zajic, D.G. Cooper, T.R. Jack and N. Kosaric (Editors), *Microbial Enhanced Oil Recovery.* PennWell, Tulsa, Okla., pp. 162–172.

Findley, J., Appleman, M.D. and Yen, T.F., 1974. Degradation of oil shale by sulfur-oxidizing bacteria. *Appl. Microbiol.*, 28(3): 460–464.

Garland, T.M., 1966. Selective plugging of water injection wells. *J. Pet. Technol.*, 18(12): 1550–1554.

Gates, G.P.L. and Parent, C.F., 1976. Water quality control presents challenge in giant Wilmington Field. *Oil Gas J.*, 74(33): 115–126.

Gaudy, A.F., Jr. and Gaudy, E.T., 1980. *Microbiology for Environmental Scientists and Engineers.* McGraw-Hill, New York, N.Y., 736 pp.

Grula, E.A., Russell, H.H., Bryant, D., Kenaga, M. and Hart, M., 1983. Isolation and screening of *Clostridium* for possible use in MEOR. In: E.C. Donaldson and J.B. Clark (Editors), *Proceedings, 1982 International Conference on Microbial Enhancement of Oil Recovery, Afton, Okla., May 16–21.* NTIS, Springfield, Va., pp. 43–47.

Grula, E.A., Russell, H.H. and Grula, M.M., 1985. Field trials in central Oklahoma using *Clostridium* strains for microbially enhanced oil recovery. In: J.E. Zajic and E.C. Donaldson (Editors), *Microbes and Oil Recovery.* Bioresources Publications, El Paso, Texas, pp. 144–150.

Grula, M.M. and Sewell, G.W., 1983. Microbial interactions with polyacrylamide polymers. In: E.C. Donaldson and J.B. Clark (Editors), *Proceedings, 1982 International Conference on Microbial Enhancement of Oil Recovery, Afton, Okla., May 16–21.* NTIS, Springfield, Va., pp. 129–134.

Hitzman, D.O., 1962. Microbiological secondary recovery of oil. *U.S. Patent*, No. 3,032,472, May 1.

Hitzman, D.O., 1972. Hydrocarbon recovery from a subterranean formation employing an aqueous medium driving fluid having increasing viscosity. *U.S. Patent*, No. 3,650,326, March 21.

Hitzman, D.O., 1983. Petroleum microbiology and the history of its role in EOR. In: E.C. Donaldson and J.B. Clark (Editors), *Proceedings, 1982 International Conference on Microbial Enhancement of Oil Recovery, Afton, Okla, May 16–21.* NTIS, Springfield, Va., pp. 162–218.

Ivanov, M.V. and Belyaev, S.S., 1983. Microbial activity in waterflooded oilfields and its possible regulation. In: E.C. Donaldson and J.B. Clark (Editors), *Proceedings, 1982 International Conference on Microbial Enhancement of Oil Recovery, Afton, Okla., May 16–21.* NTIS, Springfield, Va., pp. 48–57.

Jack, T.R. and Thompson, B.G., 1983. Patents employing microbes in oil production. In: J.E. Zajic, D.G. Cooper, T.R. Jack and N. Kosaric (Editors), *Microbial Enhanced Oil Recovery.* PennWell Books, Tulsa, Okla., pp. 14–25.

Jack, T.R., Thompson, B.G. and DiBlasio, E., 1983. The potential for use of microbes in the production of heavy oil. In: E.C. Donaldson and J.B. Clark (Editors), *Proceedings, 1982 International Conference on Microbial Enhancement of Oil Recovery, Afton, Okla., May 16–21.* NTIS, Springfield, Va., pp. 88–93.

Jang, L.K., Sharma, M.M., Findley, J.E., Chang, P.W. and Yen, T.F.., 1983. An investigation of the transport of bacteria through porous media. In: E.C. Donaldson and J.B. Clark (Editors), *Proceedings, 1982 International Conference on Microbial Enhancement of Oil Recovery, Afton, Okla, May 16–21.* NTIS, Springfield, Va., pp. 60–70.

Jang, L.K., Sharma, M.M. and Yen, T.F., 1984. The transport of bacteria in porous media and its significance in microbial EOR. *Calif. Reg. Meet., Soc. Pet. Eng. of AIME*, SPE 12770.

Jaranyi, I., Kiss, L. and Szalanczy, G., 1963. Change in some characteristics of crude oil wells through the effect of microbiological treatment. In: *Wissenschaftliche Tagung für Erdölbergbau (Scientific Meeting for Crude Oil Exploration 1963), Budapest, October 13*, pp. 633–650.

Jenneman, G.E., Knapp, R.M., Menzie, D.E., Revus, D.E., Clark, J.B. and Munnecke, D.M., 1983. Transport phenomena and plugging in Berea Sandstone cores using microorganisms. In: E.C. Donaldson and J.B. Clark (Editors), *Proceedings, 1982 International Conference on Microbial Enhancement of Oil Recovery, Afton, Okla., May 16–21.* NTIS, Springfield, Va., pp. 71–75.

Jenneman, G.E., Knapp, R.M., McInerney, M.J., Menzie, D.E. and Revus, D.E., 1984. Experimental studies in in situ microbial EOR. *Soc. Pet. Eng. J.*, 14: 33–37.

Johnson, A.C., Jr., 1979. Microbial oil release technique for enhanced recovery. In: Anonymous (Editor), *Conference on Microbiological Processes Useful in Enhanced Oil Recovery, San Diego, Calif., August 29–September 21.* NTIS, Springfield, Va., DOE Conf. 79087.

Karaskiewicz, J., 1962. Experiments in the microbiological activation of crude oil deposits. In: *Wissenschaftliche Tagung für Erdölbergbau, Budapest, October 13*, pp. 566–577.

Kianipey, S.A. and Donaldson, E.C., 1986. Mechanisms of oil displacement by microorganisms. *Annu. Conf., Soc. Pet. Eng., New Orleans, La., October 5–8*, SPE 15601, 13 pp.

Kuznetsov, S.I., Ivanov, M.V. and Lyalikova, N.N., 1963. *Introduction to Geological Microbiology.* McGraw-Hill, New York, N.Y., 252 pp.

Lazar, I., 1986. Research on the microbiology of MEOR in Romania. In: *Proceedings, 1st International MEOR Workshop, April 1–3. U.S. Dep. Energy Rep.*, DOE/BC/10852-1, pp. 124–151.

Lazar, I. and Constantinescu, P., 1985. Field trials results of microbial enhanced oil recovery. In: J.E. Zajic and E.C. Donaldson (Editors), *Microbes and Oil Recovery.* Bioresources Publications, El Paso, Texas, pp. 122–143.

Levenspiel, O., 1972. *Chemical Reaction Engineering.* John Wiley and Sons, New York, N.Y., 578 pp.

Marquis, R.E., 1983. Barobiology of deep oil formations. In: E.C. Donaldson and J.B. Clark (Editors), *Proceedings, 1982 International Conference on Microbial Enhancement of Oil Recovery, Afton, Okla., May 16–21.* NTIS, Springfield, Va., pp. 124–128.

McCoy, W.F. and Costerton, J.W., 1982. Growth of sessile *Sphaerotilus natans* in tubular recycle system. *Appl. Environ. Microbiol.*, 43: 1490–1494.

McInerney, M.J., 1983. Physiological types of microorganisms useful for enhanced oil recovery. In: E.C. Donaldson and J.B. Clark (Editors), *Proceedings, 1982 International Conference on Microbial Enhancement of Oil Recovery, Afton, Okla., May 16–21*. NTIS, Springfield, Va., pp. 38–41.

Monod, J., 1950. La technique de culture continu. Théorie et applications. *Ann. Inst. Pasteur*, 79: 390–410.

Moses, V. and Springham, D.G., 1982. *Bacteria and the Enhancement of Oil Recovery*. Applied Science Publishers, London, 178 pp.

Moses, V., Robinson, J.P. and Springham, D.G., 1983. Microbial enhancement of oil recovery in North Sea reservoirs: a requirement for anaerobic growth on crude oil. In: E.C. Donaldson and J.B. Clark (Editors), *Proceedings, 1982 International Conference on Microbial Enhancement of Oil Recovery, Afton, Okla., May 16–21*. NTIS, Springfield, Va., pp. 154–157.

Nester, E.W., Roberts, C.E., Pearsall, N.N. and McCarthy, B.J., 1978. *Microbiology*. Holt, Rinehart and Winston, New York, N.Y., 769 pp.

Peterson, J.E., 1980. *Coalinga Polymer Demonstration Project*. NTIS, Springfield, Va., DOE/SAN/1556-5, 258 pp.

Petzel, G.A. and Williams, B., 1986. Operators trim basic EOR research. *Oil. Gas J.*, 84(6): 41–45.

Pfiffner, S.M., Jenneman, G.E., Walker, G.B., Jr., Javaheri, M., McInerney, M.J. and Knapp, R.M., 1985. Isolation and characterization of halotolerant, thermophilic, anaerobic, biopolymer-producing and biosurfactant-producing bacteria. In: J.E. Zajic and E.C. Donaldson (Editors), *Microbes and Oil Recovery*. Bioresources Publications, El Paso, Texas, pp. 285–294.

Senyukov, V.M., Yulbarisov, E.M., Taldykina, N.N. and Shishenina, E.P., 1970. Mikrobiologicheskii metod obrabotki neftianoi zalezy s visokoi mineralizatsey plastovykh vod. *Mikrobiologiya*, 39: 705–710.

Singer, M.E., 1985. Microbial biosurfactants. In: J.E. Zajic and E.C. Donaldson (Editors), *Microbes and Oil Recovery*. Bioresources Publications, El Paso, Texas, pp. 19–38.

Singer, M.E., Finnerty, W.R., Bolden, P. and King, A.D., 1983. Microbial processes in the recovery of heavy petroleum. In: E.C. Donaldson and J.B. Clark (Editors), *Proceedings, 1982 International Conference on Microbial Enhancement of Oil Recovery, Afton, Okla., May 16–21*. NTIS, Springfield, Va., pp. 94–101.

Slodki, M.E. and Cadmus, M.C., 1983. Production, composition, and biodegradation of xanthan gum (polysaccharide B-1459). In: E.C. Donaldson and J.B. Clark (Editors), *Proceedings, 1982 International Conference on Microbial Enhancement of Oil Recovery, Afton, Okla., May 16–21*. NTIS, Springfield, Va., pp. 3–5.

Slodki, M.E. and Cadmus, M.C., 1985. Production and stability of xanthan gums; xanthanases and their applicability. In: J.E. Zajic and E.C. Donaldson (Editors), *Microbes and Oil Recovery*. Bioresources Publications, El Paso, Texas, pp. 190–199.

Torbati, H.M., Donaldson, E.C., Jenneman, G.E., Knapp, R.M., McInerney, M.J. and Menzie, D.E., 1985. Depth of microbial plugging and its effect on pore size distribution in Berea sandstone cores. In: J.E. Zajic and E.C. Donaldson (Editors), *Microbes and Oil Recovery*. Bioresources Publications, El Paso, Texas, pp. 213–225.

Torbati, H.M., Raiders, R.A., Donaldson, E.C., McInerney, M.J., Jenneman, G.E. and Knapp, R.M., 1986. Effect of microbial growth on pore entrance size distribution in sandstone cores. *J. Ind. Microbiol.*, 1: 227–234.

Updegraff, D.M. and Wren, G.B., 1954. The release of oil from petroleum-bearing materials by sulfate-reducing bacteria. *J. Appl. Microbiol.*, 2: 309–322.

Von Heningen, J., DeHann, A.J. and Jansen, J.D., 1958. Process for the recovery of petroleum from rocks. *The Netherlands Patent*, No. 80,580.

Yarbrough, H.F. and Coty, V.F., 1983. Microbially enhanced oil recovery from the Upper Cretaceous Nacatoch Formation, Union County, Arkansas. In: E.C. Donaldson and J.B. Clark (Editors). *Proceedings, 1982 International Conference on Microbial Enhancement of Oil Recovery, Afton, Okla., May 16–21*. NTIS, Springfield, Va., pp. 149–153.

You-Im, Chang and Yen, T.F., 1985. Effects of sodium pyrophosphate additive on the "huff and puff"/nutrient flooding MEOR process. In: J.E. Zajic and E.C. Donaldson (Editors), *Microbes and Oil Recovery*. Bioresources Publications, El Paso, Texas, pp. 247–256.

Zajic, J.E. and Donaldson, E.C., 1985. *Microbes and Oil Recovery*. Bioresources Publications, El Paso, Texas, 370 pp.

Zajic, J.E., Grula, A. and Seffens, W., 1985. Oil displacement in the Hele-Shaw model using microbes, synthetic surfactants, and polymers. In: J.E. Zajic and E.C. Donaldson (Editors), *Microbes and Oil Recovery*. Bioresources Publications, El Paso, Texas, pp. 151–152.

Zhang, Z. and Qin, T., 1983. A survey of research on the application of microbial techniques to the petroleum production in China. In: E.C. Donaldson and J.B. Clark (Editors), *Proceedings, 1982 International Conference on Microbial Enhancement of Oil Recovery, Afton, Okla., February*. NTIS, Springfield, Va., pp. 135–139.

ZoBell, C.E., 1946. Bacteriological process for treatment of fluid-bearing earth formations. *U.S. Patent*, No. 2,413,278, December 24.

ZoBell, C.E., 1947. Bacteria release of oil from sedimentary materials. *Oil Gas J.*, 46(13): 62.

ZoBell, C.E., 1953. Recovery of hydrocarbons. *U.S. Patent*, No. 2,641,566, June 9.

Chapter 15

EVALUATION AND MONITORING OF ENHANCED OIL RECOVERY PROJECTS BASED ON GEOPHYSICAL WELL LOGGING TECHNIQUES

WALTER H. FERTL and G.V. CHILINGARIAN

INTRODUCTION

When old oilfields approach the economic limits of production by secondary recovery, a re-evaluation of the saturation distribution of the producing formation and its petrophysical properties frequently leads to rejuvenation of the field or implementation of new enhanced oil recovery techniques. Properly selected and applied well logging suites and associated modern well logging interpretation techniques assist (Fertl, 1979) in the: (1) redevelopment of oilfields, (2) exploration for and recompletion in potential reservoir stringers behind casing which had been passed up earlier, (3) monitoring production behavior and reservoir depletion, (4) proper planning of enhanced oil recovery projects, (5) determination of residual oil saturation, and (6) evaluation and monitoring of infill and observation wells.

Frequently, however, only an old and/or limited (i.e., incomplete) set of well logs is available in boreholes drilled years ago. Such wellbores need to be re-evaluated by properly selected new logs.

COMPLETION CONSTRAINTS AND BOREHOLE INTEGRITY CHECKS

Before using cased wellbores for monitoring various tests and production behavior, integrity of the borehole must be checked. Tests include those for casing and cement bond integrity, and possible fluid movement behind casing due to channeling, using acoustic and/or nuclear logs, pulsed neutron, radioactive tracers, and several other techniques.

Properly run cement bond logs (CBL) provide differentiation between the presence of a microannulus and a channel behind casing. This is very important, because a channel but not the microannulus allows fluid movement behind casing. Both are easily distinguished by repeat CBL logging runs over the suspect interval with and without hydraulic pressure on the casing.

Hence, initial openhole borehole conditions (borehole rugosity, washouts, etc.), cased-hole installations (tubing, packer, valves, casing size restrictions), borehole fluids, etc., are important considerations. The key prerequisite, however, is proper isolation of zones behind pipe based on a satisfactory cement bond between casing, cement, *and* formation; otherwise, any cased-hole reservoir evaluation becomes unreliable, questionable, or totally wrong.

For about two decades, the oil industry has used wireline well logging, such as cement bond logging, to detect the presence of cement behind pipe and to evaluate the bond of the cement to both the casing and the formation.

TABLE 15-I

Cement bond evaluation (after Fertl, 1982)

Basic requirements:	fully understand gating system used, tool frequency and spacing effects Logging instrument run centered		
Interpretive conditions	**Casing signal**	**Formation signal**	**Remarks**
Free casing (uncemented)	Strong Parallel VDL pattern Chevron ("W") pattern at casing collar	None Weak if casing rests on borehole wall or sloughing shales	Amplitude decrease at collar T/R spacing indicator
Good bond to casing and formation	Weak or none in soft formation Hard (fast) formation signals may arrive at or before casing signal (VDL or scope pictures are required for analysis)	Strong, provided attenuation not too high (i.e., unconsolidated gas reservoirs)	Good acoustic coupling between casing, cement, and formation. Formation arrival time varies with depth as a function of lithology, porosity, and pore pressures in shales. Reliable analysis based on amplitude only is not possible.
Good casing bond but poor formation bond	Weak	Weak or none	Weak formation signal *also* caused by high attenuation (soft shales, unconsolidated gas sands). Reliable analysis based on amplitude only is not possible.
Microannulus [a,b]	Weak to moderately high	Moderate	Small gap between casing and cement; however, there is no vertical fluid movement.
Channeling [b]	Weak to moderately high	Moderate	Allows vertical fluid communication

[a] Caused by (1) holding pressure on casing after cementing, (2) thermal expansion during cement setting, (3) thermal contraction due to circulation, and (4) well varnish, grease, and oil-wet pipe.
[b] Repeat runs with and without casing pressure differentiate between microannulus and channeling.

The validity of cement bond log interpretation has been a subject of controversy since its introduction; and the CBL, despite its great potential, is probably one of the most abused, misused, and misunderstood logs in oilfields today (Fertl et al., 1974). Table 15-I summarizes several of the basic considerations for cement bond log evaluation which have been discussed in detail in extensive literature.

RESERVOIR POROSITY DETERMINATION

The importance of accurate, reliable porosity measurements by well logs cannot be overemphasized. Generally, modern logging suites fill these requirements very adequately.

In many old wells that are now being considered for either recompletion or for enhanced recovery techniques, however, there are no adequate logs for the required evaluation.

Similarly, when highly deviated wells are drilled, primarily for development purposes, often a complete log suite cannot be obtained in the open hole. In these circumstances, it is often still possible to evaluate adequately the wells in question with a suite of cased-hole porosity devices.

Table VII in Chapter 3 lists several through-casing porosity determination techniques which are currently available, points out the basic concepts involved, and focuses on possible constraints (Fertl, 1982b). Basically, a wide variety of neutron, acoustic, density, and several pulsed neutron logging concepts (including log-inject-log) allow through-casing porosity estimates. In addition, the same also holds true for the judicious selection and application of proper mathematical transforms.

CASED OBSERVATION WELLS

Enhanced recovery projects are frequently monitored by observation wells. The latter may be cased with steel casing or a fiberglass liner across the target interval.

Depending on well completion, type of enhanced recovery method, and reservoir characteristics, either pulsed neutron devices (particularly the salinity-independent carbon/oxygen log), neutron logs, or induction logs can be run to monitor flood profiles, movement of reservoir fluid contacts, water breakthrough, etc.

For example, high-resistivity fiberglass liners (filled with fresh water to eliminate large borehole signal effects on the induction log response) have been successfully used in several old fields to monitor on a qualitative basis any saturation changes, water breakthrough, and vertical sweep and displacement efficiencies.

TABLE 15-II

Cased-hole logging techniques (formation evaluation logs)

Special purpose logs and associated services	Acoustic Cement Bond Log (ACBL)	Cement Bond Log (CBL)
	Variable Density Log (VDL)	Variable Density Log (VDL)
Casing corrosion evaluation	Magnelog	Electromagnetic Thickness Log (ETT)
	Vertilog®	Pipe Analysis Log (PAL *)
	N.E.	Acoustic Scanner (AST)
	Casing Potential Profile (CPP)	
	Multi-Finger Caliper	
Spectral gamma ray logs	Spectralog^{sm}	Natural Gamma Ray Spectrometer (NGS)
	Carbon/Oxygen Log(C/O)	Spectrometer (GST)
	Continuous MSI-C/O Log	
Special purpose	Formation Tester(FT)	Formation Interval Tester (FIT)
	Perforating-Formation-Collar (PFC)	Perforating Depth Control (PDC)
	Photon Log(PL)	Gravel Pack Log (GPL)
Gamma ray log (natural gamma)	Gamma Ray Log (GR)	Gamma Ray Log (GR)
Neutron logs	Neutron Log (NL)	Neutron Log (NL)
	Compensated Neutron Log (CNL)	Compensated Neutron Log (CNL)
		Dual Compensated Neutron (DNP)
Pulsed neutron log	Neutron Lifetime Log®(NLL)	Thermal Decay Time Log (TDT-K, TDT-M)
	PDK-100	
Production logs	Production PCM (production equivalent of SLAM)	Simultaneous Production Logging (PLT)
	Basket Flowmeter (BF)	Fullbore Flowmeter (FBF)
	Continuous Spinner Flowmeter	Continuous Flowmeter (CFM)
	N.E.	Vortex Flowmeter (VFM)
	Basket Flowmeter (BF)	Petal Packer Flowmeter (PPF)
	Fluid Density Log (FD)	Gradiomanometer (GM)
	Fluid Density Log (FD)	Nuclear Fluid Densimeter (NFD)
	Surface-Recorded Pressure Log	HP Pressure Log (HPP)
	Surface-Recorded Pressure Log	Manometer (MT)
	Differential Temperature Log	High-Resolution Thermometer (HRT)
	Geothermal Temperature Log	High-Temperature Thermometer (HTT)
	Water Holdup Indicator Log (WHI)	Water-Holdup Meter (WHM)
	Through-Tubing Caliper Log	Through-Tubing Caliper (TTC)
	Through-Tubing Borehole Fluid Sampler	Fluid Sampler (PFS)
	Tracerlog/Flolog	Dual Radioactive Tracer Survey (DRT)
	Sonan Log	Audio Log (AUD)

TABLE 15-III

Perforating techniques

Shaped charges	Jumbo Jet II	Ultra-Jet
	Jumbo Jet®	Hyper-Jet
	Big Hole	Ultra-Pack
	Golden Jet	
Other types	Link Kone	Uni-Jet
	Bar Kone	Enerjet
	Slimkone	HyperDome
	Selective	Hyper-Select

If salinity of mixed water in the reservoir around the borehole can be established with some reliability (for example, from test samples), then this method will also provide quantitative estimates of saturation changes.

CASED-HOLE LOGGING—FORMATION EVALUATION LOGS

Inasmuch as a large variety of cased-hole logging devices available to the petroleum industry are discussed in the field cases next, both the generic classification and tradenames of these services as offered by the two major well logging service companies have been compiled as a comparative reference source in Table 15-II. Types of perforating techniques are presented in Table 15-III.

Pulsed neutron logging

In 1963, Dresser Atlas introduced the Neutron Lifetime Log (NLL) instrument, which was the first through-casing logging device that could distinguish and quantitatively evaluate oil and gas in the presence of saline formation waters. Today, the Dual Detector Neutron Lifetime Log instrument is the primary through-casing and/or through-tubing device currently in use in saline evironments. In fresh and brackish formation water environments, the carbon/oxygen log is used to evaluate potential hydrocarbon-bearing reservoirs.

The Neutron Lifetime Log® survey differs from all previous logs in that the neutron source is pulsed, which results in the ability to measure the macroscopic thermal neutron absorption capture cross-section (Σ) of the formation. Although this absorption cross-section is a nuclear measurement, the log response is similar in appearance to the induction log and can be used for many geological applications previously available only from openhole resistivity logs.

The most significant use of the Neutron Lifetime logging instrument is to discriminate among gas-, oil- and salt water-bearing formations in cased boreholes. In fact, water saturation can be calculated from the Neutron Lifetime

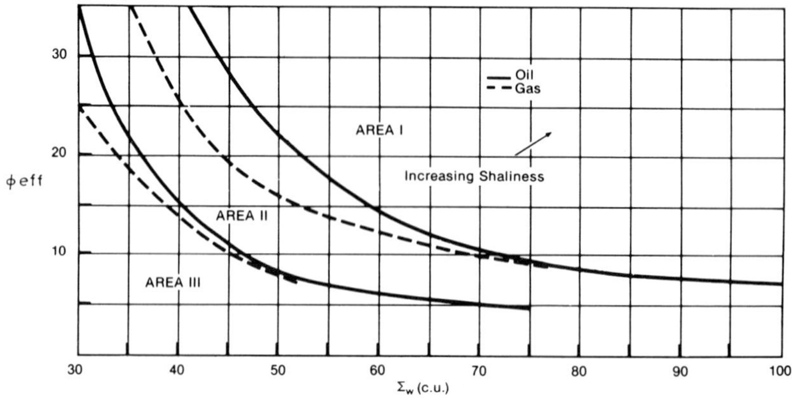

Fig. 15-1. Areas of reliable NLL water saturation calculation. *Area I:* high confidence; *Area II:* still quantitative, but may require multipass averaging; *Area III:* no quantitative interpretation.

survey, provided porosity, hydrocarbon type, matrix, and formation water salinity are known or may be estimated.

The Neutron Lifetime survey has applications in the evaluation of new hydrocarbon reserves, monitoring reservoir depletion behavior during primary and enhanced recovery, and properly planning workover operations. Gas–oil–water contacts can be located and monitored, hydrocarbon migration between zones can be detected, and recovery efficiency as well as production problems can be diagnosed using this survey. In log-inject-log operations, reliable residual oil saturation values are obtained (see Chapter 3).

As with other wireline logs, the NLL is useful only if it provides accurate, reliable and useful information about the formation of interest. The most valuable information provided by the NLL is the quantitative saturation determination within a potential reservoir. As quality and reliability of this information increases, so does the usefulness of the NLL. Fig. 15-1 provides guidelines for the conditions of greatest effectiveness for NLL interpretation, representing the combined effects of variation in the formation parameters: porosity, formation water salinity, shaliness, and hydrocarbon type. It should be noted that the accuracy of the final water saturation calculation is still dependent on an accurate knowledge of the matrix and fluid parameters and the degree of confidence in other supporting log data.

In lower formation water salinity where the NLL interpretation becomes more qualitative than quantitative, the Dresser Atlas Continuous Carbon/Oxygen Log (C/O) can be successfully used to determine water saturation. As a general rule, in formations with porosities greater than 15%, the C/O log provides a quantitative determination of water saturation regardless of formation water salinity. The C/O log provides a qualitative interpretation in rocks with porosity ranges of 10–15% and should not be used to quantitatively evaluate hydrocarbons in rocks having porosities below 10%.

DIGITAL LOG ANALYSIS (EPILOG) TECHNIQUES USING NEUTRON LIFETIME LOGGING

Over long intervals of the wellbore, the interpretation methods discussed so far can be fully exploited only through the use of digital interpretation techniques. A computer interpretation technique called the Cased Hole Exploration Service (CHES) Epilog® analysis has been developed. The CHES program combines the available open- and cased-hole data and presents the resulting digital analysis in both analog (film) and tabular formats. The CHES analysis provides detailed information on reservoir rock properties, such as porosity, shaliness, and reservoir fluid saturation. This technique also enables distinction between oil and gas pay zones, and monitoring the reservoir condition, i.e., state of depletion and hydrocarbon–water and gas–oil contacts behind pipe.

The basic format (Fig. 15-2) and the underlying concepts for both the analog presentation and the tabular listing of the computed results are described below.

Analog presentation

Track I: Gas indicator

The type of medium occupying the pore space of the reservoir rock is indicated by the comparative presentation of two curves. They are based on the values for effective reservoir porosity and the shale corrected porosity index calculated from the ratio curve of the NLL. Gas and condensate are signaled by a separation of these two curves. In the presence of oil and formation water, both curves track and exhibit no separation. Oil and water are differentiated based on the computed values for water saturation, S_w, which are presented in the other tracks on the computed log.

Track II: Fluid analysis

Water saturation. The reservoir fluid distribution is presented for the openhole and the cased-hole condition. If fluid distribution has not changed, i.e., the zone has not yet been produced in the subject well and/or been depleted by offset producers, both the openhole (R_t) and cased-hole (sigma) derived water saturation curves overlay. Depletion of hydrocarbons, however, is reflected in a separation of the water saturation curves.

Track III: Porosity and fluids—cased hole

Porosity [ϕ_e]. Porosity value presents the total volume of reservoir pore space available for reservoir fluids.

Water-filled porosity [$\phi_e S_w$]. Water-filled porosity represents the volume of pore space occupied by formation water. The area between this curve and zero porosity is the volume of pore space filled with water.

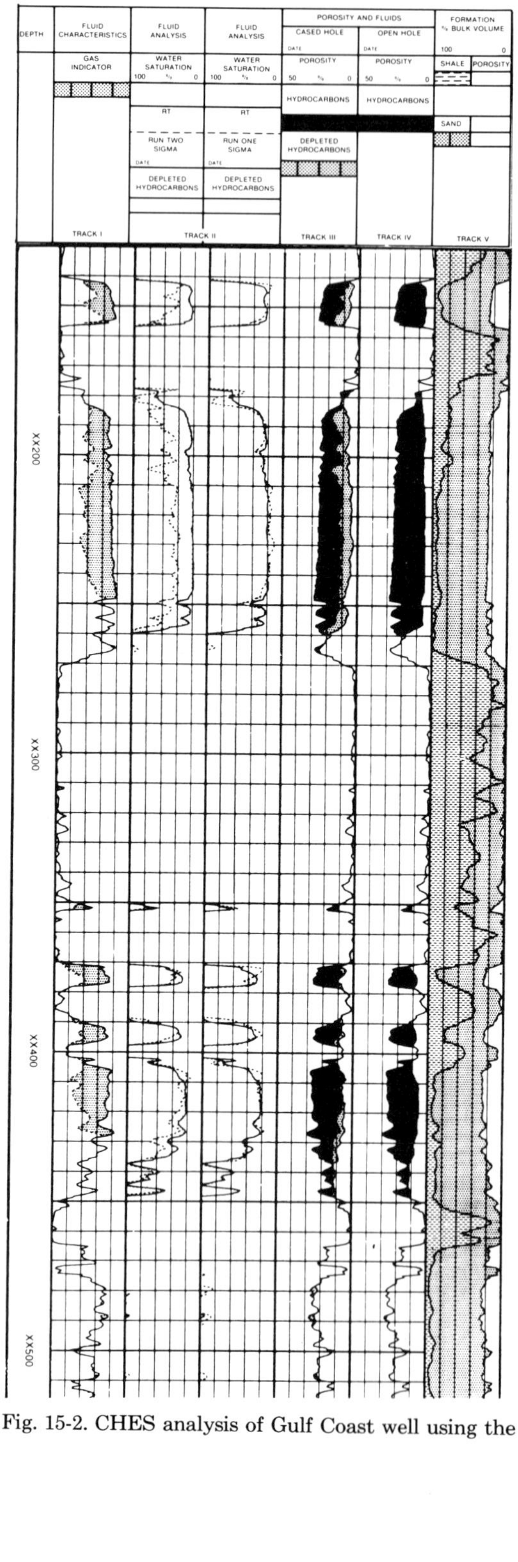

Fig. 15-2. CHES analysis of Gulf Coast well using the NLL and openhole log data.

Hydrocarbon-filled porosity [$\phi_e(1 - S_w)$]. The separation between the curves representing water-filled porosity and reservoir porosity determines the volume of pore space occupied by hydrocarbons. The fluid saturation distribution shown in Track III represents the reservoir condition at the time the Neutron Lifetime Log® survey was run.

Track IV: Porosity and fluids—open hole

Porosity [ϕ_e]. Porosity value represents the reservoir pore space occupied by reservoir fluids.

Water-filled porosity [$\phi_e S_w$]. Bulk volume curve represents the amount of pore space occupied by formation water. The area between this curve and zero porosity is the volume of pore space filled with water.

Hydrocarbon-filled porosity [$\phi_e(1 - S_w)$]. The separation between the curves representing water-filled porosity and reservoir porosity determines the volume of pore space occupied by hydrocarbons. The fluid saturation distribution shown in Track IV represents the initial reservoir condition at the time the well was drilled. Comparison of the data in Tracks III and IV shows the amount of reservoir depletion (if any), movement of hydrocarbon–water and gas–water contacts, etc.

Track V: Bulk volume analysis of the formation

The data in Track V show the percentage of formation bulk volume containing clay minerals, the volume of sand (including silt), and the reservoir porosity, which is corrected for shale and hydrocarbon effects.

Tabular listing

The tabular listing contains nine columns of data which include the following:

(1) Depth (feet or meters).

(2) Porosity (%) corrected for shale and hydrocarbon effects.

(3) Water saturation (%) in cased hole using the NLL.

(4) Water saturation (%) in open hole based on resistivity log and porosity information.

(5) Apparent fluid density (g/cm^3) of the medium occupying the pore space, such as oil, gas, and/or water.

(6) Volume of clay, in percent of bulk volume (in complex lithologies, the apparent shale volume is listed in place of clay volume).

(7) Permeability index, based on a choice of equations.

(8) Cumulative interval hydrocarbon-feet (meters), summed from top to bottom of each potential zone.

(9) Depleted cumulative hydrocarbon-feet (meters), summed from top to bottom of each potential zone. The initial cumulative hydrocarbon-feet (meters) over the zones of interest is then equal to the sum of columns (8) and (9).

TIME-LAPSE RESERVOIR MONITORING USING NEUTRON LIFETIME LOGGING

A sequence of several oil and gas zones is present in Fig. 15-3 for an old U.S. Gulf Coast well (DeVries and Fertl, 1977).

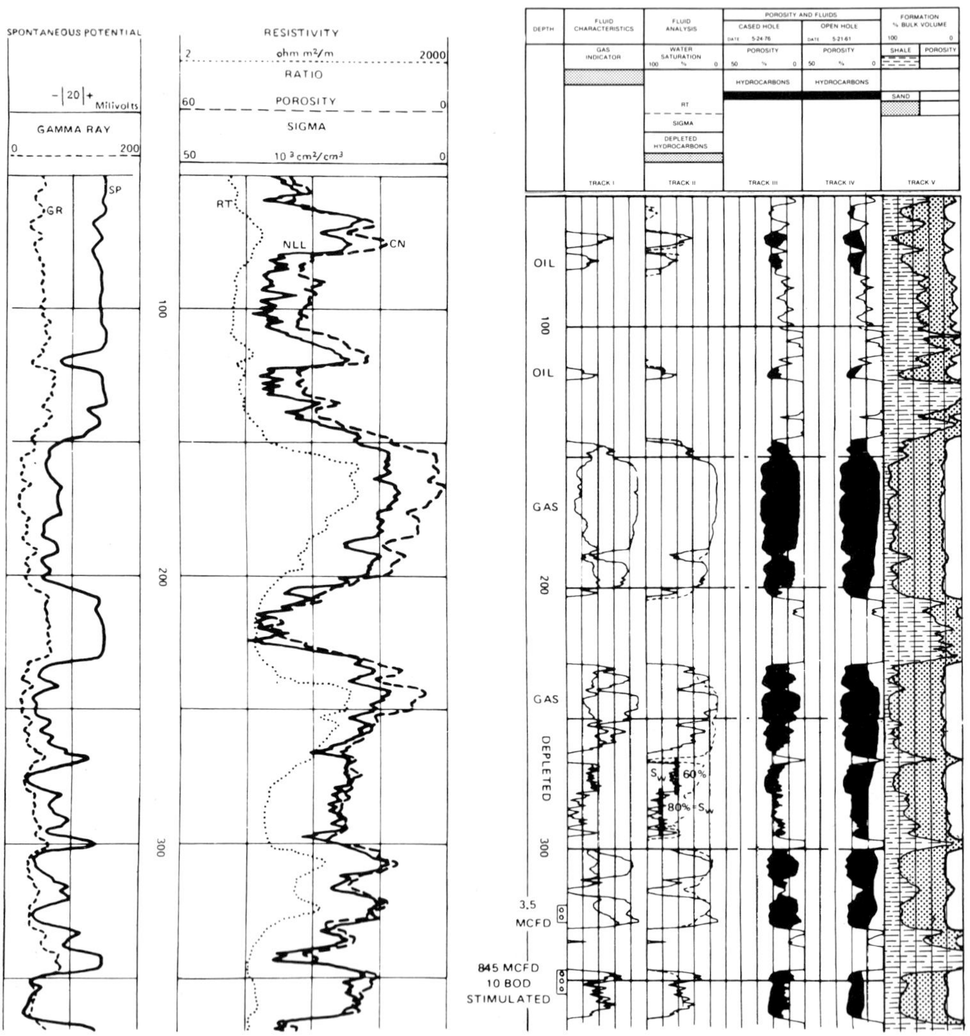

Fig. 15-3. Multipay sequence of oil and gas zones evaluated and monitored through casing with pulsed neutron (DNLL) log and compared to initial reservoir condition based on openhole logging data, U.S. Gulf Coast well. (After DeVries and Fertl, 1977, p. 13, fig. 4; courtesy of SPWLA.)

Several unproduced oil sands are present above 130 ft. A massive gas sand is present at 146–204 ft. This zone is a prolific gas producer in the area, but had not been produced in the subject well previously. Cumulative thickness of hydrocarbon zones of this interval is 11.46 ft.

A partially depleted gas zone is present from 229 to 296 ft. The zone below the shale break at 264–267 ft has been largely watered out. From 229 to 264 ft, 5.69 cumulative hydrocarbon-feet of gas are still present, with only 1.60 cumulative hydrocarbon-feet so far depleted.

Zones below 300 ft were tested after the NLL was run. The log-defined gas zone between 300 and 320 ft was perforated at 323–329 ft and tested 3.5 MMcf/day gas with no water cut.

The potential pay zone at 346–364 ft was characterized by the log analysis as oil pay with high gas/oil ratio (GOR). After the interval was perforated from 346 to 356 ft and acidized, the zone produced 845 MMcf/day of gas and 10 bbl/day of oil.

CONTINUOUS CARBON–OXYGEN (C/O) LOG

While the above pulsed neutron (NLL) devices have their optimum range of applications in saline, high-porosity reservoirs, they cannot be used to evaluate and/or monitor hydrocarbon saturation where the macroscopic cross-section of the formation water is not significantly different from that of the hydrocarbons, or in cases where salinity is constantly changing or unknown (e.g., fresh water, steamfloods, etc.). The response of the C/O log, however, is not affected by such salinity constraints.

The continuous C/O logging system has already been run successfully in many wells throughout the world under a wide variety of reservoir conditions (Table 15-IV).

Simply stated, the C/O log presents a measure of the relative abundance of carbon to oxygen derived from the detection of the gamma rays produced from both elements by the inelastic scattering of very fast neutrons. The carbon/oxygen survey is performed by taking measurements at discrete depth levels opposite the formation of interest. In order to differentiate carbon present in hydrocarbon molecules from that in the rock framework (i.e., carbonate solid matter), a Si/Ca ratio is also determined from the gamma rays resulting from neutron capture.

The source of neutrons used in C/O logging is an accelerator-type source, which produces 14 MeV neutrons by the deuterium–tritium reaction. This neutron source is pulsed at a repetition frequency of 20,000 cycles per second (20 kHz). The schematic of the instrumentation, tool specifications, and source detector operating cycle have been discussed in detail by Oliver et al. (1981).

During C/O logging, the following curves are normally recorded:

—Si: silicon curve for the capture gamma ray energy window of 2.35–5.08 MeV.

TABLE 15-IV

Evaluation of continuous carbon/oxygen (C/O) logging (after Fertl, 1982)

Applications	General remarks/constraints	Field experience/observations
Detection and quantitative evaluation of oil reserves in presence of fresh, brackish, mixed, and/or unknown formation water salinities.	Drastic salinity variations over short vertical distances may require interpretive use of Si/Ca and Ca/Si ratios with the C/O measurement.	North America (U.S.A., Canada), South America (Argentina, Venezuela), Norway and U.K., North Sea, Continental Europe (West Germany), Saudi Arabia.
Exploration for overlooked or previously not commercially attractive hydrocarbon resources in abandoned and/or workover wells.	Prerequisite is integrity of well completion (i.e., good cement job, no casing leaks, and no severe washouts).	Particularly effective in old, shallow, low-salinity, multipay oilfields, supposedly watered-out (stratified) reservoirs, etc. Excellent results obtained in U.S. Gulf Coast region, West Texas, NewMexico, Mid-Continent, Rocky Mountains, California, etc.
Evaluation of oil reserves in newly drilled wells where openhole (OH) logs are not available.	Due to mechanical or safety-related hole problems no OH logs were run.	Universally applicable in medium-to-high-porosity reservoirs. Published field cases include Glorieta and Clearfork Formations in Mitchell and Lubbock Counties, Texas.
Evaluation of newly drilled wells where OH logs give inconclusive information.	Time lag required after casing is run to eliminate possible invasion effects due to mud filtrate.	Universally accepted in medium-to-high-porosity reservoirs. Published field cases include Glorieta Formation in Lubbock County, Texas, Ector County, Texas, and multi-fault block oilfield, onshore Louisiana.
Evaluation of clean and/or shaly oil-bearing reservoirs where openhole and other cased-hole logs provide information inconsistent with reservoir performance.		
Monitoring of displacement and sweep efficiency in primary–secondary–tertiary recovery project. Reservoir performance and depletion checks.	Well completion integrity is a must.	Multi-well, field-wide monitoring of vertical and lateral reservoir performance. Input parameters in reservoir engineering studies and reservoir simulation models.
Hydrocarbon–water contacts. Detection and monitoring in presence of low or unknown formation water salinities.	Time-lapse logging. Multiple repeat runs (three or more) to improve results.	Applicable and used in many areas, domestic and overseas.

TABLE 15-IV (continued)

Applications	General remarks/constraints	Field experience/observations
Waterfloods. Monitoring of oil saturation depletion, reservoir stratification, encroachment and fresh-water breakthrough.	Time-lapse logging. Salinity index can be calculated (i.e., extent of mixing of injection and initial formation waters of contrasting salinity can be monitored on vertical and lateral reservoir basis).	Universally applicable. Salinity index has been used successfully in U.S. Mid-Continent oilfields. C/O application is dramatically superior to the requirement of special plastic-casing and induction logging monitoring.
Chemical enhanced oil recovery (micellar–polymer, polymer–caustic). Monitoring of flood efficiency, oil depletion, stratification, break-through patterns, etc.	Time-lapse logging. Well completion integrity is prerequisite.	Domestic field experiences.
Gaseous enhanced oil recovery (CO_2 miscible, hydrocarbon miscible, other gases). Monitoring of the flood efficiency, oil depletion, stratification, breakthrough patterns, etc.	Time-lapse logging. Gas saturation estimates possible in the swept zones.	Domestic field experiences.
Thermal enhanced oil recovery (cyclic steam "huff and puff" injection, steamflood, in-situ combustion). Monitoring of flood efficiency, oil depletion, stratification, break-through patterns, etc.	Time-lapse logging. Heavy oil (low API gravity) oil reservoirs, tar sands. Temperature constraints, very large casings and underreamed, gravel-pack completions.	Extensive field experiences in the U.S.A. (California, Texas, Louisiana, Mid-Continent, Rocky Mountains); Canada (Athabasca, McMurray); Europe (Emsland, West Germany); South America.
Residual oil saturation (*ROS*) based on log-inject-log techniques.	Accurate salinity control of injection fluid not critical. Assure well completion integrity. Low rate injection to avoid formation fracturing and crude oil stripping around wellbore.	Numerous domestic applications in Alaska and lower 48 states.
Oil resource evaluation in mixed and complex lithologies where formation water salinities may pose constraints on conventional formation eval-uation (carbonates, oil shales, granite wash, tuffaceous sands, etc.).	At least medium reservoir porosity. Frequently very helpful is supplementary information from natural gamma ray spectral logging.	U.S.A., Canada, Europe, South America. Carbonates (Permian basin, U.S.A.; Florida; Rocky Mtns.). Tuffaceous clastics (Argentina); granite wash (Texas, Oklahoma). Niobrara, Mowry Shale, and Phosphoria Formations (Rocky Mtns.). Oil shales (Rocky Mtns.).

TABLE 15-IV (continued)

Applications	General remarks/constraints	Field experience/observations
Distinction between gas-bearing clastic reservoirs and coal seams (i.e., lignites, etc.).	Gas sands and low-rank coals exhibit similar standard log responses. Coupled with frequent mud gas shows, this results in the testing of "gas" prospects which in fact are lignite beds.	Important in U.S. Gulf Coast region, and elsewhere, when evaluating old wells or proposed hydrocarbon prospects based on old electric logs and/or logs only.
Coal seam evaluation.	In cased (old) wells. Estimate of coal rank, Btu content. Constraints include severely washed-out seams.	Field studies in Pittsburg County, Oklahoma, based on both continuous C/O and stationary log measurements.

—Monitor: monitor curve for the reflection of source output stability.

—Si/Ca: ratio of silicon to calcium for the capture spectrum (silicon window: 3.17–4.65 MeV, calcium window: 4.86–6.62 MeV).

—C/O: ratio of carbon to oxygen for the inelastic spectrum (carbon window: 3.17–4.65 MeV).

—Ca/Si: ratio of calcium to silicon for the inelastic spectrum (calcium window: 2.50–3.30 MeV, silicon window: 1.54–1.94 MeV).

—C/I: capture/inelastic ratio, porosity curve.

—Optional: count rate data from any of the above mentioned windows may also be displayed.

The *C/O ratio* is a function of lithology, porosity and amount of oil in the reservoir, independent of salinity. Resolution of the C/O log increases with increasing porosity.

The *Ca/Si ratio* is also independent of salinity, is insignificantly affected by porosity, and provides lithology information. By combining C/O and Ca/Si data, the oil saturation can be determined by quick-look overlay techniques and/or digital analysis.

The *Si/Ca ratio* is lithology, porosity, and salinity dependent. By combining C/O and Si/Ca data, the oil saturation can be determined by quick-look overlay techniques and/or digital analysis provided salinity does not significantly change over the interval, or such changes are well known. Under such conditions, use of C/O and Ca/Si data is preferred. The Si/Ca ratio, however, still can be used to calculate a salinity index to monitor behind casing the extent of mixing of injection and initial formation waters of contrasting salinity on both a vertical and lateral basis in secondary recovery projects.

Inelastic and capture gamma ray energy spectra also can be taped digitally at the wellsite in both a continuous and stationary logging mode. The most common logging practice is taping spectral data at 1-ft intervals. These taped data can be used to retrieve count rate and ratio data for selected energy windows, including hydrogen and iron.

TABLE 15-V

Environmental effects in C/O logging (after Fertl, 1982)

Reservoir condition	Comments
Invasion	For invasion effects to disseminate, the C/O log should not be run immediately after casing has been set. Depending on the drilling operations, mud, and reservoir characteristics, a waiting period from a few days to several weeks may be advisable.
Open perforations	In a producing well killed through open perforations, the C/O interpretation will be affected over the perforated interval.
Casing type and size	Fiberglass pipe has less effect than steel casing. Very large casing sizes ($13\frac{3}{8}''$ and larger) at the most give marginal answers.
Casing fluids	Distinct, accountable shifts in C/O log ratios occur at borehole fluid interfaces.
Cement thickness	Increasing cement sheath thickness reduces all C/O log ratios: C/O and Si/Ca in similar fashion and Ca/Si to a larger degree. Thus, interpretation of C/O log data may prove pessimistic in presence of thick cement sheaths.
Completion integrity	Effectiveness of cement job is crucial. Presence of poor cement bond to casing and/or formation (particularly in severely washed-out sections) causes channeling, depletion or recharge in multi-reservoir intervals that may be at different formation pressures behind casing.
Underreamed, gravel pack –slotted liner completions	Due to constraints caused by depth of investigation and fluid distribution in massive gravel packs, information as to true reservoir conditions may be marginal.
Magnetized casing collars	Will cause spikes on the C/O log ratios, most noticeable on the Ca/Si and least on the Si/Ca ratios.
Oil density	C/O ratio values will slightly increase with decreasing API gravity of crude oils.

A standard gamma ray adapter can be run on top of the C/O device to provide an auxiliary gamma ray correlation curve.

As with any other nuclear logging instrument, statistical variations are inherent in recorded C/O logging data.

The C/O logging measurements respond to fluid saturation, porosity, and formation lithology. In high-porosity reservoirs, the depth of investigation is about 8.5 inch from the borehole wall in $6\frac{5}{8}''$ casing. Thus, environmental effects have to be considered in any quantitative C/O log analysis. Table 15-V summarizes some of the environmental effects that should be considered for the C/O log.

LOG-DERIVED EVALUATION OF THERMAL RECOVERY PROCESSES

Thermal recovery processes, such as steam stimulation (i.e., cyclic steam injection, steam soak, huff and puff), steamflooding, and in-situ combustion (i.e., fireflood), account for approximately two-thirds of the world's enhanced oil recovery. Thermal processes work best in oil reservoirs which exhibit low API gravity, high viscosity, and high porosity.

Several recent field experiences are discussed below.

Poso McVan steamflood pilot, Kern County, California

The Poso McVan heavy oil reservoir, located in Kern County, California, has been evaluated based on large-diameter whole-core and open- and cased-hole well logging suites. Comparative results for oil saturation and porosity in this reservoir have been based on core analysis, resistivity, porosity, dielectric log, and continuous C/O log.

At Poso McVan, the Etchegoin sandstones are of excellent reservoir quality, with productive intervals generally occurring at depths between 1100 and 1300 ft. Major shale breaks occur among the three reservoirs but do not affect the lateral continuity of the sand bodies.

Porosity ranges from 35% to 40%; permeability, from 500 to 50 mD; and oil gravity, from 11 to 13° API. Oil viscosity is very sensitive to temperature, ranging from 8300 cP at formation temperature of 95°F to 22 cP at the steamed reservoir temperature of 240°F. Formation water salinities are extremely low (188 ppm eq. NaCl), which causes uncertainties in the original oil saturation determination based on conventional log analysis.

For these conditions, Roux et al. (1981) developed an open- and cased-hole logging program to accurately evaluate the Poso Creek reservoir before and during steamflood operations. Conventional openhole logs in conjunction with dielectric logs were used to evaluate all new wells drilled in the project. Changes in oil saturation during the steamflood were evaluated on key wells using the continuous C/O log. Steam breakthrough was monitored using the silicon curve, capture/inelastic (C/I) ratio, and temperature log data.

Fig. 15-4 compares the oil saturation values as determined from whole-core analysis, an openhole dielectric log, and the cased-hole continuous C/O logging information. Fig. 15-5 shows the computerized continuous C/O analysis in four wells prior to steam injection. This information establishes the base logs for monitoring the efficiency and steam breakthrough patterns in the Poso McVan steamflood.

Steam-initiated gravity drainage of heavy oil reservoir, McMurray Formation, Canada

The MAISP (Mine Assisted In Situ Pilot) project, a joint venture by Petro Canada (operator), Esso Resources, Cities Service Canada, Husky, and Gulf, is

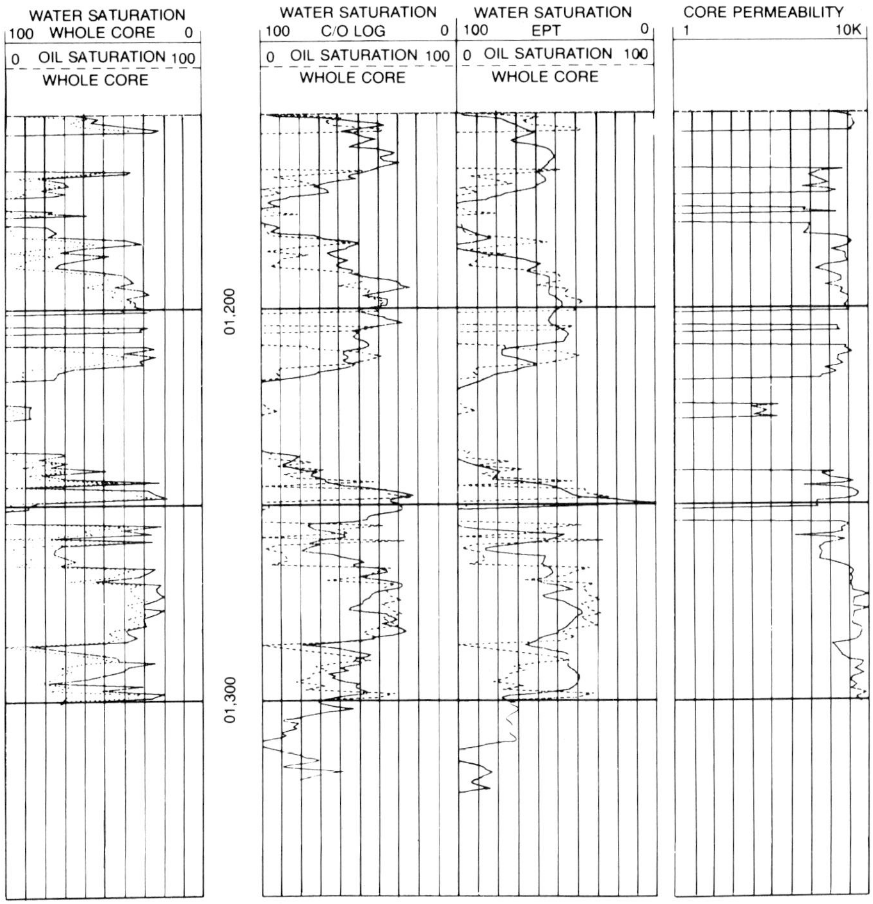

Fig. 15-4. Oil saturation distribution in a heavy oil reservoir from the Poso Creek area, Kern County, California. The whole core, C/O log, and dielectric log data are compared. (After Roux et al., 1981, p. 13, fig. 8b; courtesy of SPWLA.)

unique because the heavy oil reservoir is produced by means of a horizontal well system in which the heated oil drains by means of gravity segregation (Campbell et al., 1982). Conventional open- and cased-hole logs have been used to evaluate steam displacement and sweep efficiency of this heavy oil recovery project in the McMurray Formation, located in northeastern Alberta, Canada.

The McMurray oil sands have a great aerial extent and contain vast quantities of bitumen (as much as 900 billion barrels). The McMurray Formation of Lower Cretaceous age is a member of the Mannville Group. The upper McMurray is a fine-grained sand with shale interbedding, and exhibits lower bitumen saturations than the lower McMurray. The lower McMurray is a coarser-grained

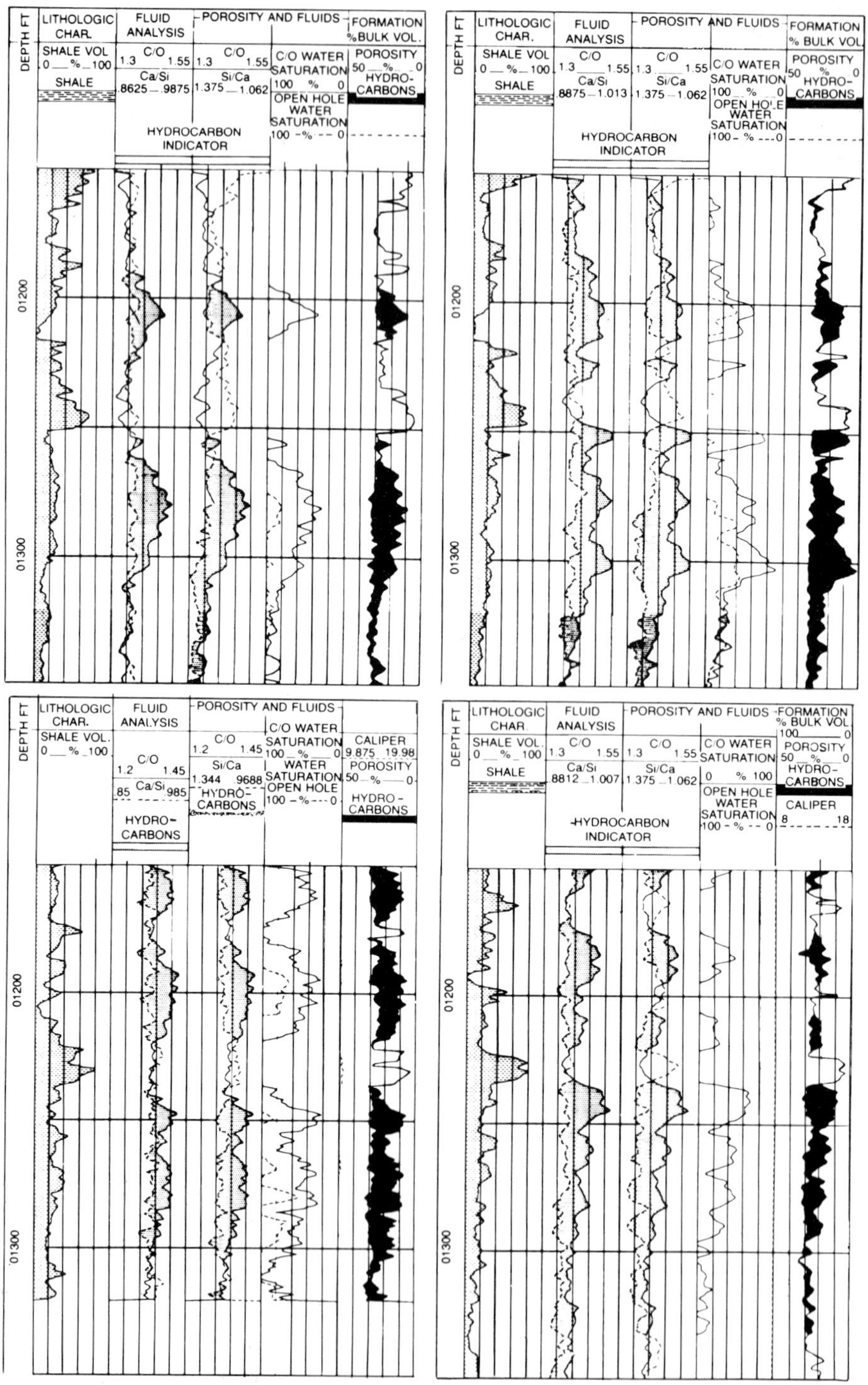

Fig. 15-5. Computerized continuous C/O log analysis from the same depth interval in four wells across the Poso Creek Reservoir, Kern County, California. The zonal correlation in the main reservoir sand—the Etchegoin Sandstone—is shown. Slightly dipping beds can be seen. (After Roux et al., 1981, p. 14, fig. 9; courtesy of SPWLA.)

sand with less shale interbedding and oil saturations varying from 0 to 90% by volume. This bottom 15-m section was drained by the pilot.

The combination logging suite of temperature and continuous C/O logs has provided a direct aid in evaluation of the reservoir bitumen recovery efficiency and in analyzing the steam communication paths through the reservoir.

The initial objective of the pilot was the drilling and completion of three horizontal wells from the underlying limestone up into the lower McMurray Formation to a distance of 310 m into the face of the hill. Six vertical observation wells drilled at the top of the hill back to the face of the hill, were used as temperature monitoring wells.

Because of the very low gravity of the oil, 7–9° API, steam was used to heat the reservoir to produce the bitumen from the bottom 15 m of the lower McMurray Formation. The pilot ran for about 16 months and steam injection was shut down in March 1981.

In April 1981, the eight post-steam boreholes were drilled on top of the hill specifically for the purpose of evaluating the reservoir after steaming. In all boreholes, temperature logs were run 24 hours after drilling. Two of the boreholes, B-1 and B-4, were cored. In five of the eight wells, the continuous C/O log was run.

The original continuous C/O logs (Fig. 15-6) and the computerized analyses (Fig. 15-7) plus the recorded temperature profiles are shown for three of the wells. Wells B-6, B-7, and B-8 are located between horizontal wells H2-A and H3-A. As with wells B-1 and B-4, only the lower sand shows signs of depletion. This is also readily seen from the temperature profiles. Well B-7 appears to have had the maximum hydrocarbon movement. Well B-8 shows total displacement in the area of 58.5–62.5 m. Whereas the C/O calculations in this area are slightly pessimistic ($ROS_{C/O} < ROS_{irr}$), the temperature log substantiates the bitumen movement in this lower section. It is interesting to note that both the upper part of the lower sand in each well, about 41.6–48 m in the B-8 well, as well as the lower sections, show bitumen displacement. This is probably due to the fact that reservoir is being drained from below; the gravity segregation is occurring as the bitumen movement progresses downward toward the horizontal drainage wells.

Cyclic steam injection pilot, Athabasca Tar Sands, Alberta, Canada

Carbon/oxygen logs are being used successfully to evaluate tar sands and the efficiency of the in-situ bitumen recovery process by steam injection (Schultz and Smith, 1979). The bitumen content, determined from the initial C/O log prior to steam injection, is in good agreement with that using core analysis. The bitumen recovery (depletion) was also monitored by subsequent C/O logging runs in several observation wells located in the pilot.

Fig. 15-8 shows the C/O log in one of these observation wells. About half of the bitumen appears to have been removed by steam injection. There is good

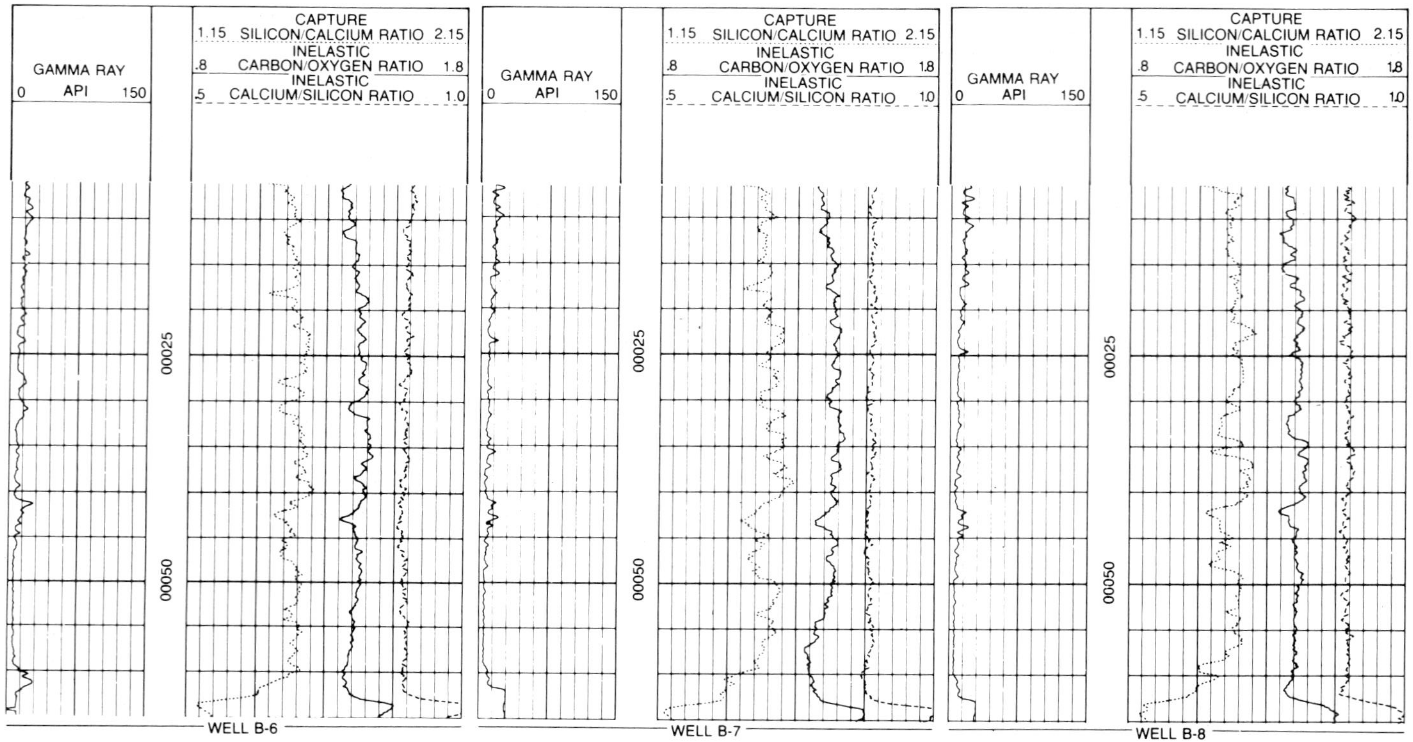

Fig. 15-6. The original C/O log data from three selected vertical wells in the MAISP Heavy Oil Evaluation Project, Alberta, Canada. This should be compared to the computerized C/O analysis for the same wells as shown in Fig. 15-7. (After Campbell et al., 1982, p. 13, fig. 10; courtesy of CIM.)

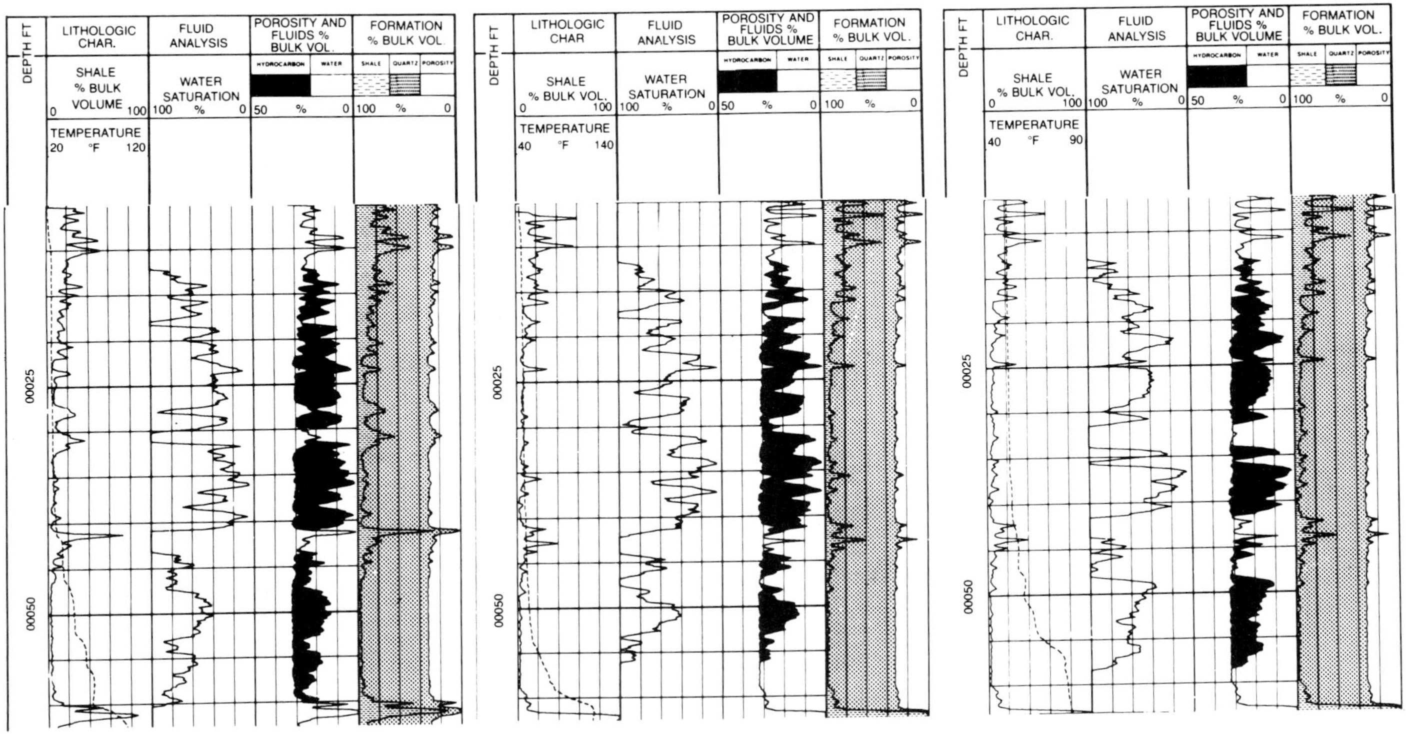

Fig. 15-7. The computerized interpretations of the C/O logs from three selected vertical wells in the MAISP Heavy Oil Evaluation Project, Alberta, Canada. This project involves gravity drainage; there is correlation between increased temperature and hydrocarbon depletion. Well B-7 shows the greatest amount of depletion from the lower sands. (After Campbell et al., 1982, p. 12, fig. 9; courtesy of CIM.)

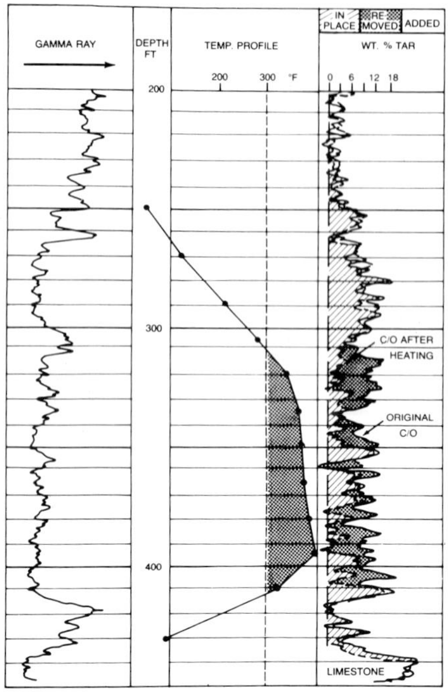

Fig. 15-8. Steam pilot evaluation in the Athabasca Tar Sands in Alberta, Canada. C/O log runs in several observation wells were able to monitor bitumen recovery (depletion) from the sands. There is good depletion from those areas heated in excess of 300°F.

correlation between this bitumen removal (recovery) and the portion of the reservoir having been heated in excess of 300°F.

Steamflood performance in Upper Conglomerate Zone, Yorba Linda Field, California

Located at the eastern edge of the Los Angeles Basin, the Yorba Linda Field produces 12–15° API gravity oil from Pliocene submarine fan channel turbidites. In this field, several silt zones exist which are discontinuous across the field. They create localized barriers to vertical permeability and greatly affect fluid and steam migration and, thus, well performance.

In this EOR project, the continuous steam injection was confined to the lower interval. The silt barriers, however, only restricted the upward migration of steam without forming a complete seal. With the steam zone continually expanding, the steam eventually worked around the discontinuous silt barrier and migrated upward to higher horizons.

The overlay of density and neutron logs, shown in Fig. 15-9, clearly indicates the buildup of steam zones (i.e., "gas, steam, air" crossover effect) beneath several silt barriers present (Cook, 1977).

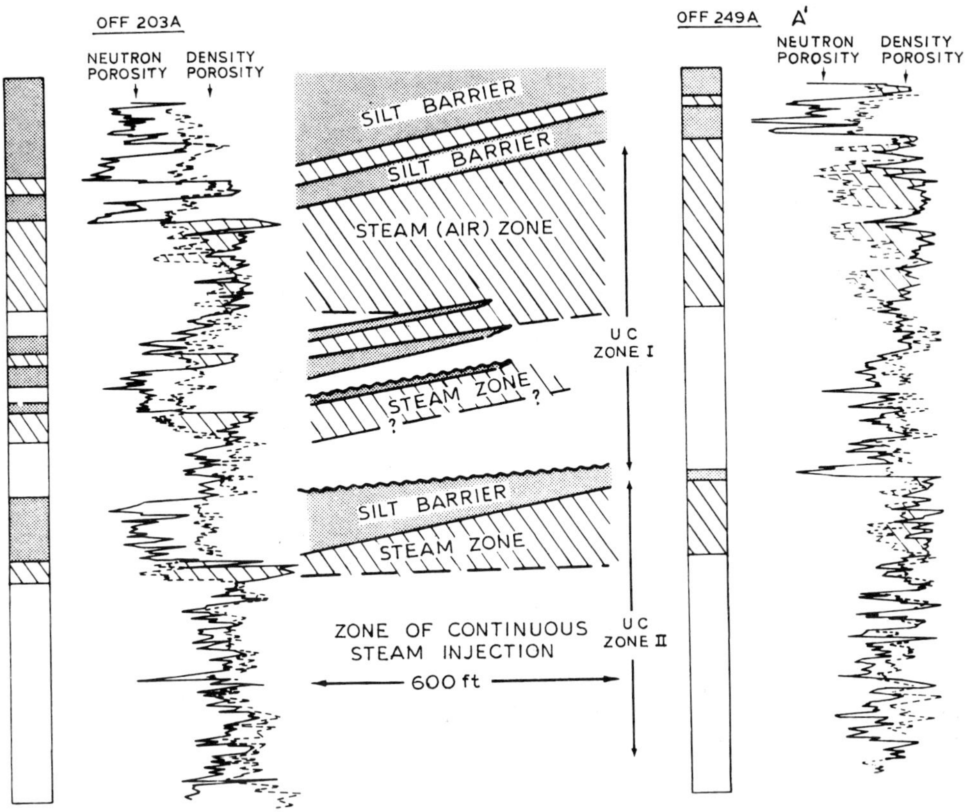

Fig. 15-9. Density–neutron response in the steam zone beneath multiple silt layers, in Yorba Linda Field, Los Angeles Basin, California. (After Cook, 1977, p. 1401, fig. 6; courtesy of Society of Petroleum Engineers of AIME.)

Use of temperature logs in monitoring steamflood and post-steam water injection, Kern River Field, California

A ten-pattern steamflood was converted to water injection after seven years of operation. Recovery of 59% of the pre-steam oil-in-place resulted, with 22% of this oil produced during the post-steam waterflood. Post-steam water injection appears to work best in reservoirs with large hot and unaffected zones remaining at the end of steam injection.

Temperature profiles run in one of the observation wells at several stages in the project's life are shown in Fig. 15-10. These profiles represent temperature (1) at original conditions, (2) during steaming of the reservoir, (3) at peak heating, and (4) at two cooling conditions during post-steam water injection (Oglesby et al., 1982).

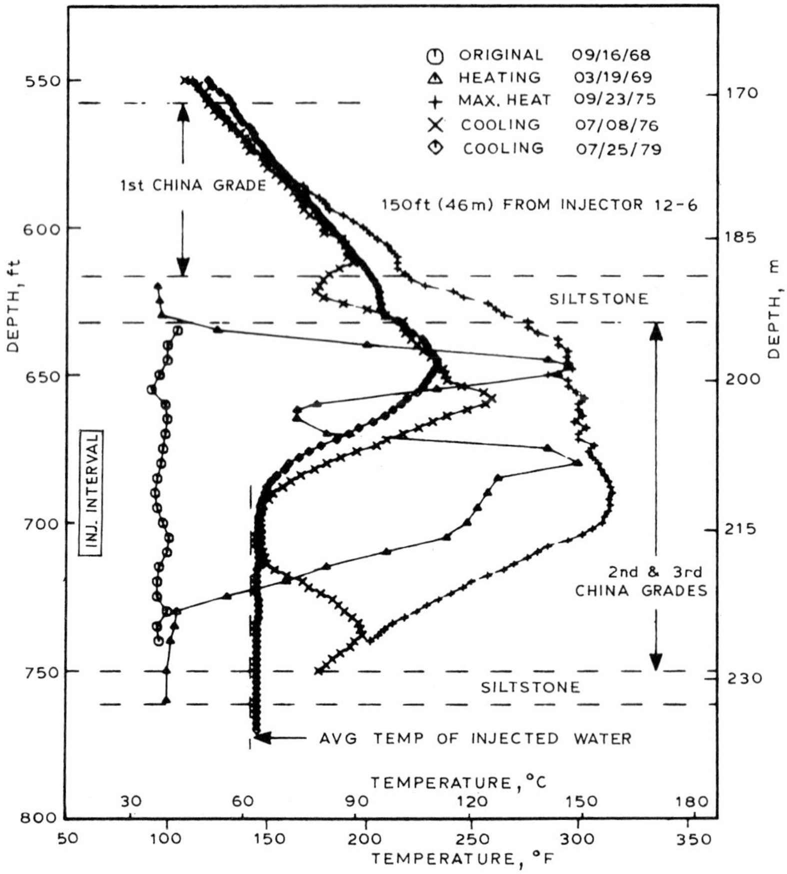

Fig. 15-10. Use of temperature logs for monitoring steamflood and post-steam water injection in Kern River Field, California. (After Oglesby et al., 1982, p. 1255, fig. 5; courtesy of Society of Petroleum Engineers of AIME.)

In-situ combustion of Permian Yates Sands, Ward County, Texas

Well log responses in a West Texas (Permian Yates sands, Ward County) fireflood pilot project are illustrated in Fig. 15-11. The subject well was drilled behind the burned front, cored, and logged to determine the direction and extent of propagation of in-situ combustion process through the reservoir and whether or not the fireflood zone had migrated to any upper sand.

The density–neutron overlay (Anthony et al., 1982), indicates a 28-ft "gas" zone from 2713 to 2741 ft within the J-2 sand. This density–neutron crossover resulted from the effect of the air sweeping of this interval behind the burned front. This 28 ft of burned zone was confirmed by conventional core analysis, which indicated zero oil saturation using the retort method.

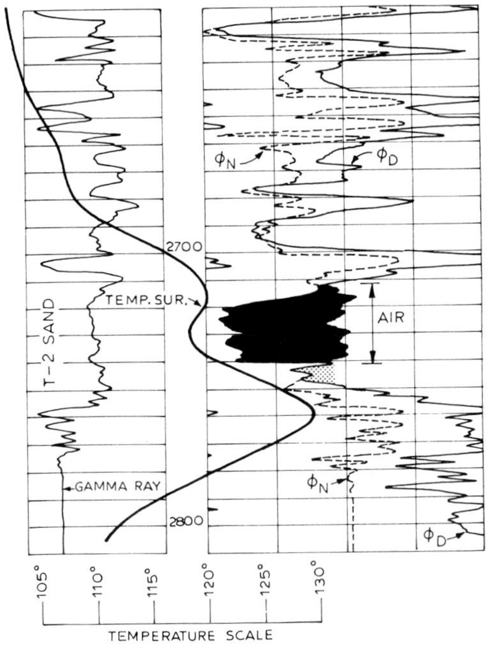

Fig. 15-11. Density–neutron response in fireflooded reservoir behind burnt front, Permian sandstone, Ward County, Texas. (After Anthony et al., 1982, p. 1247, fig. 4; courtesy of Society of Petroleum Engineers of AIME.)

The 6-ft interval below this zone was not burned due to the separation by a 1-ft, tight dolomite streak at a depth of 2741 ft. Furthermore, the subject well has been recompleted in the burned zone as air injector.

In-situ combustion in unconsolidated Nacatoch Sand, Bellevue Field, Louisiana

Joseph et al. (1983) reported on the evaluation of the combustion project in the high-porosity, Upper Cretaceous Nacatoch Sand. The Nacatoch Sand is a highly unconsolidated sand, subdivided by thin, dense, fossiliferous limestone streaks into four principal pay intervals. Crude oil has API gravity of 19°; reservoir pressure is 40 psig; average permeability of sand is 700 mD; and water saturation is 27.4%.

After termination of the air injection, a post-burn drilling program was carried out to evaluate the efficiency of the combustion process via core and logging information. The latter was used in constructing a burned-thickness isopach map.

According to Joseph et al. (1983), the reservoir "rock contains some dispersed clay minerals, mostly kaolinite with some illite and smectite, plus a few clay-rich zones. Gamma ray logs did not correlate to the amount of clay present, but did correlate to the amount of K-feldspar." Under these reservoir conditions, application of natural gamma ray spectral logging would have been warranted. Joseph et al. (1983) pointed out that "resistivity and porosity tools proved most useful in analyzing the effects of the combustion process. The EPT (electromagnetic) log, which was designed to aid in determining water saturations for formations in which the water resistivity is not known, did not provide useful data. This tool was influenced greatly by borehole irregularities." Inasmuch as the EPT is a high-frequency (1.1 GHz) pad-type device with very shallow depth of investigation, its response is significantly affected by severe borehole conditions. Hence, engineers should consider such possible constraints and pitfalls prior to commencing actual logging operations.

LOG-DERIVED EVALUATION OF MISCIBLE DISPLACEMENT PROCESSES

Miscible displacement processes include hydrocarbon, carbon dioxide (CO_2), and inert gas injection. Due to economic constraints, hydrocarbon miscible projects are on the decline, whereas inert gas injection projects are on the increase in recent years. Studies suggest that CO_2 miscible flooding could potentially recover 40% of the total EOR reserves located in the United States (Van Poollen et al., 1980).

If carbon dioxide is dense enough, it extracts hydrocarbons from the oil to form a mixture miscible with crude oil. Such a mixture can recover 95% of the oil in controlled laboratory flow settings. Heterogeneity of reservoir rocks and the low viscosity of carbon dioxide reduce the fraction of oil recovered in projects to lower but still significant levels. With the construction of three pipelines to carry naturally occurring carbon dioxide from Colorado and New Mexico to Permian Basin oilfields, large-scale implementation of enhanced oil recovery by carbon dioxide flooding is now beginning (Orr and Taber, 1984).

Miscible flood project in the South Swan Hills Field, Alberta, Canada

Results of a 15-well logging program to evaluate a miscible flood project in the South Swan Hills Field, located in Alberta, Canada, have been discussed by Anderson et al. (1980). The reservoir rock is a limestone reef in the Devonian Beaverhill Lake Formation, about 8500 ft deep. The reservoir rock consists of five distinguishable lithological zones. Zone 1, the reef-front facies of Zone 2, and Zone 3 all exhibit good porosity. Median permeabilities are 5.9 mD in Zone 1, 4.75 mD in Zone 3, 2.4 mD in all of Zone 2 (lagoonal and reef-front facies), 1.3 mD in Zone 5, and 0.8 mD in Zone 4.

Simultaneous production logging measurements were used to monitor the vertical fluid distribution (i.e., profile) in both injection and production wells under dynamic reservoir conditions (Table 15-VI). To monitor the lateral progress of the floodfront (i.e., the lateral flood distribution), several tracer materi-

TABLE 15-VI

Sensor specifications of simultaneous production logging instrumentation applied in South Swan Hills miscible flood project, Alberta, Canada (after Anderson et al., 1980)

Sensor	Accuracy (percent of full scale)	Resolution	Pressure rating	Temperature rating
Continuous flowmeter	±2	0.5 Hz	15,000 psi (103 Mpa)	350°F (177°C)
Fullbore spinner flowmeter	±2	0.08 Hz	20,000 psi (138 MPa)	390°F (200°C)
Caliper	±1.4	0.06 inch (1.5 mm)	15,000 psi (103 MPa)	350°F (177°C)
Gradiomanometer	±3	0.005 g/cm^3	15,000 psi (103 MPa)	350°F (177°C)
Manometer	±3	0.3% of full scale	15,000 psi (103 MPa)	350°F (177°C)
Thermometer	±3	0.1°F (0.056°C)	15,000 psi (103 MPa)	350°F (177°C)
Quartz pressure gauge	0.025 (percent of reading)	0.01 psi (0.07 kPa) in 1 s	12,000 psi (83 MPa)	300°F (150°C)
Dual tracer ejector	includes scintillation gamma ray detector; each chamber has 50 cm^3 capacity; ejection time selectable from 20 ms to 5 s		17,500 psi (121 MPa)	350°F (177°C)
Gamma ray	±7	≈1 APIU	17,500 psi (121 MPa)	350°F (177°C)

als were applied. The latter included tritiated water, isopropyl alcohol, and ammonium nitrate, all of which were used with the injected water, while tritium, tritiated ethane, and krypton-85 were used in the injected gas.

Besides the key objective of determining the vertical fluid distribution and the lateral variations (i.e., breakthrough patterns) within the field, the common occurrence of cross-flow and unstable flow patterns were of major concern to the operator.

The production logs in Fig. 15-12, representative of those run in this miscible flood project, have been run both up- and down-hole under dynamic (i.e., producing) reservoir conditions.

The up- and down-run flowmeter logs were normalized in the deepest interval, proven to be a static column by both flowmeter and radioactive tracer measurements. This overlay of the up and down flowmeter runs clearly indicates downflow over the distance of perforations *D* (in Zone 2) and perforations *H* (in Zone 4), which suggests crossflow from lithological Zone 2 to Zones 3 and 4. The

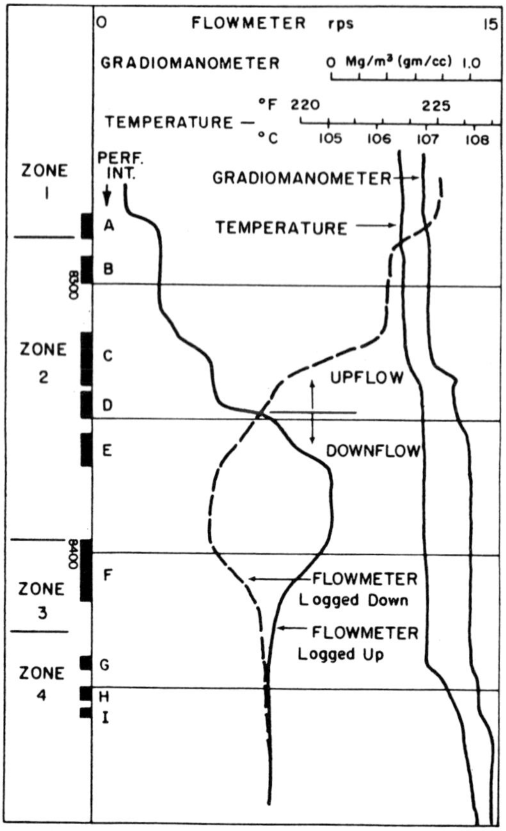

Fig. 15-12. Composite of production logs and their response as a function of changing reservoir characteristics (Zones 1–4) in a South Swan Hills Field well, Alberta, Canada. (After Anderson et al., 1980, p. 197, fig. 8; courtesy of Society of Petroleum Engineers of AIME.)

gradiomanometer shows a density of about 1.03 g/cm^3, i.e., defining the downflowing fluid as essentially all water. This downflow was verified at depths of 8430 ft and 8385 ft by injecting a water-soluble tracer material into the flow stream. The resulting flow profile for the subject well is shown in Fig. 15-13, which clearly illustrates this counterflow condition, i.e., the lithological Zones 3 and 4 are receiving fluids from Zone 2.

Gravity stable miscible CO_2 displacement in Lower Miocene Sands, Weeks Island "S" Sand, Reservoir B, Iberia Parish, Louisiana

Perry (1979) described the field test of a downward miscible CO_2 displacement in a 12,800-ft deep, steeply dipping, high-temperature and high-pressure Lower Miocene sand. Injected at the producing gas–oil contact, the slug of CO_2 is

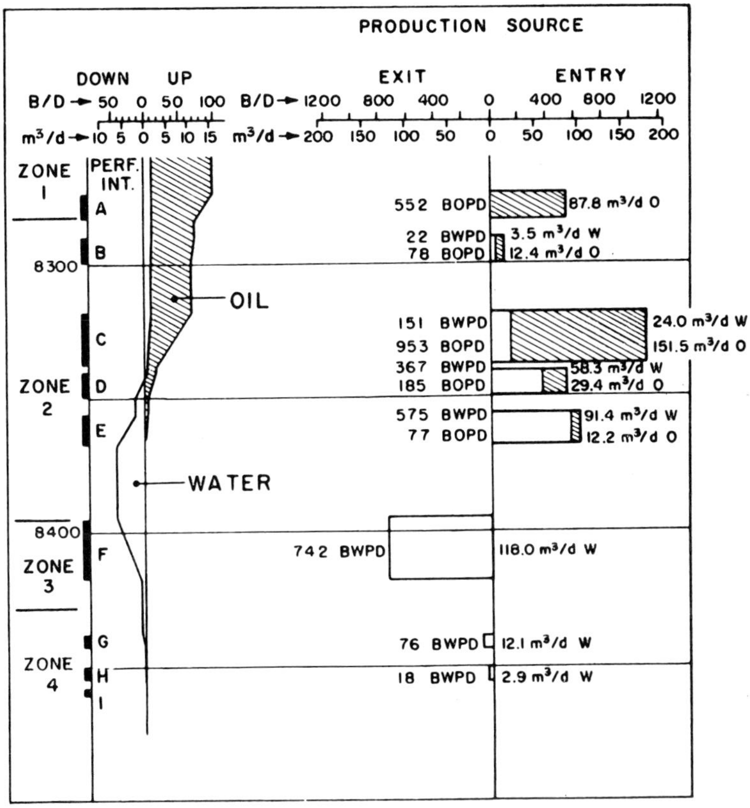

Fig. 15-13. Perforations, flow profile, and lithological zones for the well shown in Fig. 15-12, South Swan Hills Field, Alberta, Canada. (After Anderson et al., 1980, p. 197, fig. 9; courtesy of Society of Petroleum Engineers of AIME.)

designed to utilize gravity forces to stabilize displacement and increase the sweep of the injected CO_2. The latter is moved downward by production of downdip water.

Logging procedures used to monitor this flood project focused on (a) the movement of the CO_2 front down through the reservoir sand, and (b) movement of the oil bank ahead of the CO_2 slug.

Inasmuch as the hydrogen index of CO_2 is essentially zero, presence of CO_2 causes a decrease in the apparent neutron porosity. Fig. 15-14 shows time-lapse neutron porosity variations which are indicative of the CO_2 floodfront movement. Fig. 15-15 compares the Σ (capture cross-section) values, based on the pulsed neutron log and the last monitor run, which gives some indication that oil or fresh water is being pushed downdip into the saltier interval below the depth of 12,920 ft. Inasmuch as initially this reservoir had been flooded by freshwater injection, a wide variation in salinity is seen in this observation well with the

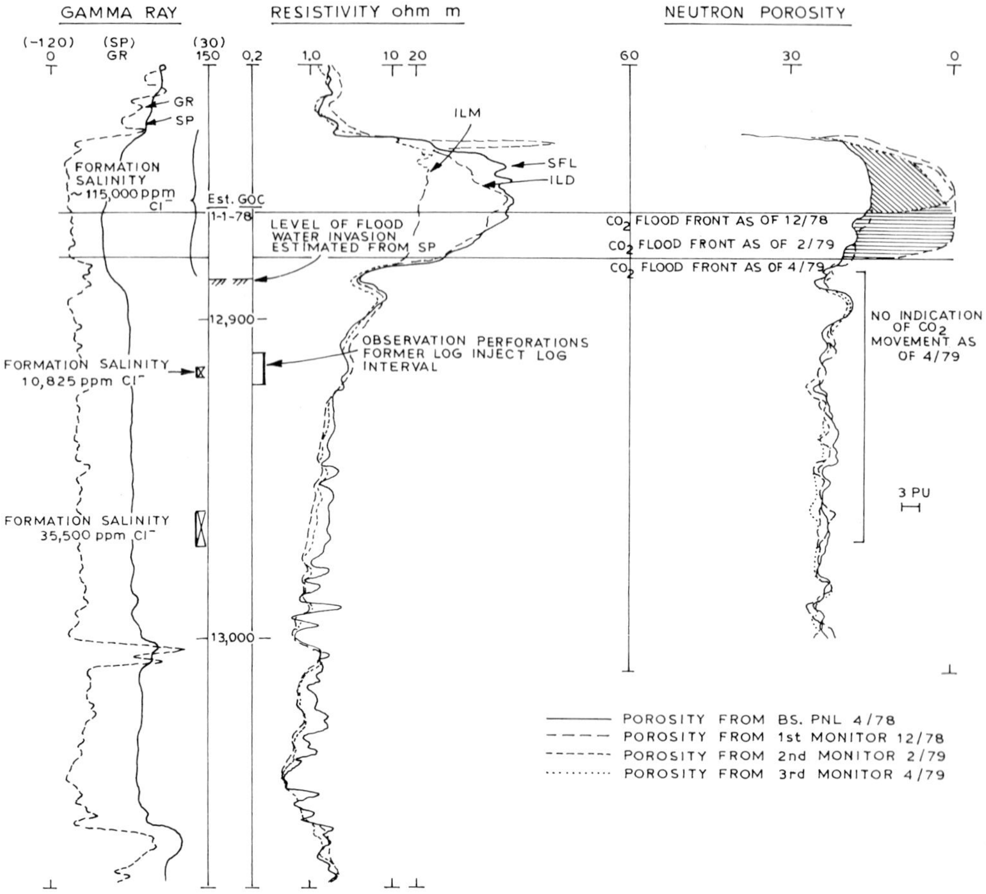

Fig. 15-14. Neutron porosity monitoring of CO_2 floodfront movement in Weeks Island S-Sand Reservoir B, Iberia Parish, Louisiana. (After Perry, 1979, p. 12, fig. 6; courtesy of the U.S. Department of Energy.)

water near the present oil–water contact expected to be virtually fresh. Oil and fresh water exhibit quite similar Σ (capture cross-section) values. The early movement of the oil bank due to CO_2 injection, therefore, is not easily recognizable.

Field project evaluation of CO_2 flooding in Wasson (San Andres) Field, Texas

Doscher and Wise (1976) showed that the San Andres reservoir of the Wasson Field has the highest potential for incremental oil production based on full-scale CO_2 flooding (Table 15-VII). The Wasson Field, one of the largest oilfields in the United States to be completely unitized, encompasses 68,500 acres in Gaines and

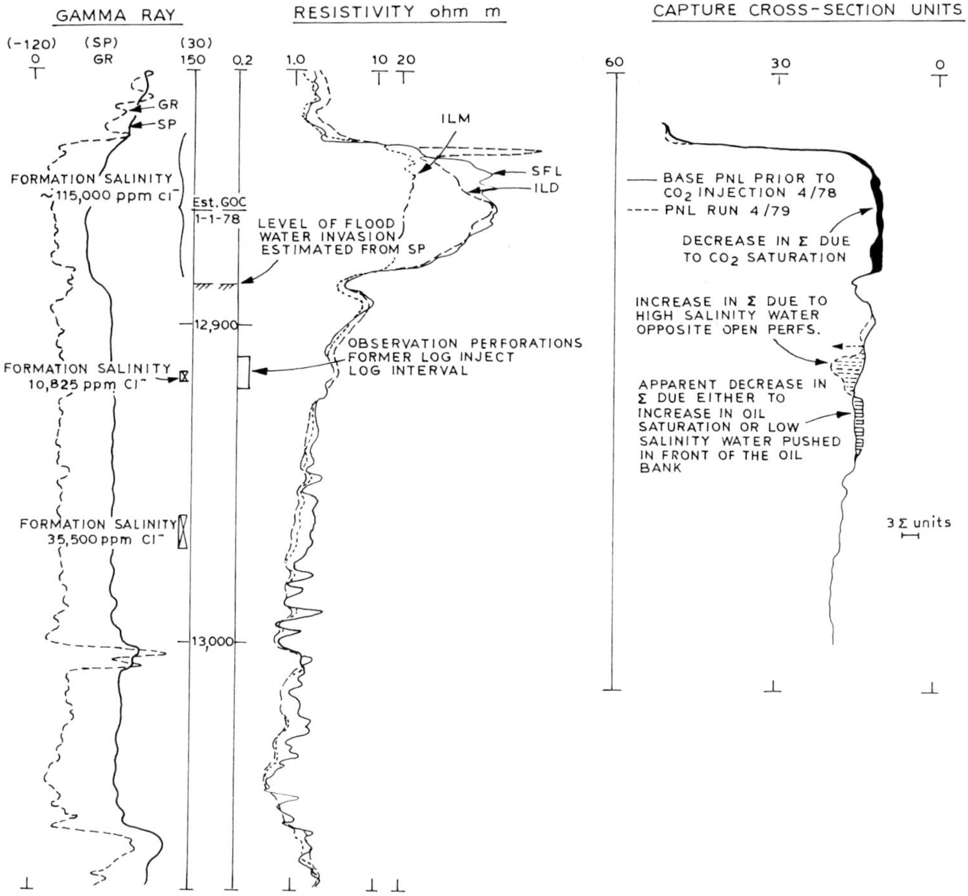

Fig. 15-15. Pulsed neutron log derived capture cross-section change due to CO_2 injection in Weeks Island S-Sand Reservoir B, Iberia Parish, Louisiana. (After Perry, 1979, p. 13, fig. 7; courtesy of the U.S. Department of Energy.)

Yoakum Counties, West Texas (Fig. 15-16). Its generalized stratigraphic section is shown in Fig. 15-17.

Swift et al. (1982) demonstrated the usefulness of a synergetic evaluation approach which integrated several petrophysical (core: pressurized and unpressurized) data, well logging data, and production information to realistically evaluate such a reservoir as a candidate for CO_2 flooding.

For example, as of February 1985, the full-scale injection of CO_2 into the waterflooded San Andres Formation in the Denver Unit of the Wasson Field near Denver City, Texas, is continuing at a rapid pace. Some 250 MMcf/day of gas are presently injected into the Denver Unit field to increase the ultimate oil

TABLE 15-VII

Oilfields proposed as candidates for CO_2 displacement flooding in the Permian Basin, West Texas (after Doscher, 1976)

Field	District/State	Major reservoir	Cumulative production to 1976 (bbl)	Hypothesized recovery (MMbbl)
Wasson	8A/Texas	San Andres	875,657,116	670
Slaughter	8A/Texas	San Andres	642,687,368	260
Levelland	8A/Texas	San Andres	242,675,781	200
Seminole	8A/Texas	San Andres	203,777,244	80
Fullerton	8/Texas	Permian	177,379,697	160
Kelly–Snyder	8A/Texas	Canyon	816,372,830	310
Diamond M	8A/Texas	Canyon Lime	196,622,305	60
Goldsmith	8/Texas	San Andres	285,990,706	130
North Cowden	8/Texas	Permian	259,005,979	130
South Cowden	8/Texas	San Andres	105,536,037	80
Foster	8/Texas	Grayburg	177,647,850	90
Howard–Glasscock	8/Texas	Yates	302,775,723	230
Ward Estes	8/Texas	Yates–Seven Rivers	315,080,915	120
Sand Hills	8/Texas	San Andres	94,933,812	220
McElroy	8/Texas	Grayburg	321,110,539	380
Yates	8/Texas	Grayburg–San Andres	619,642,206	500
Hobbs	New Mexico	San Andres	226,978,885	160
Vacuum	New Mexico	Grayburg	159,307,712	270

recovery by 280 million barrels. This will bring total recovery in this field unit to 1.2 billion barrels, or 53% of the original oil-in-place.

Porosity calibration of neutron logs in the SACROC Unit, Kelly–Snyder Field, West Texas

A major field study has developed porosity transforms for the digital geologic data base at the SACROC (Scurry Area Canyon Reef Operators Committee) Unit which covers about 50,000 acres, i.e., the major part of the giant Kelly–Snyder field, which has produced over 1 billion barrels of oil.

The reservoir is in the Canyon Limestone, a late Pennsylvanian stratigraphic reef or carbonate buildup. Today, the SACROC Unit is under CO_2 flooding, a tertiary recovery method.

To monitor fluid movement and flood efficiency, reliable porosity logs for each of more than 1500 wells in the unit are required (Swulius, 1984). Throughout the unit, a total of 350 modern porosity logs, greater than 1200 old-style neutron deflection logs, and core data in 124 wells are available. Based on these data, Swulius (1984) developed appropriate transforms, according to the particular neutron log used, for application throughout the rest of the field (Figs. 15-18 and 15-19).

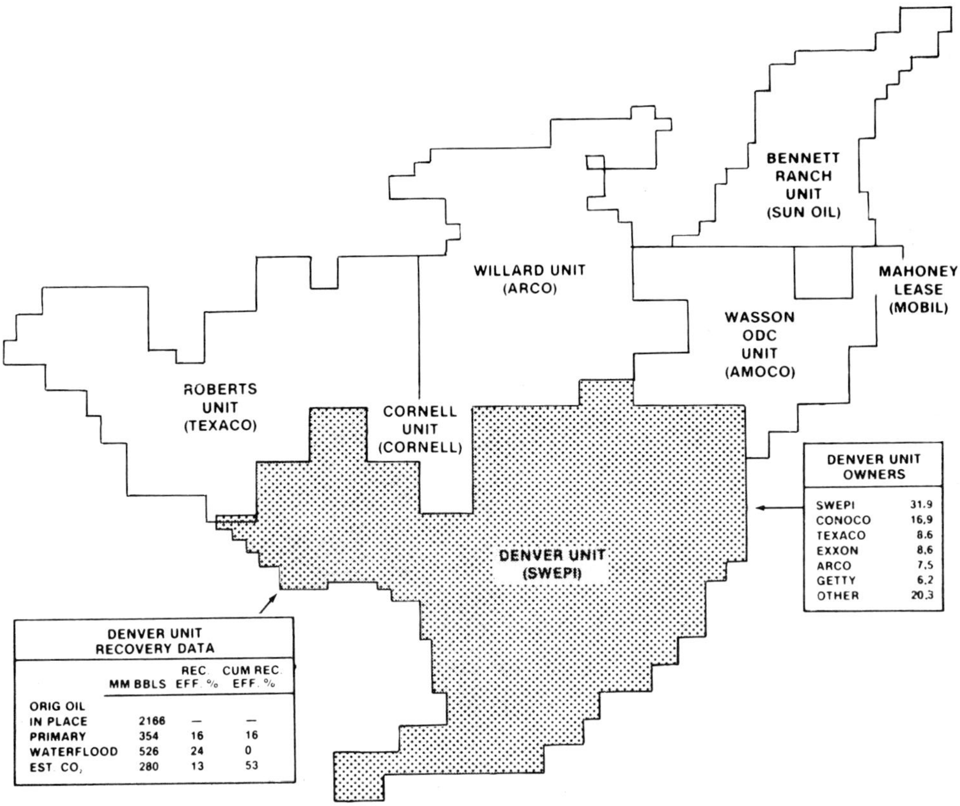

Fig. 15-16. Wasson–San Andres production units, West Texas.

Water-alternating-inert gas (N_2) tertiary recovery project in the Jay–Little Escambia Creek Fields Unit, Florida panhandle

The Jay–Little Escambia Creek (LEC) Fields Unit, located in the Florida panhandle and South Alabama, is an oil accumulation mostly in the dolomitized portion of the Smackover carbonate section at a depth below 15,000 ft. Pertinent data on rock, fluid, and reservoir properties are summarized in Table 15-VIII (Langston and Shirer, 1985).

In current operations, a total of 61 MMcf/day of N_2 and 172×10^3 bbl/day of water are being injected in a water-alternating-gas (WAG) tertiary recovery scheme. Key operational objectives to assure vertical injection conformance included (a) N_2 injection in individual wells at or near maximum rates, and (b) maintaining relatively short (1–2 weeks) injection cycles for each well.

Five radioactive tracers (krypton-85, tritiated hydrogen, methane, ethane, and propane) and one chemical tracer (sulfur hexafluoride) were selected to log the

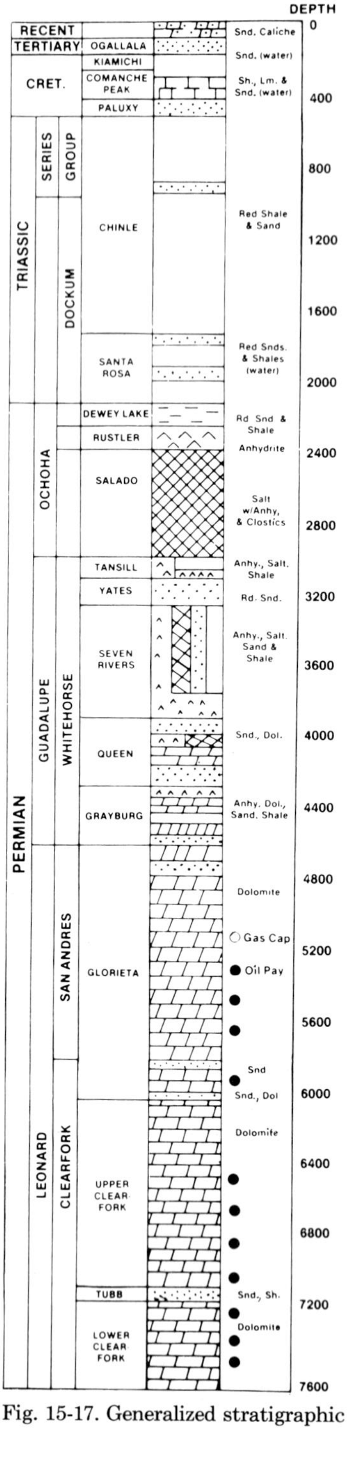

Fig. 15-17. Generalized stratigraphic section in Wasson Field, West Texas.

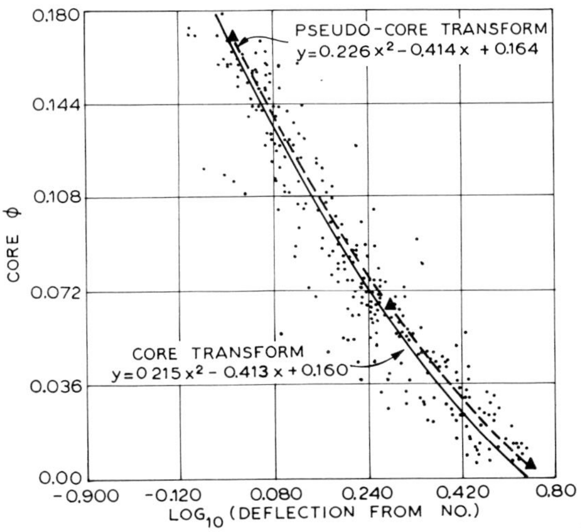

Fig. 15-18. Core transform (solid) versus pseudocore transform (dashed) in Canyon Reef Formation, SACROC Unit, Kelly-Snyder Field, West Texas. (After Swulius, 1984, p. 8, fig. 7; courtesy of Society of Petroleum Engineers of AIME.)

injected N_2. Analysis of the produced N_2 for these tracers enables determination of intervals of N_2 breakthrough.

On the other hand, spinner surveys characterize the vertical N_2 conformance in the WAG injectors. Pulsed neutron logs have been run prior to N_2 breakthrough to establish the base log response to identify and monitor intervals of N_2 breakthrough at a later stage of this WAG project.

According to Langston and Shirer (1985), the early field and well response to this tertiary operation has been positive, with an estimated 47 million barrels of incremental tertiary oil recovery being forecasted.

The WAG ratios have been increased in wellbores with deficient profiles to minimize N_2 mobility until the injection profiles are improved. A satisfactory N_2 injection profile, in which the injected gas profile corresponds to the *kh* (permeability $\times$ thickness) distribution, is shown in Fig. 15-20. Deficient N_2 injection, however, is illustrated in Fig. 15-21, with the majority of the N_2 gas entering the upper perforations of the subject interval. Furthermore, field-wide experience indicated a deterioration of all injection profiles with prolonged N_2 gas injection, resulting in an average proportion increase of 5% into the top perforations after two weeks of N_2 injection. The well shown in Fig. 15-21 clearly indicates this behavior.

These well-log-derived observations substantiated the requirement to maintain short N_2 injection cycles, e.g., one to two weeks in duration.

CHEMICAL RECOVERY PROCESSES

Chemical recovery processes, such as surfactant–polymer injection, polymer flooding and caustic flooding, appear to provide a good mechanism for recovering oil from reservoir rocks that have been successfully waterflooded but still contain significant amounts of oil.

Surfactant-flood pilot test in Louden Field, Fayette County, Illinois

Felder and Hoyer (1982) designed a logging program which is successfully used to monitor the performance of a surfactant-flood pilot test in the Louden Field, Fayette County, Illinois. Induction and carbon/oxygen (C/O) logs were run repeatedly in five observation wells to monitor the time-lapse changes in oil

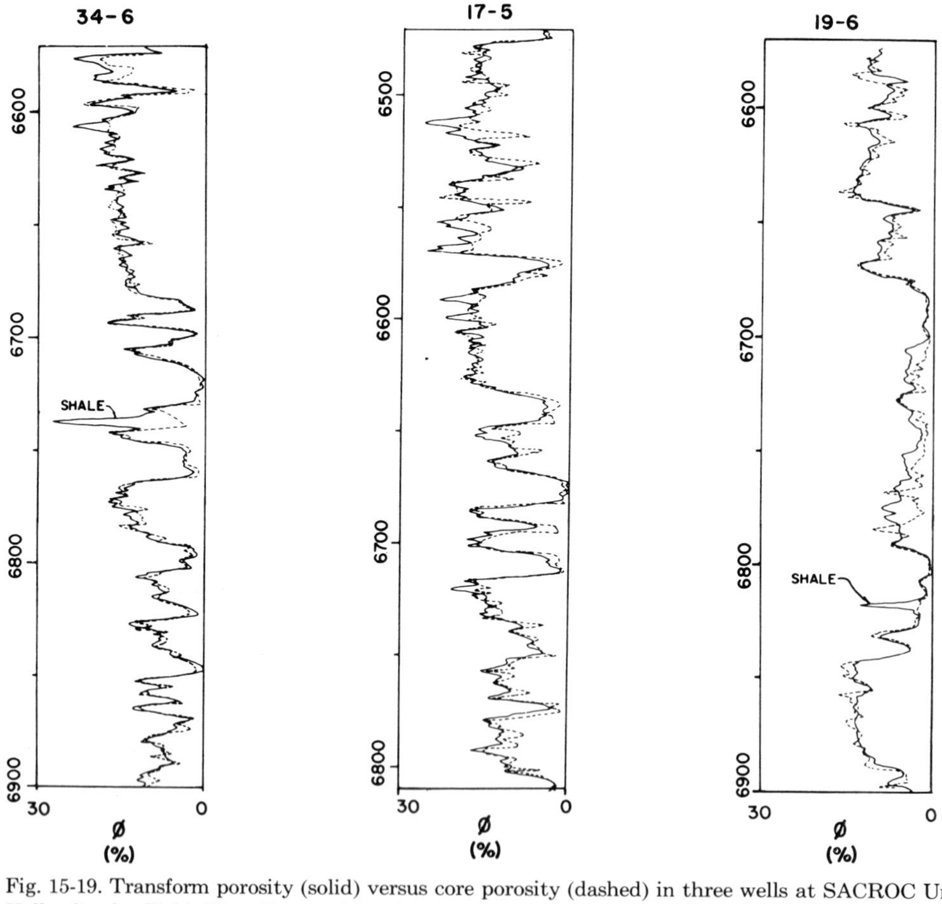

Fig. 15-19. Transform porosity (solid) versus core porosity (dashed) in three wells at SACROC Unit, Kelly–Snyder Field, West Texas. (After Swulius, 1984, p. 8, fig. 8; courtesy of Society of Petroleum Engineers of AIME.)

TABLE 15-VIII

Summary of rock and fluid properties, reservoir properties, and production–injection data, Jay–Little Escambia Creek Fields, Florida, U.S.A (after Langston and Shirer, 1985)

Rock and fluid properties	
Porosity, %	14.0
Permeability, mD	35.4
Water saturation, %	12.7
Oil formation volume factor, RB/STB	1.76
Nitrogen formation volume factor, RB/10^3 ft^3	0.66
Oil viscosity, cP	0.18
Nitrogen viscosity, cP	0.03
Oil gravity, °API	51
Solution gas/oil ratio, scf/STB	1,806
Hydrogen sulfide content, mol.%	8.8
Mobility ratio (water/oil)	0.3
N_2–oil miscibility pressure (approximate), psig	3,600
Reservoir properties	
Datum, ft subsea	15,400
Original pressure, psia	7,850
Current pressure (March 1984), psia	7,174
Saturation pressure, psia	2,830
Temperature, °F	285
Production area, acres	14,415
Net thickness, ft	95
OOIP, 10^6 STB	728
Production – Injection (July 1, 1984)	
Oil production rate, 10^3 STB/day	25
Cumulative oil production, 10^6 STB	359
Water injection rate, 10^3 bbl/day	172
Cumulative water injection, 10^6 bbl	759
Gas injection rate, 10^6 scf/day	61
Cumulative gas injection, 10^9 scf	54

saturation. All observation wells were completed with fiberglass casing to enable the use of induction logs.

Bottomhole locations of the wellbores forming the pilot test pattern are shown in Fig. 15-22.

The subsurface lithology in the pilot test area (Fig. 15-23) is represented by logs in well LR17. The interval from 1315 to 1360 ft is the water-bearing Tar Springs Sandstone. The second interval is a low-porosity limestone centered around a depth of 1420 ft, whereas the third interval from 1552 to 1565 ft is the so-called Weiler sandstone, the watered-out oil sand targeted for the pilot test.

Induction and C/O log responses measured pre-flood and post-flood oil saturations, as well as the arrival time, oil saturation, and duration of the bank of oil mobilized by the surfactant injection.

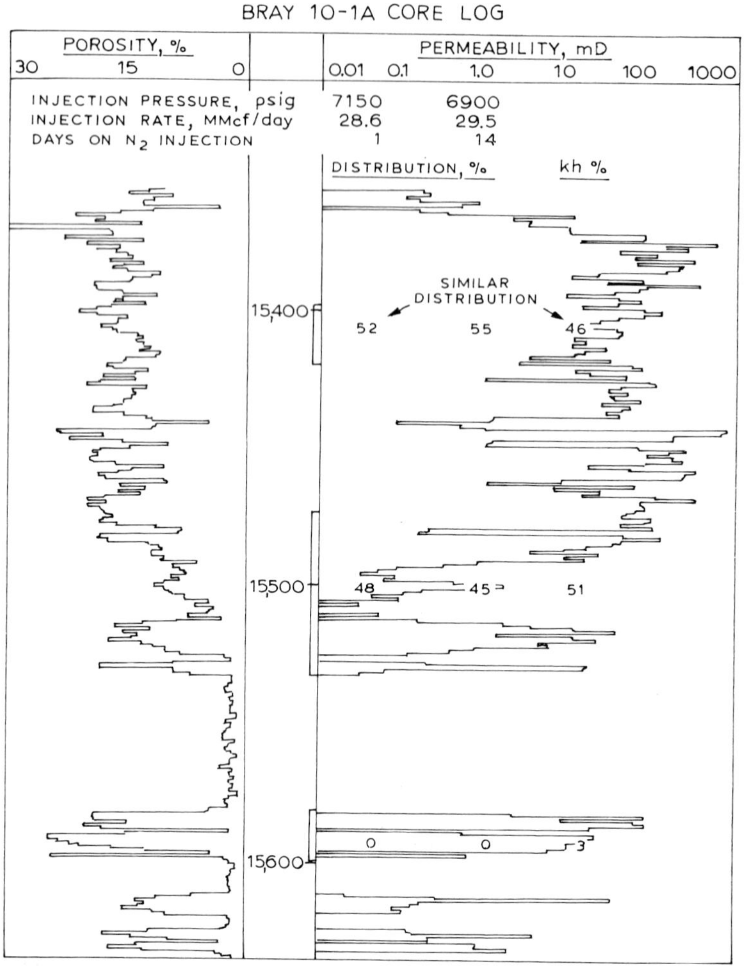

Fig. 15-20. Satisfactory N_2 injection profile in well Bray 10-1A, Jay–Little Escambia Creek Fields Unit, Florida. (After Langston and Shirer, 1985, p. 266, fig. 5; courtesy of Society of Petroleum Engineers of AIME.)

According to Felder and Hoyer (1982), an "increase in oil saturation was detected first in well LR17 (Fig. 15-24), which has a downhole location approximately 20 ft from injection well LR10 (see Fig. 15-22). The passage of an oil bank is indicated in this and other wells by dramatic saturation changes: starting at about 25%, increasing to near 50%, and then declining to below 10% (Fig. 15-25). Evidence of stratification is noted by comparing the response in the upper and lower parts of the Weiler interval, where the magnitude and timing of saturation changes are different. This suggests that vertical crossflow between zones was not highly significant." These authors also stated that the interpretation of the

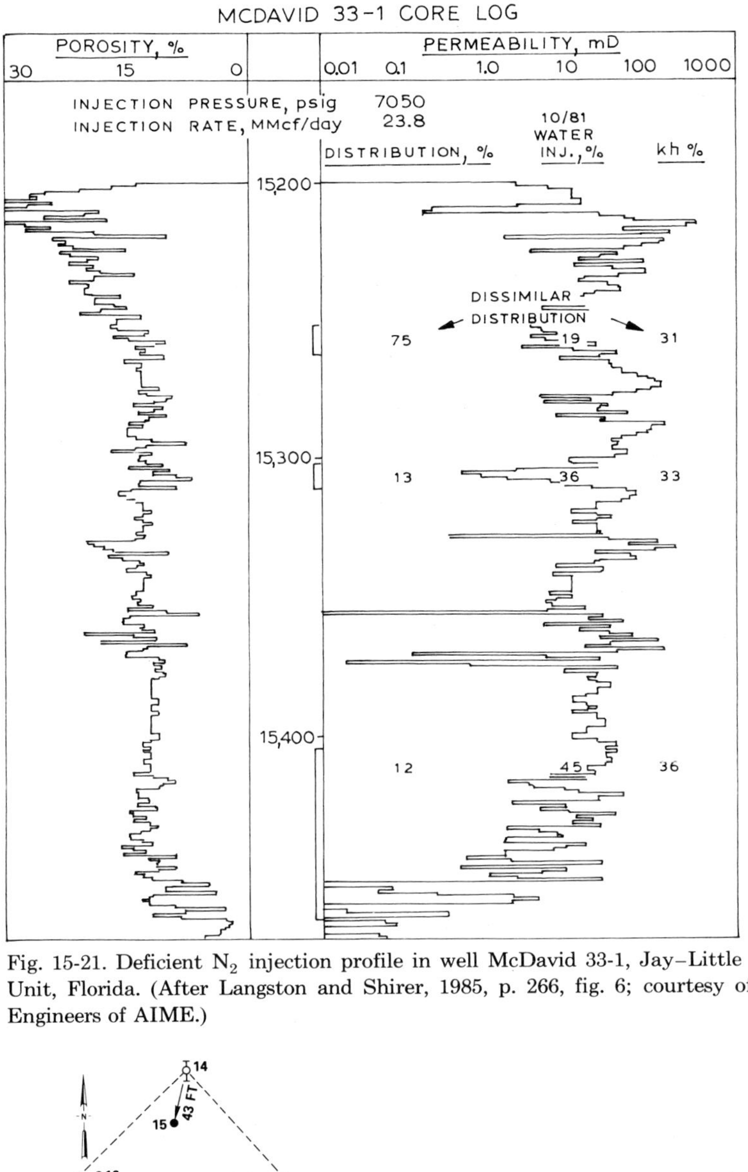

Fig. 15-21. Deficient N_2 injection profile in well McDavid 33-1, Jay–Little Escambia Creek Fields Unit, Florida. (After Langston and Shirer, 1985, p. 266, fig. 6; courtesy of Society of Petroleum Engineers of AIME.)

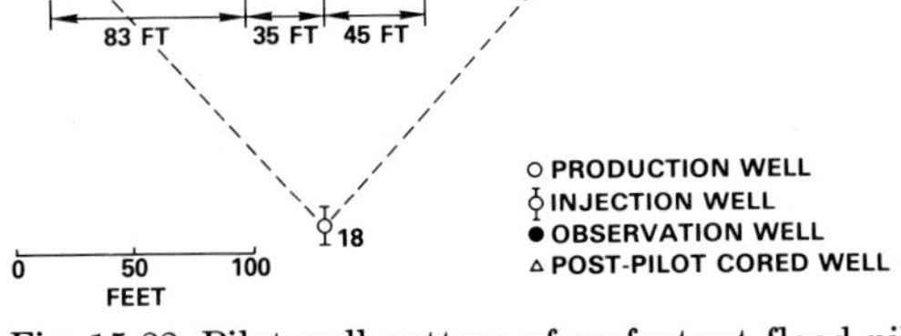

Fig. 15-22. Pilot well pattern of surfactant-flood pilot test in Louden Field, Fayette County, Illinois. (After Felder and Hoyer, 1982, p. 15, fig. 1; courtesy of Society of Petroleum Engineers of AIME.)

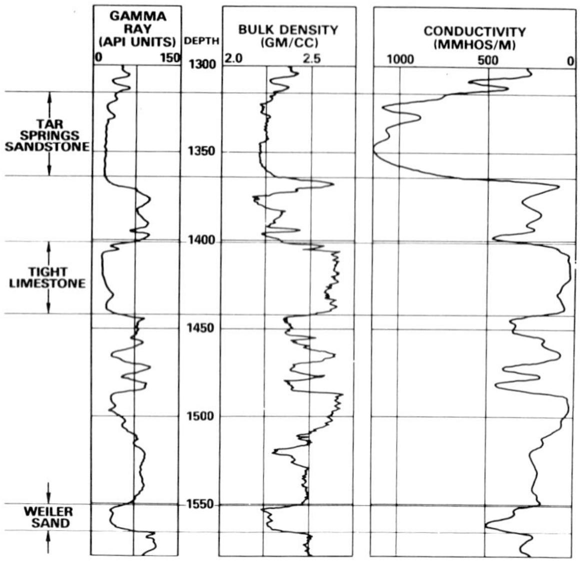

Fig. 15-23. Lithological zones in well LR17, Louden Field, Fayette County, Illinois, as defined by gamma ray, density, and induction logs. (After Felder and Hoyer, 1982, p. 15, fig. 2; courtesy of Society of Petroleum Engineers of AIME.)

surfactant-flood pilot performance obtained from logging measurements in the observation wells has been generally consistent with production data and core measurements.

Injection well logging using viscous EOR fluids in Illinois and Wyoming reservoirs

Conventional tracer tools yield acceptable results with non-viscous fluids such as water. Problems, however, have been encountered when conventional tools

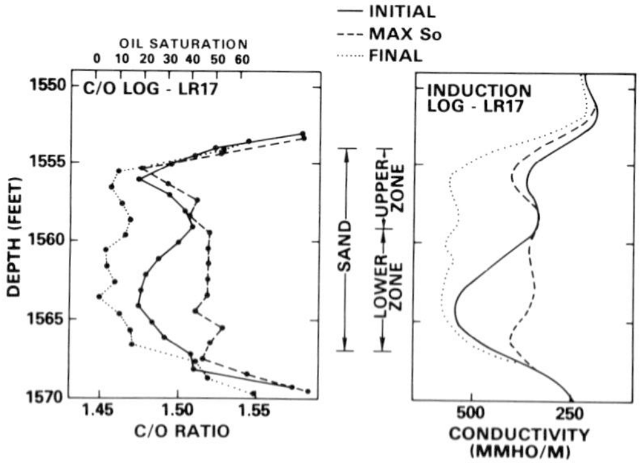

Fig. 15-24. Log responses in well LR17, Louden Field, Fayette County, Illinois. (After Felder and Hoyer, 1982, p. 19, fig. 24; courtesy of Society of Petroleum Engineers of AIME.)

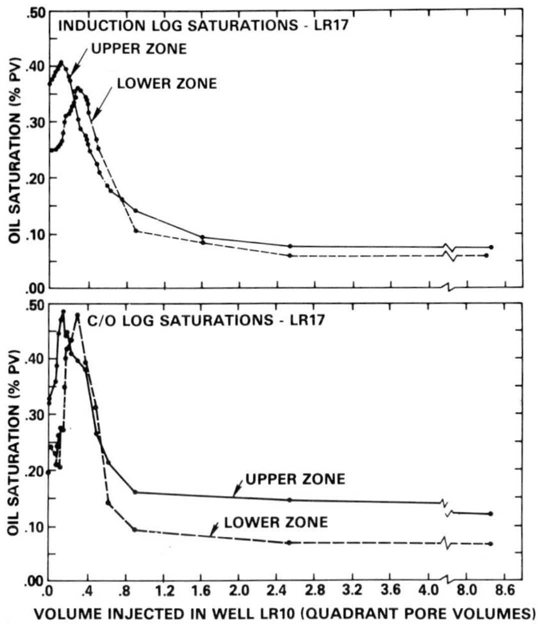

Fig. 15-25. Oil saturation changes in well LR17, Louden Field, Fayette County, Illinois. (After Felder and Hoyer, 1982, p. 23, fig. 16; courtesy of Society of Petroleum Engineers of AIME.)

were used to obtain injection profiles using viscous fluids such as surfactant and polymer solutions.

To overcome such problems, Dresser Atlas developed the Swing Arm Tracer Tool (SWAT), which ejects radioactive tracer material parallel to the direction of flow, in contrast to the perpendicular ejection feature of conventional tools.

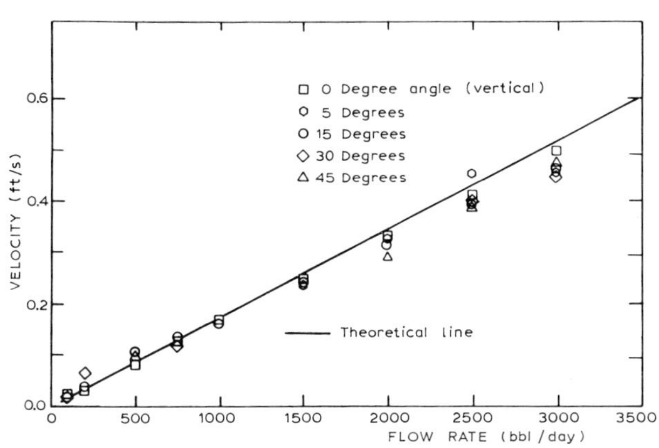

Fig. 15-26. Flow loop results of SWAT in $8\frac{1}{2}''$ casing. (After Knight and Davarzani, 1984, p. 8, fig. 7; courtesy of Society of Petroleum Engineers of AIME.)

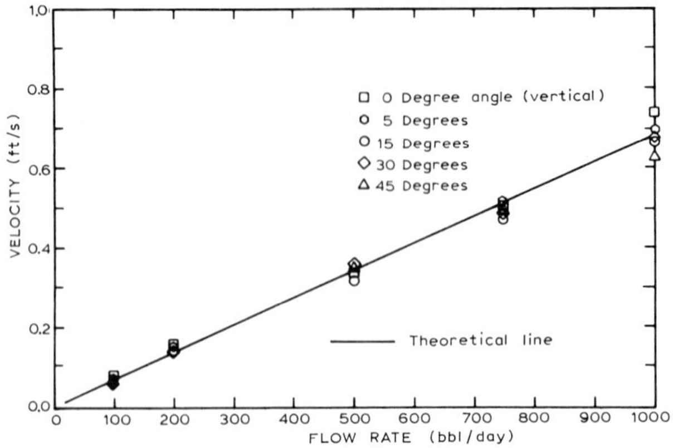

Fig. 15-27. Flow loop results of SWAT in $4\frac{1}{2}''$ casing. (After Knight and Davarzani, 1984, p. 8, fig. 8; courtesy of Society of Petroleum Engineers of AIME.)

Hence, the SWAT instrument delivers the tracer material into the region of maximum velocity to provide the best indication of flow rate.

Recently, Knight and Davarzani (1984) presented flow loop test data in $8\frac{1}{2}''$ and $4\frac{1}{2}''$ casings (Figs. 15-26 and 15-27). Reynolds numbers based on average

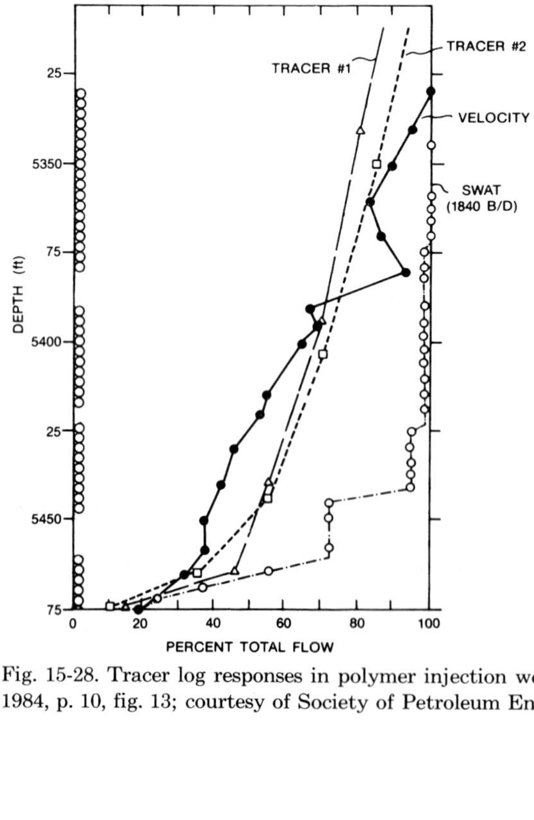

Fig. 15-28. Tracer log responses in polymer injection well, Wyoming. (After Knight and Davarzani, 1984, p. 10, fig. 13; courtesy of Society of Petroleum Engineers of AIME.)

velocity ranged from about 1100 to 33,000 in the $8\frac{1}{2}''$ casing, and from about 2200 to 22,000 in the $4\frac{1}{2}''$ casing.

Fig. 15-28 shows the polymer injection profiles based on velocity, repeat tracer runs, and the SWAT measurements. Aside from the disctinctive anomaly at 5360 ft, the conventional profile suggests uniform fluid entry. The SWAT profile, however, clearly shows that 95% of the injection fluid entered the lower 19 ft of open perforations.

REFERENCES

Anderson, R.A., Smolen, J.J., Laverdiere, L. and Davis, J.A., 1980. A production logging tool with simultaneous measurements. *J. Pet. Technol.,* 32(2): 191–198.

Anthony, M.J., Taylor, T.D. and Gallagher, B.J., 1982. Fireflooding a watered-out West Texas sandstone. *J. Pet. Technol.,* 34(10): 2244–2250.

Campbell, D., Ducheck, M., Frost, E. and Fertl, W.H., 1982. Log evaluation and monitoring of a unique heavy oil reservoir project, Alberta, Canada. *33rd Annu. Tech. Meet., Pet. Soc. of CIM, Calgary, Alta., June 6–9.*

Cook, D.J., 1977. Influence of silt zones on steam drive performance, Upper Conglomerate Zone, Yorba Linda Field, California. *J. Pet. Technol.,* 28(11): 1397–1404.

DeVries, M.R. and Fertl, W.H., 1977. Time-lapse cased hole reservoir evaluation based on the Dual-Detector Neutron Lifetime Log. *Trans. SPWLA,* Paper V.

Doscher, T.M. and Wise, F.A., 1976. Enhanced crude oil recovery potential—an estimate. *J. Pet. Technol.,* 27(5): 575–585.

Felder, R.D. and Hoyer, W.A., 1982. The use of well logs to monitor a surfactant flood pilot test. *57th Annu. Fall Meet., Soc. Pet. Eng. of AIME, New Orleans, La., September 26–29,* SPE 11147.

Fertl, W.H., 1979. Modern evaluation techniques for redeveloping old reservoirs. *Pet. Eng.,* 51(15): 48–68.

Fertl, W.H., 1982a. New developments in carbon/oxygen (C/O) logging. *Trans., 29th Annu. Southwest. Pet. Short Course, Lubbock, Texas, April 21–22.*

Fertl, W.H., 1982b. Well logging and its applications in cased holes. *Int. Pet. Exhib. Tech. Symp., Soc. Pet. Eng., Beijing, March 18–26,* SPE 10034.

Fertl, W.H., Pilkington, P.E. and Scott, J.B., 1974. A look at cement bond logs. *J. Pet. Technol.,* 25(6): 607–617.

Joseph, C., Jay, C.B., Eslinger, E.V. and Mundis, C.J., 1983. Evaluation of an in situ combustion process by postburn core and log analysis. *J. Pet. Technol.,* 35(4): 827–835.

Knight, B.L. and Davarzani, M.J., 1984. Injection well logging using viscous EOR fluids. *59th Annu. Fall Meet., Soc. Pet. Eng. of AIME, Houston, Texas, September 16–19,* SPE 13143.

Langston, E.P. and Shirer, J.A., 1985. Performance of Jay/LEC Fields unit under mature waterflood and early tertiary operations. *J. Pet. Technol.,* 36(2): 261–268.

Oglesby, K.D., Blevins, T.R., Rogers, E.E. and Johnson, W.M., 1982. Status of the 10-pattern steamflood Kern River Field, California. *J. Pet. Technol.,* 34(9): 2251–2257.

Oliver, D.W., Frost, E. and Fertl, W.H., 1981. Continuous carbon/oxygen (C/O) logging—instrumentation, interpretive concepts, and field applications. *Trans. SPWLA,* Paper TT, 28 pp.

Orr, F.M. and Taber, T.J., 1984. User of carbon dioxide in enhanced oil recovery. *Science,* 224: 563–569.

Perry, G.E., 1979. *Weeks Island "S" Sand Reservoir B-Gravity Stable Miscible CO_2 Displacement, Iberia Parish, Louisiana.* DOE Contract EF(77-C-05-5232), 15 pp.

Roux, C., Fertl, W.H., Frost, E., Stedman, D. and Elliott, D., 1981. Well logs evaluate and monitor heavy oil steam flood in Kern County, California. *Trans. SPWLA,* Paper JJ: 17 pp.

Schultz, W.E. and Smith, H.D., 1979. Carbon/oxygen log monitors Athabasca Tars and recovery. *Oil Gas J.*, 96: 165–168.

Swift, T.E., Kumar, R.M., McCoy, R.L., Wilhelm, M.H., Glascock, M.R., Goodrich, J.H. and Watts, R.J., 1982. *Field Project to Obtain Pressure Core, Wireline Log, and Production Test Data for Evaluation of CO_2 Flooding Potential, Texas Pacific Bennett Ranch Unit Well No. 310, Wasson (San Andres) Field, Yoakum, Co., Texas.* DOE Contract No. DE-AC21-79MC08341, 272 pp.

Swulius, T.M., 1984. Porosity Calibration of Neutron Deflection Logs, SACROC Unit. *59th Annu. Fall Meet., Soc. Pet. Eng. of AIME, Houston, Texas, September 16–19*, SPE 13289.

Van Poollen, H.K. and Associates, 1980. *Fundamentals of Enhanced Oil Recovery.* PennWell Books, Tulsa, Okla., 155 pp.

Chapter 16

ENVIRONMENTAL FACTORS ASSOCIATED WITH OIL RECOVERY

ERLE C. DONALDSON, TEH FU YEN and GEORGE V. CHILINGARIAN

INTRODUCTION

All field operations (activities) involving the production, storage and transportation of crude oil and petroleum products are required to meet Federal, State and local laws, regulations, and standards in the United States. In this context, the laws are those Federal, State and local government enactments which provide for regulations for the enforcement (by agencies such as the Environmental Protection Agency) of standards established under the laws.

Federal legislations that affect the EOR operation requirements include: (1) National Environmental Policy Act of 1969 (NEPA), (2) Resource Conservation and Recovery Act of 1976 (RCRA), (3) Clean Water Act of 1977, (4) Clean Air Act of 1977, (5) Occupational Safety and Health Act of 1970 (OSHA), and (6) the Transportation Safety Act of 1974. These federal legislations are the basic guidelines for all the Federal, State, and local regulations and standards.

The National Environmental Policy Act of 1969, PL 91-190, requires Federal agencies in the United States to consider the environment in all Federal actions that may significantly affect the environment, and to prepare Environmental Impact Statements. To implement the act, the Council on Environmental Quality was established to present guidelines to Federal agencies with respect to policy, responsibilities, planning and response elements, operations and coordinating instructions.

In the study of the environmental aspects of oil recovery, there are eight subject areas that are of concern for each project at each specific location. They are discussed here in the following sections.

Atmospheric emissions

The components of concern are dust, engine exhaust, off-well gases, gas-flaring combustion products, particulates, SO_2, CO, NO_x, H_2S, and hydrocarbons. In thermal processes air pollution results mainly from boiler combustion products. There is a possibility that vented steam may contain hydrocarbons and H_2S. Any other chemicals that were added to the reservoirs to enhance sweep efficiency or fluid mobility might also cause gaseous emission problems if they are vaporized and entrained in the steam. In-situ combustion processes potentially can produce H_2S as a toxic and odoriferous gas. Thermal processes need monitoring of emissions to ensure compliance with air pollution ordinances and prompt warnings of hazardous air conditions. In carbon dioxide flooding, venting of H_2S and local, short-time, blanketing concentrations of CO_2 are the major concerns. Accidental spills and equipment leaks are the other possible ways air pollution can occur. Atmospheric emissions will probably not be important in either process, but monitoring of emissions is necessary to determine their exact levels, and duration; and thus be able to provide proof of insignificant or no environmental deterioration.

Water use

Thermal steam injection processes and micellar–polymer flooding processes both need large amounts of fresh water (up to 20 barrels of water per barrel of oil recovered). There is only a short-term demand for water use in the polymer process for salinity control.

The potential constraints are subsidence of land if underground aquifers are overpumped. Surface impacts to animals and humans may occur if the surface water supply is lowered. Site specific studies are essential in order to determine if sufficient water is available for EOR processes at specific sites.

Water quality impacts

Thermal and chemical recovery processes involve the injection of various additives and chemicals into the reservoirs. Only minor quantities of these chemicals are expected to be recovered with the produced oil. The interaction of the chemicals left in the subsurface geologic formations with connate fluids, brines, and rocks are not completely understood. Nor can it be positively stated that toxic substances or hazardous materials are not produced by the decomposition of such chemicals. Such toxic or hazardous substances may be confined to the reservoir, but it cannot be positively confirmed that migration or transfer of such materials to other strata and water aquifers do not occur, unless a careful monitoring program is implemented.

Waste water effluents

Prior to initiating all the EOR processes, spent fluids (drilling muds), have to be disposed of. Brine and waters containing chemical additives, which are also produced during the operations, also must be disposed of. Leakage of these fluids into aquifers and subsurface waters may occur either during injection or during transportation to disposal sites. If suitable geologic formations are available above or below the petroleum reservoirs, these process fluids may be injected into them. If they are not available, surface storage in lined ponds is necessary. The ability to detect leakage by runoff or leaching from temporary storage ponds must be developed and made readily available to protect underground and surface waters from contamination.

Solid wastes

The most common solid wastes generated by EOR method (such as ash, if coal or coke is used to fire steam boilers) along with sludge from stack gas scrubbers are also produced resulting in disposal problems. Seepage, runoff, or leachate from solid waste piles can cause ground or surface water contamination. Airborne particulate matter blown from the waste piles can possibly contain hazardous chemicals.

Occupational safety and health

The hazards that are considered most usual in the case of thermal processes, are the ones present when high temperature and pressures are being used. The potential impact on health by chemical additives and H_2S also is present. During the CO_2 flooding process, the above hazards have to be considered, along with the possibility that if large quantities of CO_2 escape, it can blanket a local area with a smothering atmosphere for a short period of time. Various chemicals used during micellar–polymer flooding are in most instances toxic at high concentrations, and some of the chemicals may be carcinogenic. Operating personnel in all cases must be prepared to wear clothing that protects them from exposure to these chemicals.

The recent advent of microbial enhancement of oil recovery introduces a new environmental concern to the oilfield operations. Procedures and policies are necessary for proper handling of large quantities of microbial cultures which are destined for injection into the oil wells. The pathogenesis of all cultures should be understood and proper precautions taken to protect field workers and populations near the oilfield from infections by microbes used in EOR. Even the use of non-pathogenic bacteria should be handled with strict environmental controls, because aerosols of high concentrations of any microorganism can cause infections of mucous membranes. Even in isolated places, strict compliance is necessary because animals and insects that later come in contact with human population centers can transmit various types of diseases.

Physical disturbances

Land subsidence can occur when the underlying ground support is removed. This is a remote possibility in CO_2 recovery, but ground stability can be affected by the increase of pressure applied in reservoirs if there is a nearby fault. Hence, increased seismicity and a tendency for earthquakes may occur. Finally EOR processes will use more land than conventional oil recovery techniques; more wells need to be drilled and larger areas for waste disposal are necessary. These concerns require site-specific studies to determine the possible effects on the land.

Noise

The impact of noise depends on the receptors, noise, buffers, power sources, duration, and the density of drilling and injection at various sites. Permissible noise levels have to be determined for each site and for each process. Even though the duration of obnoxious noise levels may be short, they are considered a nuisance by the adjacent communities and could be a potential constraint on the commercial application of EOR processes.

Major environmental concerns of EOR oilfield operations are presented in this chapter. Companies and field managers must ensure that the operations are not shut-down due to any of the potential environmental impacts discussed. Before beginning a field operation, all environmental regulations must be complied with so that its progress will not be hindered by the legal implications of incomplete data. Second, one must ensure that adequate environmental controls are available and are applied to the field projects to mitigate any environmental deterioration. Third, technical support must be rendered in the event of environmentally damaging accidents.

Thus the purpose of this chapter is to identify environmental constraints to future EOR commercialization, and eliminate them, in the most efficient manner possible. Essentially the approach consists of the following steps:

(1) Collection of sufficient and accurate baseline environmental data for each EOR project from existing field tests, as well as from laboratory or new field tests, if such data is lacking or insufficient.

(2) Determination from these data, whether or not all the requirements of the government regulations are fulfilled for each site.

(3) Comparison of the baseline data with potential environmental impacts of EOR processes to ensure that adequate environmental control measures are applied to each project. This may simply mean that existing projects may just require preparation of an acceptable environmental assessment. One must keep abreast of improved pollution control technology in order to offer technical assistance in the event of accidental environmental damage.

It is safe to predict that the future will bring a greater population density and the spread of urban development will further encroach on oilfields. Inasmuch as standard of living depends on continuing supplies of petroleum, oilfields cannot be abandoned because of urbanization. In the future, offset wells disguised as buildings (similar to those used at Long Beach, California) may have to replace the exposed pumping units, and very tight standards of pollution control will have to be maintained for noise, vapor emissions, tank spills, and oilfield water processing. Concern for the environment requires prevention of any kind of adverse effect on air, water, or soil by EOR activities. Thus, the ecological considerations of every technological process must be considered. Government controls, economic considerations, and diminishing resources have forced a complete re-evaluation of waste effluents from all industries. Waste potential for pollution of surface waters and shallow underground fresh water must be minimized for the sake of economy and the safety of humans.

Enhanced oil recovery processes are still being developed. The transition from pilot testing to full-scale field application depends on the economic success of field research. Underlying all development and expansion activities associated with oil production is the constant need to protect the environment from pollution. Unanswered questions regarding the environmental consequences of EOR activities should be addressed during the period of pilot field testing. Environmental data can be obtained from pilot projects for design and planning of fieldwide commercial implementation of every process.

Potential adverse environmental impact is increased by the injection of large quantities of chemicals to displace oil to the producing wells. Several chemical processes for oil recovery require surface processing of large quantities of saline waters as well as chemical mixtures. The large volumes of fluids create a potential for pollution of surface waters and shallow freshwater aquifers. When the oilfield is finally abandoned, large quantities of chemicals used for oil displacement are left in the underground reservoirs. This creates a potential for environmental hazards, which depends on the rate of chemical degradation in the subsurface environment (Donaldson, 1979).

PRIMARY AND SECONDARY PRODUCTION

The general, overall estimate of the original oil-in-place that can ultimately be produced by primary (natural energy drive) and secondary (waterflood) methods of recovery is only 40%. Production of the remaining oil in the ground is the challenge of EOR. The U.S. Department of Energy has estimated this resource to be approximately 300 billion barrels of oil in the United States alone (Donaldson, 1978). On a worlwide basis, this estimate is 3700 billion barrels.

Primary production relies on gas-cap or solution-gas drive to move the oil to production wells and is generally applicable only to oils with an API gravity greater than 20° (this is generally taken as the division between "heavy oils" and "light oils"). As the reservoir energy is depleted, it becomes necessary to install pumps to create a pressure sink at the well and lift the fluids to the surface. Water production is relatively low, but facilities are required for its treatment and disposal by injection into a brine aquifer. Produced gas must be compressed and sent to a pipeline; however, in many cases the gas production is too low to make compression and transportation economically feasible and the temptation is to simply release it into the atmosphere or to burn the gas. In either case air pollution is the end result and some other alternative must be imposed, because the trend of population encroachment on the oilfields makes air pollution untenable.

The off-gas will have to be transported to a pipeline at an economical loss if necessary. Otherwise, the simple flare must grow in complexity by addition of a scrubber to remove sulfur, nitrogen and carbon oxides, followed by disposal facilities for the resulting scrubber liquids.

Although the stringent environmental demands raise the cost and increase the complexity of oil production, they result in conservation of the resource and increased ultimate recovery.

Secondary recovery is conducted by injecting water into the petroleum reservoir to displace oil to the production wells. The water is composed of various combinations of (1) brine produced from the oil reservoir, (2) brine obtained from subsurface salaquifers, and (3) water from nearby rivers or lakes.

The subsurface and surface waters have special characteristics that impose the need for surface treatment facilities. If these surface facilities are not carefully designed and operated they become a source of aesthetic blight and air, water, land, and noise pollution. When the facility is in a remote wilderness some of this can be tolerated, but the pressure of population growth now demands greater initial investment to abate each one of the obnoxious manifestations of pollution mentioned above.

Subsurface brines are aerated to complete the oxidation of the minerals and salts, which are then removed by sedimentation and filtration. Subsurface brines are in a chemically reduced state that retain certain salts in ionic form. Upon oxidation of ions to a higher valence state, when brought to the surface, salts can precipitate and rapidly plug injection wells; for example:

$$3FeS + Fe(OH)_2 + O_2 \rightarrow Fe_2O_3 + Fe_2S_3 + H_2O \quad (16\text{-}1)$$

In the above reaction, i.e., the change from ferrous oxide and sulfide to ferric forms, precipitates cause the produced brine to turn black in color. After aeration, sedimentation ponds for gradual precipitation of the oxides are required. Precipitation is followed by filtration and addition of a biocide prior to injection. The biocide is necessary because aeration enables a microbial population to develop an impermeable colony at the face of the wellbore and to plug the pores of the formation matrix upon supply of oxygen dissolved in the injection water.

If the formation contains sulfur, sulfate-reducing bacteria introduced by the injected water cause an increase of hydrogen sulfide production and the resulting air pollution problem from the produced fluids. Ivanov and Belyaev (1983) showed that the aerobic bacteria population in the vicinity of the wellbore actually provides nutrients for the anaerobic (sulfate reducers) deeper in the reservoir by producing low-molecular-weight oxygenated hydrocarbons as metabolic products (acids, alcohols, and ketones). These low-molecular-weight oxygen compounds assist in the growth and maintenance of the anaerobic bacteria, which are responsible for sour petroleum reservoirs.

Grula (1983) presented laboratory evidence that polyacrylamide used for improved waterflood at Long Beach, California, actually promoted the growth of anaerobic bacteria that were uniquely able to reduce the viscosity of the polyacrylamide and increase the production of hydrogen sulfide. Grula's tests were requested by the U.S. Department of Energy after E.C. Donaldson (senior author) surmised that the four-fold increase of hydrogen sulfide production and total lack of the anticipated increase in oil production, which were observed at a pilot injection project at Long Beach, California, was due to promotion of bacterial growth in the reservoir by polyacrylamide.

This points out that increasing demands for careful environmental controls will indeed increase the cost of oil production because microbial control, and research to develop greater insight into adverse microbial activity, have become necessary.

Some strains of indigenous microbes also promote oilfield equipment corrosion, e.g., filamentous forms of *Sphaerotilus, Crenothrix*, and the stalk-forming iron bacterium *Gallionella*.

Some chemicals, and mixing equipment, are required in secondary recovery operations for bacteria and scale control. The chemicals are stored dry and mixed at the site, which is a mundane procedure when remote from human population, but now must be considered in a different light. The potential of dust pollution from these chemicals and injury to curious individuals (illegally trespassing, or otherwise) must be included in the design and construction phases. Thus the problem of environmental control requires close scrutiny of all process procedures and equipment.

A humorous, but graphic, example of this occurred at a petrochemical plant facility near Houston, Texas, bordering on a large creek that led to the Houston ship channel. A waste treatment–sedimentation pond acquired a population of rodents (nutria) that were damaging the sealed walls of the pond by burrowing into them. Not long after the problem developed, however, the solution appeared in the guise of an alligator who took up residence in the pond with an adequate supply of food (the nutria). This solved the technical problem, until some children found their way into the plant property and teased the alligator by throwing sticks and rocks. Although the children were illegally trespassing, the alligator had to be removed because it had become an attractive nuisance and the treatment pond had to be reconstructed of cement at considerable expense in the interest of environmental control.

Secondary oil recovery

Secondary recovery is conducted by injecting treated produced brines into oil-bearing strata to displace the oil to producing wells. The brines are treated for removal of iron and suspended solids, and after filtration a small amount of sodium bisulfite is added to remove entrained oxygen. Sodium tripolyphosphate is then added to prevent scale formation in the pipes. The addition of sodium bisulfite and sodium tripolyphosphate is a simple process, and the concentrations used are so small that no environmental hazards exist from their use (Donaldson, 1978).

The accumulation and processing of large volumes of oilfield brines does present serious potential environmental problems, however, because the escape of brines can damage surface waters. Sound engineering practices prevent this from occurring (Payne et al., 1953; McClure et al., 1960). Strict enforcement of state regulations against oilfield brine pollution has mandated the best available technology in the oilfield. The conclusion of a study of freshwater aquifers in oilfields of five major oil-producing states (Dowd, 1978) was that, from the interpretation of chloride analysis, no pollution could be attributed to oil- and gas-field brine injection operations.

CHEMICAL FLOODING

The two EOR processes that employ, and are based on, the technology developed for secondary recovery are known as chemical flooding (or improved waterflooding), and the micellar–polymer process. Chemical flooding, as the name implies, involves the addition of chemicals to the brine prior to injection. At present, three types of chemical compounds are in the forefront of chemical flooding. These are caustics (sodium hydroxide, sodium silicate, ammonium hydroxide, etc.), detergents (alkyl aryl sulfonates), and polymers (polyacrylamide and polysaccharide). A large body of experience has already been accumulated in the use of polymer solutions in oilfields, inasmuch as more than 100 fields have been treated with polymers of various kinds.

Additional facilities (mixing tanks, storage tanks, pumps, pipes, etc.) are necessary to mix and process the chemical solutions at the field site; however, these additional facilities are not very different from those used in routine oilfield operations. Large quantities of chemicals are injected over a long period of time, but it is not necessary to store large inventories of chemicals at the field site. Hence, the potential for environmental damage from storage and operation of the facilities is confined to a small area of the field and is limited to spills and/or leaks in quantities less than the capacity of a railway car or truck. Precautions against health hazards to personnel using the chemicals in the field must be taken, because all the chemical compounds are irritants to mucous membranes.

Chemical compounds produced with the brine and oil are not a cause for concern because only a small fraction of the injected chemical is eventually produced. The amount of chemical injected in a chemical flood is designed to be approximately equal to the amount of material expected to be retained by adsorption, channeling, and dilution in the petroleum reservoir. Chemicals dissolved in the produced brines are recycled through the brine treatment system and reinjected into the reservoir. Chemicals dissolved in the oil are transported to the refinery to be processed as a part of the crude oil. Surplus aqueous fluids are disposed of by injection into a nonproducing brine aquifer above, or below the oil zone when they cannot be used in the brine-injection program.

The major potential environmental problems associated with chemical flooding are: (1) local air pollution by hydrogen sulfide removed from the produced brine during treatment; (2) spills or leaks of chemical additives during transportation, storage, and processing; (3) health hazards from the dry chemicals and solutions to personnel operating the field; (4) leaks from surface storage and treatment ponds for produced brine; (5) leaks from high-pressure pipes transporting mixed chemicals to the wells; (6) underground leaks into shallow aquifers from damaged or corroded wells; (7) production into shallow aquifers from improperly plugged abandoned well (possible when the pressure of the oil reservoir is raised as a result of EOR activities); (8) production into aquifers above the oil reservoir through an incompetent seal above the reservoir (fracture or fault); (9) subsidence along a fault plane caused by change of reservoir pressure; and (10) subsidence caused by chemical disaggregation of the oil reservoir rock matrix.

In 1980, the Energy Resources Company of Cambridge, Massachusetts, completed a review of the toxicity of chemical compounds that were being used, and might be used in the future, for oilfield operations. The general conclusions of the report are that (with the exception of chlorinated hydrocarbons used as biocides) the environmental hazards may be reduced to safe engineering design and operation in the field. The toxicity of many of the chlorinated hydrocarbons require careful storage and use to avoid local employee health problems and spills that could find their way into freshwater aquifers. The report, however, is a comprehensive review and is recommended for consideration prior to oilfield development (Silvestro and Desmarais, 1980).

MICELLAR–POLYMER PROCESSES

The micellar–polymer EOR process involves four separate phases of fluid injection, beginning with a "preflush" injection of fresh water or low-salinity water, necessary to adjust the salinity and pH of the reservoir. The microemulsion is composed of a hydrocarbon, brine, surfactant, cosurfactants (generally alcohols), and electrolytes. The third injected fluid is a polymer solution which increases viscosity and assists in more efficient displacement of the oil bank and the microemulsion. Brine injection follows the polymer solution as the fourth fluid-injection phase.

Although the micellar–polymer process is considerably more complex than chemical flooding, the environmental concerns are the same. The application of the micellar–polymer process increases the need for surface equipment and requires additional transportation and storage, and a large variety of chemical compounds. During the injection of chemicals (the second and third phases mentioned above), a large amount of chemicals are injected in a relatively short time in comparison to chemical flooding. Large inventories of chemicals are, therefore, required at the field site, and several environmental concerns are intensified, although no new ones are added. Health hazards to personnel become

TABLE 16-I

Common chemicals used in polymer or microemulsion flooding

Surfactants	Cosurfactants	Polymers	Bactericides
Olefin sulfonates	Amino compounds	Polyacrylamide	Sodium dichlorophenol
Alkyl aryl sulfonates	*n*-Butanol	Polysaccharides	Pentachlorophenol
Diethyleneglycol sulfate	Cyclohexanol	Polyethylene oxide	Aldehydes
Dodecyltrimethyl ammonium chloride	Isobutanol	Carboxyvinyl polymer	Alkyl amines
Polyglycerol monolaurate	Isopropanol	Ketoses	Alkyl phosphates
Sodium lauryl sulfonate	Ketones	Aldoses	
	Glutaraldehyde		

greater, because larger quantities of chemicals must be processed. The principal polymers used for mobility control (polyacrylamide and polysaccharide) are non-toxic if ingested; however, aerosols or powders taken into the lungs can result in serious pulmonary problems. Anyone entering an area containing a hopper and mixing equipment for the polymers, therefore, should be protected with a breathing mask. Any leaks from surface storage or pollution of shallow aquifers from abandoned or incompetent wells, going undetected for long periods, would result in severe damage to ground or surface streams. Recognition of intensified environmental concerns, however, should result in more stringent design specifications of equipment and closer surveillance of operations. If the presence of abandoned or improperly plugged wells is suspected, monitoring of any freshwater aquifers along the hydraulic gradient during the chemical-injection phases is necessary.

Table 16-I provides a list of some of the most common chemicals used in EOR operations. These chemicals are generally well known organic compounds and their toxicity and limiting concentrations in water have been established (Perry and Donaldson, 1980; Silvestro and Desmarais, 1980). If accidentally released from the oil sands or wells, with time and distance of traveling through underground geologic strata the chemicals will decrease in concentration by adsorption, dilution, and chemical and biological degradation. Thus, in the unlikely event that such chemicals are released, their concentrations usually will be below toxic levels. Freshwater aquifers are shallow and are situated above oil reservoirs; therefore, contamination from wells is limited to leaks from the well casing adjacent to the aquifer. This is unlikely except for very old wells which may have corroded casing.

In field tests of chemical flooding and micellar–polymer processes conducted so far, low-salinity water is used at various stages (preflush, surfactant solutions, polymer solutions, etc.). The trend, however, is to design systems less sensitive to salt content and perhaps compatible with the produced oilfield brines. If this trend does continue and chemical solutions can be prepared in brines supplied by a brine aquifer and the oil reservoir, the presently foreseen socioeconomic impact of freshwater demand will be alleviated. If it is not possible to use oilfield brines for all the chemical solutions by the time EOR processes are developed on a commercial scale, a close analysis of water availability and negotiations for allocation will be required prior to implementation.

THERMAL PROCESSES

Three types of thermal processes have currently gained widespread recognition: (1) steam soak, (2) steam drive, and (3) in-situ combustion. Processes using steam are viable commercial enterprises for viscous oils in California, where approximately 250,000 barrels of oil are produced each day by steam injection.

Steam soak

The steam-soak process involves the injection of steam for 30–120 days, followed by a shut-in period to allow for heat transfer and penetration of the steam into the reservoir. The wells are then opened for the production phase of the process, which may last as long as 6 months. The actual duration of the periods of injection, soak, and production depends on the characteristics of the oil and formation, and experience of the operator.

Steam drive

The steam-drive process employs a continuous injection of steam to enhance recovery at adjacent production wells. Oil is displaced to the production wells by a combination of in-situ distillation, solvent extraction by the advancing products of distillation, and gas expansion. Condensation of the steam produces a zone of hot water that behaves as a hot-water flood. The maximum anticipated oil recovery from the commercial projects now in progress is about 50% of the reservoir oil-in-place.

The steam processes require a constant source of fuel to generate steam along with an ample source of fresh water. Crude oil produced from the field, natural gas, processed fuel, and coal are used for steam generation on the basis of availability, economics, and air pollutant emissions. Water requirements depend on the process in use, total amount of steam required, and the ratio of the amount of injected steam (as equivalent amount of water) to the amount of oil produced, which may range from 0.1 to 5.0 or more. Steam requirements gradually increase on approaching the depletion stage of the reservoir.

The actual pollutant emissions from steam generation depend on a large amount of variables, such as the type of fuel and its chemical composition, the efficiency of heat exchange, the design of boilers and the stack, and meteorology. If produced oil is used for fuel, approximately one fourth of the production is consumed in producing steam. Emissions (in lb/hr) from a 50-million Btu/hr steam generator have been measured as: 1.3 lb/hr SO_2, 1.3 lb/hr CO_2, 0.6 lb/hr NO_x, 0.6 lb/hr unburned hydrocarbons and 0.1 lb/hr solid particles (Perry and Donaldson, 1980). One can realize from the data that steam stimulation processes add a potential or real air pollution hazard arising from direct-fired boilers. Inasmuch as the total amounts of these pollutants must be maintained below state and federal emission standards, the use of steam is constrained by the particular environmental standards involved. Perhaps this constraint could be lifted if the boiler flue gases can be compressed and added to the steam line, or injected separately into the oil-producing zone, to assist in production of the oil.

The demand for fresh water is a major consideration of steam injection processes, because water containing salts cannot be used due to the severe, irreversible damage to the boilers (Shields et al., 1983). The steam is employed to improve the mobility of viscous crude oils that cannot be produced by other known techniques. Tests also are being designed to determine the feasibility of using steam-drive processes for enhanced recovery of less viscous oils in the mid-continent region of the United States. In all cases, the demand for water must be compatible with regional water availability to municipalities and agricultural use; therefore, this is a major initial consideration for all proposed steam injection projects. Several other environmental concerns, in addition to air emission from the boilers and the water demand, are associated with steam-drive processes. Large amounts of produced water must be separated from the oil, treated, and disposed of in accordance with existing state pollution-control regulations. Produced oil emulsions are frequently difficult to break and require special treatment for resolution. The entrainment of oil from oil–water separation and emulsion-breaking systems also requires close control because of the adverse effect it has on the water-disposal system. Provisions must also be made for disposal of solid wastes from scrubbers, boilers, tanks, and flowlines.

In-situ combustion

Several fireflood pilot studies were made in the mid-continent region in the 1950's and 1960's. At that time they proved to be uneconomical. The pilot studies, however, were very important in development of theories and technology unique to this process.

The in-situ combustion processes are based on the generation of heat within the reservoir by burning some of the oil in the reservoir. One procedure simply involves the injection of air into a well followed by ignition of the oil and continued injection of air as the burning front expands radially around the well. Heat at the burning front (800–1200°F) produces steam, coke, distillation of the oil, thermal cracking of some of the heavy ends, and pressure to move the front toward peripheral production wells. This forward combustion process is not very efficient because the largest amount of heat dissipation, or transfer to the reservoir, occurs behind the burning front. The heat in the burned-out zone behind the front is of little value, but other methods have been devised to improve the heat transfer to the oil zone ahead of the burning front.

One method that has gained prominence is the wet combustion process. After the burning front reaches a predetermined size, water is injected in place of air to act as a medium to transfer heat from the burned-out zone to the oil zone ahead. After an initial transfer of heat as steam, the burned-out zone is cooled. Continued injection of water results in a hot-water drive which reduces the viscosity of the oil and furnishes the driving energy to move the oil to the production wells. Other methods involving the transfer of heat to the oil zone by water are also employed, such as alternate injection of air and water.

Air emissions from a well-designed in-situ fireflood are minimal. The thermal cracking reactions at the hot combustion front produce light hydrocarbons that may escape during production at the surface along with some carbon monoxide, carbon dioxide, and sulfur dioxide. The larger portion of these compounds, however, probably remains confined in the reservoir.

Water produced with the crude oil can contain significant quantities of metals and metal oxides dissolved in the acidic waters. These can cause contamination problems through spills and from leaks of the acid waters while they are being processed on the surface for subsurface disposal. The pH of the reservoir water may be lowered to 2.5, enough to cause severe corrosion of well casings in plugged (abandoned) wells. This may lead to contamination of other aquifers communicating with the fireflooded oil reservoir through the corroded segment of the abandoned well.

GASFLOODING

Any process employing, or producing, a gas has a potential for air pollution which may be local (confined to less than 5 acres) or regional (covering several counties, or an entire state). Gasflooding is the introduction of gases, such as light hydrocarbons and carbon dioxide, into the petroleum reservoir to improve oil-recovery efficiency through several theoretically beneficial effects: (1) the injected gas increases the reservoir pressure, furnishing energy to drive the oil to production wells; (2) by dissolving in the oil, the gas reduces the viscosity and density of the oil and improves the mobility; (3) the gas may be injected directly into a gas cap above the oil zone to furnish a reserve of displacement pressure; and (4) sufficient pressure may be maintained for liquefaction of the gas at reservoir temperature to produce a miscible-phase drive.

Before a decision to use gasflooding is made, a large constantly available source of inexpensive gas must exist near the oilfield. Examples are a producing gas field or a gas-processing or manufacturing plant producing one of the required gases as a by-product. The gas can be transported to the field by pipeline, truck, or railroad tank car.

Transportation of the gases to the field by pipe, truck, or rail is under strict regulation by both state and federal statutes. An accidental spill, or large leak, from any of these sources may cause severe local hazards, including the possibility of a fire or explosion. Dissipation and dilution in the atmosphere, however, will remove the hazard, but will present a pollution problem. One also must consider the fact that a viable economic incentive exists for not losing the valuable commodity being transported.

Water may be injected in large quantities in addition to the gas, and this will require the treating and processing of large volumes of water at the surface, similar to those in waterflooding operations (secondary recovery process). Produced water can be recycled, with additional water being obtained from a brine aquifer or nearby source of surface water. The water is generally treated to (1) remove solids, (2) adjust pH, (3) remove oxygen (to reduce corrosiveness), and (4) prevent bacterial growth.

Air emissions at the field generally originate from power facilities, such as diesel or natural gas engines used in generating electric power. Some hydrogen sulfide also may escape from the water-treating facilities. Other minor emissions may originate from storage tanks, vapor-recovery units, and heater treaters.

The potential for surface-water or shallow-groundwater pollution is the same as discussed previously for any field project processing large quantities of brine at the surface for reinjection. Stored or impounded brine may leak, field pipelines may develop leaks, and well casings may crack or corrode during injection and production operations, causing groundwater contamination. The potential of pipeline breaks and leaks has long been recognized by operators during secondary oil production. Because of strict regulations, proper equipment design, and constant monitoring, the likelihood of leaks and spills during EOR operations is considered less than for conventional operations.

Production of oil is very often accompanied by production of gases containing hydrocarbons, carbon dioxide, nitrogen, and sulfur compounds (Eaton et al., 1980).

RESEARCH

The environmental aspects of EOR require the acquisition of a considerable data base to: (1) assure compliance with laws, regulations, and environmental standards; (2) ensure that a minimum of environmental disturbance occurs as a result of the EOR program; and (3) alleviate any accidental environmental damage that might occur.

The constraints to EOR development from laws and regulations, water availability, socioeconomic impacts, environmental monitoring programs, etc., have been assessed. Some areas of research for development of specific data also have been suggested (Donaldson, 1978).

Research is recommended for determination of the rates and products of decomposition of the chemical compounds used for EOR under laboratory-simulated subsurface conditions. The data can be used to analyze the long-term viability of the large quantities of EOR compounds, which will be injected during future commercialization of EOR technology. As a companion to this work, the dispersion and chromatographic properties of these compounds also should be investigated under simulated reservoir hydrologic conditions. These data can be employed to predict the long-term migration properties of the injected compounds.

Biological degradation rates of compounds used in EOR operations by various cultures of bacteria have been studied (Grula and Grula, 1983). This work, however, should be continued because of its importance to the analysis of the pollution duration, especially of the ground surface and shallow subsurface environment in addition to discovery of microbial methods for pollution abatement.

Studies are recommended to determine the circumstances under which the injected compounds could initiate local seismic disturbances. In a related study, the reactions of these chemicals with the cementing material and supporting matrix of sedimentary rocks should be conducted. Data derived from this study can be used to determine whether long-term contact of the chemicals with sedimentary rocks will disaggregate the strata and cause land subsidence to occur.

Other research related to environmental concerns of EOR has been suggested by Wilson and Franklin (1978).

REFERENCES

Donaldson, E.C., 1978. *Environmental Compliance for Enhanced Oil Recovery.* Bartlesville Energy Technology Center, U.S. Department of Energy, Project Planning Document, FY 1979, 90 pp.

Donaldson, E.C., 1979. Environmental aspects of enhanced oil recovery. *Energy Sour.*, 3(3): 213–229.

Dowd, W.T., 1978. *Study of Fresh-Water Aquifers.* Interstate Oil Compact Commission, Oklahoma City, Okla., Press Release, August 18.

Eaton, W.S., Bush F.G., III, Coster, J., Delwiche, J.C. and Horttey, H.O., 1980. *Fugitive Hydrocarbon Emissions from Petroleum Production Operations, Volumes I and II.* American Petroleum Institute, Washington, D.C., 150 pp.

Grula, M.M., 1983. Microbial interactions with polyacrylamide polymers. In: *Proceedings, International Conference on Microbial Enhanced Oil Recovery.* NTIS, Springfield, Va., pp. 129–134.

Grula, M.M. and Grula, E.A., 1983. *Biodegradation of Materials Used in EOR. Final Report.* NTIS, Springfield, Va., DOE/BC/10304-17, 84 pp.

Ivanov, M.V. and Belyaev, S.S., 1983. Microbial activity in waterflooded oil fields and its possible regulation. In: *Proceedings, International Conference on Microbial Enhanced Oil Recovery.* NTIS, Springfield, Va., pp. 48–59.

McClure, J.C., Akins, D.W., Jr., Calhoun, G.H., Davis, W.B., Ellis, C.F., Nichols, J.R., Pierce, C.W. and Caltin, J.E. (Vocational Training Committee of API), 1960. *Subsurface Salt-Water Disposal.* American Petroleum Institute, Dallas, Texas, 92 pp.

Payne, B.W. and Staff of East Texas Salt Water Disposal Company, 1953. *Salt Water Disposal — East Texas Field.* Petroleum Extension Service, University of Texas, Austin, Texas, 116 pp.

Perry, C.W. and Donaldson, E.C., 1980. An overview of the status of EOR in the U.S., and related environment concerns. *Proc. 2nd U.S. DOE Energy Environmental Control Symp., Reston, Va., March 19,* 13 pp.

Shields, J., Koplan, E. and Royce, B.A., 1983. *Assessment of Water Issues Associated with EOR: A User's Guide*. NTIS, Springfield, Va., DOE/BC/10412-40, 187 pp.

Silvestro, E. and Desmarais, A.M., 1980. *Toxicity of Chemical Compounds Used for Enhanced Oil Recovery. Final Report.* Bartlesville Energy Technology Center, U.S. Department of Energy, Contract DE-RP 19-79BC10014, 191 pp.

Wilson, C.B. and Franklin, R., 1978. *Environmental Development Plan — Oil Supply Program*. NTIS, Springfield, Va., DOE/EDP-0024.

Appendix

FINITE ELEMENT SOLUTION FOR FLOW OF IMMISCIBLE FLUIDS IN PETROLEUM RESERVOIRS

ANANT R. KUKRETI, FARUK CIVAN and YATENDRA RAJAPAKSA

INTRODUCTION

Studying reservoirs using numerical simulators has been of continuing interest because most of the important decisions for producing the reservoir contents depend on information obtained by simulation. Numerical simulations are obtained from the solution of the reservoir model equations by some appropriate technique. Most of the conventional reservoir simulators utilize finite difference approximations for numerical solutions. The standard finite difference method, however, is best suited for regular-shaped and continuous property systems. For reservoirs involving heterogeneities, discontinuities and irregular-shaped boundaries, the finite element method is more suited. Even though this method is mathematically more sophisticated and more tedious to formulate, it is an alternative method to achieve a high quality of solutions.

In this chapter the application of the finite element method is illustrated. For this purpose the flow of immiscible and incompressible fluids through permeable media is considered. A general analytical formulation is presented considering multiple fluid phases. The media is assumed heterogeneous and anisotropic. The formulation begins with a listing of equations describing conservation of mass, as well as the set of auxiliary relations needed to mathematically describe the flow through permeable media. Then, a methodology based on finite element method is presented to numerically solve the multiple-phase flow problems. The type of boundary and initial conditions required to solve the problem are explained. The generation of the element matrices is illustrated for the case of two immiscible fluids such as oil and water. Though the formulation is carried out for a two-dimensional domain, it can be readily extended to three-dimensional systems. The methodology is automated for use on digital computers and a computer program written in FORTRAN is presented. Finally, an example is given to demonstrate the finite element method for reservoir engineering applications.

FLOW DESCRIPTION

Behavior of immiscible fluids in petroleum reservoirs is predicted from a simultaneous solution of the conservation equations for mass and momentum, because the reservoir system (rock and fluids) is assumed isothermal. Whereas a complete form of the mass conservation equation is written for each of the phases, a simplified form of the momentum equation, so called Darcy's Law, is used to express the volumetric fluxes of various phases. The pressure discontinuity across the phase interfaces are expressed in terms of the capillary pressure. Capillary pressures are correlated empirically with the fractional volumes occupied by each phase, i.e., saturations. The constitutive equations for the fluids are expressed in terms of pressure. Rock porosity is expressed as a pressure function. Rock permeability is related to phase saturations. These equations are discussed in detail elsewhere (Collins, 1961). In this section only a brief summary of these equations are given.

The basic differential equation expressing the principle of mass conservation for incompressible fluids and considering multiphase fluid flow through a permeable medium is as follows:

$$\nabla \cdot u_i + \frac{\partial}{\partial t}(\phi S_i) = 0 \text{ for } i = 1, 2, \ldots, NP \tag{A-1}$$

where i refers to phases, NP denotes the number of phases, u_i is the volumetric flux for phase i, ϕ is the porosity of the medium, and S_i is the saturation for phase i. ∇ denotes the vector gradient differential operator and $\partial/\partial t$ corresponds to differentiation with respect to time t. Equation (A-1) also accounts for the effect of compressibility of the reservoir formation (i.e., $\partial\phi/\partial t \neq 0$).

By definition of saturation:

$$\sum_{i=1}^{NP} S_i = 1. \tag{A-2}$$

The fluid volumetric flux u_i, for phase i, is expressed by Darcy's Law as follows:

$$u_i = -k\frac{k_{ri}}{\mu_i}(\nabla p_i - \rho_i g \nabla z) \tag{A-3}$$

where k is the absolute permeability of the medium, k_{ri} is the relative permeability of phase i, μ_i is the viscosity of phase i, ρ_i is the fluid density for phase i, g is the gravitational acceleration, and ∇z denotes the unit vector in the vertical direction.

The capillary pressure, p_{cij}, between two phases i and j is defined as the difference in their respective phase pressures. Choosing one phase ($j = 1$) as the reference datum, for phase i the following equation can be written:

$$p_{ci} = p_i - p_1 \tag{A-4}$$

Substitution of equation (A-2) into equation (A-1) yields:

$$-\nabla \cdot \left[k\frac{k_{ri}}{\mu_i}(\nabla p_i - \rho_i g \nabla z)\right] + \frac{\partial}{\partial t}(\phi S_i) = 0 \tag{A-5}$$

The auxiliary equations expressing the state for various fluids and rock properties are:

Rock compressibility:

$$\phi = \phi(p_i), \quad i = 1, 2, \ldots, NP \tag{A-6a}$$

Relative permeability:

$$k_{ri} = k_{ri}(S_1, S_2, \ldots, S_{NP}), \quad i = 1, 2, \ldots, NP \tag{A-6b}$$

Capillary pressures:

$$p_{ci} = p_i - p_1 = p_{ci}(S_1, S_2, \ldots, S_{NP}), \quad i = 1, 2, \ldots, NP \tag{A-6c}$$

Taking the unknown variables as $p_1, S_2, S_3, \ldots, S_{NP}$, the chain rule is used to obtain the derivatives involved in equation (A-5) using equations (A-2), (A-4), and (A-6) (the equation of state). Thus:

$$\nabla p_i = \nabla(p_1 + p_{ci}) = \nabla p_1 + \sum_{j=2}^{NP} \frac{\partial p_{ci}}{\partial S_j} \nabla S_j \tag{A-7}$$

$$\frac{\partial S_1}{\partial t} = \frac{\partial}{\partial t}\left(1 - \sum_{j=2}^{NP} S_j\right) = -\sum_{j=2}^{NP} \frac{\partial S_j}{\partial t} = -\sum_{j=2}^{NP} \dot{S}_j \tag{A-8}$$

$$\frac{\partial}{\partial t}(\phi S_1) = S_1 \frac{\partial \phi}{\partial p_1} \dot{p}_1 - \phi \sum_{j=2}^{NP} \dot{S}_j \tag{A-9}$$

and for $i > 1$ (i.e., $i = 2, 3, \ldots, NP$):

$$\frac{\partial}{\partial t}(\phi S_i) = S_i \frac{\partial \phi}{\partial p_i} \dot{p}_1 + S_i \frac{\partial \phi}{\partial p_i} \sum_{j=2}^{NP} \frac{\partial p_{ci}}{\partial S_j} \dot{S}_j + \phi \dot{S}_i \tag{A-10}$$

In equations (A-7) through (A-10) overdot denotes the time derivatives. In these equations all other derivatives are expressed in terms of the derivatives of the unknowns chosen as $p_1, S_2, S_3, \ldots, S_{NP}$.

In view of equations (A-7) through (A-10), for $i = 1$ equation (A-5) becomes:

$$-\nabla \cdot \left(k \frac{k_{r1}}{\mu_1} \nabla p_1\right) + S_1 \frac{\partial \phi}{\partial p_1} \dot{p}_1 - \phi \sum_{j=2}^{NP} \dot{S}_j + \rho_1 \nabla \cdot \left(k \frac{k_{ri}}{\mu_1} g \nabla z\right) = 0 \tag{A-11}$$

and for $i > 1$ (i.e., $i = 2, 3, \ldots, NP$):

$$-\nabla \cdot \left(k \frac{k_{ri}}{\mu_i} \nabla p_1\right) - \nabla \cdot \left(\sum_{j=2}^{NP} k \frac{k_{ri}}{\mu_i} \frac{\partial p_{ci}}{\partial S_j} \nabla S_j\right) + S_i \frac{\partial \phi}{\partial p_i} \dot{p}_1$$

$$+ S_i \frac{\partial \phi}{\partial p_i} \sum_{j=2}^{NP} \frac{\partial p_{ci}}{\partial S_j} \dot{S}_j + \phi \dot{S}_i + \rho_i \nabla \cdot \left(k \frac{k_{ri}}{\mu_i} g \nabla z\right) = 0 \tag{A-12}$$

Equation (A-11) can be rewritten in matrix form as follows (for $i=1$):

$$-\nabla\cdot\left(k\left[\frac{k_{\mathrm{r}i}}{\mu_1}\;\Big\vert\;0\right]\nabla\begin{bmatrix}p_1\\ \hline S_2\\ S_3\\ \vdots\\ S_{NP}\end{bmatrix}\right)+\left[S_1\frac{\partial\phi}{\partial p_1}\;\Big\vert\;-\phi\right]\begin{bmatrix}\dot{p}_1\\ \hline \dot{S}_2\\ \dot{S}_3\\ \vdots\\ \dot{S}_{NP}\end{bmatrix}$$

$$+\nabla\cdot\left(k\frac{k_{\mathrm{r}i}}{\mu_1}\rho_1 g\,\nabla z\right)=0 \tag{A-13}$$

and, similarly, equation (A-12) becomes (for $i>1$):

$$-\nabla\cdot\left(k\left[\frac{k_{\mathrm{r}i}}{\mu_i}\;\Big\vert\;\frac{k_{\mathrm{r}i}}{\mu_i}\frac{\partial p_{\mathrm{c}i}}{\partial S_2}\;\Big\vert\;\cdots\;\Big\vert\;\frac{k_{\mathrm{r}i}}{\mu_i}\frac{\partial p_{\mathrm{c}i}}{\partial S_{NP}}\right]\nabla\begin{bmatrix}p_1\\ \hline S_2\\ \hline \vdots\\ \hline S_{NP}\end{bmatrix}\right)$$

$$+\left[S_i\frac{\partial\phi}{\partial p_i}\;\Big\vert\;\left(S_i\frac{\partial\phi}{\partial p_i}\frac{\partial p_{\mathrm{c}i}}{\partial S_2}+\phi\right)\;\Big\vert\;\cdots\;\Big\vert\;\left(S_i\frac{\partial\phi}{\partial p_i}\frac{\partial p_{\mathrm{c}i}}{\partial S_{NP}}+\phi\right)\right]\begin{bmatrix}\dot{p}_1\\ \hline \dot{S}_2\\ \hline \vdots\\ \hline \dot{S}_{NP}\end{bmatrix}$$

$$+\nabla\cdot\left(k\frac{k_{\mathrm{r}i}}{\mu_i}\rho_i g\nabla z\right)=0 \tag{A-14}$$

Denoting:

$$\{X^{\mathrm{e}}\}=\begin{bmatrix}p_1\\ S_2\\ S_3\\ \vdots\\ S_{NP}\end{bmatrix} \tag{A-15}$$

$$k_{ij}=\begin{cases}\dfrac{k_{\mathrm{r}i}}{\mu_i} & \text{for } i=1,2,\ldots,NP \text{ and } j=1\\ 0 & \text{for } i=1 \text{ and } j=2,3,\ldots,NP\\ \dfrac{k_{\mathrm{r}i}}{\mu_i}\dfrac{\partial p_{\mathrm{c}i}}{\partial S_j} & \text{for } i=2,3,\ldots,NP \text{ and } j=2,3,\ldots,NP\end{cases} \tag{A-16}$$

$$C_{ij}=\begin{cases}-\phi & \text{for } i=1 \text{ and } j=2,3,\ldots,NP\\ S_i\dfrac{\partial\phi}{\partial p_i} & \text{for } i=1,2,\ldots,NP \text{ and } j=1\\ S_i\dfrac{\partial\phi}{\partial p_i}\dfrac{\partial p_{\mathrm{c}i}}{\partial S_j}+\phi & \text{for } i=2,3,\ldots,NP \text{ and}\\ & \quad j=2,3,\ldots,NP\end{cases} \tag{A-17}$$

$$\{q_g\} = \begin{cases} k\dfrac{k_{r1}}{\mu_1}\rho_1 g \nabla z \\ k\dfrac{k_{r2}}{\mu_2}\rho_2 g \nabla z \\ \vdots \\ k\dfrac{k_{rNP}}{\mu_{NP}}\rho_{NP} g \nabla z \end{cases} \tag{A-18}$$

Equations (A-13) and (A-14) can be combined as one differential equation expressed in matrix form as follows:

$$\{L(\{X\})\} = -\nabla \cdot k\left[k_{ij}\right]\nabla\{X\} + \left[C_{ij}\right]\{\dot{X}\} + \nabla \cdot \{q_g\} \tag{A-19}$$

The solution of equation (A-19) subject to a set of initial conditions and boundary conditions can be obtained using the finite element method as described in the next section.

The initial conditions are specified for the variables p_1, S_2, S_3, ..., S_{NP}. The boundary conditions are usually of two types: sealed or no flow boundary, and open or flow boundary, both of which are described by Darcy's Law.

FINITE ELEMENT SOLUTION

For numerical solution, one can consider a two-dimensional permeable medium (i.e., the "domain"), which is represented by an assemblage of subdivisions called "finite elements". These element are considered interconnected at points which are called "nodes" or "node points". This process is known as discretization of the continum. Various geometrically shaped finite elements (triangular, rectangular, quadrilateral, etc.) can be considered to discretize the domain. In addition, finite elements with straight edges or curved edges can be considered. Generally the nodes are selected at the corners of an element and sometimes along the element interfaces. Figure A-1 shows the discretization of a two-dimensional domain into a system of triangular finite elements. At each node of a finite element, values of pressure and saturation are treated as the unknowns, called "nodal values". Simple polynomial functions (called "shape functions") expressed in terms of spatial coordinates (x, y) are chosen to approximate the variation of the actual pressure (p) and saturation (S) over each finite element for each phase. The unknown nodal value is taken as the amplitude of the respective shape function. Thus, the domain is spanned by piecewise shape functions defined separately on each element of the domain:

$$\{X^e\} = \sum_{i=1}^{NP}\sum_{j=1}^{NN} N_j(x, y)Y_{ji}^e \tag{A-20}$$

where Y_{ji} represents the nodal value for the ith phase which is associated with the shape function N_j; x, y denote the two-dimensional cartesian spatial vari-

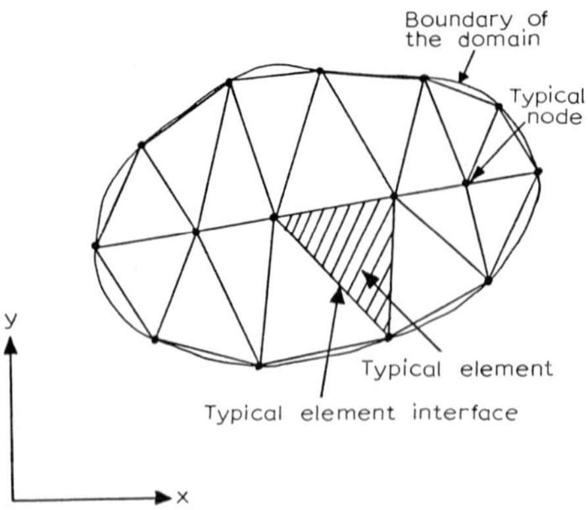

Fig. A-1. Two-dimensional domain represented as an assemblage of triangular finite elements.

ables; NN denotes the number of node points considered in the typical finite element used to discretize the domain; and superscript "e" denotes that the variable pertains to this finite element. Equation (A-20) can be written in matrix form as follows:

$$\{X^e\} = [\overline{N}]\{Z^e\} \qquad \text{(A-21)}$$

In this equation, $[\overline{N}]$ is a matrix of order $(NP \times NNP)$, where $NNP = NN \times NP$. A typical element $\overline{N}_{il}$ $(l = 1, 2, \ldots, NNP)$ of matrix $[\overline{N}]$ is equal to N_j for:

$$l = (i-1)NN + j \qquad \text{(A-22)}$$

and will be equal to zero for all other l's. It should be kept in mind that $i = 1, 2, \ldots, NP$, $j = 1, 2, \ldots, NN$, and $l = 1, 2, \ldots, NNP$. Also, vector $\{Z^e\}$ in equation (A-21) is of order $(NNP \times 1)$ and is given by:

$$\{Z^e\} = \begin{bmatrix} Y_{11} \\ Y_{21} \\ \vdots \\ Y_{NN,1} \\ \hdashline Y_{12} \\ Y_{22} \\ \vdots \\ Y_{NN,2} \\ \hdashline \vdots \\ \hdashline Y_{1,NP} \\ Y_{2,NP} \\ \vdots \\ Y_{NN,NP} \end{bmatrix} = \begin{bmatrix} \{Y^e\}_1 \\ \hdashline \{Y^e\}_2 \\ \hdashline \vdots \\ \hdashline \{Y^e\}_{NP} \end{bmatrix} \qquad \text{(A-23)}$$

where $\{Y^e\}_i$ are the finite element nodal values for the ith phase.

When equation (A-21) is substituted into equation (A-19) it will result in a residual, $\{R^e\}$, which is given by:

$$\{R^e\} = -\nabla \cdot k\left[k_{ij}\right]\nabla[\bar{N}]\{Z^e\} + \left[C_{ij}\right][\bar{N}]\{\dot{Z}^e\} + \nabla \cdot \{q_g^e\} \tag{A-24}$$

In Galerkin's method this residual is weighted by the shape functions and the integral over the finite element domain, Ω, is forced to vanish (Zienkiewicz, 1977), i.e.:

$$\int\int_\Omega [\bar{N}]^T\{R^e\}\,\mathrm{d}\Omega = 0 \tag{A-25}$$

Substituting equation (A-24) into equation (A-25), gives:

$$\int\int_\Omega \Big(-[\bar{N}]^T\nabla \cdot k\left[k_{ij}\right]\nabla[\bar{N}]\{Z^e\} + [\bar{N}]^T\left[C_{ij}\right][\bar{N}]\{\dot{Z}^e\}$$
$$+[\bar{N}]^T\;\nabla \cdot \{q_g^e\}\Big)\,\mathrm{d}\Omega = 0 \tag{A-26}$$

which after applying Green's Theorem to perform the integrations reduces to the following form:

$$\left(\int\int_\Omega \nabla[\bar{N}]^T \cdot k\left[k_{ij}\right]\nabla[\bar{N}]\,\mathrm{d}\Omega\right)\{Z^e\} + \left(\int\int_\Omega [\bar{N}]^T\left[C_{ij}\right][\bar{N}]\,\mathrm{d}\Omega\right)\{\dot{Z}^e\}$$
$$-\int\int_\Omega \nabla[\bar{N}]^T \cdot \{q_g^e\}\,\mathrm{d}\Omega - \int_{\partial\Omega}[\bar{N}]^T\left(k\left[k_{ij}\right]\nabla[\bar{N}]\{Z^e\} - \{q_g^e\}\right)\cdot \mathrm{d}n = 0 \tag{A-27}$$

where $\partial\Omega$ denotes the finite element boundaries and n denotes the unit outward normal vector to these finite element boundaries. Thus, the last term in equation (A-27) represents a line integral on the boundaries of the finite element. All other integrals appearing in equation (A-27) have to be performed over the finite element area. So they are designated with double integrals.

The boundary integral term appearing in equation (A-27) can be simplified by using Darcy's Law, given as equation (A-3). For the first phase (i.e., for $i = 1$), this equation can be written as follows:

$$u_1 = -k\left[\begin{array}{c|c} \dfrac{k_{r1}}{\mu_1} & 0 \end{array}\right]\left[\begin{array}{c} p_1 \\ \hline S_2 \\ \vdots \\ S_{NP} \end{array}\right] + k\frac{k_{r1}}{\mu_1}\rho_1 g\nabla z \tag{A-28}$$

For other phases (i.e., for $i > 1$), equation (A-3) in view of equation (A-7) can be written as follows:

$$u_i = -k\left[\begin{array}{c|c|c|c} \dfrac{k_{ri}}{\mu_i} & \dfrac{k_{ri}}{\mu_i}\dfrac{\partial p_{ci}}{\partial S_2} & \cdots & \dfrac{k_{ri}}{\mu_i}\dfrac{\partial p_{ci}}{\partial S_{NP}} \end{array}\right]\left[\begin{array}{c} p_1 \\ \hline S_2 \\ \hline \vdots \\ \hline S_{NP} \end{array}\right]$$
$$+k\frac{k_{ri}}{\mu_i}\rho_i g\nabla z \tag{A-29}$$

Thus, grouping all the phase velocities into one vector, $\{u^e\}$, for the finite element, gives:

$$\{u^e\} = \begin{bmatrix} u_1^e \\ u_2^e \\ \vdots \\ u_{NP}^e \end{bmatrix} = -\left(k\left[k_{ij}\right]\nabla\{X^e\} - \{q_g^e\}\right) \tag{A-30}$$

where $\{X^e\}$, $[k_{ij}]$ and $\{q_g^e\}$ have been defined in equations (A-15), (A-16) and (A-18), respectively.

Thus, in view of equation (A-30), equation (A-27) simplifies to:

$$[K^e]\{Z^e\} + [C^e]\{\dot{Z}^e\} = \{Q_g^e\} + \{Q_u^e\} \tag{A-31}$$

where:

$$[K^e] = \iint_\Omega \nabla[\bar{N}]^T \cdot k\left[k_{ij}\right]\nabla[\bar{N}]\,\mathrm{d}\Omega \tag{A-32}$$

$$[C^e] = \iint_\Omega [\bar{N}]^T\left[C_{ij}\right][\bar{N}]\,\mathrm{d}\Omega \tag{A-33}$$

$$\{Q_g^e\} = \iint_\Omega \nabla[\bar{N}]^T \cdot \{q_g^e\}\,\mathrm{d}\Omega \tag{A-34}$$

$$\{Q_\mathrm{u}^e\} = \int_{\partial\Omega} [\bar{N}]^T\{u^e\} \cdot \mathrm{d}n \tag{A-35}$$

It should be noted that $\{Q_u^e\}$ represents the normal outward flow through the finite element boundaries and so represents a "natural boundary" condition or Neumann-type boundary condition. Thus, if a finite element edge or boundary is impermeable, then there is "no flow" across that edge or $\{u^e\} = 0$ across this boundary.

In order to couple or assemble the individual finite elements together to give the domain or global system equations, the following has to be satisfied:

(1) The nodal values (pressures and saturations of the different phases, i.e., $\{Z^e\}$ at common nodes of the finite elements must be equal.

(2) The sum of the boundary flows from adjacent elements must equate the prescribed outflow through the node. Thus, if an interior finite element of a domain does not represent a source or a sink, then for each interior element $\{Q_u^e\} = 0$.

Satisfying these two conditions, the domain or global system equations can be constructed as

$$[K]\{Z\} + [C]\{\dot{Z}\} = \{Q_g\} + \{Q_u\} \tag{A-36}$$

by summing up contributions from the finite element matrices. In matrices $[K]$ and $[C]$, a term giving the coupling between nodal quantities i and j is formed by summing the coupling terms from the finite elements connected to those nodes. Thus, internal boundary integrals are thereby automatically eliminated.

To solve equation (A-36) in the time domain, the total time interval in which the solution is desired is divided into a sequence of time increments, $[t_n, t_{n+1}]$, and the time step (or increment) is defined as follows:

$$\Delta t = t_{n+1} - t_n \tag{A-37}$$

It is convenient to change origin at each time step by defining a new time variable τ as follows:

$$\tau = t - t_n \tag{A-38}$$

where t varies from t_n to t_{n+1} and τ varies from 0 to Δt. During a time step, $0 \leqslant \tau \leqslant \Delta t$, each element of the vector $\{Z\} = \{Z(\tau)\}$ * is assumed to vary linearly with respect to τ, i.e.:

$$\{Z(\tau)\} = \{A\} + \{B\}\tau \quad \text{for } 0 \leqslant \tau \leqslant \Delta t \tag{A-39}$$

where $\{A\}$ and $\{B\}$ are constant vectors to be determined. It will be assumed that the solution for equation (A-36) at the beginning of the time step (i.e., at $\tau = 0$) is known. Then substituting $\tau = 0$ in equation (A-39), gives:

$$\{A\} = \{Z(0)\} \tag{A-40}$$

which reduces equation (A-39) to:

$$\{Z(\tau)\} = \{Z(0)\} + \{B\}\tau \quad \text{for } 0 \leqslant \tau \leqslant \Delta t \tag{A-41}$$

Differentiating this equation with respect to time, τ, gives:

$$\{\dot{Z}(\tau)\} = \{B\} \quad \text{for } 0 \leqslant \tau \leqslant \Delta t \tag{A-42}$$

Now substituting equations (A-41) and (A-42) into equation (A-36) and solving for $\{B\}$, gives:

$$([K]\tau + [C])\{B\} = \{Q_g\} + \{Q_u\} - [K]\{Z(0)\} \quad \text{for } 0 \leqslant \tau \leqslant \Delta t \tag{A-43}$$

which can be rewritten in a simple form as:

$$[D]\{B\} = \{F\} \quad \text{for } 0 \leqslant \tau \leqslant \Delta t \tag{A-44}$$

where:

$$[D] = [K]\tau + [C] \tag{A-45}$$

$$\{F\} = \{Q_g\} + \{Q_u\} - [K]\{Z(0)\} \tag{A-46}$$

In a problem, at some of the node points in the domain, the nodal values (i.e., pressures and/or saturations) may be required to be held constant and equal to

---

* The values of pressures and saturations at each node point vary with time, which are computed as vector $\{Z\}$ in equation (A-36); therefore, in the remaining text of this section $\{Z\}$ is denoted as $\{Z(\tau)\}$.

their values at the initial time, i.e., at $t=0$. Such node points are called as constrained nodes, because the pressure and/or saturation value at these node points are constrained to a specified value at all times. For example, if the ith nodal value in the vector $\{Z(\tau)\}$ is to be held constant to a value equal to $Z_i(0)$ during the whole time domain analysis, then for this nodal value equation (A-41) requires that the ith row element, B_i, of the vector $\{B\}$ must be equal to zero at all times. In order to enforce this condition, the system assembled matrices $[D]$ and $\{F\}$ have to be modified at every time step in the following manner before solving equation (A-44):

(1) In the ith row and the jth column of matrix $[D]$, set the diagonal term equal to unity (i.e., $D(i, i)=1$) and all the off-diagonal terms equal to zero (i.e., $D(i, j)=0$ for $i \neq j$).

(2) Set the ith row element of the vector $\{F\}$ equal to zero (i.e., $F(i)=0$).

It should be remembered that in equation (A-43) the elements of matrices $[K]$ and $[C]$, and vector $\{Q_g\}$ are functions of k_{ri}, $\partial p_{ci}/\partial S_j$, and $\partial\phi/\partial p_i$ (see equations (A-18), (A-32), (A-33) and (A-34)) and have to be computed at $\tau=\Delta t$ in order to compute $\{Z(\Delta t)\}$. Since $k_{ri}=k_{ri}(S_1, S_2, \ldots, S_{NP})$, $p_{ci}=p_{ci}(S_1, S_2, \ldots, S_{NP})$ and $\phi=\phi(p_i)$, elements of $[K]$, $[C]$, and $\{Q_g\}$ can only be computed if $\{Z(\Delta t)\}$ is known. Thus, equation (A-44) will be solved for $\{B\}$ iteratively in a time step. In the following paragraphs of this section this procedure is explained.

At the beginning of any time step, i.e., at $\tau=0$, the vectors $\{Z(0)\}$ and $\{B\}_{\tau=0}$* are known. The main objective is to first solve equation (A-44) for $\{B\}_{\tau=\Delta t}$ and then to substitute the results into equation (A-41) and solve for $\{Z(\Delta t)\}$, which is given by:

$$\{Z(\Delta t)\} = \{Z(0)\} + \Delta t\{B\}_{\tau=\Delta t} \tag{A-47}$$

The steps to obtain $\{Z(\Delta t)\}$ are as follows:

(1) To start the iteration, assume that:

$$\{B\}_{\tau=\Delta t} \approx \{B\}_{\tau=0} \tag{A-48}$$

(2) Substitute equation (A-48) into equation (A-47) and compute $\{Z(\Delta t)\}$. Let this approximate solution be designated as $\{Z(\Delta t)\}^{(k)}$.

(3) Using $\{Z(\Delta t)\}^{(k)}$ and equations (A-16), (A-17) and (A-18) for each finite element, formulate the system matrices $[K]_{\tau=\Delta t}$, $[C]_{\tau=\Delta t}$, and $\{Q_g\}_{\tau=\Delta t}$.

(4) Compute $[D]_{\tau=\Delta t}$ and $\{F\}_{\tau=\Delta t}$ by substituting the results of step 3 and $\tau=\Delta t$ into equations (A-45) and (A-46), which gives:

$$[D]_{\tau=\Delta t} = \Delta t[K]_{\tau=\Delta t} + [C]_{\tau=\Delta t} \tag{A-49}$$

$$\{F\}_{\tau=\Delta t} = \{Q_g\}_{\tau=\Delta t} + \{Q_u\} - [K]_{\tau=\Delta t}\{Z(0)\} \tag{A-50}$$

* Subscript attached outside a vector or a matrix denotes the time at which it is evaluated.

(5) Modify $[D]_{\tau=\Delta t}$ and $\{F\}_{\tau=\Delta t}$ to account for specified pressure- and saturation-related boundary conditions.

(6) Now solve equation (A-44) for a new value of the vector $\{B\}_{\tau=\Delta t}$. The equation to be solved for $\{B\}_{\tau=\Delta t}$ is:

$$[D]_{\tau=\Delta t}\{B\}_{\tau=\Delta t} = \{F\}_{\tau=\Delta t} \tag{A-51}$$

(7) Substituting the results of step 6 into equation (A-47), compute the new value for vector $\{Z(\Delta t)\}$, which might be designated as $\{Z(\Delta t)\}^{(k+1)}$.

(8) Check if convergence is reached from the following condition:

$$\frac{\left|Z_i(\Delta t)^{(k+1)} - Z_i(\Delta t)^{(k)}\right|}{\left|Z_i(\Delta t)^{(k)}\right|} \leq \text{tolerance for } i = 1, 2, \ldots, NNP \tag{A-52}$$

where "tolerance" has to be specified for desired error limit; it depends on the accuracy limit desired for values of nodal pressures and saturations predicted. If convergence is not reached, then set $\{Z(\Delta t)\}^{(k)} = \{Z(\Delta t)\}^{(k+1)}$ and repeat steps 3 through 8.

The time step computations are complete when equation (A-52) is satisfied. Then set $\{Z(0)\} = \{Z(\Delta t)\}$ and $(B\}_{\tau=0} = \{B\}_{\tau=\Delta t}$ for the next time step and repeat the aforementioned eight steps. The analysis is stopped when the total time for which the solution was desired has been reached.

In the first time step, only $\{Z(0)\}$ is known, which actually contains the values of the pressure of the first phase and saturation of the other phases at each node point in the domain. These are the initial conditions for the problem and have to be specified to start the solution. Thus, in the first time step, $\{B\}_{\tau=0}$ is not known. To initiate the iteration cycle in this time step, take $\{B\}_{\tau=0} \approx 0$ and then perform the eight steps mentioned in the preceding paragraph.

FORMULATION OF ELEMENT MATRICES FOR TWO-PHASE FLOW (WATER AND OIL)

In this section, the finite element matrices, given by equations (A-31) to (A-35), are presented for two types of elements: (1) A three-node isoparametric triangular element, and (2) a four-node quadrilateral isoparametric element. The formulation is illustrated for the case of immiscible fluids such as water and oil. For the triangular element, a linear approximation for the shape functions, used to express the variation of pressures and saturations for the two phases within the element, can be adopted, whereas for a quadratic element a bilinear approximation can be chosen. The selection of the shape functions and the calculation of the element matrices has been simplified and generalized by the concept known as "isoparametric elements". In this concept, a "natural coordinate system" is adopted to define the element geometry and to express the shape functions. Thus, in order to understand this concept, it is appropriate to first define the natural coordinate system.

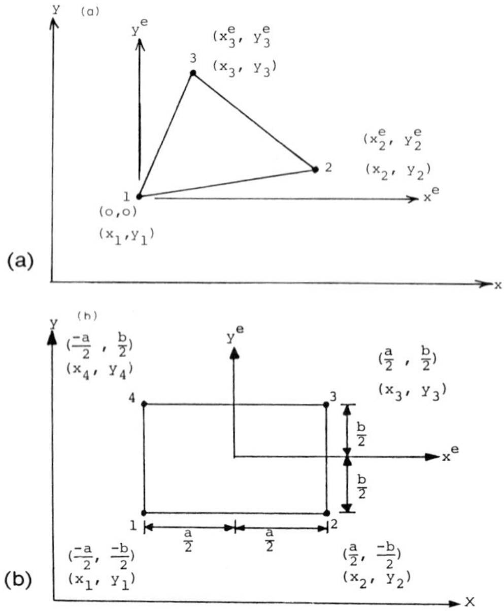

Fig. A-2. Element local coordinate system and global system: (a) three-node triangular element; (b) four-node rectangular element.

In finite element formulation, usually a "local coordinate system" is defined for a typical element. The coordinate system used to define the geometry of the domain (discretized as a finite element mesh) configuration is called "global system". For example, in Fig. A-2 the local coordinate system for a three-node triangular element and a four-node rectangular element is shown as the x^e and ye cartesian axes, whereas the global system is shown as the x and y cartesian axes. Thus, the local coordinate system changes from element to element, whereas the global system remains fixed. A "natural coordinate system" for an element is a local system which permits the coordinates of any point of the element to be expressed by a set of dimensionless numbers such that their magnitudes do not exceed unity. For example, such a choice of a natural coordinate system for a three-node triangular element is shown in Fig. A-3. Since a shape function is defined as a function which has a unity value at one nodal point and zero at the other node points of the element, the construction of shape functions for the element in its natural coordinate system becomes simpler. Also, it can be seen in Figs. A-2 and A-3 that the element coordinate system does not coincide with the global coordinate system. But to assemble the element matrices, it is essential that all elements be referenced to some fixed coordinate system, which obviously is the global system. Thus, some mapping transformation has to be developed between the two coordinate systems. The natural coordinate systems for the elements in the mesh can be so chosen, such that the natural coordinates give a unity value at the node points. If the shape functions

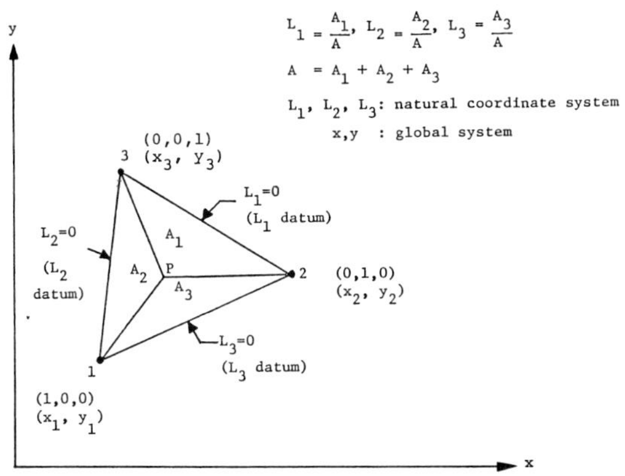

Fig. A-3. Natural coordinate system for a three-node triangular element.

developed to express the variation of pressures and saturations within the element are used to express the relationship between global coordinate system and the element natural coordinate system, then such an element is called an "isoparametric element". Thus, for such elements, the natural coordinate system not only generalizes and simplifies the formulation, but as will be seen in the two subsections that follow, it makes it easier to perform the integrations required to obtain the element matrices.

In the next two subsections the element matrices are derived for the three-node triangular and the four-node quadrilateral isoparametric elements. In the formulation presented, oil and water phases are designated as phase 1 and phase 2, respectively. In some places, the variables pertaining to oil (phase 1) and water (phase 2) are denoted by subscripts "o" and "w", respectively, so as to make it easy to understand.

There are two options which can be used to define the basic unknown variables in the problem: (1) pressure of oil and saturation of water may be chosen as the two unknown variables (i.e., X_1 and X_2, respectively, in equation (A-15)); or (2) pressure of water and saturation of water may be chosen as the two unknown variables (i.e., X_1 and X_2 in equation (A-15)). The choice as to which option would be easier and direct to use depends in which form the initial conditions for the problem are defined. For example, if pressures and saturations of water at each node point are specified at the initial time, i.e., $t = 0$, then the second option can be directly used. The element matrices are derived for both the options. The reader, however, should carefully interpret the unknown variables X_1 and X_2 correctly for the two options.

Three-node isoparametric triangular element

A typical three-node triangular element is shown in Fig. A-4, in which Y_{j1} and Y_{j2} denote the values of pressures and saturations at the jth node point (where

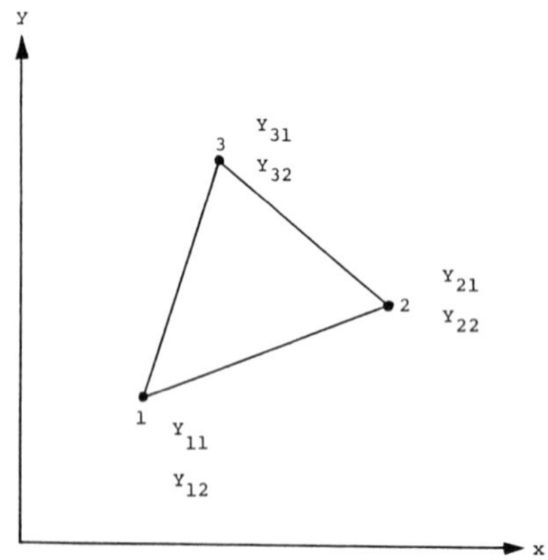

Fig. A-4. Three-node triangular element nodal values.

$j = 1, 2, 3$) of the element. These are often referred to as the "degrees of freedom" (abbreviated as d.o.f.) of the element. Thus, this element has a total of six d.o.f., with two d.o.f. at each node point. The general trial functions used to denote the variation of the unknown variables, X_1^e and X_2^e, for the element are represented in a linear form as (refer to equation (A-22)):

$$X_1^e = \sum_{j=1}^{3} N_j Y_{j1}^e = [N]\{Y^e\}_1 \tag{A-53a}$$

$$X_2^e = \sum_{j=1}^{3} N_j Y_{j2}^e = [N]\{Y^e\}_2 \tag{A-53b}$$

where:

$$[N] = [N_1 \quad N_2 \quad N_3] = \text{shape function matrix} \tag{A-54a}$$

$$\{Y^e\}_1 = \begin{bmatrix} Y_{11}^e \\ Y_{21}^e \\ Y_{31}^e \end{bmatrix} = \text{element nodal vector for phase 1}$$

$$\{Y^e\}_2 = \begin{bmatrix} Y_{12}^e \\ Y_{22}^e \\ Y_{32}^e \end{bmatrix} = \text{element nodal vector for phase 2} \tag{A-54b}$$

Defining (refer to equation (A-23)):

$$\{Z^e\} = \begin{bmatrix} \{Y^e\}_1 \\ \{Y^e\}_2 \end{bmatrix} \tag{A-55}$$

then, equations (A-53a) and (A-53b) can be grouped as:

$$\{X^e\} = \begin{bmatrix} X_1^e \\ X_2^e \end{bmatrix} = [\overline{N}]\{Z^e\} \tag{A-56}$$

where:

$$[\overline{N}]=\left[\begin{array}{ccc|ccc} N_1 & N_2 & N_3 & 0 & 0 & 0 \\ 0 & 0 & 0 & N_1 & N_2 & N_3 \end{array}\right] \tag{A-57}$$

For the three-node triangular element the shape functions (N_j, $j=1, 2, 3$) can be related to the natural coordinate system shown in Fig. A-3. These are also referred to as "area coordinates". Such coordinates are defined by taking the area subtended by any point P and its appropriate base or datum, and dividing it by the area of the whole triangle. Thus, as shown in Fig. A-3, the area coordinate L_1, is defined as follows:

$$L_1=\frac{\text{area } P23}{\text{area } 123}=\frac{A_1}{A} \tag{A-58}$$

The area coordinates L_2 and L_3 are defined similarly. Thus, the edges of the triangle (called bases) correspond to $L_1=0$, $L_2=0$ and $L_3=0$, as shown in Fig. A-3. With this definition of L_1, L_2 and L_3, it can be seen that at each node point the following is true:

$$\begin{aligned} &\text{At node point 1: } L_1=1,\ L_2=0,\ L_3=0\\ &\text{At node point 2: } L_1=0,\ L_2=1,\ L_3=0\\ &\text{At node point 3: } L_1=0,\ L_2=0,\ L_3=1 \end{aligned} \tag{A-59}$$

From Fig. A-3 and the definitions of L_1, L_2 and L_3, it can be shown that the relationship between the global cartesian coordinate system (x, y) and the local area coordinate system (L_1, L_2, L_3) is given by

$$\begin{bmatrix} x \\ y \\ 1 \end{bmatrix}=\begin{bmatrix} x_1 & x_2 & x_3 \\ y_1 & y_2 & y_3 \\ 1 & 1 & 1 \end{bmatrix}\begin{bmatrix} L_1 \\ L_2 \\ L_3 \end{bmatrix} \tag{A-60}$$

where x_j, y_j ($j=1, 2, 3$) are the x and y coordinates, respectively, of the jth node point of the element. The linear relationship of equation (A-60) shows that a polynomial in (x, y) may be expressed by a polynomial of same degree in (L_1, L_2, L_3). The advantage of a polynomial in (L_1, L_2, L_3) is that these coordinates are invariant with respect to the shape and orientation of the triangle, and that all integration formulas become remarkably simple. This advantage justifies the use of area coordinate system for development of element matrices.

Equation (A-59) shows that the area coordinates, L_1, L_2 and L_3, satisfy the definition of shape function, i.e., they give a value equal to one if evaluated at a particular node and zero if evaluated at other nodes. Thus, the shape functions, N_1, N_2 and N_3, for the triangular elements are:

$$N_1=L_1,\ N_2=L_2;\ \text{and } N_3=L_3 \tag{A-61}$$

Thus, in view of equations (A-57) and (A-61), matrix $[\overline{N}]$ can be written as follows:

$$[\overline{N}]=\left[\begin{array}{ccc|ccc} L_1 & L_2 & L_3 & 0 & 0 & 0 \\ 0 & 0 & 0 & L_1 & L_2 & L_3 \end{array}\right] \tag{A-62}$$

Inverting equation (A-60), the element natural coordinate system or the area coordinates can be expressed in terms of the global coordinates, which gives:

$$L_j = \frac{1}{2A}\left(a_j + b_j x + c_j y\right) \quad \text{for } j = 1, 2, 3 \tag{A-63}$$

where:

$$\begin{aligned} a_1 &= x_2 y_3 - x_3 y_2; \quad & b_1 &= y_2 - y_3; \quad & c_1 &= x_3 - x_2 \\ a_2 &= x_3 y_1 - x_1 y_3; \quad & b_2 &= y_3 - y_1; \quad & c_2 &= x_1 - x_3 \\ a_3 &= x_1 y_2 - x_2 y_1; \quad & b_3 &= y_1 - y_2; \quad & c_3 &= x_2 - x_1 \end{aligned} \tag{A-64}$$

and:

$$A = \frac{1}{2}\begin{vmatrix} 1 & x_1 & y_1 \\ 1 & x_2 & y_2 \\ 1 & x_3 & y_3 \end{vmatrix} \tag{A-65}$$

where | | denotes a determinant.

The evaluation of the element matrices, as given by equations (A-32) to (A-35), involve differentiation and integration with respect to global coordinates, (x, y). If area coordinates, L_1, L_2 and L_3, are used to define the shape functions, as done in equation (A-61), then all differentiation and integration have to be done with respect to the area coordinates as the variables. From equation (A-63):

$$\frac{\partial L_j}{\partial x} = \frac{b_j}{2A} \tag{A-66}$$

$$\frac{\partial L_j}{\partial y} = \frac{c_j}{2A} \tag{A-67}$$

It should be kept in mind that L_1, L_2 and L_3 cannot all be chosen as the variables. Any two can be chosen as the variables and the third can then be expressed in terms of the other two by using the following relationship:

$$L_1 + L_2 + L_3 = 1 \tag{A-68}$$

If area coordinates are used, all integrals appearing in equations (A-32) to (A-34) will be of the form $\int\int_\Omega L_1^a L_2^b L_3^c \, d\Omega$, which can be evaluated analytically by the following formula:

$$\int\int_\Omega L_1^a L_2^b L_3^c \, d\Omega = \frac{a!b!c!}{(a+b+c+2)!}(2A) \tag{A-69}$$

where A is given by equation (A-65) and ! denotes a factorial.

Defining:

$$[H] = \nabla[\bar{N}] \tag{A-70}$$

then in view of equation (A-32) the matrix $[K^e]$ for the element becomes:

$$[K^e] = \int\int_\Omega [H]^T k[k_{ij}][H] \, d\Omega \tag{A-71}$$

where the elements of matrix $[k_{ij}]$ are given by equation (A-16). For the case of two-phase flow this matrix will be of order (2×2). If the two unknown variables selected are pressure of oil and saturation of water, then:

$$k_{11} = \frac{k_{r1}}{\mu_1} = \frac{k_{ro}}{\mu_o}; \quad k_{12} = 0$$
$$k_{21} \frac{k_{r2}}{\mu_2} = \frac{k_{rw}}{\mu_w}; \quad k_{22} = \frac{k_{r2}}{\mu_2}\frac{\partial p_c}{\partial S_2} = \frac{k_{rw}}{\mu_w}\frac{\partial p_c}{\partial S_w} \tag{A-72}$$

where subscript "w" denotes quantities pertaining to water (or phase 2) and "o" denotes quantities pertaining to oil (or phase 1). On the other hand, if the two unknown variables selected are pressure of water and saturation of water, then:

$$k_{11} = \frac{k_{r2}}{\mu_2} = \frac{k_{rw}}{\mu_w}; \quad k_{12} = 0$$
$$k_{21} = \frac{k_{r1}}{\mu_1} = \frac{k_{ro}}{\mu_o}; \quad k_{22} = -\frac{k_{r1}}{\mu_1}\frac{\partial p_{c1}}{\partial S_2} = -\frac{k_{ro}}{\mu_o}\frac{\partial p_{co}}{\partial S_w} \tag{A-73}$$

Substituting equation (A-62) into equations (A-70) and (A-71) and using equations (A-66) and (A-67), gives:

$$[H] = \frac{1}{2A}\begin{bmatrix} (b_1\hat{i}+c_1\hat{j}) & (b_2\hat{i}+c_2\hat{j}) & (b_3\hat{i}+c_3\hat{j}) & 0 & 0 & 0 \\ 0 & 0 & 0 & (b_1\hat{i}+c_1\hat{j}) & (b_2\hat{i}+c_2\hat{j}) & (b_3\hat{i}+c_3\hat{j}) \end{bmatrix} \tag{A-74}$$

where $\hat{i}$ and $\hat{j}$ are unit vectors acting along the global x-axis and y-axis, respectively.

Finally, substituting equation (A-74) into equation (A-71), expanding in terms of k_{ij}'s, and performing the integration, gives:

$$[K^e] = \frac{k}{4A}\begin{bmatrix} k_{11}(b_1^2+c_1^2) & k_{11}(b_1b_2+c_1c_2) & k_{11}(b_1b_3+c_1c_3) & k_{12}(b_1^2+c_1^2) & k_{12}(b_1b_2+c_1c_2) & k_{12}(b_1b_3+c_1c_3) \\ k_{11}(b_1b_2+c_1c_2) & k_{11}(b_2^2+c_2^2) & k_{11}(b_2b_3+c_2c_3) & k_{12}(b_1b_2+c_1c_2) & k_{12}(b_2^2+c_2^2) & k_{12}(b_2b_3+c_2c_3) \\ k_{11}(b_1b_3+c_1c_3) & k_{11}(b_2b_3+c_2c_3) & k_{11}(b_3^2+c_3^2) & k_{12}(b_1b_3+c_1c_3) & k_{12}(b_2b_3+c_2c_3) & k_{12}(b_3^2+c_3^2) \\ k_{21}(b_1^2+c_1^2) & k_{21}(b_1b_2+c_1c_2) & k_{21}(b_1b_3+c_1c_3) & k_{22}(b_1^2+c_1^2) & k_{22}(b_1b_2+c_1c_2) & k_{22}(b_1b_3+c_1c_3) \\ k_{21}(b_1b_2+c_1c_2) & k_{21}(b_2^2+c_2^2) & k_{21}(b_2b_3+c_2c_3) & k_{22}(b_1b_2+c_1c_2) & k_{22}(b_2^2+c_2^2) & k_{22}(b_2b_3+c_2c_3) \\ k_{21}(b_1b_3+c_1c_3) & k_{21}(b_2b_3+c_2c_3) & k_{21}(b_3^2+c_3^2) & k_{22}(b_1b_3+c_1c_3) & k_{22}(b_2b_3+c_2c_3) & k_{22}(b_3^2+c_3^2) \end{bmatrix} \tag{A-75}$$

To formulate the elements of matrix $[C^{e}]$, equation (A-33) is used, which states:

$$[C^{e}] = \int\int_{\Omega} [\bar{N}]^{T} [C_{ij}] [\bar{N}] \, d\Omega \tag{A-76}$$

where the elements of matrix $[C_{ij}]$ are given by equation (A-17). For the case of two-phase flow, this matrix will be of order (2×2). If the two unknown variables selected are pressure of oil and saturation of water, then:

$$C_{11} = (1 - S_2)\frac{\partial \phi}{\partial p_1} = (1 - S_w)\frac{\partial \phi}{\partial p_o}$$

$$C_{12} = -\phi \tag{A-77}$$

$$C_{21} = S_2 \frac{\partial \phi}{\partial p_2} = S_w \frac{\partial \phi}{\partial p_w}$$

$$C_{22} = S_2 \frac{\partial \phi}{\partial p_2} \frac{\partial p_{c2}}{\partial S_2} + \phi = S_w \frac{\partial \phi}{\partial p_w} \frac{\partial p_{cw}}{\partial S_w} + \phi$$

If rock porosity, ϕ, is a constant then these equations reduce to:

$$C_{11} = 0; \quad C_{12} = -\phi; \quad C_{21} = 0; \quad \text{and } C_{22} = \phi \tag{A-78}$$

If the two unknown variables selected are pressure of water and saturation of water, then:

$$C_{11} = S_2 \frac{\partial \phi}{\partial p_2} = S_w \frac{\partial \phi}{\partial p_w}$$

$$C_{12} = \phi \tag{A-79}$$

$$C_{21} = S_2 \frac{\partial \phi}{\partial p_w} = S_w \frac{\partial \phi}{\partial p_w}$$

$$C_{22} = S_2 \frac{\partial \phi}{\partial p_2} \frac{\partial p_{c2}}{\partial S_2} - \phi = S_w \frac{\partial \phi}{\partial p_w} \frac{\partial p_{cw}}{\partial S_w} - \phi$$

If rock porosity, ϕ, is a constant, then for this choice of unknown variables equations similar to equation (A-78) can be obtained for the coefficient C_{ij}'s. Substituting equation (A-62) into equation (A-76), expanding in terms of C_{ij}'s and performing the integration, gives:

$$[C^{e}] = \frac{A}{12} \begin{bmatrix} 2C_{11} & C_{11} & C_{11} & 2C_{12} & C_{12} & C_{12} \\ C_{11} & 2C_{11} & C_{11} & C_{12} & 2C_{12} & C_{12} \\ C_{11} & C_{11} & 2C_{11} & C_{12} & C_{12} & 2C_{12} \\ 2C_{21} & C_{21} & C_{21} & 2C_{22} & C_{22} & C_{22} \\ C_{21} & 2C_{21} & C_{21} & C_{22} & 2C_{22} & C_{22} \\ C_{21} & C_{21} & 2C_{21} & C_{22} & C_{22} & 2C_{22} \end{bmatrix} \tag{A-80}$$

To evaluate the elements of vector $\{Q_g^e\}$, equation (A-34) is used:

$$\{Q_g^e\} = \int\int_{\Omega} \nabla[\bar{N}]^T \cdot \{q_g^e\}\, d\Omega \tag{A-81}$$

where the elements of vector $\{q_g^e\}$ are given by equation (A-18). In view of equation (A-70), equation (A-81) reduces to:

$$\{Q_g^e\} = \int\int_{\Omega} [H]^T \cdot \{q_g^e\}\, d\Omega \tag{A-82}$$

Assuming that the global y-axis is the vertical axis, the vector $g\ \nabla z = g\hat{j}$, where $\hat{j}$ is the unit vector along the y-axis. Thus substituting equations (A-18) and (A-74) into equation (A-82), expanding in terms of k_{ij}'s, and performing the integration, gives:

$$\{Q_g^e\} = \begin{bmatrix} c_1 k_{11} \rho_1 g \\ c_2 k_{11} \rho_1 g \\ c_3 k_{11} \rho_1 g \\ c_1 k_{21} \rho_1 g \\ c_2 k_{21} \rho_1 g \\ c_3 k_{21} \rho_1 g \end{bmatrix} \tag{A-83}$$

To evaluate the elements of vector $\{Q_u^e\}$, equation (A-35) is used:

$$\{Q_u^e\} = \int_{\partial\Omega} [\bar{N}]^T \{u^e\} \cdot dn \tag{A-84}$$

where:

$$\{u^e\} \cdot dn = \begin{bmatrix} u_1 \\ u_2 \end{bmatrix}$$

contains the two phase velocities (subscript 1 = oil and 2 = water) acting normal to any element boundary and expressed as velocity per unit length. Consider that the element side joining node points 2 and 3 (called side 1) has boundary phase velocities specified, as shown in Fig. A-5. Thus, for the element the boundary flow vector, $\{Q_u^e\}$, becomes:

$$\{Q_u^e\} = \int_s [\bar{N}]^T \begin{bmatrix} u_1 \\ u_2 \end{bmatrix} ds \tag{A-85}$$

where the differential boundary length ds is:

$$ds = \frac{L_{23}}{x_3 - x_2}\, dx \tag{A-86}$$

Substituting equations (A-62) and (A-86) into equation (A-85) and noting that along side 1:

$$dx = -(x_3 - x_2)\, dL_2 \tag{A-87}$$

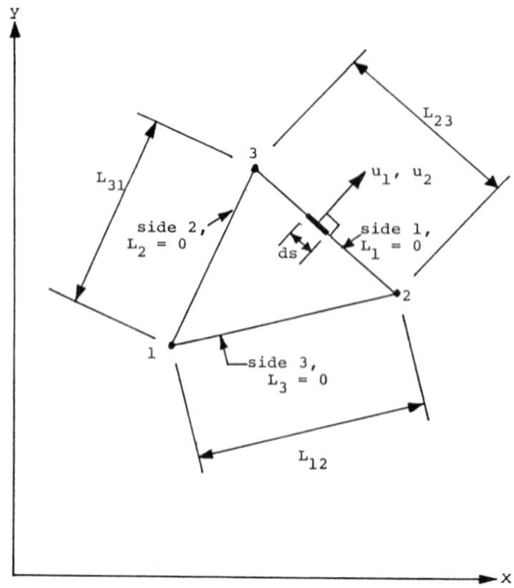

Fig. A-5. Flow velocity specified along three-node triangular element boundary.

and the limits of integration for L_2 are from 1 to 0, gives:

$$\{Q_u^e\} = \tfrac{1}{2} L_{23} \begin{bmatrix} 0 \\ u_1 \\ u_1 \\ 0 \\ u_2 \\ u_2 \end{bmatrix} \tag{A-88}$$

If all the element sides have phase velocities specified, then summing up the nodal contribution due to the flow across each side of the element, gives:

$$\{Q_u^e\} = \frac{L_{23}}{2} \begin{bmatrix} 0 \\ u_{11} \\ u_{11} \\ 0 \\ u_{21} \\ u_{21} \end{bmatrix} + \frac{L_{31}}{2} \begin{bmatrix} u_{12} \\ 0 \\ u_{12} \\ u_{22} \\ 0 \\ u_{22} \end{bmatrix} + \frac{L_{12}}{2} \begin{bmatrix} u_{13} \\ u_{13} \\ 0 \\ u_{23} \\ u_{23} \\ 0 \end{bmatrix} \tag{A-89}$$

where double subscripts have been introduced for u's to indicate which phase velocity is specified for which element side. Thus u_{ij} is the outward normal velocity specified per unit boundary length, with the first index of the subscript denoting the phase number and the second subscript denoting the element side number (see Fig. A-5).

Four-node isoparametric quadrilateral element

The shape functions for a four-node quadrilateral-shaped element are constructed from the shape functions of a four-node rectangular element. A typical four-node rectangular element is shown in Fig. A-6. This element requires that

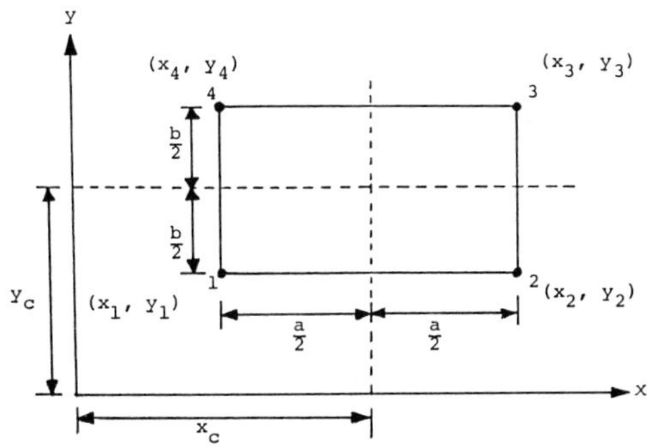

Fig. A-6. Four-node rectangular element.

the sides of the element must remain parallel to the global (x, y) coordinate axes. Defining the element natural coordinate system to be another rectangular (ξ, η) coordinate axes, which are parallel to the global (x, y) coordinate axes, satisfies the condition that for any point:

$$\xi = \frac{2}{a}(x - x_c) \tag{A-90a}$$

$$\eta = \frac{2}{b}(y - y_c) \tag{A-90b}$$

where (x_c, y_c) are the (x, y) coordinates of the centroid of the rectangular element shown in Fig. A-6. Then with respect to the (ξ, η) coordinate system, the element gets normalized to the form shown in Fig. A-7. This coordinate system is referred to as element natural coordinate system, because the coordinate variables are always within the range of ± 1. For this element the shape functions can be easily constructed as:

$$N_i = \tfrac{1}{4}(1 + \xi_i \xi)(1 + \eta_i \eta) \quad \text{for } i = 1, 2, 3, 4 \tag{A-91}$$

where (ξ_i, η_i) are the (ξ, η) coordinates of the four-corner node points of the

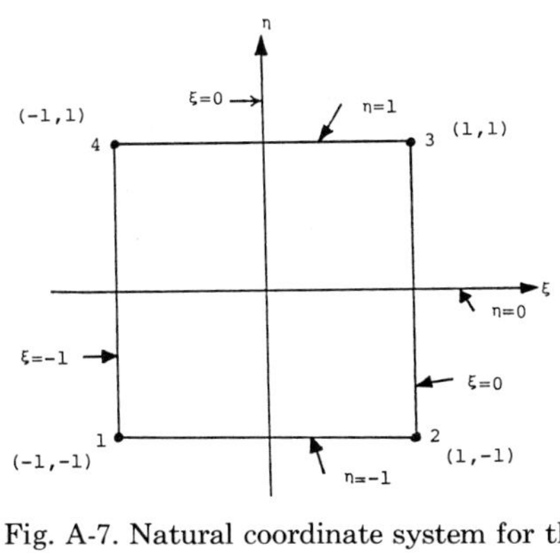

Fig. A-7. Natural coordinate system for the rectangular element.

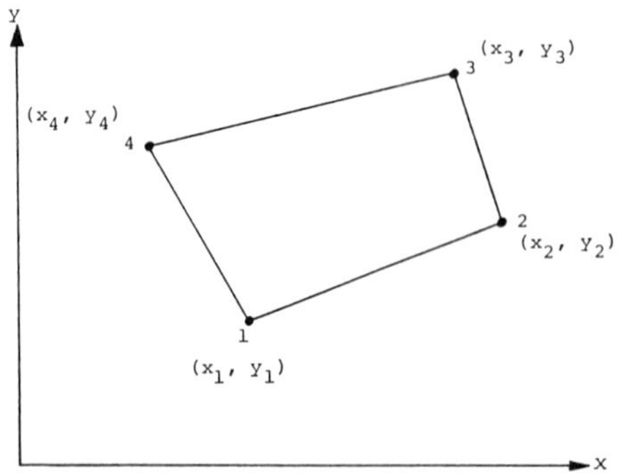

Fig. A-8. Four-node quadrilateral element.

element. From equation (A-91) it can be seen that, if N_i is evaluated at the ith node point, then $N_i = 1$ and has a zero value at all the other node points, thus satisfying the definition of a shape function.

There is no reason why the element natural coordinate system has to be a rectangular coordinate system; it can be a curvilinear coordinate system also. If the following transformation between the global rectangular (x, y) coordinate system and the element natural (ξ, η) coordinate system is used:

$$x = \sum_{i=1}^{4} N_i x_i \tag{A-92a}$$

$$y = \sum_{i=1}^{4} N_i y_i \tag{A-92b}$$

where (x_i, y_i) are the (x, y) coordinates of the ith node point and N_i are the same shape functions as defined by equation (A-91); then, a quadrilateral-shaped element in the global system (see Fig. A-8) will get normalized to an element with curvilinear (ξ, η) natural coordinate system, as shown in Fig. A-9. The origin of this coordinate system is at the intersection of the two lines that bisect the opposite sides (shown with dashed lines in Fig. A-9). The line corresponding to $\eta = 1/2$ is also shown in Fig. A-9. It should be noted that this line is not parallel to the ξ-axis but, instead, connects points a and b which are the mid-points of the left and right edges bounded by $\eta = 0$ and $\eta = 1$.

The shape functions of the quadrilateral element shown in Fig. A-9 are identical to those given in equation (A-91). These shape functions can also be used to denote the variation of the unknown variables, X_1^e and X_2^e, within the element as follows (refer to equation (A-21)):

$$X_1^e = \sum_{j=1}^{4} N_j Y_{j1}^e = [N]\{Y^e\}_1 \tag{A-93a}$$

$$X_2^e = \sum_{j=1}^{4} N_j Y_{j2}^e = [N]\{Y^e\}_2 \tag{A-93b}$$

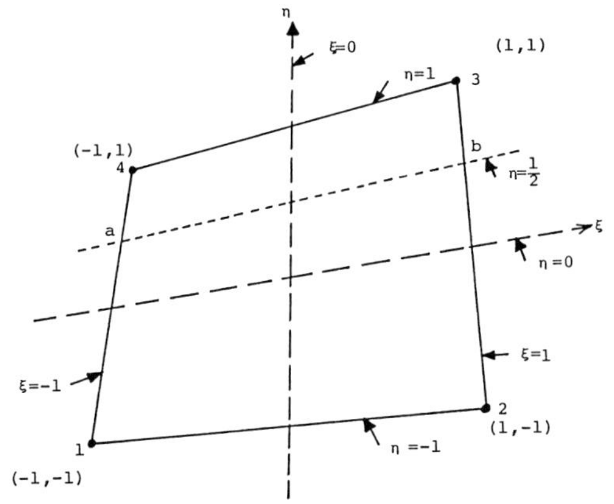

Fig. A-9. Natural coordinate system for the quadrilateral element.

where:

$$[N] = [N_1 \quad N_2 \quad N_3 \quad N_4] = \text{shape function matrix} \tag{A-94a}$$

$$\{Y^e\}_1 = \begin{bmatrix} Y^e_{11} \\ Y^e_{21} \\ Y^e_{31} \\ Y^e_{41} \end{bmatrix} = \text{element nodal vector for phase 1} \tag{A-94b}$$

$$\{Y^e\}_2 = \begin{bmatrix} Y^e_{12} \\ Y^e_{22} \\ Y^e_{32} \\ Y^e_{42} \end{bmatrix} = \text{element nodal vector for phase 2} \tag{A-94c}$$

Defining:

$$\{Z^e\} = \begin{bmatrix} \{Y^e\}_1 \\ \hline \{Y^e\}_2 \end{bmatrix} \tag{A-95}$$

then equations (A-93a) and (A-93b) can be grouped as:

$$\{X^e\} = \begin{bmatrix} X^e_1 \\ \hline X_2 \end{bmatrix} = [\bar{N}]\{Z^e\} \tag{A-96}$$

where:

$$[\bar{N}] = \left[\begin{array}{cccc|cccc} N_1 & N_2 & N_3 & N_4 & 0 & 0 & 0 & 0 \\ 0 & 0 & 0 & 0 & N_1 & N_2 & N_3 & N_4 \end{array}\right] \tag{A-97}$$

and the individual shape functions, N_i's, are defined by equation (A-91), which upon substitution of the values of the coordinates of the element node points, gives:

$$N_1 = \tfrac{1}{4}(1-\xi)(1-\eta) \tag{A-98a}$$

$$N_2 = \tfrac{1}{4}(1+\xi)(1-\eta) \tag{A-98b}$$

$$N_3 = \tfrac{1}{4}(1+\xi)(1+\eta) \tag{A-98c}$$

$$N_4 = \tfrac{1}{4}(1-\xi)(1+\eta) \tag{A-98d}$$

Because the N_i's are expressed in terms of the local coordinates, ξ, η, in order to evaluate $\nabla[\bar{N}]$ the chain rule of differentiation has to be invoked. Then the chain rule yields:

$$\frac{\partial}{\partial \xi} = \frac{\partial}{\partial x}\frac{\partial x}{\partial \xi} + \frac{\partial}{\partial y}\frac{\partial y}{\partial \xi} \tag{A-99a}$$

$$\frac{\partial}{\partial \eta} = \frac{\partial}{\partial x}\frac{\partial x}{\partial \eta} + \frac{\partial}{\partial y}\frac{\partial y}{\partial \eta} \tag{A-99b}$$

or:

$$\begin{bmatrix} \dfrac{\partial}{\partial \xi} \\ \dfrac{\partial}{\partial \eta} \end{bmatrix} = [J] \begin{bmatrix} \dfrac{\partial}{\partial x} \\ \dfrac{\partial}{\partial y} \end{bmatrix} \tag{A-100}$$

where $[J]$ is the Jacobian matrix of order (2×2) and is given by:

$$[J] = \begin{bmatrix} \dfrac{\partial x}{\partial \xi} & \dfrac{\partial y}{\partial \xi} \\ \dfrac{\partial x}{\partial \eta} & \dfrac{\partial y}{\partial \eta} \end{bmatrix} \tag{A-101}$$

Differentiation of equations (A-92a) and (A-92b) reduces this to the form:

$$[J] = \begin{bmatrix} \sum_{i=1}^{4} \dfrac{\partial N_i}{\partial \xi} x_i & \sum_{i=1}^{4} \dfrac{\partial N_i}{\partial \xi} y_i \\ \sum_{i=1}^{4} \dfrac{\partial N_i}{\partial \eta} x_i & \sum_{i=1}^{4} \dfrac{\partial N_i}{\partial \eta} y_i \end{bmatrix} = \begin{bmatrix} J_{11} & J_{12} \\ J_{21} & J_{22} \end{bmatrix} \tag{A-102}$$

The expressions for J_{ij}'s can be evaluated from equation (A-98a) to (A-98d). Inverting equation (A-100), gives:

$$\begin{bmatrix} \dfrac{\partial}{\partial x} \\ \dfrac{\partial}{\partial y} \end{bmatrix} = [J]^{-1} \begin{bmatrix} \dfrac{\partial}{\partial \xi} \\ \dfrac{\partial}{\partial \eta} \end{bmatrix} = [I] \begin{bmatrix} \dfrac{\partial}{\partial \xi} \\ \dfrac{\partial}{\partial \eta} \end{bmatrix} \tag{A-103}$$

where $[I]=[J]^{-1}$. Expanding equation (A-103), gives:

$$\frac{\partial}{\partial x}=I_{11}\frac{\partial}{\partial \xi}+I_{12}\frac{\partial}{\partial \eta} \tag{A-104a}$$

$$\frac{\partial}{\partial y}=I_{21}\frac{\partial}{\partial \xi}+I_{22}\frac{\partial}{\partial \eta} \tag{A-104b}$$

It can also be proved that:

$$\mathrm{d}x\,\mathrm{d}y=|J|\,\mathrm{d}\xi\,\mathrm{d}\eta \tag{A-105}$$

Defining $[H]$ as done in equation (A-70), equation (A-71) can be used to compute the matrix $[K^{e}]$ for the element. Substituting equations (A-97), (A-104a) and (A-104b) into equation (A-70), gives:

$$[H]=\begin{bmatrix} A_1 & A_2 & A_3 & A_4 & 0 & 0 & 0 & 0 \\ 0 & 0 & 0 & 0 & A_1 & A_2 & A_3 & A_4 \end{bmatrix} \tag{A-106}$$

where:

$$A_i=\frac{\partial N_i}{\partial x}\hat{i}+\frac{\partial N_i}{\partial y}\hat{j}=\left(I_{11}\frac{\partial N_i}{\partial \xi}+I_{12}\frac{\partial N_i}{\partial \eta}\right)\hat{i}+\left(I_{21}\frac{\partial N_i}{\partial \xi}+I_{22}\frac{\partial N_i}{\partial \eta}\right)\hat{j} \tag{A-107}$$

where $\hat{i}$ and $\hat{j}$ are unit vectors acting along the global x- and y-axis, respectively. Finally, substituting equations (A-107) and (A-105) into equation (A-71) and performing the matrix multiplications, gives:

$$[K^{e}]=\int_{-1}^{1}\int_{-1}^{1}|J|\begin{bmatrix}
k_{11}A_1\cdot A_1 & k_{11}A_1\cdot A_2 & k_{11}A_1\cdot A_3 & k_{11}A_1\cdot A_4 & k_{12}A_1\cdot A_1 & k_{12}A_1\cdot A_2 & k_{1s}A_1\cdot A_3 & k_{12}A_1\cdot A_4 \\
k_{11}A_1\cdot A_2 & k_{11}A_2\cdot A_2 & k_{11}A_2\cdot A_3 & k_{11}A_2\cdot A_4 & k_{12}A_1\cdot A_2 & k_{12}A_2\cdot A_2 & k_{12}A_2\cdot A_3 & k_{12}A_2\cdot A_4 \\
k_{11}A_1\cdot A_3 & k_{11}A_2\cdot A_3 & k_{11}A_3\cdot A_3 & k_{11}A_3\cdot A_4 & k_{12}A_1\cdot A_3 & k_{12}A_2\cdot A_3 & k_{12}A_3\cdot A_3 & k_{12}A_3\cdot A_4 \\
k_{11}A_1\cdot A_4 & k_{11}A_2\cdot A_4 & k_{11}A_3\cdot A_4 & k_{11}A_4\cdot A_4 & k_{12}A_1\cdot A_4 & k_{12}A_2\cdot A_4 & k_{12}A_3\cdot A_4 & k_{12}A_4\cdot A_4 \\
k_{21}A_1\cdot A_1 & k_{21}A_1\cdot A_2 & k_{21}A_1\cdot A_3 & k_{21}A_1\cdot A_4 & k_{22}A_1\cdot A_1 & k_{22}A_1\cdot A_2 & k_{22}A_1\cdot A_3 & k_{22}A_1\cdot A_4 \\
k_{21}A_1\cdot A_2 & k_{21}A_2\cdot A_2 & k_{21}A_2\cdot A_3 & k_{21}A_2\cdot A_4 & k_{22}A_1\cdot A_2 & k_{22}A_2\cdot A_2 & k_{22}A_2\cdot A_3 & k_{22}A_2\cdot A_4 \\
k_{21}A_3\cdot A_1 & k_{21}A_3\cdot A_2 & k_{21}A_3\cdot A_3 & k_{21}A_3\cdot A_4 & k_{22}A_3\cdot A_1 & k_{22}A_3\cdot A_2 & k_{22}A_3\cdot A_3 & k_{22}A_3\cdot A_4 \\
k_{21}A_4\cdot A_1 & k_{21}A_4\cdot A_2 & k_{21}A_4\cdot A_3 & k_{21}A_4\cdot A_4 & k_{22}A_4\cdot A_1 & k_{22}A_2\cdot A_4 & k_{22}A_3\cdot A_4 & k_{22}A_4\cdot A_4
\end{bmatrix}\mathrm{d}\xi\,\mathrm{d}\eta \tag{A-108}$$

where:

$$A_m\cdot A_n=\left(I_{11}\frac{\partial N_m}{\partial \xi}+I_{12}\frac{\partial N_m}{\partial \eta}\right)\left(I_{11}\frac{\partial N_n}{\partial \xi}+I_{12}\frac{\partial N_n}{\partial \eta}\right)+\left(I_{21}\frac{\partial N_m}{\partial \xi}+I_{22}\frac{\partial N_m}{\partial \eta}\right)\left(I_{21}\frac{\partial N_n}{\partial \xi}+I_{22}\frac{\partial N_n}{\partial \eta}\right)$$

for $m=1, 2, 3, 4$ and $n=1, 2, 3, 4$ (A-109)

and for k_{ij}'s either equation (A-72) or equation (A-73) is used, depending upon the choice of the unknown variables for the problem.

To formulate the elements of matrix $[C^e]$, equation (A-33) is used:

$$[C^e] = \iint_{\Omega} [\bar{N}]^T [C_{ij}][\bar{N}] \, d\Omega \tag{A-110}$$

where the elements of matrix $[C_{ij}]$ are given by equations (A-77) through (A-79) for the different sets of unknown variables selected for the problem. Substituting equations (A-97) and (A-105) into (A-110), gives:

$$[C^e] = \int_{-1}^{1}\int_{-1}^{1} |J| \begin{bmatrix} C_{11}N_1^2 & C_{11}N_1N_2 & C_{11}N_1N_3 & C_{11}N_1N_4 & C_{12}N_1^2 & C_{12}N_1N_2 & C_{12}N_1N_3 & C_{12}N_1N_4 \\ C_{11}N_1N_2 & C_{11}N_2^2 & C_{11}N_2N_3 & C_{11}N_2N_4 & C_{12}N_1N_2 & C_{12}N_2^2 & C_{12}N_2N_3 & C_{12}N_2N_4 \\ C_{11}N_1N_3 & C_{11}N_2N_3 & C_{11}N_3^2 & C_{11}N_3N_4 & C_{12}N_1N_3 & C_{12}N_2N_3 & C_{12}N_3^2 & C_{12}N_3N_4 \\ C_{11}N_1N_4 & C_{11}N_2N_4 & C_{11}N_3N_4 & C_{11}N_4^2 & C_{12}N_1N_4 & C_{12}N_2N_4 & C_{12}N_3N_4 & C_{12}N_4^2 \\ C_{21}N_1^2 & C_{21}N_1N_2 & C_{21}N_1N_3 & C_{21}N_1N_4 & C_{22}N_1^2 & C_{22}N_1N_2 & C_{22}N_1N_3 & C_{22}N_1N_4 \\ C_{21}N_1N_2 & C_{21}N_2^2 & C_{21}N_2N_3 & C_{21}N_2N_4 & C_{22}N_1N_2 & C_{22}N_2^2 & C_{22}N_2N_3 & C_{22}N_2N_4 \\ C_{21}N_1N_3 & C_{21}N_2N_3 & C_{21}N_3^2 & C_{21}N_3N_4 & C_{22}N_1N_3 & C_{22}N_2N_3 & C_{22}N_3^2 & C_{22}N_3N_4 \\ C_{21}N_1N_4 & C_{21}N_2N_4 & C_{21}N_3N_4 & C_{21}N_4^2 & C_{22}N_1N_4 & C_{22}N_2N_4 & C_{22}N_3N_4 & C_{22}N_4^2 \end{bmatrix} d\xi \, d\eta \tag{A-111}$$

To evaluate the element vector $\{Q_g^e\}$, equation (A-34) is used:

$$\{Q_g^e\} = \iint_{\Omega} \nabla[\bar{N}]^T \cdot \{q_g^e\} \, d\Omega \tag{A-112}$$

where the elements of vector $\{q_g^e\}$ are given by equation (A-18). In view of equation (A-70), equation (A-112) reduces to:

$$\{Q_g^e\} = \iint_{\Omega} [H]^T \cdot \{q_g^e\} \, d\Omega \tag{A-113}$$

Thus, substituting equations (A-18), (A-105) and (A-106) into equation (A-113), gives:

$$\{Q_g^e\} = \int_{-1}^{1}\int_{-1}^{1} kg\,|J| \begin{bmatrix} A_1k_{11}\rho_1g \\ A_2k_{11}\rho_1g \\ A_3k_{11}\rho_1g \\ A_4k_{11}\rho_1g \\ A_1k_{21}\rho_1g \\ A_2k_{21}\rho_1g \\ A_3k_{21}\rho_1g \\ A_4k_{21}\rho_1g \end{bmatrix} \nabla z \, d\xi \, d\eta \tag{A-114}$$

Assuming that the global y-axis is the vertical axis, the vector $\nabla z = \hat{j}$, where $\hat{j}$ is the unit vector along the y-axis. Then equation (A-114) can be further reduced as:

$$\{Q_g^e\} = \int_{-1}^{1}\int_{-1}^{1} kg\,|J| \begin{bmatrix} B_{11} \\ B_{21} \\ B_{31} \\ B_{41} \\ B_{12} \\ B_{22} \\ B_{32} \\ B_{42} \end{bmatrix} \mathrm{d}\xi\, \mathrm{d}\eta \tag{A-115}$$

where:

$$B_{ij} = \left(I_{21}\frac{\partial N_i}{\partial \xi} + I_{22}\frac{\partial N_i}{\partial \eta} \right)(k_{j1}\rho_1)$$

$$\text{for } i = 1, 2, 3, 4 \text{ and } j = 1, 2 \tag{A = 116}$$

and k_{ij}'s are given in equations (A-72) and (A-73) for the different set of unknown variables selected for the problem.

To evaluate the vector $[Q_u^e\}$, equation (A-35) is used:

$$\{Q_u^e\} = \int_{\partial\Omega} [\bar{N}]^{\mathrm{T}} \{u^e\} \cdot \mathrm{d}n \tag{A-117}$$

where:

$$\{u^e\} \cdot \mathrm{d}n = \begin{bmatrix} u_1 \\ u_2 \end{bmatrix}$$

contains the two phase velocities acting normal to any element boundary and expressed as velocity per unit length. On considering that the element side joining node points 2 and 3 (called side 2) has phase velocities specified, as shown in Fig. A-10, for the element, the boundary flow vector, $\{Q_u^e\}$, becomes:

$$\{Q_u^e\} = \int_s [\bar{N}]^{\mathrm{T}} \begin{bmatrix} u_1 \\ u_2 \end{bmatrix} \mathrm{d}s \tag{A-118}$$

in which the shape functions (N_1, N_2, N_3 and N_4) and the differential length ($\mathrm{d}s$) are given, respectively, by:

$$N_1 = 0; \quad N_2 = \tfrac{1}{2}(1-\eta); \quad N_3 = \tfrac{1}{2}(1+\eta); \quad N_4 = 0 \tag{A-119a}$$

$$\mathrm{d}s = \frac{L_{23}}{y_3 - y_2}\,\mathrm{d}y = \frac{L_{23}}{2}\,\mathrm{d}\eta \tag{A-119b}$$

Expressions given by equation (A-119a) have been obtained from equations

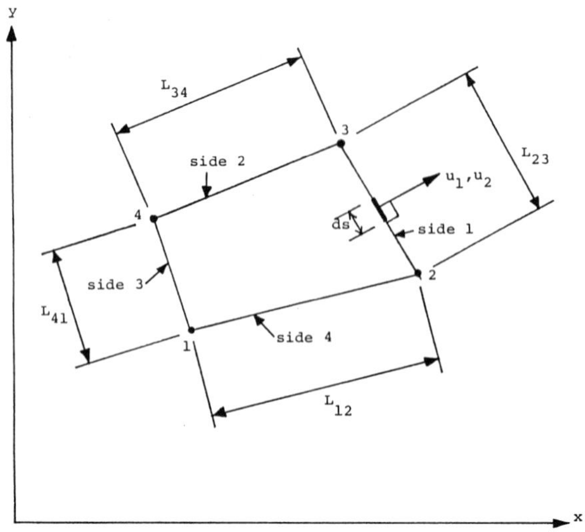

Fig. A-10. Flow velocity specified along four-node quadrilateral element.

(A-98a) through (A-98d) by substituting $\xi = 1$. Substituting equations (A-119a) and (119b) into equation (A-118) and performing the integrations, gives:

$$\{Q_u^e\} = \frac{L_{23}}{2}\begin{bmatrix} 0 \\ u_1 \\ u_1 \\ 0 \\ 0 \\ u_2 \\ u_2 \\ 0 \end{bmatrix} \tag{A-120}$$

If all the element sides have phase velocities specified, then summing up the nodal contribution due to flow across each side of the element, gives:

$$\{Q_u^e\} = \frac{L_{12}}{2}\begin{bmatrix} u_{11} \\ u_{11} \\ 0 \\ 0 \\ u_{21} \\ u_{21} \\ 0 \\ 0 \end{bmatrix} + \frac{L_{23}}{2}\begin{bmatrix} 0 \\ u_{12} \\ u_{12} \\ 0 \\ 0 \\ u_{22} \\ u_{22} \\ 0 \end{bmatrix} + \frac{L_{34}}{2}\begin{bmatrix} 0 \\ 0 \\ u_{13} \\ u_{13} \\ 0 \\ 0 \\ u_{23} \\ u_{23} \end{bmatrix} + \frac{L_{41}}{2}\begin{bmatrix} u_{14} \\ 0 \\ 0 \\ u_{14} \\ u_{24} \\ 0 \\ 0 \\ u_{24} \end{bmatrix} \tag{A-120}$$

where u_{ij} is the outward normal velocity specified per unit boundary length, with the first index of the subscript denoting the phase and the second index of the subscript denoting the element side number (see Fig. A-10).

As seen in equations (A-108) and (A-111), the matrices have to be integrated, which means that each coefficient of the matrix considered need to be integrated individually. These will be integrated numerically. This is required because the Jacobian matrix $[J]$, given in equation (A-102), is a function of ξ and η and cannot be readily evaluated because the coefficients are polynomials. Thus, $[J]^{-1} = [I]$ (refer to equation (A-103)) cannot be determined explicitly. Hence, for the derivation of the numerical integration formula, a typical coefficient of the matrices given by equations (A-108) and (A-111) is considered, which is of the form:

$$Z = \int_{-1}^{1} \int_{-1}^{1} f(\xi, \eta) \, d\xi \, d\eta \tag{A-122}$$

This integration can be evaluated using:

$$\int_{-1}^{1} \int_{-1}^{1} f(\xi, \eta) \, d\xi \, d\eta = \sum_{i=1}^{n} \sum_{j=1}^{n} H_i H_j f(\xi_i, \eta_j) \tag{A-123}$$

where H_i and H_j are called weighing coefficients, and $f(\xi_i, \eta_j)$ is the value of the function $f(\xi, \eta)$ evaluated at the sampling points specified in the arguments. A very important numerical integration procedure in which both the position of the sampling points and the weighing coefficients have been optimized is Gauss quadrature. If Gauss quadrature is used, a polynomial of order $(2n-1)$ is integrated exactly with only n function evaluations. The sampling points and weights for the interval -1 to $+1$ have been published for various values of n and are available in standard textbooks on numerical analysis, e.g., book by Carnahan et al. (1969).

The matrices to be evaluated by numerical integration are $[K^e]$ (given by equation (A-108)) and $[C^e]$ (given by equation (A-111)). The number of sampling points to be selected (called as order of the numerical integration), depends on the order of the polynomials appearing in the coefficients of these matrices. A first observation in the selection of the order of the numerical integration is that, in theory, using high-enough order, all matrices will be evaluated exactly. On the other hand, using too low an order of integration may lead to inaccurate evaluation of the matrices and, in fact, the problem solution may not be possible. Looking at the order of the polynomials appearing in the coefficients of the matrices $[K^e]$ and $[C^e]$, it is proposed to evaluate these using nine sampling points, (i.e., $n = 3$ in equation (A-123)), as shown in Fig. A-11. If ξ_i or $\eta_i =$

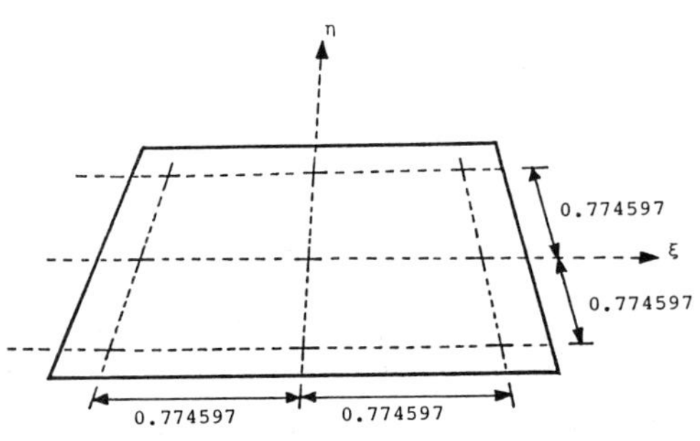

Fig. A-11. Sampling points for the numerical integration over the quadrilateral element.

± 0.774597, then the weighing coefficient $H_i = 0.55557$, and if ξ_i or $\eta_i = 0$, then $H_i = 0.88889$. This is referred to as a 3×3 order of integration.

COMPUTER PROGRAM

This section contains the description and the flow charts for the computer program formulated for predicting pressures and saturations for the case of flow of two immiscible fluids through a horizontal permeable media, which is incompressible. Thus, in its present form, the computer program presented does not incorporate gravitational flow effects and also does not consider the media porosity to be a function of phase pressures. The latter can be conveniently added to the program, however, if one desires, by adding the logic to generate the appropriate k_{ij}'s and C_{ij}'s for each element, as described in equations (A-72), (A-73), (A-77) and (A-79) before assembling the system matrices $[K]$ and $[C]$. If gravitational flow occurs, then the coding to generate the vector $\{Q_g^e\}$ for each element, as described in equations (A-83) and (A-114) for triangular and quadrilateral element, respectively, and the logic to assemble these element vectors to give the system vector $\{Q_g\}$ of equation (A-36) has to be added.

Program notation, preparation of data, and a brief description of the program organization is presented in this section. The flow chart of the main program is also presented. A complete listing of the main program and its subroutines is presented at the end of this Appendix. The application of the program to solve a numerical problem is presented in the next section.

Program notation

The notation used in the computer program is summarized in this subsection. For convenience, items are listed in alphabetical order. A symbol in the list followed by parentheses ( ) denotes a subscripted variable with a single subscript, i.e., a vector. A symbol followed by parentheses with a comma (,) denotes a subscripted variable with double subscripts, i.e., a matrix. The type of variable (integer or real) is implied by the default rule of FORTRAN, i.e., variables which begin with letters, I, J, K, L, M, and N are integers, whereas, all others are assumed to be real (i.e., with decimals).

In the computer program, the governing flow equation for the element, equation (A-31), is denoted as follows:

$$[\mathrm{EK}]\{\mathrm{EDS}\} + [\mathrm{EC}]\{\mathrm{EDDS}\} = \{\mathrm{EQU}\} \tag{A-124}$$

Similarly, the governing flow equation for the system, equation (A-36), is denoted as follows:

$$[\mathrm{SK}]\{\mathrm{DS}\} + [\mathrm{SC}]\{\mathrm{DDS}\} = \{\mathrm{SQU}\} \tag{A-125}$$

Thus, matrix names beginning with the letter E or S are used to indicate that it pertains to an element or the system, respectively.

For an element with n number of node points, the order of matrix [EK] will be $(2n \times 2n)$. If pressure nodal value for all the node points are numbered first, followed by the saturation nodal value for all the node points, then the matrix [EK], for both the triangular and quadrilateral element, can be partitioned as follows:

$$[\mathrm{EK}] = \left[\begin{array}{c|c} k_{11}[\mathrm{D}] & 0 \\ \hline k_{21}[\mathrm{D}] & k_{22}[\mathrm{D}] \end{array}\right] \tag{A-126}$$

where [D] is a square matrix of order $(n \times n)$. The k_{ij}'s have been defined earlier in equations (A-72) and (A-73) for the two optional choices of defining the nodal values, respectively. It should be noted that matrix [D] for an element will not change with time, whereas the multiplying factors k_{ij}'s change with the change in nodal saturation values. So at every time step, and even in every iteration cycle in a time step, the k_{ij}'s are updated, but, matrix [D] for every element has to be computed once. Thus, using the appropriate values of the multiplying factors k_{ij}'s and matrix [D], the updated element matrix [EK] can be computed at any instant. In the computer program, the matrix [D] for each element is generated and stored as a row in matrix [A1], in which the row number indicates the element number. Thus, in matrix [A1], the number of rows is equal to the total number of elements and the number of columns is equal to $2n$. After the first time step, knowing the element number, the element matrix [D] is extracted from matrix [A1] and is used to generate the element matrix [EK].

In the previous paragraph it can be seen that the values of the factors k_{11}, k_{21}, and k_{22} have to be computed at the instant when the element matrix [EK] is to be computed. These equations require the expressions for relative permeabilities of the two phases. The expressions for relative permeabilities used in the computer program to compute the k_{ij}'s are as follows:

$$k_{\mathrm{rw}} = S_{\mathrm{w}}^2 \tag{A-127a}$$

$$k_{\mathrm{ro}} = (1 - S_{\mathrm{w}})^2 \tag{A-127b}$$

The gradient of the capillary pressure versus water saturation curve (i.e., $\partial p_{\mathrm{c}}/\partial S_{\mathrm{w}}$) is assumed to be a constant and is treated as input data. So if the expressions for k_{ij}'s and $\partial p_{\mathrm{c}}/\partial S_{\mathrm{w}}$ change for the problem to be solved, then accordingly changes will have to be made in the computer program presented.

For an element with specified boundary flow velocity, the element number and how many sides of this element have a specified flow velocity have to be input. Then for each side, the outward normal velocity per unit length needs to be input for the two phases. Thus, some nomenclature has to be adopted to designate the element sides. In the computer program presented, the following nomenclature is adopted:

(1) For a triangular element:
   (a) Element edge joining node points 2 and 3 is called side 1.
   (b) Element edge joining node points 3 and 1 is called side 2.
   (c) Element edge joining node points 1 and 2 is called side 3.

(2) For a quadrilateral element:
   (a) Element edge joining node points 1 and 2 is called side 1.
   (b) Element edge joining node points 2 and 3 is called side 2.
   (c) Element edge joining node points 3 and 4 is called side 3.
   (d) Element edge joining node points 4 and 1 is called side 4.

The symbols used in the main program, called FLOW, are defined as follows:

Program symbol	*Definition*
ABPERM( )	Absolute permeability values of each material of the domain.
A1(,)	Stores matrix [D] for each element in a row for later use.
B( )	Vector $\{B\}$ of equation (A-44).
D(,)	Matrix used to compute the element matrix [EK] (refer to equation (A-124)).
DS( )	Nodal values predicted at the end of time step using equation (A-41).
DT	Time step in days.
EC(,)	Element matrix $[C^e]$ defined in equation (A-31).
EK(,)	Element matrix $[K^e]$ defined in equation (A-31).
EK11	Value of k_{11} as defined by equations (A-72) or (A-73).
EK21	Value of k_{21} as defined by equations (A-72) or (A-73).
EK22	Value of k_{22} as defined by equations (A-72) or (A-73).
EQU( )	Element vector $\{Q_u^e\}$ defined in equation (A-31).
ESC	Average saturation of water for the element.
F( )	Vector $\{F\}$ of equation (A-44).
ICR( )	Nodal indices or d.o.f. numbers of the element.
IT1	Number of iteration cycles performed within the current time step.
IT2	Number of time steps already performed.
MAX1	Maximum number of iteration cycles specified for a time step.
MAX2	Maximum number of time steps to cover the total time domain.
MCHECK	A counter used to check if a node point of the element has specified pressure and/or saturation which is to be held constant at a specified value.

NC(,)	Node point indices of the element. A row reserved for each element.
NCOD( )	Node point indices at which pressure and/or saturation has to be held constant at a specified value. If both pressure and saturation are specified at a node point, then that node point index is repeated twice.
ND( )	Contains indices to indicate whether pressure or saturation is specified at a node point. ND(I) = 1 if pressure is specified, and ND(I) = 2 if saturation is specified.
NDOF	Total number of unknown nodal values or d.o.f. NDOF = NUMNP * 2.
NE	Element number.
NELTYP	Code indicating the type of element. NELTYP = 3 for triangular element. NELTYP = 4 for quadrilateral element.
NF	Number of elements which have specified boundary flow velocity.
NFLOW( )	Contains element numbers which have specified boundary flow velocity.
NM( )	Contains numbers indicating material type to choose appropriate permeability and porosity value of the porous media.
NMAT	Number of material types in the porous media.
NPHASE	Code indicating whether pressure of oil or water is chosen as a nodal variable. NPHASE = 1 if oil pressure is used as a variable. NPHASE = 2 if water pressure is used as a variable.
NRSEN	Number of d.o.f. at which pressure and/or saturation is to be held constant at a specified value.
NS	Side number at which phase flow velocity is specified.
NSIDE	Total number of sides of an element for which the boundary flow velocity is specified.
NT	Total number of d.o.f. of an element.
NUMEL	Total number of elements.
NUMNP	Total number of node points.
OMU	Viscosity of oil.
P	Value of absolute permeability for the element.
PHI( )	Porosity value of each material of the porous media.
PHIEL	Value of porosity for the element.

PRESS	Value of initial water pressure at each node point.
PS	Value of the gradient of the capillary pressure versus water saturation curve (i.e., $\partial p_c / \partial S_w$).
Q(,)	Contains phase velocities normal to element boundary. A row is reserved for each element with specified boundary flow velocity. Number of columns is equal to 6 for triangular element and 8 for quadrilateral element.
SAT	Value of initial water saturation at each node point.
SC(,)	System matrix $[C]$ defined in equation (A-36).
SK(,)	System matrix $[K]$ defined in equation (A-36).
SPEC( )	Contains values of specified pressures and/or saturations which are to be held constant.
SQU( )	System vector $\{Q_u\}$ defined in equation (A-36).
TI	Time at the end of the current time step.
TOLR	Tolerance specified to check convergence for iterations in a time step.
TT	Total time in days.
U( )	Contains the phase velocities of the element with specified boundary flow. This is extracted from $\{Q\}$.
WMU	Viscosity of water.
X( )	Contains x-coordinates of the element node points.
XORD( )	Contains x-coordinates of the system node points.
XS( )	Contains values of nodal pressures and saturations at the beginning of the time step.
Y( )	Contains y-coordinates of the element node points.
YORD( )	Contains y-coordinates of the system node points.

Description of the program

The computer program presented consists of a main program, called FLOW, and eight subroutines, the names and functions of which are briefly described as follows:

Subroutine FAC generates the updated multiplying factors, k_{ij}'s, needed to multiply matrix [D] of the element to generate the element matrix [EK], shown in equation (A-125). To compute the relative permeabilities of the two phases, equations (A-127a) and (A-127b) are used. The gradient of the capillary pressure versus water saturation curve (i.e., $\partial p_c/\partial S_w$) is passed into the subroutine by the main program, where it is read as input data. Equation (A-72) or (A-73) is used to compute the k_{ij}'s depending on whether pressure of oil or water is chosen as the nodal variable. It is called by the main program.

Subroutine TRIAN generates the element matrices $[C^e]$ and $\{Q_u^e\}$ (defined as [EC] and {EQU}, respectively, in the program) as defined in equation (A-31), and matrix [D] as defined in equation (A-126) for a three-node triangular isoparametric element. It is called by the main program.

Subroutine QUAD generates the element matrices $[C^e]$ and $\{Q_u^e\}$ (defined as [EC] and {EQU}, respectively, in the program) as defined in equation (A-31), and matrix [D] as defined in equation (A-126) for a four-node quadrilateral isoparametric element. It is called by the main program.

Subroutine BMTRX generates the derivative operator needed to formulate the Jacobian matrix, $[J]$, as given in equation (A-102). It also computes the value of the determinant of the Jacobian matrix and its inverse. It is called by subroutine QUAD.

Subroutine FORMEK generates the element matrix $[K^e]$ (defined as [EK] in the program) using the appropriate k_{ij}'s and matrix [D], as shown in equation (A-126). It is called by the main program.

Subroutine ASSEM assembles the element matrix, i.e., $[K^e]$ or $[C^e]$, into the system matrix in appropriate rows and columns.

Subroutine ASSEMF assembles the element vector $\{Q_u^e\}$ (defined as {EQU} in the program) in appropriate rows.

Subroutine SOLVE solves equation (A-51) for vector $\{B\}_{\tau=\Delta t}$ using the Gauss elimination method.

In order to keep track of the value of pressure and saturation at a node point, it is convenient to arrange the vector containing the nodal values, such that successively at each node point the pressure value is followed by the saturation value. In order to achieve this, the element matrices, [EK], [EC] and {EQU}, have to be rearranged before assembling them into the system matrices, [SK], [SC] and {SQU}, respectively. For matrices [EC] and {EQU} this is done in the respective element subroutines (i.e., in TRIAN and QUAD). The matrix [EK] is rearranged in subroutine FORMEK before it is output from this subroutine.

The main program is divided into the following five steps:

Step I: Input and printing of the problem data. The data to be input is described in Table A-I. It is assumed that the data are given as a group of lines typed on a terminal.

Step II: Formulation of element matrices, $[K^e]$, $[C^e]$ and $\{Q_u^e\}$ (coded as [EK], [EC] and {EQU}). First, the average saturation of water (S_w) for the element is computed from the known node point saturation values. Using this value of S_w for the element, the factors k_{11}, k_{21} and k_{22} are computed by calling subroutine FAC. Next subroutines TRIAN and QUAD are called to formulate the element matrices [D], [EC] and {EQU}. Finally, using the computed values of k_{12}, k_{21},

TABLE A-I

Preparation of data for the problem

Data	Number of lines	Items on data lines and format
Problem heading	1	Heading describing the problem in 80 characters Format (20A4)
Control data	1	NUMNP, NUMEL, NPHASE, PRESS, SAT, NRESN, NMAT, NF Format (3I5, 2E10.2, 3I5)
Node point data	NUMNP	NP, XORD(NP), YORD(NP) Format (I5, 2F10.3)
Element data	NUMEL	NE, NM(NE), (NC(NE, J1), J1 = 1,4) Format (6I4)
Data for elements in which pressure and/or saturation is to be held constant at the node points	NRESN	NCOD(I), ND(I), SPEC(I) Format (2I5, E10.3)
Material properties of the porous media	NMAT	PHI(I), ABPERM(I) Format (2E15.4)
Data for elements with specified boundary flow velocity (these data lines are skipped if NF = 0)	NF * SIDE	NFLOW(I), NSIDE Format (2I5) NS, Q(I, 2 * NS − 1), Q(I, 2 * NS) Format (I5, 2F10.3)
Liquid properties and gradient of capillary pressure versus water saturation curve	1	OMU, WMU, PS Format (3E10.2)
Data for time iterations	1	MAX1, MAX2, DT, TT, TOLR Format (2I10, 2E10.3, 1F10.3)

k_{22} and matrix [D] for the element, subroutine FORMEK is called to compute the element matrix [EK].

Step III: Assembly of the system matrices, [K], [C] and $\{Q_u\}$ (coded as [SK], [SC] and {SQU}). The element matrices [EK] and [EC] are assembled into the system matrices [SK] and [SC], respectively, by calling the subroutine ASSEM. Then the element vector {EQU} is assembled into system vector {SQU} by calling subroutine ASSEMF. Both the subroutines use vector {ICR} for the element to locate the rows and columns of the system matrices in which the element matrices have to be stored. Steps II and III are repeated for each element.

Step IV: Solution of the system non-linear equation (i.e., equation (A-44)) iteratively. In the computer program the following steps are followed:

(1) Transform the system non-linear equation (equation (A-125)):

$$[\mathrm{SK}]\{\mathrm{DS}\} + [\mathrm{SC}]\{\mathrm{DDS}\} = \{\mathrm{SQU}\} \quad \text{(A-128)}$$

into a system of linear equations by substituting:

$$\{\mathrm{DS}\} = \{\mathrm{XS}\} + \mathrm{DT} * \{\mathrm{B}\} \quad \text{(A-129a)}$$

$$\{\mathrm{DDS}\} = \{\mathrm{B}\} \quad \text{(A-129b)}$$

where {XS} contains the known nodal pressures and saturations at the beginning of the time step. This gives:

$$(\mathrm{DT} * [\mathrm{SK}] + [\mathrm{SC}])\{\mathrm{B}\} = \{\mathrm{SQU}\} - [\mathrm{SK}]\{\mathrm{XS}\} \quad \text{(A-130)}$$

The matrix on the left side of equation (A-130) is stored as matrix [SK] and the matrix on the right side of equation (A-130) is stored as vector {F} in the computer program, which gives:

$$[\mathrm{SK}]\{\mathrm{B}\} = \{\mathrm{F}\} \quad \text{(A-131)}$$

(2) Modify [SK] and {F} to account for node points at which pressure and/or saturation values are to be held constant at a specified value throughout the solution. To do this, first a counter MCHECK is generated, using the input vector {NCOD} which gives information as to which nodal d.o.f. is to be constrained to a specified value. Say the Nth d.o.f. is to be constrained, then the matrix [SK] and vector {F} are modified as follows: SK(N,N) = 1.0; the off-diagonal elements of the Nth row and column of [SK] are made equal to zero; and F(N) = 0.0.

(3) Solve the resulting system linear equation for vector {B} by calling subroutine SOLVE.

(4) Save the previous cycle's nodal values of vector {DS} in vector {DS1}.

(5) Compute the new nodal values for vector {DS} by substituting the new values of vector {B} into equation (A-129a).

(6) Check for convergence by computing:

$$\mathrm{X2} = |(\mathrm{DS1(I)} - \mathrm{DS(I)})/\mathrm{DS(I)}| \quad \text{for } \mathrm{I} = 1, \ldots, \mathrm{NDOF} \quad \text{(A-132)}$$

If X2 is less than TOLR for each I, then vector {DS} is assumed to have converged. If not, then if the number of iterations performed (i.e., IT1) is less than MAX1, then the following steps are performed for each element:

(a) Average saturation of water (S_w) is computed.
(b) Call subroutine FAC to compute the relative permeabilities.
(c) Call subroutine FORMEK to compute updated [EK].

(d) Call subroutine ASSEM to store [EK] into appropriate rows and columns of matrix [SK].

Now formulate new {F} and [SK] using

$$\{F\} = \{SQU\} - [SK]\{XS\} \tag{A-133a}$$

$$[SK] = DT * [SK] + [SC] \tag{A-133b}$$

and repeat steps 2 to 6.

(7) Start a new time step by modifying the initial values as {XS} = {DS}. Then compute new nodal values of vector {DS} from equation (A-129a) and making use of the latest values computed for the coefficients of vector {B}. Repeat step (6) to formulate the matrix [SK]. Finally repeat steps 2 to 6 until the number of time steps (i.e., IT2) is equal to MAX2.

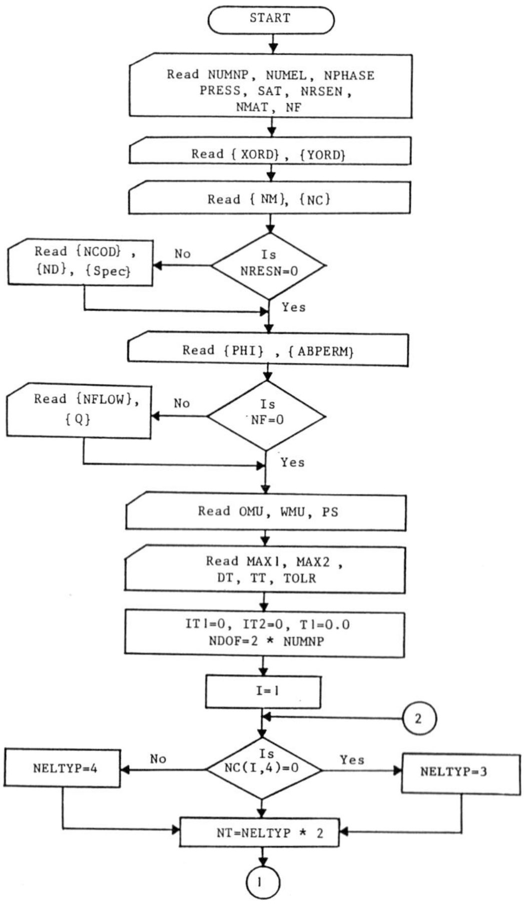

Fig. A-12. Flow chart of the main program FLOW.

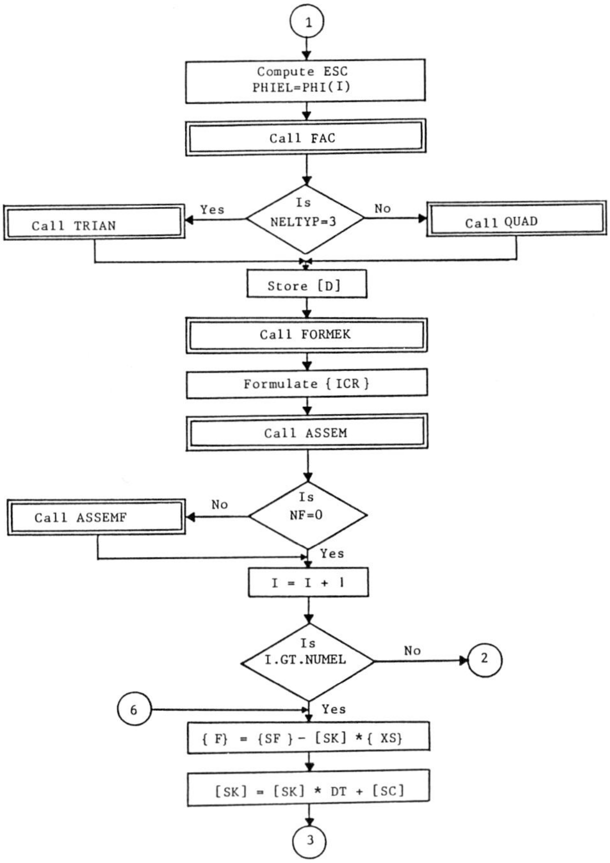

Fig. A-12 (continued).

Step V: Print results. For each node, its coordinates and values of pressure and saturation of water are printed.

The flow chart for the main computer program, FLOW, is given in Fig. A-12. The listing of this program and all its subroutines are given at the end of this Appendix.

APPLICATION

It is not uncommon to encounter reservoirs which involve complicated shapes and characteristics. Examples include, but not limited to, reservoirs containing faults, regions of relatively different material properties, boundaries through which reservoirs communicate with nearby reservoirs through a permeable zone, etc. Consequently, determining policies for efficiently producing the contents of

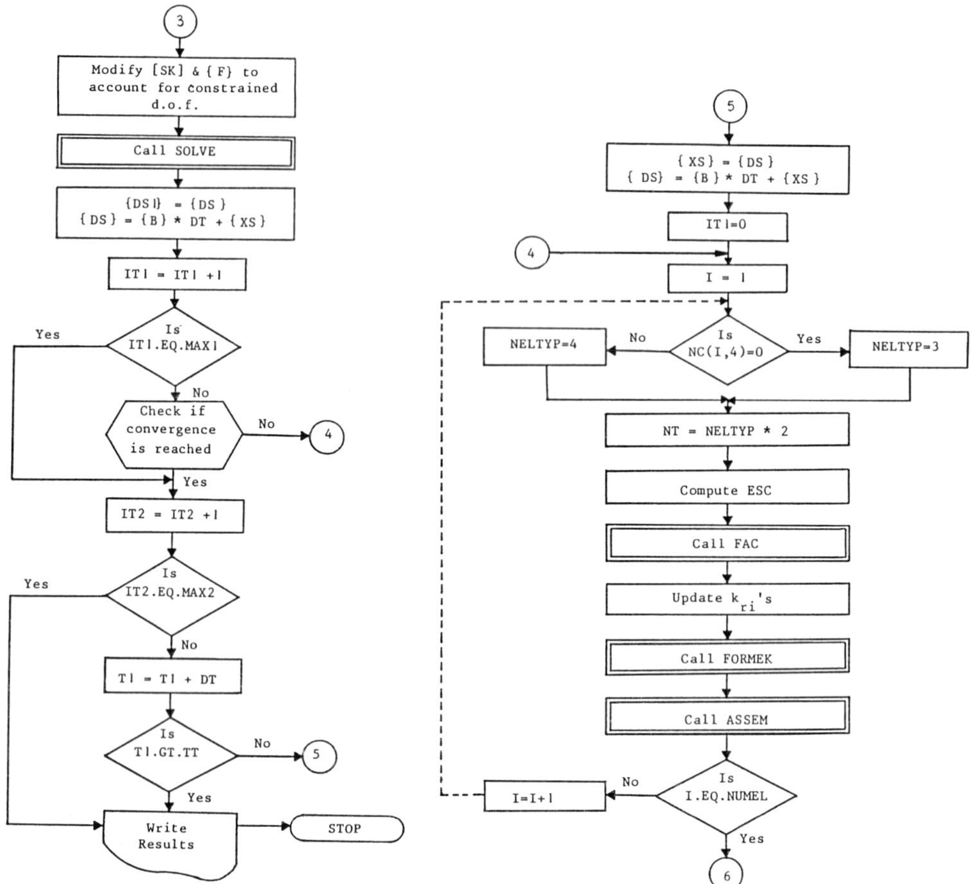

Fig. A-12 (continued).

such reservoirs is a challenging engineering problem. This type of a reservoir may be effectively simulated by the finite element method, as demonstrated in this section.

In this section, the procedure presented in the preceding sections is applied to model a reservoir which contains faults. A study is carried out to simulate a waterflood process as a way of secondary production in such a way so as to take advantage of the existence of the faults by alternating the role of the injection and production wells to improve oil recovery. A typical example of such an application has been presented by Dorsey and Brinkley (1975). For example, consider a reservoir located between two faults, as shown in Fig. A-13. Assume that the reservoir is horizontal and with dimensions and locations of the wells as shown in this figure. The reservoir data adopted for the problem is by Gulbrandsen and Wille (1985) and is shown in Table A-II. All data are assumed in darcy

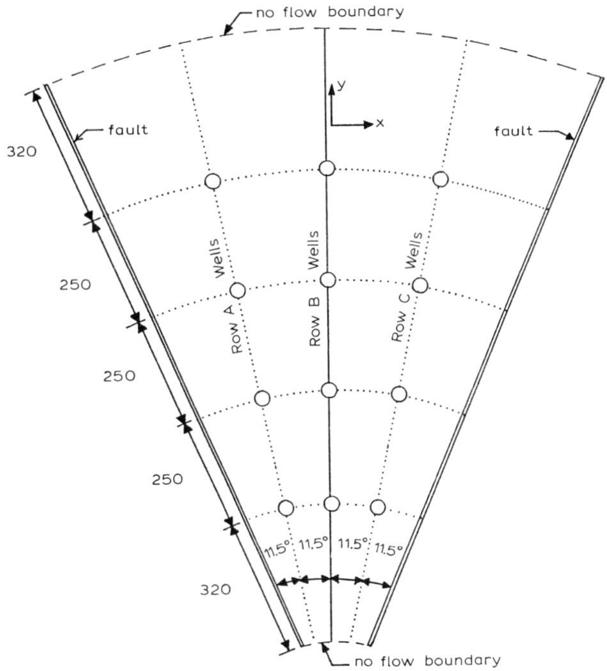

Fig. A-13. A reservoir bounded by two faults and the wells located in the reservoir.

units. The faults are assumed to be sufficiently long and the wells are located to form three rows (shown as A, B and C in Fig. A-13), each consisting of four wells. It is assumed that all the wells have been producing until the primary oil production has been completed, and the wells have been shut in. After the reservoir pressure has been stabilized, water injection is started, initially, through the wells in row B and the wells nearby the faults, i.e., the wells in rows A and C are producing oil (see Fig. A-13), as the oil is pushed by water against the faults. Following a period of production in this way, the water injection is switched over to rows A and C, while producing oil through row B wells. Operation continues by periodically alternating the role of the rows A and C wells with those of the row B wells as being injection or production wells. In the study presented the periods are taken as 120 days and the overall operation is continued by switching the locations of the production and the injection wells four times during a period. For a numerical solution, no flow boundaries are visualized as shown in Fig. A-13 by dashed lines, which are assumed to be sufficiently far from the wells (taken as 320 m away), such that the flow is not disturbed for all practical purposes. Injection or production rates are considered equal for all the wells at values given in Table A-II.

Numerical solution for the problem is obtained by using the computer program. The four-node isoparametric quadrilateral element is used to model the

TABLE A-II

Data used in the waterflood problem

Parameter	Value
Permeability, D	0.25
Porosity, fraction	0.25
Oil viscosity, cP	3
Water viscosity, cP	1.0
Initial water saturation, fraction	0.1
Initial reservoir pressure, atm	1
Injection wellbore pressure, atm	5
Production wellbore pressure, atm	1
$\partial p_c / \partial S_w$, atm	−100

Relative permeability relationships

$k_{rw} = S_w^2$

$k_{ro} = (1 - S_w)^2$

where subscript "o" refers to oil phase, and "w" to water phase

Discretization data

$\Delta t = 10$ days

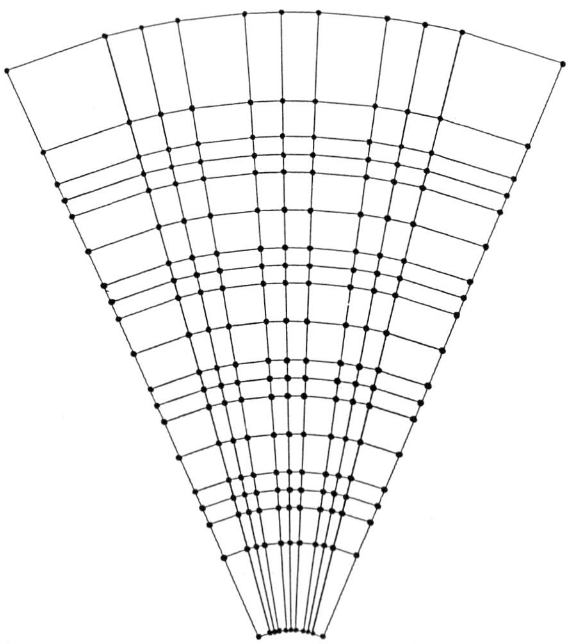

Fig. A-14. Finite element mesh of the reservoir bounded by two faults and wells located in the reservoir.

TABLE A-III

Additional statements added to the computer program given at the end of this Appendix to solve the reservoir problem

```
Before line # 158 add the following:
       IF(NCH.EQ.1) THEN

After line # 169 add
       ENDIF
       NCH = 0

After line # 203 add the following:
       M = IT2/12
       M1= IT2-M*12
       IF(M1.EQ.0) THEN
       NCH = 1
       WRITE(6,*) 'PRESSURE AND SATURATION AFTER', IT2, 'TIME STEP'
       WRITE(6,714)
       WRITE(6,716)
       DO 255 I= 1, NDOF
       N1 = I/2
       N2 = I - N1*2
       IF(N2.EQ.0) GO TO 255
       N2 = I - N1*2
       IF(N2.EQ.0) GO TO 255
       N2 = (I+1)/2
       WRITE(6,717) N2,XORD(N2),YORD(N2),DS(I),DS(I+1)
  255  CONTINUE
       ENDIF

After line # 207 add the following:
       IF(NCH.EQ.1) THEN
       READ(5,806) TITLE
  710  FORMAT(1I10)
       READ(5,710) NRESN
       DO 265 J = 1, NRESN
       READ(5,703) NCOD(J), ND(J), SPEC(J)
       MDOF = (NCOD(J)-1)*2 + ND(J)
       XS(MDOF) = SPEC(J)
       DS(MDOF) = B(MDOF)*DT + XS(MDOF)
  265  CONTINUE
       ENDIF
```

problem. The finite element mesh used is shown in Fig. A-14. To solve the problem, the program was modified to account for switching of the production wells in rows A and C with the injection wells in row B after every period. The modifications required are given in Table A-III.

For every period the program was used in cycles as follows:

Cycle 1: For the first period, the initial conditions specified in Table A-II are used, and the complete program is executed. The pressure and saturation at the node points corresponding to the well locations in row B are held constant to represent injection wells, whereas the pressure and saturation at the node points corresponding to the well locations in rows A and C are held constant to represent production wells. The results are printed every half period.

Cycle 2: For the second period, the initial conditions used are the values of nodal pressures and saturations predicted at the end of the previous cycle. The system matrices [SC], {SQU}, {DS}, and {B} obtained at the end of the previous cycle are used to start the analysis for this cycle. So these matrices do not have to be computed at the initial time of this cycle, as done in the previous cycle. The only

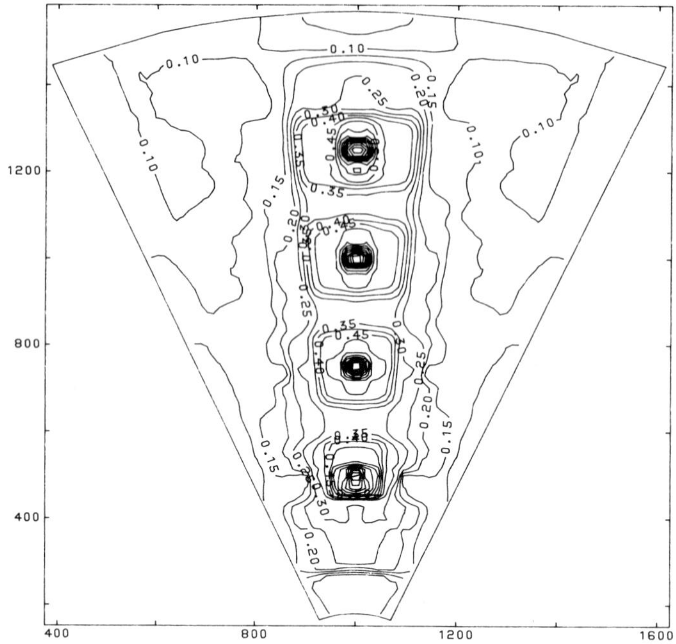

Fig. A-15. Iso-saturation lines upon the completion of the one-half period (injection through wells in row B and production through wells in rows A and C).

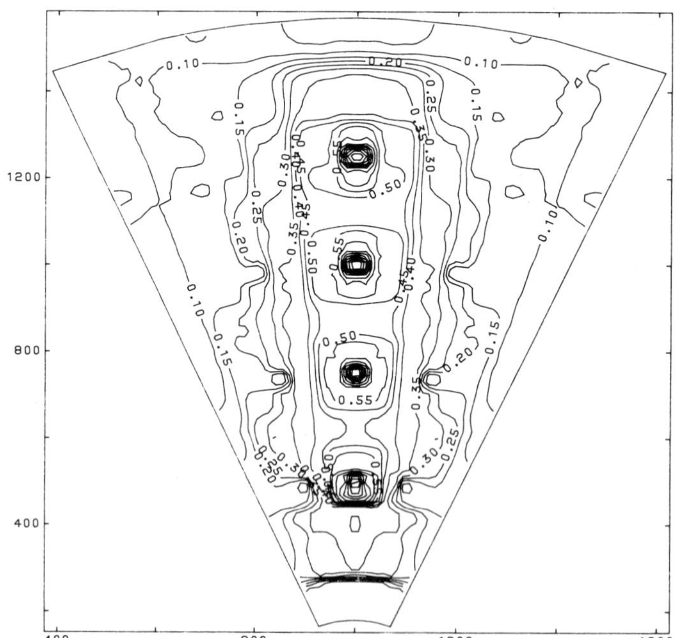

Fig. A-16. Iso-saturation lines upon the completion of the first period (injection through wells in row B and production through wells in rows A and C).

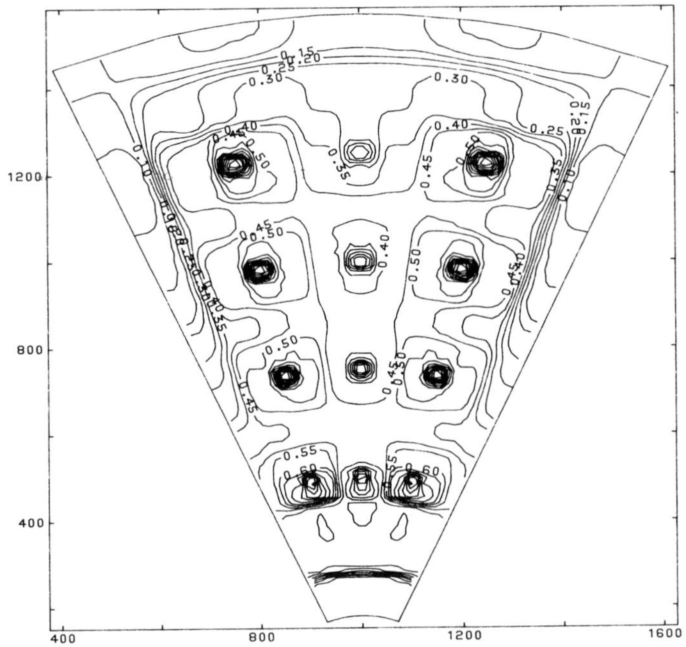

Fig. A-17. Iso-saturation lines upon the completion of the one and one-half period (injection through wells in rows A and C and production through wells in row B).

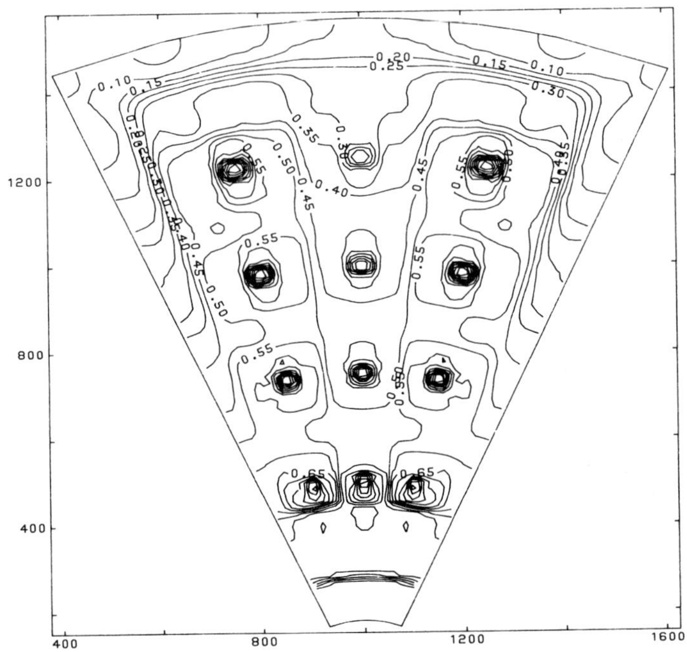

Fig. A-18. Iso-saturation lines upon the completion of the second period (injection through wells in rows A and C and production through wells in row B).

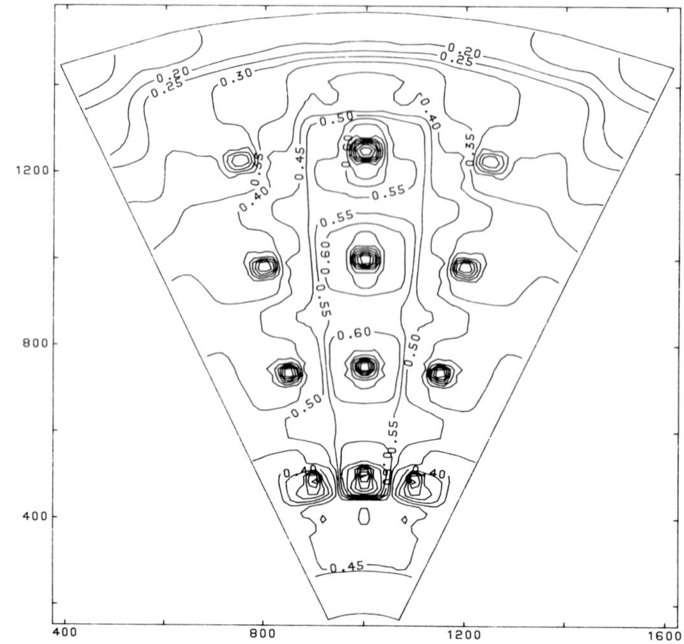

Fig. A-19. Iso-saturation lines upon the completion of the second and one-half period (injection through wells in row B and production through wells in rows A and C).

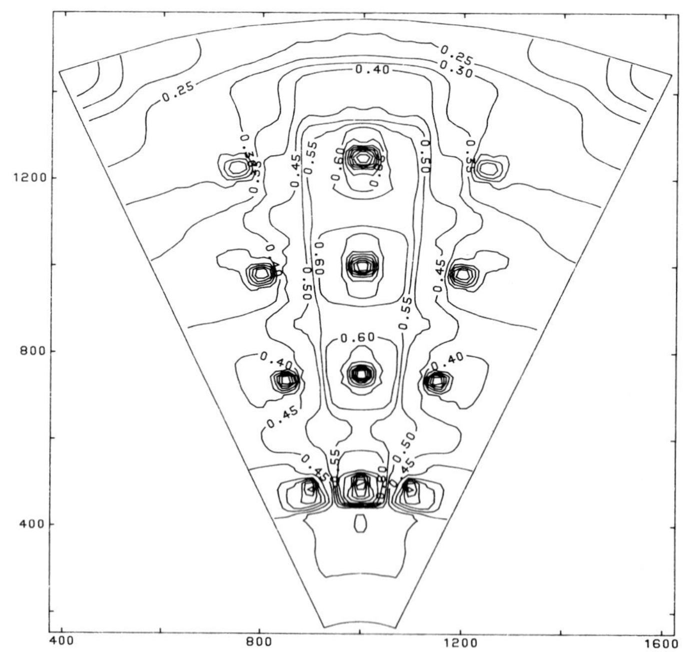

Fig. A-20. Iso-saturation lines upon the completion of the third period (injection through wells in row B and production through wells in rows A and C).

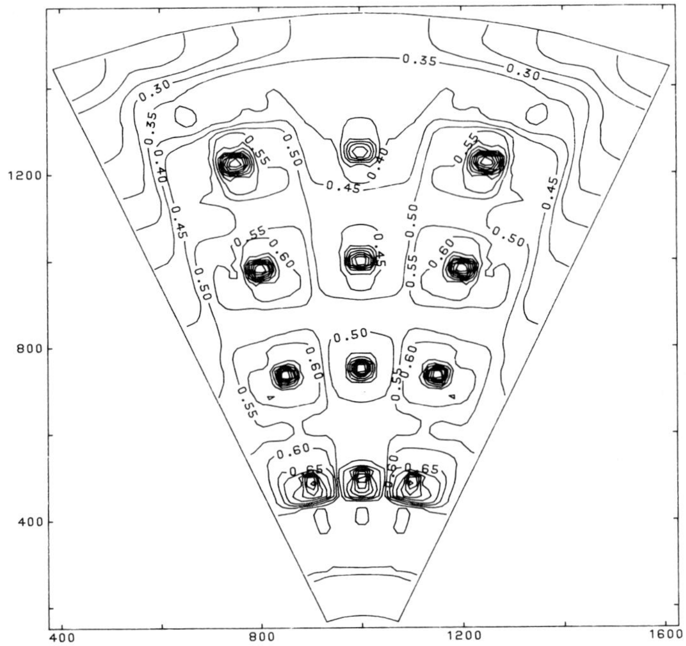

Fig. A-21. Iso-saturation lines upon the completion of the third and one-half period (injection through wells in rows A and C and production through wells in row B).

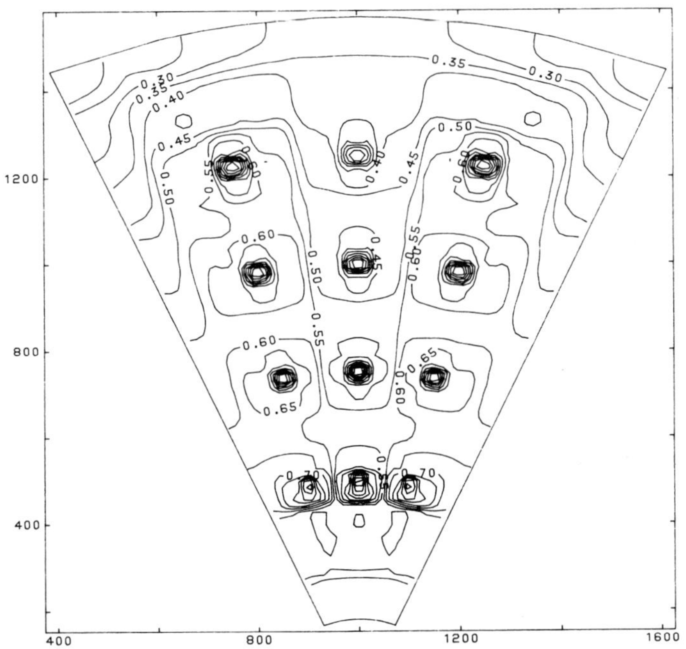

Fig. A-22. Iso-saturation lines upon the completion of the fourth period (injection through wells in rows A and C and production through wells in row B).

input data for this cycle will be the program heading, value of the parameter NRSEN, and the values of the coefficients of the vectors {NCOD}, {ND}, and {SPEC}. These have to be input to account for the situation that now the node points corresponding to the well locations in rows A and C have pressure and saturation held constant to represent injection wells, whereas the node points corresponding to the well locations in row B have pressure and saturation held constant to represent production wells. The system matrix [SK] for this cycle is formulated by performing the following steps for each element:

(1) Average saturation of water (S_w) is computed.

(2) Call subroutine FAC to compute the relative permeabilities.

(3) Call subroutine FORMEK to compute updated [EK].

(4) Call subroutine ASSEM to store [EK] into appropriate rows and columns of matrix [SK].

Now all information is known to perform the solution of the system non-linear equations iteratively for this cycle, and so the analysis is made to directly go to step IV of the computer program. The results for this cycle are printed at every half period.

Cycle 3: The logic for this cycle will be similar to the previous cycle (i.e., cycle 2). Thus, the logic of cycle 2 will be repeated for the third period taking the value of the parameters NRSEN and the vectors {NCOD}, {ND} and {SPEC} to be the same as taken for cycle 1, thus simulating the condition that the wells along row B are injection wells, whereas the wells along rows A and C are production wells.

Cycle 4: Again, the logic of cycle 2 will be repeated for the fourth period taking the value of the parameters NRSEN and the vectors {NCOD}, {ND} and {SPEC} to be the same as taken for cycle 2, thus simulating the condition that the wells along rows A and C are injection wells, whereas the wells along row B are production wells.

The solutions obtained at every half period are presented as Figs. A-15 to A-22 in the form of iso-saturation lines to cover an overall operation of four switching periods.

REFERENCES

Carnahan, B., Luther, H.A. and Wilkes, J.O., 1969. *Applied Numerical Methods*. John Wiley and Sons, New York, N.Y., 604 pp.

Collins, R.E., 1961. *Flow of Fluids Through Porous Materials*. PennWell Books, Tulsa, Okla., 270 pp.

Dorsey, J.B. and Brinkley, T.W., 1975. Performance review of Shoats Creek Unit vaporizing gas-drive project. In: *Field Case Histories, Oil and Gas Reservoirs, SPE Reprint Ser.*, 4a: 92–98.

Gulbrandsen, S. and Wille, S.O., 1985. A finite element formulation of the two-phase flow equations. *Proc. 8th SPE Symp. on Reservoir Simulation, Dallas, Texas, February 10–13, 1985, SPE Pap.*, 13516: 201–206.

Zienkiewicz, D.C., 1977. *The Finite Element Method*. McGraw-Hill, Oxford, 3rd ed., 787 pp.

PROGRAM LISTING

MAIN PROGRAM: "FLOW" LISTING

```
***********************************************************************
C
C                                    FLOW
C
***********************************************************************
C
***********************************************************************
*                                                                     *
*  THIS PROGRAM CAN BE USED TO PREDICT AT INCREMENTAL TIME STEPS      *
*  THE PRESSURES AND SATURATIONS AT DISCRETE POINTS IN THE            *
*  PERMEABLE DOMAIN FOR TWO PHASE IMMISIBLE FLUID FLOW                *
*  PROBLEMS USING FINITE ELEMENT METHOD. ONLY HORIZONTAL FLOW IS      *
*  CONSIDERED. THE FLUIDS AND ROCK ARE ASSUMED TO BE INCOMPRESSIBLE.  *
*  THE PROGRAM HAS THE OPTION OF USING EITHER TWO DIMENSIONAL 3 NODE  *
*  ISOPARAMETRIC TRIANGULAR FINITE ELEMENT OR A 4 NODE ISOPARAMETRIC  *
*  QUADRILATERAL FINITE ELEMENT.                                      *
*                                                                     *
***********************************************************************
C
      IMPLICIT REAL *8 (A-H,O-Z)
      DIMENSION XORD(100),YORD(100),X(4),Y(4),EK(8,8)
      DIMENSION EC(8,8),SK(200,201),SC(200,201),NC(100,4),ND(100)
      DIMENSION SQU(200),ICR(8),EQU(8),SPEC(100),A1(100,16),D(4,4)
      DIMENSION XS(200),F(200),DS(200),DS1(200)
      DIMENSION NCOD(100),PHI(5),ABPERM(5),NM(100)
      DIMENSION NFLOW(10),B(200),Q(10,8),U(8),TITLE(20)
C
***********************************************************************
*   THIS PROGRAM IS DIMENSIONED FOR THE FOLLOWING MAXIMUM VALUES     *
*                                                                     *
* PROGRAM          MAXIMUM VALUE                                      *
* SYMBOL                                                              *
*                                                                     *
* NUMNP            100                                                *
* NUMEL            100                                                *
* NRESN            100                                                *
* NMAT             5                                                  *
* NF               10                                                 *
***********************************************************************
C
***********************************************************************
C
C                        STEP I:  INPUT OF DATA
C
***********************************************************************
C
C
C  PRINT HEADER PAGE FOR OUTPUT
C
  806 FORMAT(20A4)
      READ(5,806) TITLE
      WRITE(6,806) TITLE
C
C  INPUT MISCELANOUS CONTROL DATA
C
      WRITE(6,800)
  800 FORMAT(1H1,1X,10HINPUT DATA/)
      READ(5,700) NUMNP,NUMEL,NPHASE,PRESS,SAT,NRESN,NMAT,NF
  700 FORMAT(3I5,2E10.2,3I5)
      WRITE(6,801) NUMNP,NUMEL,NPHASE,NRESN,NMAT,NF
```

```
  801   FORMAT (1H0,2X,6HNUMNP=,I3,2X,6HNUMEL=,I3,2X,7HNPHASE=,I3,2X,
     1  6HNRESN=,I2,2X,5HNMAT=,I2,2X,3HNF=,I2)
        WRITE(6,802)
  802   FORMAT(1H1,1X,33HNODAL DATA AND INITIAL CONDITIONS/)
        WRITE(6,714)
        WRITE(6,716)
  714   FORMAT(3X,'NODE',4X,'X',7X,'Y',10X,'PRESSURE',8X,'SATURATION')
  716   FORMAT(8X,'COORD:',3X,'COORD:')
  701   FORMAT(I5,2F10.3)
        DO 10 I=1,NUMNP
C
C  INPUT NODE POINT NUMBER AND X,Y, COORDINATES
C
        READ(5,701) NP,XORD(NP),YORD(NP)
C
C  INITIALIZE NODAL PRESSURE AND SATURATION AT TIME T=0.0
C
        XS(I*2-1)=PRESS
        XS(I*2)=SAT
C
C  PRINT NODE POINT DATA AND INITIAL NODAL PRESSURE AND SATURATION
C
        WRITE(6,803) NP,XORD(NP),YORD(NP),(XS((NP-1)*2+K),K=1,2)
  803   FORMAT(I5,2F9.3,1E15.3,5X,1F10.3)
   10   CONTINUE
        WRITE(6,804)
  804   FORMAT(1H1,1X,12HELEMENT DATA/)
C
C   FOR EACH ELEMENT READ ELEMENT NUMBER, MATERIAL PROPERTY NUMBER
C   AND NODE POINTS NUMBERS I,J,K,L IN THE COUNTER CLOCKWISE DIRECTION.
C   FOR TRIANGULAR ELEMENT L = 0
C
  702   FORMAT (6I4)
        DO 20 I=1,NUMEL
        READ(5,702) NE,NM(NE),(NC(NE,J1),J1=1,4)
C
C  PRINT ELEMENT INPUT DATA
C
        WRITE(6,702) NE,NM(NE),(NC(NE,J1),J1=1,4)
   20   CONTINUE
C
C  INPUT DATA FOR NODE POINTS AT WHICH SPECIFIED PRESSURE AND
C  SATURATION HAVE TO BE HELD CONSTANT AT ALL TIMES.
C
        IF(NRESN.GT.0) THEN
  703   FORMAT(2I5,E10.3)
C
C  FOR CONSTRAINED NODES INPUT DATA FOR NODE POINT NUMBER SPECIFY
C  IF PRESSURE AND SATURATION ARE GIVEN AND THE VALUE OF THE
C  OF THE SPECIFIED QUANTITY
C     NOTE: ND(I) = 1, IF PRESSURE IS SPECIFIED
C           ND(I) = 2, IF SATURATION IS SPECIFIED
C           IF BOTH PRESSURE AND SATURATION IS SPECIFIED THEN THAT NODE
C           POINT DATA IS REPEATED: ONCE WITH ND(I) = 1 AND SECONE TIME
C           WITH ND(I) = 2
C
        DO 30 I=1,NRESN
        READ(5,703) NCOD(I),ND(I),SPEC(I)
        WRITE(6,703) NCOD(I),ND(I),SPEC(I)
   30   CONTINUE
        ENDIF
C
C  INPUT AND PRINT POROUS MEDIA MATERIAL PROPERTIES: POROSITY AND
C  ABSOLUTE PERMEABILITY
C
```

```
  704   FORMAT(2E15.4)
  805   FORMAT(1H0,1X,16HMATERIAL NUMBER=,1I3,2X,4HPHI=,1E12.2,2X,
     1  5HPERM=,1E12.2)
        DO 40 I=1,NMAT
        READ(5,704) PHI(I),ABPERM(I)
        WRITE(6,805) I,PHI(I),ABPERM(I)
   40   CONTINUE
C
C  INPUT DATA FOR ELEMENTS WITH SPECIFIED BOUNDARY FLOW
C  FLOW OUT OF AN ELEMENT IS POSITIVE. FOR EACH ELEMENT BOUNDARY
C  READ THE FLOW OF THE TWO PHASES:
C       IF PRESSURE OF OIL IS CHOSEN AS THE UNKNOWN THEN SPECIFY FLOW
C       OF OIL AND THEN WATER, OR IF PRESSURE OF WATER IS CHOSEN AS THE
C       VARIABLE SPECIFY FLOW OF WATER AND THEN OIL
C
        IF(NF.GT.0) THEN
        DO 45 I=1,NF
        DO 45 J=1,8
        Q(I,J)=0.0
   45   CONTINUE
  706   FORMAT(2I5)
  707   FORMAT(I5,2F10.3)
        DO 50 I=1,NF
C
C  INPUT AND PRINT ELEMENT NUMBER AND NUMBER OF BOUNDARIES WHICH HAVE
C  FLOW
C
        READ(5,706) NFLOW(I), NSIDE
        WRITE(6,706) NFLOW(I), NSIDE
        DO 60 J=1,NSIDE
C
C  FOR EACH BOUNDARRY INPUT AND PRINT THE TWO PHASE VELOCITIES SPECIFIED
C    NOTE: FOR TRIANGULAR ELEMENT
C                 SIDE 1 JOINS NODE POINT 2 AND 3
C                 SIDE 2 JOINS NODE POINT 1 AND 3
C                 SIDE 3 JOINS NODE POINT 1 AND 2
C    NOTE: FOR QUADRILATERAL ELEMENT
C                 SIDE 1 JOINS NODE POINT 1 AND 2
C                 SIDE 2 JOINS NODE POINT 2 AND 3
C                 SIDE 3 JOINS NODE POINT 3 AND 4
C                 SIDE 4 JOINS NODE POINT 4 AND 1
C
        READ(5,707) NS,Q(I,2*NS-1),Q(I,2*NS)
        WRITE(5,707) NS,Q(I,2*NS-1),Q(I,2*NS)
   60   CONTINUE
   50   CONTINUE
        ENDIF
C
C  INPUT AND PRINT LIQUID PROPERTIES OF THE TWO PHASES:
C  VISCOSITY AND CAPILLARY PRESSURE GRADIENT
C
  708   FORMAT(3E10.2)
   52   READ(5,708) OMU,WMU,PS
        WRITE(6,708) OMU,WMU,PS
C
C  INPUT AND PRINT DATA REQUIRED TO PERFORM ITERATIONS: MAXIMUM NUMBER
C  OF CYCLES FOR ITERATIONS, MAXIMUM NUMBER OF TIME STEPS, TIME STEP,
C  TOTAL TIME AND TOLARANCE
C
  709   FORMAT(2I10,2E10.3,1F10.3)
        READ(5,709) MAX1,MAX2,DT,TT,TOLR
  809   FORMAT(1H0,3X,5HMAX1=,I5,3X,5HMAX2=,I5,3X,3HDT=,F8.1,3X,3HTO=,
     1  E10.3,3X,5HTOLR=,F6.3)
        WRITE(6,809) MAX1,MAX2,DT,TT,TOLR
        WRITE(6,*) 'END OF INPUT DATA'
```

```
C
C  INITIALIZE TIME AND SYSTEM MATRICES: SK, SC, SQU
C
      TI=0.0
      IT1=0.0
      IT2=0.0
      NDOF=NUMNP*2
C
C     INITIALIZE SK SC & SF MATRICES
C
      DO  80 I=1,NDOF
      SQU(I)=0.0
      DS(I)=0.0
      DO  80 J=1,NDOF
      SK(I,J)=0.0
      SC(I,J)=0.0
   80 CONTINUE
C
C *******************************************************************
C
C            STEP II: GENERATION OF ELEMENT MATRICES: EK,EC AND EQU
C
C *******************************************************************
C
      DO  90 I=1,NUMEL
C
C   CHECK FOR TYPE OF ELEMENT
C
      IF(NC(I,4).NE.0) THEN
      NELTYP=4
      ELSE
      NELTYP=3
      ENDIF
      NT=NELTYP*2
      DO 100 I1=1,8
      EQU(I1)=0.0
      DO 100 J=1,8
      EK(I1,J)=0.0
      EC(I1,J)=0.0
  100 CONTINUE
C
C   COMPUTE AVERAGE SATURATION OF THE ELEMENT
C
      ESC=0.0
      DO 110 J=1,NELTYP
      IA=NC(I,J)
      X(J)=XORD(IA)
      Y(J)=YORD(IA)
      J1=IA*2
      ESC=ESC+XS(J1)
  110 CONTINUE
      ESC=ESC/NELTYP
C
C   COMPUTE EK11,EK21,EK22: CALL SUBROUTINE FAC
C
      CALL FAC(ESC,OMU,WMU,PS,EK11,EK21,EK22,NPHASE)
C
      N=NM(I)
      PHIEL=PHI(N)
      P=ABPERM(N)
      ESC=0.0
C
C GENERATE ELEMENT VELOCITY VECTOR FOR SPECIFIED BOUNDARY FLOW ELEMENTS
C
      IF (NF.GT.0) THEN
      DO 115 J=1,NF
```

```
          IF(I.EQ.NFLOW(J)) THEN
          DO 140 K=1,NT
          U(K)=Q(J,K)
  140     CONTINUE
          ENDIF
  115     CONTINUE
          ENDIF
C
C   CALL SUBROUTINES TO COMPUTE ELEMENT MATRICES
C
          NE=I
          IF (NELTYP.EQ.3) THEN
          CALL TRIAN(X,Y,D,PHIEL,EC,NE,U,EQU,NF)
          ELSE
          CALL QUAD(X,Y,D,PHIEL,EC,NE,U,EQU,NF)
          ENDIF
          CALL FORMEK(EK11,EK21,EK22,P,D,EK,NELTYP)
C
C   STORE D MATRIX FOR LATER USE
C
          DO 120 J=1,NELTYP
          DO 120 J1=1,NELTYP
          NJ=(J-1)*NELTYP+J1
          A1(I,NJ)=D(J,J1)
  120     CONTINUE
C
C *******************************************************************
C
C         STEP III: ASSEMBLY OF SYSTEM MATRICES: SK,SC AND SQU
C
C *******************************************************************
C
          DO 150 L=1,2
          DO 150 LL=1,NELTYP
          L1=L+(LL-1)*2
          ICR(L1)=(NC(I,LL)-1)*2+L
  150     CONTINUE
          CALL ASSEM (EK,SK,NT,ICR)
          CALL ASSEM (EC,SC,NT,ICR)
          IF (NF.GT.0) THEN
          CALL ASSEMF(EQU,NT,ICR,SQU)
          ENDIF
   90     CONTINUE
          ESC=0.0
C
C ******************************************************************
C
C         STEP IV: SOLUTION OF SYSTEM NONLINEAR EQUATIONS
C
C ******************************************************************
C
C   TRANSFORM THE NONLINEAR EQUATIONS TO A SET OF LINEAR EQUATIONS
C
C   A. COMPUTE F MATRIX
C
C   (NOTE: TO TIME, T = 0, THE COMPUTED SK,SC MATRICES ARE USED.
C          THIS ASSUMES VECTOR B=0)
  156     DO 155 I=1,NDOF
          DO 155 K=1,NDOF
          F(I)=(-1)*SK(I,K)*XS(K)+F(I)+SQU(I)
  155     CONTINUE
C
C  B. COMPUTE NEW SK MATRIX (DEFINED AS MATRIX D IN THE ANALYTICAL
C     FORMULATION)
C
```

```
            DO 160 I=1,NDOF
            DO 160 J=1,NDOF
            SK(I,J)=SK(I,J)*DT+SC(I,J)
  160       CONTINUE
C
C     C. MODIFY SK AND F MATRICES TO ACCOUNT FOR CONSTRAINED D.O.F. AT
C        WHICH THE NODAL PRESSURE AND SATURATION VALUES HAVE TO BE HELD
C        CONSTANT AT ALL TIMES
C
            DO 165 I=2,NDOF,2
            DO 170 N=1,NRESN
            MCHECK=(NCOD(N)-1)*2+ND(N)
            IF(MCHECK.EQ.I) GO TO 165
  170       CONTINUE
            IF(DS(I).GT.0.95) THEN
            NRESN = NRESN + 1
            NCOD(NRESN) = I/2
            ND(NRESN)=2
            SPEC(NRESN)= 1.00
            ENDIF
  165       CONTINUE
            DO 180 I=1,NRESN
            J1=NCOD(I)
            J2=ND(I)
            K1=(J1-1)*2+J2
            DS(K1)=SPEC(I)
            XS(K1)=SPEC(I)
            DO 190 II=1,NDOF
            SK(K1,II)=0.0
            SK(II,K1)=0.0
  190       CONTINUE
            SK(K1,K1)=1.0
            F(K1)=0.0
  180       CONTINUE
C
C     D. SOLVE THE SYSTEM OF LINEAR EQUATIONS USING GAUSS ELIMINATION
C        METHOD
C
            CALL SOLVE (SK,F,B,NDOF)
C
C     E. COMPUTE NEW NODAL VALUES
C
  201       DO 200 I=1,NDOF
            F(I)=0.0
            DS1(I)=DS(I)
            DS(I)=B(I)*DT+XS(I)
            DO 200 J=1,NDOF
            SK(I,J)=0.0
  200       CONTINUE
C
C     F. CHECK IF NUMBER OF ITERATION CYCLES WITHIN TIME STEP EXCEED THE
C        MAXIMUM LIMIT SPECIFIED
C
            IT1=IT1+1
            IF(IT1.EQ.MAX1) GO TO 243
C
C     G. CHECK IF CONVERGENCE IS ACHIEVED
C
            DO 250 I=1,NDOF
            X2=(DS1(I)-DS(I))/DS(I)
            X2=ABS(X2)
            IF(X2.GT.TOLR) GO TO 251
            IF(I.EQ.NDOF) GO TO 243
  250       CONTINUE
  243       IT2=IT2+1
```

```
C
C   H. WHEN CONVERGENCE IS ACHIEVED OR WHEN MAXIMUM NUMBER OF ITEATION
C      CYCLES HAVE BEEN PERFORMED START A NEW TIME STEP
C
      IT1=0
      IF(IT2.EQ.MAX2) GO TO 261
      TI=TI+DT
      IF(TI.GT.TT) GO TO 261
      DO 260 I=1,NDOF
      XS(I)=DS(I)
      DS(I)=B(I)*DT+XS(I)
  260 CONTINUE
C
C  RE-COMPUTE ELEMENT MATRICES EK FOR NEW VALUES OF RELATIVE
C  PERMEABILITY AND ASSEMBLE NEW SYSTEM MATRIX SK
C
C   1. CHECK FOR TYPE OF ELEMENT
C
  251 DO 210 I=1,NUMEL
      IF(NC(I,4).NE.0) THEN
      NELTYP=4
      ELSE
      NELTYP=3
      ENDIF
      NT=NELTYP*2
C
C   2. COMPUTE AVERAGE SATURATION OF THE ELEMENT
C
      ESC=0.0
      DO 215 J=1,NELTYP
      IA=NC(I,J)
      M=IA*2
      ESC=DS(M)+ESC
  215 CONTINUE
      ESC=ESC/NELTYP
C
C 3. COMPUTE RELATIVE PERMEABILITIES AND CAPILLARY PRESSURE
C
      CALL FAC (ESC,OMU,WMU,PS,EK11,EK21,EK22,NPHASE)
      N=NM(I)
      P=ABPERM(N)
      ESC=0.0
C
C 4. FORMULATE D MATRIX
C
      DO 220 J=1,NELTYP
      DO 220 J1=1,NELTYP
      NJ=(J-1)*NELTYP+J1
      D(J,J1)=A1(I,NJ)
  220 CONTINUE
      DO 230 J=1,8
      DO 230 K=1,8
      EK(J,K)=0.0
  230 CONTINUE
C
C 5.  FORMULATE EK MATRIX FOR EACH ELEMENT
C
      CALL FORMEK (EK11,EK21,EK22,P,D,EK,NELTYP)
      DO 240 L=1,2
      DO 240 LL=1,NELTYP
      L1=L+(LL-1)*2
      ICR(L1)=(NC(I,LL)-1)*2+L
  240 CONTINUE
C
C 6.  STUFF EK MATRIX TO SK MATRIX
C
```

```
      CALL ASSEM (EK,SK,NT,ICR)
  210 CONTINUE
      GO TO 156
C
C **********************************************************************
C
C                 STEP 5: PRINT RESULTS
C
C **********************************************************************
C
  261 WRITE(6,810)
  810 FORMAT(1H1,1X,13HFINAL RESULTS/)
      WRITE(6,*) 'TIME STEP IT2', IT2
      WRITE (6,714)
      WRITE(6,716)
      DO 270 I=1,NDOF
      N1=I/2
      N2=I-N1*2
      IF(N2.EQ.0) GO TO 270
      N2=(I+1)/2
      WRITE(6,717) N2,XORD(N2),YORD(N2),DS(I),DS(I+1)
  717 FORMAT(I5,2F9.3,1E15.3,5X,1F10.3)
  270 CONTINUE
      STOP
      END
```

SUBROUTINE "FAC" LISTING

```
      SUBROUTINE FAC(ESC,OMU,WMU,PS,EK11,EK21,EK22,NPHASE)
C
C  THIS SUBROUTINE COMPUTES K11, K21 AND K22 USING EQUATIONS (72) AND
C  (73)
C    NOTE: THE FOLLOWING CHANGES ARE MADE IN EQUATIONS (127 A TO D)
C          FOR PROGRAM IMOLEMENTATION
C          1. ACTUALLY WKR = 0.0 WHEN ESC IS LESS THAN 0.16, BUT IN
C             THE PROGRAM IT'S VALUE IS TAKEN AS 0.0001 TO AVOID
C             SINGULARITY IN SOLUTION OF SYSTEM EQUATIONS
C          2. ACTUALLY OKR = 0.0 WHEN ESC IS GREATER THAN 0.80, BUT IN
C             THE PROGRAM IT'S VALUE IS TAKEN AS 0.0001 TO AVOID
C             SINGULARITY IN SOLUTION OF SYSTEM EQUATIONS
C
      IMPLICIT REAL *8(A-H,O-Z)
      IF (ESC.LT.0.16) THEN
      OKR=1.0
      WKR=0.0001
      ELSEIF (ESC.GT.0.8) THEN
      OKR=0.0001
      WKR=1.0
      ELSE
      OKR=((0.8-ESC)**2)/0.64
      WKR=((ESC-0.16)**2)/0.64
      ENDIF
      IF (NPHASE.EQ.1) THEN
      EK11=OKR/OMU
      EK21=WKR/WMU
      EK22=-WKR/WMU*PS
      ELSE
      EK11=WKR/WMU
      EK21=OKR/OMU
      EK22=OKR/OMU*PS
      ENDIF
      RETURN
      END
```

SUBROUTINE "TRIAN" LISTING

```
      SUBROUTINE TRIAN(X,Y,D,PHIEL,EC,MM,U,EQU,NF)
C
C  THIS SUBROUTINE FORMULATES THE ELEMENT MATRICES EK,EC, AND EQU FOR
C  3 NODE ISOPARAMETRIC TRIANGULAR ELEMENTS
C
      IMPLICIT REAL *8(A-H,O-Z)
      DIMENSION X(3),Y(3),B(3),C(3),D(4,4),U(6),C1(3,3),EC(8,8)
      DIMENSION EQU(8)
C
C  INITIALIZE THE MATRICES
C
      DO 450 I=1,3
      DO 450 J=1,3
      D(I,J) = 0.0
      C1(I,J)=0.0
  450 CONTINUE
      DO 460 I=1,8
      EQU(I)=0.0
      DO 460 J=1,8
      EC(I,J)=0.0
  460 CONTINUE
      IF(NF.GT.0) THEN
C
C  COMPUTE THE LENGTH OF EACH SIDE ,AREA AND OTHER PARAMETERS FOR
C  THE TRIANGLE
C
      L12=SQRT((X(1)-X(2))**2+(Y(1)-Y(2))**2)
      L13=SQRT((X(1)-X(3))**2+(Y(1)-Y(3))**2)
      L23=SQRT((X(2)-X(3))**2+(Y(2)-Y(3))**2)
      ENDIF
      DO 480 I=2,3
      X(I)=X(I)-X(1)
      Y(I)=Y(I)-Y(1)
  480 CONTINUE
      X(1)=0.0
      Y(1)=0.0
      AREA=(X(2)*Y(3)-Y(2)*X(3))/2
      B(1)=Y(2)-Y(3)
      C(1)=X(3)-X(2)
      B(2)=Y(3)
      C(2)=-X(3)
      B(3)=-Y(2)
      C(3)=X(2)
C
C     FORMULATES THE D MATRIX
C
      CONC=0.25/AREA
      DO 490 I=1,3
      D(I,I)=(B(I)*B(I)+C(I)*C(I))*CONC
      DO 490 J=1,3
      D(I,J)=(B(I)*B(J)+C(I)*C(J))*CONC
  490 CONTINUE
C
C     FORMULATE EC MATRIX
C
      DO 510 I=1,3
      DO 510 J=1,3
      C1(I,I)=AREA/6.0
      IF(I.EQ.J) GO TO 510
      C1(I,J)=AREA/12.0
  510 CONTINUE
      DO 520 I=1,3
      DO 520 J=1,3
```

```
      I1=2*I
      J1=2*J
      I2=I1-1
      EC(I2,J1)=C1(I,J)*PHIEL
  520 EC(I1,J1)=C1(I,J)*PHIEL*(-1)
      IF (NF.GT.0) THEN
C
C  FORMULATES EQU MATRIX
C
      EQU(1)=0.5*(L13*U(3)+L12*U(5))
      EQU(2)=0.5*(L23*U(4)+L12*U(6))
      EQU(3)=0.5*(L23*U(1)+L12*U(5))
      EQU(4)=0.5*(L13*U(2)+L12*U(6))
      EQU(5)=0.5*(L23*U(1)+L13*U(3))
      EQU(6)=0.5*(L23*U(2)+L13*U(4))
      ENDIF
      RETURN
      END
```

SUBROUTINE "QUAD" LISTING

```
      SUBROUTINE QUAD(X,Y,D,PHIEL,EC,MM,U,EQU,NF)
C
C  THIS SUBROUTINE FORMULATES THE ELEMENT MATRICES EK,EC AND EQU
C  FOR THE 4 NODE QUADRILATAAL ISOPARAMETRIC ELEMENTS
C
      IMPLICIT REAL *8(A-H,O-Z)
      DIMENSION X(4),Y(4),D(4,4),XG(3),WGT(3),HH(2,4),XX(2,4)
      DIMENSION C(4,4),EC(8,8),U(8),EQU(8),AN(4)
C
C  INITIALIZE THE MATRICAS
C
      DO 450 I=1,4
      DO 450 J=1,4
      D(I,J)=0.0
      C(I,J)=0.0
  450 CONTINUE
      DO 460 I=1,8
      EQU(I)=0.0
      DO 460 J=1,8
      EC(I,J)=0.0
  460 CONTINUE
      IF (NF.GT.0) THEN
C
C  COMPUTES THE LENGTHS OF THE ELEMENT SIDES
C
      L12=SQRT((X(1)-X(2))**2+(Y(1)-Y(2))**2)
      L23=SQRT((X(2)-X(3))**2+(Y(2)-Y(3))**2)
      L34=SQRT((X(3)-X(4))**2+(Y(3)-Y(4))**2)
      L41=SQRT((X(4)-X(1))**2+(Y(4)-Y(1))**2)
      ENDIF
C
C  USE GAUSS QUADRATURE WEIGHTING FUNCTIONS AND DATA POINTS
C
      XG(1)=-0.774597
      XG(2)=0.0
      XG(3)=0.774597
      WGT(1)=0.55557
      WGT(2)=0.88889
      WGT(3)=0.55557
      DO 5 I=1,4
      XX(1,I)=X(I)
      XX(2,I)=Y(I)
    5 CONTINUE
      DO 80 LX=1,3
      R=XG(LX)
      DO 80 LY=1,3
      S=XG(LY)
C
C EVALUATE THE DERIVATIVE OPERATOR HH AND THE DETERMINANT OF THE
C JACOBIAN METRIX
C
      CALL BMTRX(XX,HH,DETER,R,S,MM,AN)
      WT=WGT(LX)*WGT(LY)*DETER
C
C PEFURM NUMERICAL INTEGRATION TO EVALUATE THE ELEMENT MATRICES
C
      DO 40 J=1,4
      DO 40 K=J,4
      D(J,K)=(HH(1,J)*HH(1,K)+HH(2,J)*HH(2,K))*WT+D(J,K)
   40 CONTINUE

      WT=WGT(LX)*WGT(LY)*DETER
      DO 70 J=1,4
```

```
           DO 70 K=J,4
           C(J,K)=AN(J)*AN(K)*WT+C(J,K)
      70   CONTINUE
      80   CONTINUE
C
C USE SYMMETRY OF D AND C
C
           DO 50 J=1,4
           DO 50 K=J+1,4
           D(K,J)=D(J,K)
           C(K,J)=C(J,K)
      50   CONTINUE
C
C FORMULATE EC
C
           DO 520 I=1,4
           DO 520 J=1,4
           I1=2*I
           J1=2*J
           I2=I1-1
           EC(I2,J1)=C(I,J)*PHIEL
           EC(I1,J1)=C(I,J)*PHIEL*(-1)
     520   CONTINUE
           IF (NF.GT.0) THEN
C
C FORMULATE EQU MATRIX
C
           EQU(1)=0.5*(L12*U(1)+L41*U(7))
           EQU(2)=0.5*(L12*U(2)+L41*U(3))
           EQU(3)=0.5*(L12*U(5)+L23*U(4))
           EQU(4)=0.5*(L12*U(6)+L23*U(8))
           EQU(5)=0.5*(L23*U(1)+L34*U(7))
           EQU(6)=0.5*(L23*U(2)+L34*U(3))
           EQU(7)=0.5*(L34*U(5)+L41*U(4))
           EQU(8)=0.5*(L34*U(6)+L41*U(8))
           ENDIF
           RETURN
           END
```

SUBROUTINE "BMTRX" LISTING

```
      SUBROUTINE BMTRX(XX,BB,DETER,R,S,MM,AN)
C
C   THIS SUBROUTINE FORMULATE THE INTERPOLATION FUNCTIONS, THE
C   DETERMINANT OF THE JACOBIAN MATRIX AND IT'S INVERSE AND THE
C   GLOBAL DERIVATIVE OPERATOR FOR A 4 NODE QUADRILATERAL ISOPARAMETRIC
C   ELEMENT
C
      IMPLICIT REAL *8(A-H,O-Z)
      DIMENSION XX(2,4),BB(2,4),P(2,4),XJ(2,2),XJI(2,2),AN(4)
C
      RP=1.0+R
      SP=1.0+S
      RM=1.0-R
      SM=1.0-S
C
C  INTERPOLATION FUNCTIONS (SHAPE FUNCTIONS) FOR 4 NODE ISOPARAMETRIC
C  QUADRILATERAL ELEMENT
C
       AN(1)=0.25*RM*SM
       AN(2)=0.25*RP*SM
       AN(3)=0.25*RP*SP
       AN(4)=0.25*RM*SP
C
C  NATURAL CORDINATE DERIVATIVES OF THE INTERPOLATION FUNCTIONS
C
C        1. WITH RESPECT TO R
C
      P(1,1)=-0.25*SM
      P(1,2)=-P(1,1)
      P(1,3)=0.25*SP
      P(1,4)=-P(1,3)
C
C        2. WITH RESPECT TO S
C
      P(2,1)=-0.25*RM
      P(2,2)=-0.25*RP
      P(2,3)=-P(2,2)
      P(2,4)=-P(2,1)
  715 FORMAT(8F10.4)
C
C  EVALUATE THE JACOBIAN MATRIX AT POINT (R,S)
C
      DO 30 I=1,2
      DO 30 J=1,2
      DUM=0.0
      DO 20 K=1,4
   20 DUM=DUM+P(I,K)*XX(J,K)
   30 XJ(I,J)=DUM
C
C  COMPUTE DETERMINANT OF THE JACOBIAN MATRIX AT POINT (R,S)
C
      DETER=XJ(1,1)*XJ(2,2)-XJ(2,1)*XJ(1,2)
      IF(DETER.GT.0.00000001) GO TO 40
      WRITE(6,2000) MM
 2000 FORMAT(10X,' ZERO OR NEGATIVE JACOBIAN DETERMINANT FOR ELEMENT',
     1 I4,/)
      GO TO 499
C
C  COMPUTE INVERSE OF THE JACOBIAN MATRIX
C
   40 DUM=1.0/DETER
      XJI(1,1)=XJ(2,2)*DUM
      XJI(1,2)=-XJ(1,2)*DUM
```

```
      XJI(2,1)=-XJ(2,1)*DUM
      XJI(2,2)=XJ(1,1)*DUM
C
C  EVALUATE THE GLOBAL DERIVATIVE OPERATOR
      DO 60 K=1,4
      BB(1,K)=0.0
      BB(2,K)=0.0
      DO 50 I=1,2
      BB(1,K)=BB(1,K)+XJI(1,I)*P(I,K)
   50 BB(2,K)=BB(2,K)+XJI(2,I)*P(I,K)
   60 CONTINUE
  499 RETURN
      END
```

SUBROUTINE "FORMEK" LISTING

```
      SUBROUTINE FORMEK (EK11,EK21,EK22,P,D,EK,NELTYP)
C
C  THIS SUBROUTINE FORMULATES ELEMENT MATIRX EK FROM D MATRIX
C
      IMPLICIT REAL *8 (A-H,O-Z)
      DIMENSION D(4,4), EK(8,8)
      DO 450 J=1,8
      DO 450 I=1,8
      EK(I,J)=0.0
  450 CONTINUE
      DO 100 I=1,NELTYP
      DO 100 J=1,NELTYP
      I2=2*I
      J2=2*J
      EK(I2-1,J2-1)=D(I,J)*EK11*P
      EK(I2,J2-1)=D(I,J)*EK21*P
      EK(I2,J2)=D(I,J)*EK22*P
  100 CONTINUE
      RETURN
      END
```

SUBROUTINE "ASSEM" LISTING

```
      SUBROUTINE ASSEM (A,B,NT,ICR)
C
C  THIS SUBROUTINE ASSEMBLES THE ELEMENT METRICES EK AND EC INTO
C  THE SYSTEM MATRICES SK AND SC, RESPECTIVELY
C
      IMPLICIT REAL *8 (A-H,O-Z)
      DIMENSION A(8,8),B(200,201),ICR(8)
      DO 202 K=1,NT
      II=ICR(K)
      DO 202 M=1,NT
      JJ=ICR(M)
      B(II,JJ)=B(II,JJ)+A(K,M)
  202 CONTINUE
      RETURN
      END
```

SUBROUTINE "ASSEMF" LISTING

```
      SUBROUTINE ASSEMF(A,NT,ICR,B)
C
C THIS SUBROUTINE ASSEMBLES THE ELEMENT MATRIX EQU INTO SYSTEM MATRIX
C SQU
C
      IMPLICIT REAL *8 (A-H,O-Z)
      DIMENSION A(8),B(200),ICR(NT)
      DO 10 K=1,NT
      II=ICR(K)
      B(II)=B(II)+A(K)
   10 CONTINUE
      RETURN
      END
```

SUBROUTINE "SOLVE" LISTING

```
      SUBROUTINE SOLVE (A,B,D,N)
C
C  THIS SUBROUTINE SOLVES THE SYSTEM OF LINEAR SIMULTANEOUS EQUATIONS
C  USING GAUSS ELIMINATION METHOD
C
      IMPLICIT REAL *8 (A-H,O-Z)
      DIMENSION A(200,201),B(200),D(200)
      DO 5 I=1,N
      D(I)=0.0
    5 CONTINUE
      DO 10 I=1,N
      A(I,N+1)=B(I)
   10 CONTINUE
      DO 20 I=1,N
      PIVOT=1/A(I,I)
      A(I,I)=1.0
      DO 30 J=I+1,N+1
      IF(A(I,J).EQ.0) GO TO 30
      A(I,J)=A(I,J)*PIVOT
   30 CONTINUE
      DO 40 K=I+1,N
      IF(A(K,I).EQ.0) GO TO 40
      DO 50 J=I+1,N+1
      IF(A(I,J).EQ.0) GO TO 50
      A(K,J)=A(K,J)-A(K,I)*A(I,J)
   50 CONTINUE
      A(K,I)=0.0
   40 CONTINUE
   20 CONTINUE
      DO 60 I=N,1,-1
      DO 70  K=N,I+1,-1
      D(I)=D(I)+A(I,K)*D(K)
   70 CONTINUE
      D(I)=A(I,N+1)-D(I)
   60 CONTINUE
      RETURN
      END
```

REFERENCES INDEX *

* The editors are grateful to Mr. Baljit Singh for his assistance in preparation of the indices and proof reading.

SUBJECT INDEX